DEVELOPMENTS IN FOOD SCIENCE 43

FLAVOUR SCIENCE
RECENT ADVANCES AND TRENDS

DEVELOPMENTS IN FOOD SCIENCE 43

FLAVOUR SCIENCE
RECENT ADVANCES AND TRENDS

Edited by
WENDER L.P. BREDIE
Sensory Science, Department of Food Science, The Royal Veterinary and Agricultural University, Frederiksberg C, Denmark

MIKAEL AGERLIN PETERSEN
Quality and Technology, Department of Food Science, The Royal Veterinary and Agricultural University, Frederiksberg C, Denmark

ELSEVIER

Amsterdam - Boston - Heidelberg - London - New York - Oxford
Paris - San Diego - San Francisco - Singapore - Sydney - Tokyo

Elsevier
Radarweg 29, PO Box 211, 1000 AE Amsterdam, The Netherlands
The Boulevard, Langford Lane, Kidlington, Oxford OX5 1GB, UK

First edition 2006
Reprinted 2006

British Library Cataloguing in Publication Data
A catalogue record for this book is available from the British Library

Library of Congress Cataloging-in-Publication Data
A catalog record for this book is available from the Library of Congress

ISBN–13: 978-0-444-52742-4
ISBN–10: 0-444-52742-7

ISSN: 0167-4501

Printed and bound in *The Netherlands*

06 07 08 09 10 10 9 8 7 6 5 4 3 2

DEVELOPMENTS IN FOOD SCIENCE

Volume 1 J.G. Heathcote and J.R. Hibbert
Aflatoxins: Chemical and Biological Aspects

Volume 2 H. Chiba, M. Fujimaki, K. Iwai, H. Mitsuda and Y. Morita (Editors)
Proceedings of the Fifth International Congress of Food Science and Technology

Volume 3 I.D. Morton and A.J. MacLeod (Editors)
Food Flavours
Part A. Introduction
Part B. The Flavour of Beverages
Part C. The Flavour of Fruits

Volume 4 Y. Ueno (Editor)
Trichothecenes: Chemical, Biological and Toxicological Aspects

Volume 5 J. Holas and J. Kratochvil (Editors)
Progress in Cereal Chemistry and Technology. Proceedings of the VIIth World Cereal and Bread Congress, Prague, 28 june-2 July 1982

Volume 6 I. Kiss
Testing Methods in Food Microbiology

Volume 7 H. Kurata and Y. Ueno (Editors)
Toxigenic Fungi: Their Toxins and Health Hazard. Proceedings of the Mycotoxin Symposium, Tokyo, 30 August-3 September 1983

Volume 8 V. Betina (Editor)
Mycotoxins: Production, Isolation, Separation and Purification

Volume 9 J. Holló (Editor)
Food Industries and the Environment. Proceedings of the International Symposium, Budapest, Hungary, 9-11 September 1982

Volume 10 J. Adda (Editor)
Progress in Flavour Research 1984. Proceedings of the 4th Weurman Flavour Research Symposium, Dourdan, France, 9-11 May 1984

Volume 11 J. Holló (Editor)
Fat Science 1983. Proceedings of the 16th International Society for Fat Research Congress, Budapest, Hungary, 4-7 October 1983

Volume 12 G. Charalambous (Editor)
The Shelf Life of Foods and Beverages. Proceedings of the 4th International Flavor Conference, Rhodes, Greece, 23-26 July 1985

Volume 13 M. Fujimaki, M. Namiki and H. Kato (Editors)
Amino-Carbonyl Reactions in Food and Biological Systems. Proceedings of the 3rd International Symposium on the Maillard Reaction, Susuno, Shizuoka, Japan,1-5 July 1985

Volume 14 J. Škoda and H. Škodová
Molecular Genetics. An Outline for Food Chemists and Biotechnologists.

Volume 15 D.E. Kramer and J. Liston (Editors)
Seafood Quality Determination. Proceedings of the International Symposium, Anchorage, Alaska, U.S.A., 10-14 November 1986

Volume 16 R.C. Baker. P. Wong Hahn and K.R. Robbins
Fundamentals of New Food Product Development

Volume 17 G. Charalambous (Editor)
Frontiers of Flavor. Proceedings of the 5th International Flavor Conference, Porto Karras,Chalkidiki, Greece, 1-3 July 1987

Volume 18 B.M. Lawrence, B.D. Mookherjee and B.J. Willis (Editors)
Flavors and Fragrances: A World Perspective. Proceedings of the 10th International Congress of Essential Oils, Fragrances and Flavors, Washington, DC, U.S.A., 16-20 November 1986

Volume 19 G. Charalambous and G. Doxastakis (Editors)
Food Emulsifiers: Chemistry, Technology, Functional Properties and Applictations

Volume 20 B.W. Berry and K.F. Leddy
Meat Freezing. A Source Book

Volume 21	J. Davídek, J. Velišek and J. Pokorný (Editors) Chemical Changes during Food Processing
Volume 22	V. Kyzlink Principles of Food Preservation
Volume 23	H. Niewiadomski Rapeseed. Chemistry and Technology
Volume 24	G. Charalambous (Editor) Flavors and Off-flavors '89. Proceedings of the 6th International Flavor Conference, Rehymnon, Crete, Greece, 5-7 July 1989
Volume 25	R. Rouseff (Editor) Bitterness in Foods and Beverages
Volume 26	J. Chelkowski (Editor) Cereal Grain. Mycotoxins, Fungi and Quality in Drying and Storage
Volume 27	M. Verzele and D. De Keukeleire Chemistry and Analysis of Hop and Beer Bitter Acids
Volume 28	G. Charalambous (Editor) Off-Flavors in Foods and Beverages
Volume 29	G. Charalambous (Editor) Food Science and Human Nutrition
Volume 30	H.H. Huss, M. Jakobsen and J. Liston (Editors) Quality Assurance in the Fish Industry. Proceedings of an International Conference, Copenhagen, Denmark, 26-30 August 1991
Volume 31	R.A. Samson, A.D. Hocking, J.I.Pitt and A.D. King (Editors) Modern Methods in Food Mycology
Volume 32	G. Charalambous (Editor) Food Flavors, Ingredients and Composition. Proceedings of the 7th International Flavor Conference, Pythagorion, Samos, Greece, 24-26 June 1992
Volume 33	G. Charalambous (Editor) Shelf Life Studies of Foods and Beverages. Chemical, Biological, Physical and Nutritional Aspects
Volume 34	G. Charalambous (Editor) Spices, Herbs and Edible Fungi
Volume 35	H. Maarse and D.G. van der Heij (Editors) Trends in Flavour Research. Proceedings of the 7th Weurman Flavour Research Symposium, Noordwijkerhout, The Netherlands, 15-18 June 1993
Volume 36	J.J. Bimbenet, E. Dumoulin and G. Trystram (Editors) Automatic Control of Food and Biological Processes. Proceedings of the ACoFoP III Symposium, Paris, France, 25-26 October 1994
Volume 37A+B	G. Charalambous (Editor) Food Flavors: Generation, Analysis and Process Influence Proceedings of the 8th International Flavor Conference, Cos, Greece, 6-8 July 1994
Volume 38	J.B. Luten, T. Børresen and J. Oehlenschläger (Editors) Seafood from Producer to Consumer, Integrated Approach to Quality Proceedings of the International Seafood Conference on the occasion of the 25th anniversary of the WEFTA, held in Noordwijkerhout, The Netherlands, 13-16 November 1995
Volume 39	D. Wetzel and G. Charalambous † (Editors) Instrumental Methods in Food and Beverage Analysis
Volume 40	E.T. Contis, C.-T. Ho, C.J. Mussinan, T.H. Parliment, F. Shahidi and A.M. Spanier (Editors) Food Flavors: Formation, Analysis and Packaging Influences Proceedings of the 9th International Flavor Conference The George Charalambous Memorial Symposium
Volume 41	G. Doxastakis and V. Kiosseoglou (Editors) Novel Macromolecules in Food Systems
Volume 42	M. Sakaguchi (Editor) More Efficient Utilization of Fish and Fisheries Products
Volume 43	W.L.P. Bredie and M.A. Petersen Flavour Science: Recent Advances and Trends

Contents

4. Key aroma and taste components

5. Flavour changes in food production and storage

9. Advanced instrumental analyses

Preface

Flavour science is a multidisciplinary subject encompassing biochemistry, chemical and physical aspects of food science and biotechnology, the chemistry of natural products as well as the biochemistry, physiology and psychology of human perception. Flavour science is evolving from the systematic study of volatile flavour compounds in foods into a science aiming to provide an understanding of all aspects of flavour, in the food, the production chain, the perception by consumers and their contentment during and after eating.

Among the international flavour and sensory symposia, the Weurman Flavour Research Symposium is one of the few meetings addressing both the width and depth of flavour science in a comprehensive and personal way. The Weurman symposium has a long tradition in Europe and is a premier forum for scientists from academia and industry to discuss *advances* and *trends* in flavour science. Participation is by invitation and the limited number of delegates can attend all plenary sessions at a location away from daily disturbances. Dr. Cornelius Weurman had already in 1975 the vision that a successful flavour meeting needs delegates who are active in research and who actively contribute at the Symposium. He envisaged that such a meeting should encourage interaction between senior and young scientists as well as between academia and industry in an informal and open atmosphere. The spirit and ideas of Dr Weurman are still modern and alive even after 30 years.

The 11th Weurman Flavour Research Symposium, held from 21 June to 24 June, 2005 in Roskilde, Denmark, followed again the Weurman format and was attended by 166 persons from 25 countries world-wide, of which 134 attendees were from European countries. In setting up of the scientific programme of the Symposium, the Scientific Committee decided to include promising areas in genomics, structure-activity relations and biotechnology. Also, more established areas on characterisation of key flavour components, flavour generation, flavour stability in foods, and flavour release were included. Unfortunately, the connection between flavour and health did not attract a great number of contributions, but undoubtedly will win popularity in the future.

The Symposium was divided in 9 sessions with contributions on biological aspects of perception and structure-activity relationships (SAR) (10 presentations); genomics and biotechnology (9); enzymes and biological systems (19); key aroma and taste components (22); changes in food production and storage (21); flavours generated by thermal processes (14); retention and release (23); sensory - instrumental relationships (22) and advanced instrumental analysis (12). The present book reflects all of these topics and contains 142 research papers from the 110 posters and 42 plenary lectures presented at Roskilde. Besides the plenary sessions, discussions were encouraged in the workshops on “Gastronomy: the ultimate flavour science?”, “Methods for artificial perception: can machine replace man?” and “Challenges for data analysis methods in flavour science”, illustrating some of the current research and new research initiatives in

the Scandinavian countries. The proceedings gives an almost complete record of the Symposium with each section starting with a key-note contribution presented as an extended paper followed by regular research papers.
The Symposium offered also a unique opportunity to present the 2004 *Firmenich flavour and fragrance award* lecture. The award winner Dr Andrea Buettner, from the Technical University of Munich, Germany, showed her outstanding work on the chemistry and physiology of the oral and retronasal pathways of flavour sensation. The lecture will be published as a full journal paper elsewhere.
The scientific principles - both in the natural sciences and humanities - underlying the gastronomic preparation of delicious and satisfying meals are recently receiving more attention in Denmark. Although Molecular Gastronomy has longer traditions at other places in the world, the broader study of how to prepare foods and meals of high sensory quality and satiating ability, while being fairly nutritious for different types of consumers, is still a challenge for research. In this respect, flavour scientists could make use of their know-how on flavour (bio-)chemistry, flavour release, human physiology and cognitive/behavioural psychology in an even more active role in healthy product and meal design. There are still ample possibilities for future flavour research and continuation of the Weurman symposia. We look forward to the 12th Symposium in 2008 in Switzerland, which will round off two successful Weurman rotations between six European countries.
As the main organisers of the Symposium we are grateful for the generous sponsorships by the following companies and organisations: Centre for Advanced Food Studies (LMC), Denmark; Chew Tech I/S, Denmark; Chr. Hansen A/S, Denmark; Danisco A/S, Denmark; Danske Slagterier, Denmark; Firmenich SA, Switzerland; Givaudan SA, Switzerland; Ionicon Analytik GmbH, Austria; Kraft Foods GmbH, Germany; Nestlé SA, Switzerland; Quest International B.V., The Netherlands; Symrise GmbH & Co. KG, Germany; Tripos GmbH, Germany and Unilever B.V., The Netherlands. Their donations enabled us reduce the fees for 19 PhD students, organise the workshop on gastronomy and reducing some other costs at the Symposium.
We would also like to acknowledge the many people that have contributed to the organisation of the meeting. In particular the members of the Scientific Committee for their help in the reviewing process. Our thanks are also due to scientific secretary Lise Nissen, multimedia designer Robert Skammelsen Schmidt, the helping PhD students from KVL and Marianne Sjødahl from DIS congress for their invaluable assistance. We are also most grateful to scientific secretary Dorte Juncher for her excellent support throughout the Symposium and her incredible endurance and thoroughness during the editing process. Thanks are to the Mayor of Copenhagen for offering the Symposium a welcome reception at the Civic Hall. We are also thankful to LMC and The Royal Veterinary and Agricultural University (KVL) for allowing us to take the time necessary for preparing a successful Symposium and editing of the 11th Weurman proceedings.

Frederiksberg C, January 2006

Wender Bredie
Mikael Agerlin Petersen
Department of Food Science, The Royal Veterinary and Agricultural University, KVL

Scientific Committee

- Margit Dall Aaslyng (Danish Meat Research Institute, Roskilde, Denmark)
- Hans Christian Beck (Biotechnological Institute, Kolding, Denmark)
- Wender L.P. Bredie (KVL, Frederiksberg C, Denmark)
- Rasmus Bro (KVL, Frederiksberg C, Denmark)
- Per Bruun Brockhoff (The Technical University of Denmark, Lyngby, Denmark)
- Lars Porskjær Christensen (Danish Institute of Agricultural Sciences, Årslev, Denmark)
- Gunnar Hall (Swedish Institute for Food and Biotechnology (SIK), Göteborg, Sweden)
- Anne Maria Hansen (Technological Institute, Kolding, Denmark)
- Jørn Marcussen (Danisco A/S, Brabrand, Denmark)
- Magni Martens (The Norwegian Food Research Institute (Matforsk), Ås, Norway and KVL, Frederiksberg C, Denmark)
- Per Munk Nielsen (Novozymes, A/S, Bagsværd, Denmark)
- Jacob Nielsen (Danish Institute of Agricultural Sciences, Foulum, Denmark)
- Mikael Agerlin Petersen (KVL, Frederiksberg C, Denmark)
- Hanne Refsgaard (Novo Nordisk A/S, Måløv, Denmark)
- Louise Stahnke (Chr. Hansen A/S, Hørsholm, Denmark)
- Øydis Ueland (The Norwegian Food Research Institute (Matforsk), Ås, Norway)

1

Biological aspects of flavour perception and structure-activity relationships

W.L.P. Bredie and M.A. Petersen (Editors)
Flavour Science: Recent Advances and Trends

Molecular and gustatory characterisation of the impact taste compounds in black tea infusions

Thomas Hofmann[a], Susanne Scharbert[b] and Timo Stark[a]

[a]Institut für Lebensmittelchemie, Universität Münster, Corrensstrasse 45, 48149 Münster, Germany; [b]Deutsche Forschungsanstalt für Lebensmittelchemie, Lichtenbergstrasse 4, 85748 Garching, Germany

ABSTRACT

Bioresponse-guided fractionation of black tea infusions, identification of intense taste compounds by LC-MS/MS and 1D/2D-NMR spectroscopy and quantitative analysis were followed by calculation of dose-over-threshold values and biomimetic taste reconstruction. The analyses revealed that besides epigallocatechingallate, catechin, and caffeine a series of flavon-3-ol glycopyranosides are key contributors to black tea taste. Neither the high molecular thearubigens, nor the theaflavins were important tastants. Recordings of human dose-response functions in model systems demonstrated that the flavanol-3-glycosides do not only impart a velvety astringent taste sensation, but do also show a major contribution to the bitter taste by amplifying the bitterness of caffeine.

1. INTRODUCTION

For centuries, the aqueous infusion of the dried leaves and buds of *Camellia sinensis* has been consumed by humans as a highly desirable beverage. Since the taste quality is one of the key criteria used by the tea tasters to describe the quality of tea liquors, multiple attempts have been made to correlate the sensory results of the tea tasters and the molecules exhibiting the typical taste of tea infusions. The data reported so far on the key tastants is, however, very contradictory. For example, the orange coloured, low-molecular weight theaflavins as well as the red-brown polymeric thearubigins, both generated during tea fermentation upon flavan-3-ol oxidation [1,2], are believed to be responsible for the astringency of black tea infusions and have been recommended as a measure of tea quality [3]. In contradiction, other researchers could not find any statistical correlation between the overall astringent taste of tea infusions and the theaflavin concentration, but indicated a relationship between oral astringency and some flavan-3-ols such as, e.g. epigallocatechin-3-gallate [4]. Besides these phenols, 5-*N*-

ethyl-L-glutamine (theanine) is reported to exhibit sweet-brothy and/or umami-like taste quality and is believed to contribute to the taste profile of tea infusions [5].
The objectives of the present study were to identify the key taste compounds in a black tea infusion by bridging the gap between analytical chemistry and human taste perception. This was done by means of the recently developed taste dilution analysis method [6], the determination of dose-response relationships for the tastants, and, finally, to confirm their taste contribution by means of taste reconstitution experiments.

2. MATERIALS AND METHODS

The tea drug Darjeeling Gold-Auslese, TGFOP, Summer (Tee-Handelskontor, Bremen, Germany) was infused with boiling tap water (1 g/100 ml) and maintained for 4 min prior to filtration using a cellulose filter. Reference materials of theaflavins were synthesised as reported recently [7], catechins and flavon-3-ol glycosides were isolated from the tea drug [8]. Details on the ultrafiltration, taste dilution analysis (TDA), sensory analyses [8], and quantitative analysis of tastants [9] were reported recently.

3. RESULTS AND DISCUSSION

In order to evaluate the taste profile of the Darjeeling tea infusion, the trained sensory panel was asked to rate the intensity of individual taste qualities on a scale from 0 (not detectable) to 3 (strongly detectable). By far the highest scores of 2.2 and 1.6 were observed for the intensity of the astringent, mouth-drying taste quality, and the bitterness, respectively (Table 1).

Table 1. Taste profile analysis of the Darjeeling tea infusion, and the artificial taste recombinate.

	Intensities of individual taste qualities in[a]	
Taste quality	Tea infusion	Taste recombinate
Astringent, mouth-drying	2.2	2.1
Bitter	1.6	1.5
Sour	0.5	0.5
Sweet	0.4	0.4
Salty	0	0
Umami	0	0

[a]Intensities were judged on a scale from 0 (not detectable) to 3 (strongly detectable).

3.1. Identification of key taste compounds

To gain first insight into the astringent and bitter compounds, the tea infusion was separated by means of multiple-step ultrafiltration using filters with cut-offs of 10 and 1 kDa in sequence. Three fractions were obtained, a deeply brown coloured, but tasteless fraction containing thearubigene-type polymers with molecular weights above 10 kDa, a red-brown coloured, tasteless fraction containing the compounds with molecular weights between 1 and 10 kDa, and a nearly colourless fraction containing the tea's low molecular weight compounds (LMW; <1 kDa) and reflecting the typical taste profile of the tea [8]. As these experiments pointed out that the LMW compounds are the main contributors to tea taste, this fraction was further analysed by RP8-HPLC (Figure 1, left side). Aimed at rating the tea compounds in their relative taste impact, the effluent was separated into 43 fractions, which were freed from solvent and then used for the taste dilution analysis (TDA) by using the recently developed half-tongue test (Figure 1, right side). Due to their high taste dilution (TD)-factor of 8192, fractions no. 33 and 34 were evaluated with by far the highest taste impacts for astringency, closely followed by fractions no. 30-32 and 23 judged with TD factors between 1024 and 4096 (Figure 1). In comparison, fractions no. 6, 7, 10-12, and 16 were evaluated with lower taste impacts, whereas the low TD-factors determined for all the other fractions excluded major contributions to the perception of tea astringency.

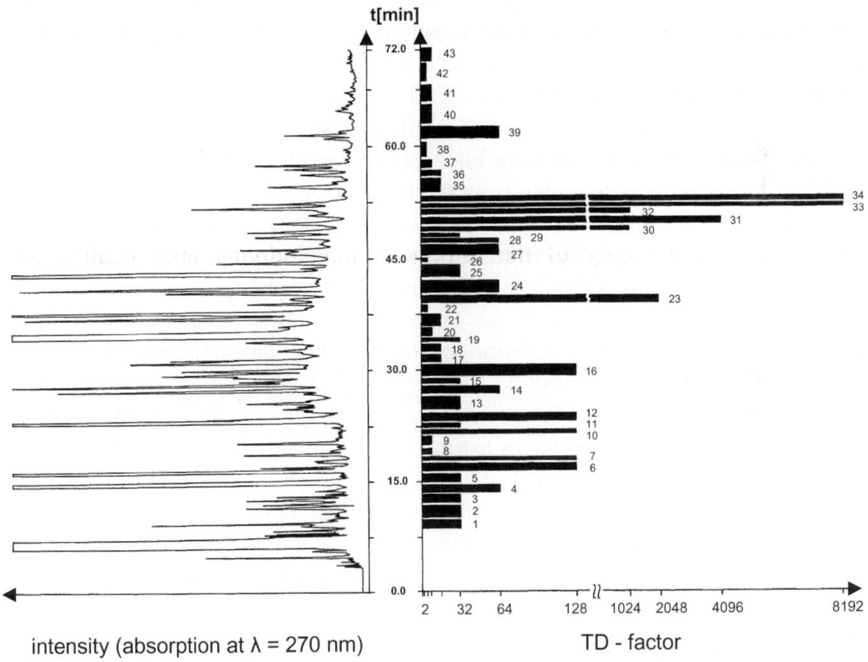

Figure 1. HPLC chromatogram (left side), and taste dilution (TD) chromatogram (for astringency) of the LMW compounds isolated from the tea infusion (Darjeeling Gold-Auslese) [8].

First, we investigated the phenolic substances which were already discussed in literature as the astringent compounds in black tea, namely the catechins and the theaflavins. Analysis of the tea infusion by means of HPLC-DAD and HPLC-MS/MS as well as co-chromatography with reference compounds led to the identification of gallocatechin (fraction 11), epigallocatechin (fraction 15), catechin (fraction 15), epigallocatechin-3-

gallate (EGCG, fraction 20), epicatechin (fraction 22), gallocatechin-3-gallate (fraction 24), epicatechin-3-gallate (fraction 25), catechin-3-gallate (fraction 26), theaflavin (fraction 37), theaflavic acid (fraction 37), theaflavin-3-gallate (fraction 38), theaflavin-3'-gallate (fraction 38), and theaflavin-3,3'-digallate (fraction 39) as the key taste compounds in the individual HPLC fractions (Figure 1) [8]. On comparison of these data with the results of the TDA (Figure 1) it was obvious that those fractions containing the catechin and theaflavin-type compounds were evaluated with TD-factors below 128, whereas the unknown compounds in fractions no. 30-34 were judged with TD factors of up to 8192 and, therefore, were expected to be of major relevance for tea taste. Isolation of the tastants from fractions no. 30-34 by means of polyamide chromatography and preparative RP-HPLC, followed by LC-MS/MS and 1D/2D-NMR spectroscopic structure determination led to the unequivocal identification of 14 highly taste-active flavon-3-ol glycosides (Table 2), amongst which quercetin-3-*O*-[α-L-rhamnopyranosyl-(1→6)-β-D-glucopyranoside] (rutin) was the most predominant [8].
In order to study the sensory impact of these tastants, the oral recognition thresholds were determined in water using the half-tongue test [8]. The oral sensation imparted by the catechins was described as astringent with threshold concentrations between 190 to 930 μmol/l, whereas the theaflavins induced a mouth-drying, rough-astringent sensation at threshold levels between 13 and 26 μmol/l (Table 2). In contrast, the flavon-3-ol glycosides induced a mouth-drying and velvety mouth-coating sensation at very low threshold concentrations spanning from 0.001 to 19.8 μmol/l (Table 2).

3.2. Quantification and calculation of dose-over-threshold (Dot)-factors

Aimed at evaluating their taste contribution, all the polyphenols, caffeine as well as theanine have been quantified in the tea infusion, and rated in their sensory impact based on the ratio of the concentration and the taste threshold of a compound [9]. Calculation of dose-over-threshold (Dot) factors revealed that from the group of catechins and theaflavins only the concentration of EGCG in the tea infusion exceeded its taste threshold concentrations for astringency by a factor of 1.7 (Table 2). In contrast, the concentration of the other catechins and theaflavins was found to be below their taste threshold concentration. Calculation of the Dot-factors for the flavonol-3 glycosides revealed high values for most of these glycosides, amongst which rutin showed the highest Dot-factor of 9652. Although theanine was present in a high concentration, its high threshold concentration of 6000 μmol/l (water) ruled out any taste contribution of this amino acid. In contrast, the bitter caffeine showed a Dot-factor of 2.0, thus demonstrating the alkaloid as a contributor to tea's bitterness (Table 2).
In order to confirm the results obtained so far, we prepared an aqueous taste reconstitute containing the 'natural' concentrations of caffeine as well as those flavonol-3-glycosides and catechins which have been evaluated with Dot-factors >0.5 (Table 2).
Theaflavins were not considered for these experiments, since recent investigations already excluded these compounds as key tastants [7,9]. The sensory analysis revealed that the taste profile of this recombinate did not differ significantly from that of the tea infusion (Table 1). In conclusion, the eight flavonol-3-glycosides evaluated with Dot-

factors >0.5, catechin, epigallocatechin-3-gallate, and caffeine have been successfully identified as the key taste compounds of the Darjeeling tea infusion [9].

Table 2. Thresholds, concentrations and dose-over-threshold (Dot) factors of tastants in black tea (Darjeeling Gold-Auslese).

Tastant	Threshold (μmol/l)	Conc. (μmol/l)	Dot-factor[a]
Group I: compounds imparting puckering astringency and rough oral sensations			
Epigallocatechin-3-gallate (EGCG)	190.0	328.0	1.7
Theaflavin	16.0	11.0	0.7
Catechin	410.0	221.0	0.5
Theaflavin-3,3'-digallate	13.0	6.7	0.5
Epicatechin-3-gallate	260.0	113.0	0.4
Theaflavin-3-gallate	15.0	6.4	0.4
Theaflavin-3'-gallate	15.0	4.3	0.3
Epigallocatechin	520.0	131.0	0.3
Gallocatechin	540.0	131.0	0.3
Epicatechin	930.0	84.0	0.1
Catechin-3-gallate	250.0	11.0	<0.1
Gallocatechin-3-gallate	390.0	11.0	<0.1
Theaflavic acid	24.0	0.009	<0.1
Group II: compounds imparting mouth-drying and velvety-like astringency			
Q-3-O-[α-L-rha-(1→6)-β-D-glc] (rutin)[b]	0.0015	11.1	9652.0
K-3-O-[α-L-rha-(1→6)-β-D-glc][b]	0.25	6.5	26.0
Q-3-O-β-D-gal[b]	0.43	5.4	12.6
Q-3-O-β-D-glc[b]	0.65	6.0	9.2
K-3-O-β-D-glc[b]	0.67	4.9	7.3
M-3-O-β-D-glc[b]	2.10	9.3	4.4
Q-3-O-[β-D-glc-(1→3)-O-α-L-rha-(1→6)-O-β-D-gal][b]	1.36	3.3	2.4
M-3-O-β–D-gal[b]	2.70	6.5	2.4
K-3-O-β–D-gal[b]	6.70	3.0	0.4
K-3-O-[β-D-glc-(1→3)-O-α-L-rha-(1→6)-O-β-D-glc][b]	19.80	8.6	0.4
Q-3-O-[β-D-glc-(1→3)-O-α-L-rha-(1→6)-O-β-D-glc][b]	18.40	7.1	0.4
A-8-C-[α-L-rha-(1→2)-β-D-glc][b]	2.80	0.9	0.3
M-3-O-[α-L-rha-(1→6)-β-D-glc][b]	10.50	2.2	0.2
K-3-O-[β-D-glc-(1→3)-O-α-L-rha-(1→6)-O-β-D-gal][b]	5.80	0.8	0.2
5N-Ethyl-L-glutamine (theanine)	6000.00	281.0	< 0.1
Group III: bitter tasting compounds			
Caffeine	500.0	990.0	2.0
Epigallocatechin-3-gallate (EGCG)	380.0	328.0	0.9

[a]The dose-over-threshold (Dot) factor is calculated as the ratio of concentration and taste threshold, [b]Abbreviations used for quercetin (Q), kaempferol (K), myricetin (M), apigenin (A), rhamnose (rha), glucose (glc), and galactose (gal).

3.3. Modulation of caffeine bitterness by flavonol-3-glycosides

Although the flavonol-3-glycosides do not exhibit any bitter taste on their own, our recent studies demonstrated that the omission of these compounds from the taste recombinate led to a reduction of the bitterness intensity by about 50% [9]. Aimed at investigating the molecular drivers for the bitterness of black tea, this prompted us to study whether flavanol-3-glycosides are able to modulate the bitter taste intensity of caffeine. Human dose-response functions were recorded for EGCG and caffeine as well as for a solution of caffeine containing rutin in its natural ratio of 90:1 (Figure 2).

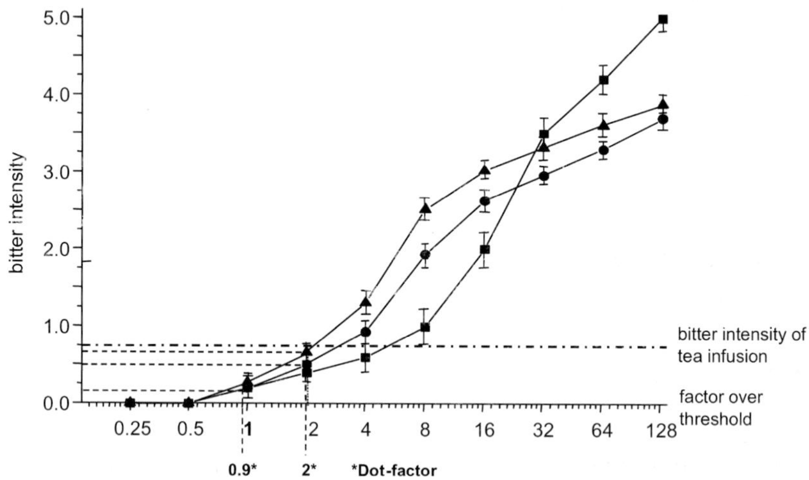

Figure 2. Human dose-response functions for bitterness of aqueous solutions of EGCG (■), caffeine (●), and caffeine + rutin (90/1; ▲).

Measurement of the bitter intensity of these solutions in 1:2 dilutions revealed that at a level of 128-fold over the threshold the EGCG exhibited the most intense bitter taste with a score of 5.0. However, at the Dot-factor of 0.9 determined for EGCG in tea, the bitter perception was just at the recognition threshold judged with an intensity score of <0.2. In comparison, the caffeine solution at the Dot-level (2.0) induced a bitter perception with an intensity of 0.45. In the presence of 'natural' amounts of rutin, the intensity was significantly increased from 0.45 to 0.70 (Figure 2). On the 5-point scale the bitterness of the caffeine/rutin solution nearly matched that of the tea infusion (0.75), thus demonstrating that the key bitter principles had been successfully identified.

References

1. E.A.H. Roberts and R.F. Smith, Analyst, 86 (1961) 94.
2. D.J. Millin, D.J. Crispin and D. Swain, J. Agric. Food Chem., 17 (1969) 717.
3. P.J. Hilton and R.Z. Ellis, J. Sci. Food Agric., 23 (1972) 227.
4. Z. Ding, S. Kuhr and U.H. Engelhardt, Z. Lebensmittel Untersuch Forsch., 195 (1992) 108.
5. K.H. Ekborg-Ott, A. Taylor and D.W. Armstrong, J. Sci. Food Agric., 45 (1997) 353.
6. O. Frank, H. Ottinger and T. Hofmann, J. Agric. Food Chem., 49 (2001) 231.
7. S. Scharbert, M. Jezussek and T. Hofmann, Eur. Food Res. Technol., 218 (2004) 442.
8. S. Scharbert, N. Holzmann and T. Hofmann, J. Agric. Food Chem., 52 (2004) 3498.
9. S. Scharbert and T. Hofmann, J. Agric. Food Chem., 53 (13) (2005) 5377.

W.L.P. Bredie and M.A. Petersen (Editors)
Flavour Science: Recent Advances and Trends

Evidence for antagonism between odorants at olfactory receptor binding in humans

G. Sanz, C. Schlegel, J.-C. Pernollet and L. Briand

Biochimie de l'Olfaction et de la Gustation, Neurobiologie de l'Olfaction et de la Prise Alimentaire, INRA, Domaine de Vilvert, Bâtiment 526, F 78352 Jouy-en-Josas Cedex France

ABSTRACT

The odorant repertoire of two human olfactory receptors (ORs) belonging to two major phylogenetic classes representing ORs from aquatic (class I) and terrestrial animals (class II) were elucidated. For this purpose, a new biomimetic screening assay based on calcium imaging on HEK293 cells expressing an OR and the promiscuous G protein $G_{\alpha 16}$ was developed. Class I OR52D1 is functional, exhibiting a narrow repertoire related to that of its orthologous murine OR, demonstrating that this class I OR is not an evolutionary relic. In contrast, class II OR1G1 was broadly tuned towards odorants of 9 to 10 carbons chain length, with diverse functional groups. The existence of antagonism between odorants at level of OR binding was demonstrated. OR1G1 antagonists were observed to be OR specific and structurally related to its agonists, with a shorter size.

1. INTRODUCTION

Humans are able to detect and discriminate myriads of structurally diverse odorants owing to approximately 350 olfactory receptors (ORs). Functional studies have demonstrated that odorant perception results in a combinatory code. One OR recognises multiple odorants and different odorants are recognised by different combinations of ORs [1,2]. Moreover, recent data revealed that, in addition to their agonist role, odorants could also inhibit or antagonise other ORs [3-5]. This dual agonist/antagonist combinatorial coding is in good agreement with psychophysical observations of mixture perception, designated as odour masking or counteraction phenomenon [6,7].

The human OR genes, as other mammals, have been classified according to two major phylogenetic classes named class I (fish-like) and class II (terrestrial-type) ORs. Class I ORs have been originally identified in fish and subsequently found in vertebrate species to be intermixed with class II ORs. Whereas human class I ORs have been first

suggested to be evolutionary relics [8,9], the pseudogene fraction among the human class I ORs (52%) is considerably lower than that observed for human class II ORs (77%) [10], suggested than human class I receptors could be functional.
Here we report the odorant repertoire of two human ORs. The class I OR52D1 was chosen because it is the orthologue of the known mouse OR S19, for which some agonists have been described [2], whereas the class II OR1G1 was studied because it has been proved to be expressed in olfactory epithelium [11].

2. MATERIALS AND METHODS

2.1. Vector constructions and cell culture

Human OR genes were amplified by PCR from human genomic DNA (Novagen) using gene-specific primers. In order to help ORs to translocate to the plasma membrane, OR52D1 and OR1G1 receptors were fused at their amino terminal end to the first 36 amino acids of bovine rhodopsin. HEK293 cells (Human Embryo Kidney cells) that stably express $G_{\alpha 16}$ and ORs were transfected and cultured as described elsewhere [12]. The G protein subunit, $G_{\alpha 16}$, was co-expressed because it has been shown to couple OR to intracellular Ca^{2+} release [13].

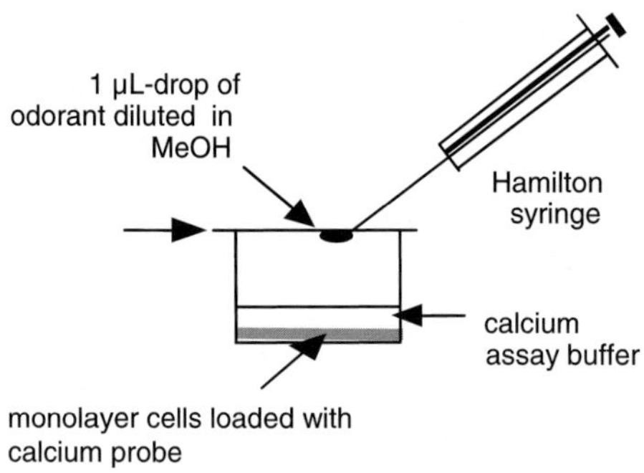

Figure 1. Cells seeded in a 96-well tissue-culture plate were loaded with Ca^{2+} sensitive Fluo-4 fluoprobe and covered with 60 µl of calcium assay buffer. Wells were sealed with a transparent adhesive plastic film. MeOH diluted odorant was applied using a 10 µl Hamilton syringe as a 1-µl drop hanging beneath the inner face of the plastic film. The MeOH drop evaporated freely in a few s leading to progressive stimulation of cells with the odorant. When antagonist odorants were screened, agonist and antagonist were mixed into MeOH and co-applied as a 1-µl drop.

2.2. Calcium imaging

HEK293 derivative cells were seeded onto a poly-L-lysine-coated 96-well tissue-culture plate. Twenty-four hours post-seeding, cells were washed and loaded 30 min at 37 °C with 2.5 µM of the Ca^{2+}-sensitive fluorescent dye Fluo-4 acetoxymethyl ester (Molecular Probes), as described [12]. Calcium imaging was carried out at 28 °C using an inverted epifluorescence microscope (CK40 Olympus) equipped with a digital camera (ORCA-ER, Hamamatsu Photonics).

2.3. Agonist and antagonist screening using VOFA

Agonist and antagonist screening was achieved using a new method of odorant application called volatile-odorant functional assay (VOFA) (Figure 1).

3. RESULTS

Using OR1G1/$G_{\alpha16}$ expressing cells, we tested 95 odorants individually by VOFA at a concentration of 10 μM in 1 μl-drop. We found that various odorants belonging to different chemical classes differently elicited OR1G1 Ca^{2+} responses. As illustrated in Figure 2, most active odorants are 8-, 9- and 10-carbon molecules, with an optimum for 9-carbon length. Among the 5 strong agonists, which exhibited aliphatic chains, we found 2 alcohols (2-ethyl-1-hexanol, 1-nonanol), 1 ester (ethyl isobutyrate), 1 lactone (γ-decalactone), and 1 aldehyde (nonanal). Medium agonists were thioesters, ketones, one aliphatic acid, and diverse cyclic molecules such as pyrazines or thiazoles.

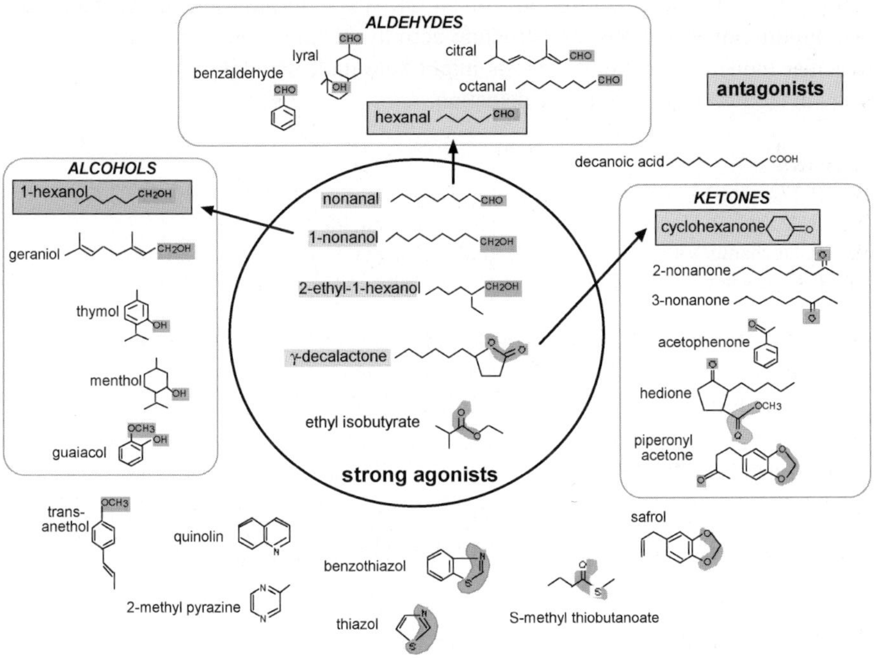

Figure 2. Structural comparison of OR1G1 ligands and antagonists. Strong agonists are located in the circle, while medium agonists are outside and gathered by chemical classes. Antagonists are shown in grey boxes. Arrows outline the chemical relationships between agonist and antagonist molecules. Peculiar structure features shared by active molecules are also highlighted in grey.

Studying OR1G1 responses, we observed that co-applications of equimolar odorant mixtures were less active than pure odorants applied at an identical concentration. By testing odorant couples, we revealed that OR1G1 antagonists were all 6-carbon molecules (Figure 2), with a functional group in common with agonists. In contrast to OR1G1, the fish-like OR52D1 receptor was observed to be functional with a more limited repertoire than the class II OR1G1, which was globally different from OR1G1 agonists (data not shown). Moreover, we observed that OR52D1 activation was not as size-dependent as observed with OR1G1, suggesting a different mode of interaction with its agonists.

4. DISCUSSION AND CONCLUSION

For the first time the odorant repertoire of two human ORs, belonging to different phylogenetic classes were identified. Interestingly, we found that fish-like OR52D1 odorant spectrum includes the reported agonists of its murine orthologous OR S19, which was shown to respond to C7 to C9 aliphatic acids and alcohols [2]. We also revealed antagonists against OR1G1 sharing common features with its agonists. While well documented in other G-protein coupled receptors, antagonists at receptor binding level were also recently reported for rodent ORs [3-5] and for human spermatozoa ORs [14]. Future investigations on structure-activity relationships of human ORs using molecular modelling and mutagenesis might help understanding how aroma perception occurs at the first level of sensory detection.

References

1. P. Ducamp-Viret, M.A. Chaput and A. Duchamp, Science, 284 (1999) 2171.
2. B. Malnic, J. Hirono, T. Sato and L.B. Buck, Cell, 96 (1999) 7137.
3. P. Duchamp-Viret, A. Duchamp and M.A. Chaput, Eur. J. Neurosci., 18 (2003) 2690.
4. R.C. Araneda, Z. Peterlin, X. Zhang, A. Chesler and S. Firestein, J. Physiol., 555 (2004) 743.
5. Y. Oka, M. Omura, H. Kataoka and K. Touhara, EMBO J., 23 (2004) 120.
6. D.G. Laing and G.W. Francis, Physiol. Behav., 46 (1989) 8.
7. J.E. Cometto-Muniz, W.S. Cain, M.H. Abraham and J.M. Gola, Physiol. Behav., 67 (1999) 269.
8. J.A. Buettner, G. Glusman, N. Ben-Arie, P. Ramos, D. Lancet and G.A. Evans, Genomics, 53 (1998) 56.
9. M. Bulger, J.H. van Doorninck, N. Saitoh, A. Telling, C. Farrell, M.A. Bender, G. Felsenfeld, R. Axel, M. Groudine and J.H. von Doorninck, Proc. Natl. Acad. Sci. USA, 96 (1999) 5129.
10. G. Glusman, I. Yanai, I. Rubin and D. Lancet, Genome Res., 11 (2001) 685.
11. V. Matarazzo, N. Zsurger, J.C. Guillemot, O. Clot-Faybesse, J.M. Botto, C. Dal Farra, M. Crowe, J. Demaille, J.P. Vincent, J. Mazella and C. Ronin, Chem. Senses, 27 (2002) 691.
12. G. Sanz, C. Schlegel, J.-C. Pernollet and L. Briand, Chem. Senses, 30 (2005) 69.
13. D. Krautwurst, K.W. Yau and R.R. Reed, Cell, 95 (1998) 917.
14. M. Spehr, G. Gisselmann, A. Poplawski, J.A. Riffel, C.H. Wetzel, R.K. Zimmer and H. Hatt, Science, 299 (2003) 2054.

W.L.P. Bredie and M.A. Petersen (Editors)
Flavour Science: Recent Advances and Trends

3D-QSAR study of ligands for a human olfactory receptor

Anne Tromelin[a], Guenhaël Sanz[b], Loïc Briand[b], Jean-Claude Pernollet[b] and Elisabeth Guichard[a]

*[a]UMRA INRA-ENESAD, 17 rue Sully, 21065 Dijon Cedex, France;
[b]Biochimie de l'Olfaction et de la Gustation, NOPA, INRA Domaine de Vilvert, 78352 Jouy-en-Josas Cedex, France*

ABSTRACT

Only about 350 olfactory receptors (OR) provide a basis for the remarkable ability of humans to recognise and discriminate a large number of odorants. A recent study reports the odorant repertoire of a human class II OR called OR1G1, including both agonists and antagonists. We used these affinity data to perform a 3D molecular modelling study of these ligands using Catalyst/HypoGen software (Catalyst v4.9.1, Accelrys Inc., San Diego, 2004) to propose alignment models for OR1G1 ligands. We obtained a triple-alignment model, which satisfactorily explained the experimental activities and was able both to predict the antagonist effects of some compounds and to identify new potent agonists, which were validated by direct experiments.

1. INTRODUCTION

Binding of an aroma molecule onto an olfactory receptor (OR) is the first step in the detection of odorants [1] and constitutes the basis of odour discrimination [2,3]. Until now, odorant repertoires have been reported for only very few ORs of human origin [4]. In a recent study, using calcium imaging and heterologous expression of a human OR (OR1G1) in HEK293 cells, both agonists and antagonists were identified and classified from a set of about one hundred odorants [5].

We used these affinity data to perform a molecular modelling study of these ligands using Catalyst/HypoGen software. Catalyst/HypoGen takes into account molecular flexibility by considering each compound as a collection of conformers. It generates models, named 'hypotheses', which describe ligands as sets of chemical functions. These hypotheses should be able to predict the activities of different compounds having the same receptor binding mechanism. In a previous work, we used Catalyst software to perform a 3D-QSAR study in order to describe the interaction between flavour compounds and beta-lactoglobulin. In this way, we separated the initial whole training set into two ligand groups that can be related to the existence of many different binding sites on beta-lactoglobulin [6]. In the present study, we employed our sorting-out approach with the aim to propose alignment models for OR1G1 ligands. The best predicted ligands were validated using functional expression of OR1G1 in mammalian cells.

2. MATERIALS AND METHODS

2.1. Biological assays

Odorants and ATP were purchased from Sigma-Aldrich, Fluka or Acros Organics (Noisy-le-Grand, France) at the highest purity available. Odorant solutions were prepared as 100 mM stocks in 100% MeOH (Spectroscopic grade, Sigma).
HEK293 cells (Human Embryo Kidney cells) were stably transfected with the $G_{\alpha16}$ protein and the human OR1G1 olfactory receptor (HEK293-$G_{\alpha16}$-OR1G1). HEK293 derivative cells were seeded onto a poly-L-lysine-coated 96-well plate at a density of $1x10^5$ cells per well. Twenty-four hours later, HEK293-$G_{\alpha16}$-OR1G1 cells were loaded with the Ca^{2+}-sensitive dye Fluo-4 (Molecular Probes) and stimulated during 10 min with different concentrations of OR1G1 agonists applied as vapour phase as previously described [5]. Calcium imaging was carried out at 28 °C using an inverted epifluorescence microscope (CK40 Olympus, Rungis, France) equipped with a digital camera (ORCA-ER, Hamamatsu Photonics, Massy, France). For antagonism studies, HEK293-$G_{\alpha16}$-OR1G1 cells were stimulated with mixtures of antagonist and strong agonist. For both agonist and antagonist activity measurement, the number of responding cells was normalised as percentage of the cells responding to 100 μM ATP.

2.2. Compounds and physicochemical data

We used activity data reported for the 95 compounds previously tested with HEK293-$G_{\alpha16}$-OR1G1 cells [5]. According to Catalyst calculation, we used the percent of non-responding cells normalised to responding cells as activity values, so that the 'Activity' equalled (100 - %cell) / %cell.

2.3. Computational Methods

The 95 compounds were built with Catalyst (Catalyst version 4.9.1 software, Accelrys Inc., San Diego, August 2004) running on a Silicon Graphics workstation (SGI-O_2). The conformers of each compound were generated using Catalyst/COMPARE module to provide the best conformational coverage and eliminate redundant conformers for a maximum number of generated conformers, defaulted to 250 in a 0-20 kcal/mol energy-range [7]. HypoGen module [8] and in some cases HypoRefine module (extension of the HypoGen algorithm that uses exclusion volumes), were used to perform automated hypothesis generation. Both HypoGen and HypoRefine automatically generated the simplest hypotheses that best correlated the estimated and experimental affinities. The statistical relevance of the various hypotheses was, therefore, assessed on the basis of their cost relative to the null hypothesis and the fixed hypothesis (the total costs should be as close as possible to the fixed cost) [9].

3. RESULTS

The first hypothesis generation, performed on the entire set, did not lead to a significant model (correl=0.2, total cost=2398, fixed cost=156, null cost=2475). Subsequently an iterative procedure to sort-out several subsets of ligands was performed. This procedure involved two steps. The first step consisted in dividing the whole initial set into subsets according to the alignment observations, and in performing hypothesis generation runs, following new selection of subsets based on alignment, and so on until obtaining statistically significant hypotheses. In this way, we obtained three core-subsets, containing 5 to 9 compounds respectively. Starting from these core-subsets, the second step consisted in building up larger groups. Hypothesis generation runs were performed on each core-subset. The obtained new hypotheses were tested on the remaining compounds, and the well-estimated molecules were added to the relevant core-subsets, in order to constitute new larger subsets. This procedure was repeated until either activity was correctly estimated in the remaining group. The subsets finally obtained were: I-33compounds (correl=0.94, total cost=154, fixed cost=65, null cost=885); II-11cmpds (correl=0.99, total cost=28, fixed cost=22, null cost=710); and III-23cmpds (correl=0.96, total cost=94, fixed cost=50, null cost=676). Hypothesis models were validated by leave-more-out procedure, removing the minimum initial core-subset of each group in order to generate new hypothesis models, which adequately satisfied to statistical requirements (suitable correl and cost values). Examination of alignments showed that vanillin and previously identified antagonists (1-hexanol, hexanal and cyclohexanone) were in very close positions (Figure 1).

(a)

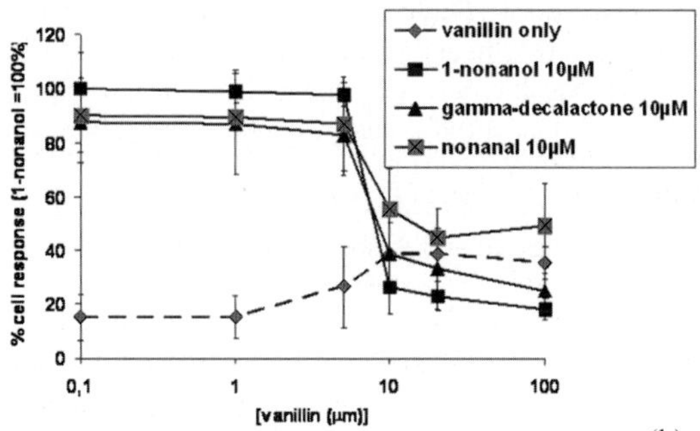

(b)

Figure 1. (a) Alignment of vanillin (dark grey) and three other antagonists: hexanal, 1-hexanol, cyclohexanone (light grey). (b) Percent of responding cells to binary mixtures of the antagonist vanillin (dashed line) with 1-nonanol, γ-decalactone and nonanal (agonists). Odorants were diluted in methanol and applied as a 1-µl drop, freely evaporating in the well containing the cells.

Analysis of HEK293-$G_{\alpha16}$-OR1G1 cell responses to binary mixtures of vanillin with each of the strong agonists, 1-nonanol, nonanal and γ-decalactone, demonstrated the antagonist activity of vanillin. Using the best significant hypotheses provided by the three groups I-33cmpds, II-11cmpds, and III-23cmpds, we performed a database screening on FlavorBase2004 [10] transposed in Catalyst environment.

Four compounds were selected as potential agonists: 2-methylundecanal, 3-methylthio-1-hexanol, tridecanal and 9-decen-1-ol. Preliminary results showed that HEK293-$G_{\alpha16}$-OR1G1 cell responses induced by 3-methylthio-1-hexanol and 2 methyl undecanal were lower than responses induced by 1-nonanol, whereas tridecanal and 9-decen-1-ol elicited stronger responses than 1-nonanol.

4. DISCUSSION AND CONCLUSION

Hypotheses generated on the whole set of 95 compounds failed to distinguish affinities and provided only inconsistent results. This suggests that the olfactory receptor OR1G1 could exhibit several binding sites. Indeed, the sorting-out selection of compounds led to three subsets, which permitted to obtain highly significant hypothesis models. The predictive power of these models was then attested by the experimental validation of vanillin as an antagonist, and also by the identification of several novel agonists. This approach allowed to propose some association of ligands and restricted choices of their conformers, which could greatly facilitate further docking tests through molecular modelling and suggest clues for the understanding of odorant mixture perception.

References

1. L. Buck and R. Axel, Cell, 65 (1991) 175.
2. D. Krautwurst, K.W. Yau and R.R. Reed, Cell, 95 (7) (1998) 917.
3. B. Malnic, J. Hirono, T. Sato and L.B. Buck, Cell, 96 (5) (1999) 713.
4. C.H. Wetzel, M. Oles, C. Wellerdieck, M. Kuczkowiak, G. Gisselmann and H. Hatt, J. Neurosci., 19 (17) (1999) 7426.
5. G. Sanz, C. Schlegel, J.C. Pernollet and L. Briand, Chem. Senses, 30 (1) (2005) 69.
6. A. Tromelin and E. Guichard, J. Agric. Food Chem., 51 (7) (2003) 1977.
7. A. Smellie, S.L. Teig and P. Tobwin, J. Comput. Chem., 16 (2) (1995) 171.
8. O.F. Güner (ed.), Pharmacophore perception, development and use in drug design, La Jolla, CA (2000) 171.
9. Y. Kurogi and O.F. Guner, Curr. Med. Chem., 8 (9) (2001) 1035.
10. Leffingwell and Associates (as accessed from www.leffingwell.com).

W.L.P. Bredie and M.A. Petersen (Editors)
Flavour Science: Recent Advances and Trends

Effect of physiology and physical chemistry on aroma delivery and perception

Andrew J. Taylor, Kris S.-K. Pearson, Mike D. Hodgson, James P. Langridge and Robert S.T. Linforth

Division of Food Sciences, School of Biosciences, University of Nottingham, Sutton Bonington Campus, Loughborough, LE12 5RD, UK

ABSTRACT

To perceive aroma compounds, they must be delivered to the olfactory receptors, a process that depends on physical chemistry and physiology. The role of these two factors was studied using a variety of delivery mechanisms, ranging from gas phase delivery to the location and form of the aroma compounds in the food matrix. The effect of mouth loading and body position was also investigated. Delivery to the olfactory receptors was monitored *in vivo* using atmospheric pressure chemical ionisation-mass spectrometry (APCI-MS) and physiological responses were monitored simultaneously. It was found that body position and mouth loading affected the opening of the velum while common aroma solvents had a significant effect on the delivery of hydrophobic compounds. Aroma delivery from droplets of pure compounds suspended in the food matrix was very intense and could not be explained by conventional partition mechanisms. A delivery mechanism based on direct volatilisation from the air-liquid interface was proposed to explain this intense release which is also seen from some commercial encapsulated materials.

1. INTRODUCTION

The various steps involved in aroma perception have been studied extensively and there is an in depth understanding of key areas like mastication, aroma release and aroma receptor structure. The challenge now is to build this knowledge into a ‘systems biology’ approach where we try to understand how the component parts of the systems work together to generate the sensation we recognise as flavour perception. The shape of the aroma profile delivered to the olfactory receptors is thought to influence the way that an aroma is perceived [1]. However, exactly how mouth/nose physiology [2] and the physical chemistry of release affect the aroma profile is not yet clear. One of the difficulties in this type of work, is the introduction of aromas under controlled conditions [3]. Intubation to deliver aroma compounds directly into the throat is one approach [4]. We have modified the environment in the food matrix to alter the release properties of the aroma compounds as well as using pulsed delivery into the mouth (via an olfactometer). In this paper, the role of the velum (which closes off the mouth from the throat), the location and physical form of the aroma compounds in foods and the role of the throat in aroma persistence are studied using *in vivo* techniques. Aroma delivery is by the lesser-studied retronasal route and the results show the interplay between physical chemistry, physiology and perception.

2. MATERIALS AND METHODS

Mastication was measured by electromyography, swallowing by electroglottography, nasal air flow via a turbine flowmeter and aroma delivery was measured *in vivo* using APCI-MS (see [5] for details). Layered chewing gum samples were prepared by rolling two layers of gum, one containing sucrose, the other a commercial strawberry aroma preparation and sucrose and then folding and re-rolling the samples to form 2, 4, 8, or 16 layers. Gelatin (6%)-sucrose gels contained one of 16 aroma compounds representing hydrophilic and hydrophobic compounds (125 mg/kg) added either as a solution in propylene glycol (0.2 g/kg in final product) or as droplets, injected into the gel centre (0.25, 0.5, 1.0 μl) just before it set. The aroma content of each set of gels was measured and the amounts were the same (within experimental error). Samples were eaten by 4 panellists (chew for 45 s then swallow), and aroma delivery to the nose measured by APCI-MS for a total time of 60 s. Sensory analysis (30 panellists; paired comparison test) of some samples was undertaken.

3. RESULTS

3.1. Role of velum in aroma delivery

The velum can seal the mouth to throat opening and, in the closed position prevents aroma transfer from mouth to nose. The effect of mouth liquid volume on the position of the velum was studied by pumping liquid containing an aroma compound into panellist's mouths who were chewing un-flavoured gum to facilitate mouth to throat aroma transfer and who were instructed not to swallow. By monitoring the presence of aroma in nasal air (using APCI-MS) the point at which the velum closed could be measured as a function of liquid volume in mouth. Typical values for 5 panellists were 7.6 ± 3.0 ml. Individuals vary in their velum behaviour due to variations in anatomy and their method of eating or drinking.

The position of the velum is also important during brain imaging experiments where the subject is usually supine. Small volumes of liquid are usually presented to the mouth to prevent discomfort to the subjects. Experiments using a typical fMRI paradigm showed that the velum was closed, presumably because subjects cannot swallow during the fMRI process and they have to stop saliva flowing into the lungs. Significant aroma transfer from the mouth only occurred when subjects swallowed the liquid.

3.2. Role of mouth in aroma delivery

3.2.1. Physiology

It is well established that the chewing action of the mouth pumps aliquots of air backwards and forwards between the mouth and throat [6]. This is manifested by the presence of small spikes of aroma which correlate with chewing events, measured by EMG [6]. The effect of mouth movements on gas phase mixing was investigated by pumping an air flow into the mouth which contained a known amount of an aroma (isopentyl acetate) either as a constant concentration or as pulses of aroma (periodicity 1 s). Panellists chewed unflavoured gum while receiving a 30 s constant stimulus, followed by 30 s of pulsed delivery and then 30 s of constant delivery. The exhaled breath was monitored for evidence of this pattern but none was seen, in contrast to orthonasal introduction of the same pattern where patterns could be identified both by in nose monitoring and by sensory time-intensity measurements [7].
The effect of mastication on mixing aromas in the food matrix was studied by asking panellists to chew layered gum which contained the same total amount of aroma but distributed in 1, 2, 4 or 8 layers (the other layers were made from sweetened but non-aromatised gum. There were no significant differences between aroma delivery from the different samples, showing that the mouth efficiently mixed the aromas from the layers in mouth and delivered a 'smoothed' profile to the nose.

3.2.2. Physical chemistry

The effect of commonly used flavour solvents (water, ethanol and propylene glycol (PG)) on aroma release was studied using aroma chemicals with physical properties that covered the hydrophobic/hydrophilic and volatility range. Aromas were incorporated into gelatin sucrose confectionery gels, eaten by panellists and aroma release monitored with APCI-MS. As expected, the *in vivo* release of hydrophilic compounds was not affected by ethanol or PG, but significant effects were noted with hydrophobic compounds.
Some of the aroma compounds were incorporated into the gels as droplets of pure compound and *in vivo* release compared with gels containing solubilised aroma. Release was dramatically different with the droplets producing very intense, short time release and the conventional gels producing much lower and more prolonged release. There was a highly significant difference in sensory intensity between the gel types. The differences between the two types of release was expressed as the ratio of the *in vivo* release peaks and values varied from 4 (hydrophilic, low vapour pressure compounds) to 2400 (hydrophobic, high vapour pressure compounds). Video studies on droplets in water demonstrated rapid movement to the air-liquid interface followed by rapid volatilisation. This observation suggests that aroma release in this case does not involve partition and also explains the 'burst' of release seen from aromas encapsulated in glass like materials where the aroma is also present as droplets in a matrix.

3.3. Persistence from the throat

Results from several research groups [8-12] has concluded that the coating of food on the throat after swallowing is a major source of aroma persistence as tidal air passes over the thin coating of food on the throat. The deposition of the food is physiological but release is governed by physical chemistry. Depletion is exponential and dependent on the aroma compound's physicochemical properties. Normand *et al.* [10] have modelled the process mathematically and tested the model experimentally.

4. DISCUSSION AND CONCLUSIONS

The results above coupled with the information available in the literature, demonstrate a basic understanding of the interaction of physiology and physical chemistry on aroma release *in vivo*. For some situations, physiology is dominant, in others physical chemistry is the driving force. The notion that release is always partition driven has been shown to be incorrect and this may explain some anomalies in the literature relating to flavour release modelling. Further work with larger numbers of panellists is needed to measure the physiological variations in the population from which we could determine whether variation followed a normal distribution or existed as 'clusters' of behaviour.

References

1. I. Baek, R.S.T Linforth, A. Blake and A.J. Taylor, Chem. Senses, 24 (1999) 155.
2. A. Buettner, A. Beer, C. Hannig, M. Settles and P. Schieberle, Food Quality Prefer., 13 (2002) 497.
3. J. Hort and T.A. Hollowood, J. Agric. Food Chem., 52 (2004) 4834.
4. T. Hummel, S. Heilman, B.N. Landis, J. Rededn and J. Frasnelli, Flavour Fragrance J., (2005) in press.
5. M. Hodgson, R.S.T. Linforth and A.J. Taylor, J. Agric. Food Chem., 51 (2003) 5052.
6. J.-L. Le Quéré and P.X. Étiévant (eds.), Flavour research at the dawn of the 21st century, proceedings of the 10th Weurman flavour research symposium, Paris, France (2003) 143.
7. J.P. Langridge, PhD Thesis, University of Nottingham, UK (2004).
8. R.S.T. Linforth and A.J. Taylor, J. Agric. Food Chem., 48 (2000) 5419.
9. A. Buettner and P. Schieberle, Food Chem., 71 (2000) 347.
10. V. Normand, S. Avison and A. Parker, Chem. Senses, 29 (2004) 235.
11. K.M. Wright, B.P. Hills, T.A. Hollowood, R.S.T. Linforth and A.J. Taylor, Int. J. Food Sci. Technol., 38 (2003) 343.
12. M. Hodgson, A. Parker, R.S.T. Linforth and A.J. Taylor, Flavour Fragrance J., 19 (2004) 470.

Structure-activity relationships of trigeminal effects for artificial and naturally occurring alkamides related to spilanthol

Jakob P. Ley, Gerhard Krammer, Jan Looft, Gerald Reinders and Heinz-Jürgen Bertram

Corporate Research & Development, Symrise GmbH & Co. KG, P.O. Box 1253, 37601 Holzminden, Germany

ABSTRACT

Twenty six variations of the structure of spilanthol ((*E*,*E*,*Z*)-2,6,8-decatrienoic acid *N*-isobutylamide) were synthesised starting from the appropriate unsaturated acids by amidation with simple aliphatic amines. The resulting alkamides were evaluated for their trigeminal sensory properties (burning, pungency, tingling, scratching, numbing, warming, mouth-watering, cooling). Spilanthol was the most active tingling and mouth-watering compound. *trans*-Pellitorine ((*E*,*E*)-2,4-decadienoic acid *N*-isobutylamide) showed mainly the same mouth-watering activity without the strong tingling effect. All other structural variations led to lower intensity of the trigeminal activity.

1. INTRODUCTION

The alkamide spilanthol ((*E*,*E*,*Z*)-2,6,8-decatrienoic acid *N*-isobutylamide), which occurs in several plants of the Compositae family, e.g. in *Spilanthes acmella*, shows a pronounced tingling and mouth-watering effect on ingestion [1]. For shogaols (2-hydroxyisobutylamides of various unsaturated fatty acids) and some other alkamides, the structure-activity of the tingling effect has been described earlier [2,3]. In a further study it was shown that the stereochemistry of the fatty acid part strongly influences the sensory properties [4]. To our knowledge, no study exists covering all important trigeminal effects on natural and synthesised alkamides.

2. MATERIALS AND METHODS

2.1. Isolation and syntheses

Spilanthol (**1**) and homospilanthol (**11**) were isolated from *Spilanthes acmella* concentrated extract (Jambu oleoresin, Robertet Flavors) by preparative HPLC. Most other amides were synthesised by amidation of the appropriate acid chloride with an amine. *cis*-Pellitorine (**9**) and derivatives **16** and **17** were synthesised by an enzymatic conversion of pear ester according to the literature [4]. The acid for dihydrospilanthol (**2**) was synthesised by a Wittig-sequence starting from acrolein and 6-bromohexanoic acid and subsequent treatment with oxalyl chloride. The acid for amide **3** was prepared by a Wittig-Horner reaction of (*Z*)-5-octenal with triethylphosphonoacetate with subsequent saponification in KOH. (*E,E*)-2,4-Decadienoic acid, (*E*)-2-decen-4-ynoic acid and (*E,E*)-2,4-undecadienoic acid were synthesised by the same procedure starting from (*E*)-2-octenal, 2-octinal and (*E*)-2-nonenal, respectively. (*E*)-2-Decenoic acid and (*E*)-3-nonenoic acid were prepared from the commercial available ethyl esters by saponification. The amides **1**, **2**, **4**, **6**, **7**, **9**, **11**, **12**, **14**, **18**, **19**, **20**, and **21** were found in nature [2,3,5].

2.1.1. Amidation

In a typical amidation reaction 10 g (50 mmol) of (*E*)-2-decenoic acid chloride (prepared from the free acid with $SOCl_2$) were dissolved in 10 ml acetone and added to a solution of 60 mmol amine in 10 ml acetone and 25 ml NaOH (9.64 g NaOH/100 ml water). In most cases a crystalline product could be obtained. If the amide separated as an oily product, a further chromatographic clean-up was performed. All amides were purified to >95% purity (GC) and fully characterised by NMR, MS and in the case of new compounds by HRMS.

2.2. Sensory testing

The test compound was dissolved in ethanol (30 or 10 mg/5 ml) and 1 ml of the solution further diluted with 11% sucrose in water (200 ml). For evaluation, the test solutions (2 to 5 ml) were sipped for 10 to 20 s and then spat out. Tasting sessions were performed by 6 to 8 fully informed and trained panellists. No more than 1 or 2 test samples were evaluated per day. Estimated intensity ratings for descriptors were 1 (low) to 9 (high).

3. RESULTS

Most of the 26 evaluated alkamides (Tables 1 and 2) showed at least a moderate trigeminal sensation, especially the tingling effect. Spilanthol (**1**) was the most active tingling and mouth-watering compound. *trans*-Pellitorine (**7**) showed mainly the same mouth-watering activity without the strong tingling effect of **1**. (*E*)-2-Decen-4-ynoic acid *N*-isobutylamide (dehydropellitorine, **10**) caused a moderate tingling and mouth-watering effect. (*E*)-2-Decenoic acid *N*-isobutylamide (dihydropellitorine, **4**) induced saliva flow but showed only a very weak tingling effect. The mouth-watering effect was strongly affected by small variations of the amide moiety as well as by the fatty acid residue. *cis*-Pellitorine ((*E*,*Z*)-2,4-decadienoic acid *N*-isobutylamide, **9**) showed high pungency and a warming effect, but no mouth-watering and tingling at all.

Table 1. Sensory profiles of N-isobutylamides and other alkamides in 11% sucrose solution.

	Structure	Name	Conc. (ppm)	Burning	Pungency	Tingling	Scratching	Numbing	Warming	Mouth wat.	Cooling
1	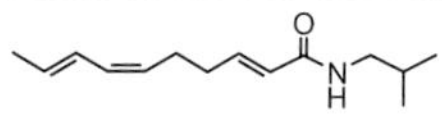	Spilanthol	30			8				7	
2	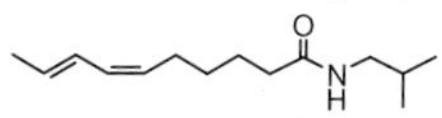	Dihydrospilanthol	10			2					
3	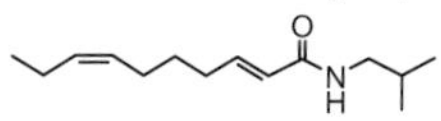	(*E*,*Z*)-2,7-Decadienoic acid *N*-isobutylamide	10	n.a.	n.a.	n.a.	n.a.	n.a.	n.a.	n.a.	n.a.
4	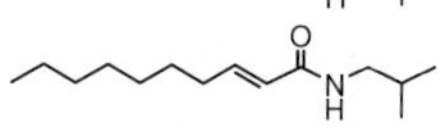	(*E*)-2-Decadienoic acid *N*-isobutylamide	10		2				4	5	
5	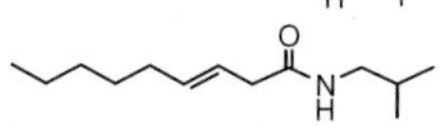	(*E*)-3-Nonadienoic acid *N*-isobutylamide	10					3			
6	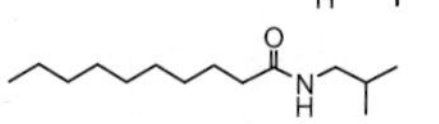	Decanoic acid *N*-isobutylamide	10			2					
7	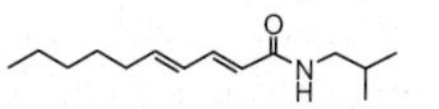	*trans*-Pellitorine	10		2	5		4		7	
8	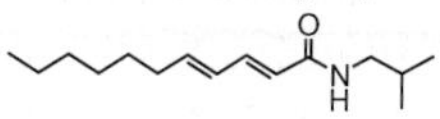	(*E*,*E*)-2,4-Undecadienoic acid *N*-isobutylamide	10					3			
9	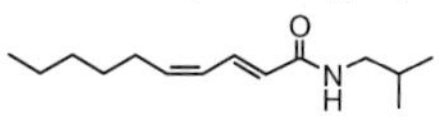	*Cis*-Pellitorine	10		5				5		
10	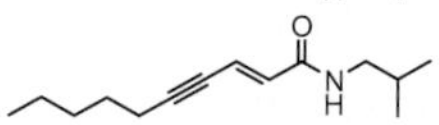	Dehydropellitorine	10			4		4		4	4
11	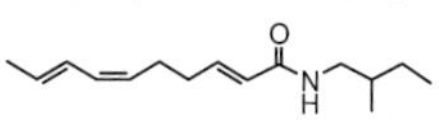	Homospilanthol	30			5		5			
12	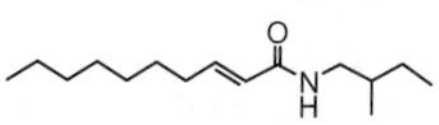	(*E*)-2-Decenoic acid *N*-(2-methylbutyl)amide	10				3	3		3	
13	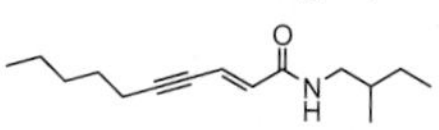	Dehydrohomopellitorine	10			4		3		4	4
14	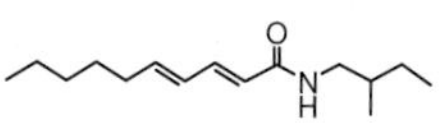	Homopellitorine	10		2	3				4	
15	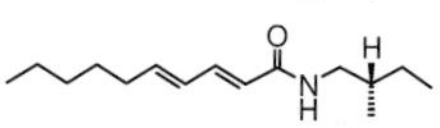	(2'*S*)-Homopellitorine	10			4		3			
16	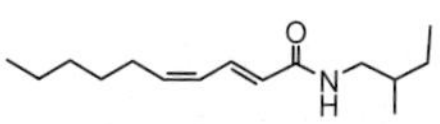	*Cis*-Homopellitorine	10	n.a.	n.a.	n.a.	n.a.	n.a.	n.a.	n.a	n.a.
17	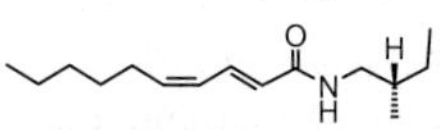	(2'*S*)-*cis*-Homopellitorine	10			3					

n.a.: not available.

Table 2. Sensory profiles of further alkamides in 11% sucrose solution.

	Structure	Name	Conc. (ppm)	Burning	Pungency	Tingling	Scratching	Numbing	Warming	Mouth wat.	Cooling
18	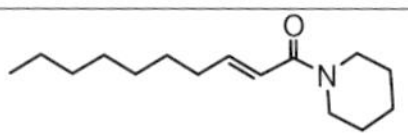	Dihydroacchilleamide	10	3	3	4					3
19	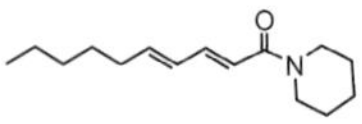	Achilleamide	10		3	5					
20	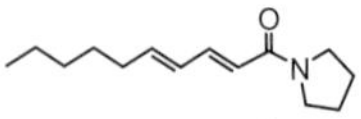	Sarmentine	10			3		4			
21	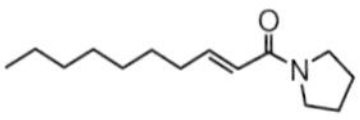	(*E*)-2-Decenoyl *N*-pyrrolidine	10					3			
22	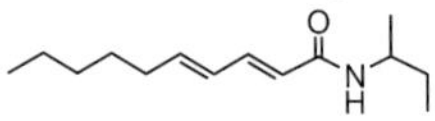	(*E*,*E*)-2,4-Decadienoic acid *N*-1'-methylpropylamide	10							4	
23	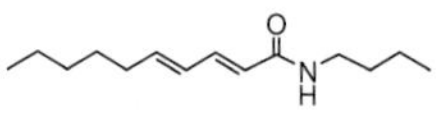	(*E*,*E*)-2,4-Decadienoic acid *N*-butylamide	10		2			3		3	
24	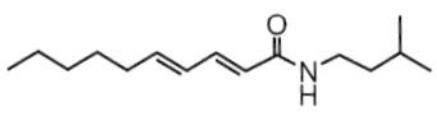	(*E*,*E*)-2,4-Decadienoic acid *N*-(3-Methylbutyl)amide	10			4				3	
25	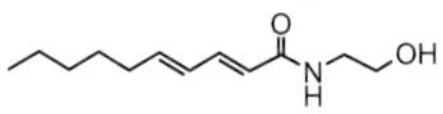	(*E*,*E*)-2,4-Decadienoic acid *N*-(2-hydroxyethyl)amide	10			4		4		4	
26	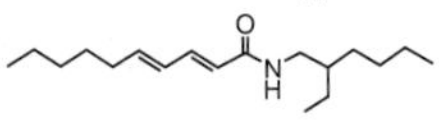	(*E*,*E*)-2,4-Decadienoic acid *N*-(2-ethylhexyl)amide	10			4		4		2	3

4. DISCUSSIONS AND CONCLUSION

The *trans* 2 double bond is necessary for the mouth-watering property, as stated earlier by Galoph *et al.* [2] and as exemplified in our study by variation **1** to **2**. The change of diene to monoene moiety (**1** to **3**) resulted in a loss of this physiologic activity. Variations in the double bond pattern between C2 and C5 (**4** to **10**) yielded interesting trigeminal activities, but the most active compounds besides spilanthol (**1**) were the 2,4-di-unsaturated amides **7**, **9** and **10**. Changing the amine moiety always causes a decrease and sometimes a change in the pattern of trigeminal activity compared to the appropriate *N*-isobutylamide.

References

1. R.S. Ramsewak, A.J. Erickson and M.G. Nair, Phytochem., 51 (1999) 729.
2. T. Hofmann, C.-H. Ho and W. Pickenhagen (eds.), Challenges in taste chemistry and biology, Washington, DC, USA (2003) 139.
3. P. Given and D. Paredes (eds.), Chemistry of taste, Washington, DC, USA (2002) 202.
4. J.P. Ley, J.-M. Hilmer, B. Weber, G. Krammer, I.L. Gatfield and H.-J. Bertram, Eur. J. Org. Chem., 24 (2004) 5135.
5. G.M. Strunz, Stud. Nat. Prod. Chem., 24 (2000) 683.

W.L.P. Bredie and M.A. Petersen (Editors)
Flavour Science: Recent Advances and Trends

Using gas chromatography-olfactometry (GCO) to measure varying odorant-specific sensory deficits (OSDs)

Katherine M. Kittel and Terry E. Acree

Department of Food Science and Technology, Cornell University, New York State Agricultural Experiment Station, 630 West North Street, Geneva, NY 14456 USA

ABSTRACT

Applications of flavour chemistry rely on the idea that people perceive flavour similarly. Odorant-specific sensory deficits (OSDs) are a severe challenge to this notion, and their effects on flavour are not fully understood. OSDs have been studied since Amoore's formative work on specific anosmia [1]. These differences in odorant thresholds for individuals with an otherwise functioning sense of smell can be divided into two categories: stable and variable OSDs. Stable OSDs are likely due to genetically determined components of the transduction system, while variable OSDs are consistent with changes in either peripheral expression or central processing. Variable OSDs are characterised by a subjects' increase in sensitivity to some compounds with time or exposure. This is a well documented phenomenon in humans and other mammals that makes it difficult to measure stable OSDs at a single time [2-5]. The use of gas chromatography-olfactometry (GCO) to investigate the phenomenon of variable OSDs will be discussed.

1. INTRODUCTION

Varying odorant-specific sensory deficits were observed by Wysocki *et al.* [2], who were able to induce the ability to perceive an odorant in initially insensitive subjects. This led to a proposed categorisation of three populations of people: the truly anosmic, the inducible, and the constitutionally sensitive. Similarly, a study by Cain and Schmidt [3] separated 30 subjects from a threshold study of glutaraldehyde into three groups, which they termed 'Non-anosmics', 'Initial anosmics', and 'Anosmics'. The initial anosmics showed a dramatic increase in sensitivity over the course of the experiment, and by the end of the study had the same sensitivity as had the initial non-anosmic group. The effect of time was examined in a related experiment, where two populations of subjects that differed in their detection threshold for *l*-carvone were retested after 70 days, and all but one of the subjects in the anosmic group improved [4]. The effect of exposure was also demonstrated by Dalton *et al.* [5], who observed a female-specific improvement in sensitivity with continuing repeated exposures of an odorant, seemingly without an end. However, these studies required many exposures to the compound in question before a threshold could be determined.

2. MATERIALS AND METHODS

2.1. Experimental design

Five subjects (4 females, 1 male; aged 23 to 61), all gas GCO novices, were tested using a 21 odorant probe in 8 sessions over 5 consecutive days. The odorant probe was composed of stimulants with a variety of chemical functional groups and produced a broad range of perceptions [6,7]. Compounds and levels in the set were modified from [8] (Table 1), and prepared in Freon 113™. Two concentration levels were used: one near threshold level (sessions 1-6, 8) and one at an increased level (81 fold) (session 7). Two internal controls, ethyl 2-methylbutyrate and *o*-aminoacetophenone, were held constant at a high level for all sessions (15 and 0.59 pMolar, respectively). Each session lasted 20 min, with a maximum of 4 sessions per day. The last two sessions (7 and 8) were held 20 min apart. The subjects were asked to signal when they detected the beginning and end of an eluting odorant, and to rate the intensity of the odorant on a category scale from 1 (weak) to 9 (strong) [9].

2.2. GCO analysis

GCO protocols used an Agilent 6890 GC (Agilent Technologies, Palo Alto, CA) modified GCO, equipped with a 30 m x 0.32 mm HP-5 column. Data was collected using PC Charmware™ (Datu Inc., Geneva, NY) software for ChemStation (Agilent Technologies). Compound response data was analysed by comparison to the known elution window. Final response and intensity data was measured against the individual central tendencies of the first six sessions.

3. RESULTS

The detection responses of the subjects were classified into anosmic, initial anosmic, and non-anosmic groups (Table 1). If the subject was unable to detect the odorant, or was only able to detect the odorant at the high dose presented in session 7, their response was categorised as anosmic. If the subject did not initially detect the odorant, but did detect the odorant during the course of the experiment, their response was categorised as initial anosmic. Non-anosmic responses detected the odorant in 6 or more sessions. If there was not enough data to classify a response into one of these categories, it was left blank. All subjects were able to detect the internal controls. Some of the non-anosmic response intensities increased with exposure, and these, along with the initial anosmic responses, may demonstrate the same increase in subjects' sensitivity as seen in Cain and Schmidt [3], indicating the presence of variable OSDs. Although the subjects showed variations in detection response, there was a net increase in compound detection over the experiment that indicated a general increase in sensitivity.

Table 1. Compounds used in the test sets with classification of the sensitivity of subjects.

Compound	Dose[a] sessions 1-6, 8	Dose session 7	S1[b]	S2	S3	S4	S5
Isoborneol	0.024	1.9	A[c]	A	A	A	A
2-Acetylpyridine	2	170	A	A	A	A	A
Phenylacetic acid	0.18	15	I	A		A	A
Isovaleric acid	3.9	320	I	A		A	A
Benzaldehyde	340	32000		A	A	A	I
Hexanal	74	6000		A	A	N	A
Benzothiazole	6.4	520	A	I		I	A
1-Octen-3-one	4.5	360	A	A		I	I
(*Z*)-4-Heptenal	0.33	27	A	N	A	I	I
2-*sec*-Butyl-3-methoxypyrazine	0.015	1.2	A	N	I	N	A
Eugenol	0.15	12	A	N	I	N	A
Phenylacetaldehyde	2	170	A	A	I	N	N
Octanal	14	1200	I	N	A	N	
Butyl acetate	530	43000	N		N	A	
Acetophenone	86	7000		I	I	N	N
l-Carvone	26	2100	N	N		N	N
β-Damascenone	0.022	1.8	N	N	N	N	
Linalool	2	170	N	N	N	N	N
cis-Rose oxide	4.8	400	N	N	N	N	N

[a]Concentration (pMolar). [b]Subject number. [c]Subject classifications for each odorant: A: anosmic; I: initial anosmic; N: non-anosmic.

4. DISCUSSION AND CONCLUSION

A difficulty with threshold studies is the inherent required exposure to a stimulus. Studies have shown that repeated exposure to a stimulus can lower thresholds [2-5]. Essentially, these studies cannot escape changing their subjects during the course of the experiment. This study indicates that GCO is a valuable tool for investigating the phenomena of stable and variable OSDs, while delivering smaller exposures to the subjects than traditional threshold determination methods.

References

1. J.E. Amoore, Nature, 214 (1967) 1095.
2. C.J. Wysocki, K.M. Dorries and G.K. Beauchamp, Proc. Nat. Acad. Sci. USA, 86 (20) (1989) 7976.
3. W.S. Cain and R. Schmidt, Chem. Senses, 27 (2002) 425.
4. H. Lawless, C. Thomas and M. Johnston, Chem. Senses, 20 (1) (1995) 9.
5. P. Dalton, N. Doolittle and P.A. Breslin, Nature Neurosci., 5 (3) (2002) 199.
6. E.T. Contis, C.-T. Ho, C.J. Mussinan, T.H. Parliment, F. Shahidi and A.M. Spanier (eds.), Food flavors: formation, analysis and packaging influences, Amsterdam, The Netherlands (1998) 27.
7. G.V. Civille and B.G. Lyon (eds.), Aroma and flavor lexicon for sensory evaluation: terms, definitions, references, and examples, ASTM International, West Conshohocken, Penn. (1996).
8. J.E. Friedrich, Thesis, Dept. Food Science and Technology, Geneva, NY, Cornell University (2000) 121.
9. H.T. Lawless and G.J. Malone, J. Sens. Stud., 1 (1986) 85.

W.L.P. Bredie and M.A. Petersen (Editors)
Flavour Science: Recent Advances and Trends

Volatile compounds of Wagyu (Japanese black cattle) beef analysed by PTR-MS

Sachiko Odake[a], Tomoko Shimamura[a], Ryozo Akuzawa[a], Akio Shimono[b] and Saskia M. Van Ruth[c]

[a]Department of Food Science and Technology, Nippon Veterinary and Animal Science University, Musashino, Tokyo, 180-0022, Japan; [b]Sanyu Plant Service CO., LTD, Sagamihara, Kanagawa, 229-1132, Japan; [c]Department of Food and Nutritional Science, University College Cork, Western Road, Cork, Ireland

ABSTRACT

Volatile compounds of highly marbled and cooked Wagyu (Japanese Black cattle) meat were analysed during mastication using proton transfer reaction mass spectrometry (PTR-MS) by collecting breath from the oral and nasal cavities. Twelve compounds were monitored. Ethyl acetate showed the highest ion counts in the oral cavity, followed by acetic acid, butanol, pentanol, and diacetyl. The counts of all lactones were very low. Ethyl acetate, butanol, and hexanol were also detected in the nasal cavity. Principal component analysis revealed that the release degree related to logP values and did not depend on the amount of aroma in the solvent extraction.

1. INTRODUCTION

Proton transfer reaction mass spectrometry (PTR-MS) has been reported as a very useful tool for simultaneously monitoring volatile organic compounds in the gas phase at very low concentrations [1]. Retronasal release of volatile compounds from red kidney beans [2] and milk of various fat contents [3] have been reported. Wagyu (Japanese black cattle) beef is famous for its high content of marble fat. The Japanese enjoy the aroma of cooked thinly sliced Wagyu beef. In this study PTR-MS was employed to investigate the released aroma volatiles of cooked Wagyu beef during mastication.

2. MATERIALS AND METHODS

2.1. Meat samples

Wagyu (Japanese black cattle) beef, the most marbled rank (A-5) produced in Matsusaka area, was purchased from a local meat shop. Thinly sliced beef of loin, 10 cm x 5 cm x 0.2 cm, was cooked for 10 s 80 °C water (named *Shabu-shabu* style in Japan). The sample was composed of water 31.8% (measured by drying at 105 °C), protein 10.3% (Kjeldahl method), and fat 52.44% (Soxhlet extraction method).

2.2. Breath analysis by PTR-MS

Breath air from the oral and nasal cavities was transferred from a single panellist to PTR-MS (Ionicon Analytik, Innsbruck, Austria) during mastication and four replicates were conducted. Twelve ions (*m/z* 57, 61, 71, 85, 87, 89, 101, 107, 111, 157, 171, and 185) were collected at a constant drift voltage of 540 V. Dwell time differed from 0.02 s to 1 s according to ions: 0.02 s per mass for *m/z* 57, 61, 71, 85, 87, and 89; 0.1 s for *m/z* 101, 107, and 111; 1 s for *m/z* 157, 171 and 185. Data collection was conducted ten times and the average value was calculated.

3. RESULTS

3.1. Ion counts measured by PTR-MS

Twelve compounds, which were reported as Wagyu beef solvent extraction aroma [4], were monitored by PTR-MS: 1-butanol (*m/z* 57), acetic acid (61), 1-pentanol (71), 1-hexanol (85), diacetyl (87), ethyl acetate (89), hexanal (101), ethylbenzene (107), 7-octen-4-ol (111), and three lactones (157, 171, and 185). The amounts of the ion counts were different for each of the compounds, arranged into three groups, and showed no dependency on the amount in the solvent extraction data [4] (Figure 1). Ethyl acetate showed the highest ion counts in the oral cavity, followed by acetic acid, butanol, pentanol, and diacetyl. Although hexanal occupied the most (half of the amount in the solvent extraction aroma), the degree of the released amount in the oral cavity was at the medium level. Ethylbenzene, hexanol, and 7-octen-4-ol belong to the medium group. Lactones whose amounts in the solvent extraction were small released at a very low level in the oral cavity. Hexanol, diacetyl, pentanol, butanol, and ethyl acetate were also detected in the nasal cavity, and the released amount of each compound was less than that in the oral cavity. Lactones (*m/z* 171 and 185) were also detected in the nasal cavity, but the level was quite low and below the limit of a signal-to-noise ratio [3].

3.2. Principal components analysis

Principal components analysis (PCA) on the released amount with loading LogP showed that the released amount was separated along the first PC axis (Figure 2), which explained 70% of the variation. The second PC axis contributed 26%. The relative locations of the oral and nasal cavities displayed closely in the first axis but were separated in the second axis. The relative direction and magnitude of logP are indicated by an arrow. The logP value of the compounds located below zero on the first axis was higher than 1, and their released amount was at the small and medium level except for pentanol. Four compounds located above zero on the first axis (ethyl acetate, butanol, acetic acid, and diacetyl) were released more than other compounds, and their logP values were less than 1. Among these four compounds, ethyl acetate and butanol were both detected in the nasal cavity at a high level, but acetic acid and diacetyl were detected only slightly.

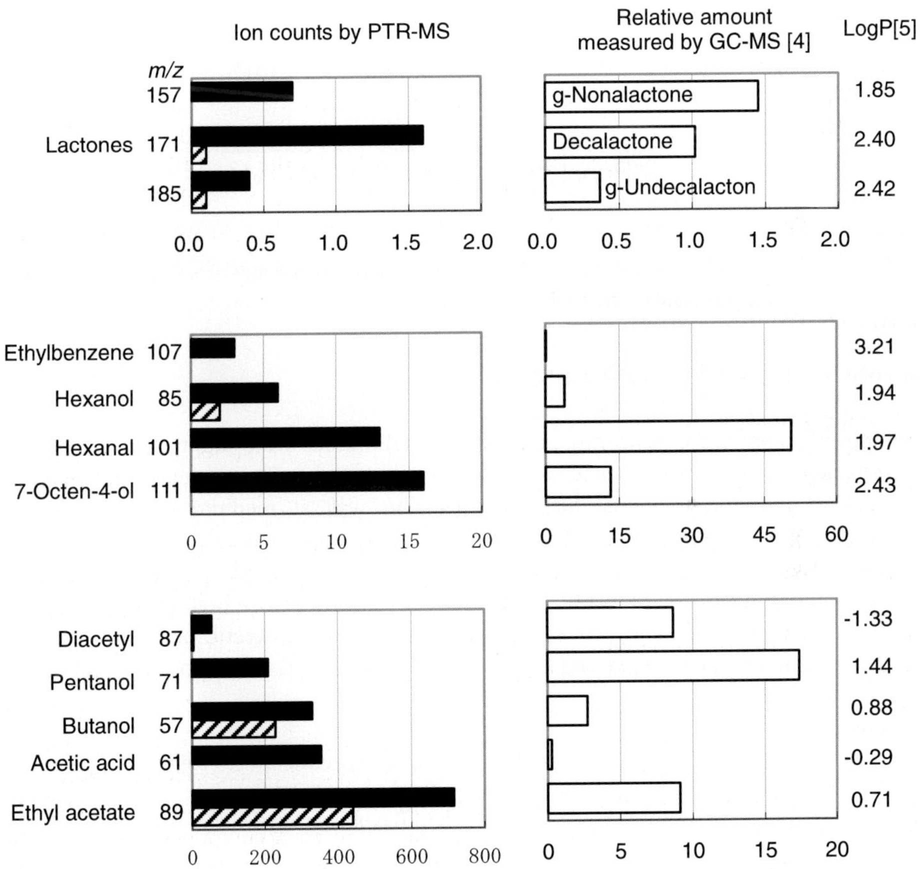

Figure 1. Ion counts measured by PTR-MS and the relative amount of compounds contained in Wagyu beef solvent extraction measured by GC-MS [4]. LogP was calculated by ACD/LogP Data Base [5]. (■) oral cavity; (▨) nasal cavity.

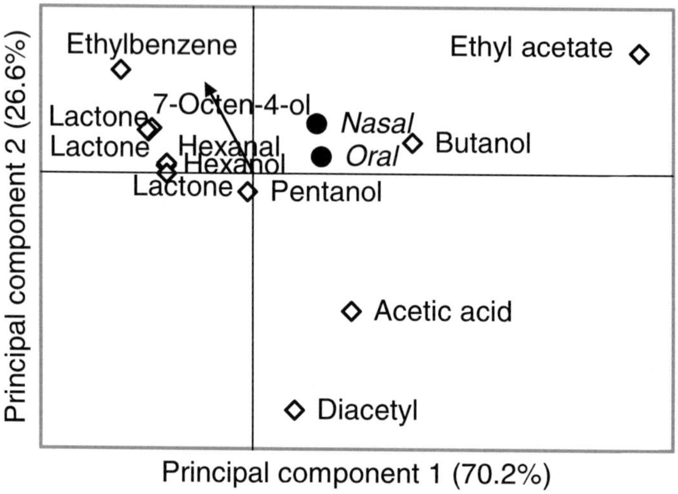

Figure 2. Principal components analysis bi-plot of released compounds from Wagyu beef. The arrow indicates the direction of increasing LogP values.

4. DISCUSSION AND CONCLUSION

Lipophilic compounds, the compounds whose logP values were higher, are considered to remain in Wagyu beef, because of its high fat content. Consequently the released amount of these compounds during mastication was at medium or small levels. Lactones were present intrinsically in very small amounts in Wagyu beef [4], and the released amounts were also very small. Comparatively hydrophilic compounds such as ethyl acetate, butanol, acetic acid, and diacetyl were released at a high level, considered to be caused by partitioning to saliva during mastication. Acetic acid and diacetyl, which are water soluble compounds, did not tend to be transferred to the nasal cavity.

References

1. W. Lindinger, A. Hansel and A. Jordan, Int. J. Mass Spectrom., 173 (1998) 191.
2. S.M. Van Ruth, L. Dings, K. Buhr and M.A. Posthumus, Food Res. Int., 37 (2004) 785.
3. K. Deibler and J. Delwiche (eds.), Handbook of flavor characterization, New York, USA (2004) 151.
4. M. Matsuishi, J. Kume, Y. Itou, M. Takahashi, M. Arai, H. Nagatomi, K. Watanabe, F. Hayase and A. Okitani, Nihon Chikusan Gakkaiho, 75 (2004) 409.
5. ACD/LogP Data Base, version 8.0, Advanced chemistry development, INC., Toronto ON, Canada.

W.L.P. Bredie and M.A. Petersen (Editors)
Flavour Science: Recent Advances and Trends

Human olfactory self-adaptation for structurally-related monoterpenes

Isabel Ovejero-López, Frans van den Berg and Wender L.P. Bredie

The Royal Veterinary and Agricultural University, Department of Food Science, Rolighedsvej 30, DK-1958 Frederiksberg C, Denmark

ABSTRACT

Olfactory adaptation influences the temporal perception of flavour. Self-adaptation and cross-adaptation phenomena are sensory effects which may explain discrepancies between the changes in stimulus concentration measured in the breath and changes in perceived sensory intensities during consumption. Olfactory self-adaptation was studied for a group of monoterpenes including (-)-menthol, thymol, (+)-carvone and (-)-carvone. The stimuli were delivered at a weakly perceptible intensity delivered by an automated dynamic olfactometer. The sensory responses were measured by the time-intensity (TI) method and subjects received a constant odour stimulation of 6 min followed by a recovery period without stimulation. Salient differences in self-adaptation were observed between the compounds and between subjects.

1. INTRODUCTION

Sensory adaptation is the reduction of sensitivity following (extensive) stimulation. Repeated or prolonged exposure to an odorant provokes a drop in sensitivity which is recovered over time in the absence of further exposure. Adaptation is an important functional mechanism preventing overflow in the central parts of the nervous system by neural activity resulting from stimuli that are either too strong or of long duration. The level of adaptation reached varies according to the nature, concentration and time of exposure of the stimulus [1-4]. Olfactory adaptation can occur at different levels in the olfactory system. Adaptation can be peripheral (receptor area) or central (central nervous system). The central adaptation (taking place somewhere from the glomeruli to the brain) tends to recover fast, whereas the recovery of cells from the receptor sites up to the glomeruli is slower [2,3]. The fast, central component of the adaptation and recovery processes play a much less predominant role at low concentrations [2,3]. Thus the lower the intensity, the smaller the relative influence of the central system.

Sensory adaptation in foods plays a role where long-lasting sensory effects are needed, such as in chewing gum. Furthermore, several studies have reported on temporal release patterns of volatile compounds in the mouth during eating and reported on temporal changes in perceived sensory intensity [1]. However, the retronasal release data are not necessarily correlated with TI sensory data. Some of the differences between sensory stimulation and sensory perception may be due to adaptation effects.
In this study, olfactory self-adaptation at a low level of stimulation was investigated for different monoterpenes. The olfactory recovery from previous exposure by means of recording changes in sensory responses was also evaluated.

2. MATERIALS AND METHODS

2.1. Odorants

The four stimuli were (-)-menthol (ChewTech A/S, Vejle, Denmark), (+)-carvone, (-)-carvone and thymol (all purchased from Sigma-Aldrich, Denmark). The odorants were placed in the odour channels of the olfactometer in their pure chemical form.

2.2. Odorant delivery system

The odorants were delivered by an automated 6 channels olfactometer at room temperature. The odour flow was presented in a stream of dry ambient air, thus by-passing the humidifier of the olfactometer. The sequences of odour stimulation were controlled by a computer program in Matlab (The MathWorks, USA).

2.3. Subjects

Five paid subjects, 3 males and 2 females (aged 24 to 64 years) were trained in odour assessments with the olfactometer. All subjects could detect the four odorants. The repetitions varied from one to three times per subject. Subjects were instructed to smell through both nostrils at a close distance to the odour port of the olfactometer. The subjects had their mouth closed during the experiment. The sensory responses were recorded on a visible scale using the FIZZ software (Biosystems, France).

2.4. Adaptation and recovery measurements

Each session in the adaptation experiment started with a detection threshold determination. This test included a short sequence of random intensities of three concentration steps around the subject's previously established threshold value. The adapting stimulus was then chosen at one step higher than the individual threshold value and delivered for 248 s. The test on olfactory recovery involved 248 s continuous stimulation, similar to the adaptation experiment. Thereafter, in consecutive intervals of 12 s, the odour was switched 'on' and 'off' for another 248 s.

2.5. Data analysis

The TI data from subjects and repetitions were for each stimulus averaged over repetitions by principal components analysis (PCA) and further averaged over subjects.

3. RESULTS AND DISCUSSION

Two examples of TI self-adaptation curves for (+)-carvone and thymol are shown in Figure 1. The curves were obtained from averaging TI responses from 5 subjects and their replicates. The self-adaptation curves showed an initial rise in perceived intensity followed by reaching a maximum and a subsequent decrease in perceived intensity. The shape of the adaptation curves varied for the different monoterpenes. (+)-Carvone showed a rapid adaptive pattern, with the onset of a falling rate of increasing intensity at 35 s. At 50 to 55 s a maximum was observed, after which a rapid decease in perceived intensity occurred. The adaptation process for (+)-carvone appeared, therefore, to be a two stage process. The first phase may relate to saturation of the subject with odorant and the onset of adaptation. The second phase may relate to adaptative processes only.

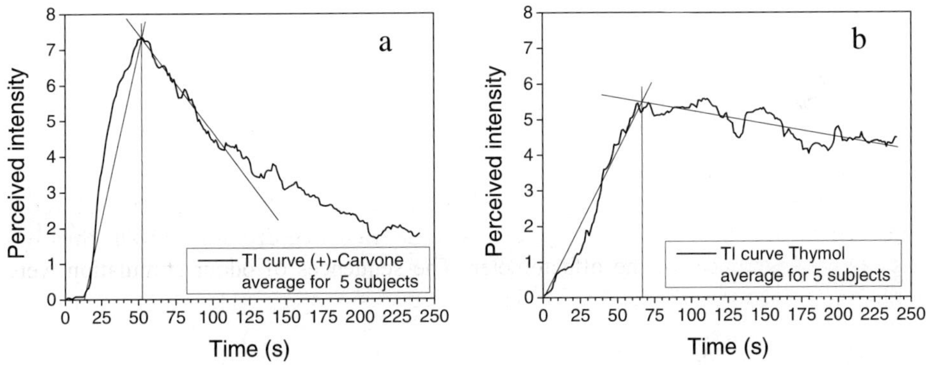

Figure 1. Adaptation TI curves for (+)-carvone (a) and for thymol (b).

The adaptation curve for thymol showed a different pattern. After an initial increase of perceived intensity a maximum was reached after 60 s. This maximum was much more persistent than for (+)-carvone. Only a small decrease was noticed over the remaining adaptive period (Figure 1). The initial part of the thymol adaptation curve may also be interpreted as a saturation/adaptation phase and the second part after reaching the maximum intensity should be related to predominantly adaptative processes. Therefore, thymol appears to give a much more persistent odour than (+)-carvone.
When comparing the sensory subjects, some differences in TI adaptation curves were noticed. There was a general agreement among subjects in the TI curves for (+)-carvone but not for thymol where subjects mainly differed in their duration before initial perception and time to reach maximum intensity.

3.1. Recovery from adaptation

The recovery experiment is illustrated by the results from subject 4 (Figure 2). The first 248 s fell into the same general pattern of the curves commented previously. Subsequently, in consecutive intervals of 12 s, the odour was switched 'on' and 'off' during 248 s. This constituted the recovery phase. The subject recovered relatively fast from adaptation to menthol, where sharp intensity peaks were obtained for perceived intensity, corresponding well with the true stimulation sequence. The intensity peaks reached the same intensity as at the beginning of the test, before getting adapted again, around 50 s into the experiment. This pattern was also observed for (-)-carvone, albeit with a lower recovery than menthol. Recovery from adaptation for (+)-carvone was not complete, while adaptation for thymol appeared to continue in the recovery sequence.

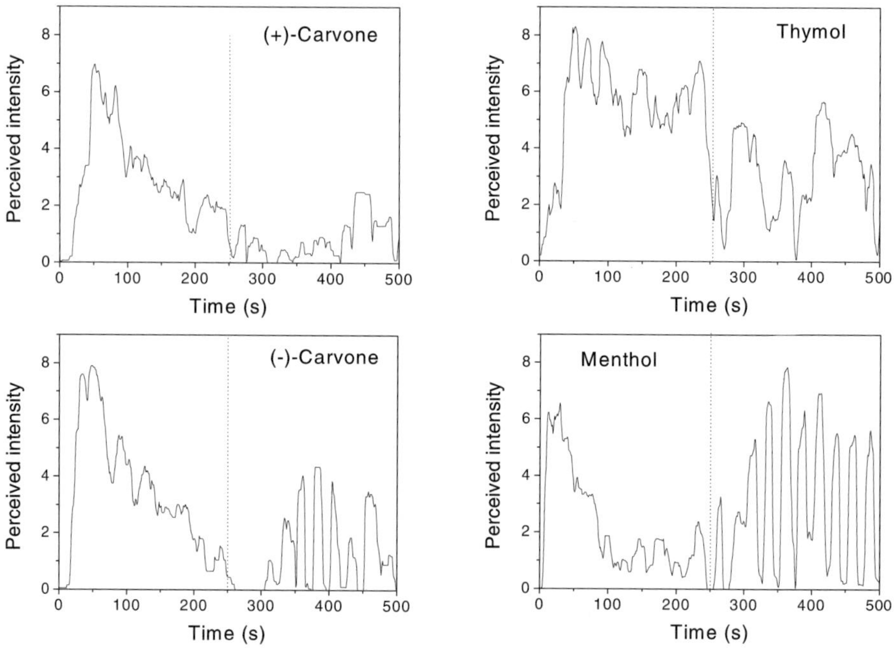

Figure 2. Average adaptation and recovery of subject 4 for each of the monoterpenes.

4. CONCLUSION

Olfactory self-adaptation was affected by the type of odorant and time of exposure. Adaptation proved to be dependent on the individual. Adaptation to (+)-carvone was relatively fast in comparison to thymol. The time necessary to start perceiving the different odour varies, as does the time to reach maximum intensity. Recovery from adaptation was also dependent on the compound and the individual.

References

1. I. Ovejero-Lopez, PhD thesis, The Royal Veterinary and Agricultural University, Denmark (2005).
2. M. Stuiver, PhD thesis, Groningen University, The Netherlands (1958).
3. E.P. Köster, PhD thesis, Utrecht University, The Netherlands (1971).
4. P. Dalton, Chem. Senses, 25 (2000) 487.

2

Genomics and biotechnology

W.L.P. Bredie and M.A. Petersen (Editors)
Flavour Science: Recent Advances and Trends

Genetic engineering of strawberry flavour

Wilfried Schwab[a], Stefan Lunkenbein[a], Elma M.J. Salentijn[b] and Asaph Aharoni[c]

[a]Biomolecular Food Technology, TU Munich, Lise-Meitner-Str. 34, 85354 Freising, Germany; [b]Plant Research International, Business Units Cell Cybernetics and Genetics & Breeding, PO Box 16, 6700 AA, Wageningen, Netherlands; [c]Department of Plant Sciences, Weizmann Institute of Science, P.O.B. 26, Rehovot 76100, Israel

ABSTRACT

Like in other fruit, a complex mixture of hundreds of compounds determine strawberry aroma. Among them 4-hydroxy-2,5-dimethyl-3(2*H*)-furanone (Furaneol®, HDMF) and its methyl ether 2,5-dimethyl-4-methoxy-3(2*H*)-furanone (DMMF) belong to the fifteen most important strawberry flavour compounds. Recently, the gene *FaOMT* (*Fragaria x ananassa* O-methyltransferase) was isolated and sequenced coding for an O-methyltransferase forming DMMF from HDMF. Heterologous expression of FaOMT and characterisation of the corresponding protein demonstrated its ability to methylate a range of substrates. In order to clarify the *in vivo* function of FaOMT we generated transgenic strawberry plants carrying a FaOMT sense or antisense construct under the control of the constitutively expressed CaMV 35S promoter. Transcript levels in transgenic and control fruit were quantified by Quantitative Real Time-PCR (QRT-PCR) and metabolite levels were determined by GC-MS and LC-ESI/MS^n. The data showed that *FaOMT* mRNA levels and concentration of DMMF were strongly down-regulated in antisense as well as some sense fruit. Fruit of several transgenic lines were devoid of DMMF. Sensory analyses of the transgenic fruit clarified the contribution of DMMF to the overall strawberry flavour. As a side effect of the reduced levels of *FaOMT* transcripts we observed the reduction of feruloyl β-D-glucose confirming the dual function of FaOMT in strawberry fruit.

1. INTRODUCTION

Combined biochemical and molecular analyses of volatile components released by fruit have demonstrated that their biogenesis forms an integral part of the ripening program. Biomolecular work on fruit ripening has been performed mainly on tomato (*Lycopersicon esculentum*) although in recent years there has been a dramatic increase in the investigations of other fruit species including melon, grape, citrus, raspberry, pear, banana and strawberry [1-4]. The majority of the studies identified genes with elevated expression during ripening and correlated their putative identity with a specific ripening process.

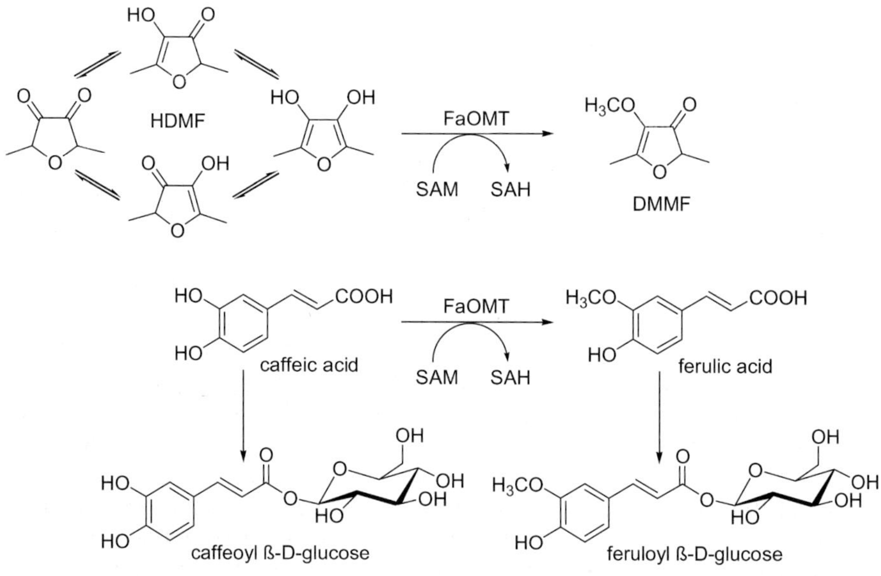

Figure 1. Substrates and products of reactions catalysed by FaOMT *in vivo* and metabolites quantified in control and transgenic fruit. HDMF: 4-hydroxy-2,5-dimethyl-3(2H)-furanone; DMMF: 2,5-dimethyl-4-methoxy-3(2H)-furanone; SAM: S-adenosyl-L-methionine, SAH: S-adenosyl-L-homocysteine.

Although strawberry fruit is derived from the flower receptacle, it has the same development and ripening characteristics of true fruit, including the degradation of chlorophyll, the accumulation of anthocyanins, the softening which is partially mediated by cell wall hydrolysing enzymes, the metabolism of sugars and organic acids and the production of flavour compounds. Compounds contributing to the flavour of strawberries have been extensively studied. More than 360 volatiles have been identified [5] but only about 15-20 of them are believed to be essential for the sensory quality of strawberries, together with the non-volatile sugars and organic acids [6]. Among these, 4-hydroxy-2,5-dimethyl-3(*2H*)-furanone (HDMF, Furaneol®) (Figure 1) is the most important because of its high concentration in strawberry fruit (up to 55 mg/kg fresh weight) [7] and low odour threshold (10 ppb in water) [6]. HDMF is

frequently accompanied by its methyl ether 2,5-dimethyl-4-methoxy-3(*2H*)-furanone (DMMF, mesifuran) (Figure 1), and HDMF-glucosides [8,9].
The quantification of HDMF and DMMF during fruit ripening indicated a rapid conversion of HDMF to DMMF and HDMF-glucoside [10] and *in vivo* feeding experiments demonstrated the incorporation of ^{14}C-label into DMMF after the application of both S-[methyl-^{14}C]-adenosyl-L-methionine (^{14}C-SAM) and ^{14}C-HDMF [11]. Recently, an O-methyltransferase (*Fragaria x ananassa* O-methyltransferase, *FaOMT*) cDNA was obtained by screening a strawberry cDNA library, cloned and heterologously expressed in *Escherichia coli*. The FaOMT protein catalyses the transfer of the methyl group from SAM not only to HDMF but also to caffeic acid, thereby forming the corresponding O-methyl ethers (Figure 1) [12]. Due to the expression pattern of FaOMT and the enzymatic activity in the different stages of fruit ripening, it was proposed that FaOMT is involved in the phenylpropanoid metabolism and in the biosynthesis of the strawberry volatile DMMF.
Most of the important molecular insights into fruit ripening have been obtained using overexpression and antisense technology in transgenic fruit [13]. In this study, by up- and downregulating FaOMT using the CaMV 35S promoter, we assessed the function of the FaOMT enzyme *in planta* and the significance of DMMF for strawberry flavour.

2. RESULTS

2.1. Transgenic strawberry plants with altered expression of FaOMT

The full-length strawberry *FaOMT* sequence in the sense and antisense orientation was placed under control of the CaMV 35S promoter in the binary vector pBINPLUS. Constructs were introduced into *Fragaria x ananassa* cv. Calypso by Agrobacterium-mediated transformation. The primary transgenic (12 FaOMT antisense (AS)-lines and 11 FaOMT sense(S)-lines) were transferred to the greenhouse. Integration of *FaOMT* was confirmed using Quantitative Real Time PCR (QRT-PCR). In several transgenic lines phenotypic changes were observed like delayed flowering and a slightly changed fruit colour in the first round of fruit production (3%). Plants with reduced growth occurred in a higher frequency (22%).

2.2. Quantitative transcript analysis by QRT-PCR

Five of the six FaOMT S-lines selected for TaqMan® analysis contained higher levels of *FaOMT* transcripts in the leaf tissue than the control plants. The one plant (FaOMT S9) that showed a lower expression level (24% of the control level) also possessed a severely altered phenotype, in that the plant and leaf sizes were significantly reduced in comparison with the control plants. This is likely to result from the known phenomenon called co-suppression [14,15]. As expected, the five antisense lines that were analysed contained lower levels of *FaOMT* mRNA ranging from 74% to 3% of the control level.

2.3. Quantification of metabolites

To determine the *in planta* function of the protein we isolated its potential substrates and products by solid phase extraction from ripe strawberries from control and transformed plants produced in the second year. Levels of HDMF and DMMF as well as concentrations of caffeoyl β-D-glucose and feruoyl β-D-glucose were quantified by GC-MS and LC-ESI/UV/MS^n, respectively in the same samples.

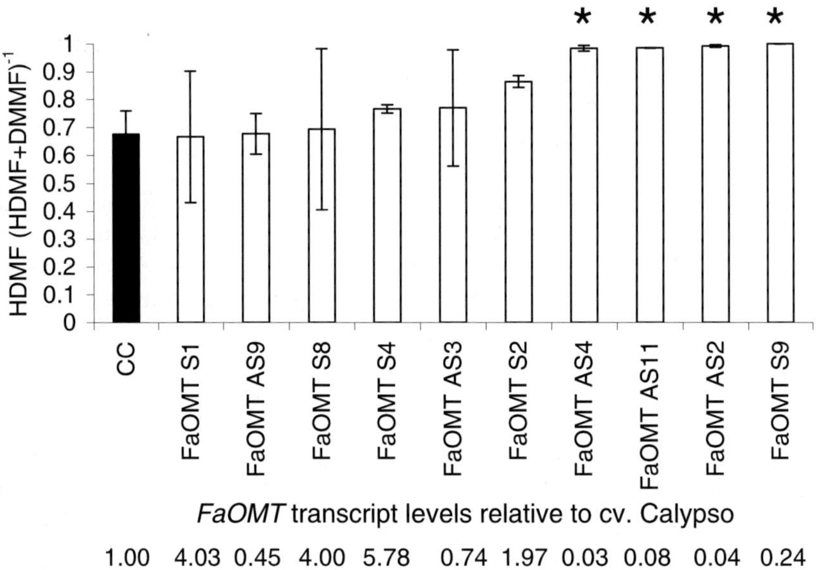

Figure 2. Normalised concentration of HDMF and DMMF expressed as the ratio of HDMF to the total amount of DMHF and DMMF (DMHF+DMMF) in control fruit (CC) and fruit of plants transformed with the sense (FaOMT S) and antisense (FaOMT AS) constructs. Asterisks (*) indicate that the data is significantly different from the data of control (CC) fruit ($p<0.01$).

Strawberries from four of the eleven transgenic lines produced significantly lower levels of the methylated product DMMF as compared to the substrate HDMF (Figure 2). Five of the six plants containing a *FaOMT* construct in the sense orientation showed overexpression of the transcript in leaves but the increase did not affect the relative levels of HDMF and DMMF in the fruit tissue. The ratio of caffeic acid expressed as caffeoyl β-D-glucose, to the combined amount of glucose esters in some selected transgenic fruit is shown in Figure 3.

As with the furanones, transgenic plants with strongly reduced levels of *FaOMT* mRNA contained the lowest levels of ferulic acid compared to caffeic acid, quantified as glucose esters. However, in none of the transgenic plants did the reduction of the transcripts result in a near total loss of feruloyl β-D-glucose as was observed for the production of DMMF. In fruit of plants FaOMT S1 and FaOMT S8 higher concentrations of the methylated product, feruloyl β-D-glucose, were detected, indicating an up-regulation of gene expression as confirmed by the *FaOMT* transcript levels in the leaf tissue.

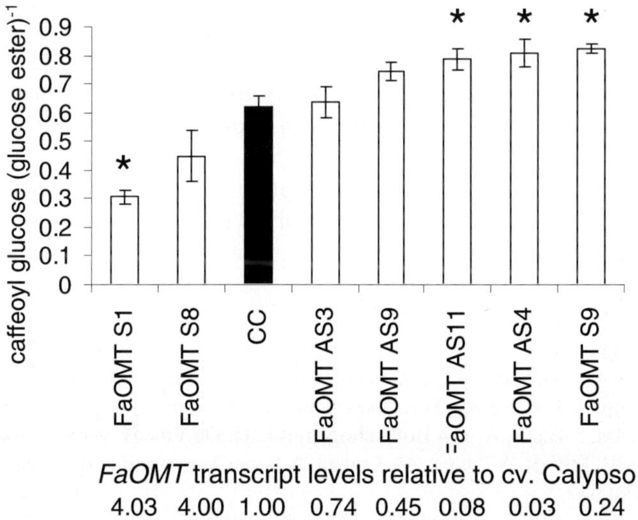

Figure 3. Normalised concentration of caffeoyl β-D-glucose and feruloyl β-D-glucose expressed as ratio of caffeoyl β-D-glucose to the total amount of glucose esters (caffeoyl β-D-glucose + feruloyl β-D-glucose) in control fruit (CC) and fruit of plants transformed with the sense (FaOMT S) and antisense (FaOMT AS) constructs. Asterisks (*) indicate that the data is significantly different from the data of control (CC) fruit ($p<0.01$).

2.4. Sensory analyses

The triangle test was employed to determine if judges could distinguish, by smelling, between homogenates prepared from control fruit and transgenic fruit. Only three out of twelve panelists were able to distinguish juice of FaOMT S9 fruits from the control. Although similar results were obtained using fruit of other plants exhibiting reduced levels of DMMF relative to HDMF (FaOMT AS4: five out of twelve; FaOMT AS11: seven out of twelve), we observed that the same three panelists who correctly distinguished the sample FaOMT S9 from controls also recognised the other transgenic fruit correctly. This implies that three individuals were able to differentiate samples with low ratios of HDMF to DMMF from those with a higher proportion of HDMF as present in the transgenic plants. In addition to the discrimination test, the three sensitive volunteers were asked to assess the flavour profiles of the samples using the descriptors 'caramel' (HDMF), 'grass-like' (*Z*-3-hexenal), 'fruity' (ethyl butanoate), 'buttery' (diacetyl), 'sour' (acetic acid), and 'sweaty' (butanoic acid) [6]. A statistical comparison of the resulting flavour profiles clearly shows that the panelists were only able to accurately perceive differences in the 'caramel' note (HDMF) in the fruit extract of FaOMT S9 and thereby identified the higher HDMF level of this transgenic line.

3. CONCLUSION

We successfully modified the ratio of the odorous furanones in strawberry fruit using an antisense approach, but only a few sensitive consumers were able to perceive the difference. Thus, in order to positively modify fruit flavour, individual sensitivities in odour perception and the impact of individual odorants on the overall aroma have to be taken into account. This implies that enhanced or reduced levels of odorants do not necessarily change the overall aroma perception of fruits. As a second effect of the reduced levels of *FaOMT* transcripts we observed the reduction of feruloyl β-D-glucose confirming the dual function of FaOMT in strawberry fruit.

References

1. A. Aharoni, A.P. Giri, F.W.A. Verstappen, C.M. Bertea, R. Sevenier, Z. Sun, M.A. Jongsma, W. Schwab and H.J. Bouwmeester, Plant Cell, 16 (2004) 3110.
2. A. Aharoni, L.C.P. Keizer, H.J. Bouwmeester, Z.K. Sun, M. Alvarez Huerta, H.A. Verhoeven, J. Blaas, A. van Houwelingen, R.C.H. De Vos, H. van der Voet, R.C. Jansen, M. Guis, J. Mol, R.W. Davis, M. Schena, A.J. van Tunen and A.P. O'Connell, Plant Cell, 12 (2000) 647.
3. A. Aharoni and F. Vorst, Plant Mol. Biol., 48 (2002) 99.
4. J. Wilkinson, M. Lanahan, T. Conner and H. Klee, Plant Mol. Biol., 27 (1995) 1097.
5. L.M. Nijssen, H. Visscher, L.C. Maarse, L.C. Willemsens and M.H. Boelens (eds.), Volatile compounds in food: qualitative and quantitative data, 7 ed., Zeist, The Netherlands (1996).
6. P. Schieberle and T.J. Hofmann, Agric. Food Chem., 45 (1997) 227.
7. M. Larsen, L. Poll and C.E. Olsen, Z. Lebensmittel Untersuch. Forsch., 195 (1992) 536.
8. F. Mayerl, R. Näf and A.F. Thomas, Phytochem., 28 (1989) 631.
9. R. Roscher, M. Herderich, J.P. Steffen, P. Schreier and W. Schwab, Phytochem., 43 (1996) 155.
10. A.G. Pérez, R. Olías, C. Sanz and J.M. Olías, J. Agric. Food Chem., 44 (1996) 3620.
11. R. Roscher, P. Schreier and W. Schwab, J. Agric. Food Chem., 45 (1997) 3202.
12. M. Wein, N. Lavid, S. Lunkenbei, E. Lewinsohn, W. Schwab and R. Kaldenhoff, Plant J., 31 (2002) 755.
13. C.G. Lu, Z. Zainal, G.A. Tucker and G.W. Lycett, Plant Cell, 13 (2001) 1819.
14. A.R. van der Krol, L.A. Mur, M. Beld, J.N. Mol and A.R. Stuitje, Plant Cell, 2 (1990) 291.
15. A.J. Hamilton and D.C. Baulcombe, Science, 286 (1999) 950.

W.L.P. Bredie and M.A. Petersen (Editors)
Flavour Science: Recent Advances and Trends

Biotechnological production of terpenoid flavour and fragrance compounds in tailored bioprocesses

Hendrik Schewe, Michael Pescheck, Dieter Sell and Jens Schrader

Biochemical Engineering, Karl-Winnacker-Institut, DECHEMA e.V., Theodor-Heuss-Allee 25, 60486 Frankfurt/Main, Germany

ABSTRACT

Integrated bioprocess concepts specific to the requirements of monoterpenes as precursors of microbial transformations were developed. *R*-Limonene and α-pinene were used as model precursors and bacteria and higher fungi served as biocatalysts. A closed-gas-loop bioreactor was set up to address the volatility of terpenoids. By this means α-terpineol was produced from *R*-limonene at the gram per litre scale with *Penicillium digitatum*. *E. coli* overexpressing a P450 BM-3 mutant served as biocatalyst to produce several hundreds of milligrams of pinene oxide, verbenol, and myrtenol from α-pinene. Here, an aqueous-organic two-phase bioreactor with diisononylphthalate as organic solvent was used for *in situ* precursor supply and product recovery.

1. INTRODUCTION

Terpenoids play a key role in the overall sensory profile of plant essential oils and thus they have a major impact on the aroma of many foods and beverages. Furthermore, they often act as natural preservatives due to their antimicrobial properties. Since the concentrations of highly bioactive oxyfunctionalised terpenoids in natural sources are usually very low, in most cases it is not economically viable to isolate them by physical means, such as extraction and distillation. Consequently, to serve the rising demand for natural food ingredients, biotechnology has become an obvious option as an alternative but still 'natural' route to these desired compounds. But there are also severe drawbacks, especially the cytotoxicity, volatility and low or non-solubility in water of the precursor and product molecules. To the authors' knowledge, only a few, biocatalytic terpene oxyfunctionalisations, which have been established on pilot or industrial scale despite the high value of certain natural terpenoid flavour compounds and the cheap and abundantly available precursors, such as *R*-limonene and α-pinene

[1]. The objective of this work was, therefore, to develop integrated bioprocess concepts specific to the requirements of monoterpene whole-cell biocatalysis.

2. MATERIALS AND METHODS

2.1. Microorganisms, culture media, blank experiments

Penicillium digitatum DSM 62840 (DSMZ, Braunschweig, Germany) was cultivated in a medium containing 10 g/l glucose, 20 g/l malt extract, 10 g/l peptone and 5 g/l yeast extract. *E. coli* BL 21 (DE3) (Novagen) expressing a pET28a^{+}-coded triple mutant of *B. megaterium* cytochrome P450 BM-3 [2] was kindly committed by the Institute of Technical Biochemistry, Stuttgart University. The strain was grown in minimal medium with 10 g/l glucose. To evaluate the potential of a non-biological formation of the target compounds, the media with precursors were incubated without microorganisms under the same conditions as in the biotransformation experiments.

2.2. Biotransformations

P. digitatum was cultivated in 2.7 l medium in a 3.7 l bioreactor (KLF 2000, Bioengineering, Switzerland) at 400 rpm, initial pH 6.3, 27 °C and 0.05 vvm [volume/(volume min)] aeration. When the glucose concentration had fallen below 1 g/l (t = 70 h in the closed-gas-loop experiment; t = 73 h in the conventional experiment), the stirrer speed was raised to 500 rpm and 2 mM limonene was added to start the biotransformation. The feeding was repeated every 24 h. With the first limonene dosage the gas-loop system was turned on. It mainly consisted of a 25 l gas holder, a compressor, a pressure vessel, a limonene-filled washing flask (for gas saturation) and stainless steel pipings (see Figure 1b). By this means additional limonene was fed via the gas phase into the bioreactor. During biotransformation the glucose concentration was re-adjusted at 2 g/l every 24 h by adding a concentrated glucose solution (500 g/l). *E. coli* was cultivated in 1.5 l medium in a 3.0 l bioreactor (KLF 2000, Bioengineering, Switzerland) at 37 °C, controlled pH 7.4 and 1.6 vvm aeration with a variable stirrer speed to keep dissolved oxygen above 50%. After consumption of initial glucose a sequential mixed feeding of glucose (700 g/l) and $MgSO_4$ (1 M) was initiated to re-adjust the glucose concentration at around 5 g/l at intervals of 0.5 h to 3 h. One hour later, the P450 BM-3 expression was induced by 0.5 mM IPTG; 3 h later, the organic phase consisting of 400 ml α-pinene and 600 ml diisononylphthalate was added (t = 0 h in Figure 2a).

2.3. Analytical methods

In the fungal experiment, the filtrated liquid culture was extracted with hexane, in the bacterial experiment, the diisononylphthalate phase was diluted 1:1 with α-pinene containing the internal standard, prior to analysis by GC-MS (17A gas chromatograph, QP5050 mass spectrometer, Shimadzu, Germany). In both experiments camphor served as internal standard. A ZB-5 Zebron column (30 m x 0.25 mm i.d.; coating thickness 0.25 μm, Phenomenex, Germany) was used.

3. RESULTS

3.1. Biotransformation of *R*-limonene in a closed-gas-loop bioreactor

We investigated whether a specific bioprocess design would allow better exploitation of the biocatalytic potential of terpene-transforming microorganisms. For this purpose a two-step bioprocess regime (1. biomass growth, 2. biotransformation) was set up for the transformation of *R*-limonene into the lilac-like fragrance and flavour compound α-terpineol by *P. digitatum* as a model fungal system. To improve the supply of limonene, the precursor was fed into the reactor via the gas phase; this feature helped circumvent any evaporative loss of the volatile limonene, which was concurrently added directly to the liquid culture. By closing the gas circuit at the same time as the biotransformation was started, any loss of limonene via the exhaust gas was avoided. In comparison with the conventional fed-batch process the closed-gas-loop bioreactor resulted in a doubling of the total α-terpineol concentration which exceeded the 1 g/l threshold (Figure 1a).

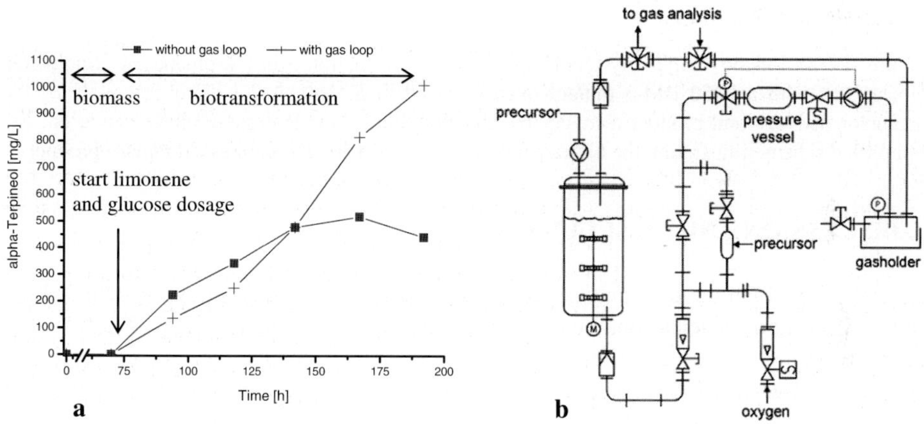

Figure 1. (a) α-Terpineol formation from *R*-limonene by *P. digitatum* DSM 62840 in a 3 l bioreactor with and without closed gas loop during the biotransformation period. (b) Flow scheme of the closed-gas-loop bioreactor.

3.2. Biotransformation of α-pinene in a two-phase bioreactor

P450 BM-3, which was exploited to oxyfunctionalise α-pinene, originally stems from *B. megaterium*. The triple mutant used in these studies had been designed by rational evolution [2]; it has a wide substrate spectrum including terpenoids. P450 BM-3 naturally acts on fatty acids and shows unusually high monooxygenase turnover frequencies of >1000 eq/min. Consequently, comparably high productivity was observed here: the three main products, pinene oxide, verbenol, and myrtenol, being important flavour compounds or their direct precursors, formed in concentrations of up to 200 mg/l within a few hours (Figure 2a). Diisononylphthalate proved to be a suitable organic solvent for *in situ* precursor supply and product extraction (Figure 2b).

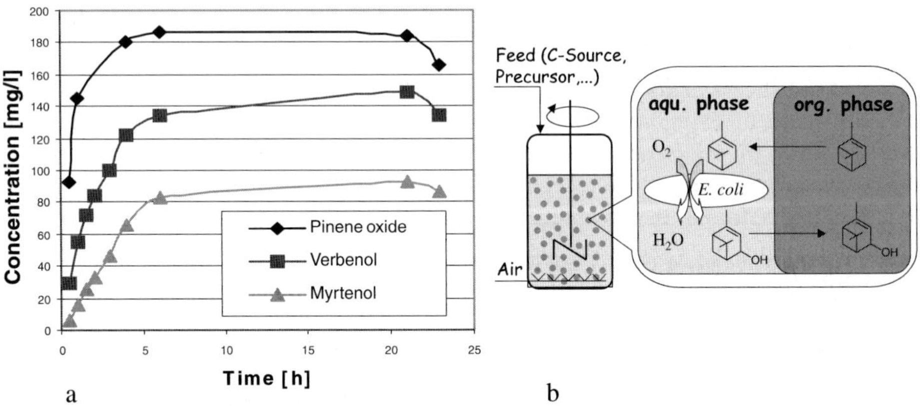

a b

Figure 2. (a) Pinene oxide, myrtenol and verbenol formation from α-pinene by whole-cell biocatalysis with a P450 BM-3 mutant overexpressed in *E. coli*. A 3 l bioreactor with *in situ* precursor delivery and product recovery by diisononylphthalate as organic phase was used. (b) Simplified scheme illustrating the biotransformation period with verbenol as exemplary product.

4. DISCUSSION AND CONCLUSION

The volatility of monoterpenes can be addressed by a closed-gas-loop system, which results in a more efficient and sustainable bioprocess. In combination with a two-step process regime, even higher fungi, usually slow transformants, can become efficient limonene-converting biocatalysts as a permanent supply of the precursor at a non-toxic level is possible. Preliminary experiments with *E. coli* overexpressing an engineered P450 BM-3 revealed that relatively high α-pinene oxyfunctionalisation rates are possible in a two-phase bioreactor containing a biocompatible precursor-supplemented organic solvent. Further improvements of microbial monoterpene biocatalysis are aimed at by high-cell-density cultures of engineered bacteria and fungi in a bioprocess combining the aforementioned two-phase and closed-gas-loop principles.

References

1. H.-J. Rehm and G. Reed (eds.), Biotechnology, Weinheim, Germany, 10 (2001) 373.
2. D. Appel, S. Lutz-Wahl, P. Fischer, U. Schwaneberg and R.D. Schmid, J. Biotechnol., 88 (2001) 167.

Identification of the gene responsible for the synthesis of volatile sulfur compounds in *Brevibacterium linens*

Mireille Yvon[a], Félix Amarita[a,b], Michèle Nardi[a], Emilie Chambellon[a], Jérôme Delettre[b] and Pascal Bonnarme[b]

[a]*Institut National de la Recherche Agronomique, Unité de Biochimie et Structures des Protéines, 78352 Jouy-en-Josas, France;* [b]*Institut National de la Recherche Agronomique, Unité Mixte de Recherches Génie et Microbiologie des Procédés Alimentaires, 78850 Thiverval-Grignon, France*

ABSTRACT

The enzymatic degradation of L-methionine and subsequent formation of volatile sulfur compounds (VSC) is believed to be essential for flavour development in cheese. L-methionine γ-lyase (MGL) can convert L-methionine to methanethiol (MTL) α-ketobutyrate and ammonia. The MGL gene encoding MGL was cloned from the type strain *Brevibacterium linens* ATCC 9175 known to produce copious amounts of MTL and related VSC. The disruption of the MGL gene, achieved in strain ATCC 9175, resulted in a 97% decrease in total VSC production in the knockout strain. Our work shows that L-methionine degradation via γ-elimination is a key step to form VSC in *B. linens*.

1. INTRODUCTION

Owing to their low detection threshold and diversity, volatile sulfur compounds (VSCs) are of prime importance in the cheese flavour, and make a significant contribution to the typical aroma. VSCs arise primarily from the degradation of L-methionine to methanethiol (MTL) by the cheese microflora. In *Brevibacterium linens*, which is an efficient VSC producer, MTL production is believed to proceed via a L-methionine-γ-elimination which is catalysed by a L-methionine-γ-lyase (MGL) [1]. The recent availability of a part of the genome sequence of *B. linens* (US Department of Energy

Joint Genome Institute), together with the publication of MGL sequences of several bacteria, gave us new tools for picking up the gene encoding MGL in *B. linens*.
The objective of our work was to identify and characterise the gene responsible for MTL production from L-methionine in *B. linens*. For this purpose, a putative MGL gene was inactivated in *B. linens* and the consequences on VSC production were investigated.

2. MATERIALS AND METHODS

All experimental procedures were as described by Amarita *et al.* (2004) [1] and Bonnarme *et al.* (2000) [2].

3. RESULTS

3.1. Sequencing of the putative MGL gene from *B. linens* strain ATCC 9175

The putative gene encoding MGL was localised in the genome of *B. linens* BL2 (Joint Genome Institute) by looking for a conserved DNA sequence (5'-T[Nx7]AANCANATGA[Nx8]GG-3', with N being any bases) of MGL gene of 5 Gram-positive bacteria. This conserved sequence was found by alignment [5] of MGL sequences of *Methanosarcina acetivorans*, *Methanosarcina mazei*, *Fusobacterium nucleatum, Oceanobacillus iheyensis* and *Bacillus halodurans.* Such a sequence (**T**CGCGAAG**AAGCAGATGA**GCGGCTTC**GG**) was found only once in the DNA sequence of the BL2 genome. It was located in an ORF of 1275 bp encoding a 425-amino-acid protein with a calculated molecular weight of 44669 Da, which was in good agreement with the MW of 43 kDa estimated for MGL by SDS-PAGE [1]. It also exhibited a pyridoxal-phosphate attachment site of Cys/Met metabolism enzymes (Prosite accession nr.° PS00868) at position 194-208, indicating that the gene really encodes a pyridoxal phosphate (PLP)-dependent enzyme.

3.2. Effect of inactivation of the putative MGL gene in *B. linens* on L-methionine conversion to VSCs

The effect of gene inactivation on L-methionine conversion to VSCs was investigated. Our results show that the inactivation of the putative MGL gene reduced the total production of VSCs from L-methionine by 97% (Figure 1). Although MTL is the reaction product of MGL on L-methionine, DMDS is, by far, the most produced VSC by the wild-type strain, since it represents 96% of total VSCs produced. This is due to the fact that MTL is a highly reactive sulfur compound that quickly reacts with itself, forming the oxidised and more stable DMDS as well as, to a lesser extent, DMTS [3]. HPLC analysis of metabolites produced from [1^{14}C]-L-methionine by the wild type strain (ATCC 9175) (Table 1) showed one major radio labeled compound (RT 12.5 min) and two other compounds only detected by UV absorbance (RT 8.1 and 11.5 min). Inactivation of the putative MGL gene totally prevented formation of the three compounds: propionate (RT 11.5 min), α-aceto-α-hydroxybutyrate (RT 12.5 min), and the compound eluting at 8.1 min. No radio-labeled metabolite was detected with the

MGL mutant (Table 1) even after 24 h of incubation. The absence of α-aceto-α-hydroxybutyrate formation – measured as formed 2,3-pentanedione - by resting-cells of the MGL mutant was confirmed by GC-MS analysis.

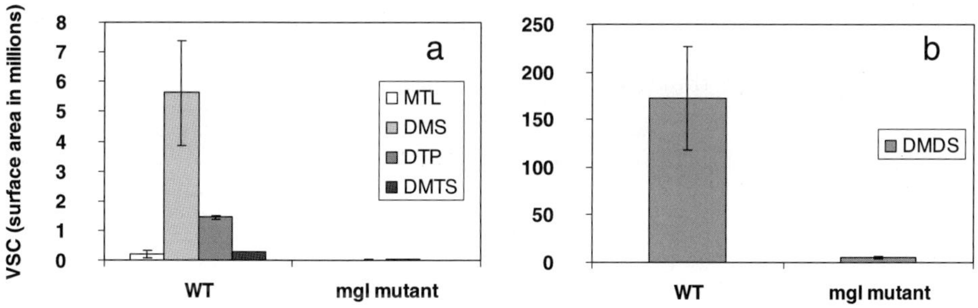

Figure 1. Production of volatile sulfur compounds by the wild type strain *B. linens* ATCC 9175 (WT) and by the MGL mutant strain TIL 872 (MGL mutant). (a) Production of methanethiol (MTL), dimethyl sulfide (DMS), dithiapentane (DTP), and dimethyl trisulfide (DMTS). (b) Production of dimethyl disulfide (DMDS).

Table 1. Peak areas of major metabolites produced from L-[1-^{14}C]-methionine by resting-cells of *B. linens* ATCC 9175 and the MGL mutant strain[a].

			Peak area for[b]					
RT			Wild type			MGL mutant		
(min)	Detection	Identification	0h	4h	24 h	0 h	4 h	24h
8.1	UV	Not identified	5 ± 2	52 ± 36	61 ± 12	0	0	0
11.5	UV	Propionate	4 ± 3	13 ± 11	80 ± 30	4 ± 1	5 ± 1	5 ± 1
12.5	Radio-activity	α–Aceto-α-hydroxybutyrate	322 ± 94	6,116 ± 777	13,569 ± 896	0	0	0

[a]Cells were incubated for 0, 4, and 24 h at 30 °C in the reaction medium at pH 8. Metabolites were analysed by ion-exclusion HPLC with both UV and radioactivity detections. Data is the means of the results from three repetitions ± standard deviations. [b]Results for UV detection are given in mV, and results for radioactivity are given in disintegrations per min.

4. DISCUSSION AND CONCLUSION

In this work, the gene encoding MGL was identified in the cheese-ripening bacterium *B. linens*. MGL is involved in L-methionine catabolism in many bacteria (including *E. coli*, *Proteus vulgaris*, *Brevibacteria* and *Pseudomonas)* and plays an essential role for the generation of metabolic energy from L-methionine via the production of α-ketobutyrate that can be degraded to propionyl-CoA and subsequently succinyl-CoA

[6]. Moreover, it is generally admitted that MGL is responsible for L-methionine conversion to MTL in *B. linens*. It is, therefore, believed to be highly involved in VSCs biosynthesis in cheese [2,4]. MGL of *B. linens* has been biochemically characterised [4], but the gene location, structure and sequence remained unknown.

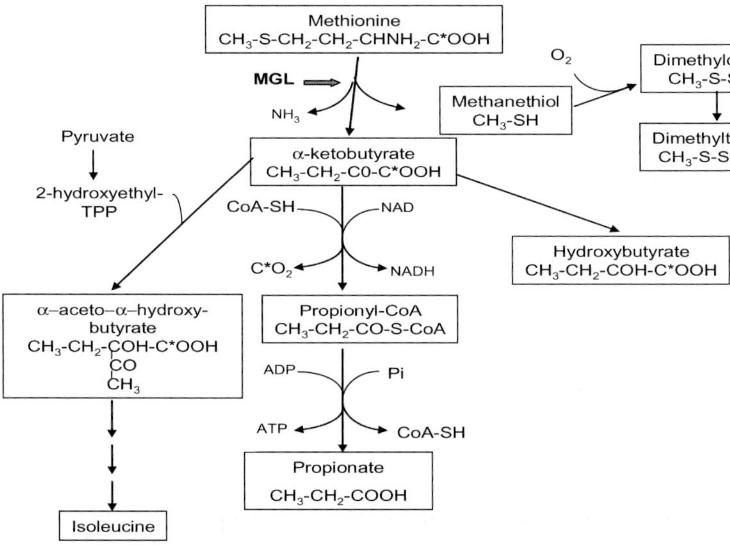

Figure 2. Putative L-Methionine catabolism pathways initiated by methionine γ-lyase (MGL).

Recently, the genome sequence of *B. linens* BL2 became available. The draft annotation did not clearly indicate the gene encoding MGL, although several genes exhibited high degrees of homology with cystathionine-synthases and cystathionine-lyases, which are two enzyme families closely related to MGL - with respect to biochemical properties.
To our knowledge, this is the first report on the gene encoding MGL in the surface cheese-ripening bacterium *B. linens*. This is of considerable interest since, until now, MGL involvement in VSC production by *B. linens* was rather speculative. In this study, we have demonstrated a clear-cut effect of MGL gene disruption on the VSC producing ability of *B. linens*. This new knowledge could represent a first step towards a better control over VSC biosynthesis in a wide variety of cheese types. Putative L-methionine catabolism pathways are presented (Figure 2).

References

1. F. Amarita, M. Yvon, M. Nardi, E. Chambellon, J. Delettre and P. Bonnarme, Appl. Environ. Microbiol., 70 (2004) 7348.
2. P. Bonnarme, L. Psoni and H.E. Spinnler, Appl. Environ. Microbiol., 66 (2000) 5514.
3. H.W. Chin and R.C. Lindsay, Food Chem., 49 (1994) 387.
4. B. Dias and B. Weimer, Appl. Environ. Microbiol., 64 (1998) 3327.
5. D.G. Higgins and P.M. Sharp, Gene, 73 (1988) 237.
6. K. Soda, H. Tanaka and N. Esaki, Trends Biochem. Sci., 8 (1983) 214.

W.L.P. Bredie and M.A. Petersen (Editors)
Flavour Science: Recent Advances and Trends

Heritability studies of aroma compounds in carrots using rapid GC methods

Detlef Ulrich[a], Thomas Nothnagel[b], Petra Straka[a], Rolf Quilitzsch[a] and Edelgard Hoberg[a]

Federal Centre for Breeding Research on Cultivated Plants (BAZ), Neuer Weg 22/23, D-06484 Quedlinburg, Germany, [a]Institute of Plant Analysis; [b]Institute of Horticultural Crops

ABSTRACT

The aim of the research is to study heritability of the aroma pattern in carrot. This paper presents the results of aroma analyses of 200 single plants from an F_2 population using headspace solid phase microextraction GC and pattern recognition for data processing. In a parallel approach molecular markers (AFLP) are developed. Together with additional analyses (sugar and carotenoids) both chemical and molecular data will be used for genome mapping.

1. INTRODUCTION

Carrot production is of economic importance for the agro-industry. Additionally, carrots are of health interest to the consumer because they are the major vegetable source of pro-vitamin A. Carrots with a desirable flavour might support an increase in consumption. The sensory quality of carrots is influenced by the sugar content, bitter compounds, and profile of aroma volatiles. The majority of volatile compounds released from raw carrots are mono- and sesquiterpenes which can comprise up to 97% of the total amount. More than 100 aroma compounds have been identified in raw carrots.
The profile of volatiles is largely influenced by the genotype [1]. Genetic variation has been found for qualitative and quantitative contents of terpenoids. This is an interesting fact for breeding plants with different sensory qualities; high terpenoid content is supposed to cause harsh flavour [2]. The authors observed that lower levels of volatiles tend to be comparable with milder flavour. This suggests that it will be useful to incorporate parents with defined volatile patterns in breeding programs. Howver, in contrast to sugar or colour analyses, the estimation of aroma compounds is not generally

implemented in carrot breeding practices. Reasons are the absence of knowledge about the inheritance and reliable analytical methods, which have to fulfill special requirements with regard to rapidity and robustness.
Automated headspace SPME-GC was used as a rapid and reliable method for analysis of carrot aroma composition in this study. Plant material in breeding experiments may be characterised by enormous qualitative and quantitative variation of volatile contents. Therefore, pattern recognition was used for data processing to prevent overlooking of qualitative changes (new compounds).
The aim of the research was to study the genetic background of the aroma profiles. In a first step parental lines with defined aromas were selected [3]. These parents were used for creation of a segregating F_2 progeny for linkage analysis. This paper presents partial results of aroma analyses of 200 single plants and the development of molecular markers (AFLP). Furthermore, first results of a genetic mapping approach are discussed.

2. MATERIALS AND METHODS

2.1. Plant material.

The plant material was obtained from the gene bank of the Horticulture Research International (HRI) in Wellesbourne in frame of the EU-project GENRES CT99-105 ('The future of European carrot'). By aroma analyses two parent lines with different terpene profiles were selected [3] to create a segregating F_2 population (n=200) for the heritability studies. The plants grew in plastic pots under greenhouse conditions.

2.2. DNA isolation and physical measurements

A leaf sample (~1 g) was harvested of each plant for DNA isolation [4]. AFLP analyses were performed according to the instruction manual of the manufacturer (Invitrogen, NL). Each carrot root was visually characterised for colour. Following, the single roots were analysed by non-invasive physical methods for colour (LAB system) and sugar content as well as carotene content (NIR spectroscopy).

2.3. Headspace-SPME GC

The details of sample preparation, SPME volatile extraction, and GC conditions as well as GC-MS for substance identification are described by Ulrich [3].

2.4. Genomic analysis

The data for each substance analysed were clustered depending on the expression in the parent lines. The distribution patterns were separated into more or less distinct binary (monogenic inheritance) or nearly continuously Gaussian-distribution (polygenic inheritance). A Chi-square test was used to test goodness-of-fit to expected ratios for monogenic inheritance in the F_2 population.

2.5. Data processing

For chemometric data processing by pattern recognition, the CHROMstat^{TM} version 2.5 software (Analyt Müllheim, Germany) was used. Data input for pattern recognition were percent reports (retention time/peak area data pairs) from the software package Chemstation^{TM} by Agilent. Using the CHROMstat^{TM} software, the chromatograms were divided in 87 time slices, of which each represents a possible peak or substance. All 87 peaks were processed by PCA. The output of pattern recognition and data export was a database comprising the areas of 87 peaks for 200 genotypes. This database was used for the linkage analyses by the software JoinMap 3.0 and MapQTL 4.0 [5].

3. RESULTS AND DISCUSSION

The presented research was a multidisciplinary approach between plant breeding, molecular biology and aroma analysis. In this context a set of multiple analyses was carried out on a single root (the single root weight differed from 1.5 g to 25 g). As a result of this complex work a database was created containing data for rating, molecular markers (AFLP), colour, sugar content, carotene content and 87 volatiles. In this paper first results for correlation of molecular data and aroma analysis are presented. Finally, all data will be used for a segregation and linkage analysis as well as a QTL (quantitative trait loci mapping) approach.

Automated headspace analysis in combination with pattern recognition as data processing is a powerful tool for analysis of aroma pattern. This technique fulfills the requirements for a rapid analysis of hundreds of small samples with a high variability of the volatiles. In Figure 1 the creation of time slices on the basis of 400 chromatograms is shown. The peak area of every time slice was the input for the PCA and the correlation with the genomic data.

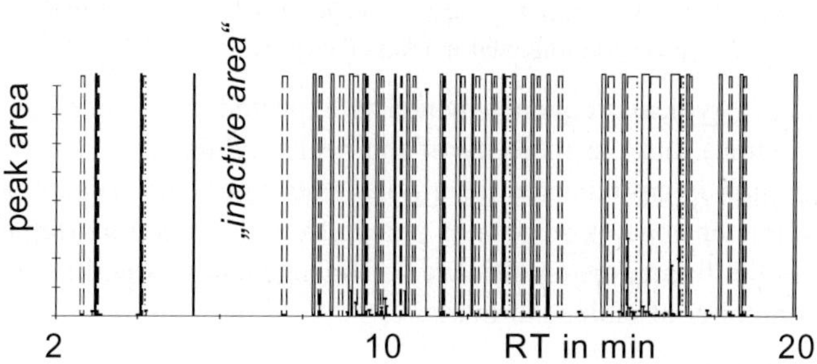

Figure 1. Time area skeleton with 87 time slices. The estimation of time slices is based on the analysis of altogether 400 chromatograms. In inactive areas no peaks (above a defined threshold) occur in all of the 400 chromatograms.

The analysis of the aroma pattern of the F_2 plants regarding the 87 peaks showed a high variability in both quality and quantity. At least 55 peaks seemed to have an informative character useable for analysis of the genetic background. When the parent lines belonged to a typical pinene type (female) and terpinolene type (male) the progeny comprised nearly all possible variants. In Figure 2 results of the frequency analyses of two peaks are presented.

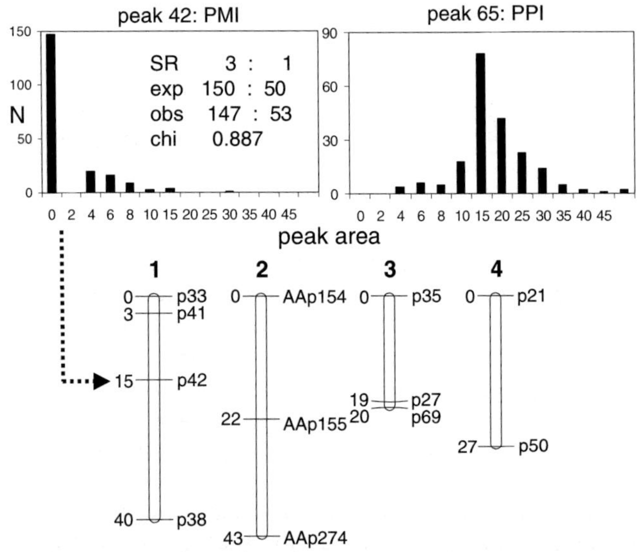

Figure 2. Results of the genetic analysis of 200 roots. *Top:* frequency analysis of aroma pattern. PMI: putative monogenic inheritance; PPI: putative polygenic inheritance (QTL); SR: segregation ratio; exp: expected distribution; obs: observed distribution; chi: chi-square test. Peak 42: unknown compound; peak 65: myristicin (CAS 607-91-0). *Bottom:* preliminary results of linkage analysis with 20 markers and 4 groups. Linkage groups: left site: genetic distance in cM; right site: volatile peaks (pii) and molecular markers (AApiii).

Peak 42 showed a typical frequency which is expected for a monogenic inheritance (Mendel distribution), but the majority of the peaks exhibited a Gaussian distribution like peak 65, which is an indicator for a polygenic inheritance (quantitative trait loci, QTL). Most character impact compounds known from the literature (terpenes) belonged to this group. In the future, the significance of linkage analysis will be improved by the use of a higher number of molecular markers.

References

1. F. Kjeldsen, L.P. Christensen and M. Edelenbos, J. Agric. Food Chem., 49 (2001) 4342.
2. H.E. Patte (ed.), Evaluation of quality of fruits and vegetables, Westport, CT (1985) 315.
3. J.-L. Le Quéré and P.X. Étiévant (eds.), Flavour research at the dawn of the 21st century, proceedings of the 10th Weurman flavour research symposium, Paris, France (2003) 440.
4. S.L. Porebski, G. Bailey and B.R. Baum, Plant Mol. Biol. Rep., 15 (1997) 8.
5. J.W. van Ooijen and R.E. Voorrips, JoinMap 3.0 software for the calculation of genetic linkage maps, Plant Research International B.V., Wageningen, The Netherlands (2001).

W.L.P. Bredie and M.A. Petersen (Editors)
Flavour Science: Recent Advances and Trends

Authentication of biotechnological flavours by isotopic analyses

Carmen Lapadatescu[a], Patrick Taillade[a] and Hervé Casabianca[b]

[a]SAF-ISIS, Zone Artisanale, 40140 Soustons, France; [b]Centre National de la Recherche Scientifique, Service Central d'Analyses, BP 22, 69390 Vernaison, France

ABSTRACT

The validation of naturalness of different flavours (acetaldehyde - 'apple' odour, acetoin - 'rancid butter' odour, 2-methyl and 3-methylbutyric acids - 'rancid cheese' odours) was made possible by developing a multi-component stable isotope picture for each compound. Flavour molecules analysed were of biotechnological or synthetic origin. The isotope results generated data that allowed for comparison of the isotopic abundance of known naturally derived or processed flavours with those of non-natural sources or processes. Our results confirmed that the biotechnological flavours had a natural origin and their isotope ratios were different from those of synthetic ones. Moreover, for a good discrimination at least two or three isotope analyses were necessary to prove the origin of flavours.

1. INTRODUCTION

The demand for natural and healthy food in the past years has also had an impact on the flavour-producing industries, as the majority of today's flavours need to be natural. Chemically synthesised flavour chemicals must under current US and European legislation be labelled 'artificial' or 'nature-identical' and cannot be used for the creation of natural flavours. Very often key flavour chemicals can, however, not be obtained from nature via extraction or distillation at reasonable prices. As an alternative, biotechnology using fermentation and enzymatic reactions can be employed for the production of so-called 'natural' aromas. Isotopic measurements can determine sources and processes by which certain flavours are derived [1]. The aim of the isotopic measurements was, therefore, to authenticate the sources or processes claimed for on the labels of flavours.

The isotope results vary as a function of origin of raw materials. Consequently, it is important to develop databases for flavour molecules from well-defined origins allowing comparisons and confirmation of authenticity. It is important to carry out these analyses since many researchers over the past decades have reported on adulteration or manipulation of synthetic materials to mimic those derived naturally because of important difference between the production costs of natural flavours versus their synthetic counterparts.

2. MATERIALS AND METHODS

2.1. Samples and techniques

Synthetic samples used in this study were supplied by Sigma-Aldrich and the natural samples came from internal SAF-ISIS production or other natural flavour suppliers. The isotopic techniques included the measurement the stable isotope ratios of carbon ($\delta^{13}C$), hydrogen (δD) and oxygen ($\delta^{18}O$). Isotopic analysis for deuterium (δD) was performed using a GV Instrument gas chromatography/pyrolysis/isotope ratio mass spectrometer (GC-P-IRMS). The pyrolysis was carried out by heating chromium at 1050 °C.

2.2. Isotopic analyses

Isotopic measurements were done on H_2 and HD, using H_2 as internal standard. The standard for hydrogen isotopes was V-SMOW (Vienne-standard mean ocean water).
Isotopic analysis of carbon ($\delta^{13}C$) was performed at 850 °C using copper oxide and. CO_2 was used as internal standard and calibrated relative to the NBS19 standard.
Isotopic analysis of oxygen ($\delta^{18}O$) was performed using a Thermofinnigan GC-P-IRMS, where the pyrolysis reaction was carried out at 1300 °C. The columns used for GC-P-IRMS were: INNOWAX 50 m, 320 μm, 0.5 μm for acids and alcohols and DB5 60 m, 320 μm, 0.25 μm for heavy molecules.
The δ value was defined as the per mil (‰) or parts per thousand deviation of the sample isotopic ratio relative to that of a standard and expressed by the equation ‰ = (R sample/R standard-1) x 10^3 where R = $^{13}C/^{12}C$ or D/H or $^{18}O/^{16}O$.

3. RESULTS AND DISCUSSION

Table 1 lists the results of the isotopic analyses ($\delta^{13}C$ and δD) of synthetic and natural acetaldehyde. Rather similar isotopic $\delta^{13}C$ values were seen for the biotechnological acetaldehyde provided by the three companies and, rather unexpected, the one which had a synthetic origin. Furthermore, the natural acetaldehyde sample of Georgia University had a $\delta^{13}C$ isotopic value quite different (-17.7 ± 4.9‰) from the biotechnological ones (-26.6/-29.3‰). The isotopic δD acetaldehyde values from SAF-ISIS, Biotech company 2 and Georgia University samples were similar, but quite different from the δD value of acetaldehyde from Biotech company 1. The δD value of Biotech company 1 sample was quiet close to that of synthetic acetaldehyde. Because of the high volatility of acetaldehyde, isotopic $\delta^{18}O$ analyses could not be performed. In conclusion, it has been demonstrated that it is difficult to distinguish between synthetic and natural samples of acetaldehyde. There are, however, indications that the samples of SAF-ISIS and Biotech company 2 have natural origin while the sample of Biotech company 1 appears to be of synthetic origin.

Table 1. Carbon and hydrogen isotopic values for acetaldehyde.

Supplier	Origin	$\delta^{13}C$ (‰)	δD (‰)	$\delta^{18}O$ (‰)
SAF-ISIS	Natural	-29.3 / -28.9	-150 / -145	n.d.
Biotech company 1	Natural	-29.4 / -24.6	-40 / -45	n.d.
Biotech company 2	Natural	-26.6 / -26.8	-175±9	n.d.
Georgia University [1]	Natural	-17.7±4.9	-202±64	n.d.
Georgia University [1]	Synthetic	-28.5±1.6	-76±6	n.d.

n.d.: not determined; SAF-ISIS samples were produced by a fermentation process. Biotech companies' samples were claimed as natural products. Georgia University samples had a synthetic or botanical origin as indicated. Values separated by a '/' are from duplicate analyses, other values are indicated with their standard deviation.

Table 2 lists the results of the isotopic analyses ($\delta^{13}C$ and δD) of synthetic and natural acetoin. We have observed an important difference between the $\delta^{13}C$ and δD isotopic values of, on the one hand SAF-ISIS and the Biotech company samples, and on the other hand the Sigma-Aldrich sample. Particularly, the δD of Sigma-Aldrich has a positive value +59‰ in opposite of the δD with negative values of SAF-ISIS (-60‰) and Biotech Company (-82‰). The quite near isotopic differences ($\delta^{13}C$ and δD) between both biotechnological companies could be due to the origin of carbon raw materials. Thus the two isotopic analyses ($\delta^{13}C$ and δD) have made it possible to conclude on the naturalness of SAF-ISIS and Biotech company samples.

Table 2. Carbon and hydrogen isotopic values for acetoin.

Supplier	Origin	$\delta^{13}C$ (‰)	δD (‰)	$\delta^{18}O$ (‰)
SAF-ISIS	Natural	-8.10 / -7.9	-60	n.d.
Biotech company	Natural	-20 / -20.1	-82	n.d.
Sigma-Aldrich	Synthetic	-25.4 / -25.6	+59	n.d.

n.d.: not determined; SAF-ISIS samples were produced by a fermentation process. Biotech company's samples were claimed as natural products. Sigma Aldrich sample had a synthetic origin. Values separated by a '/' are from duplicate analyses.

Table 3 lists the results of the isotopic analyses ($\delta^{18}O$ and δD) of synthetic and natural 2- and 3-methylbutyric acid (2- and 3-MBA). The samples of SAF-ISIS and the Chinese company were different concerning the isotopic δD values for 2-MBA and 3-MBA. Moreover, the δD values of natural 3-MBA are different from the synthetic ones (δD = +11.5‰). Additionally, we observed the SAF-ISIS and Chinese company samples were different with regard to the isotopic $\delta^{18}O$ values for 2-MBA and 3-MBA. Firstly, the

$\delta^{18}O$ differences between the two samples are less important as compared with δD values of the same ones and secondly the $\delta^{18}O$ values of the samples are very different from Sigma-Aldrich sample obtained by pure synthesis. It is also seen that for 3-MBA $\delta^{18}O$ the value of SAF-ISIS (-3‰) was identical to that of the biotechnological company. Thus, with these data, it is difficult to obtain a clear conclusion.

Table 3. Carbon, hydrogen and oxygen isotopic value for 2- or 3-methylbutyric acid.

		2-Methylbutyric acid			3-Methylbutyric acid		
Supplier	Origin	$\delta^{13}C$ (‰)	δD (‰)	$\delta^{18}O$ (‰)	$\delta^{13}C$ (‰)	δD (‰)	$\delta^{18}O$ (‰)
SAF-ISIS	Natural	n.d.	-259	-15	n.d.	-176	-3
Chinese Company	Natural	n.d.	-237	-11	n.d.	-142	-6 / -7
Biotech Company	Natural	n.d.	n.d.	n.d.	n.d.	n.d.	-3
Sigma-Aldrich	Synthetic	n.d.	n.d.	n.d.	n.d.	+11.5	+2

n.d.: not determined; SAF-ISIS samples were produced by a fermentation process. Biotech and Chinese companies' samples were claimed as natural products. Sigma Aldrich sample had a synthetic origin. The value separated by a '/' are from duplicate analyses.

In order to conclude on the naturalness of Chinese acids, we did an oxidation experiment of natural 2-MB and 3-MB alcohols by using chemical catalysts and the acids obtained were analysed by $\delta^{18}O$ isotopic measurements (data not shown). The isotopic results obtained were very surprising: the 2-MBA and 3-MBA resulted by chemical oxidation from natural alcohols had almost the same (1‰ of difference) isotopic signature as the natural acids obtained by the biotechnological pathway from natural alcohols. Using two different processes and only one isotopic analysis, we obtained similar isotopic signatures.

4. CONCLUSION

The present study showed that it was not easy to always perform a good discrimination between natural and synthetic flavour compounds. Therefore, attention must be paid to certain products claimed as natural to avoid misleading of customers. 2-MBA and 3-MBA from SAF-ISIS are known to have a natural origin due to values and biotechnological process used, but for the other samples complementary analyses are needed in order to prove their naturalness.

References

1. R.A. Culp and J.E. Noakes, J. Agric. Food Chem., 40 (1992) 1892.

W.L.P. Bredie and M.A. Petersen (Editors)
Flavour Science: Recent Advances and Trends

Exploiting natural microbial diversity for development of flavour starters

Johan E.T. van Hylckama Vlieg, Annereinou Dijkstra, Bart. A. Smit, Wim J.M. Engels, Liesbeth Rijnen, Marjo J.C. Starrenburg, Gerrit Smit and Jeroen A. Wouters

NIZO food research, P.O. Box 20, 6710 BA Ede, The Netherlands

ABSTRACT

The recent elucidation of many of the biochemical pathways involved in flavour formation in fermented food products has given an impetus to the food and ingredients industry to reshape their strain development programs. In this paper we highlight some of the latest developments and illustrate the prospects for starter culture development by exploitation of the diversity in flavour forming capacity among natural isolates.

1. INTRODUCTION

Nowadays, fermentation is not only applied as a means of food preservation but special attention is paid to flavour development. Starter cultures, especially Lactic acid bacteria (LAB), play a crucial role in flavour formation and in recent years the understanding of the physiological characteristics that determine the production of flavour compounds has increased significantly. This is exemplified by the elucidation of the metabolic pathways involved in flavour formation in dairy products. The pathway of casein breakdown leading to the development of various key flavour and off-flavour compounds has drawn particular attention. Initially, caseins are converted to large peptides by rennet and microbial proteases. Most LAB produce an extensive set of peptidases that further degrade these peptides to smaller oligopeptides and amino acids that have desired (for example 'sweet' or 'brothy') taste or an undesired (for example 'bitter') taste. Finally, volatile flavour compounds are produced from amino acids by various enzymatic and non-enzymatic conversions. Several excellent reviews are available that summarise recent advances in research on the microbiology, biochemistry and molecular biology of flavour formation by lactic acid bacteria [1-3].

An important finding has been that a large diversity in desired enzyme activities occurs among natural strains. Consequently, high-performing starter cultures can be developed

by careful selection and combination of strains with desired activities. Recent technological breakthroughs in the field of automated screening and genomics allow the efficient exploitation of this large biodiversity. In such screening programs, miniaturised fermentations are carried out in 96-well format using robotics for liquid handling, key-enzyme activity measurement with colorimetric substrates, and analysis of flavour compounds with GC-TOF or HPLC-MS. Subsequently, strains exhibiting the desired activities are tested in product model systems and pilot product trials leading to the rapid identification of high-performing strains.
The availability of automated screening platforms has given an incentive to the starter and food industries to reshape their strain development programs by targeting the key enzymes, genes and metabolites for these traits. In the current paper we will illustrate the power of this approach by highlighting an example of the development of starters that exhibit desired peptidase activities required for the removal of bitter off-flavours.

2. MATERIALS AND METHODS

Enzyme activity tests were performed on 2 ml GM17-grown cultures in 96-well 2 ml plates in quadruplet. Overnight-grown cultures were centrifuged and washed with 2 ml 50 mM sodium phosphate buffer pH 7.2. Subsequently, cells were disrupted using a mini bead-beater 96 Cell Disrupter (Merlin Diagnostic Systems, The Netherlands) with about 300 μl of 0.1 mm Zirconia / Silica (Merlin Diagnostic Systems BV, The Netherlands) and 1 ml 50 mM sodium phosphate buffer pH 7.2. The resulting suspension was centrifuged (10 min, 8300 g at 4 °C) and the supernatant containing the cell free extract was used to determine enzyme activities. All peptidase activities were determined in 96-well format at 30 °C by online monitoring of the release of *p*-nitroanilide from the following substrates: PepN, lys-*p*-nitroanilide; PepXP, H-Ala-Pro-*p*-nitroanilide; PepA, H-Glu-*p*-nitroanilide

3. RESULTS

As described above, peptidases are key enzymes in the production of flavour compounds from casein in dairy fermentation. In order to assess the diversity of peptidase activities, we have quantified the activity of three peptidases, aminopeptidase N (PepN), X-Pro dipeptidyl-peptidase (PepXP) and glutamyl-aminopeptidase (PepA) in a strain collection of Lactococci. The collection contained three groups of strains, a group of dairy isolates belonging to the subspecies *cremoris,* a group of dairy isolates belonging to the subspecies *lactis*, and a group of wild strains belonging to the subspecies *lactis*. A miniaturised and automated screening procedure performed in a 96-well plate format was used to grow the bacteria and quantify the peptidase activities in crude extracts (Figure 1). The results show that peptidase activities are highly strain dependent and that the average PepN activity in *cremoris* strains is approximately three-fold higher than the average activity in dairy isolates of the subspecies *lactis*. Within the latter subspecies the activity in dairy isolates is two-fold higher than the activity in non-dairy isolates. A similar pattern is observed with PepXP activities. PepA activities did not correlate with subspecies or isolation source. The higher levels of peptidase activities in dairy isolates may reflect their adaptation to the dairy environment where these enzymes may provide effective access to the amino acids in dairy proteins.

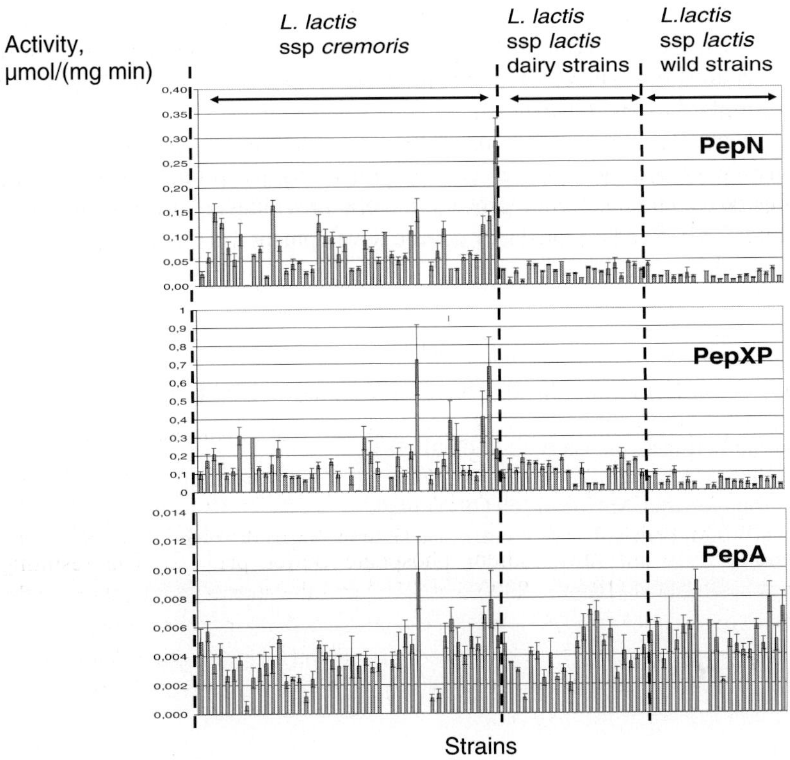

Figure 1. Diversity of aminopeptidase N (PepN), X-Pro dipeptidyl-peptidase (PepXP) and peptidaseXP and Glutamyl-aminopeptidase (PepA) peptidase activity among *L. lactis* subsp *cremoris* and *L. lactis* subsp. *lactis* grown in LM17 medium.

The power of the approach described for industrial strain development is exemplified by the development of a debittering starter culture. A bitter off-flavour may occur in many food fermentations and is often caused by unbalanced proteolysis resulting in the accumulation of certain hydrophobic peptides with a strong bitter taste. It has been shown that PepN produced by *L. lactis* is capable of degrading a bitter peptide that accumulates in bitter cheese [4]. Extensive screening programs have helped to identify strains that produce high levels of PepN. Strains that combine the high PepN activity with other beneficial characteristics have been shown to eliminate the bitter taste of certain cheese products. Some of these strains are currently marketed as commercial starter cultures.

4. CONCLUSIONS AND FUTURE OUTLOOK

In recent years the screening for flavour generating cultures has developed from a trial- and error process, limited by lack of knowledge on the available cultures, to a process that efficiently screens strains for the desired combination of flavour forming enzyme activities. By operating different analytical techniques in an automated screening platform, the functionalities and flavour compounds that can be screened for are almost unlimited. It is now feasible for example to screen large numbers of strains for the production of a specific flavour compound [5] or true flavour profiles. Moreover, performing such screenings in matrices closely mimicking the product increases the predictive value of the strain selection process, thereby providing an effective means of valorising natural LAB diversity for starter culture development.

References

1. J.E. Christensen, E.G Dudley, J.A Pederson and J.L. Steele, Antonie Van Leeuwenhoek Int. J. Gen. Molec. Microbiol., 76 (1999) 217.
2. G. Smit, J.E.T. van Hylckama Vlieg, B.A. Smit, E.H.E. Ayad and W.J.M. Engels, Aust. J. Dairy Technol., 57 (2002) 61.
3. M. Yvon and L. Rijnen, Int. Dairy J., 11 (2001) 185.
4. P.S. Tan., T.A. van Kessel, F.L. van de Veerdonk, P.F. Zuurendonk, A.P. Bruins and W.N. Koning, Appl. Environ. Microbiol., 59 (1993) 1430.
5. B.A. Smit, W.J.M. Engels, J.T.M. Wouters and G. Smit, Appl. Microbiol. Biotechnol., 64 (2004) 396.

W.L.P. Bredie and M.A. Petersen (Editors)
Flavour Science: Recent Advances and Trends

Lipase catalysed formation of methylthioesters using a continuous reactor

Hans Colstee, Marc van der Ster and Peter van der Schaft

I.F.F. (Nederland) B.V., Zevenheuvelenweg 60, P.O. Box 5021, 5004 EA Tilburg, The Netherlands

ABSTRACT

Enzymatic thioesterification has been described in literature, but no method has been described for esterification of methanethiol and carboxylic acids, which is mainly caused by the difficult handling at room temperature of the gas methanethiol. Natural methylthioesters were prepared in a solvent-free system from natural methanethiol (obtained from methionine) and carboxylic acids, by means of catalysis by immobilised lipase B from *Candida antarctica* (Novozym®435). For this purpose a laboratory scale, closed system vessel was designed consisting of a column reactor (containing the immobilised enzyme), substrate and product. In this equipment S-methyl propanethioate and S-methyl 3-methylbutanethioate could be produced during four weeks continuously. The immobilised lipase could be re-used for at least 4 additional runs.

1. INTRODUCTION

Production of foods and beverages requires the use of flavours and, especially in the Western world the use of aromas that can be designated as natural. This implies that both the raw materials and the processing are natural. In this respect, physical processes and biotransformations using enzymes or microbes are regarded as natural.

Natural methylthioesters of short-chain fatty acids three to eight carbons in length are of great interest in the flavour industry. Flavour substances like S-methyl butanethioate and S-methyl 3-methylbutanethioate are important constituents of dairy aromas, especially cheese aroma and of fruit aromas, like strawberry [1].

Enzymatic esterification of organic acids and alcohols using lipases is a well described phenomenon in literature and results in high yielding processes for the production of natural esters. These processes are quite often characterised as a solvent-free system and the lipase is immobilised on a solid support. The feasibility of enzymatic thioesterification between oleic acid and butanethiol in n-hexane with an immobilised lipase was demonstrated in the past [2]. At the same time also lipase catalysed esterification of short-chain flavour esters was demonstrated [3]. Later trans-thioesterification of fatty acid methyl esters with alkanethiols resulting in the formation of long-chain acyl thioesters was shown [4]. In addition, production of low concentrations of methylthioesters by *Geotrichum candidum* in a liquid cheese model medium was shown [5], but no method has been described so far for the efficient industrial production of natural methylthioesters for flavour use. This is mainly caused by the availability of natural methanethiol and the difficult handling of methanethiol which is a gas at room temperature. This study describes the preparation of natural methylthioesters in a solvent free system from natural methanethiol (derived from methionine) and short-chain fatty acids, by means of catalysis by immobilised lipase B from *Candida antarctica* (Novozym 435). For this purpose a laboratory scale, closed system vessel was designed consisting of a column reactor (containing the immobilised enzyme), substrate and product.

2. MATERIALS AND METHODS

2.1. Reagents

Natural methanethiol was prepared from natural methionine (internal IFF source). Natural propanoic acid and isovaleric acid were purchased on the global market. Novozym®435 (immobilised lipase B from *Candida antarctica*) was obtained from Novo Nordisk.

2.2. Analyses

Samples were analysed by means of gas-liquid chromatography (GLC) on a wax-52 type column. The structures of reaction products were identified by GC-MS.

2.3. Product recovery

After cooling the product collection vessel to room temperature, the content was collected in a flask. Applying low vacuum on the product resulted in the removal of residual methanethiol which was collected for reuse. Subsequently the flask content was fractionated by vacuum distillation at 35 mbar. Pooled selected fractions were fractionated again using vacuum distillation and finally fractions were pooled resulting in >98% pure methylthioester.

3. RESULTS AND DISCUSSION

3.1. Reaction design and conditions

A closed system vessel was designed for the production of natural methylthioesters from methanethiol and carboxylic acids (Figure 1). The enzyme column reactor (1) (volume 350 ml) is charged with 150 g Novozym 435 and placed in-line in the unit via connectors. The unit is filled with methanethiol by connecting a small autoclave containing 850 g liquid natural methanethiol (-25 °C) to the product collection vessel (2) with a volume of 5 l, which is also cooled to -25 °C and to which vacuum is applied until 450 mbar. By this action the methanethiol will be transferred to the product collection vessel. The vacuum will be relieved from the product collection vessel and the vessel will be slowly heated to 55 °C resulting in boiling of the methanethiol which will be condensed in the condenser (3) which is cooled at 8-12 °C.

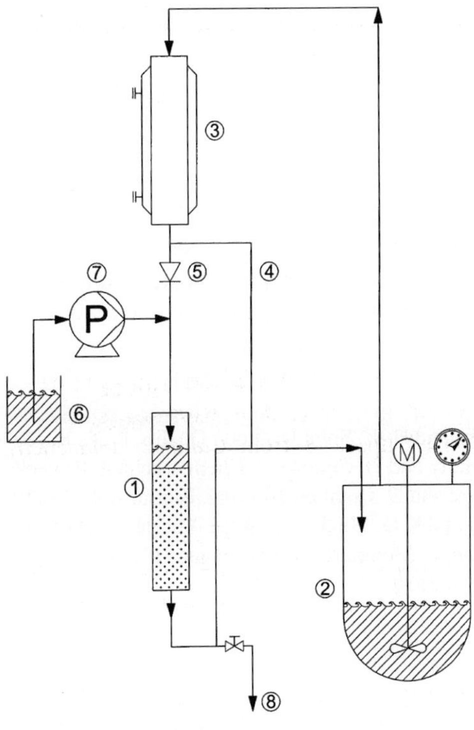

Figure 1. Closed system vessel for the production of natural methylthioesters. See text for explanation of the numbers.

Via line (4) the condensed methanethiol will flow back to the product collection vessel, so there is complete recirculation of methanethiol. The liquid methanethiol coming from condenser (3) can also be dosed to the column reactor by means of a valve (5). The column reactor (1) will be held at 50 °C. The whole system is under pressure (around 4 bar). When the enzyme column is filled with methanethiol, the methanethiol will flow to the product collection vessel (2). At this point the natural carboxylic acid can be dosed from vessel (6) through pump (7) to the enzyme column reactor, where the thioesterification reaction takes place. Methylthioester and carboxylic acid will be collected in the product collection vessel. Through valve (8) samples can be taken from the reaction mixture leaving the enzyme column reactor in order to determine the conversion.

S-methyl propanethioate and S-methyl 3-methylbutanethioate production

Natural carboxylic acid was pumped from vessel (6) into the enzyme reaction unit at about 7 g/h. Via valve (8) in-process samples were taken and the conversion was determined. The conversions ranged from 5 to 10% (w/w) for S-methyl propanethioate and from 10 to 15% (w/w) for S-methyl 3-methylbutanethioate based on the carboxylic acid. During the weekends the process was interrupted. After four weeks the process was terminated and the mixture in the product collection vessel was collected and analysed. Based on this analysis the overall conversion was 6% for S-methyl propanethioate and 11% for S-methyl 3-methylbutanethioate based on the amount of added carboxylic acid.

4. CONCLUSIONS

This closed system vessel consisting of a column reactor (containing the immobilised enzyme), substrate and product makes the efficient production of natural methyl thioesters both technically and economically feasible. In this equipment S-methyl propanethioate and S-methyl 3-methylbutanethioate could be continuously produced during four weeks and the immobilised lipase could be re-used for at least 4 additional runs.

References

1. N. Martin, V. Neelz, and H.-E. Spinnler, Food Quality Pref., 15 (2004) 247.
2. M. Caussette, A. Marty and D. Combes, J. Chem. Technol. Biotechnol., 68 (1997) 257.
3. D. Cavaille-Lefebvre and D. Combes, Biocatal. Biotransfor., 15 (1997) 265.
4. N. Weber, E. Klein and K.D. Mukherjee, Appl. Microbiol. Biotechnol., 51 (1999) 401.
5. C. Berger, J.A. Khan, P. Molimard, N. Martin and H.E. Spinnler, Appl. Environ. Microbiol., 65 (1999) 5510.

W.L.P. Bredie and M.A. Petersen (Editors)
Flavour Science: Recent Advances and Trends

Cloning and characterisation of the main intracellular esterase from *Lactobacillus rhamnosus* HN001

Marie-Laure Delabre, Julie Ng, Stephanie Wingate, Shao Q. Liu, Emily Chen, Tianli Wang, Ross Holland and Mark W. Lubbers

Fonterra Research Centre, Fonterra – Palmerston North, Private Bag 11029, Palmerston North, New Zealand

ABSTRACT

In cheese, the breakdown of milk fat into free fatty acids (FFAs) and esters by lipases and esterases contributes to flavour development. Short-chain FFAs and short-chain ethyl esters are important for the flavour of Italian-style cheeses such as Parmesan, Grana Padano, Romano and Provolone. Esterase activity has been identified in the lactic acid bacteria (LAB) used as starter bacteria and in the adjunct microflora during cheese manufacture. A novel esterase gene, designated AA7, was identified in the genome sequence of *Lactobacillus rhamnosus* HN001. AA7 was shown to be the main intracellular esterase of *L. rhamnosus* HN001. The enzyme was characterised for both hydrolytic release of FFAs and alcoholytic synthesis of ethyl esters from synthetic substrates. AA7 was shown to be particularly active on short-chain acyl substrates and had a preference for monoacylglycerol substrates. AA7 was used in cheese manufacture and catalysed an accumulation of butyric acid and ethyl esters in cheese, two of the most important flavour compounds in Italian-style cheeses.

1. INTRODUCTION

Cheese flavour is the result of the breakdown of protein, fat and carbohydrates by native milk enzymes, added enzymes, starter bacteria and the secondary microflora of cheese. Lipolysis of milk fat in ripening cheese produces FFAs, which contribute directly to cheese flavour by imparting specific fatty acid flavour notes. In dairy systems, alcoholysis of milk fat generates esters, which are important for the development of the fruity flavour in Italian-style cheeses [1]. Long-chain fatty acids (above C_{12}) may impart an undesirable 'soapy', 'tallowy' note. Therefore, selectivity of lipase/esterase for short-chain fatty acids is required to produce desirable flavour. Microbial lipases are usually non-selective and, to date, only lipases from a mammalian sources exhibit the desirable short-chain selectivity. However, the use of mammalian lipase can be undesirable based on religious, vegetarian diet or health perception reasons. Therefore, a microbial lipase/esterase with selectivity for short-chain fatty acids is highly desired by the dairy industry. *L. rhamnosus* HN001 (DR20™), a strain isolated from cheese, exhibits desirable flavour and probiotic properties. We are investigating *L. rhamnosus* HN001 enzymes involved in flavour formation. In this work, we identified, cloned and enzymatically characterised an esterase, AA7, from *L. rhamnosus* HN001. We showed a unique selectivity for short-chain FFAs and esters by AA7 in a cheese model.

2. MATERIALS AND METHODS

2.1. Enzyme preparations

The AA7 gene (GenBank accession number AR304334) was cloned by PCR and was expressed as a glutathione S-transferase (GST) fusion protein in *Escherichia coli*, purified and cleaved from its GST tag with PreScission protease™ (Amersham Biosciences, Uppsala, Sweden). The major intracellular esterase of HN001 was purified as described elsewhere [2]. The N-terminal amino acid sequence of the native and recombinant enzyme was identified by N-terminal protein sequencing. The substrate specificity of the esterases was determined using *p*-nitrophenyl (*p*-NP) esters of fatty acids C_2 to C_{10}. The standard assays for alcoholysis [3] and hydrolysis [4] were used on the recombinant enzyme (GST tag removed). *E. coli* cells expressing AA7 enzyme, or transformed with the empty vector (control), were disrupted using a French Press. Unbroken cells and membranes were removed by centrifugation.

2.2. Cheese preparation and analyses

An Italian-style cheese was manufactured from skim milk blended with cream or partially hydrolysed cream. The cream was partially hydrolysed using a commercial microbial lipase to form di- and monoacylglycerols (DAGs and MAGs), with 2% of the esterified fatty acids released from the milk fat, followed by heat to inactivate the added lipase and prevent further hydrolysis. Fifty millilitres of cell free extract CFE (17 mg/ml) or Phosphate Buffer Saline (PBS) was mixed with 1 kg of fresh curd. Ethanol was added to a final concentration of 0.1 M. The cheeses were vacuum sealed and stored for 8 weeks at 13 °C. Esters in the cheeses were then analysed as described elsewhere [5]. The method used for FFA analysis is described elsewhere [6].

3. RESULTS

3.1. Identification and purification of *L. rhamnosus* HN001 esterase AA7

AA7, an open reading frame encoding a potential esterase enzyme, was identified in the genome sequence of *L. rhamnosus* HN001 [7] and showed 89% identity to an esterase gene, designated estB from *L. casei* LILA [8]. AA7 consists of a 954 bp open reading frame encoding a putative protein of 35.7 kDa. The deduced amino acid sequence contains the characteristic GXSXG serine hydrolase motif found in most lipases and esterases. The gene did not encode an N-terminal signal sequence, which is required for extracellular localisation. The N-terminal amino acid sequence determined for the main intracellular esterase purified was identical to the sequence of the recombinant protein, except for the five additional residues GPLGS corresponding to the remaining GST tag.

3.2. Substrate selectivity of AA7 for hydrolysis of *p*-NP esters of short-chain fatty acids, and for hydrolytic and alcoholytic activity on acylglycerol

The AA7 native and recombinant esterases exhibited similar hydrolytic activities on *p*-NP (*p*-nitrophenyl) ester substrates (data not shown). AA7 was most active against *p*-NP hexanoate and was inactive on *p*-NP esters of straight-chain fatty acids greater than C_{10}. The hydrolytic and alcoholytic activities of AA7 were assayed using synthetic triacylglycerols (TAGs), DAG and MAG substrates containing fatty acids of different chain length (data not shown). Maximum hydrolytic and alcoholytic activity was found with monocaprin (C_{10}) and the activity decreased as the carbon chain length increased. Less activity was found with DAG substrate and only residual activity with TAG substrate.

3.3. Hydrolytic activity of AA7 in a cheese model

Two cheese curds were manufactured with normal (control) and partially hydrolysed cream to generate MAGs and DAGs, the preferred substrates of AA7. CFE containing AA7 had a specific activity (hydrolysis of *p*-NP butyrate) of 11.7 μmol/min/mg. PBS or CFE from *E. coli* (without AA7), had no observable impact on FFA and ester accumulation in cheese (data not shown). AA7 generated hexanoic acid and butyric acid in the control cheese; however, the amount of FFAs released was higher in a cheese made with the partially hydrolysed cream (Table 1). The concentration of longer chain FFAs (above C_{12}) was not significantly affected by the presence of AA7 (data not shown).

Table 1. Free fatty acids (FFAs) concentration (mmol/kg) in cheese.

	Hexanoic acid	Butyric acid
Control	0.03	0.01
Control + AA7	0.13	0.34
Partially hydrolysed	0.12	0.53
Partially hydrolysed + AA7	0.36	1.24

3.4. Alcoholytic activity of AA7 in a cheese model

Ethanol is required for the synthesis of ethyl esters. Without added ethanol, esters were barely detectable (data not shown). Higher amounts of esters were synthesised in the control cheese manufactured with partially hydrolysed cream, probably due to a residual ester synthesis activity from the starter or from the deactivated microbial lipase (Table 2). Nevertheless, it is clear that the presence of AA7 increased the production of esters in cheese and this occurred to a greater extent in the cheese that contained partially hydrolysed cream.

Table 2. Cheese ethyl ester composition (peak area x 10^6) in the presence of 0.1 M added ethanol.

	Ethyl hexanoate	Ethyl butyrate
Control	0	1
Control + AA7	6	9
Partially hydrolysed	4	19
Partially hydrolysed + AA7	55	46

4. DISCUSSION AND CONCLUSION

We have shown that the main intracellular esterase from *L. rhamnosus* is AA7, which exhibited a selective activity on short-chain acyl substrates and a preference for MAGs. As 95% of bovine milk fat is TAG, which are a poor substrate for AA7, it was important to generate MAG and DAG in cheese for a greater accumulation of short-chain FFAs and ethyl esters. From our study, we can conclude that cheese flavour could be finely controlled by manipulating three factors, the esterase concentration, the alcohol availability and the degree of milk fat hydrolysis.

References

1. S.-Q. Liu, R. Holland and V.L. Crow, Int. Dairy J., 14 (2004) 923.
2. S.-Q. Liu, R. Holland and V.L. Crow, Int. Dairy J., 11 (2001) 27.
3. S.-Q. Liu, R. Holland and V.L. Crow, Appl. Microbiol. Biotechnol., 63 (2003) 81.
4. L. Fernandez, M.M. Beerthuyzen, J. Brown, R.J. Siezen, T. Coolbear, R. Holland and O.P. Kuipers, Appl. Environ. Microbiol., 66 (4) (2000) 1360.
5. S.-Q. Liu, R. Holland and V.L. Crow, J. Dairy Res., 70 (2003) 359.
6. C. De Jong and H.T. Badings, J. High Resolut. Chromatogr., 13 (1990) 94.
7. T. Klaenhammer *et al.*, Antonie Van Leeuwenhoek Int. J. Gen. Molec. Microbiol., 82 (2002) 29.
8. K.M. Fenster, K.L. Parkin and J.L. Steele, J. Dairy Sci., 86 (2003) 2547.

W.L.P. Bredie and M.A. Petersen (Editors)
Flavour Science: Recent Advances and Trends

Influence of pH and carbon source on the production of vanillin from ferulic acid by *Streptomyces setonii* ATCC 39116

Nina Gunnarsson[a] and Eva Akke Palmqvist[b]

[a]Fluxome Sciences A/S, Søltofts Plads, Technical University of Denmark, Building 223, Søltofts Plads, DK-2800, Lyngby, Denmark; [b]Danisco Innovation Copenhagen, Langebrogade 1, DK-1001 Copenhagen K, Denmark

ABSTRACT

In the present study, the influence of pH and carbon source on the redox reactions involved during bioconversion of ferulic acid to vanillin, vanillic acid and vanillyl alcohol by *Streptomyces setonii* ATCC 39116 is discussed.

1. INTRODUCTION

A problem often encountered during production of aromatic aldehydes by bioconversion is the immediate oxidation or reduction of the aldehyde to products of no direct interest. Using *S. setonii*, a vanillin concentration of 6.4 g/l with a yield of 68% (mol/mol) was obtained during bioconversion of ferulic acid in shake-flask experiments at pH 7.2, and even higher vanillin concentration and yield have been reported in a bioconversion process with *S. setonii* ATCC 39116 where pH was controlled to 8.5 [1,2]. In the present study, the influence of pH and carbon source on ferulic acid conversion in *S. setonii* ATCC 39116 was investigated. Glucose and arabinose were chosen as carbon-sources, since they constitute a large fraction of sugar beet pulp, potentially a cheap raw material for ferulic acid production.

2. MATERIALS AND METHODS

The inoculum preparation and bioconversions were performed as described previously [1]. Ferulic acid, vanillin and vanillic acid were separated on a LiChrospher 100 RP-8 column (Merck, Darmstadt, Germany), operating at 25 °C with a methanol/Na-acetate buffer (pH 4.8) gradient. The proportion of methanol in the gradient was according to the following scheme (minutes, % methanol): (0, 0); (20, 50); (23, 100); (29, 100); (45, 0). Detection was performed with a diode array detector (Shimadzu, Tokyo, Japan) at 254 nm for vanillic acid, 325 nm for ferulic acid and vanillin and 280 nm for vanillyl alcohol.

3. RESULTS

A series of batch fermentations were performed. ATCC 39116 was cultivated at pH 7.2, and after 14-16 h, while the cultures were in the exponential growth-phase, ferulic acid was added to a final concentration of 60-79 mmol/l. After the addition of ferulic acid, pH was set to 7.2, 8.2 or 8.5, and then kept constant during the bioconversions. These three pH conditions were applied in cultivations with glucose or arabinose as the carbon source.

Table 1. Specific conversion rates and maximum product yields and concentrations (s: carbon-source; Glc: glucose; Ara: arabinose; Q: specific production rate [mmol/(g h)]; Y: product yield on consumed FA (mol/mol); C_{max}: maximum concentration (mmol/l); FA: ferulic acid; V: vanillin; VA: vanillic acid; VOH: vanillyl alcohol).

Fermentation (s: pH)	Q_{FA}	Q_V	C_{Vmax}	Y_{Vmax}	C_{VAmax}	Y_{VAmax}	C_{VHOmax}	Y_{VOHmax}
Glc: 7.2	-0.67	0.14	14.3	0.38	8.6	0.11	44.8	0.57
Glc: 8.2	-0.93	0.83	61.4	0.90	18.5	0.25	16.4	0.22
Glc: 8.5	-0.84	0.75	38.6	0.84	40.6	0.64	5.3	0.08
Ara: 7.2	-0.99	0.14	5.4	0.07	0.9	0.01	58.9	0.81
Ara: 8.2	-2.78	1.43	31.5	0.48	16.0	0.21	-	-
Ara: 8.5	-2.33	1.26	26.6	0.47	37.1	0.62	-	-

The results of the bioconversions are summarised in Table 1. At pH 7.2, low vanillin yields were obtained due to reduction of vanillin to vanillyl alcohol. This was more pronounced when arabinose was used as the carbon source. The highest ferulic acid conversion rates and vanillin production rates were obtained at pH 8.2, both when glucose and arabinose were used as carbon source. At pH 8.5, the vanillin yield decreased due to oxidation to vanillic acid. The ferulic acid conversion rates were several fold higher when arabinose was used as the carbon source. Also, the rates of degradation of the vanillic acid produced at pH 8.2 and 8.5 were higher when arabinose was used as the carbon source (data not shown).

4. DISCUSSION

The results of the present study demonstrate that the bioconversion of ferulic acid to vanillin, vanillic acid and vanillyl alcohol is strongly influenced by the medium pH. The intracellular pH of *Streptomyces* species has been shown to vary with extracellular pH

[3]. It is, therefore, possible that the observed influence of extracellular pH can be explained by the pH-dependency of the intracellular reactions involved in the bioconversion. The stoichiometry of vanillin oxidation and reduction depends on pH. This is due to the difference in pK_a-values at the phenolic group of vanillin (pK_a 7.38 [4]) and the products of its oxidation and reduction (pK_a 9.55 [5] and 9.95 (CAS Registry 2001), respectively). In the case of vanillin reduction when pH is well below 7.38 both vanillin and vanillyl alcohol are predominantly present in their protonated form. At pH well above 9.95, vanillin and vanillyl alcohol are both deprotonated at the phenolic group. In both cases, H^+ is not a substrate or product in the total reaction. However, at intermediate pH values, reduction of deprotonated vanillin to protonated vanillyl alcohol occurs. In this case, the stoichiometry of the total reaction is altered, and H^+ appears as a substrate in the total reaction. In the oxidation of vanillin to vanillic acid, H^+ is a product of the total reaction when pH is well below 7.38 or well above 9.55 An altered stoichiometry, with no consumption or production of protons, results in the intermediate pH range.

The energetic feasibility of the above redox reactions can be described in terms of the change in actual redox potential for the total reaction (ΔE). ΔE is dependent on the concentrations of the substrates and products of the reactions [6]. Since in these reactions, protons only occur as a substrate or a product of the total reaction in certain pH ranges, the energetic feasibility will be influenced by pH in a dual manner. This pH dependence can be assessed by calculating the redox potential differences, in terms of $\Delta E\text{-}\Delta E^0$, of the total reactions as a function of pH, while taking into account the concentrations of the protonated and deprotonated forms of the substrate and product, as defined by their pK_a–values. A positive change in ΔE corresponds to a negative change in Gibb´s free energy and, thus describes a more energetically favourable reaction. The oxidation of vanillin to vanillic acid generally becomes more energetically favourable as pH increases. In the pH region from 6 to 11, there is, however, a window of decreased energetic feasibility, with a minimum at pH 8.4. The ΔE of vanillin reduction to vanillyl alcohol similarly displays a decrease in approximately the same pH region and reaches a minimum at pH 8.7, while it is independent of pH outside of this range. As could be expected, these windows of decreased energetic feasibility are consistent with the pH ranges where vanillin is deprotonated at the phenolic group while vanillic acid and vanillyl alcohol are not. The minima of the curves correspond to the pH values where vanillin oxidation and reduction are the least energetically favoured and thus, only considering the thermodynamics of the process, the pH optimum for vanillin accumulation is in the pH range from 8.4 to 8.7.

The results from the bioconversions can be partly explained by the pH dependence of the reactions discussed above. A prominent decrease in the amount of vanillyl alcohol formed occurred when pH was increased from 7.2 to 8.2, which could indicate that this change in extracellular pH forced the intracellular pH into the region where vanillin reduction to vanillyl alcohol is less favourable. The further pH increase, from pH 8.2 to pH 8.5, resulted in even less formation of vanillyl alcohol. The oxidation of vanillin to vanillic acid was, however, more prominent at pH 8.2 than at pH 7.2, and even more prominent at pH 8.5. The equilibrium of the oxidation is, however, shifted by the

reactions leading to degradation of vanillic acid. Vanillic acid has been reported to be degraded via the β-keto-adipate pathway in *S. setonii,* using the ring-cleaving enzyme protocatechuate 3,4-dioxygenase [7]. The *in vitro* activity of protocatechuate 3,4-dioxygenase from *Streptomyces* sp. strain 2065 was shown to increase 4.5 times when the pH was increased from 6.5 to 9.5 [8], in good agreement with the pH-dependence of vanillic acid degradation shown in the present study.
The specific ferulic acid conversion rates in the experiments with arabinose as carbon source were higher than in the corresponding experiments with glucose as carbon source (Table 1). Arabinose enters the metabolism through the pentose phosphate pathway, which might lead to decreased flux through the pathway were $NADP^+$ is reduced to NADPH. The availability of $NADP^+$ co-factors and, indirectly, NAD^+ co-factors would thus increase in bacteria growing on a pentose carbon source. Ferulic acid conversion in *S. setonii* has been reported to require NAD^+ co-factors [7], and it is thus probable that the specific ferulic acid conversion rate is positively influenced by an increased availability of these co-factors when a pentose sugar is used as carbon source. With both carbon sources, the specific ferulic acid conversion rate increased when pH was increased from 7.2 to 8.2.

References

1. A. Muheim, B. Müller, T. Münch and M. Wetli, Process for the production of vanillin, European patent application, EP 0 885968 AI (1998).
2. A. Muheim and K. Lerch, Appl. Microbiol. Biotechnol., 51 (1999) 456.
3. L. Quirós and J. Salas, FEMS Microbiol. Lett., 141 (1996) 245.
4. R.A. Robinson and A.K. Kiang, Trans. Faraday Soc., 51 (1955) 1398.
5. A.V. Kurzin, A. Yu. Platonov, E.I. Evstigneev and E.D. Maiorova, Russ. J. Gen. Chem., 67 (9) (1997) 1475.
6. D. Nicholls and S. Ferguson, Bioenergetics 2, London, UK (1992) 48.
7. J.B. Sutherland, D.L. Crawford and A.L. Pometto, Can. J. Microbiol., 29 (1983) 1253.
8. S.G. Iwagami, K. Yang and J. Davies, Appl. Environ. Microbiol., 66 (4) (2000) 1499.

3

Flavours generated by enzymes and biological systems

Enzymatic conversions involved in the formation and degradation of aldehydes in fermented products

Gerrit Smit[a], Bart A. Smit[b], Wim J.M. Engels[b], Johan van Hylckama Vlieg[b], Johanneke Busch[a] and Max Batenburg[a]

[a]Food Research Centre, Unilever R&D Vlaardingen, P.O. Box 114, 3130 AC Vlaardingen, The Netherlands; [b]NIZO food research, P.O. Box 20, 6710 BA Ede, The Netherlands

ABSTRACT

The biochemical pathway leading to the formation and degradation of aldehydes, was studied with special emphasis on the identification of a branched-chain α-keto acid decarboxylase (KdcA). Using a random mutagenesis approach the gene encoding the KdcA was identified. In order to do this, a high throughput screening method was developed, based on measuring volatile metabolites by direct-inlet mass spectrometry (DI-MS). The gene of this enzyme was found to be highly homologous to the gene annotated as *ipd* in *Lactococcus lactis* IL1403 genome, which gene product is probably inactive due to a deletion at the 3' terminus of the *ipd*-gene encoding a truncated non-functional decarboxylase. The enzyme was further characterised using a KdcA over-expressing strain. Knowledge of the pathway of formation and degradation of aldehydes gives new opportunities in effectively improving the flavour profile of fermented food products, with a fermentation of soy milk by lactic acid bacteria as an example.

1. INTRODUCTION

Flavour compounds in various fermented products arise mainly from the action of enzymes from bacterial starter cultures used. In case of dairy fermentations, the formation of flavours involves various chemical and biochemical conversions of milk components. Three main pathways can be identified: the conversions of lactose (glycolysis), fat (lipolysis), and caseins (proteolysis). The predominant organisms in these starters are lactic acid bacteria (LAB).

The conversion of caseins is undoubtedly the most important biochemical pathway for flavour formation in semi-hard and hard cheeses. Degradation of caseins by the activities of rennet enzymes, and the cell-envelope proteinase and peptidases from LAB yields small peptides and free amino acids. For specific flavour development, further conversion of amino acids to various aldehydes, alcohols, acids, esters and sulfur compounds is required (see [1] for a review). A simplified scheme of the different pathways for leucine is shown in Figure 1.

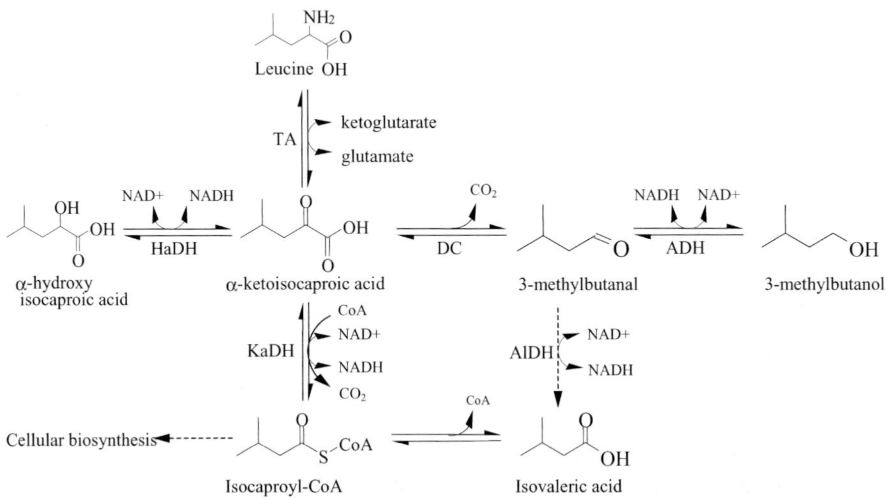

Figure 1. Reaction scheme of a simplified degradation pathway of leucine. TA: transaminase; HaDH: hydroxy acid dehydrogenase; DC: keto acid decarboxylase (KdcA); ADH: alcohol dehydrogenase; AlDH: aldehyde dehydrogenase; KaDH: keto acid dehydrogenase.

This paper focuses on pathways leading to the formation (and degradation) of aldehydes, which are key flavour components in many fermented products, as well as causing off-flavours in other products (e.g. soy based products). Special emphasis is on the characterisation of the decarboxylase enzyme, since the decarboxylating step has been described as rate-limiting in the formation of branched-chain aldehydes [2].

2. MATERIALS AND METHODS

2.1. DI-MS screening

The method is based on the injection of a small headspace sample (100 μl) from culture vials (2 ml) in 96-well format directly into a Mass Spectrometer (Direct Inlet Mass Spectrometry, DI-MS; Trace-MS, Interscience, Breda, The Netherlands). A representative mass fragment was used for identification; *m/z* 58 for 3-methylbutanal, *m/z* 72 for 2-methylpropanal and a *m/z* of 105 for benzaldehyde. The analysis of one sample took less than 1 min using this method, with a coefficient of variation for the response of less than 5%. Thus in one day over 1500 samples could be analysed. See Smit *et al.* [3] for further details.

2.2. Knock-out library, cloning and overproduction of decarboxylase

A random insertion mutant library of the decarboxylase producing strain *Lactococcus lactis* B1157 was constructed using the thermo-sensitive integration plasmid pGh9:ISS1, which was isolated from *E. coli*:pGh9:ISS*1* and transformed to *Lactococcus lactis* B1157 [4]. Over 8500 integrants were selected on LM17 agar containing 2 µg/ml erythromycin, individual colonies were organised in microtiter plates containing LM17 with erythromycin and 20% glycerol, and frozen at -80 °C. Subsequently, small scale 96-well microtiter fermentations were inoculated from these plates. The nisin controlled expression system was used to induce the enzyme in cells harbouring the cloned gene. After induction of gene expression, the cells were harvested and the cell free extracts were obtained and used for the enzymatic assay. For details on the cloning and over-expression of the decarboxylase gene, see Smit *et al.* [5].

3. RESULTS AND DISCUSSION

In order to identify the gene coding for the α-keto acid decarboxylase a high throughput screening procedure was developed based on measuring volatile metabolites by direct-inlet mass spectroscopy (DI-MS) to screen a knock-out mutant library for clones lacking the enzyme activity. Figure 2 gives an example of the profile of 3-methylbutanal injections using this method.

Decarboxylase-negative clones in the mutant library were identified by quantifying the level of 3-methylbutanal in miniaturised fermentations, in which individual mutants were grown in a 96-wells stainless steel blocks, containing 2 ml wells, filled with 1 ml LM17 medium with 2 mM α-ketoisocaproic acid (KICA). After fermentation for 24 h at 40 °C, individual headspace samples (100 µl) were analysed using DI-MS. The amount of 3-methylbutanal in the sample was calculated from the response at *m/z* 58.

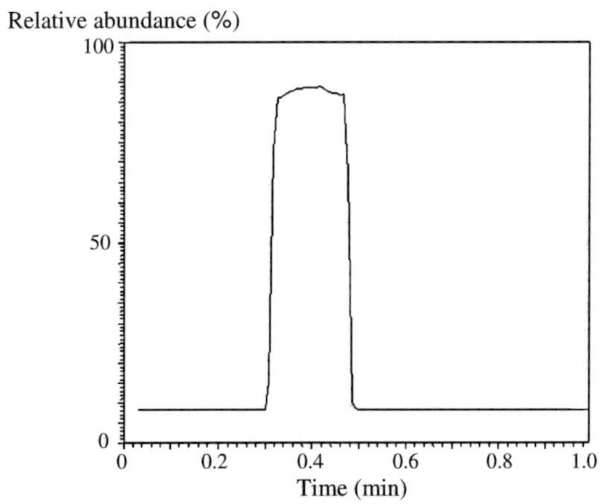

Figure 2. Example of a typical response of DI-MS, with 500 µl sample containing 100 µM 3-methylbutanal injected at 50 µl/s into the system, which was running with a pre-column pressure of 15 kPa and a split flow of 15 ml/min.

The screening of 2500 mutants resulted in the identification 2 clones lacking KdcA activity. Recovery if the pGh9:ISS1 integration sites in these clones showed that both were located in the same gene designated *kcdA*. Subsequently, the sequence analysis of the gene revealed that a major part of the gene was identical to the *ipd*-gene of *L. lactis* IL1403. The *ipd*-gene was much shorter, and seemed to lack some residues, essential for catalysis. The identified gene was over-expressed in the decarboxylase-negative *L. lactis* NZ9000 (pNZ7500). The measured activity was more than 30 times higher than in the wild type B1157 (Figure 3). In addition also the *ipd* gene of IL1403 and the corresponding region of B1157 were cloned to the same expression vector resulting in plasmids pNZ7501 and pNZ7502. Neither of these constructs resulted in the production of an active protein, which was predicted by the sequencing analysis.
Experiments with the over-expression mutant showed a 3-fold increase in 3-methylbutanal production. This increase was 30 times lower than the increased specific enzyme activity, which was measured in cell free extract of this mutant. This observation might suggest another rate-limiting step in the route, which is most likely the availability of the substrate, α-keto acid, produced by transamination (see Figure 1).

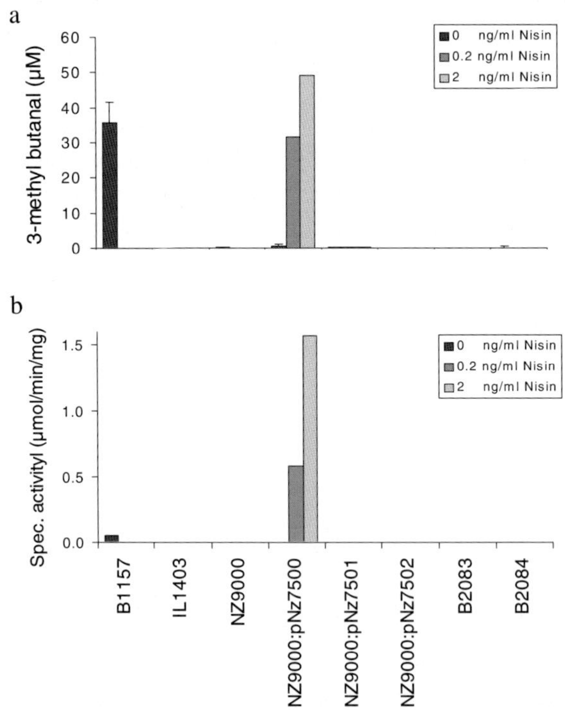

Figure 3. Flavour analysis of incubations in GM17 (panel a) and branched-chain keto acid decarboxylase (KdcA) activities (panel b) of (nisin-induced) decarboxylase over-expression mutants in relation to the wild-type strain B1157, the cloning host NZ9000 and the knock-out mutants B2083 and B2084.

Enzyme characterisation resulted in a molecular mass of a subunit of 61 kDa, and optimal activity at the pH range of 6.3 to 6.5, in the presence of thiamine pyrophosphate as a cofactor. The enzyme was highly resistant to salinity, indicating that it is active under cheese processing conditions. The activity was highest on branched-, and straight-chain α-keto acids with 4-6 carbon atoms, but also keto acids of methionine, phenylalanine and tryptophan could be converted [5]. Comparison of the enzyme with other enzymes revealed that the sequence and molecular mass were very similar to IPD and PDC, but not to a branched-chain α-keto acid decarboxylase from *Bacillus* [6]. The combination of sequence homology and the characteristics indicate that the enzyme is unique, and based on the substrate specificity it is referred to as branched chain keto acid decarboxylase (KdcA [5]).

Oxidation of a branched-chain aldehyde by aldehyde dehydrogenase (Figure 1) leads to the corresponding branched-chain organic acid. Branched chain organic acids are generally believed to be the substrates for the formation of (longer) branched-chain fatty acids. Alcohol dehydrogenase activity is identified in most LAB [7,8]. Despite the fact that the reaction equilibrium of this reaction is to the side of the alcohol, the aldehyde concentrations in many fermented products is found to be stable at relatively high concentrations [9]. This might be explained by the relatively low activity of the aldehyde dehydrogenase in LAB. The flavour intensity of aldehydes is higher than that of their corresponding alcohols, and, therefore, this conversion to alcohols might not be favourable when maximal flavour intensity is desired. However, this activity might be desirable in cases when aldehydes are reported to cause off-notes. For instance in the case of soy milk it is reported that the presence of aldehydes (e.g. hexanal) is coupled to the perceived off-notes ([10,11]; Busch and Batenburg, unpublished data).

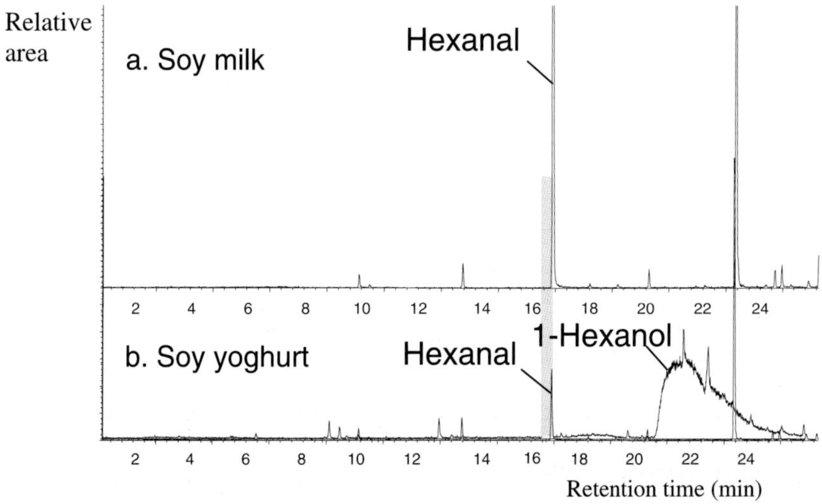

Figure 4. Conversion of hexanal to hexanol by LAB fermentation of soy milk (a) as measured by GC headspace analysis. Yoghurt cultures (b) were pre-cultured overnight from frozen stocks in 15 ml of whole soy milk at 43 °C.

Despite the fact that (branched-chain) aldehydes are reported to be desired key flavour components [9], they are also known to cause off-notes in higher concentrations as well as being key components causing the typical off-notes in soy drinks. Conversion of aldehydes such as hexanal and heptanal was observed in fermented soy milks (soy yoghurts), and the concentration of 1-hexanol was found to increase, probably due to the enzyme alcohol dehydrogenase converting hexanal to hexanol (Figure 4).
Preliminary data, also showed that desired dairy compounds such as diacetyl, 2,3-pentanedione and acetoin were formed during fermentation, and a negative correlation in sensory analyses was found with soy off-flavour (data not shown). These results indicate that both conversion of flavour compounds with negative notes and formation of more desirable flavour compounds contribute to overall flavour improvement in this case. It is expected that both effects can be further improved in order to obtain soy yoghurts of higher quality.

4. CONCLUSIONS

Based on the knowledge of formation and degradation pathways by enzymatic activity, new possibilities arise for the selection of starter cultures or combinations thereof, which give rise to desired flavour profiles in fermented products. The increasing knowledge on the amino acid converting enzymes, together with genome data, which will become available for various LAB, will expand our knowledge of flavour-forming pathways and mechanisms in different bacteria even faster. Obviously, one should also focus on other pathways (e.g. leading to flavours originating from lactose and fat), which play a role in the liking of the products. Moreover, all these enzyme activities should be present in such a manner that the ultimate flavour components formed by their activities result in a proper balance.

References

1. G. Smit (ed.), Dairy processing, improving quality, Cambridge, UK (2003) 492.
2. B.A. Smit, W.J.M. Engels, J.T.M. Wouters and G. Smit, Appl. Microbiol. Biotechnol., 64 (2004) 396.
3. B.A. Smit, W.J.M. Engels, J. Bruinsma, J.E.T. van Hylckama Vlieg, J.T.M. Wouters and G. Smit, J. Appl. Microbiol., 97 (2004) 306.
4. E. Maguin, P. Duwat, T. Hege, D. Ehrlich and A. Gruss, J. Bacteriol., 174 (1992) 5633.
5. B.A. Smit, J.E.T. van Hylckama Vlieg, W.J.M. Engels, L. Meijer, J.T.M. Wouters and G. Smit, Appl. Environ. Microbiol., 71 (2005) 303.
6. H. Oku and T. Kaneda, J. Biol. Chem., 263 (1988) 18386.
7. B. Dias and B. Weimer, Appl. Environ. Microbiol., 64 (1998) 3327.
8. M. Fernandez, M. Kleerebezem, O.P. Kuipers, R.J. Siezen and R. van Kranenburg, J. Bacteriol., 184 (2002) 82.
9. E.H.E. Ayad, A. Verheul, P.G. Bruinenberg, J.T.M. Wouters and G. Smit, Int. Dairy J., 13 (2003) 159.
10. A. Stefan and H. Steinhart, J. Agric. Food Chem., 47 (1999) 2854.
11. E.C. Ang and W.L. Boatright, J. Food Sci., 68 (2003) 388.

W.L.P. Bredie and M.A. Petersen (Editors)
Flavour Science: Recent Advances and Trends

Vitis vinifera carotenoid cleavage dioxygenase (VvCCD1): gene expression during grape berry development and cleavage of carotenoids by recombinant protein

Sandrine Mathieu[a], Nancy Terrier[b], Jérôme Procureur[a], Frédéric Bigey[a] and Ziya Günata[a]

[a]UMR IR2B, Université Montpellier II-ENSAM-INRA; [b]UMR SPO, équipe Biologie Intégrative de la Vigne et du Raisin, ENSAM-INRA, 34060 Montpellier Cedex 1, France

ABSTRACT

A Carotenoid Cleavage Dioxygenase (CCD) gene from *Vitis vinifera* was isolated and expressed in *Escherichia coli.* Recombinant VvCCD1 cleaved zeaxanthin symmetrically leading to the formation of 3-hydroxy-β-ionone, a C_{13}-norisoprenoidic compound, and a C_{14}-dialdehyde. Analysis of the gene expression during grape berries development revealed a significant induction of the gene before the onset of ripening, together with an increase in the level of C_{13}-norisoprenoids throughout the maturity.

1. INTRODUCTION

C_{13}-norisoprenoids have been detected in grape berries and leaves [1], where they are mainly found as glycoconjugates [2]. Some of them are important aroma contributors in both red and white wines and grape juices [2,3]. Carotenoids are prone to the cleavage by chemical, photochemical and oxidase-coupled mechanisms, but the cleavage is not region-specific and leads to the formation of apocarotenoids with 9, 10, 11, 13 and 15 carbon atoms [4]. Quite recently, recombinant CCDs cleaving carotenoids symmetrically at the 9,10[9',10'] bonds, resulting in the formation of C_{13}- and C_{14}-apocarotenoids, were reported in *Arabidopsis thaliana* [5] and in *Crocus sativus* [6]. Here, we report the catalytic function of a CCD gene from *Vitis vinifera*. Furthermore, changes in VvCCD1 expression and C_{13}-norisoprenoids levels in the grape berry were studied during the grape berry development of two different cultivars.

2. MATERIALS AND METHODS

2.1. In vitro assay with recombinant VvCCD1. Analysis of the metabolites

E. coli was transformed with VvCCD1 and cultivated at 37 °C. Cells were disrupted by sonication. The lysate was centrifuged and the supernatant was used for the enzymatic assay. The reaction mixture contained 35 μM of zeaxanthin (Extrasynthèse, France), 500 μl of the supernatant, 5 μM $FeSO_4$ and 1 mM DTT, and was incubated at 30 °C for 3 h. The ensemble was spiked with 3-oxo-α-ionol (synthesised in the laboratory) and extracted with pentane/dichloromethane (2:1, v/v). The organic phase was concentrated and analysed by GC-MS and HPLC. The conditions of GC-MS analysis were similar to those reported previously [1]. An HPLC equipped with a C-18 reverse phase column and an online diode array detector was used. The mobile phase was acetonitrile/water. 4,9-Dimethyldodeca-2,4,6,8,10-pentaene-1,12-dialdehyde was identified through the injection of the reference compound (BASF, Germany) and obtaining its absorption spectrum.

2.2. Grape berry material

Grape berries from *V. vinifera* L. cv. Muscat of Alexandria and Shiraz were harvested from 20 June until 10 September 2003, on the INRA-ENSAM vineyards (Montpellier, France). Six different developmental stages were chosen according to pH and the potential alcohol degree of the wine produced. Before RNA and aroma compounds extraction, samples were deseeded and powdered under liquid nitrogen.

2.3. Real-time PCR

Real-time PCR was performed on 1 μl cDNA from each cultivar at each developmental stage using a model 7700 Sequence Detection System (Applied Biosystems, Warrington, UK) and the SYBR-Green PCR Master kit (also Applied Biosystems). Data were calculated from the calibration curve and normalised using the constitutively expressed EF1 alpha gene [7].

2.4. Extraction and analysis of C_{13}-norisoprenoids from grape berries

A 70 g of powder were taken into 100 ml of H_20 milliQ containing 40 mM of gluconolactone to inhibit glycosidase activities. The mixture was stirred, centrifuged and 4-nonanol was added to the supernatant as internal standard. Subsequent conditions for the extraction and analysis of norisoprenoids were similar to those described previously [1].

3. RESULTS

3.1. Catalytic function of VvCCD1

CCD activity was tested using zeaxanthin as substrate. The analysis of the assay medium by GC-MS showed the presence of 3-hydroxy-β-ionone (Figure 1) [1]. When the assay medium was analysed by HPLC, a compound with an absorbance maxima at 396 nm and 414 nm was detected (peak 1) (Figure 2). The identity of the compound was obtained by analysing the reference compound (4,9-dimethyldodeca-2,4,6,8,10-pentaene-1,12-dialdehyde) run under the same conditions.

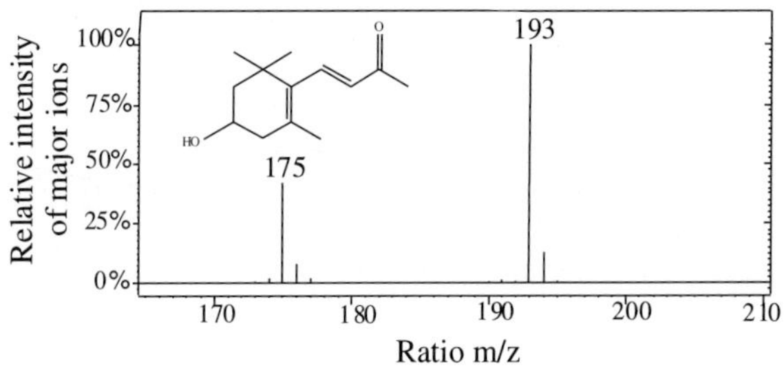

Figure 1. Mass spectrum of the VvCCD1 product 3-hydroxy-β-ionone.

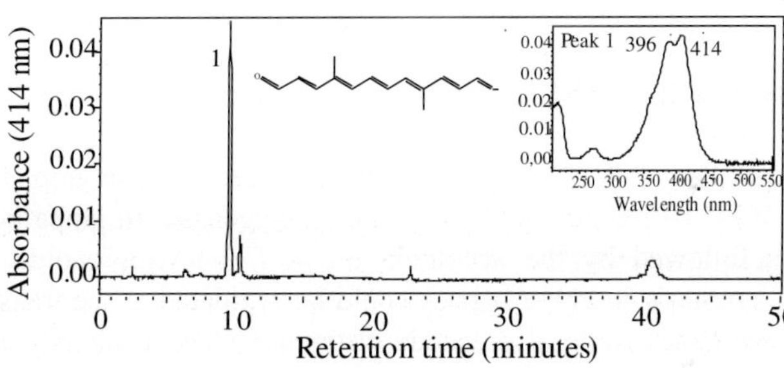

Figure 2. Identification of peak 1 as 4,9-dimethyldodeca-2,4,6,8,10-pentaene-1,12-dialdehyde (C_{14}-dialdehyde) by HPLC with online diode array detection.

3.2. Expression profile of VvCCD1 and changes in the C_{13}-norisoprenoid level during grape berries development

VvCCD1 expression pattern was expressed as the variation in the expression level of five developmental stages compared to the first stage corresponding to immature berries of Shiraz. A significant induction of VvCCD1 expression during the week preceding veraison - the onset of ripening - was observed for the two cultivars. A two-fold induction for Muscat of Alexandria and a nearly four-fold one for Shiraz was observed (Figure 3a). After veraison, the expression of the gene remained almost stable. VvCCD1 expression level was higher in Muscat of Alexandria berries than in Shiraz along grape berry development.

In both cultivars, the level of glycosylated C_{13}-norisoprenoids increased significantly after veraison (Figure 3b). Shiraz berries exhibited a progressive increase throughout ripening while a drastic increase was observed for the Muscat of Alexandria cultivar during the first week following veraison. Moreover, the total C_{13}-norisoprenoid level was nearly three times higher in the mature berries of Muscat of Alexandria than in those of Shiraz.

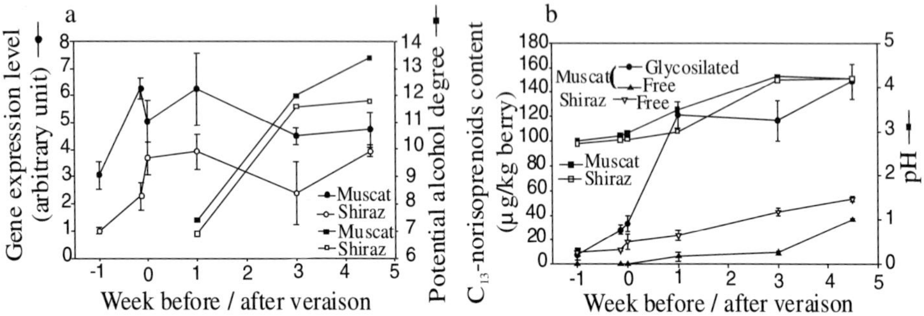

Figure 3. Analysis of grape berry development in two different cultivars, Muscat of Alexandria and Shiraz. (a) Expression profile of VvCCD1 obtained by real-time PCR and potential degree of alcohol. (b) Changes in the levels of free and glycosylated C_{13}-norisoprenoids in the corresponding cultivars and pH. Week 0 corresponded to veraison.

4. DISCUSSION AND CONCLUSION

The present work demonstrates for the first time the existence in grape berries of a gene encoding a 9,10[9',10']-Carotenoid Cleavage Dioxygenase. In grape berries, VvCCD1 induction was followed by the accumulation of C_{13}-norisoprenoids. Thus, the C_{13}-norisoprenoid synthesis in grape berries could be regulated at the transcriptional level. Other *V. vinifera* genes could also encode different CCDs, as already demonstrated for *A. thaliana* [8]. The complete grape genome sequencing will help to determine whether other CCDs are involved in the C_{13}-norisoprenoid biosynthesis.

References

1. J. Wirth, W. Guo, R.L. Baumes and Z. Günata, J. Agric. Food Chem., 49 (6) (2001) 2917.
2. P. Winterhalter and R. Rouseff (eds.), Carotenoid-derived aroma compounds: an introduction, Washington DC, USA (2002) 1.
3. G.K. Skouroumounis and P. Winterhalter, J. Agric. Food Chem., 42 (1994) 1068.
4. R. Zamora, P. Macias and J.L. Mesias, Die Nahrung, 32 (1988) 965.
5. S.H. Schwartz, X. Qin and J.A. Zeevart, J. Biol. Chem., 276 (27) (2001) 25208.
6. F. Bouvier, C. Suire, J. Mutterer and B. Camara, Plant Cell, 15 (1) (2003) 47.
7. N. Terrier, D. Glissant, J. Grimplet, F. Barrieu, P. Abbal, C. Couture, A. Ageorges, R. Atanassova, C. Léon, J.-P. Renaudin, F. Dédaldéchamp, C. Romieu, S. Delrot and S. Hamdi, Planta, September 6 (2005) 1.
8. S.H. Schwartz, X. Qin and M.C. Loewen, J. Biol. Chem., 279 (45) (2004) 46940.

W.L.P. Bredie and M.A. Petersen (Editors)
Flavour Science: Recent Advances and Trends

Labelling studies on pathways of amino acid related odorant generation by *Saccharomyces cerevisiae* in wheat bread dough

Michael Czerny and Peter Schieberle

German Research Center for Food Chemistry, Lichtenbergstr. 4, D-85748 Garching, Germany

ABSTRACT

Bio-conversion of free amino acids into important odour-active alcohols (Ehrlich-pathway) was investigated in *Saccharomyces cerevisiae* dough fermentations using $^{13}C_6$-L-leucine. Identification and quantification of the labelled 3-methylbutanol by Stable Isotope Dilution Assay demonstrated the efficacy of the Ehrlich-pathway because free L-leucine was converted to a large extent into the alcohol.

1. INTRODUCTION

Besides odorants, which stem from the flour itself, bread aroma compounds, bio-chemically formed by baker's yeast (*Saccharomyces cerevisiae*), are well-known contributors in particular to crumb aroma [1]. One important metabolic pathway leading to these odorants is the so-called Ehrlich-pathway, in which odourless amino acids present in the flour are enzymatically converted by oxidative transamination (intermediate: alpha-keto acids), decarboxylation (aldehyde) and reduction into the respective alcohols (e.g. L-leucine into 3-methylbutanol) [2,3].

Previous studies have confirmed this mechanism in fermentation experiments using ^{13}C- and ^{14}C-labelled amino acids by detecting the isotopic labelling in the corresponding alcohol [4-6], but it is still unclear, whether the Ehrlich-pathway or the *de novo* bio-synthesis is the more important pathway of odorant formation during yeast dough fermentation.

To address this question and to investigate the efficacy of the Ehrlich pathway, $^{13}C_6$-L-leucine was used in this study as a flavour precursor in yeast dough fermentation. The isotopically labelled metabolite 3-methylbutanol was quantified by a Stable Isotope Dilution Assay, which was specifically developed for this purpose.

2. MATERIALS AND METHODS

Saccharomyces cerevisiae was cultivated using a YG-medium at 26 °C for 42 h until a cell count of about 5 x 10^7 units/ml was reached. The suspension (1 ml) was mixed with wheat flour (250 g, flour type 550), tap water (750 g), and chloramphenicol (0.1 g), which was added to suppress the growth of lactic acid bacteria. The liquid dough was then fermented at 28 °C for 48 h under anaerobic conditions. After fermentation, microbiological tests confirmed that *Saccharomyces cerevisiae* was the predominant dough micro-organism. In labelling studies, $^{13}C_6$-L-leucine and L-leucine were added prior to fermentation in separate experiments.
Quantification was carried out by a Stable Isotope Dilution Assays [7] using $^2H_{10-11}$-3-methylbutanol, which was synthesised by reduction of $^2H_9$-3-methylbutanoic acid with $LiAl^2H_4$ in anhydrous diethyl ether. Wheat dough was spiked with the standard and the volatile compounds were isolated by means of the SAFE-technique [8]. The obtained extract was concentrated to about 0.1 ml and analysed by GC-MS in the CI-mode.

3. RESULTS

The metabolic property of *Saccharomyces cerevisiae* to generate 3-methylbutanol was confirmed in a preliminary experiment by fermenting wheat flour and tap water with the yeast. Quantification of the compound in the beginning and after fermentation showed that the concentration increased from 2 μmol/kg to 253 μmol/kg during fermentation (Table 1).

Table 1. Concentrations of 3-methylbutanol in *Saccharomyces cerevisiae* wheat dough.

	Concentration (μmol/kg dough)	
Odorant	Before fermentation	After fermentation[a]
3-Methylbutanol	2	253

[a]26 °C for 42 h.

To elucidate the quantitative contribution of the Ehrlich-pathway to 3-methylbutanol formation, an experiment with wheat dough spiked with $^{13}C_6$-L-leucine was performed. This amino acid does not exist in nature and differs from the natural L-leucine by the exchange of all 6 ^{12}C-atoms by ^{13}C-atoms. Because the carbon skeleton - with the exception of the loss of a ^{13}C-atom by decarboxylation - is saved in the Ehrlich-pathway, $^{13}C_5$-3-methylbutanol should be formed during fermentation. The labelled 3-methylbutanol can easily be detected by mass spectrometry and differentiated from 3-methylbutanol originating from unlabelled L-leucine.

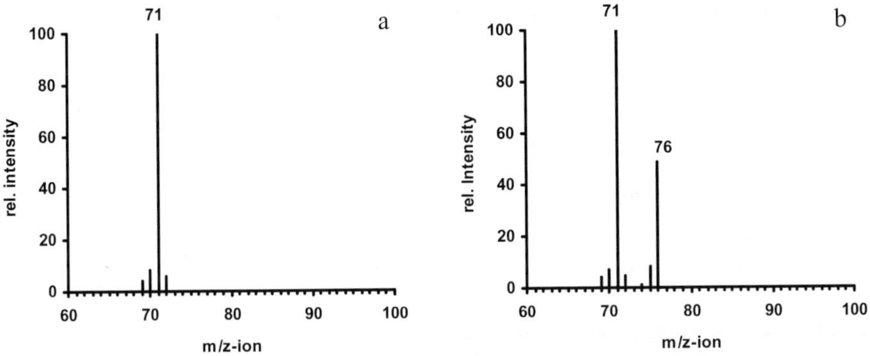

Figure 1. Mass-spectra (MS-CI) of 3-methylbutanol in extracts isolated from *Saccharomyces cerevisiae* wheat dough supplemented with L-leucine (a) and $^{13}C_6$-L-leucine (b).

Wheat dough was spiked with defined amounts of L-leucine (450 µmol/kg), fermented with *Saccharomyces cerevisiae*, and the volatiles were isolated from the dough. The obtained extract was analysed by GC-MS and the typical CI-mass spectrum for 3-methylbutanol (*m/z* 71) was obtained (Figure 1a).

In a separate experiment, $^{13}C_6$-L-leucine (450 µmol/kg) was added to another wheat dough, fermented under the same conditions and analysed in a similar way as with the unlabelled L-leucine. The CI-mass spectrum of 3-methylbutanol showed in addition to *m/z* signal of 71 a strong *m/z* signal of 76 corresponding to methylbutanol with five ^{13}C labelled carbons (Figure 1b). This mass spectrum clearly demonstrated that - besides the L-leucine present in wheat flour - the isotopically labelled amino acid was converted into the labelled alcohol.

A Stable Isotope Dilution Assay was developed to investigate the efficacy of the Ehrlich-pathway by calculating the L-leucine conversion. For this purpose, $^{2}H_{10\text{-}11}$-3-methylbutanol was synthesised and used as internal standard for the quantification of unlabelled 3-methylbutanol ($^{12}C_5$-3-methylbutanol) as well as $^{13}C_5$-3-methylbutanol in the dough supplemented with $^{13}C_6$-L-leucine.

Based on the concentration of the $^{13}C_5$-3-methylbutanol formed (377 µmol/kg) and the amount of labelled L-leucine added, an amino acid conversion rate of 84% was calculated (Table 2). The data demonstrated that bio-conversion of amino acids via the Ehrlich-pathway is very effective in odorant generation. But in consideration of the L-leucine conversion and the amount of free $^{12}C_6$-L-leucine present in the dough (33.4 µmol/kg), only 28.1 µmol/kg of the 617 µmol/kg of the unlabelled 3-methylbutanol formed during fermentation should originate from the free amino acid present in the dough (Table 2).

Table 2. Concentrations of 3-methylbutanol isotopomers and amino acid bio-conversion in *Saccharomyces cerevisiae* wheat dough supplemented with $^{13}C_6$-L-leucine[a].

Isotopomer	Conc. analysed (μmol/kg)	Bio-conversion (%)[a]	Conc. calculated (μmol/kg)[b]
$^{12}C_5$-3-Methylbutanol	617	-	28.1
$^{13}C_5$-3-Methylbutanol	377	84	-

[a]Wheat dough was spiked with $^{13}C_6$-L-leucine (450 μmol/kg). [b]Calculation is based on a bio-conversion (84%) and an amount of free L-leucine in dough of 33.4 μmol/kg.

4. DISCUSSION AND CONCLUSION

The presented approach using isotopically labelled amino acids in dough fermentation is a powerful tool to identify and balance metabolic pathways. The efficacy of the Ehrlich-pathway to bio-convert L-leucine into 3-methylbutanol was shown to be 84 of the total conversion of L-leucine%. Although the bioconversion was relatively high, the contribution of free L-leucine to 3-methylbutanol generation was relatively low. This suggested that other sources or pathways for 3-methylbutanol formation such as liberation of L-leucine from proteins by proteolysis and *de novo* synthesis essentially contribute to 3-methylbutanol formation.

References

1. P. Schieberle and W. Grosch, Z. Lebensmittel Untersuch. Forsch., 192 (1991) 130.
2. F. Ehrlich, Biochem. Z., 2 (1907) 52.
3. O. Neubauer and K. Fromherz, Hoppe-Seyler's Z. Physiol. Chem., 70 (1910) 326.
4. T. Äyräpää, J. Inst. Brew., 73 (1967) 17.
5. M.V. Wijngaarden (compiler), Proceedings of the 25th EBC congress, Oxford, UK (1995) 384.
6. C.C.-H. Chen, J. Am. Soc. Brew. Chem., 36 (1978) 39.
7. A.G. Gaonkar (ed.), Characterization of food: emerging methods, Amsterdam, The Netherlands (1995) 403.
8. W. Engel, W. Bahr and P. Schieberle, Eur. Food Res. Technol., 209 (1999) 237.

W.L.P. Bredie and M.A. Petersen (Editors)
Flavour Science: Recent Advances and Trends

Pathway analysis in horticultural crops: linalool as an example

Ellen Friel[a], Sol Green[a], Adam Matich[b], Lesley Beuning[a], Yar-Khing Yauk[a], Mindy Wang[a] and Elspeth MacRae[a]

[a]HortResearch, Mt. Albert Research Centre, Private Bag 92169, Auckland, New Zealand; [b]HortResearch, Palmerston North Research Centre, Private Bag 11030, Palmerston North, New Zealand

ABSTRACT

Linalool is an important chiral compound in the fragrance industry and is present in many products. Although, linalool has also been found in the fruit of kiwifruit and apple it is more abundant in the flowers, where it plays a key role as an intermediate to a number of interesting fragrance compounds. Three genes found to catalyse the production of linalool from geranyl diphosphate, have been mined from the HortResearch Plant EST database. The function of these genes has been proven using heterologous over-expression technologies. The similarities and differences between our genes and those already published are highlighted. Finally, we show the diversity in the fate of linalool in species of kiwifruit and apple, with discussion of the genes involved.

1. INTRODUCTION

Plants interact with their surroundings through the emission of volatile substances, often with a signal function. Plants emit these signals for a number of reasons including the attraction of pollinators or seed dispersers. Production of linalool, a floral scented monoterpene, is diurnally regulated in many flowers pollinated by bees [1], e.g. kiwifruit. Linalool can be further metabolised to produce compounds such as linalool oxide, which has been shown to elicit a strong antennal response in butterflies [2]. This onward metabolism can be enzyme mediated and the types of enzymes that have been shown to metabolise monoterpenes include dehydrogenases, heme-thiolate P450 hydroxylases [3] and alcohol acyl transferases [4]. Building a pathway map of the relationships between the many secondary metabolites and the genes responsible for catalysing the reactions that convert them, can be powerful in terms of enhancing the potential flavour attributes that can be realised from germplasms and existing horticultural commercial varieties.

2. STRATEGIES FOR FLAVOUR AND FRAGRANCE GENE DISCOVERY IN KIWIFRUIT (*ACTINIDIA)* AND APPLE (*MALUS)*

2.1. Screening the germplasm for interesting secondary metabolite profiles

Fruit, flowers and vegetative tissue from diverse kiwifruit and apple germplasm collections were screened for volatile aroma compounds using solvent extraction and/or purge and trap headspace techniques followed by gas chromatography-mass spectrometry, GC-MS [5]. Linalool is an important first step in the biosynthetic pathway of several compounds in many species of kiwifruit (*Actinidia* sp.) [5] and apple. It is found in the flowers of three commercial kiwifruit species *Actinidia deliciosa*, *A. chinensis* and *A. arguta*, but it is found in the fruit of *A. chinensis* and *A. arguta* only. Linalool is also found in the leaves, fruit and flowers of a number of apple species. Pathway mapping (Figure 1) shows that a number of linalool derived products have been identified in some of these tissues and, more interestingly, linalool derived products have been identified in tissues from which linalool has not been identified.
In nature, linalool is formed enzymatically from the monoterpene precursor geranyl diphosphate, GDP (Figure 1), which is catalysed by a linalool synthase enzyme (LIS).

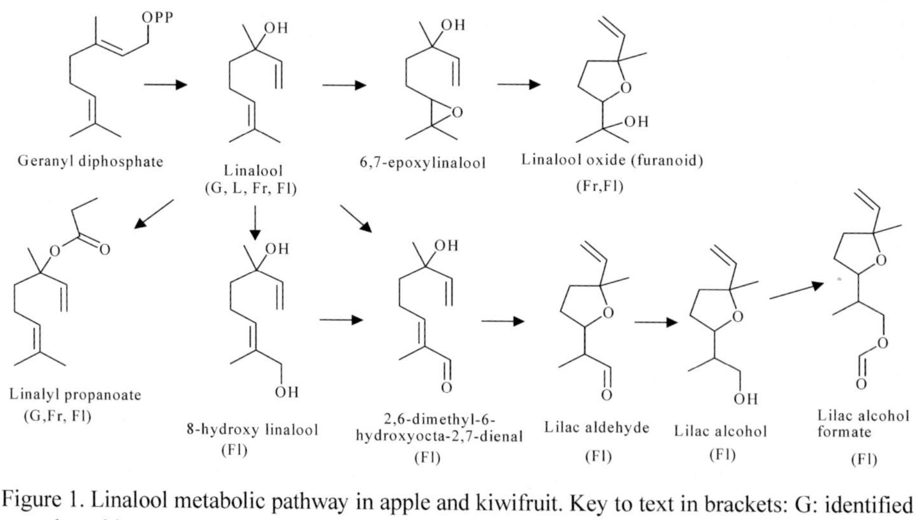

Figure 1. Linalool metabolic pathway in apple and kiwifruit. Key to text in brackets: G: identified as a glycoside; L, Fr, Fl: identified in tissues leaves, fruits and flowers respectively.

2.2. Linalool synthases from plants

To date, at least ten LIS from plants have been reported. There are three that have been isolated from the Californian native plant *Clarkia* and these have similar sequences [6].

In terms of their relationship with other terpene synthases (TPS) they form a group known as the TPS-*f* subfamily [7]. These genes are diterpene-like in that they contain a 'conifer diterpene internal sequence' domain. Other LIS which have been characterised, including one from *Arabidopsis thaliana,* show little homology to these LIS from *Clarkia* and are distributed throughout the other subfamilies *b*, *d* and *g* [7,8].
We have isolated and characterised three LIS from apple and kiwifruit plus two putative LIS from kiwifruit. Phylogenetic analysis comparing our LIS with characterised full-length terpene synthases in the public domain showed that two of our characterised LIS cluster closely together and show characteristics of both subfamily *b* and *d* (see Table 1). There were also features in common with the more recently identified TPS-*g* subfamily, characterised by myrcene and ocimene synthases from snapdragon and *Arabidopsis* LIS (AtTPS14). TPS-*g* members lack the RRX_8W motif characteristic of many other monoterpene synthases that is thought to be involved in RR-dependent isomerisation of GDP [8]. Although gene function can be predicted by homology to other known genes in the public domain and/or by grouping techniques that show similarities (e.g. phylogenetic analysis, structural modelling). It is only by expressing the gene in the appropriate systems that a reliable function can be determined.

Table 1. LIS genes including amino acid identity and similarity to *M. domestica* leaf *S*-LIS.

Compound produced (Species)	Genbank Acc. no.	Identity	Similarity	Sub-family	RRX_8W motif?
Linalool (*A.thaliana*)	AF497485	45.4	64.2	TPS-*g*	N
Linalool (*M. aquatica*)	AY083653	30.6	50.2	TPS-*b*	Y
Linalool (*P. frutescens*)	AF444798	31.4	49.4	TPS-*b*	Y
(3*R*)-Linalool (*A. annua*)	AF154124	32.5	51.2	TPS-*b*	Y
S-Linalool (*C. breweri LIS2*)	AF067603	19.9	34.3	TPS-*f*	N
Linalool (*C. concinna*)	AF067602	17.4	29.7	TPS-*f*	N
Linalool (*C. breweri LIS1*)	CBU58314	17.6	31.1	TPS-*f*	N
R-Linalool (*P. abies*)	AY473623	26.6	46.0	TPS-*d*	N
R-Linalool (*O. basilicum*)	AY693647	35.4	53.8	TPS-*d*	N
Linalool (*P. citriodora*)	AY917193	31.3	48.2	TPS-*b*	Y
S-Linalool (*M. domestica*, seeds)	-	34.9	51.8	?	N
S-Linalool (*A. arguta*, petals)	-	57.5	72.1	?	N
Putative Linalool (*A. polygama*, petals)	-	56.7	71.8	?	N
Putative Linalool (*A. chinensis*, meristems)	-	57.0	72.5	?	N

In order to prove the function of a putative enzyme, the cDNA sequence was cloned into a relevant vector and expressed as a recombinant protein in a test system (e.g. microbial or plant based systems). The headspace above bacterial cultures expressing LIS from apple leaf was sampled for four hours using solid phase micro-extraction (65 μm PDMS/DVB fibre Supelco, Australia). A peak identified as linalool, by GC-MS, (with comparison to NIST and Wiley mass spectral libraries) was found in the headspace of the sample expressing LIS incubated with GDP yet none was found in the comparable empty vector control. Subsequent analysis of purified protein extracts and transient expression in *Nicotiana benthamiana* leaves confirmed the production of linalool [9].

2.3. Building a bigger picture using pathway analysis

Mapping the biosynthetic pathways of a plant can lead to a greater understanding of its flavour and fragrance formation potential. Hypothetical pathways can be validated through labelled precursor 'feeding' experiments [10,11]. Figure 1 shows a hypothetical pathway for the metabolism of linalool in kiwifruit [5] and apple. Also shown are the tissues each linalool derived compound is found in and whether the compound has been identified as a glycoside in our germplasm. Esters such as linalyl propanoate have been identified as both free and glycosylated forms in kiwifruit thus it is likely that this plant contains both a linalool acyl transferase and a glycosyl transferase. These reactions may also be reversible, depending on the presence of other enzymes, i.e. esterases and glycosyl hydrolases respectively. In Figure 1 there are also a number of steps which may be due to P450-mediated hydroxylation of linalool, e.g. in the formation of 8-hydroxylinalool and 6,7-epoxylinalool. Relatively few P450 hydroxylase enzymes acting on monoterpenes have been published [12] and this is likely to be due to the challenging task of finding the correct combination of enzyme and substrate from the many potential enzymes and substrates. Some of the conversions in this pathway may also occur non-enzymatically. The epoxide 'opening' step may be an example of this. Finally, different parts of the pathway appear to be tissue specific. It is possible this is due to tissue specific localisation of either substrates or enzymes or both. Much of this pathway remains to be confirmed through use of labelled precursors. Although pathway mapping can give great insight into the likely enzymes involved in formation of flavour and fragrance compounds, it can also result in many more unanswered questions.

References

1. R.A. Raguso and E. Pichersky, Plant Species Biol., 14 (1999) 95.
2. S. Andersson and H.E.M. Dobson, J. Chem. Ecol., 29 (10) (2003) 2319.
3. G.W. Turner and R. Croteau, Plant Physiol., 136 (4) (2004) 4215.
4. M. Shalit, I. Guterman, H. Volpin, E. Bar, T. Tamari, N. Menda, Z. Adam, D. Zamir, A. Vainstein, D. Weiss, E. Pichersky and E. Lewinsohn, Plant Physiol., 131 (4) (2003) 1868.
5. A.J. Matich, H. Young, J.M. Allen, M.Y. Wang, S. Fielder, M.A. McNeilage and E.A. MacRae, Phytochem., 63 (3) (2003) 285.
6. L. Cseke, N. Dudareva and E. Pichersky, Mol. Biol. Evol., 15 (11) (1998) 1491.
7. J. Bohlmann, G. Meyer-Gauen and R. Croteau, Proc. Natl. Acad. Sci. USA, 95 (8) (1998) 4126.
8. N. Dudareva, D. Martin, C.M. Kish, N. Kolosova, N. Gorenstein, J. Faldt, B. Miller and J. Bohlmann, Plant Cell, 15 (5) (2003) 1227.
9. V. Allan (ed.), Fluorescence microscopy of proteins: a practical approach, Oxford, UK (1999) 163.
10. D. Burkhardt and A. Mosandl, J. Agric. Food Chem., 51 (25) (2003) 7391.
11. M. Kreck, S. Puschel, M. Wust and A. Mosandl, J. Agric. Food Chem., 51 (2) (2003) 463.
12. M. Wust, D.B. Little, M. Schalk and R. Croteau, Arch. Biochem. Biophys., 387 (1) (2001) 125.

W.L.P. Bredie and M.A. Petersen (Editors)
Flavour Science: Recent Advances and Trends

Microbial resolution of 2-methylbutyric acid and its application to several chiral flavour compounds

Toru Tachihara[a], Hiromi Hashimoto[a], Susumu Ishizaki[a], Tsuyoshi Komai[a], Akira Fujita[a], Masashi Ishikawa[a] and Takeshi Kitahara[b,c]

[a]*Technical Research Center, T. Hasegawa Co., Ltd., 335, Kariyado, Nakahara-ku, Kawasaki-shi, Kanagawa 211-0022, Japan;* [b]*Laboratory of Natural Product Chemistry Center for Basic research, The Kitasato Institute, 5-9-1 Shirokane, Minato-ku, Tokyo 108-8642, Japan;* [c]*Department of Pharmaceuticals, Teikyo Heisei University, 289-23 Uruido, Ichihara-shi, Chiba 290-0193, Japan*

ABSTRACT

Optically active 2-methylbutyric acid is a useful odorant, and an important chiral building block of other flavour compounds. We investigated microbial resolution of racemic 2-methylbutyric acid. We found a novel bacterium from soil, which utilises (*S*)-2-methylbutyric acid preferentially. This strain was identified to be *Pseudomonas* sp. by morphological observation and analysis of 16S rDNA sequences. Using this strain, we achieved an efficient preparation of (*R*)-2-methylbutyric acid for the first time. Moreover, we succeeded in the stereo-selective synthesis of stereoisomers of flavour compounds, using optically active 2-methylbutyric acid. As a result of odour evaluation of these, we found that each of the stereoisomers had different and characteristic odours.

1. INTRODUCTION

Optically active 2-methylbutyric acid is a useful flavour compound found in various natural products. The isomers of this compound have a distinct odour [1]. The (*S*)-2-methylbutyric acid is now commercially available, whereas the (*R*)-form is not yet produced commercially. In this study we sought to develop an efficient production of (*R*)-2-methylbutyric acid and investigated the microbial resolution of the racemic mixture of 2-methylbutyric acid.

2. MATERIALS AND METHODS

For the screening of microorganisms applicable to resolution of 2-methylbutyric acid, a synthetic medium with racemic 2-methylbutyric acid as the carbon source was used. It contained per liter of distilled water: 10.0 g 2-methylburyric acid, 1.0 g K_2HPO_4, 0.2 g $MgSO_4 \cdot 7H_2O$, 0.1 g NaCl, 0.1 g $CaCl_2$, 0.02 g $FeCl_2$ and 1.0 g $(NH_4)_2SO_4$. The initial pH of the medium was adjusted to 7.2 with 4 M aqueous NaOH. The medium was sterilised at 120 °C for 15 min.

Small amounts of soil or waste-water samples collected from all over Japan were suspended in 10 ml of synthetic medium and were cultivated at 30 °C with shaking at 100 oscillations per min, for 7 days. Growth of the microorganism was confirmed by visual observation and aliquots of 0.2 ml were spread on 15 g/l agar plates of the synthetic medium. The plates were incubated at 30 °C and colonies formed in 2-3 days were isolated. Each isolate was further inoculated into a test tube containing 10 ml of liquid synthetic medium, and cultivated at 30 °C, with shaking at 100 oscillations per min, for 4 days. The culture filtrate was extracted with ether. The extract was washed with brine, dried ($MgSO_4$) and concentrated *in vacuo*. The residue was analysed by gas chromatography using a CHIRAMIX^{TM} column [2].

3. RESULTS AND DISCUSSION

As a result of screening, we found a novel strain (TH-252-1) of soil-dwelling bacteria that utilises (*S*)-2-methylbutyric acid preferentially. The partial 16S rDNA sequences of TH-252-1 were 98.66% and 98.08% identical to those of *Pseudomonas alcaligenes* and *Pseudomonas pseudoalcaligenes pseudoalcaligenes*, respectively (data not shown). Therefore, we identified TH-252-1 as *Pseudomonas* sp. [3]. Optically pure (*R*)-2-methylbutyric acid was obtained (theoretical yield: 50%, enantiomeric excess: 100% ee) by microbial resolution of racemic 2-methylbutyric acid, using TH-252-1. By odour evaluation of the (*S*)-form (commercially available) and the (*R*)-form (microbial resolution) of 2-methylbutyric acid the two enantiomers were confirmed to have a distinct and different odour (Table 1).

Chiral 2-methylbutyric acid (marked **1** in the schemes) was applied for the synthesis of several chiral flavour compounds. We synthesised all the stereoisomers of the imine derivative **2** (Scheme 1) and both enantiomers of pyrrolidine derivative **3** (Scheme 2) which are found in roasted spotted shrimp. We also synthesised both enantiomers of filbertone **4** (Scheme 3), which are found in hazelnuts, and both enantiomers of ethyl 2-methylbutyrate **5** (Scheme 4), which are among others found in apple and strawberry. The synthesis of these compounds are outlined in Figure 1. The odours of the compounds were evaluated and the results are shown in Table 1 [4-6].

Table 1. Results from odour evaluations of the synthesised compounds from 2-methylbutyric acid.

Compound	Structure	Odour evaluation
2-methylbutyric acid (**1**)	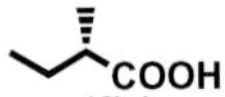 (*S*)-**1**	Fruity, sweet, tropical
	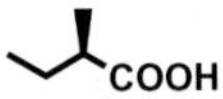 (*R*)-**1**	Cheasy, sweaty, sharp
Imine derivative (**2**)	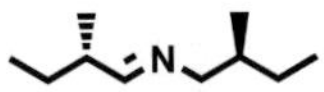 (2*S*, 2'*S*)-**2**	Green note (fruity, metallic, mild, ester-like)
	 (2*R*, 2'*R*)-**2**	Medicine-like note (fruity, phenolic)
	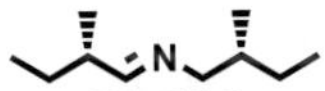 (2*R*, 2'*S*)-**2**	Fruity note (phenolic, ester-like)
	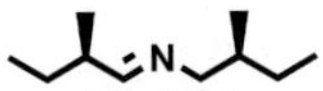 (2*S*, 2'*R*)-**2**	Fruity note (metallic, amine-like, sweet)
Pyrrolidine derivative (**3**)	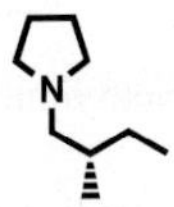 (*S*)-**3**	Seafood-like, strong, mild
	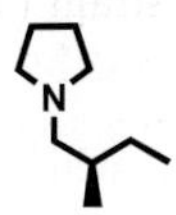 (*R*)-**3**	Roasted seafood, strong, metallic
Filbertone (**4**)	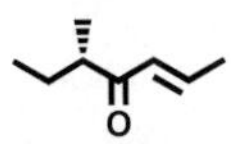 (*S*)-**4**	Sweet, fruity, fatty, strong
	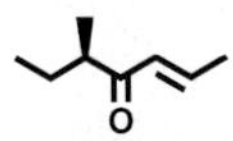 (*R*)-**4**	Metallic, earthy, weak
Ethyl 2-methylbutyrate (**5**)	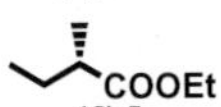 (*S*)-**5**	Sweet, fruity, natural
	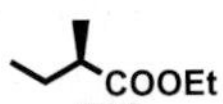 (*R*)-**5**	Heavy, oily, metallic

Scheme 1

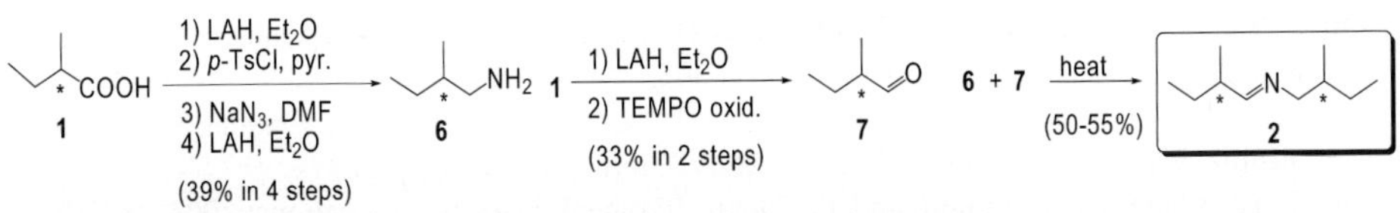

Figure 1. Synthesis of chiral derivatives from 2-methylbutyric acid.

Scheme 2

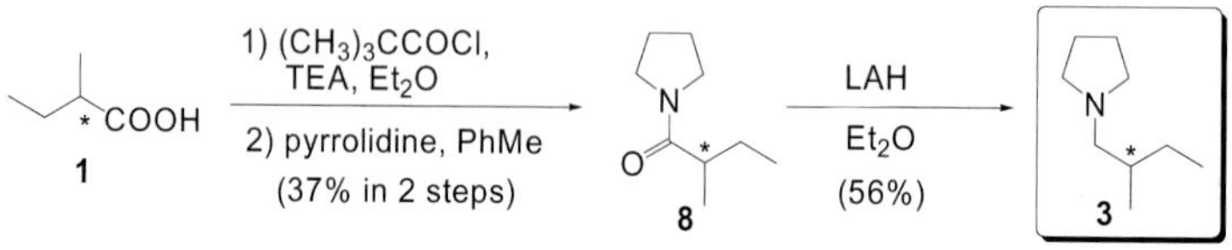

Scheme 3

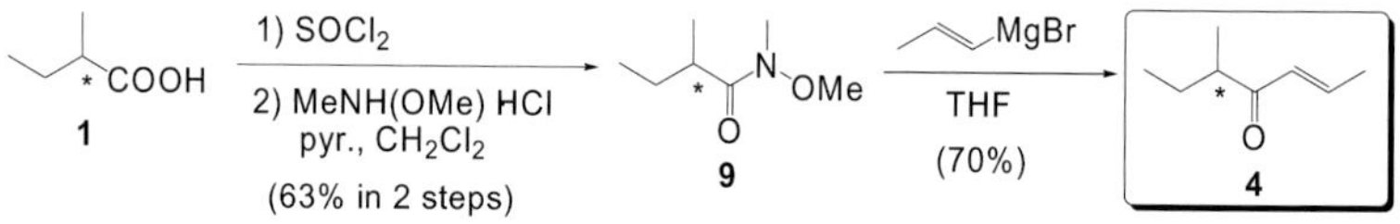

Scheme 4

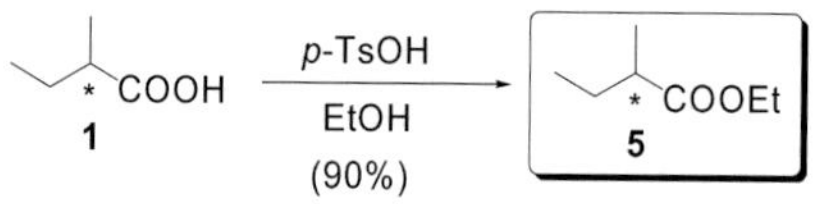

Figure 1. Synthesis of chiral derivatives from 2-methylbutyric acid (continued).

4. CONCLUSION

We found a novel bacterium from soil, which utilises (*S*)-2-methylbutyric acid preferentially. This strain was identified to be *Pseudomonas* sp. by morphological observation and analysis of 16S rDNA sequences. Using this strain, we achieved a first and efficient preparation of (*R*)-2-methylbutyric acid. Moreover, we succeeded in the stereo-selective synthesis of stereoisomers of flavour compounds, using optically active 2-methylbutyric acid as a precursor. From odour evaluations we found that each of the stereoisomers had different and characteristic odours.

References

1. K. Rittinger, C. Burschka, P. Scheeben, H. Fuchs and A. Mosandl, Tetrahedron-Asymmetr., 2 (1991) 965.
2. S. Tamogami, K. Awano, M. Amaike, Y. Takagi and T. Kitahara, Flavour Fragrance J., 16 (2001) 349.
3. T. Tachihara, H. Hashimoto and T. Komai, JP patent, Priority applications: 2004-142718 (2004).
4. T. Tachihara, S. Ishizaki, Y. Kurobayashi, H. Tamura, Y. Ikemoto, A. Onuma and T. Kitahara, Flavour Fragrance J., 18 (2003) 305.
5. T. Tachihara, S. Ishizaki, Y. Kurobayashi, H. Tamura, Y. Ikemoto, A. Onuma, K. Yoshikawa, T. Yanai and T. Kitahara, Helv. Chim. Acta, 86 (2003) 274.
6. T. Tachihara, H. Hashimoto and T. Komai, JP patent, Priority applications: 2004-287398 (2004).

W.L.P. Bredie and M.A. Petersen (Editors)
Flavour Science: Recent Advances and Trends

Effect of malolactic fermentation on the volatile aroma compounds in four sea buckthorn varieties

Katja Tiitinen[a], Marjatta Vahvaselkä[b], Mari Hakala[a], Simo Laakso[b] and Heikki Kallio[a]

[a]Department of Biochemistry and Food Chemistry, University of Turku, FI-20014 Turku; [b]Laboratory of Biochemistry and Microbiology, Helsinki University of Technology, FI-02015 TKK, Finland

ABSTRACT

Volatile compound composition of sea buckthorn juice headspace was investigated before and after malolactic fermentation of the juice. Ethyl acetate, 3-methylbutanol and 3-methylbutyl acetate were formed in abundance during fermentation, whereas concentrations of ethyl 2-methypropanoate, ethyl 3-methylbutanoate, ethyl hexanoate and ethyl octanoate decreased.

1. INTRODUCTION

Malolactic fermentation (MLF) is widely used in winemaking to reduce sourness by converting malic acid to lactic acid. It also has an effect on the aroma via formation of diacetyl and other dicarbonyl compounds in wine [1] and higher alcohols, esters, and carbonyl compounds in cider [2]. The strong, sour flavour is a characteristic of sea buckthorn, and mainly caused by malic acid [3,4]. The aroma characteristics of sea buckthorn are not well-known. The MLF has previously been applied to northern berries to decrease the sourness [5]. Our aim was to study the effect of MLF on the composition of volatile compounds in sea buckthorn juice headspace.

2. MATERIALS AND METHODS

2.1. Materials

Sea buckthorn berries of *cv.* Avgustinka, Botanicheskaya, Oranzhevaya and Chuiskaya were picked fully ripe and frozen for the analysis during summer 2003. Samples of berries were gently thawed in a microwave oven. The berries were crushed, pressed for juice and diluted 1:1 with water. The lactic acid bacteria *Oenococcus oeni* (ATCC 39401) were cultivated in a modified MRS broth and washed with 0.9% NaCl solution prior to the inoculation. The fermentation was performed over 18 h at 28 °C at a cell count of $4 \cdot 10^9$ CFU/ml. After fermentation, the juice was frozen and stored until required for analysis.

2.2. Analysis of volatiles

The juice samples were thawed and 20 g of juice was weighed in a beaker at 20 °C. The headspace was allowed to equilibrate, and volatile compounds were collected during 20 min on an SPME fibre (StableFlex divinylbenzene/carboxen/ polymethylsiloxane, 20 mm in length, 50/30 μm, Supelco, USA).

The adsorbed volatiles were analysed by a Hewlett Packard Series II 5890plus gas chromatograph coupled to an HP 5972 Series mass selective (EI) detector. A DB5 column (MS+, 30 m, ID 0.25 mm, 0.50 μm, J&W Scientific, USA) was used. The compounds were released from the fibre in the injector port (260 °C) over 10 min, collected to the upper part of the column with liquid nitrogen cooling, and then released into the column by removing the cryotrap. The column was programmed for 12 min at 35 °C, raised at a rate of 10 °C/min to 105 °C, at a rate of 1 °C/min to 135 °C, and finally at a rate of 20 °C/min to 230 °C, at which it was held for 5 min. The temperature of the detector was 270 °C. The ionisation energy was 70 eV and the detection voltage 1.8 kV. The compounds were monitored over the range *m/z* 40-350.

2.3. Identification and quantification

The compounds were identified by mass spectra (Wiley 229, HP) and the retention times of reference compounds. All peaks found at least in two of the three total ion chromatograms (TIC) of the sample were taken into account when calculating the total area of peaks (100%) and the relative areas of the volatile compounds by the absorption to the SPME fibre. In total peak areas the coefficient of variance (CV%) was less than 15% within the samples and less than 25% between the samples.

2.4. Statistical analyses

The statistical analyses were performed using SPSS v. 12.0 and Unscrambler v. 9.1.2. The effect of MLF was analysed with the nonparametric Kruskal-Wallis test. Principal components analysis (PCA) was used to interpret the data.

3. RESULTS

The changes in peak areas of volatile compounds during MLF of sea buckthorn are shown in Table 1. The compounds reported represent over 80% of the total peak area of the volatile compounds in sea buckthorn juice headspace when analysed by SPME.

During the fermentation, relative proportions of ethyl acetate and 3-methylbutanol were increased significantly ($p<0.05$). 3-Methylbutyl acetate represented up to 2% of the total peak area in the fermented juice, but in the unfermented juice it was detected only in

Chuiskaya. The contents of ethyl hexanoate and ethyl octanoate decreased during the fermentation. There was also a minor decrease in content of ethyl 2-methylpropanoate, ethyl 2-methylbutanoate and ethyl 3-methylbutanoate. The MLF was more complete in the varieties Oranzhevaya and Chuiskaya, which was noticed as larger changes in volatiles in these varieties.

Table 1. Changes in the volatile profiles of sea buckthorn varieties during MLF for 18 h at 28 °C.

ID	Compound	AVG[a]	BOT	CHU	ORA
2	Ethyl acetate	+[b]	+	+	+
5	3-Methylbutanol	+	+	+	+
6	Ethyl 2-methylpropanoate	-	n.s.	-	-
10	Ethyl 2-methylbutanoate	n.s.	n.s.	n.s.	-
11	Ethyl 3-methylbutanoate	-	n.s.	n.s.	-
12	3-Methylbutyl acetate	+	+	+	+
22	Ethyl hexanoate	-	n.s.	-	-
34	3-Methylbutyl 3-methylbutanoate	+	n.s.	+	+
46	Ethyl octanoate	n.s.	-	n.s.	-
50	3-Methylbutyl hexanoate	n.s.	+	n.s.	n.s.

[a]AVG: Avgustinka; BOT: Botanicheskaya; CHU: Chuiskaya; ORA: Oranzhevaya. [b]+ Significant ($p<0.05$) increase in proportion during MLF; - significant ($p<0.05$) decrease in proportion during MLF; n.s., no significant change during MLF.

The substrate for malolactic fermentation by *O. oeni* is malic acid, which is available in sea buckthorn juice in large quantities. Production of diacetyl, which is typical for the MLF in wine [1], was not noticed in this study probably because the content of citric acid is low in sea buckthorn.

The differences in volatile compound composition were more distinct between the varieties than due to the fermentation (Figure 1). The first two PCs explain 68% of the variance of the data. The first PC (50%) explains almost entirely the differences between samples, whereas the second PC (18%) explains both the effect of the MLF and the varieties on the volatile compound profiles. The third PC explains 17% of the data and it explains the differences between varieties (not shown in figure).

Ethyl acetate, **2**, 3-methyl butanol, **5** and 3-methylbutyl acetate, **12** describe the fermented samples of Oranzhevaya and Chuiskaya as well as the unfermented sample of Chuiskaya, which has high proportions of these compounds naturally. The compounds that decreased during the fermentation are on the positive axis of the second PC and they describe the unfermented Oranzhevaya and both the fermented and the unfermented Botanicheskaya.

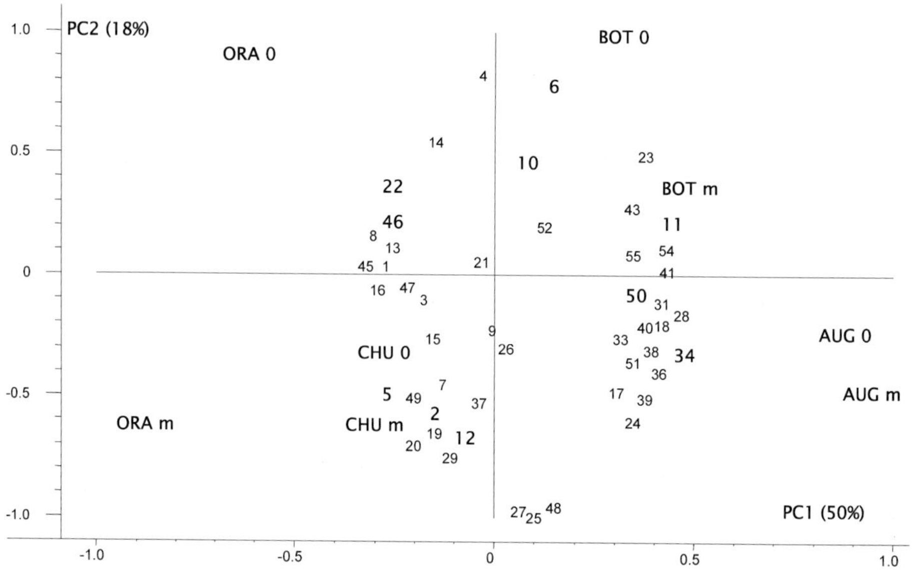

Figure 1. PCA bi-plot of the volatile compounds of sea buckthorn varieties before and after malolactic fermentation (MLF). For numbering of the compounds see Table 1. Other abbreviations used are AVG: Avgustinka; BOT: Botanicheskaya; CHU: Chuiskaya; ORA: Oranzhevaya; 0: before MLF; m: after MLF.

4. DISCUSSION AND CONCLUSION

When malolactic fermentation was applied to the sea buckthorn juice, changes in the composition of volatiles were observed. The ten compounds with changes in their relative areas in total ion chromatogram during MLF contributed with more than 80% of all of the volatiles in sea buckthorn juice headspace. The compounds formed in abundance were ethyl acetate, 3-methylbutanol and 3-methylbutyl acetate. In general, the changes due to the MLF in the composition of volatile compounds were small. The differences in the volatile profiles existed rather due to the variety than the fermentation.

References

1. G. de Revel, N. Martin, L. Pripis-Nicolau, A. Lonvaud-Funel and A. Bertrand, J. Agric. Food Chem., 47 (1999) 4003.
2. V.A. Nedovic, A. Durieux, L. Van Nedervelde, P. Rosseels, J. Vandegans, A.-M. Plaisant and J.-P. Simon, Enzyme Microb. Technol., 26 (9-10) (2000) 834.
3. K. Feng, Z. Sun, Y. Ma, W. Cao, R. Zheng and Z. Wang (eds.), Proceedings of international symposium on sea buckthorn (*H. rhamnoides L.*), Xia, China (1989) 106.
4. X. Tang, N. Kälviäinen and H. Tuorila, LWT Food Sci. Technol., 34 (2001) 102.
5. S. Viljakainen and S. Laakso, Eur. Food Res. Technol., 214 (2002) 412.

In vivo deodorisation with caffeoylquinic acid derivatives

Osamu Negishi[a] and Yukiko Negishi[b]

[a]Graduate School of Life and Environmental Sciences, University of Tsukuba, Tsukuba, Ibaraki 305-8572, Japan; [b]Institute of Nutrition Sciences, Kagawa Nutrition University, Sakado, Saitama 350-0288, Japan

ABSTRACT

Ku-ding-cha contained a large amount of the caffeoyl quinic acid (CQA) derivatives, di-CQA, 16.7% in dry weight as well as mono-CQA, 9.2%. A mixture of 3,5-DiCQA and 5-CQA with an apple acetone powder (AP), containing polyphenol oxidase, showed high capturing activities towards methanethiol (MeSH), propanethiol (PrSH), and 2-propenethiol (AllSH). Two adducts from 3,5-diCQA and methanethiol (MeSH) were identified. Furthermore, drinking Ku-ding-cha and eating apple or prune reduced the amount of allyl methyl sulfide (AMS) and AllSH gas after garlic ingestion. These results indicate the good effects of CQA derivatives on *in vivo* deodorisation.

1. INTRODUCTION

Recently we have been studying enzymatic deodorisation, a novel method involving the use of polyphenolic compounds (PPs) and polyphenol oxidases (PPOs) or peroxidases (PODs) to remove bad odours from our mouths and the environment [1]. We previously demonstrated that deodorisation with foods such as fruits, vegetables and mushrooms is achieved by multiple actions including physical and chemical interactions between volatile sulfur compounds (VSC) from *Allium* species and foods, enzymatic degradation of disulfides and addition of thiols to PPs catalysed by PPOs or PODs [2]. The formation of VSC from *Allium* species and *Allium* breath volatiles have been investigated in detail. It is well known that malodorous breath (halitosis) can originate in the mouth as well as in the gut, particularly in the case of VSC produced after ingesting *Allium* species. Furthermore, it has been demonstrated that most AMS gas originates from the gut rather than from the mouth, and this gas is likely to account for the well-known persistence of malodorous breath long after garlic ingestion. These studies suggest that Ku-ding-cha, which is generally consumed in southern China as a tea-like beverage, containing a large amount of CQA derivatives has great potential for deodorisation [3]. The purpose of this study was to investigate the effects of Ku-ding-cha and fruits on the *in vivo* removal of AMS as well as thiols.

2. MATERIALS AND METHODS

2.1. Beverages

Beverages used in this study contained very large amounts of CQA and catechin derivatives. These total CQA and catechin derivatives in Ku-ding-cha-1, Ku-ding-cha-2, green tea, black tea and oolong tea corresponded to 22.4, 90.3, 49.2, 18.9 and 27.4 mmol in 100 g of dry weight as catechol and pyrogarol moiety, respectively [3].

2.2. Measurements of thiols and AMS-capturing activities

2.2.1. Reactions with thiols

To a mixture of 10 mg apple AP (or 20 mg pear AP) and 100 μl of a 0.1% thiol aqueous suspension in a 30-ml borosilicate glass vial with open-top screw cap and Teflon/silicon disk was added 2.0 ml of a beverage that was extracted from 1.0 g tea powder with 50 ml hot water. The vial was shaken by hand at a rate of 2 strokes/s at 25 °C for 3 min. An aliquot volume of the headspace gas (2 to 6 ml) was passed through a detector tube. The effect of acetone powder and CQA was measured by mixing 2 mg apple AP, 1.0 ml of a 0.1 M acetate buffer (pH 5.0), 0.8 ml of a 2.5 mM 5-CQA or a 1.25 mM 3,5-diCQA solution, 50 μl of a 0.1% thiol aqueous suspension and 150 μl of water. The thiol-capturing activity of each material was measured in duplicate

2.2.2. Capture of AllSH and AMS in vivo

At lunch, a healthy 49 years old male with no problem with halitosis, ate 350 g of cooked rice and a soup containing 5 g of garlic paste in 10 min, rinsed his mouth with a cup of water and successively drank 150 ml of the beverage or ate 150 g of an apple (cv. Ourin) or 24 g of lyophilised prune with 130 ml of water in 5 min. He breathed deeply and stopped his breath for 20 s and then 3 l of his breath was collected in a polyester film bag with each sampling over a 5 h period. Furthermore, he did not eat anything except 100 ml of water after each sampling time for 1 to 5 h. The concentrations of AllSH and AMS gas in his breath were measured with a gas chromatograph equipped with a flame photometric detector (140 °C) and a glass column (polyphenyl ether 5 ring 10%, 3.2 mm × 3.1 m). Column temperature was held at 50 °C for 3 min, raised from 50 to 150 °C at 20 °C/min and then held at 150 °C for 14 min. Helium gas (60 ml/min) was used as a carrier gas. Retention times of MeSH, MeSMe, AllSH, AMS, $(MeS)_2$, AllSAll and $(AllS)_2$ standard samples were 1.6, 2.4, 4.1, 5.8, 7.0, 8.3 and 13.3 min, respectively. The beverages and fruits were assayed in different days and in triplicate.

3. RESULTS AND DISCUSSION

3.1. Thiol-capturing activities of CQA derivatives and beverages

The capturing activities of CQA derivatives and beverages towards thiols are shown in Table 1. No activity of 5-CQA or 3,5-diCQA alone appeared. However, by their combination with apple AP high-capturing activities were revealed towards MeSH and PrSH. From the reaction mixture between 3,5-diCQA and MeSH we obtained two kinds of 3,5-diCQA adding one SMe at the 2-position of either aromatic ring [3]. Ku-ding-cha-1 indicated extremely strong activities towards all 3 kinds of thiols in the presence of apple AP. This suggests that Ku-ding-cha exerts an effect on *in vivo* removal of bad odours with a very small amount of catalysts such as PPOs. In addition, the AllSH-capturing activity of oolong tea with pear AP was stronger than other beverages belonging to *Camellia sinensis*, which contains large amounts of catechins. This is thought to link to a low ascorbic acid content.

Table 1. Thiol-capturing activities of CQA derivatives and several beverages.

	Capturing activity (%)		
Substance and beverage	MeSH	PrSH	AllSH
5-CQA	4	9	2
5-CQA + apple AP	100	100	45
3,5-diCQA	0	14	0
3,5-diCQA + apple AP	72	83	19
Ku-ding-cha-1	3	21	7
Ku-ding-cha-1 + apple AP	100	100	100
Green tea	0	13	7
Green tea + pear AP	27	26	23
Black tea	0	4	0
Black tea + pear AP	34	30	30
Oolong tea	0	19	4
Oolong tea + pear AP	31	30	44

3.2. *In vivo* deodorisation

Figure 1 shows the time-course of AllSH and AMS decrease by drinking Ku-ding-cha and eating fruits. AllSH decreased quickly to levels that were not detected by GC during 1 h after garlic ingestion, especially in drinking Ku-ding-cha-2 or eating a prune. The amount of AMS decreased significantly 1 h after garlic ingestion, however, further decreases were slow. The AMS present early after garlic ingestion is thought to be formed in the oral cavity, the throat and the esophagus, and the AMS present later could be released from the lungs after being absorbed in the intestine. Eating apple or prune reduced the level of AMS to about 40-20 ppb and 90-30 ppb less than the control during 10-40 min after garlic ingestion, respectively. Furthermore, the amount of AMS after eating a prune was low during the experiment for 5 h. On the other hand, in Ku-ding-cha-2 the amount of AMS was about the same as in the control experiment except 10 min and 1 h after garlic ingestion. These results indicate a good effect of prune on deodorisation. Early decrease in AMS by eating an apple or prune is likely to be a result of decrease in AllSH by enzymatic deodorisation, because AMS has been reported to be formed by the methylation of AllSH. Although PPs have a weak ability to capture sulfides and disulfides [2], PPs in Ku-ding-cha probably capture large amounts of thiols produced from sulfides and disulfides in the intestine. As a result, formation of AMS gas is suppressed. The beverage was fed to a subject only once after ingestion of garlic in this investigation. If the beverage will be consumed frequently, the effect on deodorisation may be more remarkable.

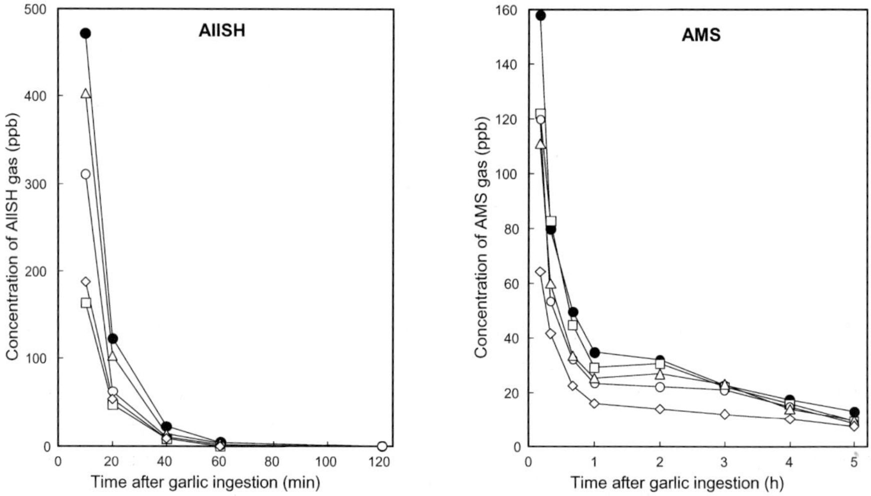

Figure 1. Effects of Ku-ding-cha-2, apple and prune on capture of AllSH and AMS *in vivo*: (●) control (water); (□) Ku-ding-cha-2; (Δ) Ku-ding-cha-2 + apple; (○) apple (cv. Ourin); (◊) prune.

4. CONCLUSION

This study indicated that bad breath after garlic ingestion can be decreased by eating prune and apple containing both PPs and PPOs, or by drinking Ku-ding-cha, which contains large amounts of PPs such as CQA derivatives.

References

1. O. Negishi, Y. Negishi, Y. Aoyagi, T. Sugahara and T. Ozawa, J. Agric. Food Chem., 49 (2001) 5509.
2. O. Negishi, Y. Negishi and T. Ozawa, J. Agric. Food Chem., 50 (2002) 3856.
3. O. Negishi, Y. Negishi, F. Yamaguchi and T. Sugahara, J. Agric. Food Chem., 52 (2004) 5513.

W.L.P. Bredie and M.A. Petersen (Editors)
Flavour Science: Recent Advances and Trends

The influence of fermentation temperature and sulfur dioxide on the volatile composition and flavour profile of cashew wine

Deborah S. Garruti[a], Fernando A.P. de Abreu[a], Maria Regina B. Franco[b] and Maria Aparecida A.P. da Silva[b]

[a]*Embrapa Agroindustria Tropical, Fortaleza, 60511-110, Brazil;* [b]*Faculdade de Engenharia de Alimentos, Unicamp, Campinas, 13081-970, Brazil*

ABSTRACT

Fresh cashew juice was inoculated with *Saccharomyces bayannus* and fermented at 18 and 30 °C. Sulfur dioxide was added at 0, 50, 100 and 200 mg/kg. The headspace volatiles were identified by gas chromatography - mass spectrometry and sniffed using the Osme technique. The flavour profiles were generated by Quantitative Descriptive Analysis. Impact volatiles included ethyl esters associated with fruit flavour. A refrigerated fermentation provided better wines, increasing the levels of compounds with cashew, fruity and sweet notes and decreasing undesirable compounds perceived as 'fermented', 'plastic' and or 'smelly'. A low level of sulfitation was required to obtain a fruity cashew wine whereas too much sulfite favoured the production of isobutanol and hexanoic acid, which could impair the sensory quality of the product.

1. INTRODUCTION

In Brazil, new products and processes are under investigation in order to reduce the high wastage of cashew apple during cashew-nut manufacturing. Therefore, the alcoholic fermentation of cashew apple juice has been investigated in an attempt to develop a novel alcoholic beverage, alternative to wine. This work aimed at completing the studies performed by Faria [1] and Dias [2] focusing, however, on the composition of volatile compounds present in the headspace of samples and on the sensory characteristics of the product. The interest in evaluating the cashew wine made at ambient temperature was particularly due to the attempt of making an alternative vinification process possible, which may be used in rural zones where refrigeration resources are scarce and costly.

2. MATERIALS AND METHODS

Fresh juice was obtained from cashew of genotype CCP 76 grown in the Northeast of Brazil (Ceará State). After clarification, the juice was added sodium metabisulfite to reach 0, 50, 100 and 200 mg/l SO_2. The musts inoculated with *S. bayannus* were incubated at 30 ºC (ambient temperature at Ceará, coded TA) and at 18 ºC (coded TR). The headspace volatiles were isolated by suction, and trapped on Porapak Q according [3] and analysed by HR-GC, GC-MS and GCO (Osme) according to [4]. Only sample TR with 100 mg/l SO_2 was submitted to GCO. A QDA [5] was performed to assess the flavour profile of the samples. Chromatographic and sensory data were submitted to ANOVA, principal components analysis and Pearson correlation.

3. RESULTS

The HR-GC analysis of the eight cashew wine treatments showed the presence of 41 volatile compounds. The aromagram of sample TR_{100} (Figure 1) showed 25 odoriferous compounds but many of them were not detected by FID and were labelled with small letters.

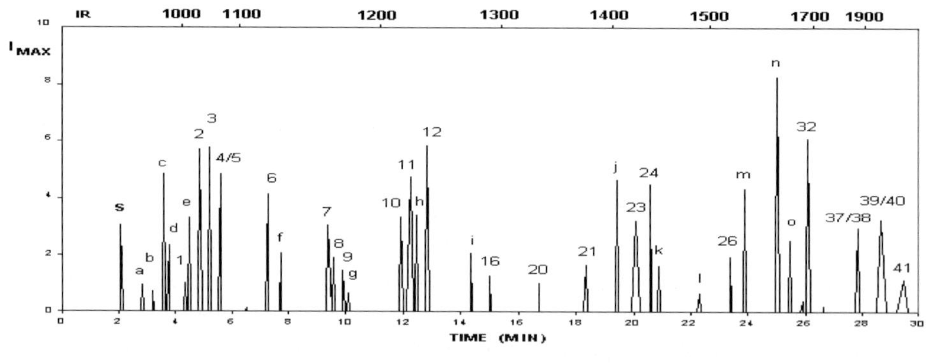

Figure 1. Osme aromagram of cashew wine fermented at 18 ºC with added 100 mg/ml SO_2.

Table 1 presents the means for the odour-active compounds only. The lowering of the fermentation temperature increased the concentration of esters with sweet, fruity and cashew-like aromas, but also increased the level of compounds that presented unpleasant aromas (isopentyl acetate, styrene and hexanoic acid). Refrigeration decreased concentration of isobutanol, described as smelly. Must sulfitation, in general, produced an increase in the esters concentration, except ethyl lactate. Styrene was not affected by sulfitation, but hexanoic acid increased significantly.

The PCA carried out with the eleven volatiles which produced a greater impact on the cashew wine aroma and flavour (Figure 2) showed that wines fermented at 18 ºC and added SO_2 presented a greater concentration of volatiles, while samples made at 30 ºC were characterised by a higher intensity of isobutanol (peak 5). Wines with no sulfitation had a smaller concentration of all odour-active compounds identified.

Table 1. Odour properties and standardised relative areas of odour-active compounds in cashew wine, as influenced by fermentation temperature and sulfur dioxide content.

Peak no	RI[a]	Compound	Odour quality	Relative peak area[b] T (°C) 30	18	SO_2 (mg/l) 0	50	100	200
1	1014	2-Methylpropyl acetate	Sweet, floral	1.41a	1.03a	0.73b	1.10ab	1.16ab	1.89a
2	1042	Ethyl butanoate	Sweet, cashew	2.07 b	5.72a	1.90c	4.28b	3.98b	5.42a
3	1057	Ethyl 2-methyl butanoate	Sweet, cashew	1.92 b	2.44a	2.08a	2.44a	1.91a	2.28a
4	1070	Ethyl 3-methylbutanoate	Sweet, cashew	5.05 b	9.78a	8.44a	7.61a	6.50a	7.10a
5	1109	2-Methyl-1-propanol	Fermented, smelly	29.03a	0.04b	16.86b	24.33ab	19.21ab	26.42a
6	1118	3-Methylbutyl acetate	Banana, plastic	19.74 b	77.55a	24.18c	51.13b	51.99b	67.25a
8	1175	1-Pentyl acetate	Sweet, cashew	0.25b	1.36a	0.24c	0.88b	0.78b	1.31a
11	1239	Ethyl hexanoate	Sweet, fruity	20.81b	43.24a	11.0b	37.69a	38.95a	40.44a
12	1253	Phenylethene	Plastic, solvent	65.10b	201.83a	139.05[a]	144.26a	105.85a	144.71a
16	1292	Unknown	Glue, mould	0.05a	0.03a	0.00b	0.04ab	0.04ab	0.08a
20	1345	Ethyl 2-hydroxy-propanoate	Acid, plastic	24.30a	12.05b	27.50a	20.95ab	12.70b	11.55b
23	1436	Ethyl octanoate	Sweet, fruity	23.33b	48.81a	14.56c	35.95b	41.36b	52.40a
37	1860	Hexanoic acid	Fermented	0.15b	0.49a	0.04b	0.27a	0.50a	0.47a
38	1888	Ethyl 3-phenyl-propanoate	Sweet, dried fruit	2.51b	3.10a	2.28b	2.81b	3.44a	2.69b
40	2050	Octanoic acid	Cashew, wine	0.45 b	0.89a	0.00b	0.60ab	0.73ab	0.88a

[a]RI: retention index. [b]Areas in a row marked with the same letter are not significantly different.

The projection of sensory descriptors determined by QDA (Figure 3) showed that wines fermented at 30 °C were characterised by 'fermented aroma', 'acid taste' and 'alcoholic flavour', whereas the samples fermented at refrigerated temperature were associated with 'fruit flavour'. The PCA also discriminated the samples as to the content of SO_2, especially on the component II. The samples with a lower SO_2 content appeared further

away from the vector corresponding to the sulfur aroma, except TR200, probably due to the perception of 'fruit flavour' being more intense.
Analysis by the Pearson test showed a significant positive linear correlation between the 'fruit flavour' of the wines and the following compounds: ethyl butyrate ($r=0.95$, $p<0.01$), ethyl 2-methylbutyrate ($r=0.90$, $p=0.01$), ethyl isovalerate ($r=0.98$, $p<0.01$), ethyl hexanoate ($r=0.99$, $p<0.01$) and styrene ($r=0.97$, $p<0.01$). 'Fruit flavour' also showed a negative correlation to the descriptors 'fermented', 'alcoholic flavour' and 'acid taste', and these descriptors correlated positively among each other.

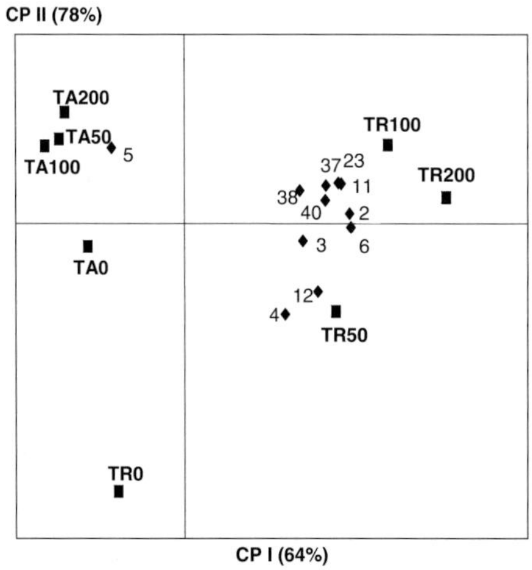

Figure 2. PCA bi-plot from GC-MS.

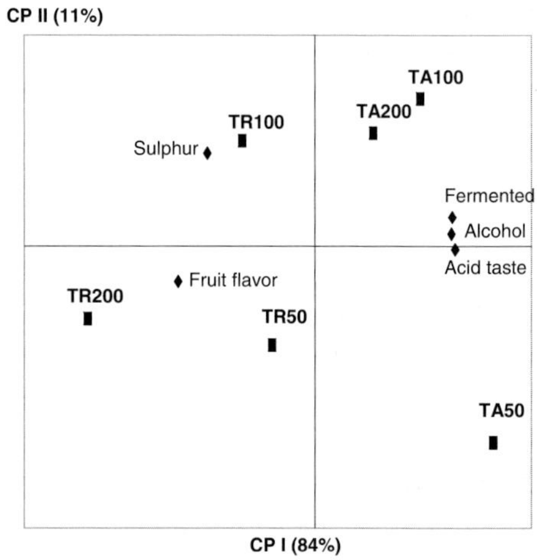

Figure 3. PCA bi-plot from sensory analysis.

4. DISCUSSION AND CONCLUSIONS

Volatiles with impact on the cashew wine sensory profile included ethyl butyrate, ethyl-2-methylbutyrate, ethyl isovalerate, ethyl hexanoate, isopentyl acetate and ethyl octanoate, most of them associated with the fruity flavour.
The fermentation temperature influenced the cashew wine volatile composition more than the variation of SO_2 dosages. Fermentation at 18 °C provided a better sensory quality wine. Refrigeration not only increased concentration of volatile compounds of desirable aroma such as cashew, fruity, and sweet, but also decreased undesirable compounds, described as fermented and smelly.

References

1. F.S.E.D.V. Faria, Master thesis, Fortaleza, Brazil (1994).
2. A.L.M. Dias, Master thesis, Fortaleza, Brazil (1996).
3. M.R.B. Franco and D.B. Rodriguez-Amaya, J. Sci. Food Agric., 34 (3) (1983) 293.
4. D.S. Garruti, M.R.B. Franco, M.A.A.P. da Silva, N.S. Janzantti and G.L. Alves, LWT Food Sci. Technol., in press.

W.L.P. Bredie and M.A. Petersen (Editors)
Flavour Science: Recent Advances and Trends

Modulation of volatile thiol and ester aromas by modified wine yeast

Jan H. Swiegers[a], Robyn Willmott[a], Alana Hill-Ling[a], Dimitra L. Capone[a], Kevin H. Pardon[a], Gordon M. Elsey[a], Kate S. Howell[a], Miguel A. de Barros Lopes[b], Mark A. Sefton[a], Mariska Lilly[c] and Isak S. Pretorius[a]

[a]The Australian Wine Research Institute, PO Box 197, Glen Osmond, Adelaide, Australia; [b]University of South Australia, North Terrace, Adelaide, Australia; [c]Institute for Wine Biotechnology, Stellenbosch University, Stellenbosch, South Africa

ABSTRACT

The volatile thiols, in particular 4-mercapto-4-methylpentan-2-one (4MMP), 3-mercaptohexan-1-ol (3MH) and 3-mercaptohexyl acetate (3MHA) are potent aroma shown to contribute strongly to the varietal aroma of Sauvignon Blanc wines. The thiols 4MMP and 3MH exist as non-volatile, aroma-inactive cysteine bound conjugates in the grape must and during fermentation the thiol is cleaved from the precursor. However, no cysteine conjugate for 3MHA has been identified. In this work we showed that 3MHA is formed from 3MH by the wine yeast *Saccharomyces cerevisiae* during fermentation. Furthermore, the alcohol acetyltransferase, Atf1p, the enzyme involved in the formation of the ester ethyl acetate, was shown to be the main enzyme responsible for the formation of 3MHA. Both a laboratory yeast and a commercial wine yeast overexpressing the *ATF1* gene produced significantly more 3MHA than the wild-type. Although an *atf1*Δ laboratory yeast strain showed reduced 3MHA formation, it was not abolished, indicating that other enzymes are also responsible for its formation. Therefore, overexpression of the *ATF1* gene in a wine yeast presents the possibility of modulating both the thiol and ester aromas in wine.

1. INTRODUCTION

Sauvignon Blanc wine has characteristic aromas described as box tree, cat urine, broom, grapefruit, blackcurrant and passion fruit [1,2]. These aromas are attributed to three potent volatile thiol compounds: i) 4-mercapto-4-methylpentan-2-one (4MMP), ii) 3-mercaptohexyl acetate (3MHA) and iii) 3-mercaptohexan-1-ol (3MH), all having an extremely low perception threshold. These thiols have also been identified in wine varieties such as Colombard, Muscat d'Alsace, Petit Manseng, Gewürztraminer, Riesling, Cabernet Sauvignon, Merlot and Semillon [1,3]. The volatile thiols 4MMP and 3MH are almost non-existent in the grapes but are released during fermentation from their non-volatile cysteine bound precursor. Although 3MHA is found in significant quantities in wine, no cysteine bound precursors could be identified [4]. Therefore, it appears that 3MHA is formed during fermentation and through the esterification of 3MH.

A large proportion of the characteristic fruity odours of wine are primarily derived from the synthesis of esters by the wine yeast, *Saccharomyces cerevisiae*. The *ATF1*- and *ATF2*-encoded alcohol acetyltransferases of this yeast are responsible for the synthesis of ethyl acetate and isopentyl acetate esters, while the *EHT1*-encoded ethanol hexanoyl transferase is responsible for synthesising ethyl caproate. However, esters such as these can be degraded by the *IAH1* encoded esterase [5-10]. The objective of this study was to investigate the possible role of alcohol acetyltransferases and esterases in the formation of 3MHA.

2. MATERIALS AND METHODS

Strains used were commercial wine yeast VIN13 (Anchor Yeast, South Africa), VIN13-ATF1-s (overexpressing *ATF1*), VIN13-ATF2-s (overexpressing *ATF2*), VIN13-EHT1-s (overexpressing *EHT1*), VIN13-IAH1-s (overexpressing *IAH1*) [6,11] and BY4742 wild type and BY4742 *atf1*Δ (EUROSCARF). Plasmid pATF1s [6] was linearised with *Apa*I and transformed into strain BY4742 using the lithium acetate procedure and confirmed by PCR.

The growth medium for fermentations consisted of 8% glucose, 0.67% YNB to which 1 mg/l of 3MH was added after autoclaving. Fermentations were carried out in 250 ml Erlenmeyer flasks equipped with an air lock. The samples were spiked with polydeuterated internal standards for stable isotope dilution analysis (SIDA). The volatile thiols were then analysed by headspace solid phase microextraction coupled with gas chromatography - mass spectrometry (HS-SPME GC-MS).

3. RESULTS

The commercial wine yeast, VIN13, was previously used to construct strains overexpressing various genes (*ATF1*, *ATF2*, *EHT1* and *IAH1*) involved in ester metabolism [6,11]. These strains were used to investigate the involvement of ester metabolism in the formation of 3MHA. A model medium was spiked with 1 mg/l of 3MH. Strains were grown overnight in YPD and the 150 ml model medium was inoculated at OD 600 of 0.05. After 2 days of fermentation at 30 °C, samples were taken and the amounts of 3MHA formed were measured using HS-SPME GC-MS (Figure 1). There was a significant increase in the concentration of 3MHA in the ferments conducted with VIN13-ATF1-s and a reduction in the concentration of 3MHA in ferments conducted with VIN13-IAH1-s. It was also shown that this was not due to chemical conversion as 3MH added to sterile fermented model media and left for 2 days at 30 °C did not result in the formation of 3MHA (data not shown).

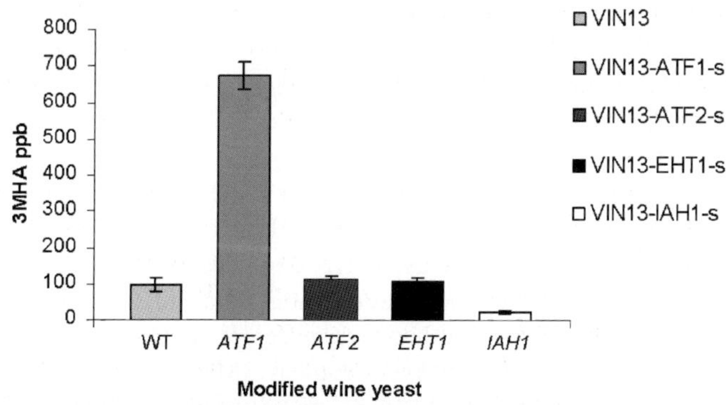

Figure 1. Conversion of 3MH to 3MHA by wild type (WT) wine yeast and wine yeast overexpressing ester metabolism related genes (*ATF1*, *ATF2*, *EHT1* and *IAH1*).

A laboratory yeast strain with deleted *ATF1* was investigated, as a deletion for the wine yeast is not available and difficult to construct. Fermentations and analyses were conducted in the same way as described above. After 2 days fermentation at 30 °C, the laboratory strain formed about 12 μg/l of 3MHA. Deletion of the *ATF1* gene did not result in the abolishment of 3MHA formation; however, 3MHA production was reduced. As a control, the *ATF1* gene was also overexpressed in the laboratory strain and large amounts of 3MHA were formed (Figure 2).

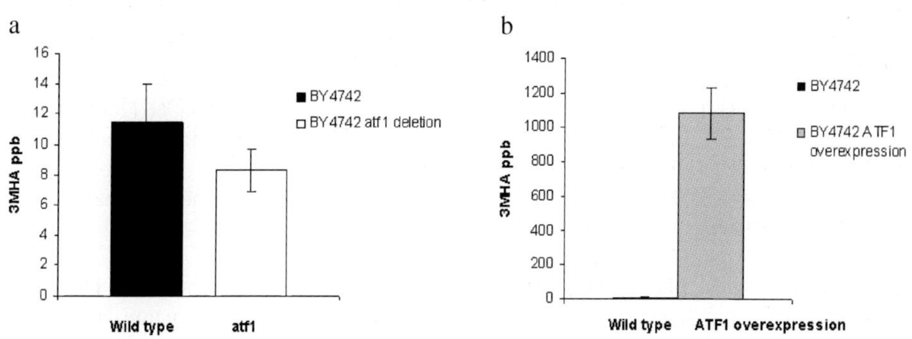

Figure 2. The bioconversion ability of laboratory yeast (BY4742) with modified *ATF1* expression. (a) A laboratory yeast with a deleted *ATF1* (atf1). (b) A laboratory yeast overexpressing *ATF1*.

Previously, the ability of wine yeast to release the thiol, 4MMP, was investigated for the same reason [12]. Fermentations and analysis were conducted in the same way as described above. After 2 days fermentation at 30 °C, analyses was done and a large variation in the ability of wine yeast to form 3MHA was observed (data not shown).

4. DISCUSSION AND CONCLUSION

In this work we have implicated the *ATF1*-encoded alcohol acetyltransferase in the formation of 3MHA. The results indicated, firstly, that wine yeast are able to bioconvert 3MH to 3MHA. Secondly, in the collection of enzymes we investigated Atf1p appears to have a major role in the conversion of 3MH to 3MHA as overexpression of *ATF1* resulted in significant increase in the amount of 3MHA formed. Thirdly, our hypothesis was supported by the fact that the esterase Iah1p, degrades 3MHA, as overexpression of *IAH1* resulted in reduced 3MHA concentration. Atf2p and Eht1p did not appear to have a role in the formation of 3MHA. Furthermore, different commercial wine yeast have a large degree of variation in their ability to convert 3MH to 3MHA. This is valuable information for winemakers, helping them to make informed choices on yeast strain selection as a tool to modulate wine flavour. Future work will entail the monitoring of expression of *ATF1* during fermentation in order to determine how winemakers can manipulate fermentation conditions to optimise 3MHA formation.

References

1. J.M. Rantz (ed.), Proceedings of the 50th anniversary annual meeting of the American society for enology and viticulture, Davis, California, USA (2001) 369.
2. T. Tominaga, C. Peyrot des Gachons and D. Dubourdieu, J. Agric. Food Chem., 46 (1998) 5215.
3. M.L. Murat, T. Tominaga and D. Dubourdieu, J. Agr. Food Chem., 49 (2001) 5412.
4. P. Darriet, T. Tominga, V. Lavigne, J. Boidron and D. Dubourdieu, Flavour Fragrance J., 10 (1995) 385.
5. T. Fujii, N. Nagasawa, A. Iwamatsu, T. Bogaki, Y. Tamai and M. Hamachi, Appl. Environ. Microbiol., 60 (1994) 2786.
6. M. Lilly, M.G. Lambrechts and I.S. Pretorius, Appl. Environ. Microbiol., 66 (2000) 744.
7. N. Nagasawa, T. Bogaki, A. Iwamatsu, M. Hamachi and C. Kumagai, Biosci. Biotechnol. Biochem., 62 (1998) 1852.
8. H. Yoshimoto, D. Fujiwara, T. Momma, C. Ito, H. Sone, Y. Kaneko and Y. Tamai, J. Ferment. Bioeng., 86 (1998) 15.
9. A.B. Mason and J.P. Dufour, Yeast, 16 (2000) 1287.
10. K.J. Verstrepen, S.D.M. Van Laere, B.M.P. Vanderhaegen, G. Derdelinckx, J.P. Dufour, I.S. Pretorius, J. Winderickx, J.M. Thevelein and F.R. Delvaux, Appl. Environ. Microbiol., 69 (2003) 5228.
11. M. Lilly, PhD Thesis, Stellenbosch University, Stellenbosch, South Africa (2004).
12. K.S. Howell, J.H. Swiegers, G.M. Elsey, T.E. Siebert, E.J. Bartowsky, G.H. Fleet, I.S. Pretorius, S. Pretorius and M.A.D. Lopes, FEMS Microbiol. Lett., 240 (2004) 125.

Heterologous expression of carotenoid-cleaving dioxygenases from plants for the production of natural flavour compounds

Frauke Patett[a], Martin Schilling[a], Dieter Sell[a], Holger Schmidt[b], Wilfried Schwab[b] and Jens Schrader[a]

[a]Biochemical Engineering, Karl-Winnacker-Institut, DECHEMA e.V., Theodor-Heuss-Allee 25, 60486 Frankfurt/Main, Germany; [b]Biomolecular Food Technology, Degussa. Stiftungsprofessur, TU München, Lise-Meitner-Str. 34, 85354 Freising, Germany

ABSTRACT

As first steps towards biochemical characterisation we investigated the heterologous production of the plant enzyme carotenoid-cleaving dioxygenase 1 from *Arabidopsis thaliana* (AtCCD1) in different *E. coli* hosts and in *P. pastoris*. An *in vitro* activity assay for carotenoid-cleaving dioxygenases based on octyl-β-D-glucopyranoside micelles of carotenoids was developed.

1. INTRODUCTION

The oxidative degradation of carotenoids leads to apocarotenoids, which exhibit various bioactivities, e.g. flavours, pigments and vitamins. C13-Norisoprenoids, e.g. ionone and damascenone, belong to the apocarotenoids. They are found in the essential oils of many plants and contribute significantly to their olfactory properties and therefore represent highly interesting flavour compounds [1]. Due to their very low natural concentrations, the extraction of pure compounds is not viable from an economic point of view. Up to now, natural apocarotenoids are only accessible by biotechnological means via unspecific cooxidation using oxidases, mostly lipoxygenases, in the presence of cooxidants such as linoleic acid [2] or using peroxidases [3].

During the last few years the first carotenoid-cleaving dioxygenases (CCD) from plants, a novel enzyme subclass, have been identified [4]. CCDs selectively cleave one double bond of the conjugated system of various carotenoid substrates. Our aim was to evaluate the potential of CCD for use as biocatalysts to produce natural flavour compounds.

Here, we compare different strategies of heterologous production using AtCCD1 from *Arabidopsis thaliana* as model CCD, which cleaves the 9,10 and 9',10' double bonds forming β-ionone from β-carotene [5].

2. MATERIALS AND METHODS

2.1. Cloning and expression of AtCCD1 in *E. coli*

Atccd1 [5] was cloned in the BamHI/NotI sites of pGEX-4T (Amersham Biosciences). The plasmid pGEX-4T-*Atccd1* was transformed in *E. coli* Rosetta strains (Novagen). Cells were grown in LB medium at 37 °C until an A_{600} of 0.4 to 0.6. Expression was induced with 0.2 mM IPTG and cultures were grown at 20 °C for an additional 3 to 5 h. Cells were harvested by centrifugation, re-suspended in PBS with nuclease and lysed by sonication. GST-AtCCD1 was purified using glutathione columns (GSTrap FF from Amersham Biosciences) according to the manufacturer's protocol.

2.2. Cloning and expression of AtCCD1 in *P. pastoris*

Atccd1 [5] was cloned in the EcoRI/ApaI sites of pPICZ (Invitrogen) and maintained in *E. coli* NovaBlue (Novagen), linearised by SacI cleavage and transformed into *P. pastoris* X-33 (Invitrogen) by electroporation. *P. pastoris* cells were grown according to Invitrogen's instructions in minimal medium containing glycerol at 30 °C until A_{600} of 4 to 6 and then transferred to minimal medium containing methanol for induction. Cells were harvested after 24 h by centrifugation and broken with glass beads.

2.3. Enzyme assay

Carotenoid micelles were prepared by mixing 400 μl of 1 mg/ml carotenoid solution in chloroform with 200 μl of 4% (w/v) octyl-β-D-glucopyranoside in chloroform followed by evaporation. Micelles were re-suspended in 600 μl enzyme solution in PBS containing 0.5 mM ascorbate. Reactions were shaken at 25 °C for 2 h in the dark.

3. RESULTS

3.1. Heterologous expression of AtCCD1 in *E. coli* and *P. pastoris*

All plant CCDs described so far were heterologously expressed in *E. coli* [4-7]. In most cases the formation of inclusion bodies was reported. Aiming at the development of a bioprocess we investigated the recombinant expression in more detail.

In *E. coli,* overexpression patterns were only observed in the insoluble fraction after cell lysis (Figure 1a). Neither an addition of betaine, iron salts or sorbitol to cultivation media nor variation of the lysis buffer composition (addition of Fe^{2+}, glycerol, salt, glutathione, cysteine, DTT, β-mercaptoethanol, Triton X-100, Tween, chloroform, DMSO) or variation of pH enhanced the yield of soluble enzyme.

Comparing different *E. coli* expression hosts, Rosetta strains (Novagen) with rare codon supplementation for eukaryotic gene expression produced the highest yields in total as well as in soluble fusion protein, as was determined by affinity chromatography of GST-AtCCD1 (data not shown). Origami strains (Novagen) facilitating disulfide bond formation showed lower expression levels. *P. pastoris* was chosen as an eukaryotic alternative. The AtCCD1 gene was successfully inserted into the host genome (Figure 1b) and soluble AtCCD1 was expressed after induction with methanol (Figure 1c).

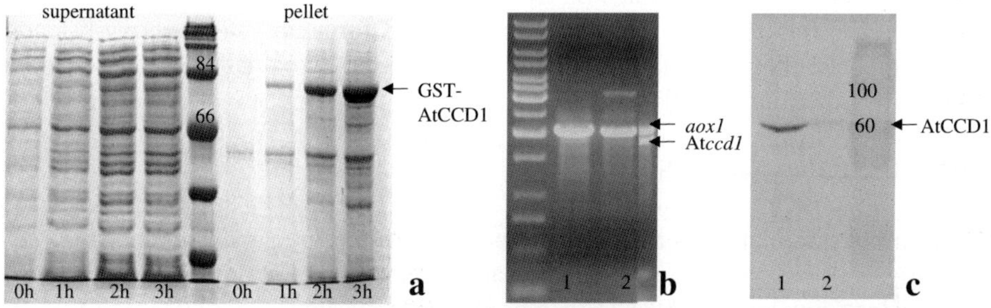

Figure 1. (a) SDS-PAGE of the expression profile of GST-AtCCD1 in *E. coli*. The supernatants and pellets after cell lysis are shown beginning from time of induction. (b) PCR amplification of genes under control of AOX1-promotor (1: untransformed *P. pastoris* X-33; 2: transformant with *Atccd1*). (c) Western blots of two *P. pastoris* X-33 transformants expressing AtCCD1.

3.2. *In vitro* cleavage of carotenoids

Carotenoid cleavage activity of gene products is often shown *in vivo* by coexpression in carotenoid-producing *E. coli* strains [4,5]. *In vitro* activity assays are complicated by the insolubility of carotenoids in aqueous systems.

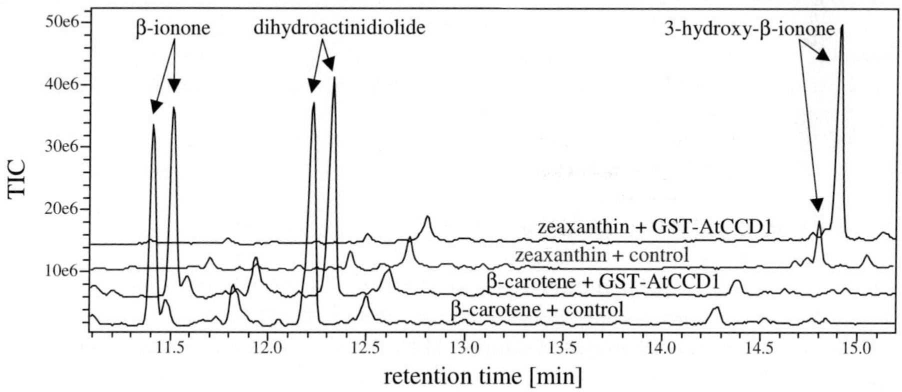

Figure 2. Cleavage experiments of β-carotene and zeaxanthin by GST-ATCCD1 containing raw cell extracts. GC-MS analyses after chloroform extraction of reaction mixtures are shown (chromatograms staggered). Controls were extracts from *E. coli* strains transformed with the empty vector.

Dissolution in organic solvents or detergent additives has mainly been applied in carotenoid transformations [4-7]. We found that preparation of carotenoid micelles with octyl-β-D-glucopyranoside before the addition of aqueous enzyme solutions led to the highest conversion yields (data not shown). The choice of carotenoid substrate proved to be even more crucial for the *in vitro* detection of CCD activity. Zeaxanthin, containing hydroxy groups, was cleaved (breakdown product 3-hydroxy-β-ionone) in contrast to β-carotene (Figure 2).
Conversion experiments (carotenoid addition in ethanol) with raw extracts from *P. pastoris* showed unspecific oxidation patterns, e.g. formation of 5,6-epoxy-β-ionone and dihydroactinidiolide. This oxidation is likely caused by oxidative species liberated from peroxisomes during cell lysis.

4. DISCUSSION AND CONCLUSION

Inclusion body formation is a common problem of heterologous expression in *E. coli*, especially with eukaryotic genes. A comparison of different expression hosts showed that the expression level of AtCCD1 can be increased by supplementing rare tRNAs and that inclusion body formation is not due to missing disulfide bonds. Aggregation in inclusion bodies may be caused by other factors, e.g. glycosylation, missing folding aids or general protein hydrophobicity. Expression in eukaryotic hosts such as *P. pastoris* is possible and might therefore prove to be a suitable alternative. Experiments with raw *P. pastoris* extracts will be repeated in the presence of radical scavengers and catalase.
The observation that only zeaxanthin was cleaved in the micellar assay is in accordance with the surface-to-core distribution in biological emulsions where xanthopylls localise in the polar surface layer and carotenoids in the apolar core [8]. Similarly, in octyl-β-D-glucopyranoside micelles zeaxanthin, but not β-carotene, seems to align near the surface thus being accessible to the enzyme in the aqueous phase.

References

1. P. Winterhalter and R.L. Rouseff (eds.), Carotenoid-derived aroma compounds, ACS Symposium Series 802, Washington, DC, USA (2002) 1.
2. D. Waldmann and P. Schreier, J. Agric. Food Chem., 43 (1995) 626.
3. H. Zorn, S. Langhoff, M. Schreibner, M. Nimtz and R.G. Berger, Biol. Chem., 384 (2003) 1049.
4. S.H. Schwartz, B.C. Tan, D.A. Gage, J.A.D. Zeevaart and D.R. McCarty, Science, 276 (1997) 1872.
5. S.H. Schwartz, X. Qin and J.A.D. Zeevaart, J. Biol. Chem., 276 (2001) 25208.
6. F. Bouvier, C. Suire, J. Mutterer and B. Camara, Plant Cell, 15 (2003) 47.
7. A.J. Simkin, S.H. Schwartz, M. Auldridge, M.G. Taylor and H.J. Klee, Plant J., 40 (2004) 882.
8. P. Borel, P. Grolier, M. Armand, A. Partier, H. Lafont, D. Lairon and V. Azais-Braesco, J. Lipid Res., 37 (1996) 250.

W.L.P. Bredie and M.A. Petersen (Editors)
Flavour Science: Recent Advances and Trends

Lilac aldehydes and lilac alcohols as metabolic by-products of fungal linalool biotransformation

Marco-Antonio Mirata[a], Matthias Wüst[b], Armin Mosandl[c], Dieter Sell[a] and Jens Schrader[a]

[a]Biochemical Engineering, Karl-Winnacker-Institut, DECHEMA e.V., Theodour-Heuss-Allee 25, 60486 Frankfurt/Main, Germany; [b]Department of Life Technologies, University of Applied Sciences Valais, Route du Rawyl 47, 1950 Sion 2, Switzerland; [c]Institute of Food Chemistry, Johann Wolfgang Goethe-University, Marie-Curie-Strasse 9, D-60439 Frankfurt/Main, Germany

ABSTRACT

Nineteen different fungi were screened to test their capability to convert (±)-linalool into lilac aldehydes and lilac alcohols. Solid phase microextraction (SPME) in combination with GC-MS was used to identify the target compounds by comparing retention times and mass spectra with chemically synthesised standards. Most of the strains showed specific and complex product spectra, often with linalool oxides and dihydrolinalool as major metabolites. *Aspergillus niger* DSM 821, *Botrytis cinerea* 5901/02 and *Botrytis cinerea* 02/FB II/2.1 generated the desired lilac fragrance compounds as metabolic by-products.

1. INTRODUCTION

Lilac aldehyde and lilac alcohol have been described as characteristic minor components of *Syringa vulgaris* L. flowers. Due to their very low odour thresholds these monoterpenoids have an important impact on the desired overall lilac odour profile. Recently, biogenetic studies with stable isotope labelled precursors have shown that *S. vulgaris* L. converts linalool into lilac aldehydes and lilac alcohols [1,2]. Due to the evidence of a plant biosynthetic pathway we hypothesised that there are also other biological systems capable of transforming linalool into the desired lilac fragrance compounds. Microorganisms, especially fungi, have been shown to be very versatile biocatalysts for the production of a wide range of flavour and fragrance compounds from terpenoid precursors [3]. Therefore, 19 different fungi, known from the literature as being able to transform terpenes, were screened for lilac aldehyde and lilac alcohol production from (±)-linalool.

2. MATERIALS AND METHODS

2.1. Microorganisms, culture media, chemicals

Botrytis cinerea 5901/2 (***A***), 5909/1, 92/lic/1, 97/4, 99/16/3, 00/II10.1, 02/FB II/2.1 (***B***), and P10 (BLWG, Veitshöchheim, Germany). *Aspergillus niger* DSM 1988, DSM 821 (***C***), *Corynespora cassiicola* DSM 62475, *Penicillium digitatum* DSM 62840 and *P. italicum* DSM 62846 (DSMZ, Braunschweig, Germany). *Geotrichum candidum* (HEVs, Sion, Switzerland). *P. digitatum* NRRL 1202 (ARS culture collection, Illinois, USA). *Saccharomyces cerevisiae* Ceppo 20, Zymaflor VL1, Uvaferm 228, SIHA Riesling n° 7 (E. Begerow GmbH & Co., Langenlonsheim, Germany). The strains were grown on MEA (malt extract 3%, soja peptone 0.3%, and agar 1.7%, pH 5.6) at 25 °C (moulds) and 30 °C (yeasts). Biotransformation experiments were performed in MYB (malt extract 3%, glucose 1%, peptone 1% and yeast extract 0.3%, pH 6.4). (±)-Linalool, (-)-α-pinene, and *tert*-butyl methyl ether (MTBE) were obtained from Fluka (Switzerland). Lilac alcohol and lilac aldehyde were prepared according to the literature procedure [4] and were used as mixtures of stereoisomers (0.01% in MTBE).

2.2. Biotransformations

Screening experiments were performed in 40 ml SPME vials filled with 15 ml MYB. After inoculation (yeast cells by loop; moulds as 0.5×10^8 spores in 500 μl) 150 μl of 0.3% (w/v) (±)-linalool (in ethanol) were added. The vials were covered with cotton stoppers and the cultures were incubated for 14 days at 220 rpm and 25 °C (moulds) or 30 °C (yeasts). Biotransformations with the selected strains were carried out for 12 days in 2l Erlenmeyer flasks filled with 500 ml MYB. After inoculation, (±)-linalool (3% w/v in ethanol) was initially added in non-toxic concentrations (A = 150 mg/l, B and C = 100 mg/l). Additional linalool (A and B = 80 mg/l, C = 50 mg/l) and glucose (A = 2 g/l, B = 0 g/l and C = 5 g/l) were fed at intervals of 1 day for B and 3 days for A and C. The cultures were incubated at 130 rpm and 25 °C (moulds) or 30 °C (yeasts). After 3 days for B and 9 days for A and C, 15 ml final culture suspension was analysed by SPME.

2.3. Blank experiment

To test the potential of a non-biological formation of the target compounds, a 500 ml flask was filled with 100 ml MYB medium. Two hundred μl linalool were added to 100 ml MYB in a 0.5 l flask and the solution was incubated at 220 rpm and 25 °C. Control of potential product formation was performed by SPME GC-MS after 9 days.

2.4. Analytical methods

For product analysis 15 ml liquid cultures were transferred into 40 ml SPME vials, the pH was adjusted to 4.0 with 1 M HCl and 3.75 g NaCl were added. The vials were immediately covered with PTFE-silicone screw caps (Supelco) and incubated for 5 min at 400 rpm and 40 °C before a 20 min SPME headspace extraction using a 75 μm CAR-PDMS coated fibre and a manual holder (Supelco, Germany) was started (α-pinene = int. standard). GC-MS analysis was performed with a GC-17 A gas chromatograph, equipped with a VB-5 Valcobond column (30 m x 0.25 mm i.d.; coating thickness 0.25 μm, ICT GmbH, Germany) and a QP5050 MS (Shimadzu, Germany): injector 250 °C, detector 280 °C, start 40 °C, hold 7 min, 40 to 280 °C at 10 K/min, hold 2 min; carrier gas (He) 1.1 ml/min; splitless injection; 5 min SPME fibre desorption; ionisation: EI 70 eV; scanned *m/z*: 35-250 (10-25 min). Lilac aldehydes and alcohols were identified by comparing their mass spectra and retention times with those of the references. Other compounds were identified by NIST mass spectral library V 2.0. The dry biomass was determined with an infrared moisture analyser (Sartorius, Germany). Glucose was analysed enzymatically (Yellow Spring Instrument, USA).

3. RESULTS

Nineteen fungal strains were screened on 15 ml scale for their ability to convert (±)-linalool into lilac aldehyde and lilac alcohol isomers. After 14 days of growth in linalool-supplemented MYB medium headspace SPME GC-MS analysis revealed seven possibly positive strains: the evaluation criterion was the detectability of at least one peak in the chromatograms of the three characteristic fragments *m/z* 111, 153 (aldehyde), and 155 (alcohol) within the time window of lilac aldehydes and alcohols which had been determined beforehand with the chemically synthesised reference compounds. The positive strains were *A. niger* DSM 1988 and DSM 821, *B. cinerea* 5901/2 and 02/FBII/2.1, *S. cerevisiae* Zymaflor VL1 and Uvaferm 228, and *C. cassiicola* DSM 62475. In subsequent experiments the selected strains were grown on 500 ml scale with sequential feeding of linalool and glucose to enhance product formation by expanding the growth and biotransformation period. SPME GC-MS analysis and comparison with a non-biological blank experiment confirmed *B. cinerea* 5901/2 and 02/FBII/2.1 and *A. niger* DSM 821 as clearly positive strains whereas the other cultures tested did not produce detectable amounts of the target compounds. Figure 1a illustrates the total ion and fragment ion chromatograms (TIC, *m/z* 111, 153, 155) by SPME GC-MS analysis of a final culture headspace after 9 days cultivation of *B. cinerea* 5901/2 in MYB fed with linalool and glucose. The mass spectra of the target compounds produced, illustrated in Figures 1b and 1c by one lilac aldehyde and one lilac alcohol isomer, showed the characteristic fragmentation pattern [1]. The stereochemistry of the isomers produced was not determined. Main transformation products were dihydrolinalool, α-terpineol and 8-hydroxylinalool. In the case of the other 'lilac-positive' strains, epoxylinalool and linalool oxide were generated as main products beside others. Blank experiments revealed that dihydrolinalool, α-terpineol and linalool oxide were also products of chemical transformation under the given incubation conditions, although at a comparably low level, but neither lilac aldehydes nor lilac alcohols were generated (data not shown).

a

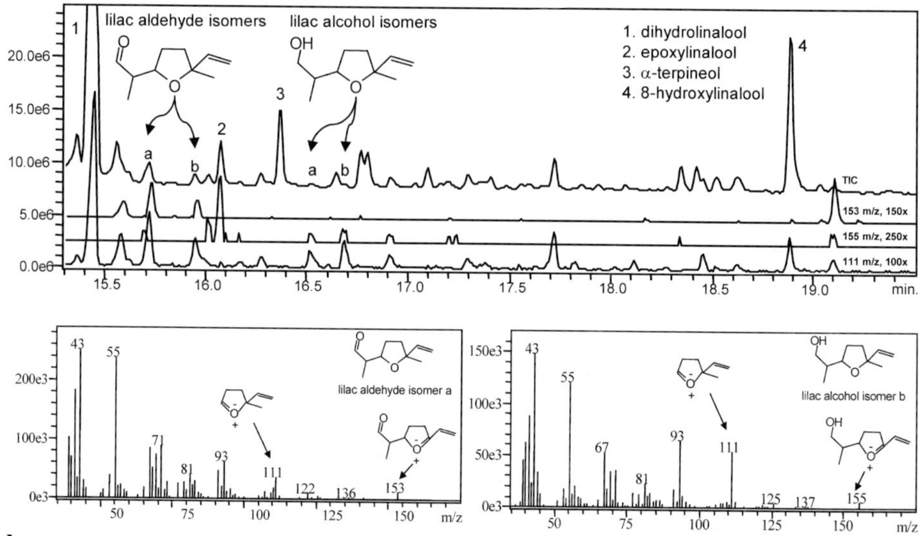

b c

Figure 1. SPME GC-MS analysis of a *B. cinerea* 5901/2 culture fed with (±)-linalool.

4. DISCUSSION AND CONCLUSION

The fungal transformation of (±)-linalool to lilac aldehyde and lilac alcohol, character-impact-compounds of *Syringa vulgaris* fragrance, was demonstrated for the first time. In cultivations with substrate and precursor feeding three undoubtedly positive fungal strains were identified. Nevertheless, low peak signals referred to only small quantities of the target compounds formed as metabolic by-products (roughly 150 μg/l). Naturally occurring lilac aldehydes and alcohols have odour thresholds of 0.2 to 4 ng as determined by GCO [4] and thus belong to the most potent terpenoid aroma compounds known at all. Due to this strong sensory impact their concentration in natural sources is very low. Given these facts, the low concentrations in the fungal biotransformations may be explained by an inherent low expression and/or activity level of the enzymes involved in lilac aldehyde and alcohol formation from linalool in nature.

References

1. M. Kreck, S. Püschel, M. Wüst and A. Mosandl, J. Agric. Food Chem., 51 (2003) 463.
2. D. Burkhardt and A. Mosandl, J. Agric. Food Chem., 51 (2003) 7391.
3. H.-J. Rehm and G. Reed (eds.), Biotechnology, Weinheim, Germany, 10 (2001) 373.
4. M. Kreck and A. Mosandl, J. Agric. Food Chem., 51 (2003) 2722.

W.L.P. Bredie and M.A. Petersen (Editors)
Flavour Science: Recent Advances and Trends

Development of a plate technique for easy and reliable detection of volatile sulfur compound-producing microorganisms

Pascal Bonnarme[a] and Hugues Guichard[b]

[a]*Institut National de la Recherche Agronomique, Unité Mixte de Recherches Génie et Microbiologie des Procédés Alimentaires, 78850 Thiverval-Grignon, France;* [b]*ADRIA Normandie, Boulevard du 13 juin 1944, BP2, F-14310 Villers Bocage, France*

ABSTRACT

Volatile sulfur compounds (VSCs) are of major importance for flavour development in foodstuffs including cheeses. They primarily result from L-methionine degradation to methanethiol (MTL), a precursor for a variety of other VSCs. A plate assay based on double layer Petri dishes containing 5,5'-dithio-bis-2-nitrobenzoic acid (DTNB) - a chemical reacting with free thiols - in the upper layer was developed. Thiol production was quantified by measuring the intensity of the yellow-orange colour resulting from DTNB reacting with thiols. *Geotrichum candidum* strains isolated from cheeses were compared by this method and colour intensity was shown to correlate to the microbial overall VSC production as measured by gas chromatography.

1. INTRODUCTION

Although generally present in very low amounts in ripened cheeses, VSCs play a major role in cheese flavour owing to their low sensory threshold values. Most VSCs – among others dimethyl disulfide (DMDS), dimethyl trisulfide (DMTS), 2,4-dithiapentane, and *S*-methylthioesters, share a common precursor, methanethiol (MTL), which arises from the degradation of the sulfur group of L-methionine. MTL production can be catalysed either by bacteria or by yeasts [1], both microbial communities being part of the cheese ecosystem. Among yeasts, *Geotrichum candidum* has been shown to produce consistent amounts of MTL together with a wide variety of other VSCs [2,3]. Due to the importance of MTL in cheese flavour and to the high volatility and reactivity of this

compound [4] it is essential to be able to compare cheese-ripening strains on the basis of their ability to produce MTL.
The main objective of this work was to develop a simple and reliable screening method for determination and comparison of *in situ* thiol-producing capability of a population of *G. candidum* strains isolated from mould-ripened cheeses. A plate colorimetric method based on the reaction of 5,5'-dithio-bis-2-nitrobenzoic acid (DTNB) with free thiols was developed on double layer Petri dishes, and subsequently used to compare 51 strains for their ability to produce thiols.

2. MATERIALS AND METHODS

To obtain a sensitive method to detect thiols, DTNB, also called the Ellman's reagent which is a thiol-specific reagent that generates a strong yellow chromophore by reaction with a thiol was used [5]. We developed a methodology consisting of two layers of media. The bottom layer (layer A) was a growth medium in which the microorganism was inoculated, and the top layer (layer B) was an agar medium containing DTNB (Figure 1). All thiols produced by the microorganism are trapped in the upper layer by reaction with DTNB. Production of this yellow chromophore resulted, through diffusion, in a homogeneous colouration of both layers of the test plate. Due to the microbial development on the surface of the plate, the measurement of the colour intensity was performed at the back side of the agar plate (Figure 1). The b colour value was the best estimate for yellow colour intensity measurements. Details about experimental procedures are described in [6].

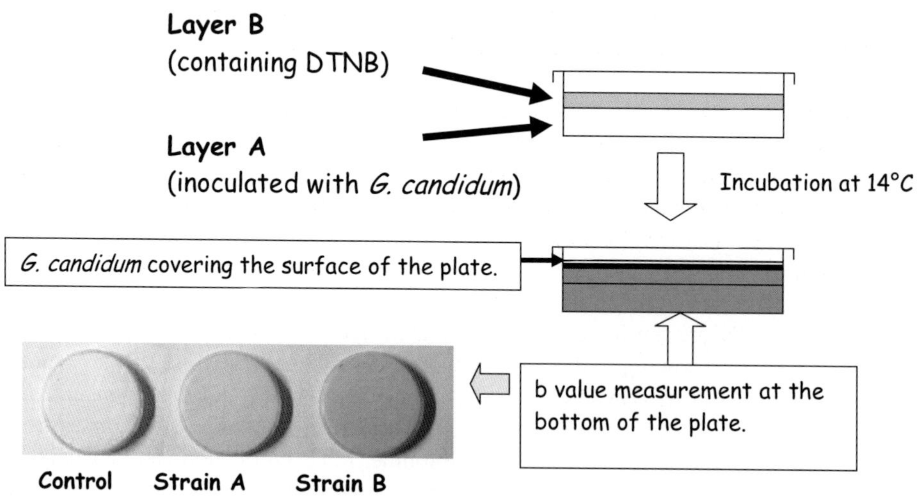

Figure 1. Detection of thiol-producing microorganisms in double-layer agar plates.

3. RESULTS AND DISCUSSION

3.1. Screening of thiol-producing strains of *G. candidum*

Time course measurements of b-values were performed for several *G. candidum* strains. The b-value remained stable in the non inoculated control plates while it steadily increased in the inoculated test plates. This showed a production of thiols which reached a maximum after around 11 days of cultivation for all strains. The screening procedure was applied to a larger population of *G. candidum* strains isolated from cheeses, which were subsequently classified according to their ability to produce thiols [6].

3.2. Production of VSCs by *G. candidum* strains under cheese-like conditions

Eighteen strains of *G. candidum* covering the whole range of thiol-producing capacities were cultivated under cheese-like conditions in a cheese curd medium (14 °C, 21 days). The production of VSCs was measured by Purge and Trap GC-MS analyses. Major volatile sulfur compounds produced were dimethyl disulfide (DMDS), methanethiol (MTL), dimethyl sulfide (DMS), *S*-methyl thioacetate (MTA), 2,4-dithiapenthane (DTP), *S*-methyl thiobutyrate (MTB), dimethyl trisulfide (DMTS). Their production levels were strongly strain dependent.

Despite MTL is a common precursor for all other produced VSCs, it is only detected in 40% of *G. candidum* strains, although all of them produced VSCs. This is most probably due to the high reactivity of MTL which can give rise, by chemical reaction, to the auto-oxidation products DMDS and DMTS [4] as well as to methylthioesters like MTA and MTB [7]. DMDS is produced by all strains tested. It is by far the most produced VSC. MTA, and DMTS are also produced by a majority of strains (94% and 78% respectively) although to a more limited extent than DMDS. In contrast, DTP and MTB were detected in few strains and synthesised in quite low amounts.

3.3. Correlation between b-values and VSC-production abilities

Total VSC production as measured by GC-MS was compared to b-value measurements (Figure 2). A good correlation (r=0.89) was obtained between b-values and total produced VSCs. This is essentially due to a strong correlation between DMDS and b (r=0.85), and between DMTS and b (r=0.81). This is consistent with the fact that both DMDS and DMTS are direct auto-oxidation products of MTL [4], the latter being specifically detected by the screening methodology.

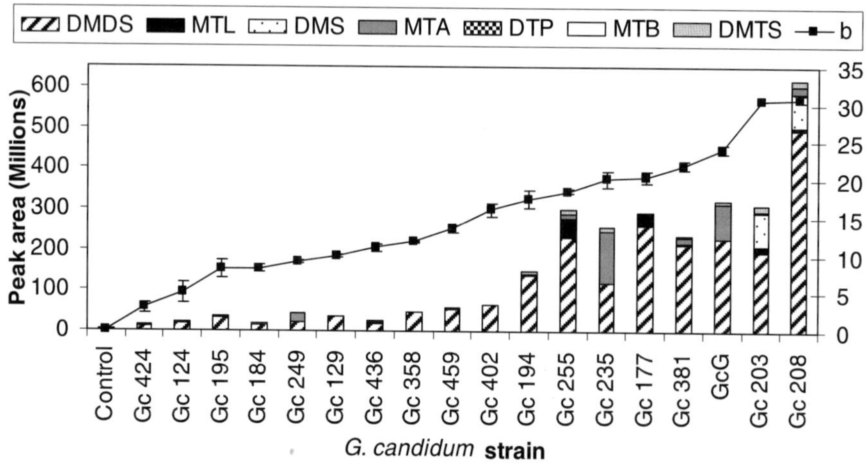

Figure 2. Relationship between total production of VSCs from 21days-old cheese-curd media and b-value measured on 11days-old test plates from *G. candidum* strains.

4. CONCLUSION

As shown in this study, GC-MS analyses provide an extensive and precise inventory of total VSCs produced by microorganisms. This can obviously give useful information for metabolic hypotheses with respect to VSCs biosynthesis. However, the method is laborious, time consuming and expensive. The double layer plate methodology provides an easy, quick and reliable method to evaluate thiol-producing abilities, which are related to VSC producing abilities. A plate method, which has also been used with success in our laboratory for screening of strictly anaerobic or aerobic bacteria, together with other cheese-ripening yeasts (data not shown), was here applied for aerobic yeasts. Furthermore, the method described could be an attractive alternative for simple and rapid screening of thiol-producing microorganisms or microbial ecosystems.

References

1. P. Bonnarme, C. Lapadatescu, M. Yvon and H.E. Spinnler, J. Dairy Res., 68 (2001) 663.
2. K. Arfi, H.E. Spinnler, R. Tâche and P. Bonnarme, Appl. Microbiol. Biotechnol., 58 (2002) 503.
3. C. Berger, J.A. Khan, P. Molimard, N. Martin and H.E. Spinnler, Appl. Environ. Microbiol., 65 (1999) 5510
4. H.W. Chin and R.C. Lindsay, Food Chem., 49 (1994) 387.
5. J. Russell and D.L. Rabenstein, Anal. Biochem., 242 (1996) 136.
6. H. Guichard and P. Bonnarme, Anal. Biochem., 338 (2005) 299.
7. S. Helinck, H.E. Spinnler, S. Parayre, M. Dame-Cahagne and P. Bonnarme, FEMS Microbiol. Lett., 193 (2000) 237.

W.L.P. Bredie and M.A. Petersen (Editors)
Flavour Science: Recent Advances and Trends

Contribution of wild strains of lactic acid bacteria to the typical aroma of an artisanal cheese

Freni Tavaria, A. César Silva-Ferreira and F. Xavier Malcata

Escola Superior de Biotecnologia, Rua Dr. António Bernardino de Almeida, P-4200-072 Porto, Portugal

ABSTRACT

Butyric acid (C_4) and 3-methylbutanoic acid (iC_5) contribute to a great extent to the overall aroma of Serra da Estrela cheese due to their abundance throughout ripening as well as to their high odour activity values. The flavour of these compounds were rated as similar ($p>0.05$). However, no significant differences could be found between 3-methylbutanoic acid alone and the reconstituted sample, suggesting a high contribution of this compound to the overall aroma perception. Wild strains of lactic acid bacteria, *Lactobacillus plantarum* ESB323 and *Lactococcus lactis* ESB331, both isolated from mature Serra da Estrela cheese were evaluated for their contribution to the volatile free fatty acid profile in cheeses manufacturing. Cheeses with the addition of *L. plantarum* ESB323 alone had higher amounts of butyric acid and 3-methylbutanoic acid than the other cheeses suggesting that inclusion of this strain may improve Serra cheese aroma.

1. INTRODUCTION

Cheeses manufactured from raw milk acquire a more intense flavour than those produced from pasteurised or heat-treated milks [1]; such realisation is mainly due to the high levels of native lactic acid bacteria present in the former [2]. Intricate biochemical reactions, involving degradation of proteins, carbohydrates and fats in the curd, lead to development of flavour and texture throughout ageing of cheese. Such degradation pathways are mainly brought about by starter and non-starter microorganisms and their enzymes, as well as by indigenous milk enzymes and coagulant enzymes [3]. A growing consumers' demand for microbiologically safer products, yet bearing the typical properties of their traditional counterparts, has provided an impetus to search for alternative processing technologies. One example is the inclusion of strains of microorganisms isolated from traditional products [4-6]. Serra da Estrela cheese is produced from raw ewe's milk without any starter addition. Odour descriptors of this cheese include 'acidic', 'sweaty' and 'sheepy-like' notes (NP-1922, 1985). These descriptors suggest that free fatty acids (FFA) play a major role in the aroma character. To evaluate the aroma impact of volatile fatty acids (VFAs) found in ripened Serra cheese, the two VFAs with the highest odour activity values (OAVs) were added to an unripened cheese matrix and the odour was evaluated. Cheeses were also manufactured from raw milk with the deliberate addition of two strains of wild LAB, isolated from Serra cheese. The VFAs as well as free amino acid profiles were analysed to assess their contribution to the overall aroma.

2. MATERIALS AND METHODS

2.1. Sensory analysis

The two FFAs with the highest OAVs, butyric acid (C_4) and 3-methylbutanoic (iC_5) acid, as well as ethyl hexanoate, believed to have no impact on the sensory perception, were used. They were added alone or in combination to 10 g of unripened cheese, at similar levels found in a 60 days old Serra da Estrela cheese (reference). The levels of added compounds were 3.18 mg/kg ethyl hexanoate, 443 mg/kg butyric acid and 601 mg/kg 3-methylbutanoic acid, respectively. The samples were evaluated for odour only, by 12 subjects using a similarity scoring test. The panel rated the odour of the samples with a similarity value (SV), which was given by comparison to the reference sample of ripened cheese, using a 0-10 discontinuous scale (where 0 means completely dissimilar and 10 means equal to the matured sample). Experiments were carried out in individual booths at 18 to 20 °C. The experiment was repeated four times.

2.2. Cheese making

Four batches of cheeses were produced according to the traditional protocol: control (CT) cheeses manufactured with indigenous flora, cheeses manufactured with *Lactobacillus plantarum* ESB323 (LB), cheeses manufactured with *Lactococcus lactis* ESB331 (LC), and cheeses with both *Lactobacillus plantarum* ESB323 and *Lactococcus lactis* ESB331 (MX). Cheeses were ripened and duly sampled at 0, 28 and 63 days after manufacture.

2.3. Volatile and free amino acid analysis

Volatile fatty acids (VFA) were analysed by GC-MS using SPME, as described elsewhere [7]. Free amino acids in cheese samples were determined using the Pico-Tag™ method (from Waters, Milford MA, USA) [8].

3. RESULTS

3.1. Sensory analysis

The sensory data presented in Figure 1, indicated that the panellists were able to detect differences between unripened (control) and reconstituted cheese (i.e. unripened cheese

+ C_4 + iC_5 + ethyl hexanoate), as well as between the latter and the traditionally ripened cheese (reference). No significant differences ($p>0.05$) were found between unripened cheese with added C_4 and unripened cheese with added iC_5 suggesting that the aroma perception of both is similar, whereas the aroma perception of ethyl hexanoate alone is poor (note that the sample added ethyl hexanoate received the same rating as the control sample, as expected). Samples containing one of the FFA plus ethyl hexanoate were also rated low. The panel found no significant differences between samples with iC_5+C4 and the reconstituted sample (C_4+iC_5+ethyl hexanoate), suggesting that iC_5 and C4 contribute to a great extent to the aroma of this cheese.

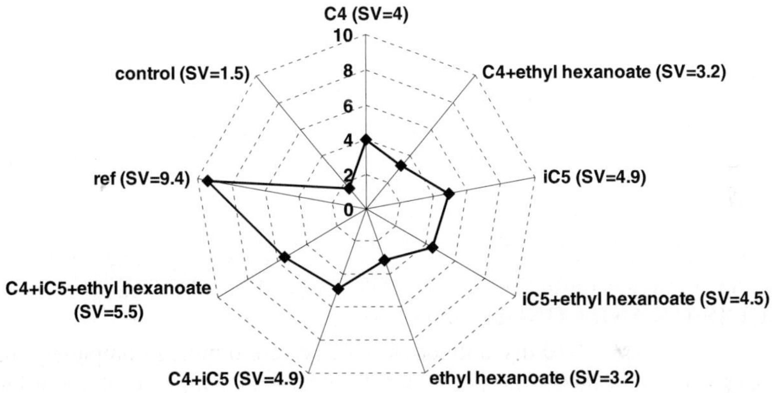

Figure 1. Sensory similarity values of the different cheese samples. Control: unripened cheese without additions; ref.: traditionally ripened cheese (60 days).

3.2. Volatiles and free amino acids

Butyric and 3-methylbutanoic acids can be related to the overall aroma of Serra da Estrela cheese, as they are the major FFA present (about 48% of the total FFA) after 60 days of ripening. Due to their OAV, they contribute to a great extent to the overall aroma. After 63 days of ripening, the amounts of the amino acids Val and Leu constituted 33.7 to 65.1% of the total free amino acid pool in the control cheeses, 9.9 to 43.7% in the MX cheeses, 27.9 to 51.8% in the LC cheeses and 11.2 to 56.1% in the LB cheeses. The amounts of Leu and Val in cheeses followed closely those of iC_5 and iso-butyric (iC_4) acids, respectively. However, no significant differences ($p>0.05$) could be observed in the amounts of Leu and Val between the various cheeses.

This was not the case for the amounts of the corresponding fatty acids. As shown in Figure 2, cheeses manufactured with *L. plantarum* had, after 63 days of maturation, approximately 9 times more iC_5 and 4 times more C_4 acid than the control cheeses, whereas cheeses manufactured with both strains had twice the amount of iC_5 and 2.5 times more C_4 than control cheeses. In addition, those manufactured with *Lc. lactis* had 6 times more iC_5 and 4 times the C_4 acid in comparison to the control cheeses. Hence suggesting that addition of one strain only leads to more efficient volatile production than when the strains are added together. Furthermore, sensory evaluation of these cheeses [9] indicated that the flavours of LB and MX cheeses are similar to that of CT cheeses, but LC cheeses were more acidic and bitter.

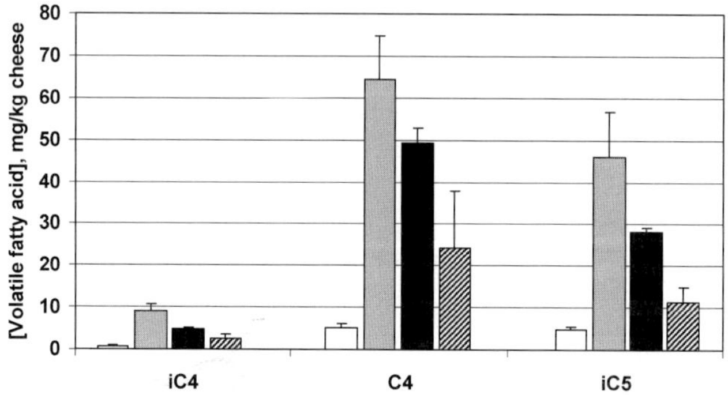

Figure 2. Concentration of iso-butyric (iC_4), butyric (C_4) and 3-methylbutanoic acids (iC_5) in cheeses manufactured without any starter addition (-□-), with added *L. plantarum* (-■-), with added *Lc. lactis* (-■-) and with both strains added (-▨-) after 63 days of ripening, respectively.

4. DISCUSSION AND CONCLUSION

From the sensory analysis, it is clear that butyric and 3-methylbutanoic acids alone do explain to a high degree the aroma of Serra da Estrela cheeses. Butyric and 3-methylbutanoic acids can be related to the overall aroma of Serra da Estrela cheese, as they are the major FFA present (about 48% of the total FFA) after 60 days of ripening. Due to their OAV, they contribute to a great extent towards the overall aroma. It is possible to produce cheeses containing high amounts of these fatty acids by inclusion of *L. plantarum*. In addition, its inclusion reduces ripening time, which is an advantage for the cheese manufacturers.

References

1. S. Buchin, V. Delague, G. Duboz, J.L. Berdagué, E. Beuvier, S. Pochet and R. Grappin, J. Dairy Sci., 81 (1998) 3097.
2. R. Grappin and E. Beuvier, Int. Dairy J., 7 (1997) 751.
3. P.F. Fox and J.M. Wallace, Adv. Appl. Microbiol., 45 (1997) 17.
4. S. Menéndez, R. Godínez, M. Hermida, J.A. Centeno and J.L. Rodríguez-Otero, Food Microbiol., 21 (2004) 97.
5. J.A Centeno, S. Menéndez, M. Hermida and J.L. Rodríguez-Otero, Int. J. Food Microbiol., 48 (1999) 97.
6. S. Menéndez, J.A. Centeno, R. Godínez and J.L. Rodríguez-Otero, Int. J. Food Microbiol., 59 (2000) 37.
7. F.K. Tavaria, A.C. Silva-Ferreira and F.X. Malcata, J. Dairy Sci., 87 (2004) 4064.
8. M.L. Alonso, A.I. Alvarez and J. Zapico, J. Liquid Chromatogr., 17 (1994) 4019.
9. A.C. Macedo, T.G. Tavares and F.X. Malcata, Food Microbiol., 21 (2004) 233.

W.L.P. Bredie and M.A. Petersen (Editors)
Flavour Science: Recent Advances and Trends

Catabolism of methionine to sulfovolatiles by lactic acid bacteria

Dattatreya S. Banavara and Scott A. Rankin

Department of Food Science, University of Wisconsin-Madison, 1605 Linden Drive, Madison, Wisconsin-53706, USA

ABSTRACT

Sulfovolatiles or volatile sulfur compounds (VSC), important odour active compounds in ripened cheeses, have been identified as catabolic products of methionine. Methionine conversion is believed to take place by at least two major pathways, one initiated by aminotransferases (ATases) and another initiated by lyases. The ATase pathway is believed to play a major role in lactococcal strains, while the pathway of VSC generation in lactobacilli is not well characterised. In this study, four lactic acid bacterial strains, *Lactobacillus helveticus* CNRZ32, *Lactobacillus casei* ATCC334, *Lactobacillus delbrueckii* ssp. *bulgaricus,* and *Lactococcus lactis* ssp. *cremoris* SK11, were studied for their ability to catabolise methionine. Cells grown in suitable media were studied in two buffers (pH 6.8 and pH 5.2 + 4% NaCl) with and without methionine, pyridoxal phosphate, α-ketoglutarate and NADH to study the effect of co-factors on VSC production. VSC production was monitored by SPME GC-MS/SIM. Results showed significant differences in the rate and extent of VSC production with the lactococcal strain producing the highest quantity. The role of co-factors and possible pathways involved are explained with the aid of ^{13}C NMR studies.

1. INTRODUCTION

Catabolism of free amino acids and peptides contributes to flavour development in cheese [1]; sulfovolatiles (VSC) can constitute a major portion of the odour active volatiles [2]. Methionine catabolism responsible for VSC formation is believed to take place by at least two major pathways, one initiated by aminotransferases (ATases) and another initiated by lyases such as methionine-γ-lyase (MGL) and cystathionine-β-lyase (CBL) [3]. ATases catalyse the transfer of an amino group from an amino acid to recipient molecules, α-keto acids. Lyases catalyse cleaving of carbon-sulfur bonds. The catabolic pathways are different for these two enzymes, however, both share a common co-factor, pyridoxal phosphate (PLP). The ATase pathway is considered dominant in lactococci [4], whereas VSC generation by lactobacilli is not well characterised [5]. The proposed VSC generation pathways are shown in Figure 1.

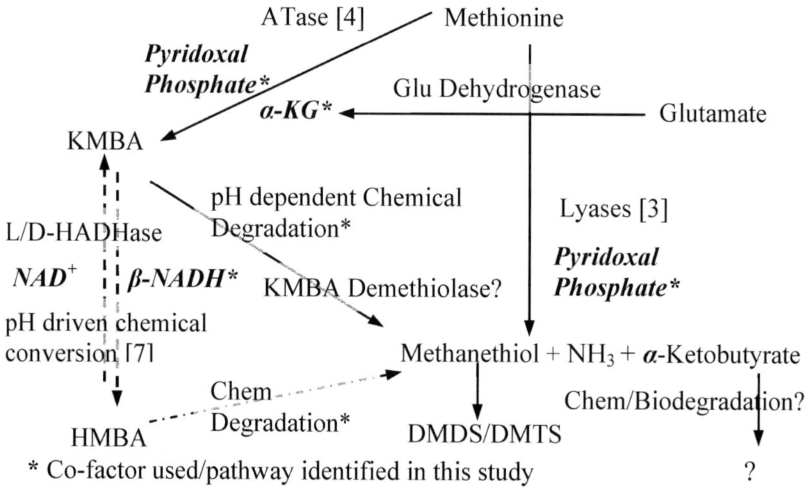

Figure 1. Enzymatic/chemical degradation pathways of methionine.

In a study with a genetically modified lactobacillus strain, substantial VSC was found with no detection of ATase pathway-derived metabolites (in the absence of keto acid) suggesting a possible lyase pathway [6]. VSC production can be largely affected by the availability of the co-factor (PLP), co-enzyme (NAD^+/NADH) and amino group acceptor molecules (keto acids). The role of these compounds on VSC production is not well known. The objective of this study was to characterise VSC production by selected lactic acid bacteria with various conditions of pH and co-factor/co-enzyme availability and to understand catabolic pathways using ^{13}C NMR.

2. MATERIALS AND METHODS

2.1. Chemicals, bacteria and cell preparation

[U]^{13}C Methionine, L-methionine, α-keto-4-(methylthio)-butanoic acid (KMBA), 2-hydroxy-4-(methylthio)-butanoic acid (HMBA), 2-Ketobutyric acid (KBA), β-NADH, Pyridoxal phosphate (PLP), α-ketoglutarate (α-KG), dimethyl disulfide (DMDS), and dimethyl trisulfide (DMTS).

Lactobacillus helveticus CNRZ32 (LH32), Lactobacillus casei ATCC334 (LC334), Lactobacillus delbrueckii ssp. bulgaricus (LDB), and Lactococcus lactis ssp. cremoris SK11 (LL11).

Cells were grown at 42 °C LH32, 37 °C LC334 & LDB, and at 30 °C LL11. The bacteria were harvested in late log phase to early stationary phase (comparable growth), washed and centrifuged to obtain whole cell suspensions at pH 5.2 + 4% NaCl and pH 6.8. Samples were prepared with and without methionine (17 mM), PLP (100 μM), NADH (100 μM), and α-KG (10 mg/5 ml).

2.2. SPME GC-MS and ^{13}C NMR

VSC formation was determined with GC-MS analysis using selected ion monitoring (SIM). A solid phase micro extraction (SPME) fiber with carboxen/ PDMS (85 μm) was used. VSCs were resolved with an RTX-Wax column. GC parameters were 35 °C for 2 min, rate 5 °C/min to 70 °C, rate 20 °C/min to 200 °C with a hold for 5 min.

^{13}C Labelled methionine was used. ^{13}C NMR spectra were obtained with a Bruker model DMX400 NMR spectrometer operating at a carbon NMR frequency of 100.6 MHz with a NMR probe of 5 mm in diameter.

3. RESULTS

Lactic acid bacteria harvested at comparable growth produced VSC with differences in rate and extent. *LL11* produced the highest quantity (up to 70 μmole/l) followed by *LC334* (up to 30 μmole/l) at 96 h (Figure 2).

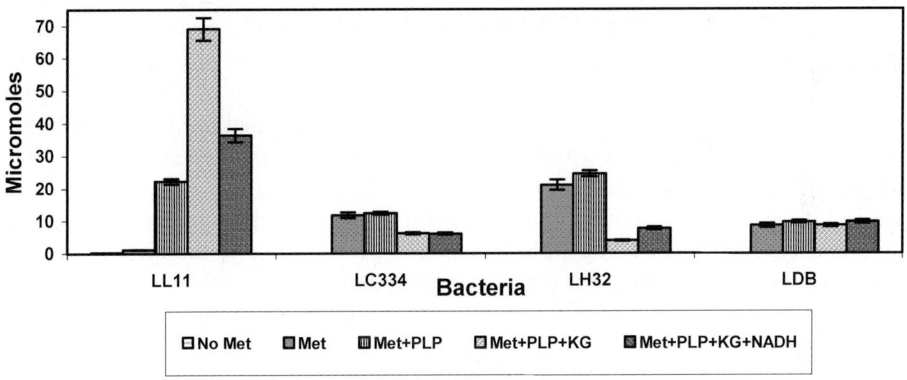

Figure 2. Production of VSC by selected LAB strains in presence of co-factors at pH 6.8.

The pH and co-factors played an important role in VSC production. All bacteria showed similar trends with respect to pH by producing highest quantities at 6.8. The abilities of these bacteria to produce VSCs under cheese-like salt and pH conditions were low. At pH 6.8, the lactobacilli showed comparable trends but in *LL11*, the addition of α-KG to PLP containing cell suspensions substantially increased VSC production (Figure 2).

NMR showed chemical shifts for KMBA as found by Gao *et al.* [4] indicating ATase dominated pathway for VSC production. In *LH32* and *LC334*, PLP alone (co-factor for both ATase and lyases) resulted in maximum VSC production. The addition of α-KG to PLP containing cells decreased VSC production. NMR studies did not show any

indication of KMBA, HMBA or KBA for lactobacilli. In *LDB*, the addition of α-KG to PLP containing cells did not show any change in VSC. Addition of NADH decreased the VSC in *LL11* indicating L/D-hydroxyacid dehydrogenase (L/D-HADHase) activity, while the changes observed in lactobacilli strains were not consistent. Under cheese-like conditions, VSC production was negligible without α-KG in all the bacteria studied. Though PLP is a co-factor for several lyases and ATases, PLP alone did not show any effect on VSC production while α-KG with PLP increased VSC production. NADH slightly increased VSC production indicating a reverse HADHase activity.
KMBA was found to be more prone to spontaneous degradation than HMBA, producing 100-fold more VSC. The conversion of KMBA to VSC was lower at lower pH. α-keto acids are known to be present in a stable hydrated form under pH conditions below their pKa and can undergo several types of tautomeric transformations and polymerisation reactions [7]. In this study, KMBA was found to be more stable at pH 5.2 than 6.8.

4. DISCUSSION

During cheese ripening, the availability of nutrients and indigenous co-factors vary largely. In this study, such co-factors affected VSC production at different pH conditions. Under conditions similar to cheese ripening (pH 5.2 + 4% NaCl), both PLP and α-KG were found to be limiting factors. PLP present in milk in variable quantities can also influence cheese aroma development significantly. At lower pH, KMBA tends to be in a stable hydrated form; addition of NADH probably increased the redox dependent HADHase activity leading to higher VSC. Overall VSC production in cheese is hindered by low enzyme activity and higher chemical stability of KMBA (if ATase pathway dominates). At higher pH, the role of α-KG was reversed in *LC334* and *LH32,* producing highest VSC levels with PLP alone as co-factor. VSC decreased with α-KG addition in the lactobacilli studied. Other studies also indicate that lyases and ATases from lactic cultures have activity optima at pH 6.0 and 8.0 and are up to 100–fold less active at cheese-like pH [8]. This phenomenon is possibly due to the competitive co-existence of both ATase and lyase pathways. NMR studies did not show chemical shifts for KMBA or HMBA in any of the lactobacilli while for *LL11* chemical shifts were observed for KMBA. The quantity and pathways of VSC production largely depends on the type of strain, pH and presence of co-factors.

References

1. A.C. Curtin and P.L.H. McSweeney, J. Dairy. Res., 70 (2003) 249.
2. T.K. Singh, M.A. Drake and K.R. Cadwallder, Compr. Rev. Food Sci. Food Safety, 2 (2003) 139.
3. B. Dias and B. Weimer, Appl. Environ. Microbiol., 64 (9) (1998) 3320.
4. S. Gao, E.S. Mooberry and J.L. Steele, Appl. Environ. Microbiol., 64 (1998) 4670.
5. L. Marilley and M.G. Casey, Int. J. Food Microbiol., 90 (2) (2004) 139.
6. K. Cadwallder (ed.), Flavor of dairy foods, proceeding of the 228th ACS Meeting, Philadelphia, PA, in press.
7. A.J.L. Cooper, J.Z. Ginos and A. Meister, Chem. Rev., 83 (3) (1983) 321.
8. B. Dias and B. Weimer, Appl. Environ. Microbiol., 64 (9) (1998) 3327.

W.L.P. Bredie and M.A. Petersen (Editors)
Flavour Science: Recent Advances and Trends

Ability of *Oenococcus oeni* to influence vanillin levels

Audrey Bloem, Aline Lonvaud, Alain Bertrand and Gilles de Revel

UMR 1219, Unité Associée INRA/Université Victor Segalen Bordeaux 2 Faculté d'œnologie, 351 cours de la Libération, F-33405 Talence cedex, France

ABSTRACT

Lactic acid bacteria (LAB) conducting the malolactic fermentation (MLF) have a significant influence on the stability and sensory quality of wine. We have investigated the role of lactic acid bacteria (LAB) with regard to their influence on vanillin formation when MLF is performed in barrels. The metabolism of *Oenococcus oeni*, the principal wine LAB species, showed to increase the levels of vanillin during MLF in presence of wood. This indicated the potential utilisation of the precursors from the oak wood and/or the production of enzymes by this species could affect vanillin formation. The precursors from the oak wood were fractionated and the action of glycosidases on the liberated compounds was studied. The release of vanillin was increased when purified β-glucosidase, α-L-arabinofuranosidase and α-L-rhamnopyranosidase were added to the wood extract.

1. INTRODUCTION

The winemaking process includes two fermentations. Firstly, the wine yeast *Saccharomyces cerevisiae* performs alcoholic fermentation and secondly, the lactic acid bacteria (LAB), especially *Oenococcus oeni*, perform malolactic fermentation (MLF). The benefits of MLF include: lowers the acidity in wine, enhances the sensory characteristics and increases microbial stability. Only certain compounds are well known for their significance in the flavour profile of wine. One of the major aroma compounds is diacetyl, which is produced by LAB and gives lactic or buttery odours.
When the process of MLF is performed in barrels, a greater increase in aromatic compounds is obtained in comparison with wines that do not undergo MLF [1]. Vanillin is one of the compounds responsible for the difference as it gives a powerful characteristic aroma. Preliminary research indicates that wine LAB enhance the release

of vanillin from wood. This fact suggests the existence of a vanillin precursor liberated in the wine in contact with wood which could be modified by LAB activity [2]. In this study, we report the preliminary results on the purification of the precursors from wood along with their release by enzymatic treatments.

2. MATERIALS AND METHODS

2.1. Lactic acid bacterial strains

Tests were performed with the *Oenococcus oeni* strain IOEB 8413 which belongs to the collection of LAB from the Faculté d'Œnologie Bordeaux. The growth was monitored by measuring the optical density (OD) at 600 nm on a 932 Uvikon spectrophotometer (Kontron, Rungis, France).

2.2. Vanillin formation in culture conditions supplemented with oak wood

The capacity of LAB to form vanillin was studied in modified MRS medium (1 g/l glucose, 10 g/l malic acid, pH 5.0), supplemented with pimaricine (5 g/l) and with oak wood chips (heated at 220 °C for 20 min, 10 g/l).

2.3. Oak wood fractionation

Wood extracts were made from oak wood chips that were heated for 48 h in a 50% hydro-alcoholic solution. Thereafter, it was filtered on 0.45 μm. Wood extracts were fractionated by solid phase extraction according to the method outlined by Dignum *et al.*, 2004 [3]. Potential precursors were eluted in 50 ml fractions following of methanol-water ratios (50:50 v/v), (60:40 v/v), (70:30 v/v), (80:20 v/v). Each fraction was air dried under a vacuum. Thereafter, the residues were dissolved in 5 ml of 0.1 M acetate-phosphate buffer (pH 5.0).

2.4. Enzymatic treatments

Commercially available preparations of β-glucosidase (from almonds, Sigma, cat no. G0395), α-L-arabinofuranosidase (Megazyme) and α-L-rhamnopyranosidase (Sigma, cat no. N1385) were added to different fractions at 10 U for each enzyme. These treatments were done for 24 h at 37 °C and were stopped by 2 ml of 1 M Na_2CO_3.

2.5. Quantification of phenolic compounds

Vanillin concentrations were performed by HPLC analysis without prior extraction using Waters (St Quentin en Yvelines, France) 600 E HPLC system and Waters 717 plus with a spectrophotometer Waters 2487 using two wavelengths. Samples were filtered through 0.45 μm filters; 20 μl were loaded on a C18 reverse-phase column (Spherisorb ODS2 C18 250 mm x 4.6 mm x 5 μm ; Interchim, Montluçon, France) and eluted at a flow rate 0.6 ml/min with a gradient of 5 ‰ (v/v) acetic acid (solvent A) and methanol with 2% of water (solvent B). The following elution programme was used: solvent B started at 0%, decreased to 80% at 40 min then to 100% at 55 min and was

returned to 0% at 60 min. The column was re-equilibrated for 10 min before the next injection. Vanillin was detected at 313 nm.
The phenolic compounds were quantified by measurement of the peak area and the concentration values extrapolated from the corresponding standard curve prepared for each compound.

3. RESULTS

To show a difference between the trials with and without MLF with oak wood, a control without inoculation but supplemented with wood chips was prepared (Figure 1). After 3 days of bacterial growth, the vanillin concentrations were similar for the inoculated medium and the control. The vanillin levels here were only due to the passive release from oak wood. However, after 7 days, it was found that the vanillin concentration was higher in presence of the LAB. This supplementary formation especially occurred during the exponential growth phase of LAB. This suggests the existence of a vanillin precursor released in medium in contact with wood which could be modified by LAB activity. This observation was consistent for all the results obtained during MLF in wine. After 7 to 15 days, these concentrations did not increase any further. This could be explained by the fact that no vanillin formation occurred during the stationary growth phase of LAB or by the fact that vanillin was utilised during this period.

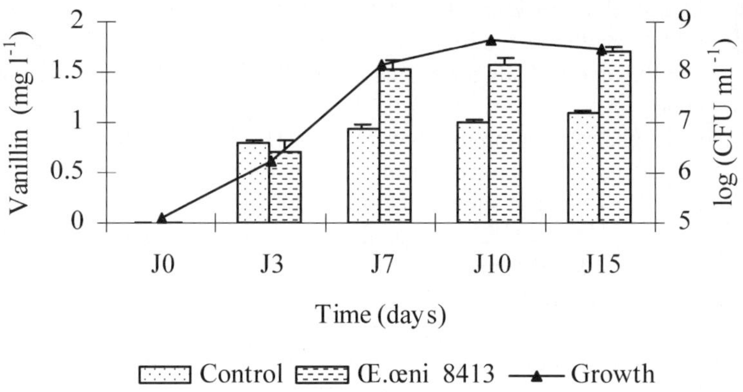

Figure 1. Vanillin formation during the growth of Oenococcus oeni 8413 in basal medium supplemented with oak wood chips.

In comparison to the control, it was found that the enzymatic treatments increased the concentrations of vanillin in all cases (Figure 2). From the figure, it can be seen that the simultaneous addition of the three purified enzymes, β-glucosidase, α-L-arabinofuranosidase and α-L-rhamnopyranosidase gave the greatest vanillin concentration with an increase of 150 μg/l. Vanillin was, also, released to a small degree by the glucosidase alone. Vanillin was mainly released in the fractions 60:40 (v/v) and 70:30 (v/v).

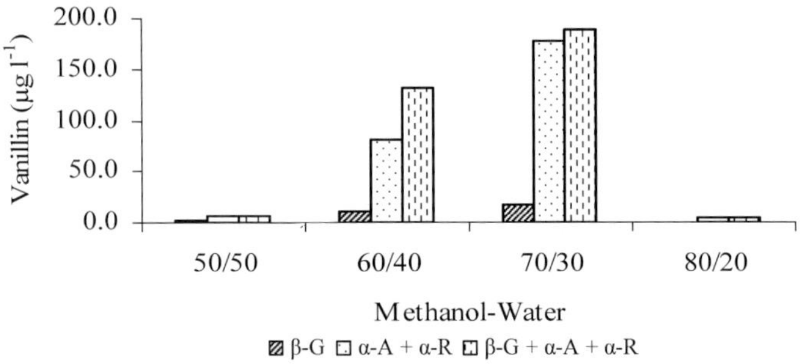

Figure 2. Vanillin released by enzymatic cleavage of the glycosides extract of oak wood. β-G: β-glucosidase; α-A: α-L-arabinofuranosidase; α-R: α-L-rhamnopyranosidase. The values shown are after deduction from the vanillin generated in the control.

4. DISCUSSION AND CONCLUSION

The impact of MLF performed in barrels on the flavour and body of wine has been accepted for many years. However, little is known about the exact mechanisms and influence of the metabolism of wine LAB.

Previous research has indicated that the concentrations of the aromatic compounds released from wood were higher in the wines after MLF compared to a wine not having undergone bacterial development [1]. In this study, the results of the bacterial development in the synthetic medium supplemented with oak wood were similar, especially for vanillin, and this suggested that LAB influences the release of compounds derived from wood.

Previously, simple phenolic compounds were considered as possible precursors, but only from ferulic acid, a small increase of vanillin level (1.7%) was observed with *Oenococcus oeni* 8413 (data not shown). During MLF in wine, the increase of vanillin levels could not be explained by this low production. Current work is being focused on other pathways responsible for vanillin production.

The action of glycosidases with oak wood fractions led to a greater release of vanillin. The increase of vanillin levels by the three enzymes indicates different types of glycones bonded to phenol groups.

The results of this study suggest the presence of glycoconjugated precursors in oak wood extract and glycosidase activities should be implicated in the vanillin release during MLF in barrels. More studies are needed to better understand the role of LAB on the hydrolysis of glycosidic flavour precursors and their identification during MLF.

References

1. G. De Revel, N. Martin, L. Pripis-Nicolau, A. Lonvaud-Funel and A. Bertrand, J. Agric. Food Chem., 47 (1999) 4003.
2. G. De Revel, A. Bloem, M. Augustin, A. Lonvaud-Funel and A. Bertrand, Food Microbiol., 22 (2005) 569.
3. M.J.W. Dignum, R. Van der Heijden, J. Kerler, C. Winkel and R. Verpoorte, Food Chem., 85 (2) (2004) 199.

W.L.P. Bredie and M.A. Petersen (Editors)
Flavour Science: Recent Advances and Trends

The biosynthesis of furaneol in strawberry: the plant cells are not alone

Ioannis Zabetakis[a], Panagiotis Koutsompogeras[a] and Adamantini Kyriacou[b]

[a]Laboratory of Food Chemistry, Department of Chemistry, University of Athens, Panepistimioupolis, Athens, 15771 Greece; [b]Department of Dietetics and Nutritional Science, Harokopio University of Athens, Athens, 176 76 Greece

ABSTRACT

In this paper, our latest results on the biosynthesis of furaneol in strawberry (*Fragaria* x *ananasa* cv. Elsanta) cells are presented. Our group has been working on the biosynthesis in strawberry of this very important flavour molecule for the past decade [1-3] and the importance of 1,2-propanediol as a putative precursor of furaneol has been described [4]. In the present work, the methylotroph *Methylobacterium extorquens* (strain with CABI registration number IMI 369321), which has been isolated from strawberry (*Fragaria x ananassa* cv. Elsanta) callus cultures [5], was grown on a mixture of methanol (0.25% v/v) and 1,2-propanediol (0.75% v/v). The microbial biotransformation of 1,2-propanediol to 2-hydroxypropanal was studied. Both the bacterial and strawberry Alcohol Dehydrogenase (ADH) enzymatic activities were assessed to define the best substrate specificity. SDS-PAGE electrophoresis experiments showed molecular weights of 45.0 kDa and 24.6 kDa for the Alcohol Dehydrogenases of the *Methylobacterium extorquens* and *Fragaria x ananassa* respectively. Our results suggest that the bacterial enzymatic activity contributes to the generation of precursors of furaneol in strawberry.

1. INTRODUCTION

Methylotrophy is defined as the ability to grow at the expense of reduced carbon compounds containing one or more carbon atoms but containing no carbon-carbon bonds [1]. Enzymes for the primary oxidation of C1 substrates such as methanol dehydrogenase have been characterised [2]. In an enlightening review on Pink

Pigmented Facultatively Methylotrophs (PPFMs) [3], the close plant-microbe interactions were described and particular emphasis has been paid to the mutual benefits of both plant and microbial cells. Bacteria are involved in many interesting *in vivo* interactions with plants [3]. Existing evidence [4] suggests a possible co-operation between the strawberries and the *Methylobacterium extorquens*, regarding their enzyme system which affects the dehydrogenation of certain alcohols. Hence the purification of the respective enzymes from the bacterium and the strawberries (ADHs) would be proved extremely valuable [5]. It is also known that a large group of metal ions or chelates (such as EDTA), deactivate the enzymes [6].

2. MATERIALS AND METHODS

2.1. Organisms, cultivation and cell extract preparations

M. extorquens isolated from strawberry callus [4] was cultivated in a medium containing 0.75% (v/v) 1,2-propanediol, 0.25% (v/v) methanol and 1.0% (w/v) Peptone [4]. The cell free extract was prepared by ultrasonic disintegration of cells. The enzyme solution from strawberries was produced by crushing 200 g of strawberries (*Fragaria x ananassa*).

2.2. Enzyme assays and chromatographic separation/purification of ADHs

The enzyme assays were based on measuring the absorption at 340 nm for NADH [6]. Chromatographic separation was carried out in a SEPHADEX chromatography column. Protein measurements were conducted by measuring the optical density (O.D.) at 280 nm. Additionally, dehydrogenation activities were measured and the most active fraction was again separated in the column. The same procedure was followed once more and the most active fraction was recollected. The procedure which was followed for the strawberries was exactly the same. Active fractions which were collected from the chromatographic separation were subsequently subjected to SDS-PAGE electrophoresis [6].

3. RESULTS

3.1. Physical and catalytic properties

NAD was used as electron acceptor, while the produced NADH was measured spectrophotometrically at 340 nm. Apparent Km values for *M. extorquens* and strawberries (*Fragaria x ananassa*) are presented in Table 1.

3.2. Chromatographic separation/purification of ADHs

Purification results for the ADHs are shown in Table 2 and Table 3. Yield (%) is the % fraction of the total activity (U) in purification step 1 towards the total activity (U) in purification step 2. Purification (fold) is the fraction of the specific activity in purification step 1 towards the specific activity in purification step 2.

Table 1. Apparent Km values of ADH of *M. extorquens* and strawberry (*F. x ananassa*).

Substrate	Km (mM) [*M. extorquens*]	Km (mM) [*F. x ananassa*]
Methanol	0.78 ± 0.00	9.69 ± 0.05
1,2-Propanediol	1.65 ± 0.01	15.84 ± 0.02
Glycerol	2.58 ± 0.02	18.60 ± 0.07
1-Propanol	3.09 ± 0.02	3.54 ± 0.01
2-Propanol	3.42 ± 0.03	11.46 ± 0.05
Formaldehyde	4.46 ± 0.03	Not detected
Ethanol	5.38 ± 0.02	6.66 ± 0.03
Benzyl alcohol	7.72 ± 0.09	19.96 ± 0.14
1-Butanol	25.78 ± 0.23	19.93 ± 0.19

Table 2. Purification (fold) for *M.extorquens*.

Purification step	Total protein (mg)	Total activity (U)	Specific activity [U/(mg protein)]	Yield (%)	Purification (fold)
1	268	226	0.8	100%	1.0
2	76	178	2.3	79%	2.8

Table 3. Purification (fold) for *F. x ananassa*.

Purification step	Total protein (mg)	Total activity (U)	Specific activity [U/(mg protein)]	Yield (%)	Purification (fold)
1	185	200	1.1	100%	1.0
2	68	171	2.5	86%	2.3

3.3. SDS - PAGE electrophoresis

The results from SDS-PAGE electrophoresis revealed a molecular weight of the ADH from *M. extorquens* of 45 kDa. The respective molecular weight of the ADH from *F. x ananassa* was 24.6 kDa.

4. DISCUSSION

In extracts of *M. extorquens*, and strawberry cells, we found that the best substrates for ADH activity were methanol and 1,2-propanediol for *M. extorquens*, while 1-propanol and ethanol were the best substrates for strawberries. There is a tendency for the ADHs to have a better specificity mostly for alcohols with 3 carbon atoms in their molecules. We obtained a purification of 2.8 fold for the bacterium ADH and 2.3 fold for the strawberry ADH. The Km value for the bacterial ADH with 1,2-propanediol is 9.6 times smaller than the respective strawberry ADH; a fact which implies the close relationship and the greater affinity-specificity for this compound by the bacterium.

5. CONCLUSION

Our research is focused on the oxidation of 1,2-propanediol to 2-hydroxypropanal. 2-Hydroxypropanal is a very important compound since it has been proved that it can be converted to 2,5-dimethyl-4-hydroxy-(2*H*)-furan-3-one (furaneol) [8]. Furaneol itself is the most important flavour compound in strawberry [5]. An enzymatic collaboration between the bacterium and the strawberry is suggested here. Our current work focuses on the interactions of *M. extorquens* and the strawberry cells in order to further highlight the role of the bacteria in the biosynthesis of furaneol. Strawberry plants containing 1,2-propanediol [8] and *M. extorquens*, which can grow on them [4], can participate in a symbiotic relationship with the fruits.

References

1. L. Chistoserdova, S.-W. Chen, A. Lapidus and M.E. Lidstrom, J. Bacteriol., 185 (2003) 2980.
2. L. Chistoserdova, M. Laukel, J.-C. Portais, J.A. Vorholt and M.E. Lidstrom, J. Bacteriol., 186 (2004) 22.
3. M.A. Holland and J.C. Polacco, Annu. Rev. Plant Physiol. Plant Mol. Biol., 45 (1994) 197.
4. I. Zabetakis, Plant Cell Tissue Organ. Cult., 50 (1997) 179.
5. K.G. Bood and I. Zabetakis, J. Food Sci., 67 (2002) 2.
6. I. Georgatsos, Enzymology, 3rd edition, Thessalonica, Greece (1991) 22; 33; 58.
7. U.K. Laemmli, Nature (London), 227 (1970) 680.
8. I. Zabetakis, P. Moutevelis-Minakakis and J.W. Gramshaw, Food Chem., 64 (1999) 311.

W.L.P. Bredie and M.A. Petersen (Editors)
Flavour Science: Recent Advances and Trends

Method for the enzymatic preparation of flavours rich in C6-C10 aldehydes

E. Kohlen, A. van der Vliet, J. Kerler, C. de Lamarliere and C. Winkel

Quest International, PO Box 2, 1400 CA Bussum, the Netherlands

ABSTRACT

Raw materials previously used as reagents for production of C6-C10 aldehydes were refined unsaturated fatty acids or fatty acid mixtures obtained by fat hydrolysis. For some oils and fats (like butter fat) undesirable compounds, like butyric acid, would be formed in the hydrolysis treatment. Therefore, a new method was developed in which aldehydes were prepared directly from oils and fats without hydrolysis. In a first step, the oil or fat was reacted with lipoxygenase in a multiphase system in the presence of air. The obtained mixture was heated, preferably under acidic conditions, resulting in an aldehyde-containing product. Aldehydes could be further isolated by distillation.

1. INTRODUCTION

C6-C10 aldehydes occur in a wide range of flavours. Their flavour profiles range from green, cucumber- or citrus-like to creamy, fatty, fried and rancid. Of particular interest are the C7-C9 unsaturated aldehydes like (*Z*)-4-heptenal, (*E*)-2-nonenal and (*E,Z*)-2,6-nonadienal. A known route to obtain these products as natural flavour ingredients is to generate the aldehyde precursor hydroxyperoxides (HPO) by unsaturated fatty acid oxidation. Next to thermal autoxidation [1], the oxidation can be carried out enzymatically using lipoxygenases (Lox). Aldehydes have been generated from purified fatty acids or hydrolysed fats using purified enzymes or lipoxygenase activity from plant material [2]. However some hydrolysed fats also contain odorous free fatty acids (e.g. butyric acid), which are detrimental to the intended flavours.

A particular benefit of enzyme-catalysed fatty acid oxidation is that the enzyme specificity enables to favour the production of certain aldehydes and, therefore, to obtain a more desirable flavour profile. It has been reported that among the 3 types of lipoxygenase activities present in soy-flour, Lox-1 activity almost exclusively catalyses

the formation of 13-hydroperoxides (13-HPO) from linoleic acid, whereas Lox-2/3 catalyses the formation of both 9- and 13-HPO [3].
Based on the knowledge that some lipoxygenases exhibit activities on triglycerides and phopholipids [4-6], a process was designed to generate especially 9-HPO derived aldehydes from non-hydrolysed fat. Enzymatic oxidation conditions were selected to enable both lipoxidation of triglycerides present in non-hydrolysed fat and generation of considerable amount of 9-HPO. In a second step, enzymatic oxidation combined with HPO degradation into aldehydes by Hock cleavage and aldehyde isolation was used to obtain a clean aldehyde-rich flavour from non-hydrolysed fat.

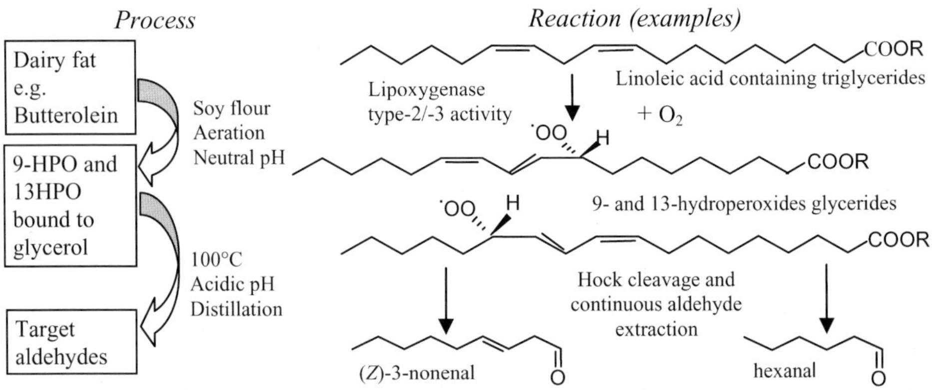

Figure 1. Principle of the method.

2. MATERIAL AND METHODS

Enzymatic assay: The polarographic method (adapted from [3]) was used to determine the Lox-activity at different pH conditions. Reagents were linoleic acid (Sigma L-1268) or trilinolein (Sigma T-6513) in ethanol/acetone solution 1:1, v/v (0.25 µl/ml), phosphate buffer (100 mM pH 5 to 7) and borate buffer 100 mM pH 8 to 10. The linoleic acid substrate solution was prepared by mixing 0.25 ml (804 µmol) linoleic acid into 5 ml of O_2-free water, adding NaOH solution until a clear solution was obtained and diluting to 100 ml with O_2-free water (final pH ± 9.5). Two lipoxygenase sources were used: crude full-fat soy flour extract (Provaflor, Cargill) and purified Lox-1 (Sigma L-8383) diluted in buffer (resp. 1:10 and 1:1000, w/w). The enzymatic assay was performed with 4.8 ml buffer, 50 µl enzyme solution and 250 µl linoleic acid (resp. 200 µl trilinolein) substrate solution. Oxygen consumption (%/min) was monitored using an Oxigraph (Model 5301B, YSI) during 5 min and used to determine the activity (5 ml of air-saturated solution contains 1.29 µmol O_2): Activity [(µmol O_2 consumed/min)/g enzyme or flour)] = (1.29 $O_{2\text{-cons/min}}$ 1000)/(100 mg enzyme or flour) [U/g enzyme or flour].
Enzymatic oxidation and aldehydic isolation: In a first step, 250 g butterolein, 90 ml 50 mM sodium phosphate buffer pH 7.0 (saturated with air) and 7.75 g of soy flour were

mixed in a 1 l reaction flask fitted with a thermometer, condensor, stirrer and a gas inlet tube. The mixture was stirred vigorously (about 1000 rpm) for 16 h while supplying air at a rate of 20 l/h. The temperature was maintained at 25 °C.
In a second step, the obtained reaction mixture was mixed with 250 g of 20 w/w % citric acid solution and subjected to simultaneous distillation-extraction using a Likens-Nickerson distillation apparatus. The distillation was carried out at 100 °C for 8 h and 75 ml di-ethylether/hexane (90:10, v/v) was used as extraction solvent. To prevent oxidation a small nitrogen flow was supplied. The organic solvent was evaporated carefully. The aldehydic residue was finally dissolved in 2.0 ml ethanol. For analysis, the product was dissolved in pentane (10% solution, v/v), injected on an HP-5 column, and analysed by GC-sniff/MS (Thermoquest, type voyager). Quantitative analysis was performed by GC-FID using octanal as external standard.

3. RESULTS AND DISCUSSION

Lipoxygenase activity in crude soy flour was compared with purified Lox-1 over a pH range (Figure 2). The soy flour lipoxygenase activity profiles varied significantly with substrate. On linoleic acid, an optimum at pH=9 was seen, but there was activity over a broad pH-range. When trilinolein was used, the optimum pH was 7 and the activity spectrum was not so broad. The profile on linoleic acid showed more similarity with the Lox-1 profile. They showed the same optimal pH but the Lox-1 activity was considerably less at lower pH. Using purified Lox-1 on trilinolein, no activity was detected. This indicates that all activity of the soybean flour on trilinolein can probably be attributed to the type-2 lipoxygenases Lox-2 and 3. On linoleic acid, Lox-1 as well as soybean flour lipoxygenases were active. Soybean flour activity (type-2 lipoxygenase) on trilinolein was considerable. At pH 7, the activity on triglycerides was 36% of the activity on the free acid (116.5 – 322.5 U/g flour).

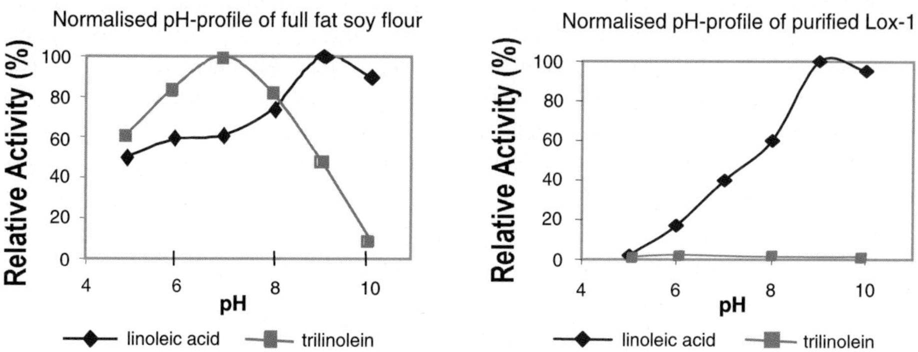

Figure 2. Normalised lipoxygenase activities of soy flour and purified Lox-1 at pH 5 to 10.

Based on the results obtained from the model enzymatic reaction, a process was designed using butterolein as a substrate subjected to lipoxidation using soy flour

followed by Hock cleavage of the peroxides in acidic conditions and simultaneous distillation-extraction of the formed aldehydes (Figure 1).
Quantitative analysis of the volatile compounds in the final flavour product (Table 1) confirmed the presence of C6-C10 aldehydes as main flavour components; (*Z*)-4-heptanal, (*E*)-2-nonenal and (*E*,*Z*)-2,6-nonadienal generated from 9-HPO, were present in high amounts. The flavour block obtained by this method was described as creamy, green, clean and particularly well suited for dairy flavouring.

Table 1. Concentrations of the aroma volatiles in the final product.

Compounds	mg/l	Compounds	mg/l
Hexanal	487	(*Z*)-2-Nonenal	<10
(*Z*)-4-Heptenal	62	(*E*,*E*)-2,6-Nonadienal	<10
Heptanal	1050	(*E*,*Z*)-2,6-Nonadienal	<10
(*E*)-2-Heptanal	124	(*E*)-2-Nonenal	1001
(*E*,*Z*)-2,4-Heptadienal	60	(*E*,*Z*)-2,4-Nonadienal	<10
Octanal	177	(*E*,*E*)-2,4-Nonadienal	72
(*E*,*E*)-2,4-Heptadienal	208	(*E*,*Z*)-2,4-Decadienal	88
(*Z*)-2-Octenal	24	(*E*,*E*)-2,4-Decadienal	202
(*E*)-2-Octenal	165	(*E*)-2-Undecenal	114
Nonanal	460	(*E*,*E*)-2,4-Undecadienal	<10
(*E*,*E*)-2,4-Octadienal	<10		

4. CONCLUSION

Based on a model reaction using trilinolein, optimum neutral pH conditions were determined at which soy flour exhibited considerable activity on triglycerides, without prior hydrolysis. Under these conditions, no type-1 lipoxygenase activity was present which means that type-2 lipoxygenase is responsible for triglycerides hydrolysis and enables the generation of 13-HPO as well as 9-HPO, which are precursors of desirable aldehydes. A two-step process, which delivers a natural aldehyde-rich block with desirable flavour characteristics was developed and patented on this basis. The process can also be applied to other type of fats.

References

1. E.N. Frankel, Lipid oxidation, Dundee, Scotland, 10 (1998).
2. J. Kerler, E. Kohlen, W. Fitz and C. Winkel, Method for the enzymatical preparation of flavours rich in C6-C10 aldehydes, world patent WO 01/39614 (2001).
3. B. Axelrod, T.M. Cheesbrough and S. Laakso, Method. Enzymol., 71 (1981) 441.
4. R.B. Koch, B. Stern and C.G. Ferrari, Arch. Biochem. Biophys., 78 (1958) 165.
5. W.R. Morrison and R. Panpaprai, J. Sci. Food Agric., 26 (1975) 1225.
6. M.P. Luquet, C. Pourplache, M. Podevin, G. Thompson and V. Larettagarde, Enzyme Microb. Technol., 15 (10) (1993) 842.

4

Key aroma and taste components

W.L.P. Bredie and M.A. Petersen (Editors)
Flavour Science: Recent Advances and Trends

Aroma compounds in black tea powders of different origins - changes induced by preparation of the infusion

Peter Schieberle and Christian Schuh

German Research Center for Food Chemistry, Lichtenbergstraße 4, D-85748 Garching, Germany

ABSTRACT

By application of the Aroma Extract Dilution Analysis on the volatiles of black tea leaves (Darjeeling Gold Selection, DG), linalool, β-ionone, 4-hydroxy-2,5-dimethyl-3(2*H*)-furanone (4-HDF), 3-hydroxy-4,5-dimethyl-2(5*H*)-furanone, phenylacetic acid and (*E*,*E*,*Z*)-2,4,6-nonatrienal were identified with the highest Flavour Dilution (FD) factors among the 25 odour-active compounds detected in the FD-factor range of 16 to 256. Quantitative measurements performed on the infusion prepared from DG indicated the same set of odorants as identified in the leaves, but the hot water treatment led to an increase of, in particular, most of the alcohols, i.e. geraniol, 4-HDF, 2-phenylethanol and (*Z*)-3-hexenol, as well as the Strecker aldehydes methylpropanal and 2-methylbutanal. Infusions prepared from four other black tea leaves of different origins also contained the same set of odorants as the infusion from DG, but the odorants identified varied significantly in their FD-factors.

1. INTRODUCTION

The infusion prepared from dried or fermented, dried leaves of the tea bush (*Camellia sinensis*) has been consumed as a beverage by mankind for more than 5700 years. Considering the amount of about 3.1 million metric tons of tea produced today, it is not surprising that many research groups have focused their attention on flavour chemistry of tea. The knowledge on this topic has been comprehensively reviewed [1-3].

For black tea production, the freshly harvested leaves are processed by four main manufacturing steps: withering, rolling, fermentation and firing, all steps inducing a lot of biochemical and chemical changes in the composition of the leaf. Because the quality of the finished product is influenced by many parameters, it is of great importance for the tea industry to comprehensively understand the changes in the concentrations of volatiles induced by the manufacturing process. Furthermore, it is also very important to know which chemical changes are caused by preparing the infusion, because the aroma of the black tea leaves and the beverage clearly differ.

Today, nearly 600 volatile compounds have been identified in tea [4] and it would for sure be a challenge for analytical chemists to monitor changes in the entire set of volatiles along with all tea processing steps in quality assessment. Fortunately, the human olfactory system 'uses' only a minor part from the bulk of volatile food compounds, because only the 'bio-active' compounds are able to interact with receptors in the human olfactory system [5]. Such compounds can be selected by combining the human olfactory system with analytical chemistry, e.g. by means of GC-Olfactometry or by calculating odour activity values based on odour thresholds in the food matrix [6]. Based on such methods, it has been shown, e.g. for the coffee beverage, that only 25 out of 1000 volatiles are needed to mimic the full aroma of coffee [7].

Surprisingly, only in two studies GC-Olfactometry has been applied to extracts of black tea powder in order to identify the most odour-active compounds [8,9] and no studies correlating volatile concentrations in tea with odour thresholds are available. The aim of the following study was, therefore, to characterise the most odour-active compounds in five commercial tea powders of different origin by application of GC-Olfactometry. Furthermore, first insights into changes in the concentrations between powder and infusion induced by the hot water treatment should be elaborated.

2. MATERIAL AND METHODS

The following five samples of black tea leaves (or powders, respectively) were purchased from the tea trade: Darjeeling Gold Selection (DG, India), Darjeeling Gold Star (DGS, India), Golden Palace (GP, Nepal), Admiral's Cup (AC, Assam, India) and Ceylon Best (CB, Sri Lanka).

For volatile isolation, the black tea leaves (50 g) were powdered in liquid nitrogen by means of a blender and stepwise extracted with solvents as described recently [10]. The extract was freed from the non-volatile material by application of the SAFE distillation [11]. After concentration of the extract to about 200 μl, the odour-active compounds were determined by GC-Olfactometry [5], and, finally, identified using reference compounds. Quantifications were performed by means of Stable Isotope Dilution Assays (SIDA) as reported recently [10].

3. RESULTS

3.1. Key odorants in black tea leaves

The volatile fraction was isolated from the Darjeeling Gold Selection (DG), because in a preliminary experiment the hot water infusion prepared from this sample was rated with the highest score by an experienced sensory panel.

By application of GC-Olfactometry, twenty-four odour-active areas were detected in the distillate covering a wide range of different odour qualities, such as flowery, violet-like, seasoning-like and, also, oat-flake like. By sniffing of serial 1+1 dilutions of the extract by means of the Aroma Extract Dilution Analysis (AEDA), seven areas showed the highest Flavour Dilution (FD) factors among the set of 25 odorants detected in the FD range of 16 to 256 [10]. The identification experiments, which were focused on these odorants, revealed vanillin, phenylacetic acid, 3-hydroxy-4,5-dimethyl-3(2*H*)-furanone, 4-hydroxy-2,5-dimethyl-2(5*H*)-furanone, β-ionone, linalool and (*E*,*E*,*Z*)-2,4,6-nonatrienal with the highest FD-factor of 256 in the extract. In particular, the (*E*,*E*,*Z*)-2,4,6-nonadienal is worth mentioning, because this odorant, showing an extremely low odour threshold, was recently identified by us for the first time as key odorant in tea aroma. The complete results of the identification will be given elsewhere [10].

3.2. Key odorants in the black tea beverage

A tea infusion was prepared by extraction of black tea leaves (12 g) with 1 l of hot water for 150 s. After filtration, the volatile fraction was isolated from the infusion, and the most odour-active compounds were characterised by application of the AEDA. A total of twenty-three odour-active areas were detected in the FD-factor range of 16 to 128 [10]. The identification of the odorants evoking the respective odour qualities showed vanillin, 4-hydroxy-2,5-dimethyl-2(5*H*)-furanone, 2-phenylethanol and (*E*,*E*,*Z*)-2,4,6-nonatrienal as most important followed by linalool, phenylacetaldehyde, β-ionone, 4,5-epoxy-(*E*)-2-decenal and 3-hydroxy-4,5-dimethyl-2(5*H*)-furanone (Figure 1).

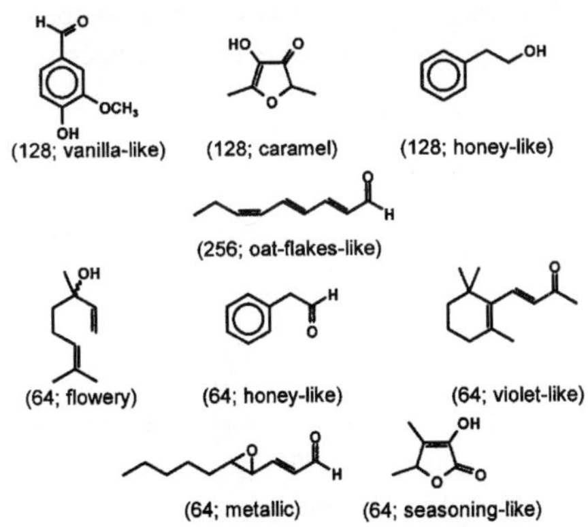

Figure 1. Key aroma compounds identified in the volatile extract/distillate from the black tea infusion (Darjeeling Gold Selection) with Flavour Dilution factors in parentheses.

If one compares the results obtained for the tea powder and the infusion prepared thereof [10], it becomes obvious that most of the odour-active compounds were identical in the leaves and the beverage prepared thereof from a qualitative point of view, however differences in the relative amounts are obvious from the FD-factors. For instance, 3-methylnonan-2,4-dione and (*Z*)-4-heptenal were detected among the key odorants in the tea leaves, but were only detected with lower FD-factors in the infusion [10]. However, because different amounts of tea leaves have been used in both experiments, the influence of the extraction process cannot directly be derived from these data, but can only be revealed based on exact quantitative data.

By using a set of quantitative data recently published by us [10], a comparison between the concentration of twenty-one aroma compounds determined directly in the black tea leaves and those determined in the infusion prepared from the same amount of leaves, but after being extracted for 150 s with hot water, was performed.

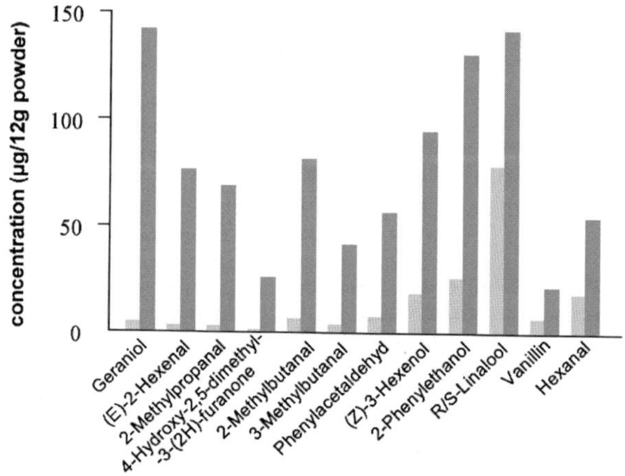

Figure 2. Comparison of the concentrations of key aroma compounds in the black tea powder (Darjeeling Gold Selection; 12 g; grey bars) and in the infusion (1 l) prepared from 12 g tea powder (black bars).

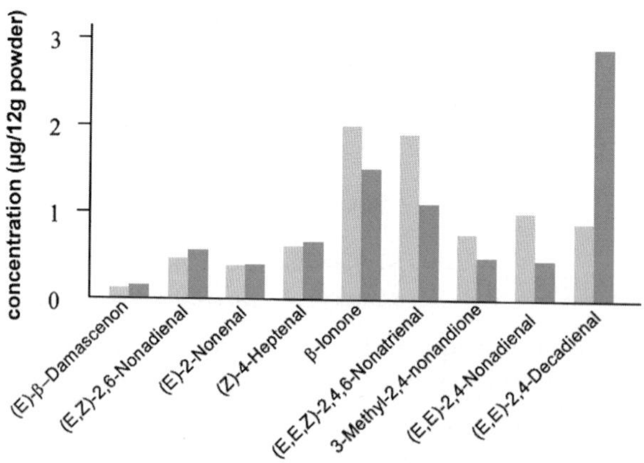

Figure 3. Comparison of the concentrations of key aroma compounds in the black tea powder (Darjeeling Gold Selection; 12 g; grey bars) and in the infusion (1 l) prepared from 12 g tea powder (black bars).

Based on the results, the aroma compounds could be allocated into two main groups: a first group of odorants being significantly increased during the hot water treatment, such as the alcohols geraniol, 2-phenylethanol and 4-hydroxy-2,5-dimethyl-3(2*H*)-furanone as well as the Strecker aldehydes 2-methylpropanal and 2-methylbutanal (Figure 2). Obviously these odorants are generated from precursors in the black tea leaves by hydrolytic cleavage of non-volatile precursors.

Table 1. Important odorants (FD≥4) in extracts isolated from five infusions of the following teas: Darjeeling Gold Selection (DG), Darjeeling Gold Star (DGS), Golden Palace (GP), Admiral's Cup (AC) and Ceylon Best (CB).

		FD-factor[a]				
Odorant	Odour quality	DG	DGS	GP	AC	CB
(*Z*)-4-Heptenal	Fishy	8	4	16	2	4
(*Z*)-1,5-Octadien-3-one	Geranium-like	4	<1	<1	2	2
(*E,E*)-2,4-Heptadienal	Fatty	16	8	16	8	8
(*E*)-2-Nonenal	Fatty, green	8	32	8	8	16
R/S-Linalool	Flowery	64	64	128	16	32
(*E,Z*)-2,6-Nonadienal	Cucumber-like	8	4	8	2	4
Phenylacetaldehyde	Honey-like	64	16	16	32	16
3-Methylbutanoic acid	Sweaty	4	4	2	2	4
(*E,E*)-2,4-Nonadienal	Fatty, green	16	8	16	8	16
3-Methyl-2,4-nonandione	Hay-like	32	32	64	64	32
(*E,E*)-2,4-Decadienal	Fatty, deep fried	16	32	16	8	16
(*E*)-β-Damascenone	Cooked apple	32	64	32	16	64
Hexanoic acid	Rancid	4	4	4	1	4
Geraniol	Rose-like	32	64	16	1	8
2-Methoxyphenol	Smoky	16	4	2	8	4
(*E,E,Z*)-2,4,6-Nonatrienal	Oat-flake-like	128	128	256	128	128
2-Phenylethanol	Honey-like	128	64	64	32	64
β-Ionone	Violet-like	64	256	64	8	32
trans-4,5-Epoxy-(*E*)-2-decenal	Metallic	64	16	16	8	8
4-Hydroxy-2,5-dimethyl-3(2*H*)-furanone	Caramel-like	128	256	128	8	128
3-Ethylphenol	Phenolic	4	2	4	1	8
3-Hydroxy-4,5-dimethyl-2(5*H*)-furanone	Seasoning-like	64	64	16	16	32
Phenylacetic acid	Honey-like	16	16	32	16	32
Vanillin	Vanilla, sweet	128	128	128	128	128

[a]Flavour dilution factor.

By contrast, a second group of odorants were either not changed in their concentrations or were decreased by the extraction process (Figure 3). Compounds belonging to this group were (*E*)-β-damascenone, β-ionone, (*E,E,Z*)-2,4,6-nonatrienal and 3-methyl-nonan-2,4-dione. It may, however, be assumed that these compounds are more or less changed due to their solubility in water, i.e. the different extraction yields, rather than by a degradation reaction. Most of these compounds are known as degradation products of carotenoids (β-ionone) or unsaturated lipids (e.g. (*E*,*E*)-2,4-nonadienal) and are, thus, formed already during tea manufacturing. It might therefore be assumed that these are simply transferred from the leaves into the beverage by the hot water treatment.

3.3. Key odorants in infusions prepared from four different black teas

It is well-known, that beverages prepared from tea leaves of different origins differ significantly in their overall aroma. To get a first insight into the differences in the aroma compounds, infusions were prepared from four further black tea samples, and the most odour-active compounds were located by application of the AEDA on the distillates obtained by solvent extraction, followed by SAFE distillation. Compounds were identified as recently reported [10]. The results of the identification experiments in combination with the FD-factors are summarised in Table 1. The data obtained for the Darjeeling Gold Selection (DG) is displayed for comparison. With the exception of (*Z*)-1,5-octadien-3-one, which was not sensory detectable in the Darjeeling Gold Star (DGS), all other aroma compounds were identified as odour-active in all five tea infusions. Only three of the most odour-active compounds (FD ≥64) namely vanillin, (*E*,*E*,*Z*)-2,4,6-nonatrienal and 3-methylnonan-2,4-dione were present with the same odour activity (i.e. the same FD-factor) in all tea beverages. Significant differences (more than a factor of 4) were observed for 4-hydroxy-2,5-dimethyl-3(2*H*)-furanone, which was much lower in the Admiral's Cup (AC) as compared to the other four tea infusions. Also β-ionone, which was highest in both Darjeeling teas (DG and DGS) was much lower in the AC and, also, in the Ceylon Best (CB). In the latter two tea infusions, also linalool and geraniol were much lower as compared to DG, DGS and Golden Palace (GP).

The data suggested that the differences in the overall aromas of the five tea infusions are undoubtedly caused by differences in the concentrations of the key aroma compounds. However, further quantitative studies must be undertaken to prove this suggestion.

References

1. S.M. Constantinides, R. Hoover, P.A. Karakoltsidis, T.C. Kelly, Y.C. Lu, M. Namiki, P. Schreier and P. Winterhalter, Food Rev. Int., 11 (1995) 371.
2. R. Teranishi, E.L. Wick and I. Hornstein (eds.), Flavor chemistry: thirty years of progress, New York, USA (1999) 135.
3. K.C. Willson and M.N. Clifford (eds.), Tea – cultivation to consumption, London, UK (1992) 603.
4. TNO, Volatile compounds in food, TNO (2005) 73A, B, E.
5. A. Goankar (ed.), Characterization of food – emerging methods, Amsterdam, The Netherlands (1995) 403.
6. W. Grosch, Flavour Fragrance J., 9 (1994) 147.
7. P. Semmelroch and W. Grosch, J. Agric. Food Chem., 44 (1996) 537.
8. H. Guth and W. Grosch, Flavour Fragrance J., 8 (1993) 173.
9. T. Parliment, C.-T. Ho and P. Schieberle (eds.), Caffeinated beverages, ACS symposium series 754, Washington DC, USA (2000) 337.
10. C. Schuh and P. Schieberle, J. Agric. Food Chem., (2005) in press.
11. W. Engel, W. Bahr and P. Schieberle, Eur. Food Res. Technol., 209 (1999) 237.

W.L.P. Bredie and M.A. Petersen (Editors)
Flavour Science: Recent Advances and Trends

Characterisation of Cheddar cheese flavour by sensory directed instrumental analysis and model studies

Keith R. Cadwallader[a], Mary Anne Drake[b], Mary E. Carunchia-Whetstine[b] and Tanoj K. Singh[a]

[a]Department of Food Science and Human Nutrition, University of Illinois at Urbana-Champaign, 1302 W. Pennsylvania Avenue, 205 ABL, Urbana IL 6180, USA; [b]Department of Food Science, Southeast Dairy Foods Research Center, North Carolina State University, Raleigh NC 27695, USA

ABSTRACT

Commercial Cheddar cheeses with and without a beefy/broth-like flavour note had a remarkable similarity in their overall GCO-AEDA and DHDA aroma compounds profiles. However, the beefy/broth-like Cheddar samples had particularly high flavour dilution factors for 3-(methylthio)propanal (methional) and 4-hydroxy-2,5-dimethyl-3(2*H*)-furanone (Furaneol). 2-Ethyl-4-hydroxy-5-methyl-3(2*H*)-furanone and 2-methyl-3-furanthiol (MFT) were identified at higher, but overall low intensities in broth-like cheeses. Sensory analysis of mild flavoured Cheddar cheese spiked with a combination of methional, Furaneol and MFT showed a more intense beefy/broth-like.

1. INTRODUCTION

Cheddar cheese flavour has been intensively studied for nearly sixty years. While much is known about the general flavour chemistry of Cheddar cheese [1], only limited studies have directly linked chemical components with descriptive sensory attributes. A Cheddar cheese flavour lexicon with standard descriptive language, definitions and standard rating scales has been developed and validated [2,3]. Components responsible for many of the sensory descriptors identified in the lexicon have been identified, e.g. [4,5]. The unambiguous linking of sensory descriptors with causative chemical components permits researchers to precisely relate sensory flavour quality with the chemistry and technology of Cheddar cheese production. The objective of this study was to identify volatile aroma compounds responsible for the beefy/broth-like flavour note in Cheddar cheese.

2. MATERIALS AND METHODS

2.1. Samples

Commercial Cheddar samples identified with beefy/broth-like flavour notes (12 samples) and control (not broth-like, 2 samples) were selected using descriptive sensory analysis as previously described [2].

2.2. Extraction of volatiles aroma compounds

Each cheese sample plus internal standards (2-ethylbutyric acid and 2-methyl-3-heptanone) was extracted with diethyl ether, followed by isolation of volatiles by solvent assisted flavour evaporation (SAFE). Aroma extracts were separated into acidic and neutral/basic fractions, see [4-6]. The analyses were performed in duplicate.

2.3. Instrumental analysis

Gas chromatography-olfactometry (GCO), including aroma extract dilution analysis (AEDA) and dynamic headspace dilution analysis (DHDA), were conducted as previously described [4-6]. Both neutral/basic and acidic fractions were analysed by GC-MS [4-6]. The selected ion monitoring (SIM) mode and use of a flame-photometric detector were also applied for the identification of trace components. Compounds were identified by comparison of GC-MS data, retention indices and odour properties.

2.4. Sensory evaluation

Selected aroma compounds, alone and in combinations, were evaluated after spiking into a mild Cheddar cheese matrix as previously described [6].

3. RESULTS

The broth-like aroma note is considered as a characteristic of aged Cheddar cheeses. But in some cases Cheddar cheeses have particularly enhanced levels of broth-like aroma notes which may be considered as a flavour defect. The broth-like flavours can sensorily be subdivided into beefy/broth-like, chicken/broth-like and mushroom/broth-like types, with the former note being most commonly encountered. Beefy/broth-like aroma notes in the Cheddar cheeses were identified by sensory evaluation. The results of two intensely beefy/broth-like and a control Cheddar samples are presented (Table 1).

There was remarkable similarity in the overall aroma compounds profiles of the beefy/broth-like and control Cheddar samples by AEDA for both acidic and neutral/basic fractions. Neutral/basic fraction of the broth-like Cheddar samples had notably higher flavour dilution factors and concentration for 3-(methylthio)propanal (methional) and 4-hydroxy-2,5-dimethyl-3(2*H*)-furanone (Furaneol). 2-Ethyl-4-

hydroxy-5-methyl-3(2*H*)-furanone (homofuraneol) and 2-methyl-3-furanthiol (MFT) were also identified in broth-like cheeses at lower intensities.

Table 1. Differentiating odorants detected in beefy/broth-like Cheddar cheeses.

Compound	RI[a]		Odour description	FD Factor					
				AEDA[b]			DHDA[c]		
	FFAP	DB-5		N[d]	B1[d]	B2[d]	N	B1	B2
2-Methyl-3-furanthiol	1312	873	Vitamin, meaty	nd	3	9	nd	nd	nd
3-(Methythio)propanal	1455	907	Potato	9	2187	2187	5	25	25
2-Methyl-(3-methyldithio)furan	1682	1170	Vitamin, meaty	nd	nd	nd	5	25	25
Maltol	1998	1175	Burnt sugar	nd	9	9	nd	nd	nd
Furaneol	2035	1058	Burnt sugar	9	729	2187	5	25	25
Homofuraneol	2090	1160	Burnt sugar	3	9	3	nd	1	1
bis-(2-Methyl-3-furyl) disulfide	2132	1542	Vitamin, meaty	nd	nd	nd	1	1	5

[a]Retention index. [b]Flavour dilution factor determined by aroma extract dilution analysis. [c]Flavour dilution factor determined by dynamic headspace dilution analysis. [d]N indicates control (not broth-like) cheese, B1 and B2 refer to beefy/broth-like cheeses.

Table 2. Sensory analysis of beefy/broth-like model cheeses.

Combination	Aroma description (intensity)[a]	Overall similarity score[b]
No spike	Not beefy/broth-like[c]	0
MFT (2 ng/g) + Furaneol (10 μg/g)	Beefy/broth-like (2)	8
MFT (4 μg/g) + Furaneol (20 μg/g)	Beefy/broth-like (3) Burnt sugar/fruity (3)	7
MFT (2 μg/g) + Furaneol (10 μg/g) + Methional (80 μg/g)	Beefy/broth-like (2)	9
MFT (4 μg/g) + Furaneol (20 μg/g) + Methional (80 μg/g)	Beefy/broth-like (3)	7

[a]Average aroma intensity on a 15-point universal scale (nine trained panellists). [b]Similarity to typical beefy/broth-like Cheddar cheese (10-point scale). [c]Cheese received the following aroma/ flavour scores: diacetyl=2, whey=4, cooked=3.5, milk fat=3.5, salty=4, sour=3, sweet=1.5.

A similar observation was also made from DHDA analyses, except that oxidised derivatives of MFT (e.g. 2-methyl-(3-methyldithio) furan and bis-(2-methyl-3-furyl) disulfide) were detected instead of MFT. Results of the sensory analysis of spiked mild flavoured Cheddar cheese indicated that the beefy/broth-like note could most accurately be produced by a combination of methional, Furaneol and MFT. However, the Furaneol and MFT combinations produced recognisable beefy/broth-like notes. Methional alone produced a potato-like note in the cheese.

Table 3. Compounds contributing specific flavour/aroma notes in Cheddar cheese.

Note	Compound	Reference
Broth-like	Methional	Present study
	Furaneol	
	2-Methyl-3-furanthiol (MFT)	
Cowy/phenolic	*p*-Cresol	[4]
Earthy/bell pepper	2-Propyl-3-methoxypyrazine	[4]
	2-Isobutyl-3-methoxypyrazine	
Nutty	2/3-Methylbutanal	[5]
	2-Methylpropanal	
Rosy/floral	Phenylacetaldehyde	[6]
	Phenylacetic acid	

4. DISCUSSION AND CONCLUSION

It is generally accepted that Cheddar cheese flavour is due to the proper balance of numerous volatile components. While some of these compounds contribute to background flavour, some play predominant roles in the overall flavour. The origin of the compounds found to be responsible for the beefy/broth-like aroma note can be explained by catabolic activity of the starter bacteria and/or Maillard reaction.
Some other specific aroma notes in Cheddar cheeses were also identified in our previous studies (Table 3). Causative chemicals identified by the instrumental analysis were further evaluated and their importance confirmed by addition or spiking into cheese models or control cheeses lacking these particular aroma notes.

References

1. T.K. Singh, M.A. Drake and K.R. Cadwallader, Compr. Rev. Food Sci. Food Safety, 2 (2003) 139.
2. M.A. Drake, S.C. McInvale, P.D. Gerard, K.R. Cadwallader and G.V. Civille, J. Food Sci., 66 (2001) 1422.
3. M.A. Drake, P.D. Gerard, S. Wright, K.R. Cadwallader and G.V. Civille, J. Sens. Stud., 17 (2002) 215.
4. O. Suriyaphan, M.A. Drake, X.Q. Chen and K.R. Cadwallader, J. Agric. Food Chem., 49 (2001) 1382.
5. Y.K. Avsar, Y. Karagul-Yuceer, M.A. Drake, T.K. Singh, Y. Yoon and K.R. Cadwallader, J. Dairy Sci., 87 (2004) 1999.
6. M.E. Carunthia-Whetstine, K.R. Cadwallader and M.A. Drake, J. Agric. Food Chem., 53 (2005) 3126.

W.L.P. Bredie and M.A. Petersen (Editors)
Flavour Science: Recent Advances and Trends

The analysis of volatiles in Tahitian vanilla (*Vanilla tahitensis*) including novel compounds

Neil C. Da Costa and Michael Pantini

International Flavors & Fragrances Inc., 1515 Highway 36, Union Beach, New Jersey 07735, USA

ABSTRACT

Tahitian vanilla was analysed as part of IFF's ongoing Generessence® programme. This programme involves the in-depth analysis of a natural product to ultimately produce a new truly nature identical flavour and to discover new flavour molecules. Selected Tahitian vanilla beans were imported from French Polynesia and extracted by percolation. The ethanolic extract underwent further liquid/liquid extraction, and sampling by a sorptive stir bar. The volatiles of the beans were analysed by dynamic headspace and GC-MS. A total of 276 components were identified through these 3 analyses. Amongst them were novel compounds not previously reported in Tahitian vanilla [1] These novel compounds were either purchased or synthesised for confirmation. The major semi-volatiles were analysed by HPLC. In addition the novel compounds were evaluated by an expert panel.

1. INTRODUCTION

Vanilla is an important flavouring material and spice, cultivated from a plant of the orchid species. Originating in South America it was first cultivated by the Aztecs and brought to Europe by the Spanish. In 1848 it was introduced to Tahiti and a vanilla-producing industry was established. The fully ripened fruits are beans which are harvested, fermented and cured. This cultivation and processing makes them an expensive and prised ingredient [2].

2. MATERIALS AND METHODS

2.1. Extraction and sampling

Select chopped beans (1.4 kg) were percolated twice with 3 l of 45% ethanol/water for 16 h and 9 h respectively at 48.9 °C, with draining for 6 h each time. The first most concentrated percolate was used for further extraction.

2.1.1. Liquid/ liquid extraction

Two litres of percolated bean extract were transferred to a separating funnel and extracted twice with 200 ml of dichloromethane. The combined extracts were dried over anhydrous sodium sulphate, filtered and reduced under rotary evaporation, followed by further concentration under nitrogen to 2 ml. This extract was analysed by GC-flame ionisation detector (GC-FID) and GC-MS on polar and non-polar columns.

2.1.2. Sorptive stir bar extraction (Twister™)

Approximately 20 ml of the aqueous/ethanol solution were poured into pairs of 30 ml glass vials on stirrer plate at 1800 rpm. Stir bars were added and volatiles adsorbed for 1 h and 4 h respectively. The stir bars were rinsed with distilled water, patted dry and thermally desorbed onto GC-FID and GC-MS systems.

2.1.3. Dynamic headspace

Approximately 30 g of cured beans were chopped and placed in a headspace jar and the headspace sampled using Tenax™ TA traps (Supelco) fitted with a reduced pressure pump (flow rate: 25 ml/min). The headspace was sampled for 1 h and 4 h respectively and the traps thermally desorbed onto GC-FID and GC-MS systems.

2.2. Analysis

2.2.1. Gas-Chromatography

The liquid/liquid extract was analysed using an HP6890 GC fitted with an OV1 capillary column (50 m x 0.32 mm i.d., 0.5 μm film thickness) used in split mode (split ratio 40:1). The helium carrier gas flow rate was 1.0 ml/min. Injection port temperature was 250 °C and the detector temperature was 320 °C. The column temperature programme was from 40 °C to 270 °C at 2 °C/min held for 10 min at the top temperature. For a polar gas chromatogram a HP6890 GC was fitted with a Carbowax capillary column (50 m x 0.32 mm i.d., 0.3 μm film thickness). Injection and detection techniques were the same as above. The temperature programme started at 60 °C held for 10 min, ramped at 2 °C/min to a final temperature of 220 °C held for 20 min.
Tenax™ TA traps and stir bars were thermally desorbed onto an HP6890 GC using a thermal desorber Model TDS 2 (Gerstel Inc.). Desorption time was 5 min at 250 °C. An OV-1 capillary column was used and the analysis performed in splitless mode. The injector temperature was programmed from -150 °C (held for 5 min during the thermal desorption) to 250 °C and the detector temperature was 320 °C.

2.2.2. *Mass Spectrometry*

The sample was injected onto an HP5890 GC. The chromatographic conditions for the OV-1 column were the same as described for GC analysis. The mass spectrometer was a Micromass Prospec high resolution, double-focusing, magnetic sector instrument, operated in the electron ionisation mode 70 eV, scanning from *m/z* 450 to *m/z* 33 at 0.3 s per decade. Analysis on a polar phase (Carbowax) was via a HP5890 GC into a Kratos Profile mass spectrometer. GC oven and mass spectrometer conditions were the same as outlined above. Spectra obtained were analysed using an in-house library and commercial libraries. The identification of flavour components was confirmed by MS match or interpretation and GC relative retention indices.

2.2.3. *HPLC*

HPLC determination of the main vanilla components was as follows, (all concentrations are in mg/100 ml), 4-hydroxybenzoic acid (146.7), 4-hydroxybenzaldehyde (54.0), vanillic acid (13.7) and vanillin (149.0). Conditions: 1% acetic acid in water/acetonitrile gradient at 35 °C, Alltima C18 column, 10 μl injection, 1.0 ml/min, 280 nm UV detection.

3. RESULTS AND CONCLUSIONS

The GC-MS analysis of the 3 sets of vanilla samples identified 276 compounds. Individually the liquid extract gave 142 compounds, the stir bar 152 compounds and the headspace 79 compounds, respectively. These included many components of interest, some not previously reported in vanilla (Table 1).

Table 1. Components of interest in Tahitian Vanilla.

Component	Concentration ppt[a]	Component	Concentration ppt
Anisyl alcohol	225.8	Caffeine	0.1
Anisic acid	87.4	Theobromine	0.1
Anisaldehyde	25.0	α-Ionone	0.4
Dianisyl ether	3.1	β-Ionone	0.4
Anisyl ethyl ether	15.0	Dihydroactinidiolide	0.2
Anisyl methyl ether	0.8	Vitispirane	0.3
Anisyl anisate	6.6	Anisyl 4-hydroxybenzoate	7.4
Anisyl *trans*-cinnamate	0.5	Anisyl *cis*-cinnamate	0.2

[a]Note that ppt is parts per thousand in the liquid extract by GC.

Three new anisyl esters were proposed for synthesis from the GC-MS analysis of the liquid extract and confirmed as being present (Figure 1). They were also evaluated for sensory characteristics in a sugar solution by a panel of internal experts.

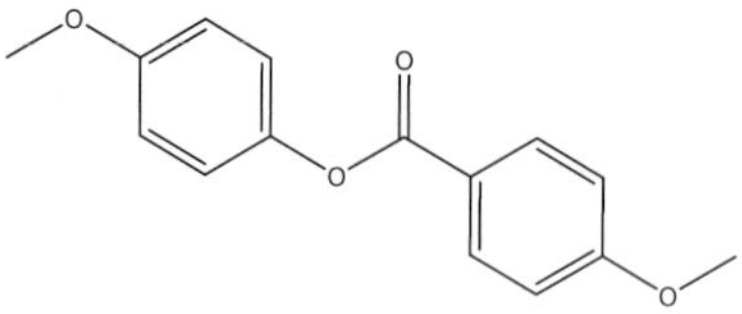

Anisyl anisate (1)

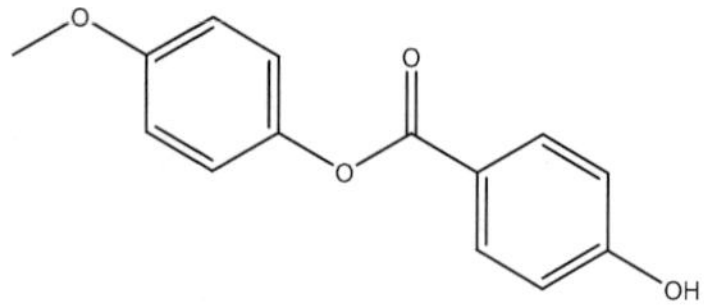

Anisyl 4-hydroxybenzoate (2)

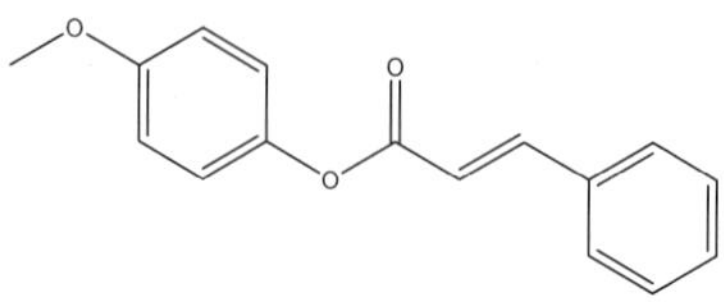

Anisyl trans-cinnamate (3)

Figure 1. Novel compounds identified in Tahitian Vanilla.

The chemical and sensory data of the three new compounds identified in Tahitian vanilla are presented below with their routes of synthesis. Synthesis from anisyl alcohol and trans-cinnamic acid in xylene with the elimination of water. *Anisyl trans-cinnamate* (3) $C_{17}H_{16}O_3$, MW 268, *m/z* 121 (100), 268 (14), 223 (8), 131 (14), 122 (9), 103 (16), 91 (11), 78 (14), 77 (31), 51 (10). ^{1}H-NMR (CDCl$_3$) δ 3.78 (s, 3H), 5.17 (s, 2H), 6.45 (d, 1H, J=16.0Hz), 6.90 (d, 2H, J=8.5Hz), 7.34-7.35 (m, 5H), 7.48 (m, 2H), 7.70 (d, 1H, J=16.0Hz), ^{13}C-NMR (CDCl$_3$) δ 55.2, 66.1, 113.9 (2), 118.0, 128.0 (2), 128.1, 128.8 (2), 130.1 (2), 130.2, 134.4, 144.9, 159.6, 166.7. Flavour descriptors: weak, sweet balsamic, slightly floral, vanillin, cherry, anise, French vanilla.

Synthesis from anisyl acid chloride and anisyl alcohol in dichloromethane with triethylamine base. *Anisyl anisate* (1) $C_{16}H_{16}O_4$, MW 272, *m/z* 121 (100), 272 (20), 135 (27), 122 (12), 92 (6), 91 (5), 78 (8), 77 (13), 76 (6). ^{1}H-NMR (CDCl$_3$) δ 3.79 (s, 3H), 3.82 (s, 3H), 5.26 (s, 2H), 6.89 (dd, 4H, J=8.77, 8.11Hz), 7.37 (d, 2H, J=8.45Hz) 8.00 (d, 2H, J=8.82Hz). Flavour descriptors: cherry, vanilla, cream, floral, coumarin, sweet, liquorice, sugar, fennel, caramelic, anise, lactone, balsamic, marshmallow.

Synthesis from anisyl alcohol and 4-hydroxybenzoic acid in the presence of thionyl chloride and triethylamine base gave a solid mixture. Component identified from mass spectrum by fragmentation interpretation, as the same in the vanilla. *Anisyl 4-hydroxybenzoate* (2) $C_{15}H_{14}O_4$, MW 258, *m/z* 121 (100), 258 (20), 122 (12), 91 (5), 78 (7), 77 (8), 65 (8), 39 (5).

Tahitian vanilla is generally characterised by a lower concentration of vanillin than the other main vanilla varieties (Bourbon and Indonesian) and a high concentration of anisyl components not reported in the other types of vanilla.

References

1. D. Fraisse, F. Maquin, D. Stahl, K. Suon and J.C. Tabet, Analysis, 12 (2) (1984) 63.
2. J.W. Purseglove, D.G. Brown, C.L. Green and S.R.J. Robbins, Spices, Harlow, UK (1981) 2, 644.

W.L.P. Bredie and M.A. Petersen (Editors)
Flavour Science: Recent Advances and Trends

Screening and identification of bitter compounds in roasted coffee brew by taste dilution analysis

Oliver Frank[a], Gerhard Zehentbauer[b] and Thomas Hofmann[a]

[a]Institut für Lebensmittelchemie, Universität Münster, Corrensstrasse 45, 48149 Münster, Germany; [b]The Procter & Gamble Company, Miami Valley Innovation Center, 11810 E. Miami River Road, Cincinnati OH45252, USA

ABSTRACT

Intense bitter taste compounds were identified in roasted coffee brew using sensory-guided fractionation, LC-MS/MS and 1D/2D-NMR spectroscopy, syntheses, and model roasting experiments with potential precursors. The intense bitter tastants of coffee were 3-*O*-caffeoyl-γ-quinide (**1**), 4-*O*-caffeoyl-γ-quinide (**2**), 4-*O*-caffeoyl-*muco*-γ-quinide (**3**), 5-*O*-caffeoyl-*muco*-γ-quinide (**4**), 5-*O*-caffeoyl-*epi*-δ-quinide (**5**), 3-*O*-feruloyl-γ-quinide (**6**), 4-*O*-feruloyl-γ-quinide (**7**) 3,4-*O*-dicaffeoyl-γ-quinide (**8**), 4,5-*O*-dicaffeoyl-*muco*-γ-quinide (**9**) and 3,5-*O*-dicaffeoyl-*epi*-δ-quinide (**10**). Determination of the bitter taste recognition thresholds showed that, depending on their chemical structure, the bitter threshold concentrations ranged between 9.8 and 180 μmol/l (water). Quantification and determination of the dose over threshold factors for the individual bitter compounds revealed that approximately 80% of the bitterness of the decaffeinated coffee beverage could be related to these 10 quinides.

1. INTRODUCTION

More than 20 years ago, Chen [1] reported that caffeine contributed to approximately 30% and trigonelline to a maximum of 1% to the overall bitterness of coffee beverage. The structures of the compounds evoking the typical bitter taste of coffee, however, remained unknown. Recently, bitter tasting diketopiperazines [2] and chlorogenic acid lactones [3] have been identified in roasted coffee, but neither quantitative data, nor human bitter taste thresholds of purified compounds have been reported. Therefore, identification of the key bitter-tasting compounds of coffee beverages needs further investigation. In order to answer the question as which of the non-volatile tastants are responsible for the bitter taste of the coffee beverage, we have recently developed the so-called taste dilution analysis as a screening procedure for taste-active non-volatiles in foods [4]. The objective of this study was to screen a coffee beverage for its key bitter taste compounds by application of taste dilution techniques, to determine the chemical structure of the compounds inducing the most intense bitter taste, and to evaluate their sensory impact on the basis of their threshold concentrations.

2. MATERIALS AND METHODS

Raw and roasted, decaffeinated Arabica Columbia was obtained from the coffee industry. After grinding the coffee beans by means of an ultracentrifuge mill equipped with a sieve (2 mm pore diameter), aliquots (54 g) of coffee powder placed in a coffee filter (No.4, Melitta, Germany) were percolated with 1.1 l hot water at 80 °C. The resulting coffee beverage (1 l) was rapidly cooled to room temperature and then used for sensory and chemical analyses. Reference material of the bitter compounds 1-5, 6-7 and 8-10 were isolated from the coffee beverage or from model roasting experiments [5]. The details on the procedure for isolation and identification of the bitter compounds, the sensory analyses and the data of the quantitative analysis are reported elsewhere [5].

3. RESULTS AND DISCUSSION

To gain insight into the volatility and hydrophobicity of the compounds imparting the bitter taste, the coffee beverage was fractionated by high vacuum distillation into a volatile and a non-volatile fraction. The distillate was separated by a sequential solvent extraction affording a pentane soluble fraction, an ethyl acetate soluble fraction, a chloroform soluble fraction, and the remaining aqueous phase. The sensory analysis of these fractions revealed that the bitter tastants were exclusively located in the non-volatile fraction and extractable with ethylacetate.

3.1. Separation and identification of key taste compounds

In order to differentiate the bitter compounds on their taste impact a taste dilution analysis was performed [4]. The ethyl acetate extract was separated by RP-HPLC and the effluent collected into 15 fractions. The fractions were freeze-dried and made up with water (pH 5.0) to the same volume. Each fraction was stepwise diluted one-to-one with water and, then, presented in order of increasing concentrations to the sensory panellists who were asked to evaluate the taste quality and to determine the detection threshold in a triangle test. As this so-called taste dilution (TD)-factor is related to its taste activity in water [4], the 15 HPLC fractions were rated according to their relative taste impact. As reflected by their high TD-factors of 256, fractions 5, 8, and 15 were evaluated with the highest bitter taste impact (Figure 1). The fractions 12 or 4, 6, and 13 were judged with a somewhat lower TD factor of 128 or 64, respectively. All the other fractions were evaluated with lower taste impacts. Therefore, the identification experiments were focused on the tastants in HPLC fractions no. 4-6, 8, 12, 13, and 15.

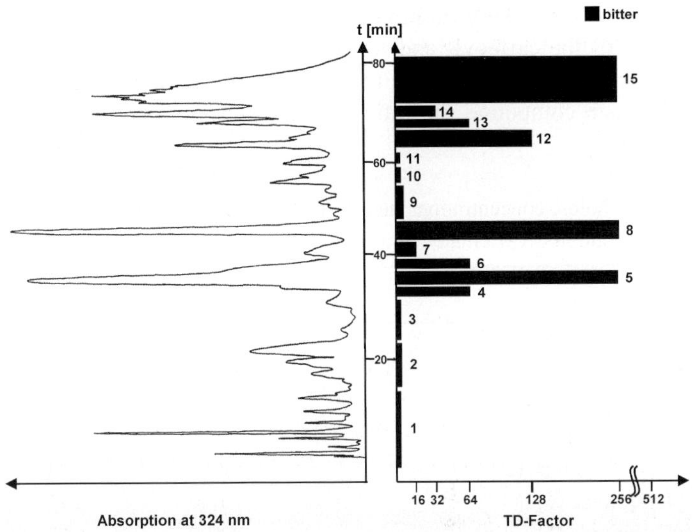

Figure 1. HPLC chromatogram (left side), and taste dilution (TD) chromatogram (for bitterness) of the solvent extractable compounds of a freshly prepared decaffeinated coffee beverage [5].

UV-vis spectra, LC-MS- and NMR analysis revealed that the bitter tasting fractions of the fractions 4-8 may be related to the hydroxycinammic acids. Isolation of these compounds and model roasting experiments with caffeoyl-, feruloyl- and dicaffeoylquinic acids led to the identification of these bitter compounds (Figure 2). In the different fractions were identified 3-*O*-caffeoyl-γ-quinide (**1,** no. 5), 4-*O*-caffeoyl-γ-quinide (**2,** no. 8), 5-*O*-caffeoyl-*epi*-δ-quinide (**5,** no. 7), 4-*O*-caffeoyl-*muco*-γ-quinide (**3,** no. 6) and 5-*O*-caffeoyl-*muco*-γ-quinide (**4,** no. 1), in the fractions 12/13 the 3-*O*-feruloyl- (**6**) and 4-*O*-feruloyl-γ-quinide (**7**) could be identified as bitter principals. In addition, 3,4-*O*-dicaffeoyl-γ-quinide (**8**), 4,5-*O*-dicaffeoyl-*muco*-γ-quinide (**9**) and 3,5-*O*-dicaffeoyl-*epi*-δ-quinide (**10**) were identified as bitter constituents of fraction 15.

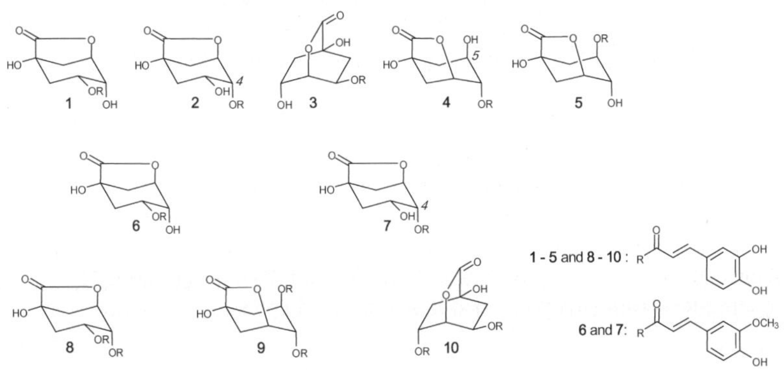

Figure 2. Structures of identified bitter tasting *O*-caffeoyl, *O*-feruloyl- and *O*-dicaffeoyl quinides.

3.2. Quantification and calculation of dose-over-threshold (Dot)-factors

The contribution of the caffeoyl- and feruloylquinides to the bitter taste of the coffee beverage was estimated by determining their concentration, taste threshold and calculating for each compound the ratio of the concentration and the taste threshold (Table 1).

Table 1. Taste thresholds, concentrations and dose-over-threshold (Dot) factors of the identified bitter tasting compound in decaffeinated Arabica Columbia coffee.

Bitter taste compounds	Threshold (mg/l)	Conc. (mg/l)	Dot factor
3-*O*-Caffeoyl-γ-quinide (**1**)	13.4	33.15	2.5
4-*O*-Caffeoyl-γ-quinide (**2**)	12.1	19.68	1.6
4-*O*-Caffeoyl-*muco*-γ-quinide (**3**)	11.2	8.27	0.7
5-*O*-Caffeoyl-*muco*-γ-quinide (**4**)	9.7	6.12	0.6
5-*O*-Caffeoyl-*epi*-δ-quinide (**5**)	60.5	3.28	0.05
3-*O*-Feruloyl-γ-quinide (**6**)	13.7	6.75	0.5
4-*O*-Feruloyl-γ-quinide (**7**)	13.7	3.03	0.2
3,4-*O*-Dicaffeoyl-γ-quinide (**8**)	4.9	5.40	1.1
4,5-*O*-Dicaffeoyl-*muco*-γ-quinide (**9**)	4.9	1.65	0.3
3,5-*O*-Dicaffeoyl-*epi*-δ-quinide (**10**)	24.9	0.80	0.03
ΣQL_{total}			7.6
Coffee beverage			9.0[a]

[a]Dot-factor was calculated as the ratio of the dry weight and the taste detection threshold of the coffee beverage.

These analyses and calculation of the dose-over-threshold (Dot) factors revealed that the concentration of 3-*O*-caffeoyl-γ-quinide (**1**), 4-*O*-caffeoyl-γ-quinide (**2**) and 3,4-*O*-dicaffeoyl-γ-quinide (**8**) in the coffee beverage exceeded the bitter threshold concentrations by a factor of 2.5, 1.6 and 1.1 respectively (Table 1). Sensory experiments with purified quinides revealed that the compounds 3-7, 9 and 10 although present below their taste threshold still contributed to the bitter taste of coffee. These compounds gave and additive effect on the bitter taste of the coffee beverage.

References

1. W.C. Chen, Doctoral Thesis, University Munich, Germany (1979).
2. M.Ginz and U.H. Engelhardt, Eur. Food Res. Technol., 213 (2001) 8.
3. M.Ginz and U.H. Engelhardt, Proceedings to the ASIC symposium, Trieste, Italy (2001) 248.
4. O. Frank, H. Ottinger and T. Hofmann, J. Agric. Food Chem., 49 (2001) 231.
5. O. Frank and T. Hofmann, Eur. Food Res. Technol., in press.

W.L.P. Bredie and M.A. Petersen (Editors)
Flavour Science: Recent Advances and Trends

The flavour chemistry of culinary *Allium* preparations

Gerhard Krammer, Christopher Sabater, Stefan Brennecke, Margit Liebig, Kathrin Freiherr, Frank Ott, Jakob P. Ley, Berthold Weber, Detlef Stöckigt, Michael Roloff, Claus Oliver Schmidt, Ian Gatfield and Heinz-Jürgen Bertram

Symrise GmbH & Co. KG, Research & Development, P.O.Box 1253, 37601 Holzminden, Germany

ABSTRACT

In culinary preparations cooked, roasted and fried *Allium* products play a major role in the taste and the aroma of the final product. In this study the correlation of various culinary preparation techniques such as cooking and frying and the resulting differences in the flavour chemistry (volatiles and non-volatiles) are demonstrated. Different preparation techniques performed on shallots, garlic and/or scallions influence the profile of volatiles and non-volatiles in culinary preparations significantly.

1. INTRODUCTION

Since ancient times *Allium ssp.* are used as a valuable taste and aroma contributor in the different cuisines of the world. The sulfur chemistry of various *Allium* species such as onion *(A. cepa L.)*, shallot *(A. ascalonicum)*, scallion *(A. fistulorum)* and garlic *(A. sativum)* has been subjected to numerous scientific studies [1,2]. In the past the interaction of lipids and amino acids was studied in detail [3]. Volatile sulfur compounds are formed during cutting by the rapid enzymatic degradation of cysteine sulfoxides followed by a cascade of chemical reactions [4-7].

The direct link, however, between the sensory profile of culinary preparations and the pattern of *Allium*-derived flavour compounds is not fully understood yet. In this study a set of Asian recipes for the preparation of different *Allium* species (shallots, garlic, scallions) has been selected and evaluated at different culinary preparations.

2. MATERIALS AND METHODS

Shallots (*A. ascalonicum*), garlic (*A. sativum*) and scallions (*A. fistalorum*) as well as palm oil and pork lard were obtained from the local market in Singapore. All details for GC-MS, GCO and NMR analysis are described earlier [8]. [1,1,2-3D]-1-octen-3-one, dimethyl disulfide-6D, and 2-nonanol were used as internal standards. HPLC/MS was performed using the Agilent 1100 HPLC/Finnigan LCQ System.

An 800 g of shallow-fried *Allium* material were blended for 3 min in a blender (Grindomix). After addition of 1 l diethyl ether and 1 ppm internal standard the mixture was subjected to a phase separation using a high performance centrifugal separation system (CINC Deutschland, Brakel, Germany). The organic phase was pooled with the organic phase derived from the solid/liquid extract of the residue with diethyl ether. The combined organic phases were pre-concentrated using a solvent assisted flavour evaporation system (SAFE). The condensed organic phase was concentrated using a Turbo vap II system (Caliper Life Science, Hopkinton, MA, USA) and subjected to GCO and GC-MS analysis. The sensory profiles of culinary targets and flavour extracts were evaluated by a group of expert flavourists.

The residue of the sample preparation for the analysis of volatiles was freeze-dried and fractionated using solid phase extraction (SPE) on a Lewatit resin (Lanxess, Germany) using an ethanol/water mixture as eluent. Fraction 1 was used for the analysis of citric acid, malic acid, oxalic acid, sucrose, fructose and amino acids. In addition a Multilayer Coil Counter Current chromatography system (P.C. Inc., Buffalo, NY, USA) was employed for the fractionation of the lyophilisate. Selected fractions were subjected to further purification via preparative HPLC and subsequent structure elucidation.

1-Propenyl sulfides were synthesised by isomerisation of allyl sulfides which were obtained starting from allyl bromide and alkyl or alkenyl sulfides. Disulfides were synthesised by oxidation of thiols and trisulfides were the products of the reaction of thiols and sulfur dichloride. Pyrazines according to Table 1 were prepared by reaction of the appropriate α-diamines and α-diketones followed by dehydrogenation.

3. RESULTS AND DISCUSSION

The following different culinary targets were studied and compared: shallots in palm oil at 120 °C versus 160 °C, shallots at 160 °C in palm oil versus pork lard, and shallots *versus* shallots and garlic versus shallots and scallion in palm oil at 120 °C. For the comparison of flavour-active compounds in the selected culinary targets an aroma extract dilution analysis (AEDA) was performed up to flavour dilution factor (FD factor) 512.

For shallots in palm oil at different temperatures (cf. Table 1) higher amounts of (*E*,*E*)-2,4-decadienal from lipid degradation and 2,5-dimethyl-4-hydroxy-3(2*H*)-furanone from carbohydrate degradation were found, and these give rise to stronger fatty and sweet, fruity notes.

Table 1. Volatiles detected (ppm) upon frying shallots in palm oil at 120 °C and at 160 °C.

	Compound selected via AEDA FD 512	120 °C	160 °C
1	Methyl 1-propenyl sulfide	<0.01	0.03
2	Methyl (*E*)-1-propenyl disulfide	0.05	2.02
3	6D-Dimethyl trisulfide (standard for quantification)	1.00	1.00
4	Dimethyl trisulfide	<0.01 (FD 128)	2.93
5	(*E*)-1-Propenyl propyl disulfide	0.32 (FD 128)	1.28
6	2-Ethyl-3,6-dimethylpyrazine	n.d.	6.5
7	2-Ethyl-3,5-dimethylpyrazine	n.d.	0.74
8	unknown: *m/z* 87, 116, 45, 39, 53, containing S	n.d.	0.49
9	tentatively: 2-Ethyl-3,5,6-trimethylpyrazine	<0.01	0.66
10	Methyl propyl trisulfide	0.09	2.05
11	2-Nonanol (standard for quantification)	1.00	1.00
12	2-Isobutyl-3,5-dimethyl-pyrazine	n.d.	<0.3
13	tentatively: Methyl 1-propenyl trisulfide	<0.02	approx.1.5
14	5-Methyl-6,7-dihydro-(5*H*)-cyclopentapyrazine	<0.01	0.44
15	(*E,E*)-2,4-Decadienal	<0.01	0.6
16	4-Hydroxy-2,5-dimethyl-(2*H*)-furan-3-one	0.01	1.28

Table 2. Volatiles detected (ppm) upon frying shallots in palm oil 160 °C and pork lard at 160 °C.

	Compound selected via AEDA FD 512	Oil 160 °C	Lard 160 °C
1	6D-Dimethyl trisulfide (standard for quantification)	1.00	1.00
2	(*E,E*)-2,4-Heptadienal	<0.1	0.25
3	2-Nonanol (standard for quantification)	1.00	1.00
4	(*E*)-2-Nonenal	<0.01 (FD 128)	0.30
5	tentatively: Methyl 1-propenyl trisulfide	approx. 1.5	approx. 3.0
6	(*E,Z*)-2,4-Decadienal	0.17 (FD 128)	0.29
7	(*E,E*)-2,4-Decadienal	0.6	1.29

Table 3. Volatiles detected (ppm) upon frying shallots (s) in palm oil at 120 °C versus shallots + garlic (g) in palm oil at 120 °C and shallots + scallions (sc) in palm oil at 120 °C.

	Compound selected via AEDA FD 512	s	s + g	s+ sc
1	Methyl (*E*)-1-propenyl disulfide	0.05	0.15	0.03 (FD 128)
2	6D-Dimethyl trisulfide (standard for quantitation)	1.00	1.00	1.00
3	tentatively: Di-1-propenyl disulfide	n.d.	0.27	n.d.
4	2-Nonanol (standard for quantification)	1.00	1.00	1.00
5	Diallyl trisulfide	n.d.	1.68	n.d.
6	2-Vinyl-(4*H*)-1,3-dithiine	<0.01 (FD 128)	0.45	n.d.

n.d.: not detected.

In addition, significantly higher amounts of methyl propenyl sulfide with a smooth character of cooked onion and garlic as well as dimethyl trisulfide with onion, cabbage-like notes and methyl propyl trisulfide with pungent green character were observed. Among the series of detected pyrazines the increase of the roasty, cocoa-like 3,6-dimethyl-2-ethylpyrazine and of 3,5-dimethyl-2-ethylpyrazine with roasty and nutty notes are prominent examples. Frying of shallots in palm oil and pork lard at 160 °C results (cf. Table 2) in higher amounts of (*E*,*Z*)-2,4-decadienal with a strong fatty, aldehyde-like character and (*E*,*E*)-2,4-decadienal with deep-dry notes for the pork lard application. In addition, increasing amounts of (*E*)-2-nonenal with a tallowy character and of propenyl propyl disulfide with raw onion-like notes and methyl (*E*)-1-propenyl disulfide characterised by sulfury onion-like notes were found. In general, the addition of garlic to shallots results (cf. Table 3) in characteristic amounts of the allicin degradation product 2-vinyl-(4*H*)-1,3-dithiine with a green, pungent garlic-like character together with diallyl trisulfide, characterised by a leek-like, garlic-like impact. On the other hand methyl propenyl sulfide, which was found for the shallot and scallion preparation, is missing in products with garlic addition.

The analysis of non-volatiles was performed on 2 levels: i) determination of the sensory relevance of amino acids, organic acids, minerals and sugars, ii) identification of additional taste-active compounds and the correlation with perceived sensory effects. Alliin was identified in lyophilisate of shallots in oil at 160 °C via HPLC/MS. Obviously this compound was not fully degraded during the preparation of the culinary target. Sensory studies showed that among other compounds alliin is responsible for the lingering, sulfury, onion-like after-taste of food prepared with *Allium* material. In addition a quercetin glycoside has been tentatively identified and tasting experiments showed that this compound contributes to the astringent notes of the culinary profile. As a direct result of the culinary preparations *N*-(1-desoxy-D-fructosyl)-isoleucine has been identified. This Amadori product is obtained from fructose and isoleucine during the heating process and is responsible for dry, dusty, and slightly meaty notes.

References

1. E. Block, Angew. Chem., 104 (1992) 1158.
2. G. Freeman and R.J. Whenham, J. Sci. Food. Agric., 26 (1975) 1869.
3. J.-L. Le Quéré and P.X. Étiévant (eds.), Flavour research at the dawn of the 21st century, proceedings of the 10th Weurman flavour research symposium, Paris, France (2003) krammer.pdf on CD-ROM.
4. H. Maarse and P.J. Groenen (eds.), Proceedings of the international symposium on aroma research, Wageningen, The Netherlands (1975) 42.
5. H. Maarse (ed.), Volatile compounds in foods and beverages, Amsterdam, The Netherlands (1981) 203.
6. J.F. Carson, Food Rev. Int., 3 (1987) 71.
7. B. Schmitt and M. Keusgen, Lebensmittelchem., 57 (2003), 148.
8. C.O. Schmidt, G.E. Krammer, B. Weber, D. Stöckigt, K. Herbrand, F. Ott, G. Kindel, S. Brennecke, I.L. Gatfield and H.-J. Bertram, Perfum. Flavor., 30 (March/April) (2005) 28.

W.L.P. Bredie and M.A. Petersen (Editors)
Flavour Science: Recent Advances and Trends

4'-Hydroxyflavanones are the bitter-masking principles of Herba Santa

Jakob P. Ley, Gerhard Krammer, Günter Kindel, Ian L. Gatfield and Heiz-Jürgen Bertram

Symrise GmbH & Co. KG, Research & Development, P.O. Box 1253, D-37601 Holzminden, Germany

ABSTRACT

In sensory studies the flavanones homoeriodictyol, its sodium salt, and eriodictyol, which occur in Herba Santa (*Eriodictyon californicum* (H. & A.) Torr.) could significantly decrease the bitter taste of caffeine. Further investigations on homoeriodictyol sodium salt elicited a broad masking activity between 10 and 40% towards different chemical classes of bitter molecules (e.g. salicin, amarogentin, paracetamol, quinine) but not towards bitter linoleic acid emulsions.

1. INTRODUCTION

Many pharmaceuticals and food ingredients from plant sources exhibit a bitter taste. For improved compliance of orally administered pharmaceuticals and palatability of food, it is of great importance to mask undesirable tastes [1]. In food, lower amount of bitterness is well tolerated or necessary, but higher bitter ratings cause rejection by consumers. Especially bitter off-tastes produced by fortification (e.g. plant extracts, minerals etc.) are not tolerated. For decreasing bitterness in food applications mostly sodium salts, sugar and/or acids are used so far. But for many applications such solutions are only partially successful due to the high amounts needed.

In our efforts to find compounds which can suppress bitter taste without influencing other taste qualities, we have started from the known, very aromatic Herba Santa (*Eriodictyon californicum* (H. & A.) Torr.) liquid extract which was described as a bitter masking agent for quinine [2]. In earlier studies it was shown that the most important single constituents of the plant dry material are the flavanones homoeriodictyol, hesperetin and eriodictyol [3]. Therefore, 4′-hydroxyflavanones were also evaluated for their bitter masking potency (Figure 1).

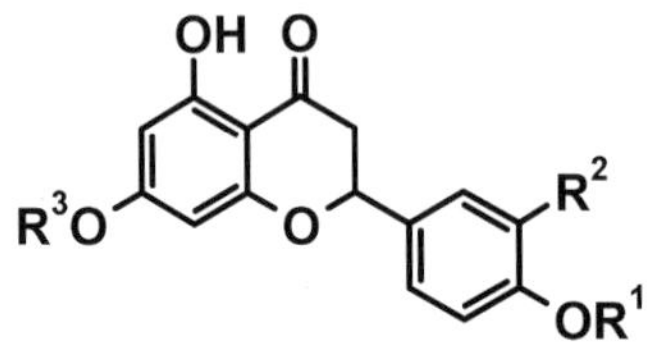

homoeriodictyol: $R^1 = R^3 = H$, $R^2 = OCH_3$
hesperetin: $R^2 = OH$, $R^3 = H$, $R^1 = CH_3$
eriodictyol: $R^1 = R^3 = H$, $R^2 = OH$
naringenin: $R^1 = R^2 = R^3 = H$

Figure 1. Structures of the investigated flavanones.

2. MATERIALS AND METHODS

2.1. Stimuli and sample preparation

Authentic samples of homoeriodictyol and eriodictyol were obtained from C. Roth (Karlsruhe, Germany), naringenin, hesperitin, caffeine, salicin and all other chemicals were from Sigma-Aldrich (Deisenhofen, Germany). Dried Herba Santa (*Eriodictyon californicum*) was from Alfred Galke GmbH (Gittelde, Germany). Fractionation was performed according to Figure 2. Homoeriodictyol sodium salt was isolated in pure form in a similar way as originally described by Geissmann [4]. For screening of bitter masking the test compounds were dissolved in a small amount of ethanol and added to an aqueous solution of the appropriate bitter compound. In the case of linoleic acid, a system containing 0.4 g free fatty acid, 0.02 g sucrose stearate and the test substance in 100 ml water was emulsified.

2.2. Sensory evaluation and data analysis

Panellists (healthy adults, no known tasting problems) were trained on caffeine as bitter standard. A minimum of 8 subjects was used in the descriptive test. The bitterness was rated against the standard dilutions of caffeine (50, 100, 150, 200, 250, 300, 350, 400, 450 and 500 ppm) or on a scale of 1 (no bitterness) to 10 (very strong bitterness). Mean rating of a 500 ppm caffeine solution ranged between 6 and 7. For other bitter tastants concentrations were used at which all persons could perceive a pronounced bitterness (rating about 7). For all experiments the test solutions were coded and in the case of colour or cloudiness they were covered. Panellists were advised to test the randomly presented samples in the given order by the sip and spit method. The raw sensory data were analysed by the Student's matched pair t-test to calculate of the level of significance of the masking effects.

3. RESULTS AND DISCUSSION

The non-volatile part of Herba Santa was fractionated and the various fractions evaluated for bitter-masking activity using a trained panel with caffeine as reference substance (Figure 2). In addition, the most important flavanones were tested for their bitter masking ability as pure compounds (cf. Figure 3, naringenin was added as structural analogue).

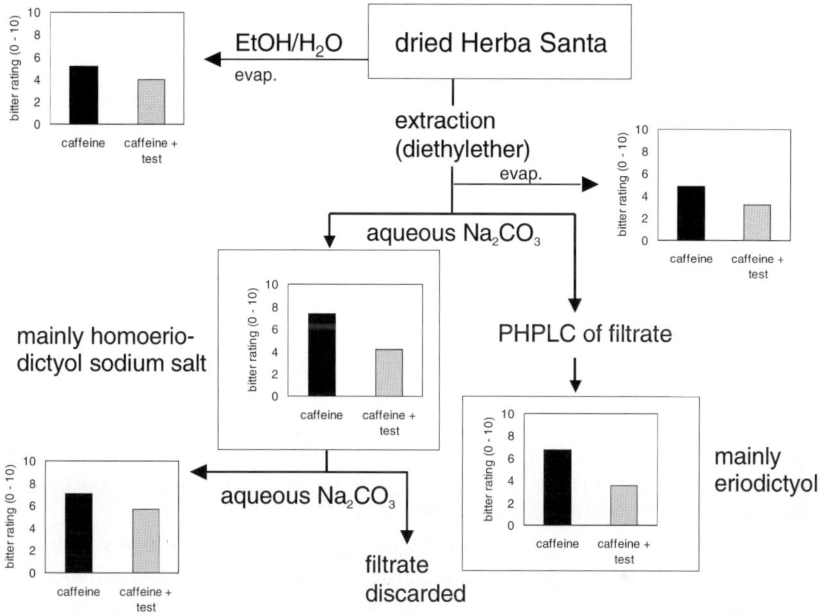

Figure 2. Fractionation of Herba Santa and evaluation of bitter masking properties of fractions.

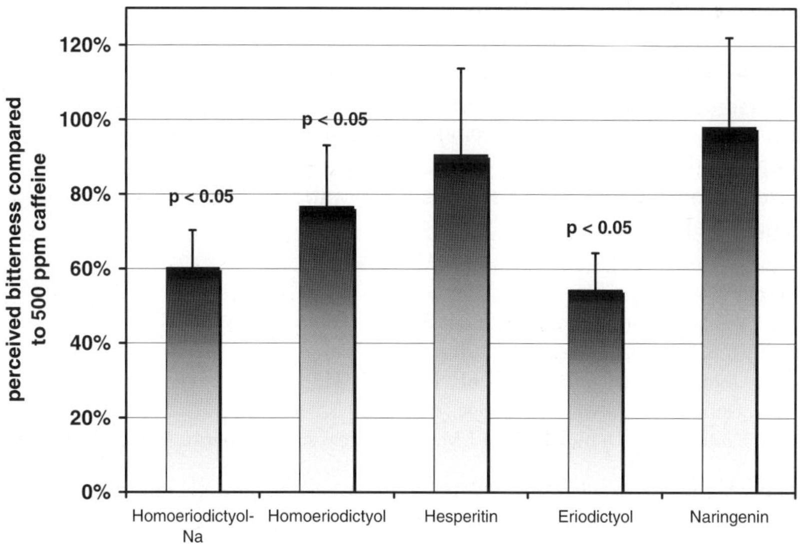

Figure 3. Masking effects of pure flavanones (100 ppm) against caffeine (500 ppm) in water.

Table 1. Activity of homoeriodictyol sodium salt (HEDNa) against various bitter tastants.

Bitter tastant	Conc. (ppm)	HEDNa (ppm)	Bitter rating without HEDNa	Bitter rating with HEDNa	Reduction (%)	Significance
Caffeine	500	100	7.4 ± 1.6	4.2 ± 1.6	43	< 0.005
Guaifenenesin	13000	200	7.0 ± 1.0	4.0 ± 1.0	43	n.d.
Paracetamol	2500	250	9.0 ± 1.0	4.0 ± 1.0	56	n.d.
Linoleic acid	4000	100	7.5 ± 1.0	8.0 ± 1.0	-7	n.d.
Quinine	5	100	4.8 ± 2.5	3.3 ± 1.5	31	< 0.005
Denatonium benzoate	0.050	100	9.2 ± 0.9	7.3 ± 2.0	21	< 0.005
Salicin	100	100	7.0 ± 1.9	4.9 ± 1.9	30	< 0.005
Amarogentin	0.030	100	9.3 ± 1.5	6.9 ± 2.3	26	< 0.05

In contrast to older studies published in the literature [5], the two leading flavanones eriodictyol and homoeriodictyol (HED) and its corresponding sodium salt could be identified as bitter masking compounds.

Other taste qualities were not influenced by the mostly tasteless homoeriodictyol. Eriodictyol exhibited nearly the same activity as the homoeriodictyol sodium salt. The nearest relative, hesperitin, showed a modest and naringenin only a very small activity against caffeine bitterness. Homoeriodictyol and its monosodium salt showed masking activities between 20 and 50% for several bitter compounds (caffeine, salicin, quinine, amarogentin, denatonium benzoate, guafenesine, and paracetamol) but not for the bitter tasting linoleic acid emulsion (cf. Table 1). Further studies to clarify the structure-activity relationships will be performed.

4. CONCLUSION

In conclusion, the flavanones homoeriodictyol, its sodium salt, and eriodictyol occurring in Herba Santa show remarkable bitter masking effects. Especially homoeriodictyol sodium salt seems to be a very interesting new taste modifier for food applications and pharmaceuticals.

References

1. D. Psczola, Food Technol., 58 (2004) 56.
2. L. Lewin, Berl. Klin. Wchnschr., 28 (1894) 644.
3. N.D. Johnson, Biochem. Syst. Ecol., 11 (1983) 211.
4. T.A. Geissmann, J. Am. Chem. Soc., 62 (1940) 3258.
5. J.B. Fantus, H.A. Dyniewicz and J.M. Dyniewicz, J. Am. Pharm. Assoc., 22 (1933) 323.

The consumption of damascenone during early wine maturation

Merran A. Daniel[a,b,c], Gordon M. Elsey[a,b,c], Michael V. Perkins[b,c] and Mark A. Sefton[a,c]

[a]The Australian Wine Research Institute, P.O. Box 197, Glen Osmond, S. A., 5064, Australia; [b]School of Chemistry, Physics and Earth Sciences, Flinders University, P.O. Box 2100, Adelaide, S. A., 5001, Australia; [c]Cooperative Research Centre for Viticulture, PO Box 154, Glen Osmond, South Australia 5064, Australia

ABSTRACT

Damascenone has been shown to undergo reaction with common wine components. Following the action of acid and heat alone, two bicyclic compounds were isolated. Two C_9 adducts, resulting from the addition of water and ethanol were also observed. Their formation was shown to be reversible. When treated with SO_2 (introduced as its aqueous equivalent, bisulfite) the concentration of damascenone rapidly decreased to zero. A C_9 sulfonic acid derivative was isolated and characterised, and its formation from damascenone shown to be irreversible in the presence of acetaldehyde.

1. INTRODUCTION

β-Damascenone (damascenone) (**1**, Figure 1) is an important flavour compound and is found ubiquitously in nature. Its flavour detection threshold has been measured at 50 ng/l in model wine [1], and 2 ng/l in water [2]. Damascenone is characterised by an intense 'stewed apple' odour, and has been found to contribute positively to the fruity aromas of red wine varieties.

Damascenone is formed in wine in relatively high concentrations during fermentation, but its concentration decreases by up to 75% during the first few months of maturation [3]. Previous investigations into the fate of damascenone concerned the role of acid-catalysed degradation, and led to the isolation of two species, termed bicyclodamascenones **2** and **3**. However, these experiments were conducted under conditions normally not utilised in wine production. Thus it was unclear whether they would form in wine itself. Damascenone which possesses unsaturated carbonyl functionalities could also react with nucleophilic species already present in wine.

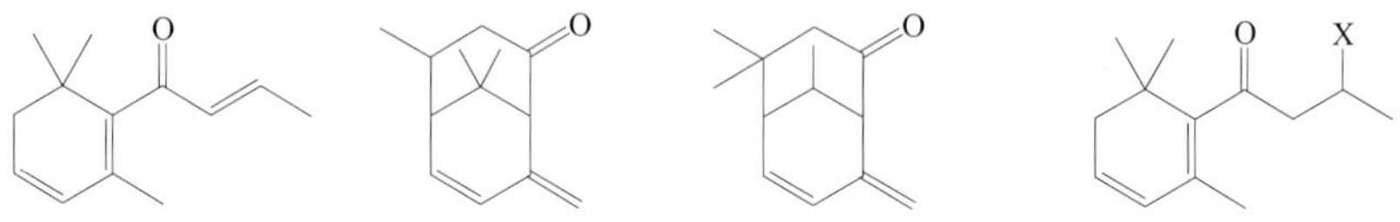

1 **2** **3** **4**, X=OH
5, X=OEt
6, X=SO_3Na

Figure 1. Damascenone derivatives prepared during this study.

2. MATERIALS AND METHODS

2.1. Chemicals

9-Hydroxydamascenone (**4**) and 9-ethoxydamascenone (**5**) were prepared as described by Daniel *et al.* [4]. Damascenone was quantified as reported elsewhere [4]. The sulfonic acid derivative (**6**) was prepared from a mixture of damascenone (0.5 mmol) and Na2S2O5 (150 mg) in dimethyl formamide (DMF, 5 ml) was heated at 45 °C for 3 days. The DMF was removed, the product dissolved in H2O, washed with EtOAc and extracted with acetonitrile/EtOAc (5:1) to yield, after drying and concentration, a yellow solid (51 mg, 35%). Spectral data: ^{1}H NMR (300 MHz, D_2O): 6.03-5.87 (2H, m, $H_{3,4}$); 3.47 (1H, ddq *J* 9.3, 3.0 and 6.7, H_9); 3.35 (1H, dd *J* 19.1 and 3.0, H_{8a}); 2.94 (1H, dd *J* 19.1 and 9.3, H_{8b}); 2.14 (2H, dd *J* 4.1 and 1.5, H_2); 1.72 (3H, s, H_{13}); 1.37 (3H, d *J* 6.7, H_{10}); 1.09 (6H, 2xs, $H_{11,12}$). ^{13}C NMR (75 MHz, D_2O): 212.2 (C_7), 140.8 (C_6), 129.3 (C_3), 129.0 (C_5), 127.7 (C_4), 50.7 (C_9), 47.3 (C_8), 39.3 (C_2), 33.3 (C_1), 25.6 (C_{11}), 25.6 (C_{12}), 18.4 (C_{13}), 15.2 (C_{10}). LC-MS (negative ion mode) (*m/z*): 271 $(M\text{-}Na^+)^-$

2.2. Heating of damascenone

Acidic conditions: Solutions of damascenone (at an initial concentration of 100 mg/l) in model wine (pH 2.8) were sealed in glass ampoules and heated at room temperature, 45 °C or 100 °C. After prescribed times, ampoules were opened and the damascenone concentration determined.

With added SO_2: Damascenone (1 mg/l) was added to buffered aqueous ethanol at either pH 3.2 or pH 3.4. Sodium metabisulfite was added to each to give sulfur dioxide concentrations of either 80 ppm (120 mg/l $Na_2S_2O_5$) or 200 ppm (300 mg/l). Aliquots of each were sealed in glass ampoules and stored at either room temperature or 45 °C. After storage the ampoules were opened, and the damascenone quantified.

2.3. Formation of the sulfonic acid derivative (6) under wine conditions

Damascenone (0.5 mmol) and sodium metabisulfite (100 mg, 0.5 mmol, equivalent to 1.0 mmol SO_2) were added to model wine (500 ml, pH 3.2) and stirred at room temperature for four weeks. The mixture was extracted first with ethyl acetate to remove unreacted damascenone, and then the products were extracted with acetonitrile and ethyl acetate (5:2), and the solvent removed.

2.4. Reversibility experiments

Solutions of either **5** in model wine (pH 3.2, room temperature) or **6** in model wine (pH 3.2, 25 °C or 45 °C, with and without added acetaldehyde (5.0 equivalents)) were sealed in glass ampoules. After prescribed times, the ampoules were opened and the concentration of damascenone measured.

3. RESULTS

3.1. Degradation of damascenone

When heated in model wine (pH 2.8) the decrease in damascenone was observed to be highly temperature dependent (40% remaining after 14 days at 100 °C, 73% remaining after 30 days at 45 °C, 88% remaining after 61 days at 25 °C). The products of acidic consumption were isolated and identified by spectroscopic methods and/or confirmed by synthesis. Two major products were assigned as the isomeric bicyclodamascenones **2** and **3**. Another two minor products proved to be the C_9 adducts of damascenone, **4** and **5**, which arise from addition of water and ethanol, respectively, across the C_8-C_9 alkene. When treated with sulfur dioxide (introduced as the aqueous equivalent, bisulfite) the levels of damascenone decreased rapidly to zero (limit of detection ~ 0.5% of initial concentration). The amounts of damascenone remaining after 14 days at 45 °C, and after 28 days at 25 °C are shown in Table 1. The consumption was essentially independent of pH, with the concentration of added SO_2 being the limiting factor. The major degradation product was identified as the sulfonic acid damascenone derivative **6**.

Table 1. Residual damascenone (μg/l) of an initial level of 1000 μg/l after treatment with SO_2.

	25 °C (28 days)		45 °C (14 days)	
	SO_2, 80 mg/l	SO_2, 200 mg/l	SO_2, 80 mg/l	SO_2, 200 mg/l
pH 3.2	502.2	168.0	89.2	0
pH 3.4	468.9	139.6	98.3	0

3.2. Reversibility of additions

The reversibility of the additions to the side chain enone functionality was investigated. A solution of the ether adduct **5** in a model wine (pH 3.2) was sealed in ampoules and stored at room temperature. After three months, analysis revealed that the drop in concentration of **5** (6%) was mirrored by an almost exactly equal increase in damascenone concentration. It was noted also that the sum of the concentrations of **5** and **1** remained constant over the duration of the experiment. Under accelerated conditions (pH 2.8, 100 °C) the level of **5** had fallen by 70% after 48 h, again matched closely by the concomitant increase in damascenone concentration. These data clearly demonstrate the reversible nature of the addition of ethanol.

Similarly, solutions of **6** in model wine (5 mg/l, pH 3.2) were sealed and stored at both room temperature and 45 °C. After four months at both temperatures, no more than trace amounts of damascenone (< 1 μg/l) were measurable. When the experiment was conducted in the presence of 5.0 equivalents of acetaldehyde, the levels of measured damascenone were essentially unchanged, confirming the irreversible nature of this addition. Analysis of the aqueous component (after removal of the solvent under vacuum) by LC-MS revealed that **6** was still present.

The losses observed due to acidic consumption of damascenone, particularly at room temperature, are insufficient to account for the reported decrease in damascenone concentration over the early months of wine maturation [3]. The addition of ethanol across the alkene was shown to be reversible, and favour formation of damascenone. Therefore **5** (and by analogy, **4**) are likely to be minor components.

The consumption of damascenone by SO_2, on the other hand was very rapid. Sulfur dioxide is known to react with carbonyl compounds to produce bisulfite adducts. Accordingly, sodium metabisulfite was reacted with damascenone in DMF and the product extracted with acetonitrile/ethyl acetate. Spectral analysis revealed that the product isolated was in fact the sulfonic acid derivative **6**, where the sulfonate functionality had been formed by 1,4-addition to the α,β-unsaturated enone, rather than 1,2-addition across the ketone. The regiochemistry of addition was established from the NMR spectra. In contrast to ether adduct **5**, the product **6** was completely stable, even in the presence of acetaldehyde, a known scavenger of SO_2. Thus, damascenone lost by reaction with SO_2 in young wines is unlikely to be regenerated by the presence of acetaldehyde formed during wine aging.

4. CONCLUSION

Decreases in damascenone content during wine maturation can be attributed to several factors, but it appears to be that it is the interaction with sulfur dioxide that is the most salient process accounting for such losses. These results also explain recent observations [5] where the loss of damascenone in white wine over time is reduced when stored under an air headspace (with resultant consumption of SO_2) rather than anaerobic.

References

1. H. Guth, J. Agric. Food Chem., 45 (1997) 3027.
2. R.G. Buttery, R. Teranishi and L. Ling, Chem. Ind., (1988) 238.
3. A.L. Waterhouse and S.E. Ebeler (eds.), The chemistry of wine flavour (ACS symposium series 714), Washington, DC, USA (1999) 39.
4. M.A. Daniel, G.M. Elsey, D.L. Capone, M.V. Perkins and M.A. Sefton, J. Agric. Food Chem., 52 (2004) 8127.
5. R.J. Blair and I.S. Pretorius (eds.), Proceedings of the 12th Australian wine industry technical conference, Adelaide, Australia (2005) 215.

W.L.P. Bredie and M.A. Petersen (Editors)
Flavour Science: Recent Advances and Trends

Sensory and structural characterisation of an umami enhancing compound in green tea (mat-cha)

Shu Kaneko[a], Kenji Kumazawa[a], Hideki Masuda[a], Andrea Henze[b] and Thomas Hofmann[b]

[a]Material R & D Laboratory, Ogawa & Co. Ltd., Chidori 15-7, 279-0032, Urayasu, Chiba, Japan; [b]Institute for Food Chemistry, University of Muenster, Corrensstrasse 45, D-48149, Muenster, Germany

ABSTRACT

Aimed at defining the key components for the umami taste of mat-cha a series of analysis were carried out on freshly prepared samples. The analyses included a bioactivity-guided fractionation using solvent extraction, solvent precipitation, preparative chromatographic separations and human psychophysical experiments. The LC-MS/MS and 1D/2D-NMR studies on isolated fractions led to the identification of (1R,2R,3R,5S)-5-carboxy-2,3,5-trihydroxy-cyclohexyl 3,4,5-trihydroxybenzoate known as theogallin, an umami enhancing compound in mat-cha. Sensory analysis showed that this compound can increase the umami intensity of sodium L-glutamate proportionally.

1. INTRODUCTION

The mat-cha, literally ‘powdered tea’, is a special type of green tea; a precious, jewel-green powder made from hand picked, high grade Japanese tea. It is whisked with hot water in a special bowl to make a creamy, frothy, healthy beverage exhibiting a typical umami-like taste besides fresh vegetable-like and green odour notes. Although various investigations have been focused on amino acids and nucleotides [1,2], the key drivers for the umami taste are still not fully defined on a molecular level.
Aimed at discovering the compounds contributing to the umami taste of the mat-cha, preparative separation techniques and instrumental analysis were combined with human psychophysical tools [3], and the gustatory activities of the compounds identified were verified by means of a bio-mimetic taste reconstitution experiment.

2. MATERIALS AND METHODS

2.1. Materials

High-grade mat-cha leaves were purchased from a Japanese tea company. All chemicals were purchased from commercial sources.

2.2. Sensory analysis

The taste intensities were rated by trained sensory panels with a five point scoring method. Bottled water was used as a solvent. The pH value of the solution was adjusted to 6.0 by adding aqueous formic acid solution (0.1 mM). The umami enhancing effect of the tea fractions and compounds, respectively, was evaluated by comparing the umami intensity of a 3 mM aqueous sodium L-glutamate (MSG) solution and a binary aqueous mixture of 3 mM MSG and the fraction under investigation.

2.3. Separation methods of the components in mat-cha

2.3.1. Water extraction of the mat-cha leaves and liquid-liquid extraction

The mat-cha leaves were extracted with 10 times the amount of distilled water at 80 °C for 4 min, and the extract was cooled rapidly to 20 °C in an ice-bath. The suspension was centrifuged (3000 rpm x 10 min), filtered, and freeze-dried *in vacuo*. The mat-cha extract was dissolved in 10 times the amount of water and was then sequentially extracted with dichloromethane (2 x 500 ml) and ethyl acetate (2 x 500 ml). The umami taste reminiscent aqueous layer was freeze-dried again.

2.3.2. Solvent precipitation, adsorption to the resin, and preparative HPLC separation

The umami taste reminiscent aqueous fraction was suspended in 50% aqueous methanol. The precipitate formed was removed by filtration and the filtrate was freeze-dried and subsequently dissolved in 10 times the amount of an aqueous solution of formic acid (pH 4.0). After application of the material onto the top of an ODS resin (C-18, Phenomenex), the column was flushed with water (fraction I) and then with methanol (fraction II). The non-adsorbed aqueous fraction I was concentrated *in vacuo* and then separated by RP-HPLC.

3. RESULTS

3.1. Bioactivity-guided screening for umami enhancing compounds in mat-cha

To locate the key drivers of the umami taste for mat-cha, the taste-active aqueous fraction containing the hydrophilic compounds was isolated from the mat-cha by means of solvent extraction, solvent precipitation, followed by chromatography on an ODS resin. This aqueous fraction was separated into 22 sub-fractions by means of preparative RP-HPLC (Figure 1). After freeze-drying and addition of a 3 mM MSG solution, the materials were evaluated for their umami enhancing activity.

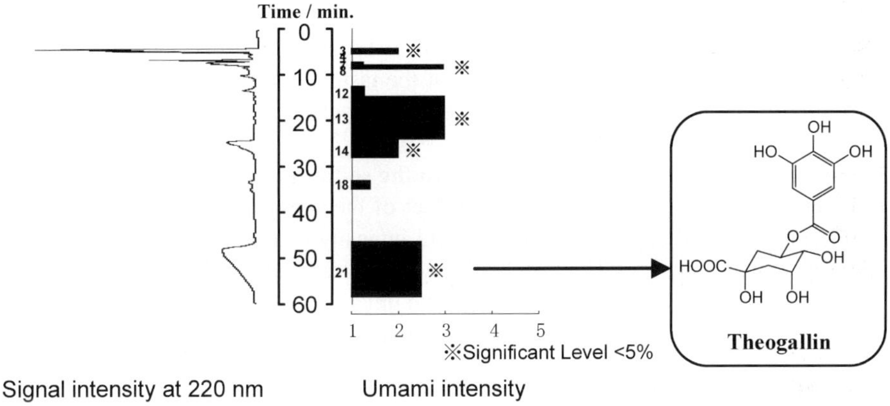

Figure 1. HPLC chromatogram (left side) and umami intensity of a 3 mM MSG solution spiked with the individual fractions collected (right side).

Addition of the tasteless or weakly astringent fractions no. 8, 13, 14, and 21 to the 3 mM MSG solution resulted in an increased umami taste. After isolation and purification, the umami-enhancing compounds in fraction 21 could be successfully identified as (1R,2R,3R,5S)-5-carboxy-2,3,5-trihydroxycyclohexyl 3,4,5-trihydroxybenzoate also known as theogallin. The data from LC-MS/MS, 1D/2D-NMR experiments and comparison of the spectroscopic data with those reported earlier [4] confirmed the identity of the compound.

3.2. Influence of theogallin on the umami taste of mat-cha

To verify the contribution of theogallin to the umami taste of the mat-cha extract, theogallin (0.14 mM) was added to a biomimetic taste reconstitute containing L-glutamic acid (1.5 mM), L-theanine (3.7 mM), L-aspartic acid (1.3 mM), succinic acid (0.11 mM), AMP (0.32 μM), GMP (0.60 μM), Na^+ (0.012 mM), K^+ (19 mM), Mg^{2+} (0.62 mM), and Cl^- (0.025 mM) in 'natural' concentrations.

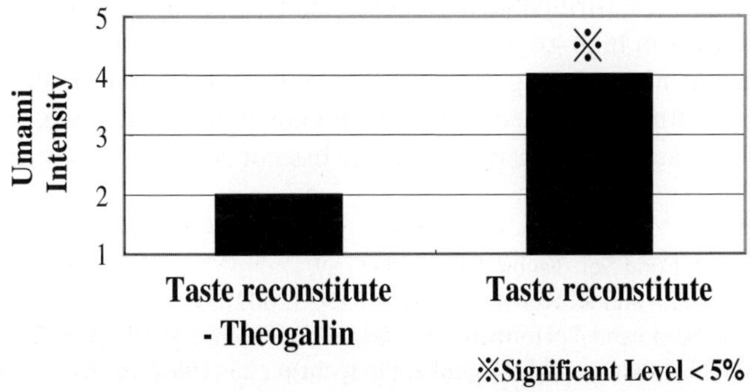

Figure 2. Influence of theogallin (0.14 mM) on the umami intensity of a biomimetic mat-cha reconstitute.

The biomimetic taste reconstitute was compared to the reconstitute lacking the theogallin by means of a duo test. As shown in Figure 2, the presence of theogallin significantly increased the umami intensity of the taste reconstitute, thus demonstrating its key importance for the umami-like taste of the mat-cha.

3.3. Umami enhancing effect of theogallin to the sodium L-glutamate

To further confirm the umami enhancing effect of theogallin, the relationship between the concentration of theogallin and the umami intensity of a 3 mM MSG solution was verified by sensory analysis (Figure 3). Theogallin itself exhibited only astringency at the threshold concentration of 0.08 mM, but the umami intensity of MSG was increased proportionally even below its astringency threshold.

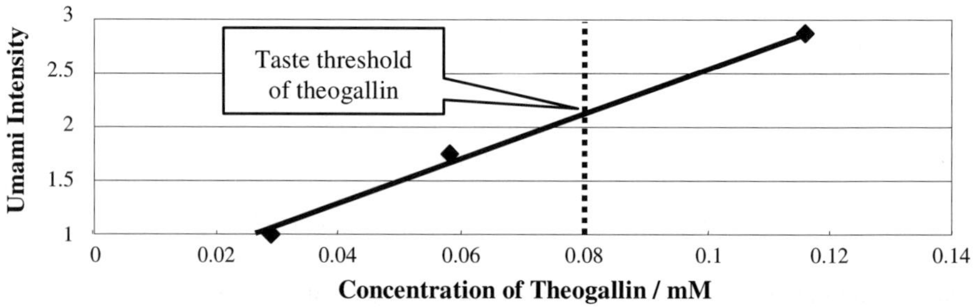

Figure 3. Influence of theogallin on the umami intensity of a 3 mM MSG solution. The umami intensity of the 3mM MSG solution was set to 1 as default.

4. DISCUSSION AND CONCLUSION

Application of a bioresponse-guided fractionation of the mat-cha revealed that the mat-cha extract contained not only umami-like taste compounds but also umami enhancing compounds. This is the first report demonstrating (1R,2R,3R,5S)-5-carboxy-2,3,5-trihydroxycyclohexyl 3,4,5-trihydroxybenzoate, known as theogallin, as one of the umami taste enhancers in high-grade mat-cha. Theogallin enhanced the umami intensity of MSG but the umami enhancing effect did not relate to the threshold of theogallin itself. Although the structure of theogallin is well known in the literature, to the best of our knowledge its umami taste enhancing activity has not been previously reported.

References

1. M. Nakagawa, J. Food Sci. Technol., 22 (1975) 59.
2. H. Horie, T. Ujihara and K. Kohata, Tea Res. J., 93 (2002) 55.
3. H. Ottinger, T. Soldo and T. Hofmann, J. Agric. Food Chem., 51 (2003) 1035.
4. H. Nishimura, G. Nonaka and I. Nishioka, Phytochem., 23 (1984) 2621.

W.L.P. Bredie and M.A. Petersen (Editors)
Flavour Science: Recent Advances and Trends

Optimisation and validation of a taste dilution analysis to characterise wine taste

Ricardo López, Laura Mateo-Vivaracho, Juan Cacho and Vicente Ferreira

Department of Analytical Chemistry, Faculty of Sciences, Universidad de Zaragoza, 50009 Zaragoza, Spain

ABSTRACT

Taste dilution analysis (TDA) has been applied to the characterisation of tastants in wine. The wines were concentrated under vacuum at 20 °C and with a stream of N_2 to obtain a completely dry extract without aroma compounds. The dry extracts were re-dissolved in water and injected in a semi-preparative C18 HPLC column. A thorough study to determine reproducibility and maximum load in the HPLC system was carried out. The effluent was separated in fractions, which were collected and concentrated to dryness under the same conditions that wine extracts, and then, again dissolved in bottled water. Different dilutions of the fractions were tasted by a sensory panel to assess the intensity of basic tastes and sensations in the mouth. Fractions were also submitted to HPLC-MS analysis to determine the identity of the tastants.

1. INTRODUCTION

A fundamental part of wine flavour quality is related with taste and mouth feel sensations, but when compared to aroma compounds, relative little attention has been paid to the non-volatile compounds in wine. TDA [1] has proved to be a powerful tool to characterise important tastants in foods and beverages. In the present work this technique has been applied to characterisation of taste components in wine.

2. MATERIALS AND METHODS

A set of five wines with marked technological and taste differences were selected. The wines included a young mono-varietal Chardonnay wine; a mono-varietal Chardonnay wine fermented in barrels; a young mono-varietal red wine *cv. Tempranillo* with marked astringency; a 10 years old red wine (1995 vintage) with light body and little astringency and a 5 years old red wine (2000 vintage) with full body and well balanced structure.

Wine aliquots (50 ml) were concentrated in a 20 °C water bath under vacuum to obtain a dry extract, free from aroma compounds. The concentrate was then re-dissolved in 1 ml of the aqueous mobile phase of an HPLC.

An HPLC-DAD system from Varian was used. One ml of re-dissolved extract was injected in a semi-preparative 250 x 10 mm i.d., 5 μm C18 HPLC column. The elution was carried out with a gradient program. A mixture of two solvents was used: 2% formic acid in water and acetonitrile UV-visible spectra were recorded from 250 to 600 nm. The effluent was separated in 9 fractions that were concentrated under the same conditions as the wine samples. Each dry extract was re-dissolved in 2 ml of bottled water. Relevant fractions were also analysed by HPLC-ESI(-)-MS using a 150x2.1 mm i.d., 3.5 μm XTerra C18 HPLC column.

The sensory analysis of the fractions was carried out in standard tasting booths. Black tulip-shaped standardised cups for wine tasting were used. Assessors were trained by tasting aqueous solutions of standard tastants.

Wine's astringency was perceived by tasters as different from astringency produced by aluminium sulphate and tannic acid therefore, tasters were trained by using different dilutions of a very astringent wine; the ethanol was removed from wine to avoid its influence in astringency perception and to achieve the same conditions in the training and in the tasting of the fractions.

The sensory evaluation of the fractions was carried out by application of TDA [1]. The sensory panel was composed by four trained tasters who assessed the presence or absence of basic tastes and oral sensations. The categories assessed were: sweetness, saltiness, sourness, bitterness and astringency.

Wine TDA analysis was performed sequentially. Wine fractions were tasted randomly by the panel, but once a fraction TDA was started it was conducted until completion to prevent oxidation or polymerisation in the course of the experiment. Dilutions were presented in order of descending concentrations. If the effect was significant, another dilution was tasted, following the dilution process until the tasters perceived no effect in any category. For each category the dilution at which a category could just be detected, was defined as Taste Dilution (TD) factor.

3. RESULTS AND DISCUSSION

The nine fractions obtained by HPLC from each wine were assessed by a sensory panel. The panel reproducibility was assessed in two ways. The values of TD achieved by the panel in the bitterness and acidity categories did not differ from more than one dilution step, however in the astringency category the TD values were the same in the 3 replicates.

Representation of TD factors against the fraction number, showed important fraction differences both in quality and intensity in astringency, sourness and bitterness categories. Sweetness and saltiness did not produce significant results and are not discussed.

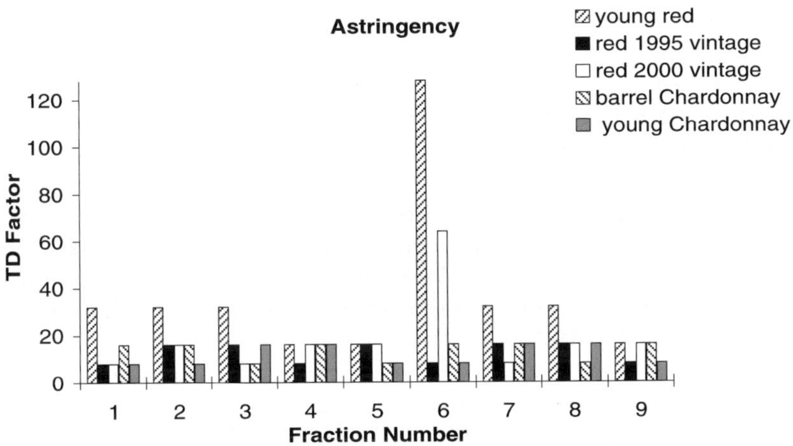

Figure 1. Taste dilution chromatograms for astringency tasting fractions of the wines.

Astringency was mainly located in fraction 6 (Figure 1). Young red wine scored the maximum TD factor in this category with 128, followed by red wine from 2000 vintage with a TD factor of 64. The 10 years old red wine and Chardonnays scored similar lower TD factors. Intensity of astringency evaluated in fraction 6 fit well with overall astringency in original wines, as young red wine was by far the most astringent and both Chardonnays and 10 years old red wine had barely a noticeable astringency. HPLC-MS analysis of fraction 6 revealed the presence of proanthocyanidins oligomers up to four unities. Tetramers were only found in young red wine, while in 5 years old wine trimers were the largest oligomers found. In Chardonnays and in 10 years old red wine only catechin and epicatechin monomers were found. Ellagitannins were absent in fraction 6.

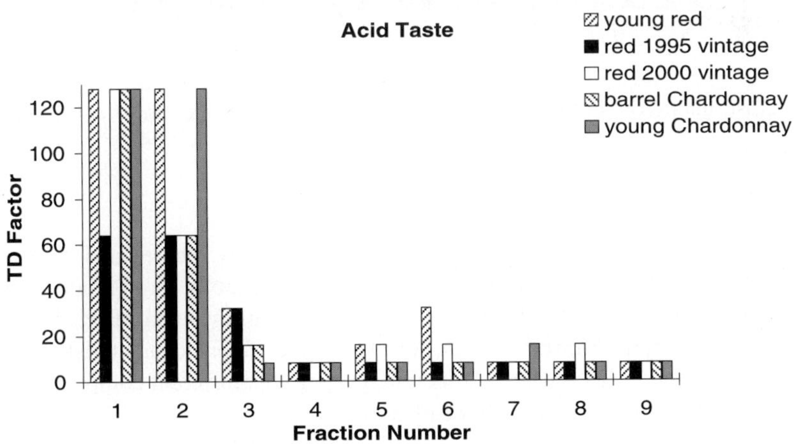

Figure 2. Taste dilution chromatograms for acid tasting fractions of the wines.

As can be noticed in Figure 2, acid taste appears at the beginning of the chromatogram, in fractions 1 and 2. All wines analysed, apart from 10 years old wine, scored TD factors of 128 in at least one of these fractions. No significant difference was found between white and red wines fractions in terms of TD factor suggesting that greater acidity perceived tasting white wines should be related with overall flavour perception rather than with greater organic acids content. As could be expected organic acids of wine were found in fractions 1 and 2. Presence of tartaric, citric, malic, lactic, succinic and oxalic acids was confirmed in all fractions 1 and 2 by HPLC-MS analysis.
In the case of the bitter taste (Figure 3) young Chardonnay and red wine from 2000 vintage scored highest TD factors in fractions 1 and 2, respectively. On the other hand, young red wine scored its highest TD factor for bitterness in fraction 6. This indicates that bitterness in wine is elicited by different groups of compounds. In the first 2 fractions bitterness might be caused by acids or ions. In fraction 6 proanthocyanidins monomers or, less likely, by hydroxycinnamic acids [2] might elute.

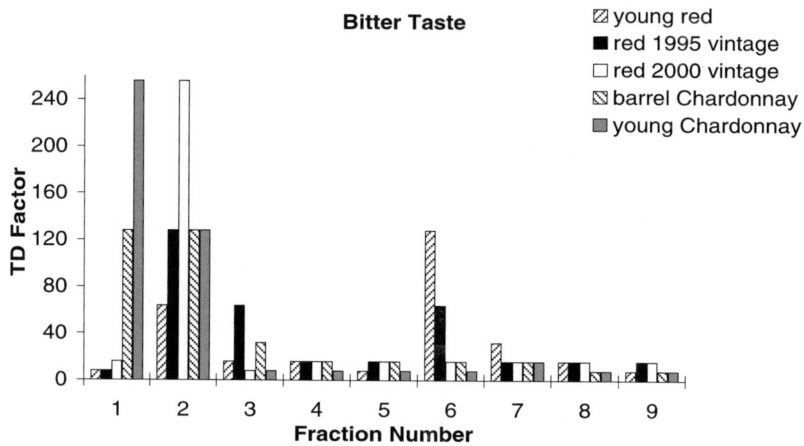

Figure 3. Taste dilution chromatograms for bitter tasting fractions of the wines.

4. CONCLUSION

Astringency, bitterness and sourness have been successfully located and evaluated by applying TDA in wine. Astringency has been found mainly in fraction 6 and its intensity is related with overall wine astringency. Regarding acid taste, no dramatic differences were found between wines despite their different characteristics. Finally, bitter taste appeared in two different zones suggesting that different compounds elicited bitterness in wine depending on the type of wine.

References

1. O. Frank, H. Ottinger and T. Hofmann, J. Agric. Food Chem., 49 (1) (2001) 231.
2. R. Gawel, Aust. J. Grape Wine R., 4 (1998) 74.

W.L.P. Bredie and M.A. Petersen (Editors)
Flavour Science: Recent Advances and Trends

Key aroma compounds in apple juice - changes during juice concentration

Martin Steinhaus[a], Johanna Bogen[a] and Peter Schieberle[a,b]

[a]German Research Centre for Food Chemistry, Lichtenbergstr. 4, D-85748 Garching, Germany; [b]Technical University of Munich, Department of Chemistry, Chair for Food Chemistry, Lichtenbergstr. 4, D-85748 Garching, Germany

ABSTRACT

Potent apple juice odorants, including 1-butanol, β-damascenone, dimethyl sulfide, ethyl butanoate, ethyl 2-methylbutanoate, hexanal, 1-hexanol, (*E*)-2-hexenal and 1-octen-3-one were quantified using stable isotope dilution assays in juice and in the aroma fractions thereof obtained by an industrial juice concentration process. Results showed a distinct loss of the major juice aroma component β-damascenone during juice concentration. High losses were also found for dimethyl sulfide and 1-octen-3-one.

1. INTRODUCTION

The consumption of fruit juices in the EU increases continuously. In 2004, the average consumption amounted to 25 l per capita. Although in most European countries orange juice is number one among the fruit juices consumed, in Germany apple juice (13 l) is clearly preferred to orange juice (9 l) [1]. By removing water, a major part of apple juice is processed to juice concentrate and aroma, which can be stored and shipped more cost-saving. On demand, the apple juice is then reconstituted from concentrate and aroma fraction by addition of water.

The decisive factor for the sensory quality of apple juice is its typical aroma, caused by a limited number of aroma-active substances, previously identified following the odour activity value concept [2]. The aim of this study was to evaluate the fate of these key odorants in apple juice prepared from a single variety (Golden Delicious) during a commercial juice concentration process. For this purpose, selected key odorants were quantified by stable isotope dilution assays [3].

2. MATERIALS AND METHODS

Apples (*Malus domestica Borkh.* cv. Golden Delicious) grown in South Tyrol (Italy) were processed in a large scale industrial juice production plant following the scheme given in Figure 1. Raw apple juice and aroma fractions 1 and 2 were collected from the same batch.

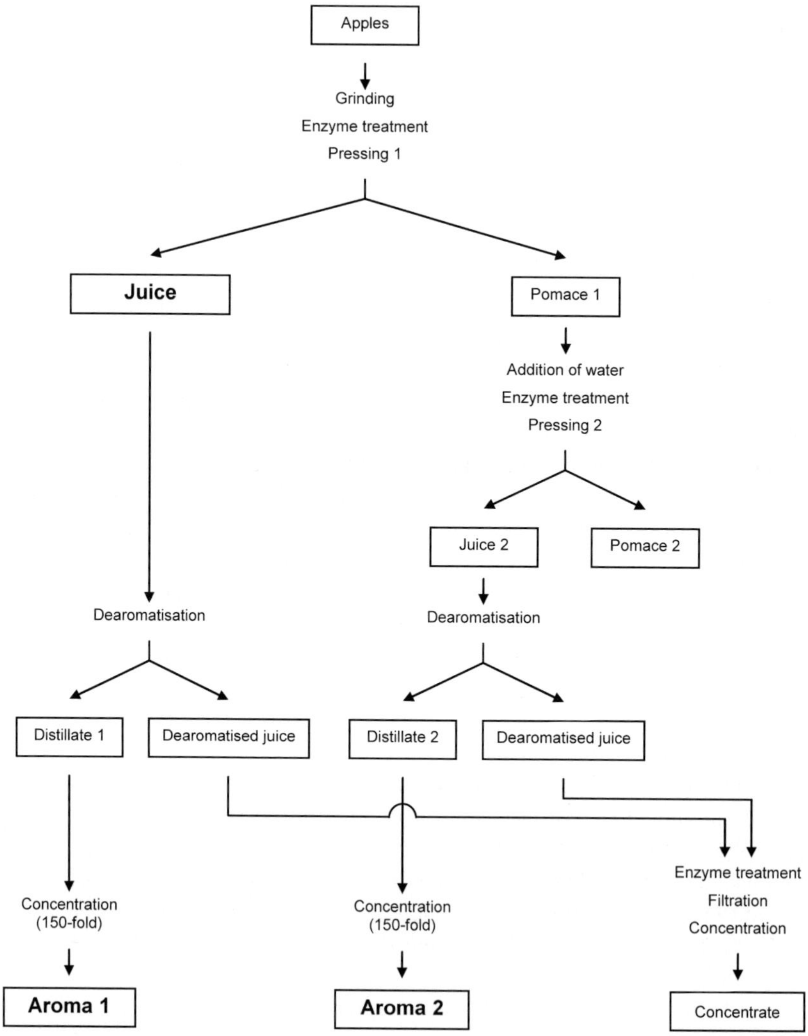

Figure 1. Scheme of the industrial apple juice processing.

For quantification known amounts of deuterium labelled analogues of the target compounds were added and the mixture was vigorously stirred, followed by extraction with dichloromethane. The extracts were concentrated by SAFE distillation [4] and fractionated by column chromatography using silica gel and pentane/diethylether mixtures (100+0; 90+10; 70+30; 0+100) as eluents. Fractions were concentrated using Vigreux columns and analysed by GC-GC/MS. For that purpose, aliquots of the concentrates were applied by cold-on-column injection onto the column in first dimension (FFAP) and heart cuts taken for each analyte, were further separated on a DB-1701 column connected to a mass spectrometer using the system described in [5]. From the area counts obtained from the mass chromatograms individual concentrations were calculated as detailed earlier [5].

3. RESULTS

3.1. Raw juice versus aroma 1

Concentrations measured for odorants in aroma 1 were divided by the concentration factor (150) to be directly comparable to the amounts present in the unprocessed juice on the basis of odour activity values (OAV = concentration divided by odour threshold; Figure 2). Results showed high losses of the baked apple-like smelling β-damascenone, exhibiting the highest OAV in the unprocessed juice, during concentration, as well as for 1-octen-3-one (mushroom-like) and dimethyl sulfide (asparagus-like). Distinct losses were also found for hexanal (grassy), ethyl 2-methylbutanoate (fruity) and 1-butanol (malty), whereas (*E*)-2-hexenal (apple-like) slightly increased.

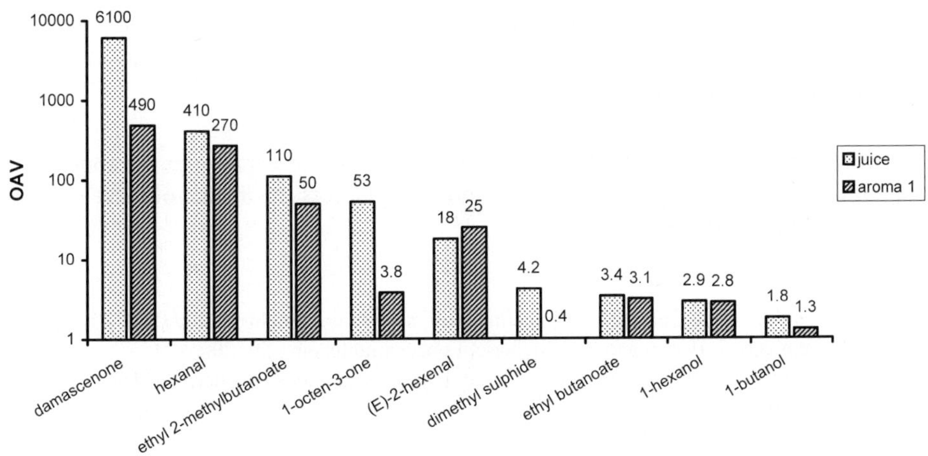

Figure 2. Odour activity values of selected apple juice odorants in the unprocessed juice and aroma 1.

3.2. Aroma 1 versus aroma 2

During the common industrial procedure of apple juice manufacturing (see Figure 1) a second aroma fraction (aroma 2) is obtained from the primary pomace after an additional enzymatic incubation (endogenous and added enzymes). This aroma fraction is estimated lower in the aroma quality as compared to aroma 1. Quantitative data (Figure 3) accordingly showed distinct differences in major odorants. Especially the important apple juice odorants hexanal and (*E*)-2-hexenal were present in smaller amounts in aroma 2.

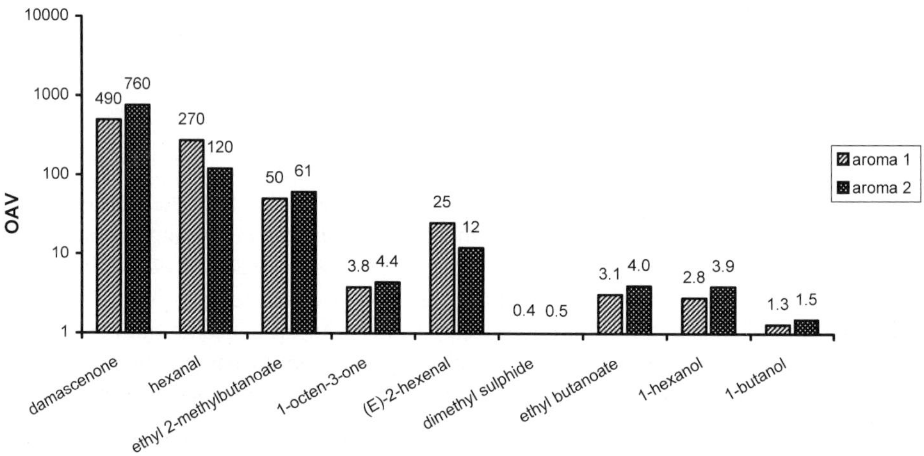

Figure 3. Odour activity values of selected apple juice odorants in aroma 1 and aroma 2.

4. CONCLUSIONS

The data showed that industrial juice concentration leads to changes in the contents of important apple juice odorants. Further efforts will address the impact of these changes on the sensory quality of apple juice.

References

1. Association of the German fruit juice industry, http://www.fruchtsaft.net/ (Nov 2005).
2. M. Steinhaus, J. Bogen and P. Schieberle, Lebensmittelchem., 59 (2005) 91.
3. A. Gaonkar (ed.), Characterization of food: emerging methods, Amsterdam, The Netherlands (1995) 403.
4. W. Engel, W. Bahr and P. Schieberle, Eur. Food Res. Technol., 209 (1999) 237.
5. M. Steinhaus, H. Fritsch and P. Schieberle, J. Agric. Food Chem., 51 (2003) 7100.

W.L.P. Bredie and M.A. Petersen (Editors)
Flavour Science: Recent Advances and Trends

Characterisation of the odour volatiles in *Citrus aurantifolia* Persa lime oil from Vietnam

Nguyen Thi Lan Phi[a], Nguyen Thi Minh Tu[b], Chieko Nishiyama[a] and Masayoshi Sawamura[a]

[a]Department of Bioresources Science, Faculty of Agriculture, Kochi University, B-200 Monobe, Nankoku, Kochi 783-8502, Japan; [b]Hanoi University of Technology, Institute of Biological and Food Technology, 1 Dai Co Viet, Hanoi, Vietnam

ABSTRACT

Essential oil of *C. aurantifolia* Persa (lime) from Vietnam was isolated by a cold-pressing method with a yield of 0.02% of the fresh fruit. The odour-active constituents were investigated by capillary GC, GC-MS, GCO and AEDA. Ninety-six compounds were detected and 92 compounds were identified in the lime oil. Limonene (73.5%) was the major component, followed by geranial (8.4%), neral (4.9%), myrcene (2.1%) and β-bisabolene (1.6%). Eight compounds, limonene, geranial, neral, myrcene, geranyl acetate, α-pinene, α-terpineol and β-bisabolene were evaluated as the odour-active compounds, showing high FD values (FD-factor > 6). From the GC-sniffing results, neryl acetate, β-bisabolene, l-carvone, geranyl acetate, α- and β-citronellol, cumin aldehyde, perillaldehyde, nerol, tridecanal, germacrene B, geraniol, dodecyl acetate, caryophyllene oxide and perillyl alcohol were suggested as contributors to the characteristic aroma of this Vietnamese lime.

1. INTRODUCTION

Lime is one of the principal citrus fruits in the world. The two species of lime that have been studied mostly are *Citrus aurantifolia* Swingle (Mexican, Key or West Indian lime) and *Citrus latifolia* Tanaka (Tahiti, Persian or Bears lime). The composition of lime oil has been reported by many researchers [1,4-9]. *C. aurantifolia* Persa (lime) is one kind of sour citrus fruits, very important and popular in Vietnam, as well as Yuzu (*C. junos* Sieb. ex Tanaka) in Japan or lemon in other countries in the world. It has been used as a folk medicine in curing sore throat, cough or hair care [2]. As a result of different geographical origins, climate and ancient habits and practices in planting, lime species has become very abundant. However, there have been very few studies reporting on the composition and no studies have been undertaken to characterise the aroma volatiles of Vietnamese lime. The purpose of this study was to elucidate the characteristic components of the essential oil from Vietnam lime, and to properly evaluate the quality, origin or genuineness.

2. MATERIALS AND METHODS

2.1. Sample preparation

Lime (*Citrus aurantifolia* Persa) fruits were obtained in September 2004 from Institute of Fruit and Vegetable Research, Hanoi, Vietnam. The peel oil was extracted by a hand pressing method as previously reported [3].

2.2. Essential oil GC analysis

The analysis of the oil composition was carried out using a Shimadzu GC-17A equipped with an FID and a DB-Wax column (60 m × 0.25 mm i.d., film thickness of 0.25 μm, Folsom, CA). The column temperature was programmed from 70 °C (2 min) to 230 °C (20 min) at 2 °C/min. The peel oil of 0.5 μl were injected directly and the split ratio of the injector being 1:50. The quantitative analysis was done in triplicate using 1-hexanol and nonadecane as internal standards. A Shimadzu GC-17A linked to a Shimadzu QP-5050 MS was used for identifying the detected volatiles.

2.3. Essential oil GCO analysis

A Shimadzu GC-17A equipped with a DB-Wax column (60 m × 0.53 mm i.d., film thickness 1 μm, Folsom, CA) connected to a humidifier ODO II (SGE, Japan) was used for GC-olfactometry (GCO). The cold-pressed lime oil was stepwise diluted three-fold with acetone based on AEDA technique which was developed by Grosch [10].

2.4. Similarity testing of reconstructed essential oil

Aroma model samples of the lime were prepared using authentic compounds on the basis of the results of chemical and sensory analyses. Propylene glycol was used as a solvent for the aroma models. A sensory test was performed to evaluate the similarity between the scents of cold-pressed lime oil and the model samples.

3. RESULTS AND DISCUSSIONS

The yield of the essential oil was 0.02% of the fresh fruit. GC-MS identified 92 compounds (some compounds are shown in Table 1) among the 96 compounds detected by GC-FID which constituted approximately 97.9% of the lime oil. The quantitative analyses were done in triplicate. There was no significant difference between replicated datasets ($p<0.05$).

Table 1. Volatile aroma components of *Citrus aurantifolia* Persa (lime) essential oil.

Peak no.	Compound	RI[a]	w/w[b] (%)	Odour description[c]	N[d]	RFA[e]	Identification[f]
1	α-Pinene	1029	0.3	Green, woody	7	6.6	RI, MS
3	β-Pinene	1117	**	Green, herbal	5	11.0	RI, MS
4	Sabinene	1127	0.1	Woody, fruity	6	8.5	RI, MS
6	Myrcene	1162	2.1	Citrus, balsamic	8	2.6	RI, MS
9	Limonene	1214	73.5	Herbal, citrus, minty	10	0.6	RI, MS
10	β-Phellandrene	1219	0.2	Citrus, minty	6	6.0	RI, MS
12	*cis*-β-Ocimene	1235	0.6	Herbal, woody	6	3.7	RI, MS
13	γ-Terpinene	1252	0.9	Fresh, mushroom	5	2.6	RI, CoGC
15	Octanal	1289	0.3	Grassy, flowery	6	5.1	RI, MS
17	Nonanal	1392	0.4	Green, waxy	5	3.8	RI, MS
22	(*E*)-Limonene oxide	1462	*	Grassy, floral	4	27.8	RI, CoGC
24	δ-Elemene	1472	0.1	Sweet, spicy	5	6.4	RI, MS
25	Citronellal	1478	0.1	Fresh, floral	5	6.4	RI, MS
26	Decanal	1496	0.6	Fatty, floral	6	3.6	RI, MS
29	Linalool	1545	0.5	Floral, spicy, pungent	6	4.0	RI, MS
39	*cis*-Verbenol[g]	1628	**	Grassy, dusty	4	22.4	MS
49	Neral	1682	4.9	Citrus, floral	9	1.9	RI, MS
53	α-Terpineol	1696	0.1	Citrus, floral, bitter	7	12	RI, MS, CoGC
55	Germacrene D	1709	0.1	Grassy, citrus	6	11.1	RI, MS
56	Neryl acetate	1726	**	Lime-like, pungent	4	9.0	RI, MS
57	β-Bisabolene	1731	1.6	Lime-like, pungent	7	2.6	RI, MS, CoGC
58	Geranial	1734	8.4	Citrus, grassy	10	1.6	RI, MS
59	*l*-Carvone	1739	*	Cool, lime-like	4	61.1	RI, CoGC
60	Geranyl acetate	1753	0.1	Lime-like	8	13.1	RI, MS
61	α-Citronellol	1755	*	Lime-like	3	51.1	RI, MS, CoGC
62	β-Citronellol	1764	*	Lime-like	4	36.5	RI, MS, CoGC
63	β-Selinene[g]	1771	**	Grassy, citrus	5	14.2	MS
64	Cumin aldehyde	1779	**	Lime-like, pungent	4	26.1	RI
65	Perillaldehyde	1783	**	Lime-like, pungent	4	12.0	RI, CoGC
66	Nerol	1797	**	Lime-like, fresh	5	12.5	RI, MS
67	Tridecanal	1811	**	Lime-like, sour	5	12.1	RI, MS
68	Germacrene B	1830	**	Lime-like, floral	5	14.5	RI, MS
69	Geraniol	1845	**	Lime-like, fresh	5	12.4	RI, MS, CoGC
70	Dodecyl acetate	1882	*	Lime-like	3	34.7	RI, MS
71	Tetradecanal	1918	**	Citrus, herbal, cool	6	14.0	RI, MS
73	Caryophyllene oxide	1989	**	Lime-like, cool, floral	4	16.3	RI, MS
74	Perillyl alcohol	2004	*	Lime-like, floral	4	28.5	RI, CoGC
75	Pentadecanal[g]	2024	0.1	Cool, sour, citrus	5	10.6	MS
84	Cedryl acetate	2170	**	Green, citrus	5	13.5	RI, CoGC
87	α-Bisabolol	2216	0.1	Lemon-like, flowery	4	6.4	RI, MS
92	Cinnamyl alcohol	2283	**	Citrus, oily	4	8.8	RI, MS, CoGC

[a]Retention index on DB-Wax column. [b]*Weight percent less than 0.005%; **Weight percent between 0.005% and 0.05%. [c]Odour description at the GC-sniffing port during GCO. [d]N: log3(FD-factor). [e]RFA: log (FD-factor)/$S^{1/2}$, where S is the weight percentage. [f]Identification method: retention index (RI), mass spectrum (MS) and co-injection with authentic standards (CoGC). [g]Tentatively identified.

Hydrocarbons including 13 monoterpenes (77.8%) and 19 sesquiterpenes (3.3%), 19 aldehydes (15.4%), 25 alcohols (1.2%), 9 acids and esters (0.2%), 3 ketones and 4 oxides in trace amount were determined. The most abundant compound was limonene (73.5%), followed by geranial (8.4%), neral (4.9%), myrcene (2.1%) and β-bisabolene (1.6%). A series of the 12 aliphatic aldehydes ranging from heptanal to nonadecanal, except heptadecanal, were also identified. Germacrene B and D were detected in this lime oil at the concentration of 0.03% and 0.1%, respectively.
The lime oil aroma was evaluated as fresh, sour and green-citrus by the assessors. As summarised in Table 1, limonene, geranial, neral, myrcene, geranyl acetate, α-pinene, α-terpineol and β-bisabolene were the higher odour-active compounds (FD factor > 6). l-Carvone in a relatively small amount, showed the highest RFA value (61.1). Geranial and neral were described as having citrus and grassy/floral aroma. The description from GC-olfactometry revealed that neryl acetate, β-bisabolene, l-carvone, geranyl acetate, α- and β-citronellol, cumin aldehyde, perillaldehyde, nerol, tridecanal, germacrene B, geraniol and dodecyl acetate would contribute to the characteristic aroma of lime.
A subsequent study looked at the sensory effects of blending of pure odorants. The aroma models were compared with the original lime oil in respect to their sensory characteristics. Various combinations and concentrations of authentic compounds were examined. The final blend, which had the odour rather similar to that of lime had a composition (%) as follows: limonene (50.00), neral (2.25), geranial (7.75), myrcene (0.33), geranyl acetate (0.67), α-citronellol (0.32), β-citronellol (0.32), nerol (0.06), tridecanal (0.07), geraniol (0.07), dodecyl acetate (0.04), β-bisabolene (0.67), neryl acetate (0.27), cumin aldehyde (0.13), caryophyllene oxide (0.07), perillaldehyde (0.13), perillyl alcohol (0.04) and l-carvone (0.13). The similarity of this aroma model to the real peel oil from the scoring test was 6.8±0.7.

References

1. B. Lawrence, Perfum. Flavor., 21 (1996) 62.
2. D.T. Loi, Vietnamese medicinal plants, Hanoi, Vietnam (1986) (in Vietnamese).
3. M. Sawamura, K. Shichiri, Y. Ootani and X. H. Zheng, Agric. Biol. Chem., 55 (1991) 2571.
4. M. Sawamura, J. Agric. Food Chem., 4 (2000) 131.
5. S.M. Njoroge, H. Ukeda, H. Kusunose and M. Sawamura, Flavour Fragrance J., 11 (1996) 25.
6. I.D. Morton and A.J. MacLeod (eds.), Food flavours, Amsterdam, The Netherlands (1990) 93.
7. N.T. Minh Tu, L.X. Thanh, A. Une, H. Ukeda and M. Sawamura, Flavour Fragrance J., 17 (2002) 169.
8. W. Feger, H. Brandauer and H. Ziegler, Flavour Fragrance J., 15 (2000) 81.
9. P. Dugo, L. Mondello, G. Lamonica and G. Dugo, J. Agric. Food Chem., 45 (1997) 3608.
10. H. Guth and W. Grosch, Flavour Fragrance J., 8 (1993) 173.

W.L.P. Bredie and M.A. Petersen (Editors)
Flavour Science: Recent Advances and Trends

Identification of character impact odorants in coriander and wild coriander leaves using GC-olfactometry and GC x GC-TOFMS

Graham Eyres[a], Jean-Pierre Dufour[a], Gabrielle Hallifax[a], Subramaniam Sotheeswaran[b] and Philip J. Marriott[c]

[a]Department of Food Science, University of Otago, PO Box 56, Dunedin, New Zealand; [b]Department of Chemistry, University of the South Pacific, PO Box 1168, Suva, Republic of the Fiji Islands; [c]Department of Applied Chemistry, Royal Melbourne Institute of Technology, GPO Box 2476V, Melbourne, Victoria 3001, Australia

ABSTRACT

The character impact odorants in the essential oil of coriander leaves (*Coriandrum sativum*) and wild coriander leaves (*Eryngium foetidum*) were determined using gas chromatography-olfactometry (GCO) and CharmAnalysis™. The main aroma compounds in *C. sativum* were (*Z*)-2-decenal, a co-eluting odour-cluster ((*E*)-2-dodecenal, (*E*)-2-dodecen-1-ol, and 1-dodecanol), β-ionone and eugenol. The most intense odorant in *E. foetidum* was (*E*)-2-dodecenal followed by eugenol, β-ionone and (*Z*)-4-dodecenal. Of the important odorants identified, (*Z*)-2-decenal, β-ionone, eugenol and (*Z*)-4-dodecenal have not previously been reported in *C. sativum* or *E. foetidum*.

1. INTRODUCTION

The leaves of the immature coriander plant (*C. sativum*) harvested prior to flowering or fruit development are a popular culinary herb with a distinctive pungent, fatty and aldehydic aroma [1,2]. The leaves of wild coriander (*E. foetidum*) have an intense coriander-like aroma and are often used interchangeably with *C. sativum* [3].
It is well known that compounds have very different odour perception thresholds and so their impact can not be determined by their concentration alone. GC-olfactometry provides a specific technique to determine the compounds responsible for a sample's distinctive aroma profile, known as the 'character-impact odorants'. A limitation of

GCO is that peak co-elution makes identification of the compound responsible for an odour challenging. Co-elution of odour-active compounds results in the perception of 'odour clusters' during GCO analysis. One possible solution for identifying character-impact odorants where co-elution occurs is to use comprehensive two-dimensional gas chromatography (GC x GC).

GC x GC consists of two 'orthogonal' columns connected in series with a cryogenic modulator at the interface. Sequentially trapping and pulsing compounds from the first column to the second creates a two-dimensional separation plane based on two different compound properties [4]. For example, two co-eluting compounds with similar boiling points on a primary column may be separated on the second dimension if they differ in their polarity. The greater peak capacity, resolution and sensitivity of GC x GC provides superior analyses compared with conventional GC analysis. Hyphenating GC x GC to time-of-flight mass spectrometry (TOFMS) presents researchers with a very powerful identification tool. For more information on the development and operation of GC x GC and TOFMS the reader is directed to several comprehensive reviews [4-6].

2. MATERIALS AND METHODS

The essential oils of *C. sativum* and *E. foetidum* leaves grown in Fiji were obtained by steam distillation. GCO analyses were performed by two trained panellists using an Hewlett Packard HP5890 Series II plus gas chromatograph with a single column (25 m BPX5 stationary phase) connected to an olfactometry port (Datu Technology, Geneva, USA) [7]. Serial dilutions (factor of 2) were analysed until no odours were detected and the results were integrated using CharmAnalysis™ [7]. Details of the analytical conditions are reported elsewhere [8]. The compounds responsible for character impact odorants located with GCO were identified using GC x GC-TOFMS. Separations were performed using an Agilent 6890 gas chromatograph retrofitted with an Everest Model Longitudinally Modulated Cryogenic System (LMCS; Chromatography Concepts, Doncaster, Australia). The column set consisted of a BPX5 first dimension (30 m) coupled in series to a BP20 second dimension (0.8 m). The operation of the LMCS and the details of the analytical conditions used for GC x GC are reported elsewhere [8,9].

3. RESULTS AND DISCUSSION

The most intense odorant identified in the *C. sativum* sample using CharmAnalysis™ was (*Z*)-2-decenal (Table 1). The second most important peak was a co-eluting odour cluster consisting of (*E*)-2-dodecenal, (*E*)-2-dodecen-1-ol and 1-dodecanol. Reference compounds of (*E*)-2-dodecenal and (*E*)-2-dodecen-1-ol were analysed separately using GCO and both eluted at the same retention index with the same odour descriptor. Therefore, both compounds potentially contribute to the odour cluster but it is not known in what proportions. β-Ionone was responsible for the third largest peak in CharmAnalysis™. Notably, β-ionone could not be identified using conventional GC-MS and has not previously been reported in *C. sativum*. Eugenol was responsible for the fourth largest peak and has also not been reported in the literature.

Table 1. Main odorants identified in *C. sativum*.

No.	Compound	Odour description	% TCV[a]	% Area[b]
1	(*Z*)-2-Decenal	Coriander/aldehydic/pungent, spicy	18.3	0.16
2	(*E*)-2-Dodecenal	Coriander/floral/pungent		5.37
	(*E*)-2-Dodecen-1-ol		14.0	4.60
	1-Dodecanol			0.19
3	β-Ionone	Floral - rose/violet	12.0	0.02
4	Eugenol	Medicinal/clove-like	7.95	< 0.01

[a]Percent of the Total Charm Value for both sniffers. [b]Percent peak area of the total ion current in GC x GC-TOFMS.

Table 2 . Main odorants identified in *E. foetidum*.

No.	Compound	Odour description	% TCV[a]	% Area[b]
1	(*E*)-2-Dodecenal	Coriander/pungent, spicy	52.9	63.5
2	Eugenol	Medicinal/clove-like	22.8	< 0.01
3	β-Ionone	Floral - rose/violet	3.9	< 0.01
4	(*Z*)-4-Dodecenal	Citronellol/fruity/pungent	3.6	0.36

[a]Percent of the Total Charm Value for both sniffers. [b]Percent peak area of the total ion current in GC x GC-TOFMS.

The most important odorant in *E. foetidum* was (*E*)-2-dodecenal (Table 2). The second most important odorant found was eugenol, followed by β-ionone and (*Z*)-4-dodecenal, all three of which have not been reported previously in *E. foetidum*.

In addition to (*E*)-2-dodecenal and (*E*)-2-dodecen-1-ol the C11, C13 and C14 (*E*)-2-alkenals and (*E*)-2-alkenols co-eluted to give odour clusters during GCO analysis of *C. sativum*. Figure 1 demonstrates the resolution obtained for the C12 odour cluster using GC x GC compared with conventional GC. Figure 1 also shows the multiple slices for each compound produced by the cryogenic modulator and the resultant contour plot (bottom inset) where the second column retention time (y-axis) is plotted against first column retention time (x-axis).

A possible solution to resolve odour clusters is to use a multidimensional GCO system incorporating the LMCS and a second orthogonal column [10]. Such a system should allow odorants to be evaluated individually without co-elution. In conjunction with the resolution and identification capabilities of GC x GC-TOFMS, these techniques should facilitate the identification of character-impact odorants in more complex essential oil samples.

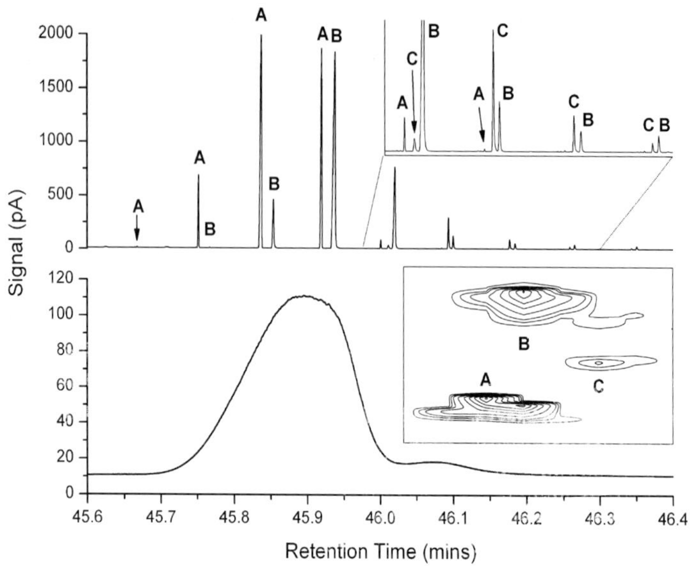

Figure 1. Resolution of C12 odour cluster obtained with GC x GC (top; lower inset – contour plot) compared with conventional GC (bottom). A: (*E*)-2-dodecenal; B: (*E*)-2-dodecen-1-ol; C: 1-dodecanol.

4. CONCLUSION

GC x GC-TOFMS allowed the identification of important odour compounds not previously identified in *C. sativum* and *E. foetidum*. Further work is underway to employ a multidimensional GCO system to resolve co-eluting odour clusters and evaluate the odour impact of each compound individually.

References

1. A.J. MacLeod and R. Islam, J. Sci. Food. Agric., 27 (1976) 721.
2. T.L. Potter, J. Agric. Food Chem., 44 (1996) 1824.
3. P.A. Leclercq, D. Nguyen Xuan, L. Vu Ngoc and T. Nguyen Van, J. Essent. Oil Res., 4 (1992) 423.
4. L. Mondello, A.C. Lewis and K.D. Bartle (eds.), Multidimensional chromatography, Chichester, England (2002) 77.
5. J.B. Phillips and J. Beens, J. Chromatogr. A, 856 (1999) 331.
6. R.A. Shellie and P.J. Marriott, Flavour Fragrance J., 18 (2003) 179.
7. T.E. Acree and R. Teranishi (eds.), Flavor science: sensible principles and techniques, Washington, DC, USA (1993) 1.
8. G. Eyres, J-P. Dufour, G. Hallifax, S. Sotheeswaran and P.J. Marriott, J. Sep. Sci., 28 (2005) 1061.
9. R.M. Kinghorn, P.J. Marriott and P.A. Dawes, J. High Resolut. Chromatogr., 23 (2000) 245.
10. F. Begnaud and A. Chaintreau, J. Chromtogr. A, 1071 (2005) 13.

W.L.P. Bredie and M.A. Petersen (Editors)
Flavour Science: Recent Advances and Trends

The astonishing sensory and coagulative properties of methylcyclopolysiloxanes

Laura Culleré, Vicente Ferreira and Juan F. Cacho

Laboratory for Flavour Analysis and Enology, Analytical Chemistry, Faculty of Sciences, University of Zaragoza, 50009 Zaragoza, Spain

ABSTRACT

The sensory properties of some methylcyclopolysiloxanes (hexamethylcyclotrisiloxane, octamethylcyclotetrasiloxane and decamethylcyclopentasiloxane) are reported. These odourless compounds are able to cause different sensory effects usually linked to the concept of astringency, such as dryness, glue-like feeling, harsh, aggressive and anaesthesia. Different tests have demonstrated that they induce the coagulation of salivary proteins even if present at extremely low concentration (below 0.1 μg/l). As common constituents of many consumer applications they may exert an impact on the sensory properties of foods.

1. INTRODUCTION

Methylcyclopolysiloxanes are widely distributed compounds. They are theoretically inert chemicals and are widely used to deactivate surfaces in glass or silica based systems and are also common constituents of detergents, closures and other consumer applications including fillings for surgical implants. The smallest members of this family are hexamethylcyclotrisiloxane (D3), octamethylcyclotetrasiloxane (D4), decamethylcyclopentasiloxane (D5) and dodecamethylcyclohexasiloxane (D6). All of them are volatile compounds and are always present in small amounts in silicone oils. Since humans are potentially exposed to these compounds, studies on their toxicity have increased more attention. At present their effects on health are unclear; although several researchers have demonstrated that they can interact and even denaturalise some proteins [1,2]. In the present paper we explore the sensory properties of methylcyclopolysiloxanes related to the concept of astringency.

2. MATERIALS AND METHODS

2.1. Sensory panel training

The sensory panel was composed by eight trained individuals belonging to the laboratory staff. They were first trained to recognise the kind of sensory effect caused by the siloxanes. This was done by tasting still mineral water containing progressively lower amounts of the compounds. The maximum concentration assessed was 10 μg/l. At least 3 h must pass after a test and each taster can only taste two samples per day.

2.2. Oral sensitivity testing

A sensory test, taking into account the difficulties and risks of measuring the perception of these compounds, was designed and applied for measuring oral sensitivity. The subjects were presented first with a sample of water and they were informed of the content of the sample (blank reference). They had to taste it and to keep it in the mouth during 5 s while spreading the liquid through the mucosa, spit it out and then keep the tongue and the mouth at rest for 30 s memorising the different perceptions. This period was extended without more restrictions for one or two minutes more. After this, they were given a coded sample containing either water or water spiked with siloxanes. They had to taste the sample as indicated earlier, to note the different buccal perceptions, and to decide if such sample was able to induce the previously learned sensory effect.

2.3. Protein precipitation

A solution containing 0.02% bovine muccine in a citric acid-phosphate butter at pH 3.5 was prepared. Volumes of 2 ml of such solutions were spiked with different amounts of siloxanes diluted in water. Tubes were mixed and incubated at room temperature for 1 h. In a second experiment, small volumes of natural human saliva (about 2 ml) were spiked with small amounts of D4 (0.1, 0.5 and 50 ng). The precipitation of salivary proteins was evaluated.

3. RESULTS AND DISCUSSION

3.1. Sensory properties of siloxanes

The three siloxanes studied (D3, D4 and D5) showed closed sensory properties related to the concept of astringency. Solutions containing small amounts of these components ($C < 1$ μg/l) caused dryness in different areas of the mouth, particularly in the area in the gum between the superior teeth and the lips. The sensory effect was defined as close to anesthesia and could last for more than half an hour. The effect needed some time to become perceptible, and usually required more than 30 s before the stimulus could be recognised. If accidentally or during the training, concentrations superior to 1 μg/l were tested, the effect was so intense that the taster could not taste anything else for several hours, developing a feeling similar to having taken a too-hot spoon of soup. Other terms used to describe such perception were glue-like, harsh and aggressive.

If the compounds are added to a liquid product, such as wine, the above described sensory effect became even more noticeable. In addition, a clear change in the overall flavour of the wine was noted. This last effect was what not trained tasters more easily noted. Trained wine tasters found it very difficult to clearly define this effect, and usually different (and contradictory) descriptive terms were used by these professionals, such as dryness, chemical taste, acidity, astringency and bitterness. None of the naïve wine tasters were able to define the effect of the supplemented siloxanes, although it became clear that a harsh dryness in their mouth remained after the consumption of the spiked samples.

3.2. Taste thresholds of siloxanes

The particular sensory properties of these compounds forced us to develop a simple sensory test. In this test, it was of important to avoid the taster to compare any sample with a reference other than the blank, to retest any sample or to move the tongue too much. The taster has to rely on his/her memory and to make a decision. This seems to be the easiest way to avoid memory effects which are particularly important with these long-lasting and ill-defined perceptions. The final decision on the sensory effect of the sample on trial is based on chi-square statistics by comparing the number of correct responses with the number of false detections of a blank.

The sensory thresholds determined in this way were astonishingly low, and our impression is that if more experiments were carried out, a greater sensitivity of the sensory panel may develop. At the beginning of the task, the sensory panel could barely detect 0.5 μg/l, but after three months of regularly tasting the panel could consistently detect the amounts shown in Table 1.

Table 1. Sensory thresholds of different siloxanes determined in water and wine.

Compound	D3	D4	D5	Mixture D3+D4+D5
Water	<1 μg/l	<5 ng/l	<0.4 μg/l	<<1 ng/l
Wine		0.5 pg/l		0.1 pg/l

The results in the table were obtained consistently, even though an explanation for the extremely low threshold concentrations remains to be found.

3.3. Protein coagulation

The addition of very low amounts of these compounds to solutions containing salivary proteins (human or bovine) provoked the precipitation of such proteins. The photograph in Figure 1 illustrates this effect. The tube on the left contains a reference solution of human saliva; the tube in the middle was spiked with 0.5 ng of D4, and that on the right with 50 ng. The figure clearly shows the ability of this compound to induce the precipitation of these proteins even if present at very low concentration.

Figure 1. Precipitation of human salivary proteins after addition of 0.5 ng (middle) and 50 ng (right) siloxane D4. The left sample is the reference without addition of D4.

3.4. Other physico-chemical and sensory properties

These compounds also showed other oddities. They are soluble in pentane and other non-polar organic solvents, but they are extremely hygroscopic and are poorly retained in reversed-phased sorbents and rp-HPLC columns. They are also very volatile, even if they have very high molecular weights, but they cannot be eluted out of Carbowax-based columns. Siloxanes are, therefore, molecules exhibiting surprising chemical, physical and sensory properties.

We believe that such properties derive from their particular structure which gives them an extremely dual character. One half of the molecule forms a hydrophobic shield, and the other half is an extremely hydrophilic 'pocket'. In hydrophilic environments, they are probably forming clusters or micelles with other non-polar substances. Presumably, they interact with the hydrophobic areas of the proteins, provoking conformational changes with significant sensory effects.

4. CONCLUSION

Siloxanes are part of our environment and they are most likely responsible for some of the perceptions related to dryness, harshness or astringency noted when consuming foods, water or even sometimes when we identify an environment as 'dusty'.

References

1. L. Sun, H. Alexander, N. Lattarulo, N.C. Blumenthal, J.L. Ricci and G. Chen, Biomaterials, 18 (1997) 1593.
2. J.J. Nicholson, S.L. Hill, C.G. Frondoza and N.R. Rose, J. Biomed. Mater. Res., 31 (1996) 345.

W.L.P. Bredie and M.A. Petersen (Editors)
Flavour Science: Recent Advances and Trends

Synergic, additive and antagonistic effects between odorants with similar odour properties

Idoia Jarauta[b], Vicente Ferreira[a] and Juan F. Cacho[a]

[a]*Laboratory for Flavour Analysis and Enology, Analytical Chemistry Faculty of Sciences, University of Zaragoza, 50009 Zaragoza, Spain;*
[b]*Bodegas Guelbenzu, Cascante ,31520 Navarra, Spain*

ABSTRACT

The existence of synergic, additive or antagonistic effects between mixtures of compounds showing similar chemical structures and/or odours has been studied. The odour threshold of a mixture of aliphatic γ-lactones is four times lower than that expected from the individual thresholds, which evidences a strong synergic effect. In the case of compounds with burnt sugar notes, the odour threshold of the mixture is exactly that expected from the individual thresholds (additive effect). In the case of vanillin-related compounds an antagonistic effect was observed.

1. INTRODUCTION

Complex food products often contain many different odorants which are formed through similar metabolic or chemical pathways. These compounds can have similar odour characteristics, which suggest that their contribution to the final sensory perception could be the result of the addition of the effects caused by the individual components. Such addition, if exists, could take different forms: a purely additive effect, in which the effect of the mixture is exactly the addition of the individual effects; a synergic effect, in which the effect of the mixture is higher than the addition of the individual effects, or an antagonistic effect In any case, the likely existence of these interactions poses a challenge to the ranking of odorants using Odour Activity Values (OAV) or sequential GC-olfactometry methods. The results of such analyses could be biased and the potential importance of some compounds could be over- or underestimated.

The objective of the study was to characterise the sensory properties of four different families of odorants found in wine: linear aliphatic lactones, burnt sugar compounds, vanillin-related compounds and ethyl-phenols.

2. MATERIALS AND METHODS

2.1. Chemicals and sample preparation

All the chemicals used were of the maximum purity available and were purchased from Lancaster, Aldrich or Fluka. The synthetic wine media used to prepare the samples contained a 10% (v/v) ethanol in water, 5 g/l of tartaric acid, and the pH was adjusted to 3.2 with 0.1 N NaOH. The concentrations and mixtures assessed by a sensory panel were taken from analytical results from Spanish aged red wines [1].

2.2. Sensory analysis

The sensory panel was composed by 12 trained tasters. Sensory tests were carried out under controlled conditions for sensory evaluation. Samples were presented to the tasters in coded black cups covered on a Petri dish. Most of the trials involved triangle tests for determining odour thresholds and recognition thresholds. All evaluations were made orthonasally.

3. RESULTS

3.1. γ-lactones

Wine contains at least five linear aliphatic γ-lactones (γ-octa, nona, deca, undeca and dodecalactones) at the concentrations shown in Table 1. The orthonasal odour thresholds of these compounds in a synthetic wine are also given in the table. As can be seen, in all the cases, the average concentrations of these compounds are well below their corresponding odour thresholds. Therefore, a first conclusion should be that these compounds are not important wine odorants. The last column of Table 1, gives the OAV of the odorants in a wine containing the average concentration. Even if all of them are considered together the total OAV is below 0.4. However, the OAV of the mixture containing the five lactones at the average concentrations found in wine is 1.33. This means that there is a strong synergism in the perception of these compounds.

Table 1. Average concentrations of γ-lactones found in wine and their sensory impact.

Compound	Average concentration[a] (C_{av})	Threshold[a] (in synthetic wine)	Average OAV
γ-Octalactone	2.3	700	0.003
γ-Nonalactone	13.4	76	0.18
γ-Decalactone	0.5	793	0.0006
γ-Undecalactone	1.2	30	0.04
γ-Dodecalactone	4.6	29	0.16
			ΣOAV_i=0.38

[a]in μg/l.

In another set of experiments, the three lactones (γ-octa, deca and undecalactone) showing smallest individual OAV (see Table 1), were removed from the mixture to determine whether the least powerful lactones have any sensory significance. The results demonstrated that they are indispensable, because the OAV of the simplified mixture decreased by a factor 4.

3.2. Compounds with burnt sugar notes

Wine contains different compounds with typical burnt sugar notes, such as sotolon, furaneol, homofuraneol and maltol. In this case, the compounds share a common odour feature, but they have different chemical structures and biochemical origins. The strategy followed is quite similar to that explained in the case of γ-lactones. Table 2 gives the concentrations at which these compounds are usually found in aged red wines.

Table 2. Average concentrations in wine and odour thresholds of some compounds with burnt sugar note.

Compound	Average C (C_{av})	Threshold (in synthetic wine)	OAV of average mixture
Sotolon	5.7	9	0.63
Furaneol	140	700	0.2
Homofuraneol	19	125	0.15
Maltol	100	5000	0.02
			$\Sigma OAV_i=1.0$

Again, the threshold values are not reached, although in this case one of the compounds has an OAV higher than 0.6, and all compounds together make 1.0 OAV. The OAV experimentally determined for the mixture of these compounds at average concentration was exactly 1.0. In this case, there appears to be an additive effect.

3.3. Vanillin-related compounds

In this case, the wine contains at least four different molecules sharing biochemical origin and a part of the structure. These compounds are vanillin, acetovanillone, ethyl vanillate and methyl vanillate. The concentrations in aged red wines and odour thresholds for these compounds can be seen in Table 3.

The maximum concentrations found in the set of wines were selected to prepare the synthetic mixture, since the odour thresholds are very high for all the members of this category. As can be seen, vanillin seems to be the single compound in this group to exert any sensory influence. Remarkably, the OAV of the mixture was found to be 0.28, two times smaller than that calculated by addition of the individual OAVs. In these mixtures there appears to be a suppression or an antagonistic effect.

Table 3. Maximum concentrations in wine and odour thresholds of some vanillin-related compounds.

Compound	Max C (C_{max})	Threshold (in synthetic wine)	OAV of average mixture
Vanillin	485	995	0.48
Acetovanillone	165	5000	0.03
Ethyl vanillate	251	6000	0.04
Methyl vanillate	64	3000	0.02
			ΣOAV_i=0.57

3.4. Ethylphenols

Results for the two ethyl phenols contained in red wine are shown in Table 4. In this case, both compounds have concentrations near or above the threshold. The experimentally calculated OAV for the mixture containing average amounts of these compounds was 3.0, virtually identical to that calculated by the addition of the two individual OAVs.

Table 4. Maximum concentrations in wine and odour thresholds of some vanillin-related compounds.

Compound	Average C (C_{av})	Threshold (in synthetic wine)	OAV of average Mixture
4-ethylphenol	96	100	0.96
4-vinylphenol	12	6	2
			ΣOAV_i=2.96

4. DISCUSSION AND CONCLUSION

Compounds with similar odours interact in complex mixtures so that the potency of the odour of the mixture, measured as OAV, can be higher, equal to or lower than that estimated by summation of individual OAVs. This fact makes necessary to work not only with single odorants but with families of odorants and to take care when the odorants of a given product are ranked according to their OAV.

The existence of these effects has also some important consequences for wine flavour chemistry. Wine γ-lactones are important odorants particularly relevant to the flavour of some red wines because of the existence of a strong synergic effect. In the case of ethyl phenols and compounds with burnt sugar notes, there are additive effects between compounds so that the importance of both groups increases. This was not observed in the case of vanillin-related compounds.

References

1. I. Jarauta, PhD thesis, University of Zaragoza, Spain (2004).

W.L.P. Bredie and M.A. Petersen (Editors)
Flavour Science: Recent Advances and Trends

Preparation of the enantiomeric forms of wine lactone, epi-wine lactone, dill ether and epi-dill ether

Stefano Serra and Claudio Fuganti

C.N.R. Istituto di Chimica del Riconoscimento Molecolare, presso Dipartimento di Chimica, Materiali ed Ingegneria Chimica del Politecnico, Via Mancinelli 7, 20133 Milano, Italy

ABSTRACT

A concise synthesis of the enantiomeric forms of wine lactone, *epi*-wine lactone, dill ether and *epi*-dill ether has been accomplished starting from the enantiomeric forms of *p*-mentha-1,8(10)-diene-3,9-diol.

1. INTRODUCTION

Recently, we have reported [1] the enzyme-mediated preparation of a number of enantiomerically pure *p*-menthan-3,9-diols and their use for the synthesis of some *p*-menthane lactones and ethers. In connection with our continuing interest in the developing of synthetic approaches to enantiopure flavour and fragrances, we extended our studies on the preparation of *p*-mentha-1-ene lactones and ethers showing *cis* stereochemistry at 3,4 position of the *p*-menthane framework [2]. The latter kind of compounds was found in Nature and different researches have demonstrated that both character and intensity of their odour are related to their relative and absolute stereochemistry [3]. This is the case of wine lactone **1**, which was first identified as an animal metabolite [4] and then recognised as a key flavour compound of different white wines [5]. The olfactory evaluation [6] revealed that natural **1** is the more powerful isomer with an odour threshold <0.04 pg/l whereas the weakest isomers show a threshold > 10^6 pg/l. A similar case is that of the dill ether **2** which is the most important constituent of dill essential oil from a sensory standpoint [7]. It is noteworthy that **1** and **2** show the same absolute configuration and *cis* stereochemistry at 3,4 position. Moreover, wine lactone and dill ether share the difficult accessibility by chemical synthesis, particularly in enantiomerically pure form.

Although different enatioselective synthesis as been reported, it seemed desirable to develop a new synthetic method to the enantiomerically pure forms of **1** and **2** by means of a procedure not involving troublesome separation, low yielding steps or the use of expensive catalyst and enantiopure starting materials.
According to the retrosynthetic analysis (Figure 1), diol **5** could be a useful building block for the preparation of both wine lactone and dill ether. **1** was obtained in high enantiomeric purity by lipase-mediated acetylation of the racemic material [1].

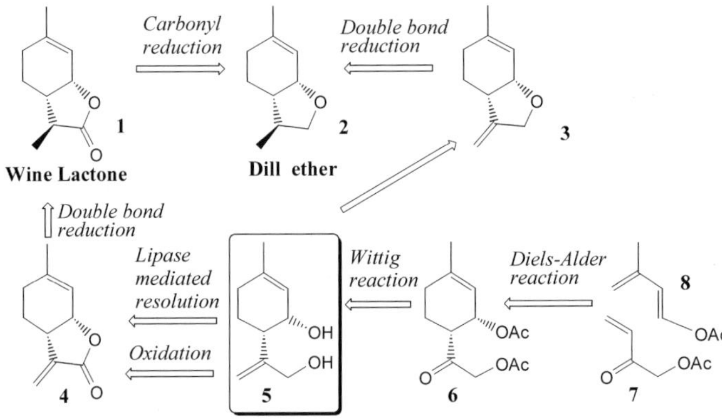

Figure 1. Retrosynthetic analysis of wine lactone **1** and dill ether **2**.

2. RESULTS

Oxidation. The preparation of wine lactone **1** by our synthetic pathway required the preparation of α,β-unsaturated lactone **4**. The chemioselective oxidation of **5** is an intriguing synthetic step since both hydroxyl functional groups are allylic (Figure 2).
We found that employing transition metal reagents, the reaction afforded a mixture of lactone **4** and keto-aldehyde **10** without formation of the lactols **8** and **9**. Pyridinium chlorochromate (PCC) gave mainly the compound **10** whilst MnO_2 and Ag_2CO_3 on celite showed reverse selectivity.
Otherwise, hypervalent iodine oxidants demonstrated a considerably different behaviour over the previous mentioned reagents. The use of *o*-iodoxybenzoic acid (IBX) in DMSO proceeded without regiochemical control affording predominantly a mixture of lactols **8** and **9** close to the keto-aldehyde **10**, unreacted **5** and a trace of lactone **4**. Further, the use of *Dess-Martin* periodinane (DMP) gave selectively **10** free of any lactol isomers. Surprisingly enough, we found that the use of catalytic TEMPO and bis-acetoxyiodobenzene (BAIB) as the co-oxidant was the best methodology for the preparation of **4**. The wanted product was obtained in yield up to 97% with complete transformation of the starting diol.

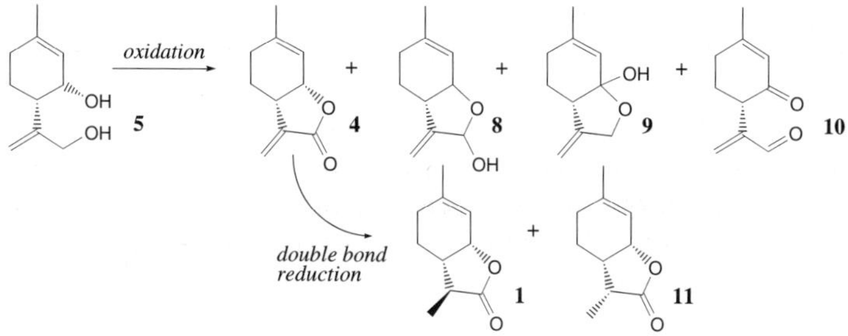

Figure 2. Oxidation of diol **5** and reduction of lactone **4**.

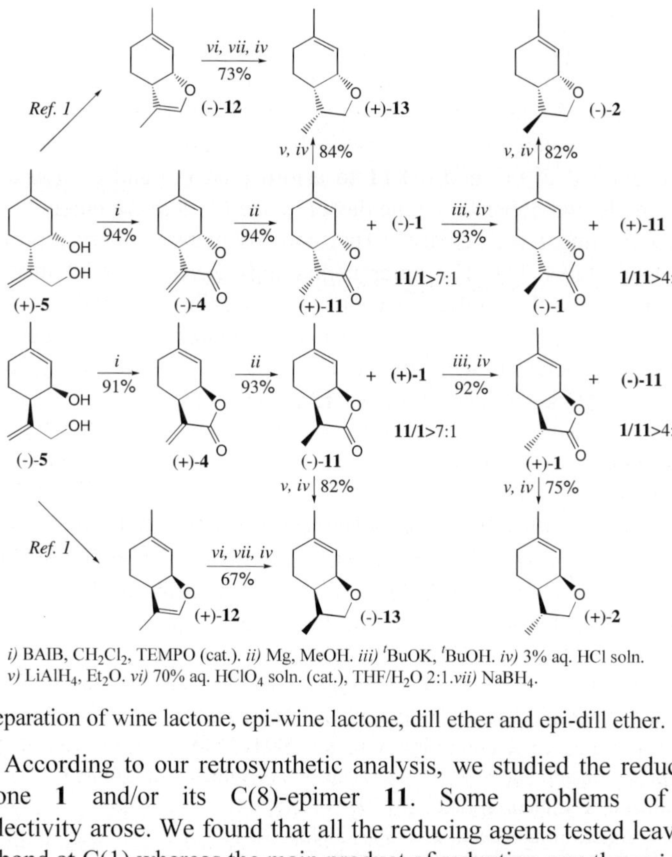

i) BAIB, CH_2Cl_2, TEMPO (cat.). *ii)* Mg, MeOH. *iii)* ^{t}BuOK, ^{t}BuOH. *iv)* 3% aq. HCl soln. *v)* $LiAlH_4$, Et_2O. *vi)* 70% aq. $HClO_4$ soln. (cat.), THF/H_2O 2:1.*vii)* $NaBH_4$.

Figure 3. Preparation of wine lactone, epi-wine lactone, dill ether and epi-dill ether.

Reduction. According to our retrosynthetic analysis, we studied the reduction of **4** to wine lactone **1** and/or its C(8)-epimer **11**. Some problems of regio and diastereoselectivity arose. We found that all the reducing agents tested leave unaffected the double bond at C(1) whereas the main product of reduction was the *epi*-wine lactone **11**. The use of magnesium turnings in methanol was the more effective procedure since all the starting **4** was reduced with the smallest amount of unidentified side products (<1%). Different results have been obtained using the hydrides as reducing agents. Either $NaBH_4$ in ethanol and in situ formed copper hydride (Bu_3SnH/LiCl/CuI) in THF reduced efficiently **4**. Otherwise, the reduction with silicon and tin hydrides afforded, closed to the lactone **11** and **1**, a considerable amount of isomerised products.

Preparation of the isomeric forms of wine lactones and dill ethers. The enantiopure (+) and (–)-**5** were oxidised with BAIB/TEMPO to afford enantiopure (–) and (+)-**4** respectively (Figure 3). The selective reduction of the latter lactones by mean of magnesium in methanol afforded epi-wine lactones (+) and (–)-**11** respectively close to a small amount of isomeric (–) and (+)-**1** respectively. The enantiomeric forms of **11** were easily separated from their epimers by chromatography. Furthermore, we found that treatment of either pure **11** or the mixture of **11/1** (>7:1) obtained as above with ^{t}BuOK in ^{t}BuOH gave a new mixture of **1/11** (>4:1) where **1** was the main component. By mean of the latter synthetic sequence natural (–)-wine lactone **1** and lactone (+)-**1** were prepared from (–) and (+)-**4** respectively in enantiopure form and with an overall yield of about 55%. The $LiAlH_4$ reduction of the lactones (–) and (+)-**1** affords smoothly the corresponding diastereoisomeric diols that were cyclised by treatment with diluted aq. HCl to give the natural (–)-dill ether **2** and its enantiomer (+)-**2**, respectively. This ring closure affords stereospecifically the bicyclic ethers with *cis* configuration at the 3,4 position in chemically and isomerically pure form. Analogously, the same procedure was applied to (+) and (–)-**11** to afford pure (+) and (–)-*epi* dill ether **13** respectively. Finally, we transformed the diols (+) and (–)-**5** in the ethers (–) and (+)-**12** respectively by treatment with catalytic HCl and following isomerisation by mean of rhodium hydride catalyst [1]. The latter compounds were dissolved in a THF/water mixture and treated with catalytic $HClO_4$. The starting materials were smoothly converted in the corresponding lactols, which were treated with an excess of $NaBH_4$. The mixtures were quenched with diluted HCl and the isolation procedure afforded diastereoselectively (+) and (–) *epi*-dill ether **13** respectively.

3. CONCLUSION

A new enantiospecific approach to the isomeric forms of natural wine lactone and dill ether is described in this work. The enantiomeric forms of *p*-mentha-1,8(10)-diene-3,9-diol were the common building blocks for this divergent synthesis. The key steps were a number of chemio and diastereoselective reactions that we have studied and optimised.

References

1. S. Serra and C. Fuganti, Helv. Chim. Acta, 85 (2002) 2489.
2. S. Serra and C. Fuganti, Helv. Chim. Acta, 87 (2004) 2100.
3. E. Brenna, C. Fuganti and S. Serra, Tetrahedron-Asymmetr., 14 (2003) 1.
4. I.A. Southwell, Tetrahedron Lett., 24 (1975) 1885.
5. H. Guth, J. Agric. Food Chem., 45 (1997) 3022.
6. H. Guth, Helv. Chim. Acta, 79 (1996) 1559.
7. M. Wüst and A. Mosandl, Eur. Food Res. Technol., 209 (1999) 3.

W.L.P. Bredie and M.A. Petersen (Editors)
Flavour Science: Recent Advances and Trends

Hierarchy and identification of additional important wine odorants

Eva M[a] Campo, Ricardo López and Vicente Ferreira

Laboratory for Flavour Analysis and Enology, Analytical Chemistry, Faculty of Sciences, Universidad de Zaragoza 50009, Zaragoza, Spain

ABSTRACT

GCO analysis has been applied to study the aroma profiles of wines from Madeira, Pedro Ximénez, Sherry, Cava and Sauternes. Wine extracts were prepared by dynamic headspace and the most relevant odorants, according to their GCO score, were ranked. Linalool, sotolon and 4-ethylguaiacol were found to be potentially important odorants when characterising some of these wines. The study also revealed the presence of three ethyl esters, not previously reported in wines. Eight additional odorants showing high GCO scores and not detected in normal dry wines remained still unidentified.

1. INTRODUCTION

The aroma of wine has been extensively studied in recent years, and there is now a wide knowledge on its chemical composition [1-3]. However, there are several particular or very specific wines remaining where little is known about their aroma volatiles. This is the case of wines made by following complex elaboration processes such as Madeira, Pedro Ximénez, Sherry, Cava or Sauternes. Such processes affect the aroma and flavour composition and lead to the formation of their typical and characteristic bouquet. It is expected that those wines present some odorants not previously identified in wine which could still play some role in the different aroma nuances of other wines.

2. MATERIALS AND METHODS

2.1. Wines

Five different wines were selected for this study; a Pedro Ximénez and a Sherry (Fino) wine both matured and blended in the *solera* system; a fortified Madeira wine, which followed an oxidation step known as *estufagem*; a sparkling, barrel fermented Cava and finally, a wine from Sauternes, this last elaborated with grapes infected with noble rot *(Botrytis Cinerea)*.

2.2. Extract preparation and GC-Olfactometry

Extracts for analysis were obtained by a dynamic headspace sampling technique at conditions close to retronasal olfaction. A controlled flow of N_2 (100 ml/min) passed through a mixture of 80 ml of wine and 20 ml of artificial saliva kept at 37 °C for 200 min. Volatiles are first trapped in a 400 mg bed packed with LiChrolut-EN resins and then eluted with 3.2 ml of dichloromethane. The extract was finally concentrated to 200 μl, of which 1 μl was injected into the GC for GCO analysis.

GCO was carried out by a trained panel of eight sniffers, who noted down retention time, intensity (using a 7 point category scale) and a description for any detected odour. The identification of the odorants was carried out by comparison of their odours, the chromatographic retention index on both DB-WAX and DB-5 columns, and MS spectra with those of authentic compounds.

In order to make a ranking of the most relevant odorants, both intensity and frequency of detection were normalised to % scale by the following formula: %MF = (%Fr * %Int)1/2, where %MF was the modified frequency (expressed as %) and %Fr and %Int were the frequency of citation (in %) and intensity of the odorant, respectively.

2.3. Identification of unknown compounds

An extract enriched in the target compounds was prepared using the same headspace method explained in the former paragraph but extracting this time 300 ml of wine, in four consecutive purges. Volatiles were retained in a single SPE bed packed with 1 g resin. Prior to the elution with 8 ml of dichloromethane, the cartridge was washed with 200 ml of a mixture water/methanol (60:40, v:v) with 1% $NaHCO_3$ to eliminate alcohols and fatty acids. The extract was concentrated up to 100 μl. From this enriched extract, 90 μl were used for GC-GC analysis.

The off-line multidimensional system employed for identification purposes consisted of two independent chromatographs, fitted to a FID-ODO and a DB-WAX column and to a MS-ODO and DB-5 column, respectively. Both were equipped with manually operating switching valves and a cold trap as interface. The fraction of interest was isolated in the first chromatograph, being the heart-cutting window of 10 s over the start and final sniffing elution times of the target compound, cryo-trapped and finally desorbed in the second chromatograph by raising the temperature.

3. RESULTS

3.1. Ranking of the odorants

More than 100 different odorants were detected for every single wine analysed by GCO. For simplicity, only the odorants reaching a minimum MF value of 50% were considered. This led to a final list of 36 odorants. Table 1 shows the ranking of the odorants reaching at least a 50% of MF in each of the wines.

Table 1. Ranking of the odorants with a MF of greater than 50% in GCO analysis.

Madeira	Pedro Ximénez	Sherry	Sauternes	Cava
3-Met.-1-butanol	Et. isobutyrate	[ni] 995	3-Met.-1-butanol	Et. hexanoate
2,3-Butanedione	3-Met.-1-butanol	Et. isobutyrate	Et. isobutyrate	Et. butyrate
Et. butyrate	Linalool	Et. isovalerate	β-Phenylethanol	3-Met.-1-butanol
Isovaleric acid	2,3-Butanedione	3-Met.-1-butanol	Et. hexanoate	Et. 2-met.butyrate
[ni]1427	Et. isovalerate	[ni] 1541	Isovaleric acid	Et. isovalerate
Guaiacol	Acetylpyrazine	β-Phenylethanol	Et. butyrate	[ni] 973
Et. hexanoate	[ni] 1059	Acetylpyrazine	Et. 2-met.butyrate	MFT
Et. 4-met.pentan.	[ni] 973	[ni] 1059	Acetylpyrazine	Acetylpyrazine
(*Z*)-3-hexenol	Isovaleric acid	Isovaleric acid	Isopentyl acetate	[ni] 1427
Et. isovalerate	Et. isobutyrate	Et. isobutyrate	(*Z*)-3-hexenol	2,3-Butanedione
Et. isobutyrate	Et. butyrate	Et. hexanoate	Et. isovalerate	β-Phenylethanol
[ni] 1794	β-Damascenone	Et. butyrate	MFT	β-Damascenone
[ni] 1815	[ni] 995	[ni] 973	3-IP-2-Methox	Isobutyl acetate
β-Damascenone	[ni] 1427	Isobutyl acetate	Acetic acid	Isovaleric acid
[ni] 1059	MFT	Isopentyl acetate	Isobutyl acetate	3-IP-2-Methox
Et. 2-met.pentan.	Guaiacol	[ni] 1059	3-IB-2-Methox	[ni] 995
3-IP-2-Methox	Isobutyl acetate	4-Ethylguaiacol	Isobutanol	
Acetic acid	[ni] 1541	β-Damascenone	[ni] 973	
Sotolon	Et. hexanoate	2,3-Butanedione	Linalool	
(*Z*)-Whiskylact.	(*Z*)-Whiskylact.	Et. 4-met.pentan.	4-Vinylguaiacol	
Isobutyl acetate	Isopentyl acetate	(*Z*)-Whiskylact.	[ni] 1427	
Hexanal		[ni] 1427	Sotolon	
1-Octen-3-one				

Abbreviations: et.: ethyl; met.: methyl; MFT: 2-methyl-3-furanthiol; pentan.: pentanoate; lact.: lactone; 3-IP-2-Methox: 3-isopropyl-2-methoxypyrazine; 3-IB-2-Methox: 3-isobutyl-2-methoxypyrazine. [ni]Not identified compound with LRI value for odour on a DB-WAX column.

3.2. Identified compounds

The enriched extract, clean from alcohols and fatty acids, yielded sufficient quantities of three target compounds for further analysis by GC-GC and GCO-MS. The identified compounds were three ethyl esters: ethyl 2-methylpentanoate, ethyl 3-methylpentanoate and ethyl 4-methylpentanoate. The identification was carried out by comparison of their odours, chromatographic retention index in both DB-WAX and DB-5 columns and MS spectra with those of authentic reference compounds.

4. DISCUSSION AND CONCLUSION

A first remark about this set of white wines is that all of them represent very different tendencies and styles. They have been made from different grape varieties and using different wine making strategies and maturation procedures, often profoundly linked to the tradition of the region of origin. The exceptional flavour and bouquet of these dessert wines is far from the traditional concept of dry wine. However, as Table 1 shows, they all share a general common aroma background with the rest of the wines, constituted by wood extractable compounds; by-products of alcoholic fermentation and β-damascenone.

On the other hand, some remarkable differences can be found between this group of wines and dry white wines. First of all, the number of odorants reaching % MF values above 50 is much higher than those obtained in the analysis of normal wines using the same GCO strategy, which is reflects their aroma complexity. Additionally, GCO revealed the presence of eight unknown compounds, showing diverse odours, not previously detected by olfactometry in wines [1-3]. Some of them are present in more than one wine as it can be seen in Table 1. Another important common fact of these wines is the presence of three ethyl esters not previously identified in wine: ethyl 2-methylpentanoate, ethyl 3-methylpentanoate and ethyl 4-methylpentanoate.

The most particular aroma profile, out of the five studied in this work, is that of fortified Madeira wine, probably as a consequence of the extreme oxidation it suffers after fermentation. The absence of linalool and 3-mercaptohexyl acetate as well as the presence of compounds like sotolon, hexanal and 1-octen-3-one is particularly outstanding. The Pedro Ximenez white is mainly characterised by its high content in linalool, being this compound the potentially most important odorant to introduce differences between samples. On the other hand, the most interesting result in the Sherry GCO profile is related to 4-ethylguaiacol, due to the rare presence of this phenol in white wines. Concerning the wine from Sauternes, the important presence of β-phenylethanol as well as both linalool and sotolon is noteworthy. Finally, the Cava showed a lower content in relevant odorants when compared to the rest of the wines.

References

1. A. Escudero, B. Gogorza, M. Melús, N. Ortín, J.Cacho and V. Ferreira, J. Agric. Food Chem., 52 (11) (2004) 3516.
2. L. Cullere, A. Escudero, V. Ferreira and J. Cacho, J. Agric. Food Chem., 52 (6) (2004) 1653.
3. R. López, N. Ortín, J.P. Pérez and V. Ferreira, J. Agric. Food Chem., 51 (11) (2003) 3419.

W.L.P. Bredie and M.A. Petersen (Editors)
Flavour Science: Recent Advances and Trends

Identification of key odorants related with high quality Touriga Nacional wine

A.C. Silva Ferreira[a], E. Falqué[b], M. Castro[a], H. Oliveira e Silva[a], B. Machado[a] and P. Guedes de Pinho[a]

[a]Escola Superior de Biotecnologia, Universidade Católica Portuguesa, Rua Dr. Antonio Bernardino de Almeida, PT-4700-072 Porto, Portugal; [b]Faculdad de Ciencias, Universidad de Vigo, Spain

ABSTRACT

High quality Touriga Nacional (TN) wines are characterised by a fruity-citric aroma described as sweet and of fresh citrus, evoking the bergamot fruit (*Citrus bergamia*). By sensory analysis the identification of the most important descriptors were found. Among 18 descriptors 3 were selected: 'bergamot-like' aroma, 'orange-like' and 'violet'. A GCO of an extract of a typical TN wine allowed the identification of three related odorant zones ZO1, ZO2 and ZO3 showing bergamot-like odours. By AEDA the importance of ZO2 was confirmed, this odorant zone corresponded to the presence of linalool and linalyl acetate. Results from a similarity test showed that the highest value was observed when linalool alone was added. Results obtained from the analysis of several red wines from different varieties show that terpenols are present in higher amounts in wines from the TN variety. These compounds are most likely the indicator/trigger of the varietal aroma of TN wines.

1. INTRODUCTION

High quality Touriga Nacional (TN) wines are characterised by a fruity-citric aroma described as sweet and of fresh citrus, evoking the bergamot fruit (*Citrus bergamia*) [1]. In fact, a 'bergamot like' descriptor is currently employed to rate high quality TN wines. Previous work has shown that TN wines have higher levels of terpenol compounds [2,3] compared with other wines made from other red wine varieties. This cultivar is one of the most important red wine varieties. Wines produced with this variety have higher commercial prices. The fact, that the aroma of TN wines is mainly related to the terpene composition can be an important source of information for enologists, as they can better control the levels of these compounds in grapes, by correct viticulture practices. The terpenes content can also be influenced by changing some technological parameters during wine production. The aim of this work was to determine the sensory quality descriptors and chemical compounds responsible for the TN red wines determined by experts as high quality.

2. MATERIAL AND METHODS

2.1. Wines

Seven TN wines were selected by an industry panel as corresponding to TN high quality wines with the typical 'bergamot-like' aroma. Other monovarietal red wines (n=75) were from Tinta Roriz (TR), Tinto Cão (TC), Tinta Franca (TF), Tinta Barroca (TB).

2.2. Sensory testing

Two sensory panels were used; one composed of winemaking producers, who selected the high quality TN wines. A second one composed of 12 graduate students, who were selected on the basis of their sensory performances [4]. Tests were performed individually, using tulip glasses containing 30 ml of wine in a room at a constant temperature of 20 ºC.

The identification of the most important descriptors, related to the bergamot-floral characteristics of TN high quality wines was performed according the AFNOR 09-021 procedure [5] by a trained panel (the second of the two panels) [8]. Typical high quality TN wines were selected by the industry panel. Four free-choice profiling sessions were performed. In each session two different TN typical wines were analysed. The initial vocabulary of 18 preliminary terms were after discussion reduced and the most appropriate terms defining the wine aroma were selected for the formal descriptive analysis. In two later sessions a TN wine presented before (TN) was again presented to the panel, and the panellists were asked whether or not the 18 attributes were detected (presence or absence). Two sessions were conducted with wines that had 2 or 3 attributes from the initial 18 added to them.

The similarity value (SV) of each sample with the TN high quality was determined by a comparison test using a discontinuous scale from 0 to 10.

2.3. GC-Olfactometry and GC-MS analysis

GCO screening analysis was employed on dichloromethane extracts of TN wines in order to determine odorant zones related to the bergamot aroma. The extracts of bergamot oil were also analysed to determine the Dilution Factors of the most important odorant zones. This GCO analysis was carried out with three trained panellists and was performed twice. Flavour dilution (FD) factors of the odour-active compounds were determined. Aroma dilution extract analysis (AEDA) was performed according to [6].

The quantification of α-pinene, linalool, terpineol, geraniol, nerol, citronellol, linalyl acetate, limonene was done by GC-MS as earlier described [7].

3. RESULTS

3.1. Terpenols in red wines

Among 75 red wines analysed, the levels of terpenols ((linalool, terpineol, geraniol, nerol and citronellol) were much higher in TN wines than in the other wines tested. In fact wines from TN are considered to have floral characteristics, which could be associated with the presence of terpenol compounds.

3.2. Identification of key sensory descriptors

Comparing TN wine selected by the industry panel with a non-typical wine spiked with reference materials for 3 combine attributes showed that the wine spiked with bergamot tea, scraped mandarin skin and violet was most similar to TN wine. Following this, six sessions were held, in each of which 3 wines were presented with 13 reference standards to help panellists identify and remember sensory attributes found in the evaluated wine samples [9]. It was the consensus that bergamot is the note that best described the typical aroma of TN wine.

3.3. Determination of compounds related to sensory descriptors

The second step of this work was to determine from the global aroma contributors of bergamot essential oil. Hence, an AEDA of a DCM extract of bergamot diluted oil was performed. The major aroma contributors were: α-pinene (FD=1024), linalool (FD=512) and linalyl acetate, γ-terpinene, (*E*)-β-ocimene (all with FD=256). The first one, which exhibited a pine-like aroma, was found to be the strongest aroma contributor to this essential oil, since it had the highest FD factor. The presence of linalool and linalyl acetate was confirmed, and their quantification by GC-MS showed that these two volatile compounds were present in the highest concentration. This suggested their important role in the flavour of bergamot oil. γ-terpinene, (*E*)-β-ocimene and β-phellandrene, tentatively identified according to the Kovats retention index cited in the bibliography, were also important for the overall aroma property of bergamot essential oil [10].

GCO analysis was simultaneously performed with the same TN wine used in the AFNOR analysis. Three odorant zones (OZ) were identified with an aroma related to the bergamot-like aroma. A first odorant zone OZ1 (RI=1023, with a FD=4) with an aroma described as pineapple/pine/fruity, a second one, OZ2 (RI=1560, FD=4) with a floral, early grey aroma descriptor: and finally a third odorant zone, OZ3 (RI=1940, FD=32) described as floral.

A GCO of a DCM extract of a non-TN wine was also performed. The aromatic intensities of these zones from the non-TN wine were much less. Among the 3 chromatographic odorant zones related to floral aroma the most similar to the bergamot-like aroma was OZ2. This OZ, corresponded to the presence of linalool and linalyl acetate identified by GC-MS. In order to investigate the contribution of these molecules to the 'bergamot-like' aroma of TN wines, these two compounds were added separately or in combinations to a TR wine in concentrations naturally found in TN wines. This TR wine was selected as it had a very low concentration of terpenols. The similarity values obtained for each pair comparison test between TR wine and the spiked samples are given in Table 1.

Table 1. Results obtained from sensory analysis (12 persons).

Added compounds to TR	Similarity value (SV)	Standard deviation
TN	9.7	1.1
TR + Linalool	5.9	1.9
TR + Linalyl acetate	3.3	2.1
TR + Linalool + Linalyl acetate	5.5	2.6
TR	2.8	2.4

The highest similarity value was observed when linalool was added to TR wine. (SV=5.9), the addition of linalyl acetate has a small impact (SV=3.3). The ANOVA calculations for the data showed differences between samples ($p<0.001$) and no significant differences between assessors. All pair additions contributed in a high degree to TN aroma perception, SV ranged from 5.5 to 5.9.

4. CONCLUSION

This work aimed to correlate the characteristic descriptors of TN wines such as floral, bergamot-like aroma with the presence of specific compounds. Three relevant odour zones from GCO analysis related to such descriptors. AEDA was employed to evaluate the relative importance of each of these zones. Linalool and linalyl acetate were identified as important odorants in the aroma of TN typical wines.

References

1. B. Lawrence, Perfum. Flavor., Oct/Nov (1982) 43.
2. A.C. Silva Ferreira and P. Guedes de Pinho, Anal. Chim. Acta., 513 (2004) 169.
3. Tec-Doc (ed.), 7th international symposium of enology, Paris, France (2003) 584.
4. S. Issanchou, I. Lesschaeve and E.P. Köster, J. Sens. Stud., 10 (1995) 349.
5. AFNOR NFV09-021, Recueil des normes françaises, contrôle de qualité des produits alimentaires, analyses sensorielle, 4ª edition (1991).
6. H. Maarse and D.G. van der Heij (eds.), Trends in flavour research, proceedings of the 7th Weurman flavour research symposium, Amsterdam, The Netherlands, (1995) 179.
7. A.C. Silva Ferreira, T. Hogg and P. Guedes de Pinho, J. Agric. Food Chem., 51 (5) (2003) 1373.
8. E. Falqué, A.C. Silva Ferreira, T. Hogg and P. Guedes de Pinho, Flavour Fragrance J., 19 (2004) 298.
9. B. Rainey, J. Sens. Stud., 1 (1986) 149.
10. A. Verzera, A. Trozzi and A. Cotroneo, J. Agric. Food Chem., 51 (2003) 206.

W.L.P. Bredie and M.A. Petersen (Editors)
Flavour Science: Recent Advances and Trends

Spiking as a method for quantification of aroma compounds in semi-hard cheeses

Mikael Agerlin Petersen[a], Adel Ali Tammam[b] and Ylva Ardö[a]

[a]*Department of Food Science, Royal Veterinary and Agricultural University, Centre for Advanced Food Studies, Frederiksberg, Denmark;* [b]*Department of Dairy Science, Faculty of Agriculture, Assiut University, Assiut, Egypt*

ABSTRACT

A dynamic headspace method for analysis of aroma compounds in semi-hard cheeses with different fat contents (13 and 24%) was developed. The method included homogenisation of the cheese sample with water and an internal standard, purging with 200 ml nitrogen/min for 60 min at 30 °C followed by analysis of the volatiles by GC-MS. For absolute quantification, standard curves describing the relation between absolute concentration and area in chromatograms were established by spiking with 30 aroma compounds at three levels. The relationships obtained were very different. Some compounds behaved similarly, independently of fat content in the cheese, while others were clearly affected, and for some compounds reliable relationships could not be directly established.

1. INTRODUCTION

Dynamic headspace analysis is used for a multitude of products. Data for relative concentrations are easily obtained (as in [1]), but if concentrations are to be compared with thresholds, or if kinetics of formation of aroma compounds is considered, absolute quantifications have to be carried out (see e.g. [2]). No matter which quantification method is used, standards must be added either to the product itself or to a model system. This is problematic especially in solid products, because the addition of standards to a complex solid matrix is not straightforward, as well as the release of volatile compounds is very variable, depending on the nature of the matrix and the individual compounds. In this work, cheese samples were homogenised with water in order to obtain homogenous liquid samples. This should enable reliable addition of internal standard and spiking with selected aroma compounds.

2. MATERIALS AND METHODS

Samples of two semi-hard cheeses (Riberhus: 27% fat and Cheasy: 13% fat) were homogenised with water in order to obtain homogenous suspensions, which were added an internal standard, and were spiked with selected aroma compounds at three levels (0.31, 0.63 and 0.94 ppm). The homogenised samples were purged with nitrogen, while the aroma were trapped on Tenax TA. The traps were thermally desorbed using a Perkin Elmer ATD 400 and the volatiles analysed using an Agilent G1800 GC-MS. All calculations were based on relative areas, i.e. area of peak divided by area of internal standard. Concentrations in samples were determined by extrapolation to relative area = 0 (Figure 1). The logP values were estimated using an interactive logP predictor [3].

3. RESULTS AND DISCUSSION

In Table 1, compounds are grouped after r^2 of the standard curves obtained. It appears that an important factor determining the level of r^2 is the breakthrough volume. The samples were purged with 48 l N_2/g Tenax. Because it is commonly considered safe to purge with half the breakthrough volume [4], breakthrough volumes less than 96 l/g are potentially critical.

Compounds in group A (Table 1) had $r^2 > 0.9$, and except for 1-butanol they all had high breakthrough volumes indicating efficient trapping. The breakthrough volume for 1-butanol is, however, still higher than 48 l/g, and the standard curves exhibited high correlations ($r^2 \approx 0.99$). The logP values varied from 0.33 to 2.45, indicating that logP itself was not critical for the behaviour of a given compound in the analysis.

The three compounds in group B had some deflection of the standard curves (leading to lower r^2-values), and the breakthrough volumes were all in the critical area.

The three compounds in group C had low correlations and generally small peaks, due to very low breakthrough volumes and thus excessive losses during purging.

Group D consisted of acetoin and 2-butanone which naturally occur in very high concentrations in the cheeses. The spiking levels were, therefore, low compared to the natural content. A low correlation was seen between added amount and areas in chromatograms. It was not possible to spike with higher levels because the peaks already tended to overload in the unspiked samples.

Group E consists of four acids which all yielded very small peaks with no relation to the amount added during spiking. This is most probably due to dissociation since all the acids have pK_a values that are considerably lower than the pH of the cheeses. The dissociated form, which has low volatility, has therefore, been a dominant factor.

For some of the compounds in group A the slopes and intercepts of the standard curves were rather independent of fat content in the cheese. 2,3-Pentanedione had a much steeper curve in the low fat cheese than in the full fat cheese and the opposite was the case for dimethyl trisulfide, 3-methyl-2-pentanone, hexanal, 1-hexanol and 2-heptanone.

Table 1. Results from spiking of two cheeses at three levels of standards (0.3, 0.6 and 0.9 ppm). Breakthrough volumes, 1-octanol-water partition coefficient (logP), r^2 for calibration curves and extrapolated concentrations in cheeses are shown.

	Compound	Break-through volume [a)]	logP [b)]	Riberhus r^2	Riberhus Conc. in cheese, ppm IV	Riberhus Conc. in cheese, ppm II	Cheasy r^2	Cheasy Conc. in cheese, ppm IV	Cheasy Conc. in cheese, ppm II
A	4-Methyl-2-pentanone	1000[c)]	1.8	0.95	0.067	0.025	0.95	0.082	0.001
	2,3-Pentanedione		0.3	0.997	0.020	0.006	0.998	0.017	0.001
	2-Hexanone	1000	1.8	0.996	0.026	0.004	0.991	0.024	0.001
	1-Butanol	**56**	1.0	0.997	0.061	0.056	0.986	0.039	0.004
	3-Methyl-1-butanol	158	1.4	0.99	0.342	0.286	0.99	0.042	0.006
	Ethyl hexanoate		2.4	0.99	0.023	0.003	0.998	0.004	0.001
	Dimethyl trisulfide		1.5	0.97	0.000	0.009	0.99	0.011	0.002
	3-Methyl-2-pentanone	1000[c)]	1.8	0.96	0.102	0.020	0.91	0.095	0.002
	Toluene	400	2.5	0.95	0.050	0.004	0.96	0.078	0.004
	Dimethyl disulfide		1.3	0.97	0.158	0.106	0.93	0.087	0.004
	Hexanal	500	1.7	0.97	0.050	0.001	0.99	0.042	0.003
	3-Methyl-2-buten-1-ol		1.4	0.99	0.042	0.011	0.97	0.049	0.015
	1-Hexanol	1800	1.9	0.997	0.004	0.004	0.99	0.005	0.003
	2-Heptanone	5000	2.3	0.998	0.040	0.026	0.996	0.111	0.089
B	Ethyl acetate	**34**	0.6	0.95	0.103	0.019	0.84	0.151	0.016
	3-Methylbutanal	**67**[d)]	1.0	0.77	0.147	0.007	0.80	0.161	0.014
	2-Methyl-1-propanol	**20**	0.9	0.91	0.190	0.107	0.67	0.280	0.032
C	2-Propanone	**6**	0.5	0.25			0.85		
	Acetic acid	**5.6**	-0.3	0.75			0.18		
	1-Propanethiol		1.5	0.62			0.58		
D	Acetoin		0.2	0.37			0.04		
	2-Butanone	**40**	1.0	0.42			0.76		
E	Propanoic acid	**32**	0.1	0.25			0.01		
	2-Methylpropanoic acid	140[e)]	0.5	0.58			0.54		
	Pentanoic acid	700	1.0	0.44			0.10		
	Hexanoic acid	3100	1.4	0.06			0.00		

[a]From http://www.sisweb.com/index/referenc/resins.htm. [b]From http://www.molinspiration.com /services/logp.html. [c]Value for 2-hexanone. [d]Value for 2-methylbutanal. [e]Value for butyric acid. IV: Using all four levels for regression. II: Only using 0-addition and lowest added level for regression.

These differences can not be explained by differences in fat content, since 2,3-pentanedione is far the most polar of the compounds mentioned, and, therefore, is expected to have a higher affinity for a low-fat matrix. The cheese slurries did, however, contain considerable amounts of both water and fat and would thus dissolve a broad

range of compounds. Other factors as protein structures, are known to be able to bind aroma molecules [5], and this could also explain some of the effects seen.
When spiking is used for quantification, the standard curve will mostly be used outside the interval actually measured (extrapolation). For reasons of accuracy, extrapolations should be as short as possible. On the other hand, some spacing is needed between the spiking levels to obtain an accurate standard curve. Especially if the standard curve is not exactly linear, excessive deviation may occur. In Table 1 large differences are seen between IV and II, also when standard curves exhibit high correlations. An example of this is shown in Figure 1. Even though the curve fitting is good ($r^2 > 0.99$), the limited curvature at the higher levels leads to essential errors when the curve is used at the lower levels, for instance for determination of the natural level in the spiked cheese, e.g. for Riberhus the intercept with the x-axis is 0.040 ppm using all points but 0.026 using the two lowest levels.

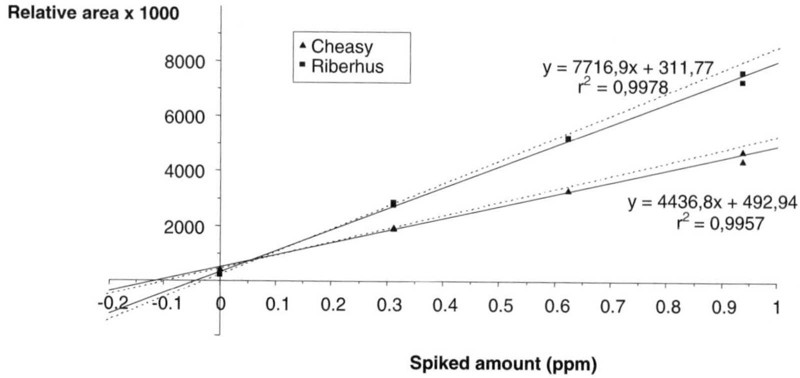

Figure 1. Standard curves for 2-heptanone. Solid lines: regression using all points. Dashed lines: regression using points from two lowest levels.

4. CONCLUSION

Spiking is shown to be a reliable method for quantification of aroma volatiles in cheeses, because standards are mixed with the original matrix or a slurry thereof. The spiking levels must be carefully chosen, i.e. close to the natural level, but should still givie enough variation to obtain a reliable and preferably linear standard curve.

References

1. M. Juric, G. Bertelsen, G. Mortensen and M.A. Petersen, Int. Dairy J., 13 (2003) 239.
2. B.V. Thage, M.L. Broe, M.H. Petersen, M.A. Petersen, M. Bennedsen and Y. Ardö, Int. Dairy J., 15 (2005) 805.
3. Molinspiration Cheminformatics, http://www.molinspiration.com/services/logp.html (2005).
4. Scientific Instrument Services Inc, http://www.sisweb.com/index/referenc/resins.htm (2005).
5. A.M. Seuvre, M.A. Espinosa Diaz and A. Voilley, Food Chem., 77 (2001) 421.

Modification of bread crust flavour with enzymes and flavour precursors

Wender L.P. Bredie[a], Marinke Boesveld[a†], Magni Martens[ac] and Lone Dybdal[b]

[a]Department of Food Science, The Royal Veterinary and Agricultural University, DK-1958 Frederiksberg C, Denmark; [b]Novozymes, DK-2880 Bagsværd, Denmark; [†]Present address: Kerry Bio-Science, NL-1411 GP Naarden, The Netherlands; [c]Matforsk, N-1430 Aas, Norway

ABSTRACT

The modification of bread crust flavour by enzymes and proline as a model for specific proteases was investigated. Several desirable compounds such as 2-acetyl-1-pyrroline and 6-acetyltetrahydropyridine could be produced in some of these small-scale bread systems. However, also sensory less desirable bitter-tasting and burnt-smelling compounds such as pyrrolizines and azepinones were identified. These compounds were predominantly generated in breads containing relatively high levels of proline.

1. INTRODUCTION

The flavour of bread originates from many volatile and non-volatile components derived from Maillard reactions and metabolites from yeast activity, which also play a role in Maillard reactions. Bread crust aroma is associated with the key odorants 2-acetyl-1-pyrroline (AP) and isomers of 6-acetyltetrahydropyridine (ATHP) [1]. Important precursors for AP and ATHPs are proline, ornithine and citrulline. During baking, Strecker degradation of ornithine leads to effective precursors for AP and alkylpyrazines in reactions with carbonyl compounds [2]. In contrast, proline is an important source for ATHPs, but also produces bitter tasting compounds in model Maillard reactions [3]. This study reports on the modification of bread crust flavour by: 1) arginase which hydrolyses arginine to ornithine; 2) an enzyme preparation from *Humicola insolens* with various hemicellulase activities, used to improve bread volume and crumb structure and 3) proline as a model for the theoretical action of proline-releasing exopeptidases.

2. MATERIALS AND METHODS

2.1. Bread baking

Bread dough consisted of wheat flour, water, yeast, salt, ascorbic acid and sucrose, and was mixed with the appropriate level of precursors. Arginase (Sigma) and Pentopan 500 BG (Novo Nordisk) were added at 10,000 and 7,650 units/kg flour, respectively. After a standard protocol for relaxing, sheeting, rolling and moulding, the dough was fermented for 45 min at 32 °C and baked in pans at 240 °C for 25 min. After cooling, the crust was removed, packed in aluminium-PE laminated bags and stored at -18 °C. Chemicals used in bread baking were of high purity and where necessary certified for human consumption.

2.2. Crust extraction and analysis

A 50 g bread crust was ground and suspended in 200 ml buffer (pH 7.0) and extracted with three volumes of 50 ml diethylether/pentane (2:1). The internal standard (IS) 1,2-dichlorobenzene was added before extract concentration under a stream of nitrogen to a final volume of 0.2 ml. Extracts were analysed by GC-MS on a 30 m DB-wax column with Helium as carrier gas. Spectra were recorded in the EI mode with an ionisation voltage of 70 eV. The GC oven temperature was ramped at 40 °C for 10 min and raised at 3 °C/min to 240 °C with a final hold of 30 min. Compounds were identified by interpretation of mass spectra and comparison to reference spectra and where available linear retention indices (LRI). Concentrations were estimated relative to the IS. All crust samples were analysed in duplicate.

2.3. Sensory evaluations

A trained sensory panel (n=8) profiled the odour characteristics of bread crusts from different combinations of flavour precursors and enzyme additions, in total 7 samples, in 3 replicates. The odour and flavour (in mouth assessment) changes in both crust and crumb by addition of different levels of proline (0, 0.3 and 1.0 g/kg flour) with or without glucose (10 g/kg flour) were evaluated by another sensory panel (n=9) in 2 replicates. Nine untrained assessors participated in a preliminary affective test.

3. RESULTS AND DISCUSSION

More than 80 compounds were identified in crust extracts from the different model breads, however, no sulfur-containing compounds and only few secondary lipid oxidation products were found. The crust aroma compounds, AP and ATHPs, were identified in trace amounts in breads with added proline. The ornithine-enriched bread also contained AP, but ATHPs could not be detected. ATHP, however, was present in breads without precursor additions and breads with added Pentopan, arginine or arginine/arginase; AP was not detected in these breads (Table 1). The addition of arginine or ornithine/fructose increased the level of pyrroles and pyrazines in the crust.

A similar increase was observed for the arginine/arginase bread (Figure 1a), indicating a low activity of arginase (see also Table 1).

In breads with Pentopan, the total amount of alcohols increased which was mainly due to 2-methoxy-4-vinylphenol and vanillin. These phenols may derive from thermal degradation of ferulic acid liberated from hemicellulose by esterases in Pentopan.

Addition of proline (1.0 g/kg) showed significant increases in pyrroles, pyrrolizines and azepinones. Their quantities increased further in presence of Pentopan (Figure 1b). Pentopan most likely generated pentoses that participated in the Maillard reaction with proline which explained such a synergism.

Table 1. Generation of 2-acetyl-1-pyrroline (AP) and 6-acetyl-1,2,3,4-tetrahydropyridine (ATHP) in breads with different enzyme and precursor additions to the dough.

Precursors added	AP	ATHP	ATHP[b]
No addition (reference)	-[a]	+	-
Pentopan	-	+	-
Arginine (5.0 g/kg)	-	+	-
Arginine (5.0 g/kg) + arginase	-	+	-
Ornithine (5.0 g/kg) + fructose (5.0 g/kg)	+	+	-
Proline (1.0 g/kg)	+	+	+
Proline (1.0 g/kg) + Pentopan	+	+	+

[a]- Not detected; + Trace, confirmed by MS and LRI. [b]Mixture of unstable isomers.

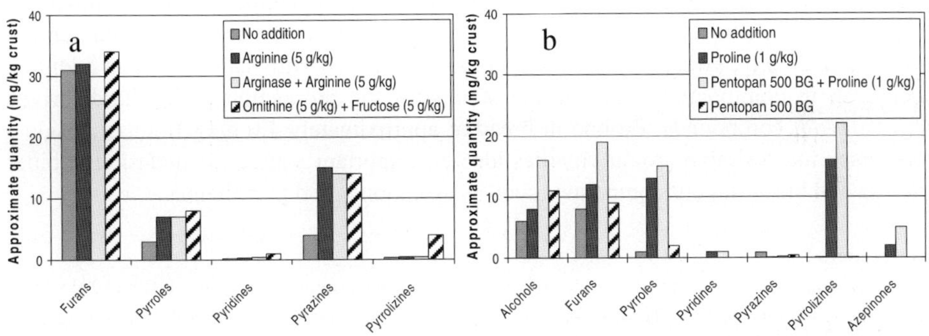

Figure 1. Influence of arginase, arginine and ornithine on the formation of Maillard volatiles (a) and Pentopan and proline on the formation of alcohols and Maillard volatiles (b) in bread crust.

Seven 2,3-dihydro-1*H*-pyrrolizines and two azepinones were identified in the bread crusts. Both classes of compounds were predominantly identified in the proline-enriched breads. Only traces of azepinones were present in the ornithine/fructose bread. Azepinones were not detected in the reference and breads with added arginine or Pentopan alone. Pyrrolizines have earlier been shown in heated proline sugar systems and some in beer [4]. The pyrrolizines have been described by 'smoky' and other odours [5], but their taste properties have not been reported. Two azepinones were

identified in relatively large quantities in proline-enriched breads. 7-Methyl-2,3,6,7-tetrahydrocyclopent(b)azepin-8(1*H*)-one has also been reported in roasted malt and beer with a bitter taste at >10 ppm in water [3].
The crust odour profiling data from breads with arginine, ornithine or Pentopan were analysed by ANOVA and principal components analysis (PCA). The arginine/arginase bread showed a tendency to increase toasted and cracker-like odours compared to the reference bread and the bread with arginine alone. Pentopan addition gave more pronounced fatty, sweet and burnt odours, whereas the ornithine/fructose bread increased cracker-like, burnt and popcorn-like odours compared to the reference.
Sensory profiling of crust odour and flavour in breads with added proline and/or glucose showed that the crust flavour better discriminated the samples than the crust odour. Bitter taste was mostly associated with the 1.0 g/kg proline addition. Addition of glucose (10 g/kg) to this system also increased the perceived burnt flavour. A lower level of proline addition (0.3 g/kg) had little effect on the odour and flavour profile. However, when glucose was added the burnt flavour, and chemical-like and burnt odours increased. A preliminary affective test with proline-enriched breads indicated that bread crust flavour preference could be explained by variation in sweet taste and fresh bread-like flavour and odour. Bitter taste and burnt flavour negatively correlated with crust flavour preference.

4. CONCLUSIONS

Addition of arginase and arginine to bread has little effect on the crust odour profile and shows no important changes in the level of flavour components. Bread with relatively high levels of Pentopan gives large increases in 2-methoxy-4-vinylphenol and vanillin in the crust. Fatty, sweet and burnt odours are associated with such Pentopan addition. Pentopan in combination with proline stimulates the production of Maillard-derived crust flavour compounds. Proline in bread at approximately 1.0 g/kg flour is, beside a precursor for desirable aroma volatiles also an important source for undesirable bitter tasting and burnt flavour compounds such as azepinones and pyrrolizines.

References

1. H. Maarse (ed.), Volatile compounds in foods and beverages, New York, USA (1991) 41.
2. A.J. Taylor and D.S. Mottram (eds.), Flavour science: recent developments, Cambridge, UK (1996) 221.
3. R. Tressl, B. Helak, H. Köppler and D. Rewicki, J. Agric. Food Chem., 33 (1985) 1132.
4. R. Tressl, D. Rewicki, B. Helak, H. Kamperschröer and N. Martin, J. Agric. Food Chem., 33 (1985) 919.
5. G.R. Waller and M.S. Feather (eds.), The Maillard reaction in foods and nutrition, Washington, USA (1983) 185.

The spectator role of potassium hydroxide in the isomerisation of eugenol to isoeugenol

Christophe C. Galopin[a], Cristian Bologa[b] and William B. DeVoe[a]

[a]*Givaudan Flavors Corporation, Research and Development, Organic Synthesis Lab, 1199 Edison Drive, Cincinnati, OH 45216, USA;*
[b]*Univiversity of New Mexico, School of Medicine, Division of Biocomputing, 2703 Frontier NE, Albuquerque, NM, 87131, USA*

ABSTRACT

The role of potassium hydroxide in the isomerisation of eugenol to isoeugenol was investigated. The theoretical value of the enthalpy of activation of an isomerisation mechanism of eugenol involving a hydroxide was calculated to be between 38.8 and 44.8 kcal/mol. The experimental values of the enthalpy and entropy of activation of the isomerisation were 32.3 kcal/mol and 0.2 cal/(K mol), respectively. The experimental value of the enthalpy suggests that the isomerisation is going through a less energy-demanding path than the one involving the hydroxide. The low value of the entropy suggests that the mechanism involves a highly concentrated species, such as water.

1. INTRODUCTION

Isoeugenol is an ingredient with a sweet spicy, clove-like odour and a woody nuance; it is described as finer and fuller in odour than eugenol with a balsamic, amber quality [1]. While isoeugenol is only present in Nature in small quantities, it can be easily obtained by isomerisation of eugenol from cloves in the presence of potassium hydroxyde and heat [2]. The isomerisation process is dependent on the concentration of KOH [3]. This observation is considered as proof that KOH participated in the isomerisation by abstracting a proton from eugenol [4]. However, these allylic rearrangements are unlikely to proceed via a simple proton abstraction [5]. In the case of eugenol the isomerisation occurs on its potassium salt which has an electron-rich allylic system. This electronic effect reduces the instability of the allylic proton because the obtained system has many unfavourable mesomeric forms (Figure 1).

It is suggested that KOH, in addition to being a key factor in dissolving eugenol in water, may only have a physical impact - such as setting the right ionic strength for the reaction to occur [6]. Several authors [7-9] have noticed that neutral species such as KCl or KF could also affect the kinetics of the isomerisation while other hydroxides such as NaOH [10] had a much more limited effect than KOH.

In this study we compared the measured experimental enthalpy of activation, $\Delta H^{\neq}$, of the isomerisation of eugenol, in the presence of KOH, with the theoretical value assuming that the hydroxide participates actively in the reaction (Figure 2).

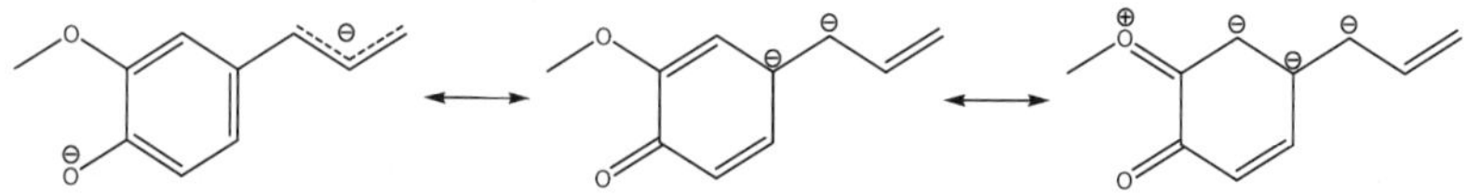

Figure 1. Examples of unfavourable mesomeric forms of eugenol.

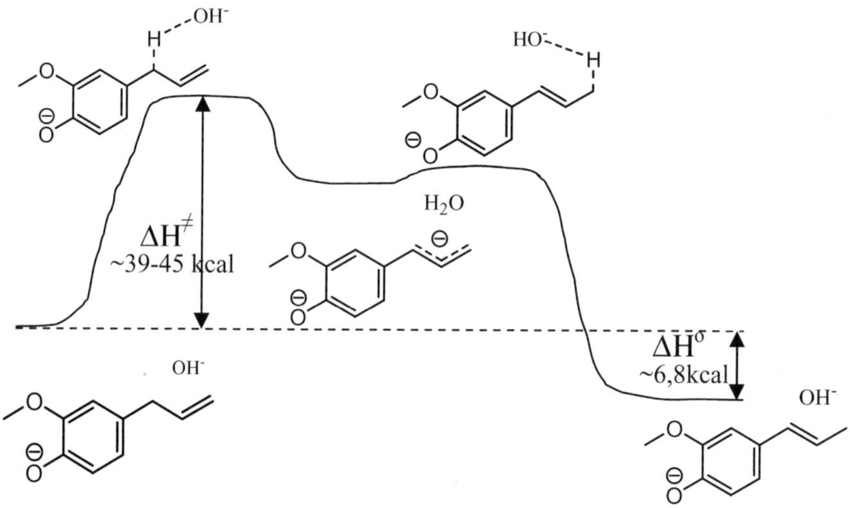

Figure 2. Energy diagram of the isomerisation of eugenol involving a hydroxide.

2. METHODS AND PROCEDURES

The value of $\Delta H^{\neq}$ can be measured experimentally thanks to Eyring's equation which relates the energy of activation, $\Delta G^{\neq} = \Delta H^{\neq} - T\Delta S^{\neq}$, to k, the reaction kinetic parameter:

$$k = \frac{k_B T}{h} e^{-\frac{\Delta G^{\neq}}{RT}} \quad \textit{also written as} \quad R \ln\left(\frac{k_B T}{k h}\right) = \Delta H^{\neq} \frac{1}{T} - \Delta S^{\neq}$$

The second form of the equation is a line with a slope equal to $\Delta H^{\neq}$, the activation enthalpy, and a y-intercept equal to $-\Delta S^{\neq}$. Assuming that this isomerisation involves a rate-determining elemental step of the first order, k can be obtained, at each temperature, from the slope of the function: *ln([eugenol]₀/[eugenol]) = kt.*

To determine the kinetic parameters, k, the reaction was run at eight different time intervals at six different temperatures. Each data point was obtained, in triplicate, by

heating 1 ml of an aqueous solution, containing 0.1 g of eugenol and 0.4 g of KOH 85%, in an Emrys[TM] Optimizer microwave (Personal Chemistry, Uppsala, Sweden). The instrument was set up so that the timer only started once the solution had reached its set temperature. After cooling down to room temperature, the reaction mixture was quenched with 1 ml of aqueous H_2SO_4 25% and extracted with 3 ml of MtBE. The MtBE layer was injected in a 6890N GC (Agilent Technologies, Wilmington, DE). The ratio [eugenol]/[eugenol]$_o$ was calculated by dividing the GC area of eugenol by the sum of the GC areas of eugenol, *cis*-isoeugenol and *trans*-isoeugenol.
The theoretical value of $\Delta H^{\neq}$ was calculated by optimising the geometries of all species in Figure 2 using the semi-empirical methods AM1 [11] and PM3 [12]. A transition state search was performed at the same levels of theory using a quadratic synchronous transit method [13] in order to find the coordinates of the two transition states, which were further refined using the eigenvector following method [14].

3. RESULTS AND DISCUSSION

The slopes of the kinetic curves at different temperatures (Figure 3) and the Eyring's plot allowed the determination of the experimental values of $\Delta H^{\neq}$ and $\Delta S^{\neq}$ to be 32.3 ± 0.6 kcal/mol and 0.2 ± 1.4 cal/(K mol), respectively. The experimental value of $\Delta H^{\neq}$ is significantly lower than either theoretical value of 38.8 (AM1) or 44.8 (PM3) kcal/mol. This kinetic experiment demonstrates that the isomerisation of eugenol proceeds through a step that is more thermodynamically favourable than the abstraction of one of the allylic protons by KOH. Moreover the low value of the entropy of activation, $\Delta S^{\neq}$, indicates that formation of the transition state from the starting material only involves small perturbations. This observation suggests that the mechanism of the isomerisation is either intramolecular, or, intermolecular and involving a species in very high concentration such as water.

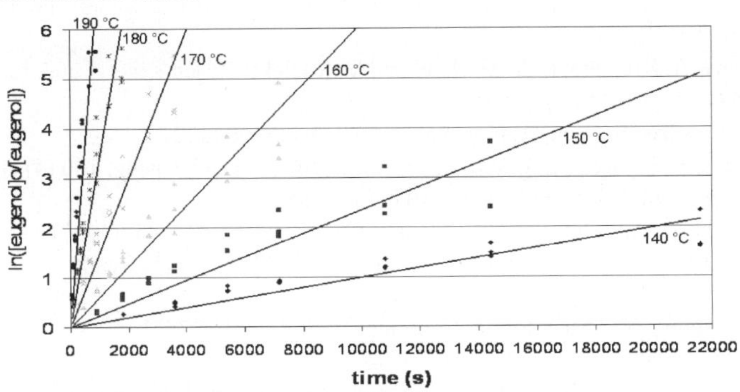

Figure 3. Kinetic equations at different temperatures.

While an intramolecular mechanism, such as a 1,3-proton shift, is thermally disallowed [15], an ene-type [4+2] sigmatropic rearrangement involving water seems like a plausible mechanism (Figure 4), even though the highly structured transition state would suggest a negative entropy of activation.

Although it is difficult to write a reaction mechanism explaining all the data, it is clear that neither the value of the enthalpy of activation, nor that of the entropy of activation support the theory of a bimolecular interaction of eugenol with a hydroxide. The reported relationship [1] between the rate of isomerisation and the concentration of KOH must, therefore, be due to a physical effect rather than a chemical effect.

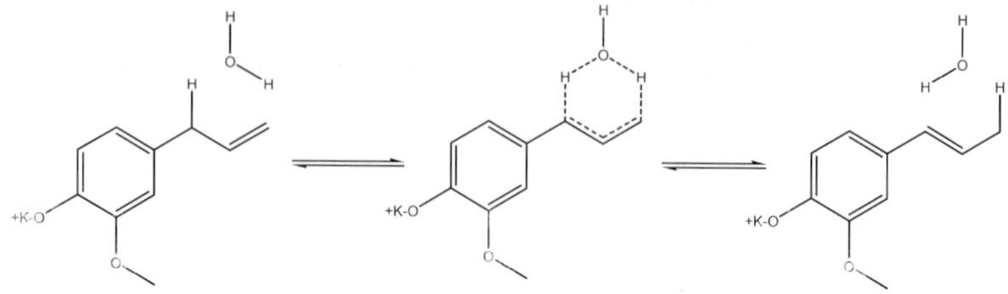

Figure 4. Example of an intermolecular mechanism involving water.

4. CONCLUSION

The experimental value of the $\Delta H^{\neq}$ of the isomerisation of eugenol to isoeugenol in the presence of KOH does not correlate with the theoretical value of a mechanism involving abstraction of one allylic proton by KOH. The low value (0.2 ± 1.4 cal/(K mol)) of $\Delta S^{\neq}$ also indicates that the mechanism is unlikely to be a bimolecular interaction with KOH. This experimental work supports the theory [4] that the role of KOH in the isomerisation of eugenol is not chemical.

References

1. Flavor-Base Pro, Leffingwells and Associates, FEMA# 2468 (2001).
2. J. Ehrlich, Propenyl derivatives of aromatic hydrocarbons such as isoeugenol, US Patent No. 19301230 (1930).
3. L. Cerveny, A. Krejcikova, A. Marhoul and V. Ruzicka, React. Kinet. Catal. Lett., 33 (2) (1987) 471.
4. D. Kishore and S. Kannan, Appl. Catal. A, 270 (1-2) (2004) 227.
5. V. Kobychev, N. Vitkovskaya and B. Trofimov, Int. J. Quant. Chem., 100 (4) (2004) 367.
6. R. Horiuchi, Bull. Chem. Soc. Japan, 10 (1935) 314.
7. G. Salmoria, E. Dall'Oglio and C. Zucco, Synth. Commun., 27 (24) (1997) 4335.
8. A. Radhakrishna, S. Suri, K. Rao, K. Sivaprakash and B.B. Singh, Synth. Commun., 20 (3) (1990) 345.
9. A. Loupy and L.-N. Thach, Synth. Commun., 23 (18) (1993) 2571.
10. Unpublished results of Givaudan Flavors R&D.
11. M.J.S. Dewar, E.G. Zoebisch, E.F. Healy and J.J.P. Stewart, J. Am. Chem. Soc., 107 (1985) 3902.
12. J.J.P. Stewart, J. Comput. Chem., 10 (2) (1989) 209.
13. C. Peng and H.B. Schlegel, Israel J. Chem., 33 (1993) 449.
14. J. Baker, J. Comput. Chem., 7 (4) (1986) 385.
15. R.B. Woodward and R. Hoffman, J. Am. Chem. Soc., 78 (11) (1965) 2511.

W.L.P. Bredie and M.A. Petersen (Editors)
Flavour Science: Recent Advances and Trends

Characterisation of key odorant compounds in creams from different origins with distinct flavours

Estelle Pionnier and Daniel Hugelshofer

Nestlé Product Technology Centre Konolfingen, Nestlé-Strasse 3, 3510 Konolfingen, Switzerland

ABSTRACT

Key odorants from four different creams were analysed by Headspace Sorptive Extraction (HSSE) and Gas Chromatography-Olfactometry-Mass Spectrometry (GCO-MS) using a detection frequency methodology. Among the odour peaks detected in one or more of the four creams, 32 aroma compounds were identified such as ketones, acids, lactones and sulfur compounds. Nine key odorants seemed to contribute actively to the 'yoghurt' cream flavour (diacetyl, acetoin, dimethyl trisulfide, 2-nonanone, butanoic acid, acetic acid, dimethyl sulfide, 2-butanone and one unknown). The aroma compounds 2-pentanone, dimethyl trisulfide, 2-nonanone and 2 unknown compounds were the major contributors to the 'animalic' cream flavour. The 'sterilised' cream flavour was predominantly due to the presence of dimethyl trisulfide, 2-nonanone, 2-pentanone, 2-heptanone, 2-furfural and 2-furanmethanol. Finally, none of the key odorants in the 'milky' flavour cream seemed to play a major role for its global aroma except three unknown compounds detected by 7 judges.

1. INTRODUCTION

Flavour quality is one of the most important factors to achieve consumer acceptance and preference. In this context, numerous studies have dealt with the flavour composition of dairy products [1-3]. This study focused on a better understanding of cream flavours depending on their origin. Thirty-nine creams from 12 different countries, made by different processes (pasteurised, sterilised or UHT) and containing different fat levels (15-40% fat) were tasted by 9 panellists. The samples were grouped according to their flavour characteristics defined as: 'yoghurt', 'animalic', 'sterilised' and 'milky'. One cream per group was selected for further studies to identify compounds responsible for the different odours and to compare the aroma composition of the four creams.

2. MATERIALS AND METHODS

2.1. Creams

Four creams were selected according to their flavour characteristics: cream 1 'yoghurt' (15% fat, pasteurised), cream 2 'animalic' (40% fat, UHT treated), cream 3 'sterilised' (25% fat, sterilised), cream 4 'milky' (35% fat, pasteurised).

2.2. Aroma extract preparation

Cream aroma extracts were obtained from Headspace Sorptive Extraction (HSSE) using stir bars coated with polydimethylsiloxane (Gerstel, Germany). The stir bar was placed in a perforated glass capsule positioned in the headspace of a 1l conical flask filled with 500 ml of stirred cream. The extraction lasted 5.30 h at 42 °C. Stir bars were preconditioned by thermal desorption before each extraction.

2.3. GCO-MS analysis

An Agilent 6890 gas chromatograph equipped with a sniffing port supplied with humidified air, coupled with a mass selective detector (MSD 5973N, Agilent Technologies) was used to perform GCO analyses. The volatiles were thermally desorbed in splitless mode (oven of the thermodesorption unit (TDU) programmed from 20 °C to 240 °C (1 min) at 40 °C/min), and cryofocused in a CIS-4 PTV injector at -23 °C. Helium was used as carrier gas. The injection of the cryofocused analytes on the column was done in solvent vent mode, by rapid heating of the injector from -23 °C to 260 °C (3 min) at 12 °C/s. A HP-FFAP column (50 m x 0.2 mm ID, 0.3 μm film thickness; Agilent Technologies) was used with He at 1.5 ml/min. The oven program was: from 40 °C to 70 °C at 5 °C/min, then from 70 °C to 240 °C at 3 °C/min. The GC effluent was split 1:1 between the mass spectrometer and the sniffing port.

Frequency of detection was used to perform the olfactometry study [4] with a trained panel of 10 people. A complete sniffing analysis lasted 62 min and was performed by two judges who each sniffed twice during 15 min to stay alert. The panellists were asked to delimit each odorant zone and to describe the perceived odours. The aromagrammes from the 10 subjects were joined. To be considered as a key odorant, the molecule of interest had to be detected at the same retention time at least by four people. The identification of the potent odorants was based on comparison of GC retention indices (RI), mass spectra and odour properties. Linear retention indices (RI) of the compounds were calculated using a series of n-alkanes (C8-C32) injected under the same chromatographic conditions.

3. RESULTS AND DISCUSSION

A total of 32 odorants were positively identified and belonged to various chemical classes such as ketones, acids, lactones, sulfur compounds and others (Table 1).

Table 1. Key odorant compounds identified in the four creams (C1, C2, C3, and C4).

Peak no.	RI	Compound	Odour description	Detection frequency C 1	C 2	C 3	C 4
1	897	**Dimethyl sulfide**	Sulfury, rancid, cheese	**6**	4	< 4	4
2	914	2-Propanone	No common description	< 4	4	< 4	< 4
3	949	**2-Butanone**	Perfume, milk	**6**	4	< 4	< 4
4	992	**Diacetyl**	Buttery	**9**	< 4	< 4	< 4
5	994	**2-Pentanone**	Caramel, cream, milk, burnt	< 4	**9**	**7**	< 4
6	1022	α-Pinene	Fruit	< 4	< 4	< 4	4
7	1079	2-Hexanone	Floral, medicinal	< 4	< 4	4	< 4
8	1190	**2-Heptanone**	Dairy, fruity	< 4	5	**6**	< 4
9	1289/1293	2-Octanone/Octanal	Cooked milk, floral	4	5	< 4	< 4
10	1304	**Acetoin + unknown**	Buttery, creamy + moldy	**9**	< 4	< 4	< 4
11	1393/1396	**Dimethyl trisulfide+ 2-Nonanone**	Rancid, cabbage, hot milk, floral	**8**	**8**	**9**	4
12	1453	1-Octen-3-ol	Cardboard, plastic, floral	< 4	5	< 4	< 4
13	1462	**Acetic acid**	Acidic, parmesan	**7**	< 4	< 4	< 4
14	1487	**2-Furfural**	Caramel, bread, milk	< 4	5	**6**	< 4
15	1564	1-Octanol	Floral, green	4	< 4	< 4	< 4
16	1606	2-Undecanone	Floral ,milky	< 4	< 4	4	< 4
17	1637	**Butanoic acid**	Cheese, rancid, acidic	**8**	< 4	< 4	< 4
18	1676	**2-Furanmethanol**	Burnt, floral, sweet	4	5	**6**	< 4
19	1819	2-Tridecanone	Cardboard, cheese, soapy	4	< 4	< 4	5
20	1824	δ-Hexalactone	Milky, cheesy, coffee	< 4	< 4	< 4	5
21	1855	Hexanoic acid	Fermented, acidic	< 4	< 4	4	5
22	1931	Dimethyl sulfone	Cooked milk, flower	< 4	5	4	< 4
23	1995	δ-Octalactone	Caramel, vanilla	< 4	< 4	< 4	4
24	2055	γ-Nonalactone	No common description	< 4	< 4	< 4	4
25	2171	γ-Decalactone	Caramel, milky	< 4	< 4	4	< 4
26	2171/2177	Nonanoic acid+ γ-Decalactone	Cooked milk, soapy	< 4	< 4	< 4	4
27	2220	δ-Decalactone	Milky, vanilla, cheese	< 4	< 4	< 4	4
28	2284	n-Decanoic acid	Burnt, buttery	< 4	< 4	4	< 4
29	2340	δ-Undecalactone	Vanilla, caramel	< 4	< 4	< 4	4
30	2414	δ-Dodecalactone	Old milk, floral, fruit	< 4	5	< 4	< 4
31	2423	γ-(*Z*)-6-Dodecenolactone	Milky, floral	4	< 4	< 4	< 4
32	2536	5-Hydroxymethyl-2-furfural	Milky, vanilla, caramel	< 4	4	4	< 4

Fifteen key odorants were detected in cream 1 by 4 or more judges. Twelve compounds were positively identified and three compounds responsible for 'oxidised, cardboard' (RI=1108, 7 judges), 'milk, floral' (RI=1167, 5 judges), 'nutty, bread' (RI=1358, 5 judges) odour notes could not be assigned any compound. Eight of the identified compounds and one unknown odorant seemed to be the major contributors to the 'yoghurt' cream flavour as they were detected by at least 6 judges: diacetyl, acetoin, dimethyl trisulfide, 2-nonanone, butanoic acid, acetic acid, dimethyl sulfide, 2-butanone and the compound responsible for 'oxidised, cardboard' odours.

Eighteen key odorant compounds were detected in cream 2. Thirteen could be identified and the 5 remaining unknown compounds were responsible for 'milky' (RI=1072, 4 judges), 'green, metallic, cellar' (RI=1106, 6 judges), 'mushroom, burnt' (RI=1312, 7 judges), 'cereal' (RI=1358, 4 judges), and 'bouillon, milky' (RI=1704, 5 judges) odours. The typical 'animalic' cream flavour seemed to be predominantly due to the presence of 2-pentanone, dimethyl trisulfide, 2-nonanone and the two unknown compounds with RI values of 1106 and 1312.

Seventeen molecules were detected as key odorants in cream 3. Among them, 5 compounds could not be identified and were described with 'floral, milky' (RI=1250, 5 judges), 'floral, cardboard' (RI=1312, 5 judges), 'nutty, bread' (RI=1358, 4 judges), 'milky, burnt' (RI=1657, 4 judges), and 'vanilla, milky, medicinal' (RI=2144, 5 judges) odorant notes. The frequency detection results showed that the major contributors for the 'sterilised' cream flavour were dimethyl trisulfide, 2-nonanone, 2-pentanone, 2-heptanone, furfural and 2-furanmethanol.

Nineteen compounds were detected as key odorants in cream 4. Eight compounds are still unknown and were described with the following attributes: 'no common attribute' (RI=1054, 4 judges), 'milky, rubbery, solvent' (RI=1105, 7 judges), 'dairy, burnt' (RI=1250, 5 judges), 'cardboard, floral, oxidised' (RI=1317, 4 judges), 'popcorn, cereals' (RI=1359, 7 judges), 'burnt, earth, floral, bread' (RI=1654, 7 judges), 'milky, solvent, lemon' (RI=2190, 4 judges), and 'milky, cardboard' (RI=2320, 4 judges). Except three unknown compounds detected by 7 judges with RI values of 1105, 1359 and 1654, no odorant identified in this cream was shown to be a major contributor for this 'milky' flavour.

4. CONCLUSION

The differences in the flavour of four dairy creams could be explained by some odour-active compounds uniquely present in the specific cream and by concentration differences within a common set of flavour compounds between the creams. More aroma isolation experiments should further identify the compounds for some of the 'unknown' odour-active GC regions. Finally, the quantification of key odorants will be performed for each cream in order to finalise this investigation with sensory analysis of reconstituted aroma mixtures.

References

1. L. Moio, P. Etievant, D. Langlois, J. Dekimpe and F. Addeo, J. Dairy Res., 49 (1994) 4825.
2. J. G. Bendall, J. Agric. Food. Chem., 49 (2001) 4825.
3. Y. Karagül-Yücer, M. Drake and K.R. Cadwallader, J. Sens. Stud., 19 (2004) 1.
4. S. Le Guen, C. Prost and M. Demaimay, J. Agric. Food. Chem., 48 (2000) 1307.

5

Flavour changes in food production and storage

W.L.P. Bredie and M.A. Petersen (Editors)
Flavour Science: Recent Advances and Trends

Aroma changes from raw to processed products in fruits and vegetables

Leif Poll, Ghita S. Nielsen, Camilla Varming and Mikael A. Petersen

The Royal Veterinary and Agricultural University, Department of Food Science and The Centre for Advanced Food Studies, Rolighedsvej 30, 1958 Frederiksberg C, Denmark

ABSTRACT

The present paper deals with aroma changes in the production and storage of precooked potatoes, frozen leeks and fruit juices (apple and blackcurrant). In precooked potatoes and frozen leeks development of aliphatic aldehydes results in off-flavours, however, blanching (leeks) and exclusion of oxygen reduce off-flavour formation. In fruit juices high losses of esters and terpenes in the different processing steps are seen and result in juices of poor aroma quality with losses of 80 to 90% of the aroma volatiles. The aroma losses and changes are results of enzyme activity, thermal influence, evaporation and removal by water or waste.

1. INTRODUCTION

The consumption of fruits and vegetables by the Danish population is lacking behind the recommendations from the health authorities. In order to successfully promote a higher consumption of these commodities, two factors are of importance namely, convenience and an appealing, characteristic flavour.

For many fruit and vegetable products it is believed that the best sensory quality is obtained if the aroma composition is close to that of the raw material. Processing often results in a more convenient and stable product compared to fresh fruits and vegetables, but processing also changes the aroma composition, often by a loss of aroma compounds but rearrangements and formation of new compounds (off-flavours) can also be important. Depending on the fruit and vegetable species and on the product ready for sale, possible processing methods include peeling, slicing, blanching or heating, freezing, pressing, pasteurisation and concentration or drying.

Four main reactions or processes are important for the aroma composition in processed products: *enzyme activity, thermal influence, evaporation and removal by water or waste.* In this paper four examples of aroma changes during processing of fruits and vegetables are presented, namely precooked potatoes, frozen leeks, apple juice and blackcurrant juice.

2. MATERIALS AND METHODS

Dynamic headspace analysis using Tenax traps and thermal desorption has been used in most of the studies presented [1,2], but solvent extraction with ether-pentane has also been used [3]. The quantification of aroma compounds was carried out using model systems or spiking. To point out sensory important aroma compounds GC-sniffing has been an important tool, the Nasal Impact Frequency Profiling [1,2,4] has been found very useful. Threshold values have also been used to indicate important compounds [5].

3. RESULTS AND DISCUSSION

3.1. Precooked potatoes

In the catering sector, precooked potatoes are important convenience products for meal preparations. Important aroma compounds in cooked potatoes are methional, methoxypyrazines and C5-C10 aldehydes. The industrial precooking process has the following steps: peeling (steam), vacuum packaging, cooking, cooling and storage (up to 28 days at 2-5 °C). It has been shown that the industrially precooked potatoes develop a rancid, cardboard-like off-flavour after approximately 7 days. The same off-flavour is developed in peeled cooked (not vacuum packed) potatoes 7 h after cooking [3]. Aroma analysis of fresh boiled and boiled stored (24 h) potatoes showed a substantial increase in the aldehyde concentration during storage (Table 1).

Table 1. Concentration of aldehydes in fresh boiled and boiled stored potatoes, sniff intensity and odour threshold values in water and potato mash for the same aldehydes. Modified after [3,5].

	Concentration (ppb)		Odour threshold (ppb)		Sniff intensity (0-5)[b]	
	Fresh boiled	Stored[a] boiled	Water	Potatoes	Fresh boiled	Stored boiled
Pentanal	28	120	22	-[c]	0.1	0.6
Hexanal	43	1300	2.5	47	2.3	3.1
Nonanal	0	6	-	-	0.5	1.8
(*E*)-2-Octenal	2	45	0.34	600	1.5	0.8
(*E*)-2-Nonenal	1	10	0.23	-	3.1	4.4
(*E*,*E*)-2,4-Heptadienal	1	35	56	-	0	0.8
(*E*,*E*)-2,4-Nonadienal	1	38	0.0017	-	0	4.5
(*E*,*E*)-2,4-Decadienal	1	72	0.045	63	0.4	1.0

[a]For 24 h. [b]Five judges evaluated the intensity. [c]Value not determined.

The odour threshold values (Table 1) in water for some aldehydes are very low, for a few of the aldehydes the thresholds in potato mash were also determined, and as expected these values were much higher. GC-sniffing (5 judges) also indicated that the odour intensity of aldehydes in stored potatoes was higher than in the fresh boiled samples. Especially (*E*,*E*)-2,4-nonadienal, which has an almond, paint, rancid-like odour seems to be important for the developed off-flavour. The aldehydes originate from fatty acids. It is hypothesised that lipoxygenase activity in the disrupted cells after peeling produce hydroperoxides, which again produce aldehydes by autoxidation. The cooking process could also, for a shorter time, result in high lipoxygenase activity.

3.2. Frozen leek

The aroma of freshly cut leek mainly consists of numerous sulfur-containing volatile compounds originating from the alliinase catalysed decomposition of cysteine sulfoxides. However, the lipoxygenase pathway also contributes to the formation of aroma in fresh and especially in frozen, stored leek slices by the production of mainly aldehydes. As illustrated in Figure 1, the amount of sulfur compounds decreases substantially (illustrated by dipropyl disulfide) whereas the amount of aldehydes increases (illustrated by (*E*,*E*)-2,4-nonadienal) during frozen storage of leek slices.

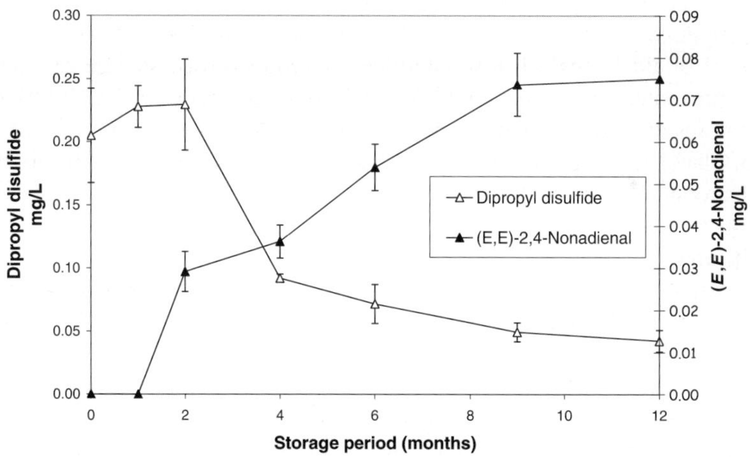

Figure 1. Development of selected aroma compounds in leek slices during frozen storage. Vertical bars indicate standard deviation. Modified after [6].

Production of frozen vegetables often includes a blanching step mainly to inactivate enzymes and thus prevent off-flavour formation but for leek slices the loss of texture, a side-effect of the blanching procedure, is unfavourable to the quality. Another consequence of blanching is the leaching out of aroma compounds to the blanching water, which for some aroma compounds is considerable [7]. Table 2 illustrates the effect of blanching prior to frozen storage as well as the impact of storing the leek slices in 100% nitrogen. The unblanched samples stored for 12 months in atmospheric air (UNB 21% O_2) showed a decline in sulfur compounds (represented by methyl propenyl disulfide and dipropyl disulfide) and an increase in aldehydes (represented by hexanal and (*E*,*E*)-2,4-decadienal). This development was partly prevented by the blanching procedure as the decline of sulfur compounds was less and the formation of aldehydes was prevented effectively.

Table 2. Concentration of selected aroma compounds[a] found in fresh leek (0 M) and in leek stored frozen for 12 month (12 M). Modified after [8].

	Hexanal	(*E*,*E*)-2,4-Decadienal	Methyl propenyl disulfide	Dipropyl disulfide
UNB[b] 0 M	0.31±0.09	0.0±0.0	0.17±0.06	0.62±0.18
B[c] 0 M	0.027±0.003	0.0±0.0	0.06±0.01	0.46±0.22
UNB 21% O_2[d] 12 M	4.7±1.4	0.23±0.49	0.009±0.003	0.39±0.11
B 21% O_2 12 M	0.78±0.2	0.0±0.0	0.090±0.009	0.50±0.06
UNB 0% O_2[e] 12 M	3.0±0.7	0.049±0.035	0.029±0.005	1.0±0.1
B 0% O_2 12 M	0.11±0.02	0.0±0.0	0.081±0.002	0.52±0.09

[a]All values are in mg/l ± standard deviation. [b]Unblanched slices. [c]Blanched slices. [d]Atmospheric air. [e]100% nitrogen.

Nitrogen packaging prior to frozen storage had a very positive effect on the keeping of sulfur compounds both in the blanched and the unblanched samples and the development of aldehydes was also significantly influenced by storage in nitrogen. As blanching prevented the formation of hexanal and (*E*,*E*)-2,4-decadienal during frozen storage, the results indicate that the production was due to enzymatic activity. Frozen storage of blanched samples in 100% nitrogen obtained an aroma profile closest to the fresh leek aroma.

3.3. Fruit juice

Juice processing is basically a very simple process. The fruits are crushed (the pulp) and by pressing, the fluid material from the vacuoles (the juice) is separated from the cell wall material (the press cake). This process will result in a cloudy juice, where the sensorily important compounds are sugars, acids, phenolics and aroma compounds. The pressing process from pulp to juice results in a reduction of esters in the juice depending on their carbon number and polarity (Table 3). Long chained esters are retained in the cell wall material to a larger extend than short chained esters. Aldehydes and alcohols are transferred to the juice to the same degree as short chained esters. During juice processing aroma compounds can be lost by evaporation and enzymatic activity. Table 4 shows that in pasteurised apple pulp 71-74% of the ester content was retained after 1.3% weight loss by evaporation (1 h 20 °C) whereas only 41-44% of the esters was retained in the unpasteurised pulp. The difference is most likely due to esterase activity. Often the juice production process involves a short-time heating of the pulp for inactivation of enzymes and microorganisms, pectinolytic enzyme treatment, and pasteurisation of the juice to inactivate microorganisms.

Table 3. Transport of aroma compounds from apple pulp to juice after cold pressing determined by dynamic headspace analysis.

Compounds	Transfer from pulp to juice
C5 Esters	70%
C6 Esters	68-72%
C7 Esters	61%
C8 Esters	52-54%
C10 Esters	43%
C4 Alcohols	75%
C6 Alcohols	66%
C6 Aldehydes	78-79%

Table 4. Aroma retained in apple pulp after 1 h holding time at 20 °C. Weight loss due to evaporation was 1.3%. Results from dynamic headspace analysis.

Compounds	Apple pulp	Pasteurised apple pulp
Butanol	61%	77%
Ethyl butanoate	41%	74%
Hexyl acetate	44%	71%

In Table 5 the influence of these processes on selected aroma compounds in blackcurrants is shown. From raw material (thawed berries) to pasteurised juice rather high losses of aroma compounds (except for damascenone) are seen. An interesting increase of pinene and cineol was seen in the crushing and heating step which could be due to release of these compounds from their corresponding glucosides or terpene rearrangements. The pasteurisation step also causes a decrease of aroma compounds which could be due to evaporation. The blackcurrant juice was produced in a laboratory scale, but the same extent of loss was seen in an industrial plant [9].

Most of the juice on the market has been concentrated meaning that most of the water from the juice has been evaporated and this process results in a juice concentrate and an aroma distillate.

Table 5. Relative levels of aroma volatiles at different steps of blackcurrant juice processing [9].

	Ethyl butanoate	α-Pinene	Cineol	4-Terpineol	β-Damascenone
Thawed berries	100	100	100	100	100
Crushing	91	140	124	95	74
Heating[a]	61	182	126	87	134
Enzyme treatment[b]	45	160	82	80	126
Pressing juice[c]	31	37	57	70	78
Pasteurisation	26	5	38	44	156

[a]75 °C for 2 min. [b]50 °C for 2 h. [c]400 bar, 40-50 °C.

If the process runs perfectly the total amount of aroma is recovered in the distillate. Experiments on an industrial concentration plant [10] shoved that approximately 50% of the terpene hydrocarbons, 30% of terpene alcohols and 20% of the esters were lost in the concentration process. On the other hand benzene and furan derivates were recovered by more than 100%. Juices which originate from concentrates and aroma distillates have a less fruity and more caramel-like flavour. The industrial concentration process involves a heating process (5-10 min at 90-100 °C) and a distillation process, leading to aroma losses by evaporation as well as rearrangements of terpenes [10,11].

4. CONCLUSION

Aldehydes and sulfur compounds are produced by enzyme activity during peeling or slicing. Especially aliphatic aldehydes are important as off-flavours in precooked stored potatoes and in frozen stored leeks. Aldehydes in lower concentrations contribute to the fresh leek flavour but become off-flavours in higher concentrations. Blanching and short time heating will inactivate enzymes but can possibly for a short time accelerate enzyme activity. The blanching process can leach out aroma compounds from the leeks to the blanching water.

When fruits or berries are crushed the enzymes and substrates are mixed resulting in enzyme activity, which can lower the ester concentration and liberate terpenes from glucosides. Even approximately 1% evaporation of water by weight can lead to a high losses of the ester content in the juice process. In the concentration process evaporation of volatiles also takes place. Short time heating like a pasteurisation does not change the aroma composition in apple juice, whereas heating for a longer time results in decrease of some terpenes and increase of others. Furan compounds and damascenone are increased by heating. The cell wall material (press cake) retains substantial amounts of aroma compounds depending partly on chain length of the compounds.

References

1. G.S. Nielsen, L.M. Larsen and L. Poll, J. Agric. Food Chem., 52 (6) (2004) 1642.
2. C. Varming, M.A. Petersen and L. Poll, J. Agric. Food Chem., 52 (6) (2004) 1647.
3. M.A. Petersen, L. Poll and L.M. Larsen, Food Sci. Technol., 32 (1999) 32.
4. P. Pollien, A. Ott, F. Montigon, M. Baumgartner, R. Munõs-Box and A. Chaintreau, J. Agric. Food Chem., 45 (7) (1997) 2630.
5. K. Jensen, M.A. Petersen, L. Poll and P.B. Brockhoff, J. Agric. Food Chem., 47 (3) (1999) 1145.
6. G.S. Nielsen, L.M. Larsen and L. Poll, J. Agric. Food Chem., 51 (7) (2003) 1970.
7. J.-L. Le Quéré and P.X. Étiévant (eds.), Flavour research at the dawn of the 21st century, proceedings of the 10th Weurman flavour research symposium, Paris, France (2003) 678.
8. G.S. Nielsen, L.M. Larsen and L. Poll, J. Agric. Food Chem., 52 (15) (2004) 4844.
9. B.B. Mikkelsen and L. Poll, J. Food Sci., 67 (9) (2002) 3447.
10. J.-L. Le Quéré and P.X. Étiévant (eds.), Flavour research at the dawn of the 21st century, proceedings of the 10th Weurman flavour research symposium, Paris, France (2003) 741.
11. C. Varming, M.L. Andersen and L. Poll, J. Agric. Food Chem., 52 (25) (2004) 7628.

W.L.P. Bredie and M.A. Petersen (Editors)
Flavour Science: Recent Advances and Trends

Occurrence of polyfunctional thiols in fresh and aged lager beers

Catherine Vermeulen, Sabine Bailly and Sonia Collin

Unité de Brasserie et des Industries Alimentaires, Faculté d'ingénierie biologique, agronomique et environnementale, Université catholique de Louvain, Croix du Sud, 2 bte 7, B-1348 Louvain-la-Neuve, Belgium

ABSTRACT

In the present work, fresh Belgian lager beers have been specifically extracted by *p*HMB (*p*-hydroxymercuribenzoic acid), according to Tominaga *et al.* [1]. GC-MS, GCO and GC-PFPD analyses of the extracts revealed the presence of potent odorants such as 2-mercaptoethanol (roasted, garbage), 3-mercaptopropanol (roasted, potato), 3-mercaptohexanol (rhubarb, exotic fruit), 3-mercapto-3-methylbutanol (onion, vegetable), 2-mercapto-3-methylbutanol (onion), 1-mercapto-3-pentanone (burned), 4-mercapto-4-methyl-2-pentanone (blackcurrant, box tree), 3-mercaptopropyl acetate (roasted meat), 2-mercaptoethyl acetate (toasted, roasted, burned), and 2-methyl-3-furanthiol (broth, meaty). 3-Methyl-2-buten-1-thiol, known, however, as the skunky off-flavour of samples exposed to sunlight, was also detected in fresh light-protected beers. Possible pathways for the formation of these thiols are discussed.

1. INTRODUCTION

Around 80 different thiols have been identified in foods and beverages [2]. These sulfur compounds are in most cases only present as traces, highly reactive (oxidation, heat, light), and commercially unavailable. Therefore, their occurrence in beer is most probably underestimated. Until now, many papers have been devoted to the aromatic composition of fresh and aged beers [3-6]. Nevertheless, little is still known about beer thiols, and especially polyfunctional thiols. Yet, spiking of copper (trap for thiol) clearly demonstrates their huge sensory impact in this fermented beverage [7].

Our first contribution to this field was the synthesis and the sensory characterisation of more than sixty thiols potentially present in foods: mercaptoaldehydes and mercaptoketones (coming from the reaction between H_2S or thioacetic acid and α,β-unsaturated aldehydes or ketones) as well as mercaptoalcohol and mercaptoacetate

analogues (resulting from reduction and esterification of the former). The combinatorial chemistry methodology hyphenated to the use of an equimolar SCD detector and/or mass spectrometer allowed us to build the library [8-11].
All those data were further used to identify new thiols in four lager beer *p*HMB extracts, which were analysed by GCO, GC-PFPD and GC-MS. In order to find potential precursors, a similar extraction was applied on wort and aged beer (5 days at 40 °C).

2. MATERIALS AND METHODS

2.1. Extraction of polyfunctional thiols by pHMB

The protocol is the same as the one proposed by Tominaga *et al.* for wines [1] except that only one liquid-liquid extraction was applied (500 ml of beer with 200 ml of CH_2Cl_2) and that two centrifugation steps were needed to break the emulsion. An external standard (thiazole) was added just before concentration under nitrogen.

2.2. Gas chromatography olfactomety (GCO) and GC-MS

The analyses were carried out as described in [8-11].

2.3. Gas chromatography hyphenated to a pulsed flame photometric detector

The injections on the gas chromatography hyphenated to the pulsed flame photometric detector (GC-PFPD) were carried out in the splitless mode at 250 °C, the split being turned on after 0.5 min. At the detector, the following parameters were selected: 210 °C as temperature, 600 V as voltage, 18 ms as gate width, 6 ms as gate delay, 600 mV as trigger level and 3.45 Hz as pulse frequency. The flows of H_2, Air1, and Air2 were respectively of 8.5 ml/min, 8.7 ml/min, and 10.8 ml/min. The chromatographic column was the same as the one used in GCO and the temperature program was similar to the previous one (last rate was set at 5 °C/min).

3. RESULTS

As depicted in Table 1, eleven polyfunctional thiols were identified in lager beers. Most of them were found in the four commercial samples. 2-Mercapto-3-methylbutanol seemed to be more specific of one brand probably characterised by a lower antioxidative ability of this beer. Among those compounds, only 2-methyl-3-furanthiol [5], 3-methyl-2-buten-1-thiol [12], 3-mercaptohexanol [7], 2-mercapto-3-methylbutanol [3], and 4-mercapto-4-methyl-2-pentanone [7] were previously described as natural beer constituents in the literature. None of these compounds were found in the *p*HMB wort extract, suggesting a key role of fermentation. On the other hand, thiol concentration revealed to be still higher in beer I after accelerated ageing (5 days at 40 °C).

Table 1. Retention indices, odour attributes and structures of polyfunctional thiols identified in fresh and aged beers.

Thiols	Structure	GC detection method	Present in *p*HMB extract
2-Methyl-3-furanthiol $Ir_{CP\text{-}Sil\ 5\ CB} = 847$, $Ir_{FFAP} = 1306$ Broth, meaty	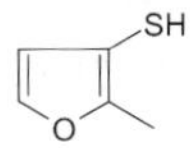	GCO/PFPD/MS	Fresh lager beers I,II,III,IV Aged lager beer I
2-Mercaptoethanol $Ir_{CP\text{-}Sil\ 5\ CB} = 717$, $Ir_{FFAP} = 1501$ Roasted, garbage	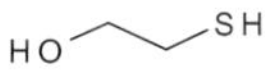	GCO/PFPD	Fresh lager beers I,II,III,IV Aged lager beer I
3-Mercaptopropanol $Ir_{CP\text{-}Sil\ 5\ CB} = 840$, $Ir_{FFAP} = 1665$ Roasted, broth, potato		GCO/PFPD	Fresh lager beers I,II,III,IV Aged lager beer I
3-Mercapto-3-methylbutanol $Ir_{CP\text{-}Sil\ 5\ CB} = 944$, $Ir_{FFAP} = 1671$ Broth, onion, sweat, vegetable	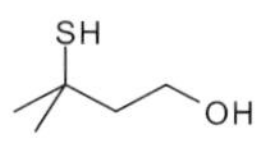	GCO/PFPD/MS	Fresh lager beers I,II,III,IV Aged lager beer I
2-Mercapto-3-methylbutanol $Ir_{CP\text{-}Sil\ 5\ CB} = 964$, $Ir_{FFAP} = 1656$ Onion	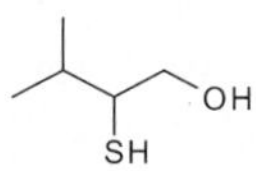	GCO/PFPD/MS	Fresh lager beers IV
1-Mercapto-3-pentanol $Ir_{CP\text{-}Sil\ 5\ CB} = 981$, $Ir_{FFAP} = 1698$ Nettle, burned	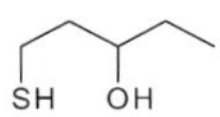	GCO/PFPD/MS	Fresh lager beers I,II,III,IV Aged lager beer I
3-Mercaptohexanol $Ir_{CP\text{-}Sil\ 5\ CB} = 1095$, $Ir_{FFAP} = 1853$ Rhubarb, fruity	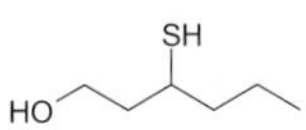	GCO	Fresh lager beers I,II,III,IV Aged lager beer I
3-Methyl-2-buten-1-thiol $Ir_{CP\text{-}Sil\ 5\ CB} = 808$, $Ir_{FFAP} = 1112$ Plastic, old beer, pungent, skunky	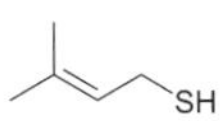	GCO/MS	Fresh lager beers I,II,III,IV Aged lager beer I
4-Mercapto-4-methyl-2-pentanone $Ir_{CP\text{-}Sil\ 5\ CB} = 915$, $Ir_{FFAP} = 1382$ Catty, blackcurrant, box tree	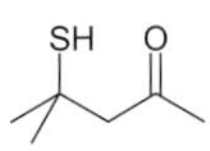	GCO	Fresh lager beers I,II,III,IV Aged lager beer I
2-Mercaptoethyl acetate $Ir_{CP\text{-}Sil\ 5\ CB} = 887$, $Ir_{FFAP} = 1444$ Toasted, burned	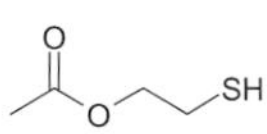	GCO/PFPD/MS	Fresh lager beers I,II,III,IV Aged lager beer I
3-Mercaptopropyl acetate $Ir_{CP\text{-}Sil\ 5\ CB} = 992$, $Ir_{FFAP} = 1565$ Roasted meat, burned	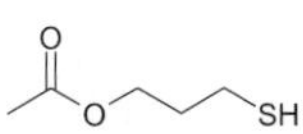	GCO/PFPD/MS	Fresh lager beers I,II,III,IV Aged lager beer I

4. CONCLUSION

The *p*HMB extraction technique hyphenated to GCO-PFPD/MS analyses allowed us to identify 6 new polyfunctional thiols in fresh lager beers. In order to explain the formation of these compounds, the following mechanisms may be suggested. Hydrogen sulfide excreted by yeast during fermentation could be able to:

- substitute allylic alcohol of hop (synthesis of 3-methyl-2-buten-1-thiol)
- achieve a 1,4 addition to α,β-unsaturated aldehydes or ketones in the wort (synthesis of 1-mercapto-3-pentanol, 3-mercaptohexanol, 4-mercapto-4-methyl-2-pentanone)
- carry out either an electrophilic addition or a radical anti-Markovnikov addition on a double bond (synthesis of 3-mercapto-3-methylbutanol and 2-mercapto-3-methylbutanol, respectively)
- interact with Maillard reaction intermediates (synthesis of 2-methyl-3-furanthiol).

Other beer thiols were likely derived from amino acid Ehrlich degradation followed by a reduction and an esterification (synthesis of 2-mercaptoethanol, 3-mercaptopropanol and their corresponding acetates).

References

1. T. Tominaga, M.-L. Murat and D. Dubourdieu, J. Agric. Food Chem., 46 (1998) 1044.
2. C. Vermeulen, L. Gijs and S. Collin, Food Rev. Int., 21 (2005) 69.
3. A. Olsen, Carlsberg Res. Commun., 53 (1988) 1.
4. P. Schieberle, Z. Lebensmittel Untersuch. Forsch., 193 (1991) 558.
5. G. Lermusieau, M. Bulens and S. Collin, J. Agric. Food Chem., 49 (2001) 3867.
6. L. Gijs, F. Chevance, V. Jerkovic and S. Collin, J. Agric. Food Chem., 50 (2002) 5612.
7. C. Vermeulen, S. Lecocq and S. Collin, Polyfunctional thiols and drinkability of beer, proceedings of the 29th EBC Congress, Nürnberg, Germany (2003) 91/1-91/11.
8. C. Vermeulen, J. Pellaud, L. Gijs and S. Collin, J. Agric. Food Chem., 49 (2001) 5445.
9. C. Vermeulen and S. Collin, J. Agric. Food Chem., 50 (2002) 5654.
10. C. Vermeulen, C. Guyot-Declerck and S. Collin, J. Agric. Food Chem., 51 (2003) 3623.
11. C. Vermeulen and S. Collin, J. Agric. Food Chem., 51 (2003) 3618.
12. G. Lermusieau and S. Collin, J. Am. Soc. Brew. Chem., 61 (2003) 109.

W.L.P. Bredie and M.A. Petersen (Editors)
Flavour Science: Recent Advances and Trends

Flavour and health-promoting compounds in broccoli and cauliflower - an inconsistency?

Angelika Krumbein, Ilona Schonhof and Bernhard Brückner

Institute of Vegetables and Ornamental Crops, Theodor-Echtermeyer Weg 1, D-14979 Grossbeeren, Germany

ABSTRACT

The most potent odour-active volatiles in raw broccoli and cauliflower were found to have green and sulfur odour notes, using GCO. The different cultivars could be differentiated by the sensory attributes bitter, pungent, green/grassy, leek, cauliflower and broccoli, and correlated aroma volatiles. Some breakdown products of glucosinolates, with anticarcinogenic potential, may contribute to these sensory properties. The bitter taste was closely associated with 3-butenylisothiocyanate. A consumer test showed that cultivars with bitter taste were rejected. A compromise has to be found between positive and negative flavours and health-promoting components.

1. INTRODUCTION

Consumer acceptability tends to be based on appearance and flavour rather than on knowledge of health-promoting substances in vegetables. Breakdown products of glucosinolates are not only potential anticarcinogens, but they can also contribute to the flavour [1-3]. The objectives of the present study were (1) to investigate the aroma volatiles responsible for sensory properties in raw broccoli and cauliflower, and (2) to evaluate how flavour compounds with health-promoting effects influence the acceptance of these vegetables by consumers.

2. MATERIALS AND METHODS

Four broccoli types (‘Emperor’, ‘Shogun’, ‘Marathon’, ‘Viola’), Chinese broccoli, and four cauliflower types (‘Marine’, ‘Minarett’, ‘Rosalind’, ‘Alverda’) were grown in a field in Grossbeeren over three consecutive years.
For dynamic headspace sampling fresh florets (300 g) were blended with 250 ml distilled water for 30 s and held for 180 s; 400 ml saturated calcium chloride solution

was added, and the mixture was re-blended for 10 s. The mixture was placed in a 3 l flask containing a magnetic stirrer, and purified air (150 ml/min) was passed through the mixture and out of the flask through a Tenax trap (200 mg Tenax TA). The sampling was continued for 150 min, the trap was removed, and volatiles were extracted with acetone.

GC-MS was carried out using an HP5890 Series II plus MSD 5972A; splitless, injector temperature 250 °C; Supelcowax 10 column of 30 m x 0.25 mm i.d. and 0.25 μm film thickness; 1 ml helium/min; temperature programme: 3 min at 40 °C; from 40 °C at 1 °C/min to 60 °C, 2 min at 60 °C; from 60 °C at 5 °C/min to 180 °C, 10 min at 180 °C. GCO was carried out using an HP6890 with FID and an SGE ODO-1 sniffing port. Sniffing was performed by five trained assessors.

Ten trained panellists established a sensory profile of uncooked products by evaluating odour, flavour and taste attributes on unstructured line scales with the anchor points 0 - not perceptible and 100 - strongly perceptible. A consumer panel (100 housewives) rated the acceptability of the products.

3. RESULTS AND DISCUSSION

The descriptive analysis showed, that the cultivars could be differentiated by the odour attributes ‘cabbage’, ‘cauliflower’, ‘broccoli’, and ‘green/grassy’; the flavour attributes ‘broccoli’, ‘cauliflower’, ‘leek’, ‘spicy’, ‘pungent’, ‘green/grassy’, and ‘mouldy’; and the taste attributes ‘sweet’ and ‘bitter’ (Figure 1).

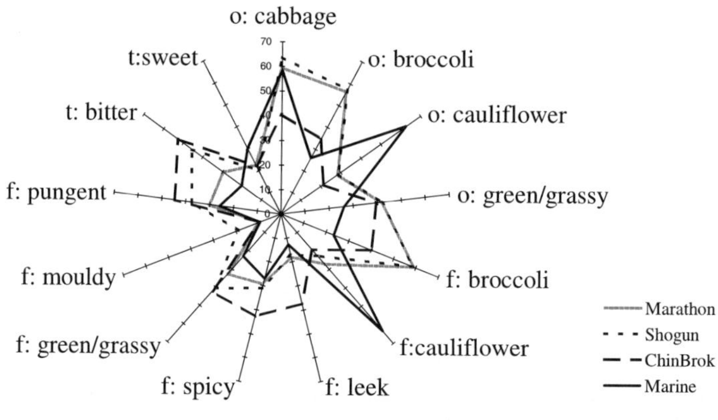

Figure 1. Sensory profile of broccoli and cauliflower by odour (o), flavour (f), and taste (t).

The most odour-active volatiles had green and sulfur odours (Table 1). Besides allylisothiocyanate, a breakdown product of the glucosinolate sinigrin, further breakdown products were identified in a few of the other cultivars (P16 to P18). GCO analysis showed 3-butenylisothiocyanate (P16) to coincide with a pungent odour and 5-methylthiopentanonitrile (P17) with a sulfur odour in *B.* rapa (data not shown). 4-Methylthiobutanonitrile (P18) with an odour threshold of 82 ppb [1] may also contribute to the aroma.

A principal components analysis (PCA) was carried out to describe the relationships between sensory attributes and odour-active compounds. PC1 (26.6%) was characterised by high loadings of the attribute cauliflower (Figure 2). The aroma volatiles dimethyl disulfide (P3), 4-methylthiobutanonitrile (P18), and allylisothiocyanate (P11) were associated with the attribute cauliflower. The sensory attributes bitter, pungent, and green/grassy were negatively related to PC1. The aroma volatiles 3-butenylisothiocyanate (P16) and 5-methylthiopentanonitrile (P17) were closely related to the attributes bitter and pungent. The volatiles (*Z*)-3-hexenyl acetate (P10), 2-ethylthiophene (P6), and dimethyl trisulfide (P12) were associated with the attributes green and leek. PC2 (17.9%) was dominated by the volatiles hexanal (P4), (*Z*)-3-hexenal (P5), (*E*)-2-hexenal (P8), 1-penten-3-ol (P7), and (*E*,*Z*)-2,4-heptadienal (P15). These were associated with the attribute broccoli.

Table 1. Odour-active compounds in the broccoli cultivars 'Marathon' and 'Marine'.

Peak no.	Compound	FD-factor 'Marathon'	FD-factor 'Marine'	Odour description
1	1-Penten-3-one	32	64	Green
2	2,3-Pentandione	1	n.d.	Sweet
3	Dimethyl disulfide	n.d.[a]	1	Sulfur
4	Hexanal	64	128	Green, grassy
5	(*Z*)-3-Hexenal	32	512	Fresh green, sweet
6	2-Ethylthiophene	1	n.d.	Mouldy
7	1-Penten-3-ol	1	n.d.	Grassy
8	(*E*)-2-Eexenal	16	4	Green
9	Methylthiocyanate	1	n.d.	Mouldy
10	(*Z*)-3-Hexenyl acetate	2	n.d.	Fruit candy
11	Allylisothiocyanate	n.d.	1	Sulfur
12	Dimethyl trisulfide	32	16	Sulfur
13	(*Z*)-3-Hexenol	64	16	Green, grassy
14	(*E*)-2-Octenal	16	n.d.	Earthy
15	(*E*,*Z*)-2,4-Heptadienal	2	8	Herbaceous
16	3-Butenylisothiocyanate	n.d.	n.d.	Pungent
17	5-Methylthiopentanonitrile	n.d.	n.d.	Sulfur
18	4-Methylthiobutanonitrile	n.d.	n.d.	n.d.

[a]n.d.: not detected.

Consumers evaluated eight cultivars (excluding Chinese broccoli). The highest acceptance in flavour was given to the cauliflower cultivar 'Marine' (acceptance value: 84) whereas the lowest acceptance was given to the cultivar 'Shogun' (acceptance value: 62). The consumer acceptance in flavour and overall liking was negatively correlated with bitter taste (Spearman's rank correlation R = -0.70 and -0.71 for the two traits, respectively). Consumers rejected cultivars with a high content of glucosinolates (progoitrin, sinigrin, gluconapin, glucobrassicin and neoglucobrassicin) [4].

Results from the PCA showed a close relationship between bitter taste and the relatively high content of the breakdown product of gluconapin, 3-butenylisothiocyanate. The relatively low content of allylisothiocyanate, a breakdown product of sinigrin, was more related to the attribute 'cauliflower' than to 'bitter'. The bitter-tasting sinigrin could be detected by the consumers at about 106 mg/l, this was higher than earlier reported [5].

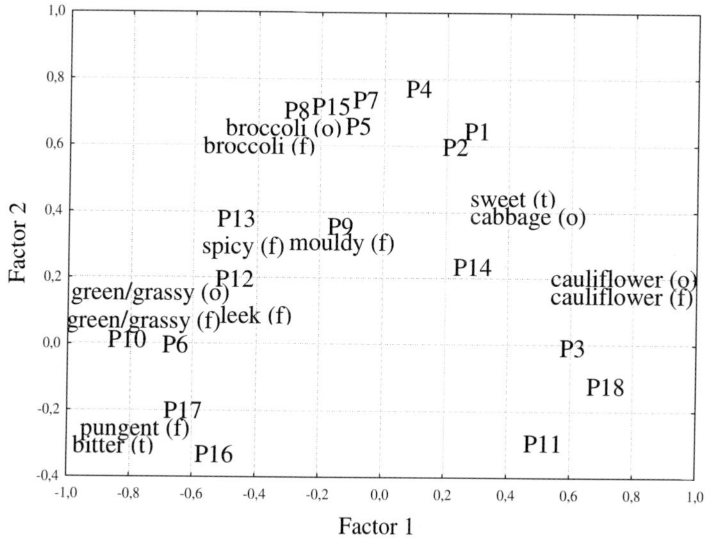

Figure 2. PCA bi-plot of odour (o), flavour (f) and taste (t) attributes together with the odour-active compounds (cf. Table 1) of the broccoli and cauliflower cultivars.

4. CONCLUSION

This study showed that consumer acceptance of fresh broccoli and cauliflower was negatively correlated with bitter taste. Since bitter glucosinolates have been proven to have anticarcinogenic properties, a compromise has to be found to optimise the concentration of desired glucosinolates with acceptance by consumers in order to give appropriate recommendations for breeders and growers of *Brassicacea*.

References

1. R.G. Buttery, D.G. Guadagni, L.C. Ling, R.M. Seifert and W. Lipton, J. Agric. Food Chem., 24 (1976) 829.
2. E. Engel, C. Baty, D. le Corre, I. Succhon and N. Martine, J. Agric. Food Chem., 50 (2002) 6459.
3. H.-P. Kruse and M. Rothe (eds.), Flavour perception, aroma evaluation, Potsdam, Germany (1997) 504.
4. I. Schonhof, A. Krumbein and B. Brückner, Nahrung, 48 (1) (2004) 25.
5. A. Drewnovski and C. Gomez-Carneros, Am. J. Clin. Nutr., 72 (2000) 1424.

W.L.P. Bredie and M.A. Petersen (Editors)
Flavour Science: Recent Advances and Trends

Assessment of fresh salmon quality under different storage conditions using solid phase micro-extraction

Jean-Pierre Dufour[a], Rana Wierda[a], Erwan Pierre[a] and Graham Fletcher[b]

[a]Department of Food Science, University of Otago, PO Box 56, Dunedin, New Zealand; [b]New Zealand Institute for Crop & Food Research Ltd, Private Bag 92169, Auckland, New Zealand

ABSTRACT

Research on technologies to extend the shelf life of fresh salmon is complicated by the need to employ taste panels, which are time-consuming and costly to assess the sensory quality of the product. This research sought to develop an instrumental method of analysis that correlated with sensory results. A method for the analysis of volatiles in fresh salmon during storage was developed using solid phase microextraction (SPME) and gas chromatography. Salmon samples were stored under air or under modified atmosphere. The SPME technique enabled the identification of more than 100 volatile compounds, which were grouped according to whether they increased, decreased, or did not change during storage. There were clear differences in the rate of change of some volatile compounds between different conditions of storage. By comparing trends of the volatiles that changed during storage under different atmospheric conditions it was possible to identify a number of compounds that correlated with sensory changes in the salmon. It is suggested that these compounds could be used in quality control testing as markers for the evolution of salmon freshness during storage.

1. INTRODUCTION

There is a lot of interest in the development of shelf life extension methods for fresh salmon, largely because fresh quality salmon is in high demand, particularly for the lucrative export market. Currently such research requires the use of sensory analysis by trained taste panels to accurately monitor quality changes occurring during storage. These taste panels take a lot of time to set up and require significant financial input. For this reason, various other methods have been investigated to determine their suitability for monitoring salmon quality during storage. Some of the methods that have been investigated in the past include measurement of microbiological changes, texture changes, and chemical changes [1-3]. While these methods are useful indicators of fish quality, they only measure an aspect of quality rather than giving the holistic picture that can be obtained using sensory evaluation.

The current research looked at using solid phase microextraction (SPME) headspace analysis combined with gas chromatography to monitor changes in fresh salmon quality during storage. SPME is now a well-established method for volatile analysis and has been previously used to measure volatiles in seafood products [4-9]. The main objectives of this research were to develop an SPME headspace analysis method for the measurement of salmon volatiles that can be used to monitor changes in the volatile profile of the salmon during storage and to correlate these changes with sensory changes as determined by a trained sensory taste panel.

2. MATERIALS AND METHODS

2.1. Sensory analysis

The details for fish and sensory analysis used for correlation of sensory data with SPME results were described by Fletcher *et al.* [3].

2.2. Sample preparation for SPME

Salmon portions were stored in ambient air at 0 °C or 9 °C, or at 0 °C in a 40:60 (v/v) CO_2:N_2 mix. The salmon portions were then kept at -80 °C until further preparation. Samples (in 4 cm x 4 cm x 2 cm portions) were defrosted at 4 °C overnight and were then kept on ice until cooking. Samples were removed from the foil bags and wrapped in aluminum foil parcels (foil size 16 cm x 16 cm) for oven cooking (fan bake, 180 °C, 14 min). Four samples were cooked at once, with samples being placed 6 cm apart and each sample being 3 cm from the centre of the oven tray. The cooked samples were vacuum-sealed in new foil bags and cooled on ice before freezing at -80 °C. Frozen cooked salmon samples were dipped in liquid N_2 until it stopped bubbling. The sample was then placed in a pre-cooled (with liquid N_2) stainless steel Waring blender and blended into a fine powder (20-30 s) of which 8 g was poured into each of two 20 ml headspace vials through a funnel (pre-cooled with liquid N_2). Headspace vials were sealed with an aluminum seal (with a 10 mm center hole) fitted with a 20 mm chlorobutyl septum (Chromacol, Herts, UK) and stored at -80 °C until analysis.

2.3. SPME GC-FID/MS analysis

Unless stated otherwise, SPME sampling was carried out on duplicate vials using a 75 μm carboxen-polydimethylsiloxane (CAR-PDMS) coated fibre (Supelco, Bellafonte, PA, USA). The samples were held at room temperature for 30 min prior to incubating with the SPME fibre exposed to the sample headspace for 90 min at 60 °C. The sample volatiles were desorbed from the CAR-PDMS coating by placing the fibre for 5 min in the injector (200 °C, splitless 2 min) of a Fisons 8000 series GC (Carlo Erba Instruments, Milan, Italy) fitted with a flame ionisation detector (FID). Separation was achieved on a CP-Wax 52CB column (50 m, 0.32 mm id, 0.45μm film thickness, 5 m retention gap) (Chrompak, Middelburg, The Netherlands) using helium (constant flow: 23 cm/s at 40 °C) as the carrier gas. The oven temperature was increased from 40 °C to 250 °C at 3 °C/min. The detector temperature was 250 °C. For volatile identification purposes, a number of samples were analysed using gas chromatography – mass spectrometry using a Fisons 8000 Top GC (Carlo Erba Instruments, Milan, Italy) coupled to a Finnigan MAT MD 1000 mass detector (Finnigan Instruments, Manchester, UK) using the same column, carrier gas, and temperature settings as described for the GC-FID analysis. The mass spectrometer was operated in the electron impact ionisation mode (70 eV). Source and interface temperatures were 195 °C and 250 °C, respectively. Detector voltage was 250 V; mass range was from 35 to 400 amu; scan rate was 0.9 scan/s.

3. RESULTS

Initial SPME method development involved the determination of the best SPME fibre coating and incubation conditions for the analysis of salmon volatiles (Figure 1).

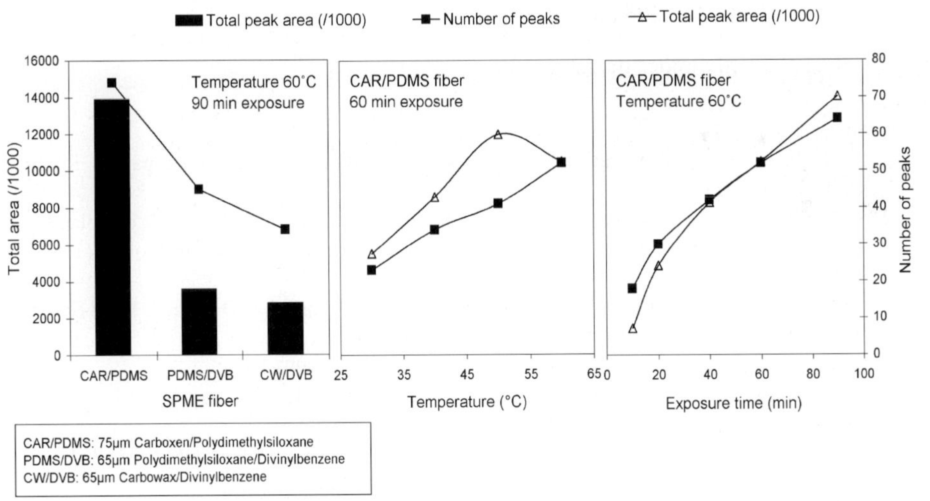

Figure 1. Effect of fibre coating and incubation conditions on sampling of salmon volatiles.

More than 100 volatile compounds (including aldehydes, alcohols, sulfur compounds, and short-chain fatty acids) were identified in the salmon samples using the SPME method. A comparison of the chromatographic peak responses of compounds at the different stages of storage showed that several volatile compounds changed (increased or decreased) during storage. The changes in peak area observed for a number of compounds (acetoin, cyclopentanol, 1-penten-3-ol) were found to correlate well with changes in the sensory quality index score of the fresh salmon (Table 1), with negative correlation coefficient values indicating markers of freshness and positive values indicating markers of spoilage.

Table 1. Correlation of volatile compounds (peak area) with sensory analysis (quality index) [3].

	Correlation coefficients		
Compound	40:60 CO_2:N_2, 0 °C	Air, 0 °C	Air, 9 °C
1-Penten-3-ol	-0.96	-0.63	-1.00
Cyclopentanol	-0.91	-0.69	-0.99
Acetoin	0.93	0.78	0.96

4. DISCUSSION AND CONCLUSION

The current research has identified cyclopentanol and 1-penten-3-ol as potential markers of freshness and acetoin as a potential marker of spoilage in fresh stored salmon. The SPME method described above could easily be standardised using the level (peak area) of marker compounds in the fresh salmon sample as a reference. A comparison of the level of the marker compounds in the fresh salmon and in the stored sample could subsequently be used to determine cut-off ratios for sensory acceptability. This approach is presently under investigation.

References

1. H.H. Huss (ed.), Quality and changes in fresh fish, FAO fisheries technical papers, Rome, Italy, 348 (1995).
2. G. Olafsdottir, E. Martinsdottir, J. Oehlenschlager, P. Dalgaard, B. Jensen, I. Undeland, I.M. Mackie, G. Henehan, J. Nielsen and H. Nilsen, Trends Food Sci. Technol., 8 (1997) 258.
3. G.C. Fletcher, G. Summers, V. Corrigan, S. Cumarasamy and J.P. Dufour, J. Food Sci., 67 (6) (2002) 2362.
4. S.W. Lloyd and C.C. Grimm, J. Agric. Food Chem., 47 (1999) 164.
5. M. Linder and R.G. Ackman, J. Food Sci., 67 (6) (2002) 2032.
6. T. Serot and C. Lafficher, Food Chem., 82 (2003) 513.
7. M.A. Mansur, A. Bhadra, H. Takamura and T. Matoba, Fisheries Sci., 69 (2003) 864.
8. R. Triqui and N. Bouchriti, J. Agric. Food Chem., 51 (2003) 7540.
9. X. Li, Z. Zeng, J. Zhou, S. Gong, W. Wang and Y. Chen, J. Chromatogr. A, 1041 (2004) 1.

W.L.P. Bredie and M.A. Petersen (Editors)
Flavour Science: Recent Advances and Trends

Varietal differences in the aroma compound profile of blackcurrant berries

Lars P. Christensen[a] and Hanne L. Pedersen[b]

[a]Department of Food Science, Danish Institute of Agricultural Sciences, Research Centre Aarslev, Kirstinebjergvej 10, DK-5792 Aarslev, Denmark; [b]Department of Horticulture, Danish Institute of Agricultural Sciences, Research Centre Aarslev, Kirstinebjergvej 10, DK-5792 Aarslev, Denmark

ABSTRACT

The content of volatile compounds of blackcurrant berries from 13 varieties grown under organic field conditions was determined. Volatile compounds were collected by dynamic headspace technique and a total of 45 volatile compounds were quantified and identified by GC-FID and GC-MS, respectively. Only minor qualitative differences were observed between berries of the different varieties, whereas large quantitative differences were found for important flavour compounds such as aliphatic esters (methyl- and ethyl butanoate, methyl- and ethyl hexanoate) and monoterpenes (sabinene, 3-carene, γ-terpinene, terpinolene, (*Z*)-β-ocimene, and 1,8-cineole). Different soil cultivation and nitrogen supply had only a minor effect on the production of volatile compounds of blackcurrant berries and the effects were non-significant. The results of the present study clearly show that improvement and/or alterations of the flavour of blackcurrant berries should be based on genetic resources rather than on optimising growing conditions, although the latter may have an effect on bush performance and hence the yield of blackcurrant berries.

1. INTRODUCTION

Blackcurrant (*Ribes nigrum* L.) berries is an important raw material for the food industry due to their characteristic colour and excellent flavour. The berries are used to prepare jams, ice cream, and especially juice. In addition to numerous terpenoids, blackcurrant berries contain aliphatic esters, carbonyl compounds, and alcohols [1-5]. The esters such as methyl butanoate, ethyl butanoate, ethyl hexanoate and methyl acetate are considered to be responsible for the fruity notes of blackcurrant berries, whereas some terpenoids have been associated with the typical blackcurrant aroma (e.g., 1,8-cineole, β-damascenone) [1-5]. The berries of blackcurrant have been extensively investigated for aroma compounds. However, only little information exist upon the varietal differences affecting the aroma profiles of the berries from cultivars grown under similar field conditions. Furthermore, the effect of soil cultivation and nitrogen is very important in organic production where only organic fertilisers are allowed and needs further clarification in relation to generation of aroma volatiles.

2. MATERIALS AND METHODS

2.1. Plant material

Berries from 13 blackcurrant varieties [Ben Lomond (BLom), Ben Loyal (BLoy), Ben Nevis (BN), Cie5-32, Cie5-63, Haakon, Kristin, Otelo, P8-5-24, Ben Rua (BR), Sunniva, Tenah, Tsema] were grown under organic field conditions [6] for investigation of varietal differences of volatile compounds. The effect of soil cultivation and nitrogen supply on the content of volatile compounds of the berries was investigated [6] in five varieties [Ben Alder (BA), BLom, Farleigh, Intercontinental and Titania]. Berries were frozen immediately after harvest and packed in airtight aluminium foil pouches.

2.2. Dynamic headspace sampling (purge and trap)

Blackcurrant berries (55 g) were thawed for 30 min after which 55 ml distilled water was added. The mixture of blackcurrant berries and water was blended for 30 s and 100 g of the mash was transferred to a 250 ml conical bottle and 20 μl 4-methyl-1-pentanol (814 ng/μl) was added as internal standard (IS). Samples were placed in a in a thermostated chamber and purged with N_2 (90 min, 150 ml/min) under constant stirring at 25 °C for dynamic headspace sampling. The volatile compounds were trapped on absorbent traps (glass tubes, 4 mm i.d. 180 mm length) filled with 200 mg Porapak Q 50-80 mesh. No breakthrough of volatiles was detected with this trap size. The Porapak columns were eluted with 2 ml methylene chloride and the headspace samples carefully concentrated to 100 μl under a flow of N_2 (50 ml/min) prior to GC analyses.

2.3. Analysis of volatile compounds

Volatile compounds were quantified by GC-FID on a Hewlett-Packard 5890 series II Plus GC equipped with a FID detector (230 °C) and identified by GC-MS (Varian Saturn 2000 ion trap MS). Compounds were separated on a Chrompack WCOT fused silica capillary column (50 m × 0.25 mm i.d., DF = 0.2 μm liquid phase: CP-Wax 52CB) using the following temperature program: 32 °C for 1.5 min, followed by 3 °C/min to 40 °C, isothermal for 10 min, and 3 °C/min to 220 °C followed by constant temperature for 10 min. Carrier gas: helium (flow 1.1 ml/min). Injection volume: 1 μl. Quantification of volatile compounds was done relative to the IS. Identification of volatile compounds was done by probability-based matching with mass spectra in the NIST database (NIST (1998), version 6.0), and by comparison with mass spectra and retention indices of authentic compounds.

3. RESULTS

A total of 45 volatile compounds were identified and quantified by GC-MS and GC-FID in headspace samples from berries of 13 organically grown blackcurrant cultivars. The dominant groups were aliphatic esters and monoterpenes and large quantitative differences in these compounds amongst blackcurrant varieties were found (Table 1).

Table 1. Approximate concentration (ng/g) of the major volatile compounds in selected varieties of blackcurrant berries. The varieties were selected among 13 varieties and represents some of the largest quantitative differences among the major volatile compounds.[a]

	Variety						
Volatile compounds	Bloy	BN	Cie5-32	Cie5-63	Kristin	Sunniva	Tsema
Ethyl acetate	250	320	150	390	190	260	270
Methyl butanoate	17100	96	140	19800	690	17800	22400
α-Pinene	430	310	170	410	540	410	520
α-Thjuene	640	82	64	94	610	290	98
Ethyl butanoate	2680	110	3700	26200	1240	21900	19500
Hexanal	700	860	390	1180	790	670	780
β-Pinene	390	190	26	150	290	310	180
Sabinene	9900	500	770	350	9700	3730	480
3-Carene	9	2900	6	2010	5040	2050	2200
Myrcene	600	190	34	200	780	390	220
α-Terpinene	790	89	34	92	480	270	160
Methyl hexanoate	680	14	4	800	47	540	1200
Limonene	310	120	18	120	420	240	490
1,8-Cineole	74	59	96	960	170	210	300
β-Phellandrene	500	130	28	120	620	280	1080
Ethyl hexanoate	71	4	42	660	110	400	710
(*Z*)-β-Ocimene	520	370	46	480	1120	370	260
γ-Terpinene	1520	80	60	93	780	430	76
(*E*)-β-Ocimene	330	210	32	280	650	220	160
p-Cymene	100	140	18	200	200	550	140
Terpinolene	260	970	15	1050	2330	870	900
Methyl octanoate	52	5	0	55	9	46	47
β-Caryophyllene	91	65	20	70	240	140	99
4-Terpineol	430	28	32	29	260	170	17

[a]Mean of triplicates with coefficient of variation (CV) less than < 10% in all cases.

The largest variation among the major volatile compounds was observed for methyl butanoate (96–22400 ng/g), ethyl butanoate (110–26200 ng/g), methyl hexanoate (4–1200 ng/g), and ethyl hexanoate (4–710 ng/g) and the terpenes sabinene (38–9900 ng/g), 3-carene (6–5040 ng/g), 1,8-cineole (45–960 ng/g), β-phellandrene (28–1080 ng/g), (*Z*)-β-ocimene (46–1120 ng/g), γ-terpinene (11–1520 ng/g), and terpinolene (15–2330 ng/g) (Table 1 and data not shown in the table).

Different soil cultivation and nitrogen supply had only a minor effect on the production of volatile compounds of blackcurrant berries and the effects were non-significant (selected data shown for the variety Farleigh in Table 2).

Table 2. The effect of soil cultivation and nitrogen supply on the approximate concentration of selected major volatile compounds (ng/g blackcurrant berries) for the variety Farleigh.[a]

	Cultivation method[b]		
Volatile compounds	Clover grass	Clover grass + mulching	Clover grass + slurry
Ethyl acetate	188	207	210
α-Pinene	840	780	910
Hexanal	560	540	580
β-Pinene	880	820	1010
3-Carene	350	390	400
Limonene	640	610	740
β-Phellandrene	1140	1390	1210
Terpinolene	140	170	160

[a]Mean of three replicates. Each replicate represents berries taken from different plants within the same treatment. CV less than 15% in all cases. [b]Differences non-significant.

4. DISCUSSION AND CONCLUSION

The major volatile compounds found in blackcurrant berries are in accordance with other investigations using headspace technique [5]. The large genotypic variation in the content of aroma compounds of blackcurrant berries indicates possibilities to improve the flavour of blackcurrant food products, and further that improvement and/or alterations of the flavour of the berries should be based on genetic resources rather than on optimising growing conditions. In particular the large concentration differences of aliphatic esters (methyl- and ethyl butanoate as well as methyl- and ethyl hexanoate) between blackcurrant berries is interesting. These aliphatic esters are known to be important contributors to the fruity flavour of blackcurrant berries [2-5] and hence may have a significant effect on the sensory quality of the berries.

References

1. M.L.R. del Castillo and G. Dobson, J. Sci. Food Agric., 82 (13) (2002) 1510.
2. A. Latrasse, J. Rigaud and J. Sarris, Sci. Aliment., 2 (1982) 145.
3. C. Varming, M.A. Petersen and L. Poll, J. Agric. Food Chem., 52 (6) (2004) 1647.
4. B.B. Mikkelsen and L. Poll, J. Food Sci., 67 (9) (2002) 3447.
5. C.K. Iversen, H.B. Jakobsen and C.-E. Olsen, J. Agric. Food Chem., 46 (3) (1998) 1132.
6. H.L. Pedersen, Acta Hort., 585 (2002) 633.

W.L.P. Bredie and M.A. Petersen (Editors)
Flavour Science: Recent Advances and Trends

Effect of development stage at harvest on the composition and yield of essential oils from thyme and oregano

Lars P. Christensen[a] and Kai Grevsen[b]

[a]Department of Food Science, Danish Institute of Agricultural Sciences, Research Centre Aarslev, Kirstinebjergvej 10, DK-5792 Aarslev, Denmark; [b]Department of Horticulture, Danish Institute of Agricultural Sciences, Research Centre Aarslev, Kirstinebjergvej 10, DK-5792 Aarslev, Denmark

ABSTRACT

The aim of the present study was to quantify essential oil components in thyme (*Thymus vulgaris*) and oregano (*Origanum vulgare* ssp. *hirtum*) and to investigate how the development stage at harvest (harvest time) and re-harvesting in the same growing season affects the chemical composition and yield of essential oils of these plants. Yields of essential oils varied from 3 to 15 kg/ha in thyme and from 75 to 165 kg/ha in oregano. Significant differences in the concentrations of the major essential oil components were observed during the different development stages. In particular re-harvesting of thyme in September just before flower opening affected the percent composition of thymol and its biosynthetic precursors *p*-cymene and γ-terpinene compared to harvesting at the flowering stage in June. The results of the present study showed that it is possible to optimise the yield and the quality of the essential oils of thyme and oregano by harvesting the plant material at the right development stage.

1. INTRODUCTION

Aromatic plants are known for their excellent odour and flavour and for their anti-microbial activity and other medicinal effects, and also antioxidant activity. These effects of aromatic plants are due to the production of large amounts of volatile terpenes that constitute the major part of their essential oils [1-4]. Consequently, aromatic plants and their essential oils are widely used as flavourings and food preservative agents as well as in cosmetics and medicines. The yield and chemical composition of essential oils depend on a number of factors, such as environmental and growing conditions, development stage at harvest and genotypes. Some of the most commonly used aromatic plants in foods are thyme (*Thymus vulgaris*) and oregano (*Origanum vulgare*), and the most important constituents in the essential oil of these species are the highly anti-microbial terpenes thymol and carvacrol [1-4]. These terpenes are also the major contributors to the pleasant odour of the oils. However, very little information exists on how the development stage at harvest and re-harvesting in the same growing season affects the chemical composition and yield of the essential oil in thyme and oregano.

2. MATERIALS AND METHODS

2.1. Plant material

In a field experiment with five harvest times, samples of thyme (2 × 300 g) and oregano (2 × 300 g) were collected at each harvest time in triplicates for determination of dry matter content and for the analysis of essential oils. Samples were kept cooled after harvest and samples (30 g) for essential oil analysis were vacuum packed in aluminium foil pouches the same day and stored at -25 °C until analysis. At each harvest time also the fresh weight yield per plot and development stage of the plants were recorded. In autumn the plots harvested as 2nd and 3rd harvest times were re-harvested before flower opening.

2.2. Extraction of essential oil components

Essential oils were extracted from aerial parts of thyme (0.5 g) and oregano (0.5 g) with dichloromethane (5 ml) for 20 h at 5 °C. After filtration 10 μl 4-methyl-1-pentanol (8140 ng/μl) was added as internal standard. All extractions were made in triplicates.

2.3. Analysis of plant extracts by GC and GC-MS

A Hewlett-Packard 5890 series II Plus GC equipped with a FID detector (230 °C) was used. Essential oil components were separated on a Chrompack WCOT fused silica capillary column (50 m × 0.25 mm i.d., DF = 0.2 μm liquid phase; CP-Wax 52 CB) using the following temperature program. 33 °C for 1 min, followed by 2 °C/min to 130 °C, 3.5 °C/min to 160 °C, 8 °C/min to 220 °C followed by constant temperature for 10 min. Carrier gas: helium (1.4 ml/min). Column head pressure: 20 psi. Injection volume: 1 μl. Compounds were quantified from GC-FID peaks areas relative to the int. std. Identification of essential oil components was performed by GC-MS (Varian Saturn 2000 ion trap MS). Compounds suggested by NIST (NIST (1998), version 6.0) were verified by comparison with mass spectra and retention indices of authentic compounds.

3. RESULTS

Thyme and oregano were grown under similar field conditions and plant material harvested at five different development stages during the growing season from June to August, and two times in September and October (re-harvest) in both 2003 and 2004.

Essential oils were extracted with dichloromethane and analysed by GC and GC-MS. The profile of the oil components of thyme was similar to that of oregano in the major compounds but at different concentrations. The major compounds in the essential oil of thyme (T) and oregano (O), near the flowering stage were thymol (T: 61.9%; O: 1.4%), carvacrol (T: 2.8%; O: 75.8%), *p*-cymene (T: 12.1%; O: 4.7%), γ-terpinene (T: 10.2%; O: 8.0%), α-terpinene (T: 0.4%; O: 1.0%), myrcene (T: 2.3%; O: 2.5%), α-pinene (T: 1.8%; O: 1.8%), linalool (T: 3.0%; O: 0%), and β-caryophyllene (T: 4.0%; O: 1.9%) (Figure 1). Yields of essential oils varied from 3 to 15 kg/ha in thyme and from 75 to 165 kg/ha in oregano. The qualitative profile of the essential oils did not change during the different development stages. However, significant differences in the concentration of the major essential oil components during the growing season with maximum yields at the flowering stage were observed (Figures 1 and 2). A similar distribution of major volatile compounds was observed in the essential oils of thyme and oregano harvested in 2004 (data not shown).

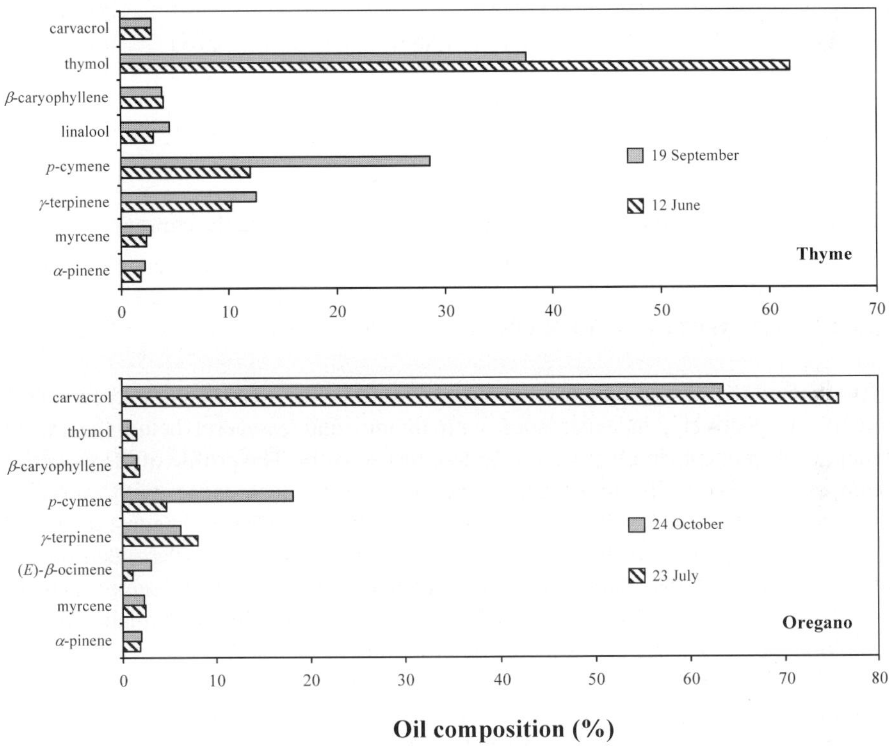

Figure 1. Distribution of major volatile compounds in the essential oils of thyme and oregano harvested near the flowering stage in June (75% open flowers) and July (50% open flowers) 2003, respectively, and at re-harvest in September for thyme and October for oregano. Values are mean of triplicates (CV < 5%).

In particular re-harvest of thyme in September just before flower opening affected the percent composition of thymol and its biosynthetic precursors *p*-cymene and γ-terpinene compared to harvesting at the flowering stage in June.
The yields of essential oil per area in re-harvested thyme were not significantly affected, because the concentration of essential oils was much higher in re-harvested material (Figure 2). Repetition of the experiments in 2004 gave almost the same results, although the yields at re-harvest was lower (data not shown).

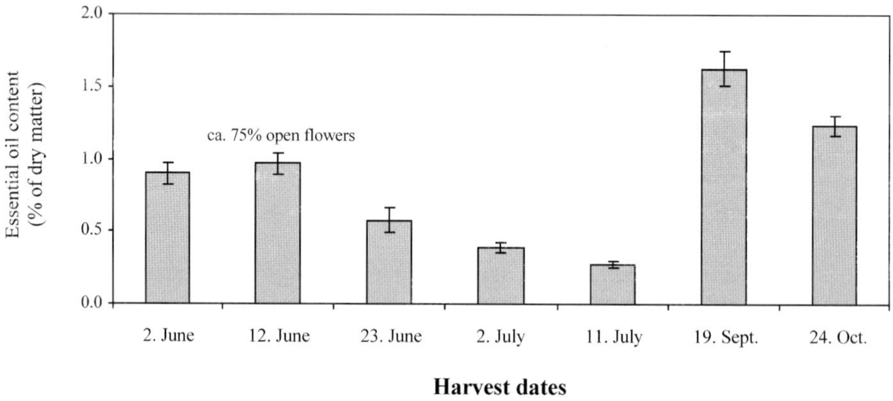

Figure 2. Essential oil content in thyme at different harvest times during the growing season and at re-harvest in the year 2003.

4. DISCUSSION AND CONCLUSION

The investigated thyme and oregano varieties were classified as a thymol and carvacrol chemotype, respectively, in accordance with thymol and carvacrol being the major essential oil components in thyme and oregano, respectively. The profile of the essential oil components found in thyme and oregano was in accordance with previous investigations [3,4]. The results of the present investigation clearly demonstrated that the yield and composition of the essential oils of thyme and oregano depended on the development stage of the plants (harvest time) and re-harvest in the same growing season. This information may for example be used to obtain the best quality of the essential oil whether it is for use as medicine or in food products.

References

1. V. de FeO, M. Bruno, B. Tahiri, F. Napolitano and F. Senatore, J. Agric. Food Chem., 51 (13) (2003) 3849.
2. I. Rasooli and S.A. Mirmostafa, J. Agric. Food Chem., 51 (8) (2003) 2200.
3. D.J. Daferera, B.N. Ziogas and M.G. Polissiou, J. Agric. Food Chem., 48 (6) (2000) 2576.
4. M. Russo, G.C. Galletti, P. Bocchini and A. Carnacini, J. Agric. Food Chem., 46 (9) (1998) 3741.

W.L.P. Bredie and M.A. Petersen (Editors)
Flavour Science: Recent Advances and Trends

Deceleration of beer ageing by amino acid and Strecker aldehyde monitoring over the brewing process

Andreas Stephan, Helge Fritsch and Georg Stettner

Bitburger Brauerei Th. Simon GmbH, Römermauer 3, 54634 Bitburg, Germany

ABSTRACT

In the pilot plant of the Bitburger brewery, studies on Strecker aldehyde generation during the brewing process were carried out. For the first time verification experiments of single amino acid addition to a standard mash were performed in a 20 hl scale. Changes in concentration of the free amino acids and their corresponding Strecker aldehydes were monitored across the total process from the mash up to the filled and 6-months stored bottle. The examinations revealed a couple of starting points to decelerate the Strecker aldehyde generation in the final product by controlling precursor concentrations at key process steps.

1. INTRODUCTION

Traditionally brewed beers according to the German purity law of 1516 are influenced during storage mainly by temperature and oxygen. The typical flavour of fresh beer changes into an oxidised flavour described as 'papery', 'bread-like' and after long-time ageing even as 'sherry-like'. In the past, the substantial cause of this change in flavour perception was attributed to the papery-smelling (*E*)-2-nonenal [1]. Present analytical and sensory studies have shown that the main odour-active components of the aged flavour are caused by Strecker degradation. The compounds 2-methylbutanal, 3-methylbutanal, methional and 2-phenylethanal are mainly responsible for the perceived ageing flavour [2,3].

Knowledge of these correlations is an important point, but for industrial production the translation of these basic facts into improved technological processes is the key. Some aspects of our studies about the impact of the amino acid precursors and the controllability of the generated corresponding aldehydes over the total process are shown below.

2. MATERIALS AND METHODS

2.1. Materials

The monitoring studies were carried out in duplicate in a 20 hl pilot plant at the Bitburger brewery. A standard 2-mash decoction procedure was used to produce Pilsener type beers from 300 kg barley malt, variety Scarlett. In contrast to the standard worts, the test worts were spiked with 250 g of methionine, leucine and phenylalanine (VWR, Darmstadt, Germany), respectively at the beginning of boiling. Boiling time was 70 min at 100 °C. Fermentation temperature was 10.5 °C.

2.2. Instrumental analysis

Odorants and standards needed for quantification (2- and 3-methylbutanal, methional, phenylacetaldehyde, pentanal and benzaldehyde were obtained from Aldrich, Steinheim, Germany. The GC-MS method used for quantification was described previously [3]. Amino acids were quantified after derivatisation with *o*-phthalaldehyde by fluorescence detection.

2.3. Sensory analysis

Beer samples were tested freshly and after 3 months by at least 10 trained test persons in the Bitburger sensory laboratory [4].

3. RESULTS

The addition of leucine and phenylalanine resulted in a duplication of their natural amounts, methionine addition resulted in a fivefold increase. Isoleucine was not added. Consequently, isoleucine and the corresponding Strecker aldehyde 2-methylbutanal acted as intrinsic control compounds in the trial brew.

The analytical data of the process steps wort boiling, wort cooling, fermentation up to 14 days, freshly bottled and stored beer (3 and 6 months at 28 °C) are shown in Tables 1-3. Both amino acids and corresponding Strecker aldehydes were significantly reduced after wort cooling during the fermentation period up to the bottled beer. During storage a marked increase of Strecker aldehydes was observed.

Aldehyde generation during wort boiling correlated approximately linear with amino acid spiking. The reduction process during fermentation and maturation led for the spiked as well as for the non-spiked trial to nearly the same aldehyde concentrations in freshly bottled beers except for the methional (Table 3).

During storage of the final products the generation of Strecker aldehydes diverged. Beers produced of wort with higher free amino acid amounts resulted in a forced aldehyde generation associated with an intensified perception of flavours associated with ageing (Figure 1).

Table 1. Amino acid (mg/l) and Strecker aldehyde (μg/l) amounts during brewhouse process (STD: standard; AAM: amino acids added mixture).

	Wort before boiling and addition	Wort after boiling and cooling	
	STD	STD	AAM
Amino acids			
Methionine	26	26	135
Lucine	118	115	223
Phenylalanine	90	88	172
Isoleucine	51	51	53
Aldehydes			
Methional	39.2	22.8	92.8
3-Methylbutanal	127	44.6	80.0
Phenylacetaldehyde	64.8	68.0	107
2-Methylbutanal	78.8	19.6	20.8

Table 2. Amino acid (mg/l) degradation during fermenting process (STD: standard; AAM: amino acids added mixture).

	2 days		4 days		8 days		14 days	
Fermentation	STD	AAM	STD	AAM	STD	AAM	STD	AAM
Amino acids								
Methionine	3	97	6	74	3	72	5	65
Leucine	87	170	50	160	43	151	31	134
Phenylalanine	85	143	67	145	65	148	49	129
Isoleucine	39	43	23	36	19	37	14	38

Table 3. Amino acid (mg/l) amounts in fresh beer and Strecker aldehyde (μg/l) generation during storage at 28 °C (STD: standard; AAM: amino acid added mixture).

	Freshly bottled		Stored 3 months		Stored 6 months	
Beer storage	STD	AAM	STD	AAM	STD	AAM
Amino acids						
Methionine	5	67				
Leucine	40	90				
Phenylalanine	91	168				
Isoleucine	29	29				
Aldehydes						
Methional	0.6	3.1	1.8	10.9	3.4	14.6
3-Methylbutanal	2.5	2.7	5.8	10.2	12.0	18.1
Phenylacetaldehyde	3.0	3.5	8.3	13.7	12.5	18.8
2-Methylbutanal	2.5	2.5	4.7	5.5	10.9	9.2

The difference of the sensory perceptions between the freshly bottled standard and trial beers is mainly based on the varying methional amounts. The aged flavour of the freshly bottled spiked sample is described as 'papery' with a slight sense of a 'potato-like' note. The flavour perception of the 3 months aged spiked sample develops into a significant 'bread-like' note in contrast to the standard.

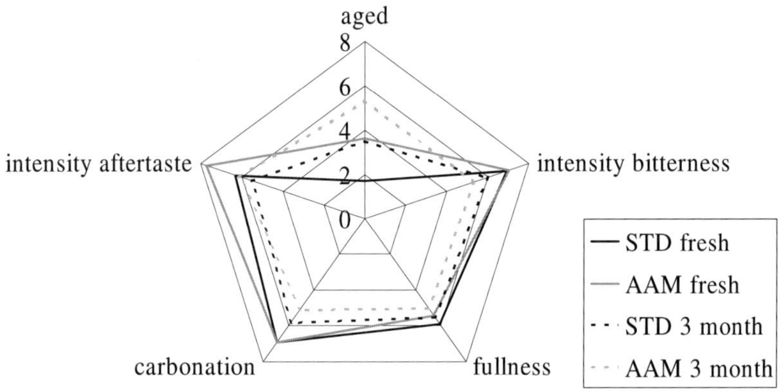

Figure 1. Results of the sensory evaluation of freshly bottled and 3 months stored samples (STD: standard; AAM: amino acid added mixture).

4. DISCUSSION AND CONCLUSION

Although boiled worts with and without amino acid addition differed significantly according to their content of Strecker aldehydes, fermentation resulted in products with nearly the same content of those compounds. However, after removing the yeast adequate amounts of free amino acids remain in the freshly bottled AAM sample to ignite a forced aldehyde generation during storage.

In conclusion, the results of these examinations open up new vistas to decelerate the Strecker aldehyde generation in the final product by controlling the precursor concentrations at each process step.

References

1. A.M. Jamieson and J.E.A. van Gheluwe, Proc. Amer. Soc. Brew. Chem., 29 (1970) 192.
2. K.R. Cadwallader and H.P. Weenen (eds.), Freshness and shelf life of foods, ACS symposium series 836, Washington DC, USA (2002) 70.
3. European Brewery Convention (ed.), Proceedings of the 29th EBC Congress, Nürnberg, Germany (2003) 732.
4. European Brewery Convention (ed.), Proceedings of the 30th EBC Congress, Nürnberg, Germany (2005) 947.

W.L.P. Bredie and M.A. Petersen (Editors)
Flavour Science: Recent Advances and Trends

Packaging material and formulation of flavoured yoghurts: how to choose the kind of polymer in accordance with the yoghurt composition?

Anne Saint-Eve[a], Cécile Levy[a], Marine Le Moigne[a], Solenn Coic[b], Violette Ducruet[b] and Isabelle Souchon[a]

[a]UMR GMPA, INRA, Thiveral Grignon, France; [b]UMR SCALE, INRA, 1 Avenue des Olympiades, Massy, Cedex, France

ABSTRACT

The influence of the packaging polymer (polypropylene or polystyrene) on the sensory and physicochemical characteristics of flavoured stirred yoghurts was investigated during 28 days of storage at 4 °C. The yoghurts with either 0% or 4% of fat content were flavoured with a strawberry aroma. A higher impact of packaging on sensory and physicochemical characteristics was observed for the fat free yoghurts compared to the 4% fat yoghurts. A fast loss of the fruity notes during the first 14 days was obtained in accordance with the decrease of the level of aroma compounds. A relative stabilisation of the flavour in the case of polypropylene was observed. However, polystyrene packaging seemed to be preferable for fat free yoghurts.

1. INTRODUCTION

Packaging materials are essential to protect the food product and to preserve its inherent quality during the storage. The interactions between food constituents and packaging materials may alter the sensory quality of the product [1]. The interactions occurring during storage are mainly (i) sorption phenomena from food to packaging materials, and (ii) migration phenomena from packaging to the food. The sorption of flavour compounds is influenced by the properties of the polymers and the flavour molecules, and by the external conditions of storage (time, temperature, etc.) [2].

Knowledge of the behaviour of yoghurt during storage is important because changes of quality characteristics may appear. Indeed, the time of storage can cause a fall in the acidity of fresh dairy products and an increase in yoghurt viscosity [3-4].

There arre only few studies addressing the influence of the packaging type on fresh dairy products. A few studies are based on model systems. In this study, the influence of the packaging (polypropylene (PP) or polystyrene (PS)) on the sensory characteristics of flavoured stirred yoghurts was investigated during 28 days of storage at 4 °C. The yoghurts flavoured with a strawberry aroma had a fat content of either 0% or 4%.

2. MATERIALS AND METHODS

2.1. Product preparation

Fermented milk products were either 0% or 4% fat stirred flavoured yoghurts [5]. Yoghurts were flavoured with a strawberry aroma containing 17 volatile compounds. Their concentrations ranged from 1.01 to 32.53 mg/kg in the final product. The products (Table 1) were conditioned in three different kinds of packaging: glass used as reference (G), polystyrene (PS) and polypropylene (PP).

Table 1. Experimental design of the yoghurts, packaging and storage conditions.

Product code	Fat content (%)	Packaging	Days of storage	Product code	Fat content (%)	Packaging	Days of storage
0gt2	0	Glass	2	4gt2	4	Glass	2
0pst2	0	PS	2	4pst2	4	PS	2
0ppt2	0	PP	2	4ppt2	4	PP	2
0gt14	0	Glass	14	4gt14	4	Glass	14
0pst14	0	PS	14	4pst14	4	PS	14
0ppt14	0	PP	14	4ppt14	4	PP	14
0gt28	0	Glass	28	4gt28	4	Glass	28
0pst28	0	PS	28	4pst28	4	PS	28
0ppt28	0	PP	28	4ppt28	4	PP	28

2.2. Physicochemical measurements

The complex viscosity of the three yoghurt matrices was measured at 10 °C, in the harmonic regime with a rheometer [5]. The pH of the products was measured at 4 °C. The release of volatile compounds at 4 °C was measured by solid-phase microextraction using a gas chromatograph combined with an automatic headspace sampler CombiPal.

2.3. Sensory profiling

A descriptive sensory analysis of the yoghurts was performed by 15 trained subjects. The sensory analyses were done after 2, 14 and 28 days of storage and were based on 30 descriptors related to visual texture with a spoon (V), texture in mouth (M), odour (N) (orthonasal olfactory perception), aroma (M) (retronasal olfactory perception) and taste. The subjects evaluated the intensity perceived for each descriptor on an unstructured scale anchored with the terms 'very weak' and 'very intense'. A replication was performed for each product.

3. RESULTS AND DISCUSSION

3.1. Evolution during storage of the yoghurts

A principal components analysis (PCA) based on the correlation matrix was performed on the yoghurt samples, using sensory and physicochemical variables, which significantly varied according to the product. A 3-ways analysis of variance (time, packaging and subject) with interaction on the sensory data set, and 2-ways ANOVA (time, packaging) with interaction on the physicochemical data set was performed. In total 17 sensory descriptors and 12 physicochemical parameters varied significantly according to the time of storage, the packaging or the interaction time of storage*packaging ($p<0.05$). The first factorial plot of PCA showed that fruity notes perception and the quantities of aroma compounds decreased from the day 2 to day 14 of storage, contrary to the thick perception, which increased during the storage. Indeed, perceived intensities of the odour notes 'kiwi', 'apple' and 'solvent' declined. In parallel, several aroma compounds had a significant and similar evolution during the storage: ethyl acetate, furaneol and limonene, which decreased during storage (53.2% for the limonene from day 2 to day 28 of storage). This finding corroborated with an earlier study [4], which also observed a decrease of some flavour compounds during the storage of flavoured yoghurts in plastic packaging.

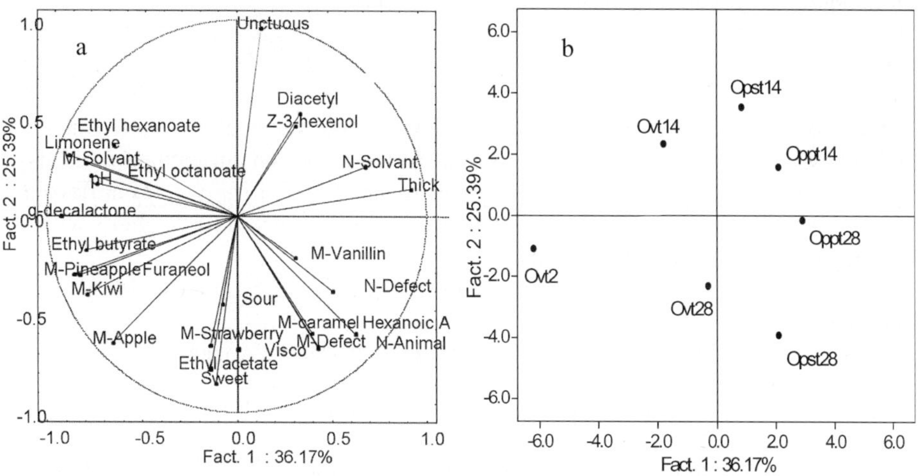

Figure 1. Principal components analysis of the descriptive sensory analysis and analysis of aroma volatile of the 0% fat yoghurts. (a) Loadings plot. (b) Scores plot. Sample codes are from Table 1.

The quantity of hexanoic acid increased with more than a factor 2 between day 14 and day 28, which was statistically significant ($p<0.02$). This increase could be related to the increase of 'animal' odour and 'defect' flavour perception, especially between day 14 and day 28 of storage. Moreover, the pH decreased during ageing from 4.30 to 4.10.

The PC2 illustrated mainly the evolution of yoghurt between day 14 and day 28 of storage. Concerning the texture attributes, yoghurts were perceived thicker from the day 14. Rheological measurements confirmed this observation: an increase of the complex viscosity of the low fat yoghurts of 71% was observed. Higher viscosities were attributed to exopolysaccharide production due to bacterial activity, confirmed by the decrease of pH during the storage [4-6]. According to [4], the changes in rheological properties observed during ageing could partly explain the results in flavour release.

3.2. Effect of type of packaging during storage

In general, the time effect dominated the packaging effect, but differences between the 3 packaging materials appeared. Yoghurts stored in PS and glass materials had a similar evolution. Defect notes appeared at the end of the storage time, mainly for yoghurts stored in polystyrene, and probably due to the migration of small molecules from the packaging to the food. Beyond 14 days, a stabilisation of the composition and the perception of the product flavour was observed only when the latter was packed in a polypropylene container. Moreover, the quantity of yoghurt aroma compounds in the 3 packagings decreased during the storage, but yoghurt stored in the glass jar showed the lowest decrease. The properties of the packaging polymers (higher density of polystyrene) could explain the evolution of the products by kinetics of sorption, i.e. the sorption being slower on polystyrene than on polypropylene. However, in a fat free yoghurt, the flavour compounds, and in particular the esters, exhibited a greater affinity for polypropylene than for polystyrene.

4. CONCLUSION

The sensory and physicochemical analyses of the low fat yoghurts stored at 4 °C highlighted for all the three packaging materials a fast evolution during the first 14 days. Polystyrene packaging seemed to be preferable to avoid loss of fruity notes in flavoured yoghurts used in this study. Less significant effects were observed for yoghurts with 4% fat content than with fat free yoghurts.

References

1. R. Van Willige, D. Schoolmeester, A. van Ooij, J. Linssen and A. Voragen, J. Food Sci., 67 (6) (2001) 2023.
2. V. Ducruet, N. Fournier, P. Saillard, A. Feigenbaum and E. Guichard, J. Agric. Food Chem., 49 (5) (2001) 2290.
3. N. Martin, J. Skokanova, E. Latrille, C. Beal and G. Corrieu, J. Sens. Stud., 14 (1999) 139.
4. S. Lubbers, N. Decourcelle, N. Vallet and E. Guichard, J. Agric. Food Chem., 52 (10) (2004) 3077.
5. E. Paçi Kora, E. Latrille, I. Souchon and N. Martin, J. Sens. Stud., 18 (5) (2003) 367.
6. Z.B. Guzel-Seydim, E. Sezgin and A. Seydim, Food Control, 16 (3) (2005) 205.
7. M.S. Brauss, S.T. Linforth, I. Cayeux, B. Harvey and A.J. Taylor, J. Agric. Food Chem., 47 (1999) 2055.

W.L.P. Bredie and M.A. Petersen (Editors)
Flavour Science: Recent Advances and Trends

Light-induced off-flavour in cloudy apple juice

Midori Hashizume[a,b], Tamotsu Okugawa[b], Michael H. Gordon[a] and Donald S. Mottram[a]

[a]School of Food Biosciences, The University of Reading, Whiteknights, Reading, RG6 6AP, UK; [b]Glico Dairy Products Co., Ltd., Institute for Technical Research, 2-14-1 Musashino, Akishima-shi, Tokyo, 1960021, Japan

ABSTRACT

Cloudy apple juice develops a metallic flavour when it is exposed to light. Off-flavour induced by fluorescent lights in cloudy apple juice was analysed by SPME-GC/MS. 1-Octen-3-one, 2-heptenal and (*E*)-2-octenal were identified in light-exposed cloudy apple juice. These compounds are known to be derived from lipid oxidation.

1. INTRODUCTION

Lipid oxidation induced by fluorescent light is known to be one of the main causes of deterioration of milk and milk products [1,2]. On the other hand, fruit juices such as apple, orange and grapefruit contain a low concentration of lipids, and nutritional information for consumers reports that they are free from fat. However, cloudy juices contain a trace of fatty acids derived from plant tissues in pulp fragments. Unpleasant metallic and mushroom flavours are detected by sensory evaluation after cloudy apple juice is exposed to fluorescent light for 24 h. Such beverages can be adversely affected by strong lighting in supermarkets. In this study, the light-induced off-flavour in cloudy apple juice was analysed by SPME GC-MS, and the possible effects on flavour changes are discussed.

2. MATERIALS AND METHODS

2.1. Samples

Commercial products of cloudy apple juice were imported under proper storage temperature from Glico Dairy Products Co., Ltd in Japan to the University of Reading in the UK. Juice concentrates (40 °Brix), to which vitamin C had been added for the inhibition of browning, were reconstituted and flavoured to produce cloudy apple juice (12 °Brix) in the factories prior to transportation. Samples were stored in glass bottles, exposed to 3,000 lux fluorescent light and kept at 8 °C from 24 h to 48 h. Simultaneously, equivalent samples were stored in the dark as a reference control.

2.2. Analysis of volatiles

Solid phase microextraction (SPME) was performed with a 75 μm Carboxen/ polydimethylsiloxane (PDMS) fibre. Twenty grams of cloudy apple juice, an internal standard (1.25 μg bromobenzene) and a stirring bar were placed in a 40 ml vial and capped with a PTFE septum. The SPME fibre was exposed to the sample headspace for 45 min. Each sample was analysed in triplicate. A Hewlett-Packard 5890 Series II Gas Chromatograph and a 5972 Series Mass Selective Detector were used for analysis. The column was a 60 m VF-5ms column (i.d. 0.25 mm, 0.25 μm, Varian, Inc.). The oven temperature was kept at 40 °C for 2 min and programmed to increase at 4 °C per min. The final temperature was 350 °C and held for a further 10 min. The temperature of the injector was 250 °C and that of the detector was 280 °C. The GC linear retention indices (LRI) of the isolated compounds were determined externally with a mixture of n-alkanes (C_6-C_{25}) which were run under the same conditions.

2.3. GC-Olfactometry analysis

GC-Olfactometry (GCO) evaluation was performed on an HP5890 gas chromatograph (Agilent, USA) fitted with an ODO II odour port (SGE International Pty Ltd., UK) using 60 m Rtx®-5MS column (i.d. 0.25 mm, 0.25 μm, RESTEC Co., Bellfone, PA).

2.4. Fatty acid analysis

Fatty acid methyl esters (FAME) were analysed as follows: 100 ml cloudy apple juice was extracted with chloroform-methanol (2:1) and transformed into methyl esters using 0.5 M sodium methoxide in methanol at room temperature. The resulting compounds were analysed by GC-FID (HP6890 gas chromatograph; Agilent, USA). Pentadecanoic acid was used as internal standard.

3. RESULTS

3.1. Flavour analysis

There were obvious differences between the 24 h light-induced sample and the equivalent reference. Description of the flavour of the light-exposed sample was 'metallic', 'mushroom-like' and 'oily'. More compounds were found in the light-induced sample than those in the reference (Table 1). The main compounds identified were esters. Thirteen esters were identified and the amount of these compounds did not change significantly after exposure to light. Most of the esters have been reported previously [3].

Table 1. Volatile compounds found in cloudy apple juice products.

Compound	LRI[b]	Relative peak area[a] A	Relative peak area[a] B	ID[c]
2-Methylpropanoic acid methyl ester	681	7	3	MS+LRI
2-Pentanone	684	15	13	MS+LRI
3-Pentanone	697	n.d.	8	MS+LRI
Pentanal	699	15	35	MS+LRI
Propanoic acid ethyl ester	709	19	10	MS+LRI
Acetic acid propyl ester	712	11	7	MS+LRI
Butanoic acid methyl ester	720	21	11	MS+LRI
2-Methyl-1-butanol	739	8	5	MS+LRI
2-Methylpropanoic acid ethyl ester	755	3	2	MS+LRI
Toluene	766	9	7	MS+LRI
2-Methyl-1-penten-3-one	768	n.d.	20	MS
Acetic acid 2-methylpropyl ester	771	30	22	MS+LRI
2-Methylbutanoic acid methyl ester	774	138	88	MS+LRI
Octane	799	n.d.	8	MS+LRI
Hexanal	802	241	388	MS+LRI
Acetic acid butyl ester	814	111	103	MS+LRI
2-Methylbutanoic acid ethyl ester	848	86	69	MS+LRI
3-Methylbutanoic acid ethyl ester	853	180	145	MS+LRI
(*E*)-2-Hexenal	856	39	59	MS+LRI
Ethylbenzene	863	7	8	MS+LRI
1-Hexanol	870	28	47	MS
Acetic acid 3-methylbutyl ester	876	72	64	MS+LRI
Acetic acid 2-methylbutyl ester	878	53	54	MS+LRI
Styrene	896	3	3	MS+LRI
Heptanal	904	n.d.	8	MS+LRI
2-Heptenal (*Z* or *E*)	960	n.d.	124	MS+LRI
1-Octen-3-one	979	n.d.	125	MS+LRI
Octanal	1006	n.d.	11	MS+LRI
Acetic acid hexyl ester	1012	9	18	MS+LRI
(*E*)-2-Hexenyl acetate	1015	n.d.	14	MS+LRI
(*E*)-2-Octenal	1063	n.d.	20	MS+LRI
Undecane	1099	7	9	MS+LRI
Nonanal	1108	n.d.	4	MS+LRI

[a]Relative percentage: (compound peak area/internal standard peak area)x 100, mean value of three replicates; n.d.: less than 2%. A: standard sample; B: sample exposed to fluorescent light for 48 h at 8 °C. [b]LRI: linear retention index on a VF-5MS column. [c]MS+LRI: mass spectrum and LRI agree with those of authentic compound; MS: mass spectrum agrees with mass spectral database.

On the other hand, many ketones and aldehydes were detected only from the light-induced sample, including 2-heptenal and (*E*)-2-octenal which are known to be linoleic acid derivatives [4]. An unpleasant strong metallic flavour was identified as 1-octen-3-one by GCO. While the area where hexanal eluted in GCO was described as having a pleasant apple aroma, other aldehydes such as (*E*)-2-hexenal, heptanal and (*E*)-2-heptenal were described with both pleasant and unpleasant odours.

3.2. Lipid analysis

Cloudy apple juice contains enough polyunsaturated fatty acids (PUFAs) such as linoleic acid and linolenic acid to form off-flavours (Table 2). Although arachidonic acid is known as one of precursors of 1-octen-3-one detected in light-induced samples, it was not detected by GC-FID. The origin of PUFAs could not be identified, since sodium methoxide randomly transesterifies ester derivatives of fatty acids.

Table 2. Fatty acids present in acylglycerols of the cloudy apple juice.

Fatty acids	C16:0	C18:0	C18:1	C18:2	C18:3	C20:0	C20:4	C22:0
(mg/kg)	27.4	4.1	10.7	96.2	17.5	1.0	n.d.	0.6

4. DISCUSSION AND CONCLUSION

1-Octen-3-one, (*E*)-2-octenal and (*E*)-2-heptenal were detected by GC-MS in cloudy apple juice after exposure to light, and GCO indicated that 1-octen-3-one caused mushroom and metallic off-flavours. These volatile compounds are formed from the oxidation of polyunsaturated fatty acids (PUFAs). The flavour threshold value of 1-octen-3-one in water is 0.0001 ppm [4]. Although PUFAs in cloudy apple juice are present at levels of about 200 mg/kg, it is suggested that small amounts of PUFAs are enough to cause light-induced off-flavour in cloudy apple juice. Therefore, 1-octen-3-one is useful for monitoring the quality of apple juice. In future work, lipids in cloudy apple juices will be analysed to determine the origin of PUFAs. Possible sources are wax in peel, triacylglycerols in seeds or phospholipids in plant cells.

References

1. P.S. Dimick, Milk Food Technol., 36 (7) (1973) 383.
2. M. Shiota, N. Takahashi, H. Konishi and T. Yoshioka, J. Am. Oil Chem. Soc., 81 (5) (2004) 455.
3. P.S. Dimick and J.C. Hoskin, CRC Crit. Rev. Food Sci., 18 (4) (1983) 387.
4. D.A. Forss, Prog. Chem. Fats Other Lipids, 13 (1971) 181.

W.L.P. Bredie and M.A. Petersen (Editors)
Flavour Science: Recent Advances and Trends

Formation and determination of microbially-derived off-flavour in apple juice

Barbara Siegmund, Barbara Zierler and Werner Pfannhauser

Graz University of Technology, Institute for Food Chemistry and Technology, Petersgasse 12/II, A8010 Graz, Austria

ABSTRACT

The role of two bacteria (namely *Alicyclobacillus acidoterrestris* and *Streptomyces griseus griseus*) on the off-flavour formation in apple juice was investigated. The strains were able to form a number of off-flavour compounds causing medicinal and/or earthy-musty notes in the juice under conditions that were common for the storage of shelf-stable products.

1. INTRODUCTION

The rejection of foods due to the occurrence of off-flavour has become one of the most frequent consumers' complaints. Microorganisms of various species do not only account for spoilage, but may also be responsible for off-flavour formation in the respective food [1]. In apple juice Cerny *et al.* [2] were the very first to identify a B*acillus* strain being responsible for off-flavour formation. Since the 1990s, various incidents of microbially derived off-flavour in fruit juices have been reported. These reports showed that various microorganisms (so-called thermoacidophilic bacteria) may survive under the common conditions for pasteurisation, even at the low pH values present in fruit juices. They are able to survive in shelf-stable products and may also germinate in the product at temperatures between 25 to 30 °C. The determined off-flavours have been described either as 'medicinal', 'chemical' or 'disinfectant' (e.g. caused by compounds like guaiacol or 2,6-dibromophenol) or as 'musty', 'earthy' (caused by compounds like 2-methylisoborneol, geosmin or 2-isopropyl-3-methoxypyrazine). The most extensively studied strain in this context is *Alicyclobacillus acidoterrestris*. Recently, other strains - mainly subspecies of the *Alicyclobacillus* genus [3], but also others like *Penicillium expansum* [4] - were described to be responsible for off-flavour formation in juices. It is not unlikely that even more strains from different species might be involved in this off-flavour problem.

In this study special emphasis was given to the growth behaviour and formation of volatile metabolites of two types of bacteria in apple juice, namely *Alicyclobacillus acidoterrestris* and a *Actinomyces spp.* (i.e. *Streptomyces griseus griseus*). The *Streptomyces* strain was chosen based on its ubiquitous character in the rural environment (e.g. water and soil) and the high probability of contamination via such sources. The qualitative and quantitative determination of 17 relevant off-flavour compounds formed by these bacteria was performed by GC-MS after headspace solid-phase microextraction (SPME) of the juice samples. The selection of the compounds was based on data found in literature [5]. To be able to judge the sensory relevance, odour thresholds of the compounds in an apple juice matrix were determined.

2. MATERIALS AND METHODS

2.1. Microbial growth experiments

Controlled growth of the investigated strains was performed in commercially available, aseptically filtered apple juice. The juice samples were inoculated with defined amounts of the strains *Alicyclobacills acidoterrestris* and *Streptomyces griseus griseus* (DSMZ, Braunschweig, Germany), either one strain per sample or both strains simultaneously. The bacteria were cultivated under different conditions regarding temperature and oxygen supply. Growth behaviour was controlled via microscopy as well as via inoculation of cell cultures on agar plates to prove for the absence of any undesired organisms. Growth rates were determined by calculating the colony forming units (CFU) per ml.

2.2. Sample preparation and GC-MS measurements

Headspace solid-phase microextraction (HS-SPME) was chosen for the extraction of the volatile metabolites from the juice sample. A DVB/CAR/PDMS fibre (film thickness 50/30 μm; fibre length 2 cm; Supelco) was used. Na_2SO_4 was added to the juice to improve the extraction yield of the sample. Extraction was performed at a sampling temperature of 60 °C. For the GC-MS analysis a Hewlett Packard System (HP G1800A GCD system) was used. The analytical column was an HP5 (cross-linked 5% phenyl methyl siloxane, 30 m x 0.25 mm x 1 μm). The analytical methods were fully validated; limits of detection and limits of quantification were determined for each of the 17 compounds. Details on the analytical method are given in [5].

2.3. Sensory evaluation

The sensory test panel consisted of 16 trained members. Training was performed in a number of training sessions, including training and tests on the basic tastes, odour recognition, as well as specific training on off-flavour compounds and off-flavour in apple juice. The odour thresholds were determined as group BET (best estimate threshold) according to [6]. Every value was determined at least twice using a minimum of 12 panellists per experiment.

3. RESULTS AND DISCUSSION

Several experiments were performed to investigate the growth behaviour of the two strains in apple juice. The concentrations and odour activities values (OAV) of all 17 compounds were determined in each experiment. In the following the results are discussed considering guaiacol as metabolite of *A. acidoterrestris* and [(1S)-endo]-(-)-borneol (called 'borneol' in the following) as metabolite of the *Streptomyces spp.* (see Table 1 for the odour thresholds in apple juice).

Table 1. Odour thresholds of off-flavour compounds (group BET) in apple juice.

Compounds	Odour threshold (μg/l)	± S.D.[a] (μg/l)	Number of trials[b]
Guaiacol	0.58	0.09	3 (14)
Borneol	47.7	7.7	2 (13)

[a]S.D.: standard deviation; [b]The number of subjects per trial is given in parenthesis.

Growth experiments at different temperatures showed increasing growth rates with increasing temperature. Accordingly, the concentrations of the volatile compounds rose with the temperature (Figures 1a and 1b). This holds true for all investigated compounds. Not all compounds could be detected at 4 °C (e.g. borneol, Figure 1b). On the contrary, a temperature of 22 °C was sufficient to gain fairly high concentrations of all the off-flavour compounds. The influence of the compounds on the off-flavour varied strongly from compound to compound. Guaiacol, for example, showed low concentrations at a growth temperature of 4 °C (OAV<1), whereas similar concentration curves were obtained at both elevated temperatures (OAV>1 after 10 days at 22 °C and 5 days at 30 °C). For borneol an OAV>1 could not be reached under the investigated conditions. Other compounds (e.g. geosmin, results not shown) showed large differences in concentrations at 22 °C and 30 °C, some of them with very high OAVs.

Growth behaviour of the strains was also strongly influenced by the available oxygen. Both strains showed higher growth rates with higher oxygen supply. In contrast, borneol and guaiacol were formed in higher concentrations in the experiment with limited oxygen (Figure 2b). Again, this behaviour did not hold true for all compounds of interest. The investigated pyrazines, for example, did not show different concentration curves depending on the oxygen in the system; others like 1-octen-3-ol were preferably formed in the oxygen-rich experiment.

Co-inoculation of the apple juice with both strains led to significantly reduced growth rates for both microorganisms (results not shown). The production of the volatile metabolites was diminished in the same dimension. Antibiotic production of the *Streptomyces spp.* was supposed to be one of the reasons for this behaviour.

These investigations showed that microbially-derived off-flavour in apple juice appeared to be a very complex issue, especially when more than one strain is involved. Our results indicated fairly different formation mechanisms for the compounds of interest. In order to get this off-flavour problem under control, further knowledge about the formation mechanisms and the involved enzyme systems is required.

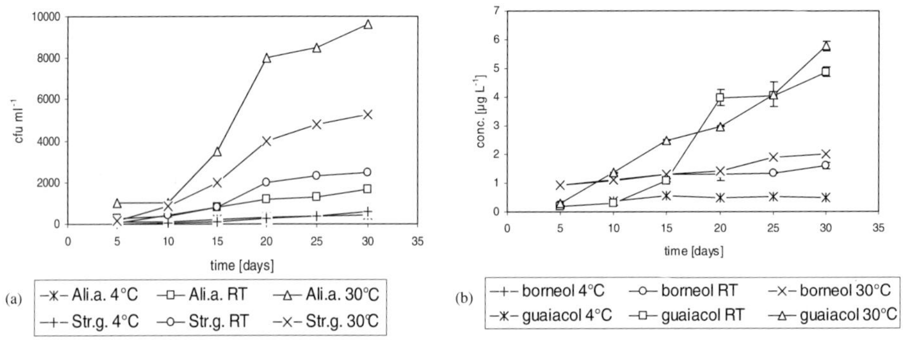

Figure 1. (a) Growth of the strains under different temperatures, one strain per sample (Ali.a. *Alicyclobacillus acidoterrestris*, Str.g. *Streptomyces spp.*); RT: room temperature (22 °C). (b) Formation of borneol and guaiacol in the respective experiments.

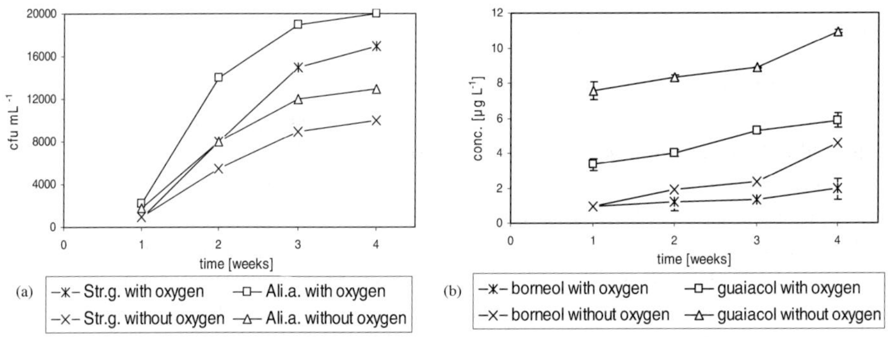

Figure 2. (a) Growth of the strains with and without oxgen supply under optimum growth temperatures; one strain per sample (Ali.a. *Alicyclobacillus acidoterrestris*, Str.g. *Streptomyces spp.*). (b) Formation of borneol and guaiacol in the respective experiments.

References

1. B. Baigrie (ed.), Taints and off-flavours in food, Cambridge, UK (2003) 112.
2. G. Cerny, W. Hennlich and K. Poralla, Z. Lebensmittel Untersuch. Forsch., 179 (1984) 224.
3. S.S. Chang and D.H. Kang, Crit. Rev. Microbiol., 20 (2) (2004) 55.
4. J.P. Mattheis and R.G. Roberts, Appl. Environ. Microbiol., 58 (9) (1992) 3170.
5. B. Zierler, B. Siegmund and W. Pfannhauser, Anal. Chim. Acta, 520 (2004) 3.
6. M. Meilgaard, G.V. Civille and B.T. Carr, Sensory evaluation techniques, Boca Raton, Fl. (1999).

W.L.P. Bredie and M.A. Petersen (Editors)
Flavour Science: Recent Advances and Trends

Off-flavours of soy ingredients: astringency - sensory perception, key molecules and masking strategies

Mickaël Labbe and Mark Springett

Danone Vitapole Centre de Recherche Daniel Carasso, Route Départementale 128, 91767 Palaiseau Cedex, France

ABSTRACT

The effect of type of soy ingredient and yoghurt cultures on astringency of soy-based yoghurts was determined. Soy-based yoghurts were prepared by mixing four different soy ingredients at a protein content of 5.5% prior to fermentation. Four soy ingredients (a soymilk powder, two soy protein isolates and full fat soy flour) and five pure or mixed bacterial strains were studied in a full factorial experiment. A trained sensory panel evaluated the soy products. Soy ingredients had a strong effect on astringency and bitterness. Bacterial strains were shown to decrease soy off-flavours, and at the same time, to increase the sweetness and creamy flavour. These results indicate that the choice of soy ingredient and starter cultures could play a crucial role in formulating soy-based yoghurt with reduced astringency.

1. INTRODUCTION

Traditionally, soybeans are consumed as either non-fermented or fermented products. The reasons for the popularity of soybeans are ease of production worldwide, low cost, good nutritional value and health benefits. Recently, studies have revealed the strong potential of soy foods in lowering blood cholesterol levels and the incidence of heart disease, cancer and other chronic diseases [1,2]. Despite the many advantages of soybeans, they have not been widely accepted by Western consumers because of perceived off-flavours characterised by a beany note and an objectionable aftertaste [3,4] described as bitter and astringent. Research has focused on the problem of beany flavour which arises due to lipid oxidation and bitter notes which are mainly due to the isoflavones and certain peptides [5,6]. However, there have been few studies focusing on the flavour defect astringency, which has been detected in foods including persimmons, wine, tea, unripe fruits, ciders and especially in soy products.

To date, data is insufficient to draw any conclusion on the impact of lactic acid bacteria on astringency. Previous studies [7,8] have evaluated the microbiological aspects of the growth of yoghurt cultures and their ability to ferment soybean carbohydrates. The study reported here was conducted to investigate the impact of type of soy ingredient and yoghurt cultures on the astringency of fermented soy products.

2. MATERIALS AND METHODS

2.1. Ingredients

The two Soy Protein Isolates (SPI1, SPI2), Soy Milk Powder (SMP) and Soy full fat Flour (SF) were obtained from commercial suppliers. The brief composition of the soy ingredients is summarised in the Table 1.

Table 1. Composition of soy ingredients.

	SMP	SPI 1	SPI 2	SF
Moisture, g/100g	5.0	4.5	5.0	10.0
Protein (N x 6.25, as is), g/100g	47.0	81.0	79.0	39.0
Fat total (acid hydrolysis), g/100g	25.0	4.0	5.0	20.0
Carbohydrates, g/100g	25.0	< 1	0.0	25.0
Ash, g/100	7.0	10.5	11.0	4.0
Isoflavones total, mg/g protein	2.0	-	2.0	2.0

2.2. Microorganisms and enumeration

Cultures of *St. thermophilus* (St1, St2, St3) and *Lb. delbruecki ssp. bulgaricus* (Lb1, Lb2), from the Danone culture collection were used. Bacteria were enumerated on MRS and M17 agars and results reported as colony forming units (cfu)/g.

2.3. Soy-based yoghurt manufacture and experimental design

To evaluate the impact of soy ingredients type (SMP, SPI1, SPI2, SF) and strains (St1/Lb1, St1/Lb2, St2/Lb1, St2/Lb2, St3) on the astringency, a full 4 x 5 factorial experiment was made. Four mixes at 5.5% protein (3.4% protein from SMP and 2.1% protein from SMP, SPI1, SPI2 or SF) were fermented with 2% cultures until a pH value of 4.6.

2.4. Sensory and statistical evaluation

To evaluate the flavour attributes (sweet, bitter, astringent, acid, creamy flavour and beany flavour) of soy-based yoghurts, judges were familiarised with a brief description of each attribute prior to sensory evaluation tests, using the standards listed in Table 2. Each soy-based yoghurt was coded and evaluated, in triplicate, by 8 panellists in random order (14 days after production), on a 10 cm unstructured line-scale. These means were separated using the Duncan multiple test.

Table 2. Standards used for evaluation of taste and flavour attributes.

Attribute	Reference standard	Standard concentration (g/l)
Sweet	Sucrose	Low int. = 1; High int. = 10
Bitter	Quinine chloride monohydrate	Low int. = 0.008; High int. = 0.03
Acid	Citric acid	Low int. = 0.25; High int. = 1
Astringent	Alum	Low int. = 0.1; High int. = 0.6
Beany flavour	Water/soymilk ratios (v/v)	Low int. = 10:1; High int. = 0:1
Creamy flavour	Water/whole cow milk ratios (v/v)	Low int. = 10:1; High int. = 0:1

3. RESULTS

3.1. Growth and survival of lactic acid bacteria in soy-based yoghurt

In general, the pH decreased as the fermentation time increased for all the soy ingredients and all the strains. Among the soy ingredients, the best acidification kinetics in terms of short lag phase and time to reach final pH value, were obtained with the mix SPI1. Among the strains, the best acidification kinetics were obtained with the strains St2 and St3 which reached a pH value below 4.6 in <5 h. Regarding the growth of strains, the best results were obtained with *St. thermophilus* strains and the lowest final number was obtained for *Lb. bulgaricus* strains due to their low ability to metabolise sucrose, raffinose and stachyose. In this study, we did not find any decline in the numbers of bacteria during storage and therefore, the impact of strains on astringency was not associated with a difference in inoculation efficiency but more to a difference in metabolism of each starter culture.

3.2. Impact of soy ingredients on astringency

The impact of soy ingredients was significant on astringency ($p=0.008$), acidity ($p=0.013$) and beany flavour ($p=0.031$) but not significant on bitterness ($p=0.209$), sweetness ($p=0.185$) and creamy flavour ($p=0.282$) (Table 3). The most astringent product (SF) was also the most bitter, the most acid, the most beany and the least creamy and sweet. This indicated that astringency was positively correlated with bitterness, acidity and beany flavour and negatively with sweetness and creamy flavour. This positive correlation between astringency and bitterness could be explained by a common molecular origin such as isoflavones and peptides.

Table 3. Impact of soy ingredients on sensory attributes of soy-based yoghurts.

Mix	Astringency	Bitterness	Sweetness	Acidity	Creamy	Beany
SF	6.09 a	2.83 a	0.82 b	4.43 a	1.36 a	6.86 a
SMP	4.43 b	2.33 ab	1.22 ab	3.97 a	2.23 a	5.60 ab
SPI 1	4.36 b	2.16 ab	1.41 ab	2.03 b	2.43 a	5.33 b
SPI 2	4.07 b	1.87 b	1.70 a	2.00 b	2.30 a	4.89 b

Values in the same column with different letters were significantly different ($p<0.05$).

However, the isoflavone content can not be the only cause of astringency because the most neutral product was produced from an isoflavone-rich ingredient (SPI2). SPI2 was also stronger in creamy and sweetness, which may have masked the astringency.

3.3. Impact of strains on astringency

The impact of strains (Table 4) was significant on the astringency ($p=0.044$), bitterness ($p=0.006$), sweetness ($p=0.004$), creamy flavour ($p=0.030$), acidity ($p=0.050$) and beany flavour ($p=0.049$). As previously, astringency and bitterness were positively correlated and clearly strain-dependent.

Table 4. Impact of the strains on sensory attributes of soy-based yoghurts.

Strains	Astringency	Bitterness	Sweetness	Acidity	Creamy	Beany
St1Lb1	4.54 ab	2.08 bc	1.64 ab	3.97 a	2.16 abc	5.21 b
St1Lb2	5.54 a	3.00 a	0.72 c	3.71 a	1.16 c	5.16 b
St2Lb1	3.79 b	1.37 c	2.07 a	2.71 b	3.08 a	5.54 ab
St2Lb2	5.51 a	2.62 ab	0.87 c	2.75 b	1.66 bc	5.41 ab
St3	4.29 ab	2.41 ab	1.15 bc	2.39 b	2.33 ab	7.04 a

Values in the same column with different letters were significantly different ($p<0.05$).

Indeed, the perceived astringency and bitterness were correlated with the presence of the strain Lb2. This result could be explained by a peptidolytic activity. St3 had the highest beany note. There was, however, not an increase in astringency. This indicated that lipid oxidation was not the mechanism of formation for astringency. The best strain association was St2-Lb1 with a weak impression of astringency and a possible masking effect from the higher creamy flavour and the sweetness taste.

4. DISCUSSION AND CONCLUSION

The best yoghurt obtained was with the soy ingredient SPI2 and fermented with strains St2 and Lb1. Soy ingredients and starter cultures clearly had an impact on astringency, making their selection a key step in the development of soy-based yoghurts without astringency. Further reduction of astringency should consider the peptidolytic acitivity, post acidification characteristics and flavour generation capabilities of the strains.

References

1. M.J. Messina, V. Persky, K.D.R. Setchell and S. Barnes, Nutr. Cancer, 21 (2) (1994) 113.
2. K.K. Carroll and E.M. Kurowska, J. Nutr., 125 (35) (1995) 594S.
3. W.J. Wolf and J.C. Coward (eds.), Soybeans as a food source, Cleveland, USA (1975) 82.
4. T. Watanabe and A. Kishi (eds.), The book of soybeans, New York, USA (1984) 31.
5. A.S. Huang, O.A.L. Hsieh and S.S. Chang, J. Food Sci., 47 (1) (1982) 19.
6. K. Okubo, M. Iijima, Y. Kobayashi, M. Yoshikoshi, T. Uchida and S. Kudou, Biosci. Biotechnol. Biochem., 56 (1) (1992) 99.
7. H.L. Wang, L. Kraidej and C.W. Hesseltine, J. Milk Food Technol., 37 (2) (1974) 71.
8. Y.J. Cheng, L.D. Thompson and H.C. Brittin, J. Food Sci., 55 (4) (1990) 1178.

W.L.P. Bredie and M.A. Petersen (Editors)
Flavour Science: Recent Advances and Trends

Optimising soy sauce quality by linking flavour composition with consumer preference

Max Batenburg, Joop Wesdorp, Frank Meijer, Wilma den Hoed, Pieter Musters, Mikkel Suijker and Gerrit Smit

Food Research Centre, Unilever R&D Vlaardingen, P.O. Box 114, 3130AC Vlaardingen, The Netherlands

ABSTRACT

The molecular basis of preference was investigated for Chinese soy sauce by combining identification of key flavour compounds with sensory and consumer studies. Based on the results, routes were explored to avoid off-flavour generation during fermentation and to optimise the composition of the final blend.

1. INTRODUCTION

Soy sauce is produced by solid state mould fermentation ('koji') of the raw materials, soy and wheat, followed by a submerged high-salt ripening phase ('moromi') [1]. During the latter phase, proteins and carbohydrates are hydrolysed to amino acids and sugars, and subsequent Maillard reactions lead to the generation of a range of savoury aroma compounds [2]. Depending on the conditions, lactic acid bacteria and yeasts may, also contribute to flavour generation, but in the short Chinese 'jiang pei' process their role is limited. In China, the fermentation product is usually marketed as a mixture with other ingredients, like monosodium glutamate and hydrolysed vegetable protein.

The aim of the present study was to improve the consumer acceptance of the product by identification of the key sensory drivers of liking, and translation of these sensory attributes into the product composition. This approach requires collaboration of consumer and sensory research, flavour science, and finally biotechnology to establish the conditions needed to achieve the desired flavour profile.

2. MATERIALS AND METHODS

2.1. Aroma analysis

Soy sauce, standardised in pH and saturated with NaCl, was shaken for 2 h with half a part (w/w) of methyl formate with internal standards. For GC-Olfactometry, 2 μl of the solvent layer, obtained by centrifugation and $MgSO_4$-drying, was injected on an HP5890, equipped with a 50 m Carbowax 58CB (0.32 mm/0.2 μm) column, an FID- and a sniffing port. AEDA (aroma extract dilution analysis) was performed with 3 assessors sniffing at 6 levels of dilution. GC-MS identification was performed off-line.

2.2. Consumer study

Based on a Takuchi factorial design, 24 soy sauces were prepared with the same soy sauce recipe but with different additions affecting flavour or appearance. Seven compounds and ingredients, suggested to be potential drivers of liking, were added at three concentration levels in various combinations to a relatively neutral soy sauce base. A standardised pork dish was prepared with all 24 sauces, and in sets of eight products ranked and scored for liking on a 7-point scale by 100 Shanghai consumers in an in-hall test.

2.3. Microbiological challenge test

The 125 g koji substrate (soy grits, wheat bran, and water) was sterilised in 250 ml Scott flasks. The amount of water added to the raw materials was varied as well as the temperature during further incubation. A small inoculum (10^2 CFU/g) of a *B. subtilis* was added, previously isolated from a factory koji, and the outgrowth and resulting formation of 3-methylbutyric acid was measured after 24 and 48 h. Conditions for isolation and analysis of the volatiles were as described above.

3. RESULTS AND DISCUSSION

Key aroma compounds were identified via GC-Olfactometry, quantification and odour activity determination, and finally flavour recombination. Table 1 shows the results of AEDA and odour activity value (OAV) determination. Dominating in the aroma are hydroxyfuranones, phenolic compounds and short-chain fatty acids. In contrast to Japanese and Chinese 'jiang lao' soy sauce, but in line with the production route, homofuraneol, esters and other fermentation products are of low importance, also according to additional headspace GCO analyses. The putative key components as derived from GCO were quantified via GC-MS using isotope dilution analysis or standard addition methodology. Odour activity values (OAVs) obtained from comparison with published flavour thresholds [3,4], essentially confirmed the GCO data. As a final check of the completeness of our knowledge, a synthetic match of the soy sauce was prepared, containing 31 aroma compounds and 39 tastants (minerals, organic acids, amino acids, sugars and sweeteners), dosed according to their measured concentration. In a comparison with batches of real soy sauce the match was not discriminated by naive consumers, and was apparently within the natural variation.

Table 1. Flavour dilution factors from aroma extract dilution analysis (AEDA), measured concentrations (in ppm), and odour activity values of soy sauce key aroma compounds.

	FD factor	Concentration	OAV
Acetic acid (acetic)	5	800.00	13.4
Methional (potato)	200	0.04	200.0
Butyric acid (cheesy)	30	15.90	64.0
2-Phenylacetaldehyde (floral)	1	0.71	177.0
3-Methylbutyric acid (sweaty)	200	23.20	23.2
Cyclotene (bouillon)	5	3.50	11.7
Guaiacol (smoky)	500	0.75	83.0
2-Phenylethanol (floral)	200	0.67	0.6
Maltol (sweet)	100	6.40	2.5
Furaneol (sweet)	500	6.20	200.0
Unknown (urine)	30	-	-
Homo-furaneol (sweet)	1	0.64	14.9
Sotolon (bouillon, maggi)	500	0.59	1970.0
Abhexon (bouillon, maggi)	30	-	-
2,6-Dimethoxyphenol (smoky)	100	0.41	16.6
Phenylacetic acid (honey,stable)	200	3.56	0.4

Combination of in-home consumer preference studies and quantitative descriptive analysis of commercial soy sauces suggested several attributes influencing consumer liking. Sensory attributes positively correlating with liking included sweet taste, dark and preferably reddish colour, 'treacle' aroma and 'table seasoning' aroma. A strong 'animalic/sweaty' aroma on the other hand was suggested to affect consumer liking in a negative way. To investigate whether these correlations are indeed based on causal relations, a new in-hall consumer preference study was designed, based on deliberate variation of these potential key drivers of liking. The added compounds and ingredients were selected on the basis of their unimodal sensory effect in the QDA panel. Sweetness with least side effects was introduced, e.g. via artificial sweeteners, better than with sucrose. On the basis of our flavour knowledge 'animalic' and 'seasoning-like' could be translated to 3-methylbutyric acid and abhexon respectively. For 'treacle' a pure compound could not be identified, but cane molasses was found effective to introduce this note.

From the results (Figure 1) it can be concluded that sweetness is a dominating factor with respect to liking. The factors affecting appearance have small effects in spite of the relatively high dose of e.g. the starch; the standard level of caramel colorant of 3% is preferred over very high dosages. Aroma additions, introducing 'treacle' and 'seasoning' notes, do have significant positive effects on liking; high levels of 'animalic' fatty acids have a negative effect. For the 'seasoning' aroma of abhexon a significant positive effect is observed only at higher sweetener concentrations; a negative effect was seen when 'sweet' was set at its lowest level (data not shown). No further significant interactions were observed in this set of parameters. The data indicated that flavour composition and liking were indeed based on causal relations.

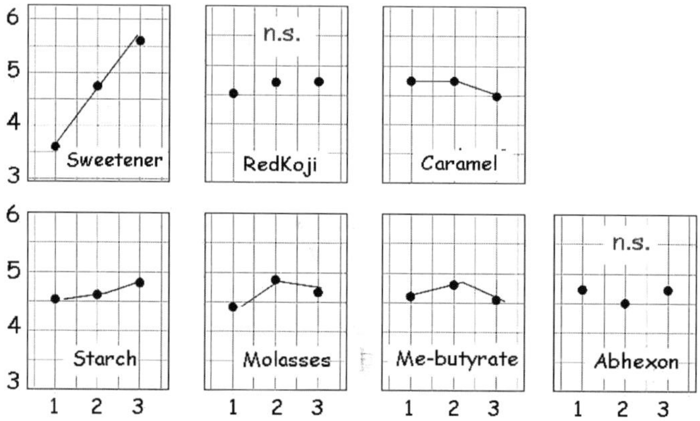

Figure 1. Effects of potential drivers on 'liking' in a pork dish recipe prepared with soy sauce. Average scores per dosage level; n.s., not significant. Dose in soy sauce: cyclamate + saccharin: 0+0%, 0.05+0.025%, 0.06+0.06%; red koji: 0, 0.02, 0.035%; caramel: 3, 6, 9%; starch: 0, 6, 9%; molasses: 0, 2.5, 7.5%; 3-methylbutyric acid: 0, 12, 72 ppm; abhexon 0, 0.1, 1.0 ppm.

High levels of volatile acids like 3-methyl butyric acid appear undesirable. Moreover, a large batch-to-batch variability was observed in the volatile acid content, sometimes exceeding 100 ppm. In a lab-scale soy sauce process the moisture and temperature conditions during the koji phase were therefore investigated to avoid the Bacillus growth causing the flavour defect. The results (not shown) stressed the importance of a good control of these process parameters. Additional protection might be obtained from application of the natural antibiotic nisin which appeared to be an effective inhibitor under practical conditions. Nisin-producing lactic acid bacteria, which were found to grow out well under koji conditions, may offer a more cost-effective solution.

Based on the data presented, a prototype soy sauce could be composed, containing only sweeteners, molasses and abhexon flavouring as extra components, that clearly outperformed the original product in an in-home, sequential monadic, preference test. To our knowledge this is one of very few published examples where product improvement was obtained via direct linking of preference data, sensory characterisation, and molecular composition.

References

1. K.H. Steinkraus (ed.), Handbook of indigenous fermented foods, New York, USA (1996).
2. C.O. Chicester, E.M. Mark and B.S. Schweigert (eds.), Advantages in food research, Orlando, USA, 30 (1986) 195.
3. R.G. Buttery, R. Teranishi and J.G. Turnbaugh, J. Agric. Food Chem., 38 (1990) 336.
4. R.G. Buttery, L.C. Ling and D.J. Stern, J. Agric. Food. Chem., 45 (1997) 837.

W.L.P. Bredie and M.A. Petersen (Editors)
Flavour Science: Recent Advances and Trends

Analysis of Gruyere-type cheeses by purge and trap GC-MS and solvent assisted flavour evaporation GCO/MS

Hedwig Schlichtherle-Cerny, Roland Gauch and Miroslava Imhof

Agroscope Liebefeld-Posieux, Swiss Federal Research Station for Animal Production and Dairy Products (ALP), Schwarzenburgstrasse 161, CH-3003 Berne, Switzerland

ABSTRACT

The aim of this study was to characterise the volatile flavour compounds in a Gruyere-type cheese manufactured from pasteurised milk using different strains of *Lactobacillus casei*. These strains were used as non-starter cultures to manufacture the different cheeses. A dynamic headspace extraction technique, purge and trap, coupled to gas chromatography-mass spectrometry (GC-MS), and solvent assisted flavour evaporation (SAFE) in combination with GC-MS and GC-olfactometry (GCO) were used to determine the volatile aroma compounds formed after 180 days of ripening. The purge and trap method revealed aldehydes and ketones, such as 2-methylpropanal, 2-methylbutanal, 3-methylbutanal, and 2,3-butanedione, as well as dimethyl disulfide and dimethyl trisulfide among other compounds, which have been described earlier as key aroma compounds of Gruyere and Cheddar cheeses. Free fatty acids were not detected by the purge and trap extraction method. On the other hand, the analysis of the SAFE extract revealed intense signals for propanoic, butanoic, 2- and 3-methylbutanoic acid as well as pentanoic acid and hexanoic acid. Less volatile, more polar odorants, such as the caramel-like 2,5-dimethyl-4-hydroxy-3(2*H*)-furanone and 2-ethyl-4-hydroxy-5-methyl-3(2*H*)-furanone, but also the mushroom-like smelling 1-octen-3-one were detected in the SAFE extract by GCO, however, not using purge and trap GC-MS.

1. INTRODUCTION

Microbial enzymes are essential for flavour development in cheese. Non-starter lactic acid bacteria (NSLAB) originate from the factory environment or are already present in the raw milk. Selected NSLAB strains are often added as adjunct cultures during cheese

production, since they have been reported to be crucial for flavour and aroma development in cheeses. NSLAB are able to convert amino acids produced from the starter cultures by milk protein degradation into aroma compounds [1]. Methional, 2- and 3-methylbutanal are examples of key aroma compounds in Gruyere cheese [2]. They originate from the amino acids methionine, isoleucine and leucine, respectively.
Facultatively heterofermentative lactic acid bacilli (FHL) are indigenous to raw milk and belong also to NSLAB. They inhibit growth of propionic acid bacteria and are used in Switzerland to prevent late fermentation defects caused by the latter. FHL ferment hexoses mostly to lactic acid and grow in cheese metabolising citrate into formic acid, acetic acid and CO_2 [3].
The aim of the present study was to compare the aroma compounds formed in Gruyere-type hard cheeses manufactured with different *Lactobacillus casei* cultures by different extraction methods.

2. MATERIALS AND METHODS

Gruyere-type hard cheeses were produced from pasteurised milk using three different FHL non starter *Lactobacillus casei* strains (FAM 3228, FAM 6161, and FAM 8407, Agroscope Liebefeld-Posieux) and ripened for 180 days. The control cheese was ripened without adjunct cultures.
The cheeses (5 g) were suspended in distilled water (20 ml), and 10 g of the suspension were extracted by dynamic headspace (purge and trap) using a Tekmar 3100 instrument. The effluent was trapped on a Tenax trap (No. 1). GC-MS was conducted using a SPB-1 sulfur capillary column (Supelco, 30 m × 0.32 mm × 4 μm film thickness) with helium as carrier gas (55 kPa) applying the following temperature program: 45 °C (13 min) to 240 °C at 5 °C/min. Mass spectra were obtained in the EI mode at 70 eV and a scan range from *m/z* 26-250. Purge and trap analyses were carried out in triplicate. The software Systat 11 (Systat Software Inc.) was employed for the statistical data analysis of the purge and trap data.
For solvent assisted flavour evaporation (SAFE) the cheeses (50 g) were extracted with diethyl ether/pentane (2:1), and the extracts were submitted to SAFE according to [4]. GCO/MS was performed with 1 μl aroma extract using an Optima 1701 capillary column (Macherey-Nagel, 30 m × 0.25 mm × 0.25 μm film thickness) at a constant helium carrier gas flow of 55 ml/min. The oven temperature was 35 °C for 5 min, and increased by 5 °C/min to 280 °C. The column effluent was assessed for odours by 4 trained panellists. Mass spectra were obtained in the EI mode at 70 eV and a scan range from *m/z* 29-400. Odorants were identified on the basis of identical GC linear retention indices and mass spectra with authentic reference compounds.

3. RESULTS AND DISCUSSION

The purge and trap profiles of the Gruyere-type cheeses revealed volatile aldehydes and ketones which have been described as aroma compounds in Gruyere and other cheeses, such as 2- and 3-methylbutanal, 2,3-butanedione, dimethyl disulfide and dimethyl trisulfide [3]. Free fatty acids were not detected in the purge and trap chromatogram, probably partly due to dissociation of the acids present in the aqueous suspension of the cheese. The comparison of the peak heights obtained by purge and trap GC-MS showed that in particular with the FHL strains FAM 6161 and FAM 8407 more dimethyl disulfide and dimethyl trisulfide was formed (Figure 1) as compared to the control and to FAM 3228, respectively. FAM 3228 followed by FAM 8407 produced significantly greater levels of 2-methylpropanal as well as for 2- and 3-methylbutanal as compared to the control and to FAM 6161.

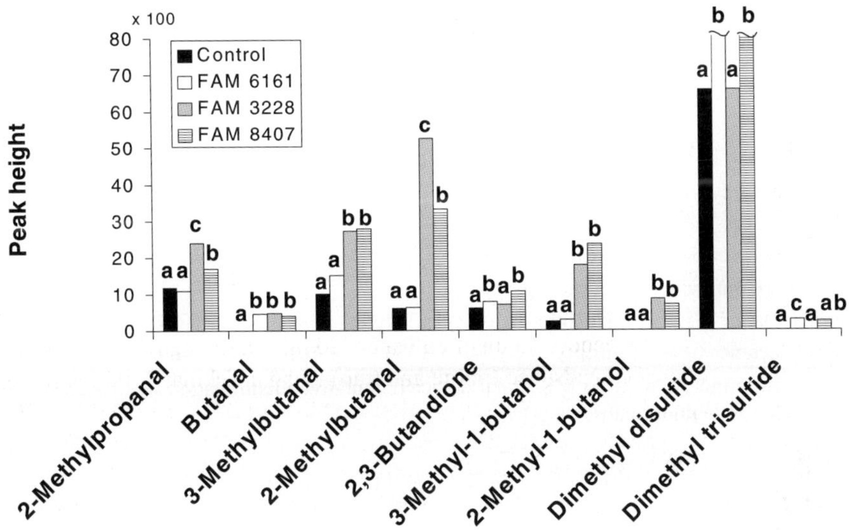

Figure 1. Selected volatiles produced by different lactic acid bacteria strains in Gruyere-type cheese after 180 days of ripening (purge and trap GC-MS). Peak heights showing different letters for a compound are significantly different ($p<0.05$).

Solvent assisted flavour evaporation (SAFE) in combination with GC-MS and GCO/MS was used in comparison to the purge and trap extraction method. Table 1 compares the peak heights for selected odorants. The extracts revealed very intense signals for free fatty acids, such as acetic, butanoic, 2-methylbutanoic acid and 3-methylbutanoic acid. Butanoic acid together with 2- and 3-methylbutanoic acid were described among the character impact compounds of Gruyere cheese [2]. The less volatile, more polar caramel-like odorants 2,5-dimethyl-4-hydroxy-3(2*H*)-furanone and 2-ethyl-4-hydroxy-3(2*H*)-furanone were detected by SAFE-GCO/MS, however, not by purge and trap GC-MS. These two compounds were described as character impact compounds in Emmental cheese [5]. The mushroom-like compound 1-octen-3-one, as well as the coconut and fruity smelling δ-deca- and δ-dodecalactones were also found by SAFE-GCO/MS only.

Table 1. Profiles of selected odorants analysed by SAFE GCO/MS.

			Peak height/1000		
Compound	Odour quality	RI[a]	FAM8407	FAM6161	FAM3228
2,3-Butandione	Buttery, creamy	675		n.d.[b]	
Acetic acid	Pungent	765	21873	20108	22083
Dimethyl disulfide	Onion, sulfury	802	180	165	153
Propanoic acid	Pungent, rancid	904	5817	6246	4899
2-Methylpropanoic acid	Goat-like	978	5285	4390	2653
Butanoic acid	Rancid, sweaty	1009	15338	13895	12367
Methional	Cooked potato	1030		n.d.[b]	
3-Methylbutanoic acid	Rancid, sweaty	1035	4473	3674	3888
2-Methylbutanoic acid	Sweaty, fatty	1054	8808	8000	6520
1-Octen-3-one	Mushroom-like	1066		n.d.[b]	
Hexanoic acid	Rancid, cheesy	1173	5862	6150	5854
2-Nonanone	Cheesy	1181	511	703	581
2,5-Dimethyl-4-hydroxy-3-(2*H*)-furanone	Sweet, caramel-like	1237	183	143	266
2-Ethyl-4-hydroxy-5-methyl-3-(2*H*)-furanone	Sweet, caramel, burnt sugar	1336	130	62	-
δ –Decalactone	Coconut-like	1714	680	624	1017
δ –Dodecalactone	Fruity	1934	501	469	787

[a]Linear retention index. [b]n.d.: no MS signal detected, tentative identification was based on linear retention index and odour quality.

4. CONCLUSION

The results showed that purge and trap GC-MS retrieved highly volatile odorants, whereas SAFE-GCO/MS revealed more polar, less volatile compounds from Gruyere-type cheeses. Both methods can be used as complementary techniques. Facultatively heterofermentative lactic acid bacillus strains were shown to produce important aroma compounds, such as sulfides and aldehydes.

References

1. A. Kieronczyk, S. Skeie, T. Langsrud and M. Yvon, Appl. Environ. Microbiol., 69 (2) (2003) 734.
2. M. Rychlik and J.O. Bosset, Int. Dairy J., 11 (2001) 895.
3. P.F. Fox, P.L.H. McSweeney, T.M. Cogan and T.P. Guinee (eds.), Cheese: chemistry, physics, and microbiology, Amsterdam, The Netherlands (2004) 140.
4. W. Engel, W. Bahr and P. Schieberle, Eur. Food Res. Technol., 209 (1999) 237.
5. M. Preininger and W. Grosch, Food Sci. Technol. Lebensm. Wiss. Technol., 27 (1994) 237.

W.L.P. Bredie and M.A. Petersen (Editors)
Flavour Science: Recent Advances and Trends

Influence of pasteurisation and pulp amount on partition coefficients of aroma compounds in orange juice

Cécilia Berlinet[a], Pierre Brat[b], Cédric Plessis[a] and Violette Ducruet[a]

[a]INRA UMR SCALE (ENSIA/CNAM/INRA), 1 avenue des Olympiades, 91744 Massy Cedex, France; [b]CIRAD-FLHOR, TA 50/16, 78 rue Jean-François Breton, 34398 Montpellier Cedex 5, France

ABSTRACT

The influence of flash-pasteurisation and pulp amount (10 and 20% pulp) on vapour/liquid partition coefficients of aroma compounds (ethyl butyrate, hexanal, β-myrcene, limonene and octanal) were investigated in processed orange juice. Before pasteurisation, the highest retention of aroma compounds was obtained with 20% pulp, whereas after pasteurisation, it was obtained with 10% pulp. The particle size distribution of the 10% pulp pasteurised orange juice differed markedly from the corresponding fresh juice and could explain the higher retention observed after pasteurisation.

1. INTRODUCTION

Orange juice is a multiphase system which is formed by an aqueous phase, the serum, containing soluble compounds and a water insoluble phase including pulp and cloud. Eighty percent of total juice volatiles are within these insoluble particles of which 90% are in the pulp and 10% are in the cloud [1]. Numerous studies of the effect of pasteurisation on the aromatic composition of orange juice are reported in the literature. Only a few deal with the effect of insoluble solids on the aroma composition. Jordan *et al.* [2] showed that the headspace of freshly extracted orange juice contained less aldehydes, terpenes and alcohols after a reduction of the insoluble solids. More recently, Rega *et al.* [3] revealed that the pasteurisation process decreased the headspace amounts of important compounds like acetaldehyde, ethyl butyrate and hexanal. These studies were conducted by SPME and underline the headspace evolution.

When measuring the partition coefficients of volatile compounds in orange juice insight into the molecular interactions between insoluble solids and volatiles explaining retention or release, can be gained. In this work, the influence of pasteurisation and pulp amount (10 and 20% pulp) on vapour/liquid partition coefficients of the aroma compounds ethyl butyrate, hexanal, β-myrcene, limonene and octanal in processed orange juice was studied. The phase ration Variation (PRV) method was used because of the lack of calibration which is useful for food matrix analysis [4].

2. MATERIALS AND METHODS

2.1. Juice samples

Orange juices were purchased from a citrus factory located in the region of Valencia, Spain. Fresh juices containing 20% (A1) and 10% (B1) of pulp were flash-pasteurised for 20 s at 98 °C, thus giving A2 and B2, respectively. Aliquots of each juice were immediately frozen at -20 °C before sending to France.

2.2. Flavour analysis

For extraction of aroma compounds 10 g of orange juice was added 80 μg of butyl valerate (internal standard) and extracted by dichloromethane as reported previously [1]. Volatile compounds were separated on a DB-WAX column (30 m x 0.25 mm i.d. x 0.25 μm, J&W Scientific). Compounds were identified from linear retention indices and EI mass spectra (Wiley 275 library).

2.3. Determinatin of partition coefficients

Partition coefficients were determined by the PRV method which is based on the relationship between the vapour/liquid partition coefficient K and the phase ratio of volumes of headspace and liquid phase [4]. Increasing volumes (0.5; 1; 2; 3 and 4 ml) of orange juice were introduced into headspace vials (22 ml) representing phase ratios of 43, 21, 10, 6.3 and 4.5. Analyses were carried out in duplicate. After equilibration for 3 h at 45 °C headspace samples of 1.5 ml were taken by a Multi Purpose Sampler 2 (Gerstel) before injection on a DB-FFAP column (30 m x 0.32 mm i.d. x 1 μm film thickness, Agilent). Three sets of increasing volumes from 0.5 to 4 ml were tested.

2.4. Particle size distribution

The particle size distribution of orange juice was established by laser light scattering at room temperature, using a Mastersizer MS 1000 (Malvern Instruments, Orsay, France) The measurement was made before and after pasteurisation for each content of pulp.

3. RESULTS

The results of the liquid-liquid extractions of aroma compounds and the standard deviations (in parenthesis) are given in Table 1. The concentrations of the aroma compounds in (A1) and (B1), and (A2) and (B2) were not significantly different. Table

2 shows the partition coefficients of aroma compounds in the fresh juices A1 and B1). All the vapour/liquid partition coefficients were higher in (B1) than in (A1), suggesting more retention in (A1). The difference was not significant for ethyl butyrate, hexanal and octanal, which are the less hydrophobic compounds. Concerning ethyl butyrate, the slight difference is coherent with the results obtained by Rega *et al.* [3], showing that an increase in viscosity and texture could affect diffusion of ethyl butyrate and could increase its retention into the juice matrix [5]. Concerning terpenes, the difference is significant ($p<0.05$). It is known that the more hydrophobic the compound is, the more it is retained by pulp. This phenomenon involves molecular interactions between pulp and hydrophobic aroma compounds.

Table 1. Concentration of some volatile compounds (ppm) in the 20% pulp juice before (A1) and after pasteurisation (A2) and in the 10% pulp juice before (B1) and after pasteurisation (B2).

Aroma compound	A1	A2	B1	B2
Ethyl butyrate (4%)	1.56	1.36	1.64	1.12
Hexanal (16%)	0.13	0.16	0.12	0.13
β-Myrcene (3%)	9.68	9.52	9.75	9.15
Limonene (3%)	357	336	364	340
Octanal (4%)	1.39	1.85	1.52	1.82

Table 2. Partition coefficients ($x10^{-2}$) in fresh juices with 20% pulp (A1) and 10% pulp (B1).

Aroma compound	A1	B1	Statistical test[a]	LogP [b]
Ethyl butyrate	10.3 (18%)	13.7 (7%)	NS	1.7
Hexanal	5.3 (14%)	6.7 (15%)	NS	1.78
β-Myrcene	29.2 (12%)	78.8 (6%)	S	4.17
Limonene	18.8 (12%)	29.4 (1%)	S	4.2
Octanal	8.9 (9%)	11.5 (0.2%)	NS	2.78

[a]S: significantly different and NS: not significantly different at ($p<0.05$); [b]Estimated octanol-water partition coefficient.

Figure 1 compares the partition coefficients of the juices before and after pasteurisation. Pasteurisation had different effects on the partition coefficients in juices with 10 and 20% pulp. A comparison between (A1) and (A2) shows that the partition coefficients increased during pasteurisation of the 20% pulp juice except for ethyl butyrate. Pasteurisation of the 10% pulp juice led to decreasing partition coefficients for all substances. In order to understand this phenomenon, particle size distributions were determined in the two different juices. Table 3 gives the percentage of particles <10 μm and <100 μm in the juices (A1), (A2), (B1) and (B2).

Table 3. Percentage of particles <10 μm and 100 μm in the particle size distribution of the 20% pulp juice before (A1) and after (A2) pasteurisation, and in the 10% pulp juice (B1) and (B2).

	A1	A2	B1	B2
< 10 μm	2.5	5.8	3.1	24.9
< 100 μm	21.9	21.1	33.5	59

The particle size of (B1) was strongly affected by pasteurisation with considerable increases of both smaller (<10 μm) and bigger (<100 μm) particles. On the contrary, the 20% pulp juice was not effected to any higher degree; its content of bigger particles remained stable. It is hypothesised that the increase of the proportion of smaller particles of pulp in (B2) could facilitate molecular interactions between pulp and aroma compounds, and thus could increase the retention observed in Figure 1.

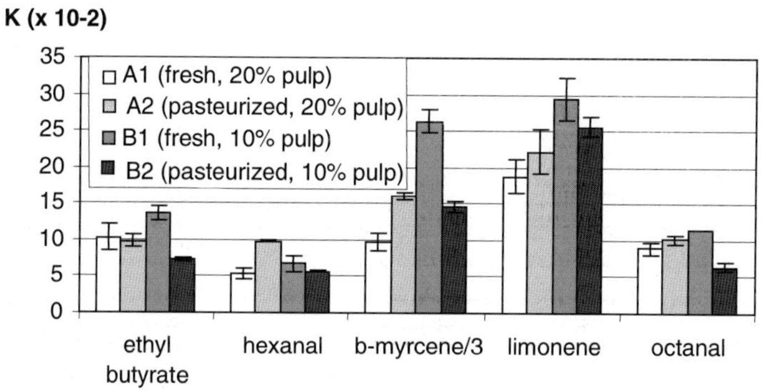

Figure 1. Partition coefficients of aroma compounds in fresh juices (A1), (B1) and pasteurised juices (A2) and (B2) n=3.

4. CONCLUSION

Concerning the fresh juices, the highest retention of aroma compounds was observed in juice with 20% pulp. After pasteurisation, the retention behaviour was reversed, the 10% pulp showing the highest retention. Olfactometry analysis will be conducted with these juices in order to confirm the sensory significance of the retention effects.

References

1. P. Brat, B. Rega, P. Alter, M. Reynes and J.-M. Brillouet, J. Agric. Food Chem., 51 (2003) 3442.
2. M.J. Jordan, T.N. Tillman, B. Mucci and J. Laencina, Lebensm. Wiss. Technol. Food Sci. Technol., 34 (2001) 244.
3. B. Rega, N. Fournier and E. Guichard, poster presented at 7th Wartburg symposium on flavor chemistry and biology, Eisenach, Germany (2004).
4. C. Jouquand, V. Ducruet and P. Giampaoli, Food Chem., 85 (2004) 467.
5. B. Rega, N. Fournier, S. Nicklaus and E. Guichard, J. Agric. Food Chem., 52 (2004) 4204.

Comparison of cold pressed and essence orange oil oxidative stability using TI GCO and GC-MS

Ozan Gurbuz[a,b], Brenda Odor[c] and Russell Rouseff[a]

[a]University of Florida, IFAS, Citrus Research and Education Center, USA; [b]Uludag University, Bursa, Turkey; [c]Brenda L. Odor Consultants, Inc., USA

ABSTRACT

Undiluted, commercial orange oils were exposed to oxygen and stored in darkness for up to three months at room temperature to compare their relative oxidative stability. Sensory evaluations indicated that in essence oil, EO, aroma had been altered more than in the cold pressed oil, CPO, characterised primarily by the loss of citrus character rather than pronounced off-flavours. MS analyses indicated increased levels of terpene oxides and ketones, but greatly reduced levels of terpene aldehydes. Major increases of limonene oxides (8 to 9 fold) and *cis* and *trans* *p*-2,8-menthadiene-1-ol (3 to 4 fold) were observed in both oil types, but initial and final levels were always higher in EO. Common acid catalysed hydration products of limonene such as terpene-4-ol and α-terpineol decreased approximately 10% in both types during storage. Decanal and octanal decreased 37% and 38% respectively in EO, whereas only 26-27% was lost in CPO. Losses of citronellal, dodecanal and β-sinsensal ranged from 52 to 63% in EO but only 27 to 31% in CPO. Putative off-flavour compounds such as carvone and carveol increased 3 to 6 fold during storage but were always twice as high in EO. Time-intensity GCO confirmed the loss of critical aldehydes, but also demonstrated that the aroma impact of produced carveol and carvone were minimal. Observed sensory changes were due to losses of potent aldehydes and ketones. The total carotenoid concentration in the CP oil decreased by 54% during 12 weeks storage suggesting that they may act as sacrificial antioxidants.

1. INTRODUCTION

Orange oils are products of significant commercial value but like all natural products they vary with fruit cultivar, maturity, and processing techniques. Commercial products are therefore usually blended to improve uniformity. The chemical composition of orange oils has been well studied, but only limited information is available on orange oil stability [1,2]. In these studies, orange oils were emulsified in acidic media and subjected to both oxygen and UV light. The purpose of this study was to compare the relative stability of the undiluted oil after exposure to oxygen but without exposure to light as oils are rarely exposed to light during storage. Orange essence oil, EO, is a condensate from distilled juice whereas cold pressed peel oil is obtained directly from the peel and contains peel pigments (carotenoids). The relative oxidative stability of these two products was examined using both time-intensity (TI) GCO and GC-MS.

2. MATERIALS AND METHODS

Orange oils: Florida orange essence oil and cold pressed orange oil were saturated with oxygen and stored at room temperature for 12 weeks. Controls were untreated and stored cold. Neat injections of the oils were used for all analyses.
GC-MS analysis: Perkin-Elmer Clarus 500 GC/quadropole mass spectrometer; columns: DB-5MS 60 m, 0.25 mm ID, 0.25 μm and DB-WAX 60 m, 0.25 mm ID, 0.5 μm; helium carrier gas flow = 1.5 ml/min; 0.2 μl injection splitless mode; temperature program: 40-240 °C at 7 °C/min, hold for 8 min (wax), 40 - 290 °C at 7 °C/min, hold for 5 min (DB-5).
GC-olfactometry: Hewlett-Packard 5890 gas chromatograph with ZB-5 capillary column, 30 m x 0.32 mm, 0.5 μm; carrier gas is helium; 0.1-0.2 μl splitless injection; Time-intensity GCO and other details as in Bazemore and co-workers [3]. Two trained assessors were used for the TI GCO analyses and each sample was analysed in duplicate by each assessor.
Carotenoid analysis: Total carotenoids were analysed by measuring their absorbance at 420 nm. Measuring cells were blanked with pure limonene and a calibration curve established using a 200 ppm β-carotene standard dissolved in pure limonene.

3. RESULTS

The major changes due to oxidation of the samples were, as expected, decreased aldehyde levels and increased terpene oxidation products. Figure 1 shows the relative loss of five aldehydes (octanal, citronellal, decanal, dodecanal, and β-sinensal) in the two oils after 12 weeks storage at ambient temperature. Important flavour components decreased in both oils but losses were greatest in EO. Although initial levels of all five aldehydes were similar in both oils, losses of citronellal, dodecanal and β-sinensal were twice as great in EO compared to CPO. Figure 2 compares the levels of seven oxidation products in the two oils. The levels of *p*-cymene, *cis*-limonene oxide, *trans*-limonene oxide, *trans-p*-2,8-menthadiene-1-ol, *cis-p*-2,8-menthadiene-1-ol, *trans*-carveol, and *cis*-carveol were always greatest in EO. The two isomers of *p*-2,8-menthadien-1-ol had higher percentage increase in the CPO compared to the EO only because initial levels were much lower. Levels of *p*-cymene were surprisingly low even after 12 weeks storage. Carvone (not shown) increased 3 to 4 fold from initial concentrations. Initial and final carvone concentrations in CPO were one third to half of that in EO. Linalool decreased slightly (14% in EO and 9% in CPO) during 12 weeks storage. Acid catalysed hydration products of limonene such as terpene-4-ol and α-terpineol decreased approximately 10% in both types during storage. Decanoic acid was not observed in either oil at time zero, but was found in both 12 week oxidised samples. However, levels were 20 times higher in EO than in CPO.

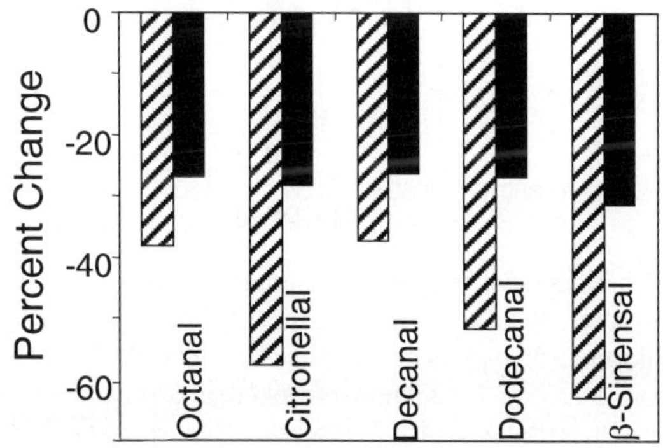

Figure 1. Comparative loss of aldehydes during 12 weeks of storage at ambient temperature. Cross hatched bar: EO (essence oil); solid bar: CPO (cold pressed oil).

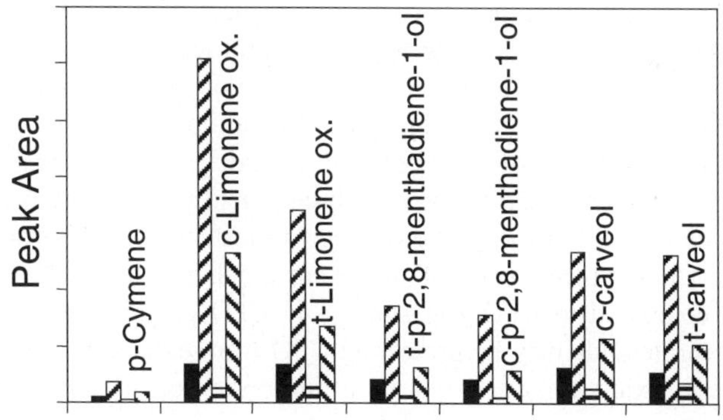

Figure 2. Relative change in oxidation products. First two bars are EO (essence oil) initial and after 12 weeks, last two bars are CPO (cold pressed oil) initial and after 12 weeks storage.

Thirty-five aroma active components were observed in the initial essence oil which is similar to what had been found in previous reports [4,5]. As shown in the lower part of Figure 3, there were over 50 aroma peaks observed in the 12 weeks EO sample. Peaks 5, 8 and 9 have been reported to be major off-flavours in previous studies. However, it can be seen from the aroma intensities in Figure 3, that these well known off-flavours are not major aroma impact compounds in orange essence. The same conclusion is reached for CPO (data not shown).

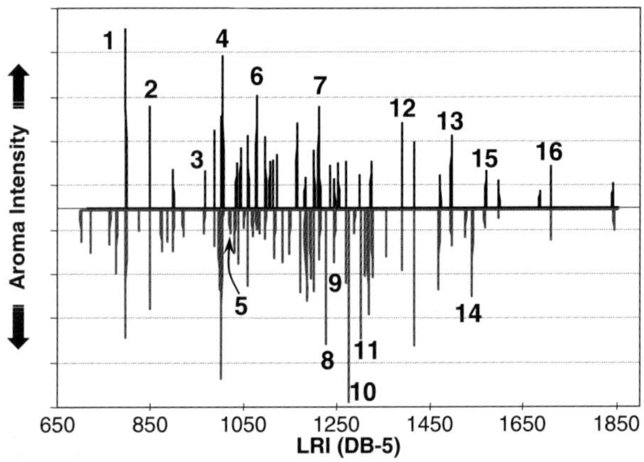

Figure 3. Essence oil aromagrams. Upper aromagram: Initial. Lower aromagram: 12 weeks storage. 1: coeluted ethyl butanoate and hexanal; 2: ethyl-2-methyl butanoate; 3: 6-methylheptanal; 4: limonene; 5: *p*-cymene; 6: 6-methyl octanal; 7: decanal; 8: *cis*-carveol; 9: carvone; 10: unknown (spearmint); 11: unknown (cooked); 12: β-damascenone; 13: β-ionone; 14: unknown (soapy); 15: 8-tetradecanal; 16: β-sinensal. The compounds have been tentatively identified based on linear retention index value (DB-5) and aroma descriptors.

4. DISCUSSION AND CONCLUSION

Oxidation of citrus oils in the absence of light produces some of the same compounds observed in other studies [1,6] where light was employed. However, the number and amounts of oxidative products formed in the absence of light was very different. There were significant losses of important aldehydes. In addition to the five aldehydes shown in Figure 1, there were losses of ethyl butanoate/hexanal, aldehyde peaks 3, 6 and 15 in addition to β-damascenone, and β-ionone. Three potent off-flavours (peaks 10, 11 and 14) remain unidentified. Sensorily, the orange EO lost its desirable 'orange' character faster than the CPO. Since losses of important aldehydes and ketones were always greater in EO compared to CPO and since only CPO contains entrained carotenoids from the peel, carotenoids may be protecting CPO from oxidation. The overall sensory impression due to oxidation was primarily a diminished or missing 'orange' character rather than the introduction of pronounced off-flavours.

References

1. L. Buckholz Jr. and H. Daun, J. Food Sci., 43 (2) (1978) 535.
2. M. Ziegler, H. Brandauer, E. Ziegler and G. Ziegler, J. Essent. Oil Res., 3 (4) (1991) 209.
3. R. Bazemore, K. Goodner and R. Rouseff, J. Food Sci., 64 (5) (1999) 800.
4. J.-L. Le Quéré and P.X. Étiévant (eds.), Flavour research at the dawn of the 21st century, proceedings of the 10th Weurman flavour research symposium, Paris, France (2003) 289.
5. A. Hognadottir and R.L. Rouseff, J. Chromatogr. A, 998 (1-2) (2003) 201.
6. P. Schieberle and W. Grosch, Z. Lebensmittel Untersuch. Forsch., 189 (1) (1989) 26.

W.L.P. Bredie and M.A. Petersen (Editors)
Flavour Science: Recent Advances and Trends

Flavour quality of organic tomatoes grown in different systems

Merete Edelenbos, Anette K. Thybo and Lars P. Christensen

Department of Food Science, Danish Institute of Agricultural Sciences, Research Centre Aarslev, Kirstinebjergvej 10, DK-5792 Aarslev, Denmark

ABSTRACT

Tomatoes were grown organically in a greenhouse in soil in an open system (control), on compost beds in a confined system and on compost beds with holes for root penetration in a combined system. Tomatoes were harvested in the spring (May) and in the autumn (October) at comparable maturity and analysed for flavour compounds and sensory quality. A total of 31 volatile compounds were collected from blended tomatoes using a dynamic headspace technique and identified and quantified by GC and GC-MS. The aroma profile of organically grown tomatoes was in agreement with published data on conventional tomatoes. The content of volatiles, soluble solids, citric acid and dry matter was almost the same regardless of growing system. Sensory analysis revealed that tomatoes harvested in the spring from the confined system scored higher in tomato flavour and sourness than those from the combined system. In the autumn there was no effect of growing system on sweetness, sourness and tomato flavour but on the content of hexanal, (*Z*)-3-hexenal and (*Z*)-3-hexen-1-ol. Principal components analysis (PCA) revealed that system accounted for 15% and harvest time for 67% of the variation in the volatile data indicating that growing system had only a minor effect on the chemical composition and sensory quality of tomatoes harvested at comparable maturity. Growers of organic tomatoes can therefore switch to a confined or combined growing system without affecting the flavour quality of tomatoes.

1. INTRODUCTION

Flavour and firmness are important quality criteria for tomatoes [1]. Tomato flavour is attributed to the content of sugars, organic acids and volatile compounds. Many factors determine tomato flavour e.g. cultivar, growing system, maturity, harvest time and post-harvest treatments [1,2]. In Denmark, organic tomatoes are grown directly in the soil. It

is difficult, however, to provide enough organic manure, to prevent leaching out of nutrients and to control soil-born diseases in a soil system. Recently, Sørensen and Thorup-Kristensen [3] reported results on two alternative systems, a confined system with tomatoes grown on compost beds and a combined system with tomatoes grown on compost beds with holes for root penetration. Nutrients and water were re-circulated in these systems, and compost was replaced every year.
The aim of the present study was to determine the effect of growing system (open, confined and combined systems) on selected flavour compounds and sensory quality of tomatoes harvested in the spring and autumn.

2. MATERIALS AND METHODS

Tomatoes (*Lycopersicon esculentum* Mill. cv. 'Aromata' grafted on cv. 'Beaufort') were grown in a greenhouse on open, confined and combined beds (Figure 1) in a Latin square design with three replicates. First class, red tomatoes were harvested at comparable maturity in May (spring) and October (autumn) and analysed for soluble solids, total acidity expressed as citric acid [4], dry matter and volatile compounds.

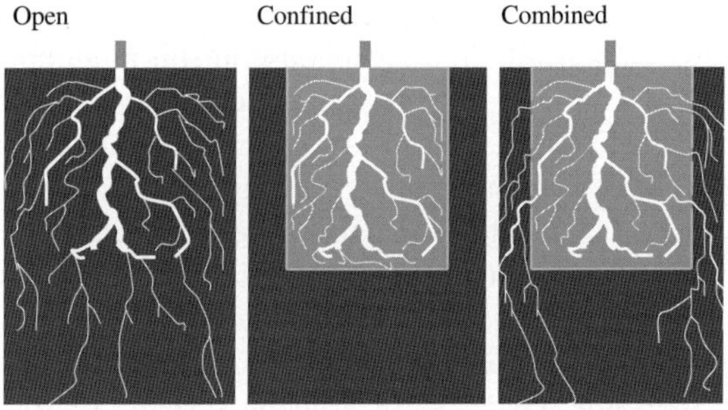

Figure 1. Development of the tomato roots in different growing systems [3].

Volatiles were collected from blended tomatoes (30 s, then 180 s holding time before inactivation of enzymes with $CaCl_2$) by dynamic headspace sampling using a modified method of [1] and quantified by GC. Volatiles were separated on a CP-Wax 52CB column (50 m × 0.25 mm i.d.; DF = 0.2 μm) by isothermal for 3 min at 40 °C, followed by 1 °C/min to 60 °C, isothermal for 2 min, then 5 °C/min to 180 °C, isothermal for 10 min, then 10°C/min to 220°C, followed by constant temperature for 10 min. Volatile compounds were identified by GC-MS (EI 70 eV). Compounds suggested by the mass spectral database (NIST (1998), version 6.0) were verified by comparison with mass spectra and retention indices of authentic reference compounds.
Tomatoes were analysed by descriptive sensory analysis by a trained panel of ten assessors using a 15 cm unstructured line scale from 0 (none) to 15 (very much) [5]. Half a tomato from each sample was served in randomised order on a white plate at room temperature and evaluated twice for sensory quality at each harvest time.
Data were analysed by analysis of variance and principal components analysis (PCA).

3. RESULTS

A total of 31 volatile compounds were quantified and identified in organic tomatoes (Figure 2) including compounds of importance for tomato flavour e.g. 1-penten-3-one, hexanal, (*Z*)-3-hexenal, 2- and 3-methyl-1-butanol, (*E*)-2-hexenal, (*E*)-2-heptenal, 6-methyl-5-hepten-2-one, (*Z*)-3-hexen-1-ol, geranyl acetone and β-ionone [6].

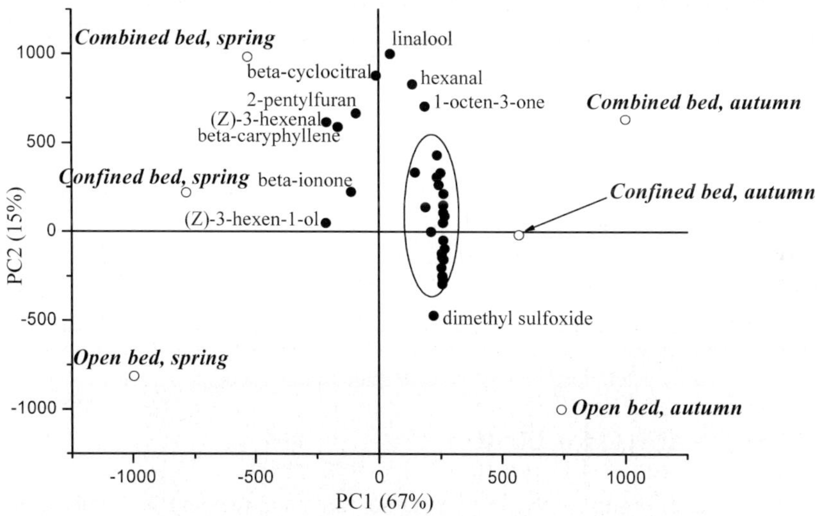

Figure 2. Variation in the volatile compounds in organic tomatoes harvested in the spring and autumn from an open, confined and combined growing system. The circle represents a group of correlated volatiles (1-penten-3-one, 2-methyl-2-butenal, 2-methyl-4-pentanal, 1-penten-3-ol, 2/3-methyl butanol, (*E*)-2-hexenal, (*E*)-2-heptenal, 1-nitro-3-methylbutane, 6-methyl-5-hepten-2-one, (*E,E*)-2,4-hexadienal, nonanal, (*E*)-2-octenal, (*E*)-3-hexen-1-ol, 1-octen-3-ol, 6-methyl-5-hepten-2-ol, camphor, benzaldehyde, α-humulene, geraniol, geranyl acetone, 2-phenylethanol).

Regardless of growing system, the content of volatiles, soluble solids, citric acid and dry matter was almost the same with higher values of soluble solids, citric acid, dry matter and hexanal in the spring than in the autumn (Table 1). Sensory analysis revealed that tomatoes harvested in the spring from the confined system scored higher in tomato flavour and sourness than those from the combined system. However, the variation in the content of volatile compounds and other chemical constituents could not explain these differences (Table 1). In the autumn there was no effect of growing system on tomato flavour, sourness and sweetness, however, analysis of volatile compounds revealed significant differences in the content of hexanal, (*Z*)-3-hexenal and (*Z*)-3-hexen-1-ol (Table 1). Harvest time (PC1) accounted for 67% and growing systems (PC2) for 15% of the variation in the volatile data by PCA (Figure 2). (*Z*)-3-Hexen-1-ol, (*Z*)-3-hexenal and β-caryophyllene were present in higher concentration in the autumn than in the spring, whereas it was opposite for the other volatiles (Figure 2).

Tomatoes from the combined system had higher contents of linalool, hexanal, β-cyclocitral, 1-octen-3-one, 2-pentylfuran, β-caryophyllene and (*Z*)-3-hexenal than tomatoes from the open system (Figure 2; Table 1).

Table 1. Effect of growing system on chemical composition and sensory quality of organic grown tomatoes harvested in the spring and autumn 2002.[a]

	Spring harvest			Autumn harvest		
	Open	Confined	Combined	Open	Confined	Combined
Soluble solids, %	4.5	5.0	4.8	3.9	4.0	3.9
Total acidity, %	0.43	0.46	0.43	0.35	0.34	0.35
Dry matter, %	5.6	6.0	5.9	5.2	5.3	5.2
Hexanal[b]	283	304	308	257b	297a	302a
(*Z*)-3-Hexenal[b]	402	445	424	471c	526b	556a
(*Z*)-3-Hexen-1-ol[b]	1.2	2.4	1.8	3.0a	2.6ab	2.1b
Sourness[c]	5.2ab	5.6a	4.7b	8.4	8.0	7.7
Sweetness[c]	5.3	5.8	5.9	3.7	3.7	3.8
Tomato flavour[c]	5.7ab	6.2a	5.6b	5.7	5.6	6.4

[a]Significant differences are indicated within rows at each harvest time at $p=0.05$. [b]Content in ng/g fresh weight. [c]Intensity determined on unstructured scale from 0 (none) to 15 (very much).

4. DISCUSSION AND CONCLUSION

The aroma profile of organically grown tomatoes was in agreement with published data on conventional tomatoes [1]. The volatiles represented two major areas of tomato flavour: those with an earthy, musty, green flavour and those with a fruity, floral, ripe tomato and sweet tomato flavour [1,6,7]. Aliphatic C-6 volatiles in combination with sugars and organic acids in a balanced ratio have been suggested as the most important contributors to tomato flavour and consumer acceptance [4]. Growing system had only a little effect on the sensory scores for sourness and tomato flavour in the spring and on the content of hexanal, (*Z*)-3-hexenal and (*Z*)-3-hexen-1-ol in the autumn. Growers of organic tomatoes can, therefore, switch to a confined or combined growing system without affecting the flavour quality of tomatoes.

References

1. A. Krumbein and H. Auerswald, Nahrung, 42 (6) (1998) 395.
2. H. Auerswald, D. Schwarz, C. Kornelson, A. Krumbein and B. Brückner, Sci. Hortic., 82 (3-4) (1999) 227.
3. J.N. Sørensen and K. Thorup-Kristensen, Gartner Tidende, 119 (9) (2003) 34.
4. J.J. Ruiz, A. Alonso, S. Garcia-Martinez, M. Valero, P. Blasco and F. Ruiz-Bevia, J. Sci. Food Agric., 85 (1) (2005) 54.
5. A.K. Thybo, I.E. Bechmann and K. Brandt, J. Sci. Food Agric., 85 (13) (2005) 2289.
6. E.A. Baldwin, K. Goodner, A. Plotto, K. Pritchett and M. Einstein, J. Food Sci., 69 (8) (2004) S310.
7. M. Edelenbos, A. Thybo and M. Nielsen, Gartner Tidende, 119 (11) (2003) 14.

1-Ethoxy-1-(1-ethoxy-ethoxy)-ethane: a new acetaldehyde precursor

Klaus Gassenmeier[a], Andrew Daniher[b] and Stefan Furrer[b]

[a]Givaudan Schweiz AG, Ueberlandstr. 138, 8600 Dübendorf, Switzerland; [b]Givaudan Flavours Corp., 1199 Edison Drive, Cincinnati, Ohio 45216, USA

ABSTRACT

In a natural acetaldehyde rich fraction of orange oil 1-ethoxy-1-(1-ethoxy-ethoxy)-ethane was identified. To the best of our knowledge this is the first identification of 1-ethoxy-1-(1-ethoxy-ethoxy)-ethane in a natural product. The release of acetaldehyde from 1-ethoxy-1-(1-ethoxy-ethoxy)-ethane at various pH values was measured and compared to acetaldehyde release from acetaldehyde diethyl acetal and aldemax.

1. INTRODUCTION

Acetaldehyde is a valuable ingredient in a wide variety of fresh and prepared food products. It adds a fresh, juicy character to products like orange, passion fruit, raspberry and yoghurt [1]. Often it is used in conjunction with ethyl butyrate to fortify orange juice flavours as these two components contribute to the characteristic 'juicy' notes. It is also used in Liqueur formulations (e.g., Peach Schnapps) to help 'lift' and add top note to the aroma. Typical use levels in finished products are 4-25 mg/kg [2].
However the use of acetaldehyde can be difficult due to its low boiling point of 20 °C and its chemical reactivity. Concentrated liquid flavouring preparations containing acetaldehyde are difficult to handle because of their high volatility and inflammability, and vapours may form explosive mixtures with air. For safety reasons such formulations are not desirable, and frequently spray dried acetaldehyde, which is easier to handle is preferred. Due to the high temperature of spray drying of 150 - 200 °C compared to the boiling point of 20 °C for acetaldehyde huge losses occur during the process. Alternative attempts to produce dry acetaldehyde including freeze drying [3] result in smaller losses. However the total amount of encapsulated acetaldehyde in the various encapsulation systems does not exceed 2% [4]. Another problem of acetaldehyde is its

instability. It may polymerise, oxidise or combine with other materials in the presence of an acid or a base (Figure 1). Some of the formed products exhibit an off-flavour.

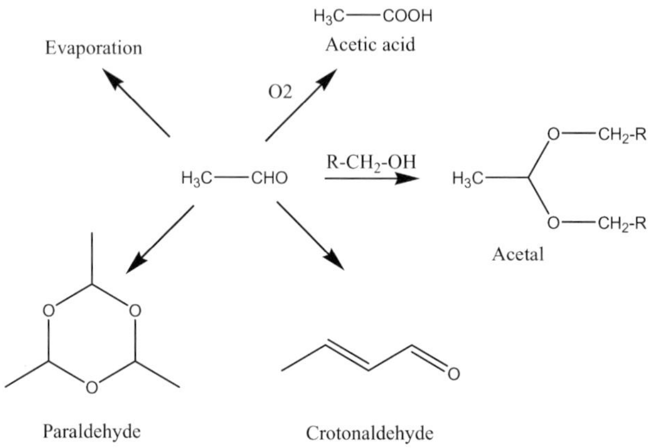

Figure 1. Mechanisms of loss and degradation of acetaldehyde.

To overcome some of the problems, nature identical and artificial precursors were proposed and commercialised (e.g. 1,2-di[(1'-ethoxy)ethoxy]propane; Aldemax) [5] and acetaldehyde diethylacetal. The acetaldehyde release from these precursors is acid catalysed. Aldemax forms upon hydrolysis 2 mol acetaldehyde per mol Aldemax. One of the drawbacks is that it also forms propylene glycol, which is limiting the application of Aldemax in some countries. Furthermore Aldemax is artificial. Acetaldehyde diethyl acetal yields ethanol and acetaldehyde upon hydrolysis. In instant beverage applications its use is problematic because it exhibits an unwanted choking alcoholic-like, whisky flavour which is considered as an off-flavour in soft drinks.

2. DISCOVERY OF A NEW ACETALDEHYDE PRECURSOR

In many countries only flavours derived from the named fruit are allowed to add back to concentrated juice in order to reconstitute the original juice aroma. Since acetaldehyde plays a very important role, the flavour industry has developed specific acetaldehyde-rich fractions that were obtained during distillation of orange oils. In the course of a quality and stability check of an acetaldehyde-rich fraction from orange oil we obtained a mass spectrum of an unknown compound (Figure 2). Since it eluted much later than the predominant acetaldehyde and ethanol, the compound was enriched by distilling of the acetaldehyde and ethanol and purified by preparative GC. NMR structure elucidation of a purified sample was tried in $CDCl_3$ first. However the sample was already decomposed. In a second trial the purified compound was dissolved in deuterobenzene and the structure of the unknown was proposed to be 1-ethoxy-(1-ethoxy-ethoxy)-ethane (Figure 3).

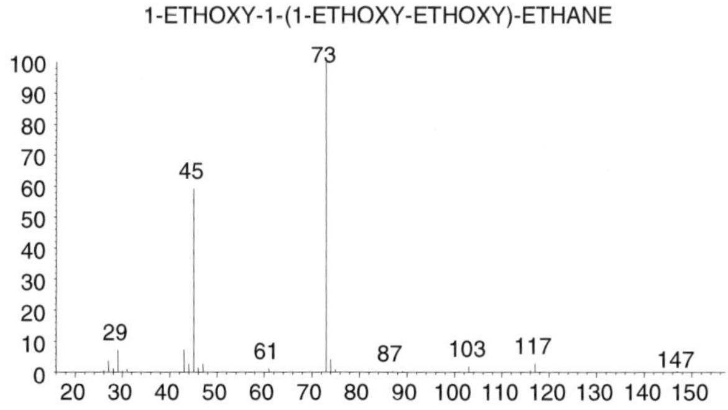

Figure 2. Mass spectrum of the unknown in an acetaldehyde rich fraction from orange oil.

A reference sample was synthesised according to the procedure of Oon *et al.* [6]. The synthetic compound co-eluted with the unknown compound on a DB-1 and DB-WAX column and showed an identical mass spectrum. Retention indices were 936 on DB-1-column and 1177 on a DB-WAX column.

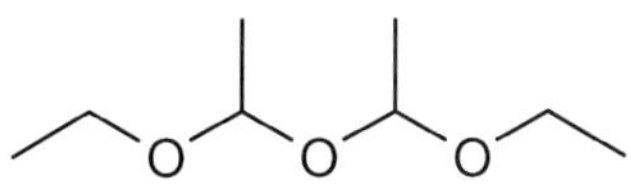

Figure 3. Structure of 1-ethoxy-(1-ethoxy-ethoxy)-ethane.

During purification, unexpected instabilities of the target molecule were found. It was assumed, that due to humidity and/or traces of acid 1-ethoxy-1-(1-ethoxy-ethoxy)-ethane was hydrolysed. Therefore, it might have use as acetaldehyde precursor. To test this assumption the release of acetaldehyde from 1-ethoxy-1-(1-ethoxy-ethoxy)-ethane was monitored at different pHs, and rate constants and half life times were calculated. The absorbance of acetaldehyde, which was formed by hydrolysis, was measured in a photometer at 280 nm. Measurements were done in a quartz cuvette (V = 1.5 ml) with 5 mg precursor. The cuvette was closed with Parafilm. Every 5 s the absorbance was measured. A typical curve is displayed in Figure 4. Release of acetaldehyde followed a first order reaction and was strongly dependent on pH as seen in Table 1.

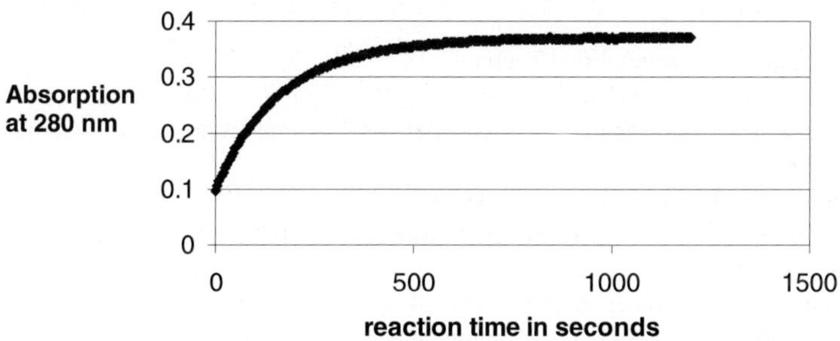

Figure 4. Release of acetaldehyde from 1-ethoxy-1-(1-ethoxy-ethoxy)-ethane at pH 2.8.

Table 1. Rate constants and half life time for the formation of acetaldehyde from 1-ethoxy-1-(1-ethoxy-ethoxy)-ethane at different pH values.

pH value	2.8	3.81	4.14
Rate constants (1/s)	0.005966	0.00069	0.000382
Half life time (s)	116	1004	1813

Table 2. Half life time of acetaldehyde precursors in acidic media.

	Half life time in (s)	
	pH 3.5	pH 3.0
1-Ethoxy-1-(1-ethoxy-ethoxy)-ethane	966	306
Acetaldehyde diethyl acetal	2493	579
Aldemax	3906	625

The data shows a very fast release of acetaldehyde from the precursor. At pH 2.8 it takes less than 2 min to release 50% of the acetaldehyde present. This quick release makes 1-ethoxy-1-(1-ethoxy-ethoxy)-ethane suitable for the use as acetaldehyde precursor even in instant beverages. To compare the newly identified compound with already existing materials, the release rates of Aldemax and acetaldehyde diethyl acetal were also recorded (results displayed in Table 2). 1-Ethoxy-1-(1-ethoxy-ethoxy)-ethane released acetaldehyde twice as fast as the known products aldemax and acetaldehyde diethyl acetal. Experiments showing the application of 1-ethoxy-1-(1-ethoxy-ethoxy)-ethane were described in [7].

3. CONCLUSION

In a natural acetaldehyde rich fraction of orange oil 1-ethoxy-1-(1-ethoxy-ethoxy)-ethane was identified. To the best of our knowledge this is the first identification of 1-ethoxy-1-(1-ethoxy-ethoxy)-ethane in a natural product. The compound releases acetaldehyde faster than ready available acetaldehyde precursors like aldemax and acetaldehyde diethyl acetal.

References

1. B. Byrne and G. Sherman, Food Technol., 38 (1984) 57.
2. S. Arctander, Perfume and flavour chemicals, Monograph 3, Carol Stream, USA (1994).
3. B. Byrne, Process for preparing dry acetaldehyde and product produced thereby, EP0372760 (1990).
4. R.S. DeSimone, Perfum. Flavor., 11 (1986) 15.
5. R.S. DeSimone, Aldehyde generators and foodstuff containing such generators, US 4,280,011 (1991).
6. S.M. Oon and D.G. Kubler, J. Org. Chem., 47 (1982) 1166.
7. K. Gassenmeier, J.P. Nelissen, A. Daniher and S.M. Furrer, Flavour and fragrance compounds, WO2004/048305 (2004).

W.L.P. Bredie and M.A. Petersen (Editors)
Flavour Science: Recent Advances and Trends

Formation of methyl (methylthio)methyl disulfide in broccoli (*Brassica oleracea (L.)* var. *italica)*

Jean-Claude Spadone, Walter Matthey-Doret and Imre Blank

Nestlé Research Center, Vers-chez-les-Blanc, 1000 Lausanne 26, Switzerland

ABSTRACT

Vapour cooking of fresh broccoli did not generate 2,3,5-trithiahexane (TTH) but cutting broccoli into pieces followed by cooking in water allowed the detection of this odorant. The formation of TTH was enhanced upon frozen storage of cut broccoli pieces for a few weeks. Cutting broccoli activates cysteine sulfoxide lyase transforming methylcysteine sulfoxide into methylsulfenic acid, which upon heating gives rise to dimethyl sulfide and dimethyl trisulfide that react to TTH. Thus, enzymatic reactions combined with thermal treatment may favour TTH formation.

1. INTRODUCTION

Vegetables belonging to the *Brassica* species (e.g. white cabbage and broccoli) are widely used in fresh or frozen form in home-style culinary preparations and industrial products. Methyl (methylthio)methyl disulfide, also referred to as 2,3,5-trithiahexane (TTH), has been reported as an unusual constituent of broccoli, cabbage, and cauliflower [1,2], besides methanethiol (MT) and dimethyl trisulfide (DMTS) as well-known volatile compounds of *Brassicaceae*. The sensory properties of TTH were described as garlic-like, alliaceous, sulfury, and cooked cabbage-like [2-4] with a recognition odour threshold of 40 ppb in water [5]. TTH has been suggested as a character impact compound significantly contributing to the typical flavour of various plants such as such as *Brassica pekinensis* [2], *Hua gabonii* ('garlic plant'), [3], *Marasmius alliaceus* [4] and *Scorodocarpus borneensis* ('wood garlic') [5]. In the present work, the formation of TTH in broccoli was investigated, in particular the role of enzymatic reactions and thermal treatment in developing the freshly cooked broccoli character. The sensory significance of TTH was also addressed.

2. MATERIALS AND METHODS

2.1. Materials

Fresh broccoli was purchased from a local market and kept at 4 °C for less than 3 days prior to analysis or treatment. Most of the chemicals used in this study were commercially available. 2,3,5-Trithiahexane was obtained from Givaudan (Dübendorf, CH). (*Z*)-1,5-Octadien-3-one [6] and 2-acetyl-1-pyrroline [7] were synthesised.

2.2. Solid-phase micro-extraction

Raw broccoli: fresh broccoli florets (2 g) were introduced in a SPME flask (20 ml), allowed equilibrating at 20 °C for 15 min. The volatiles present in the headspace were then adsorbed on a PDMS/DVB fibre (65 μm, Supelco, Bellefonte, USA) for 30 min. The fibre was part of a syringe attached to an automatic sampler for subsequent analysis. *Vapour-cooked broccoli:* fresh broccoli (100 g) was cooked for 5 min in a pressurised cooker. After cooling, vapour cooked broccoli (2 g) was introduced in a SPME flask (20 ml) for equilibration and analysis.

2.3. Isolation of volatiles by SDE

Cutting + boiling broccoli: fresh broccoli (100 g) was cut in small cubes (2-3 cm^3 each), which were allowed to stand at air for 15 min. They were then introduced in a round-bottom flask (1 l), to which distilled water (400 ml) was added. In parallel, redistilled diethyl ether (50 ml) was poured into a round-bottom flask (100 ml). Both flasks were attached to a Likens-Nickerson apparatus [8] and heated to boiling. The simultaneous distillation-extraction process was run for 30 min, once a constant reflux was established in each part. After cooling, the organic phase was dried over anhydrous sodium sulphate, concentrated to 1 ml on a Vigreux column and kept at -20 °C prior to analysis. *Cutting + frozen storage + boiling broccoli:* fresh broccoli (100 g) was cut in small pieces (2-3 cm^3 each), allowed to stand at air for 15 min. These cut portions were kept frozen at -20 °C for 6 weeks. They were then boiled in water using the SDE procedure run under the same conditions as described above.

2.4. Gas chromatography-olfactometry

The analysis was carried out on a Hewlett-Packard (HP) 5890 Series II gas chromatograph (Agilent, Palo Alto, USA) equipped with a 60 m x 0.25 mm (i.d.) DB-Wax capillary column (film thickness d_f 0.25 μm, J&W Scientific, Folsom, USA). The oven temperature was programmed from 35 °C (3 min) to 240 °C (10 min) at 6 °C/min (SPME) or from 20 °C (0.5 min) to 60 °C at 70 °C/min and from 60 °C to 240 °C (10 min) at 6 °C/min (SDE extracts). Helium was used as the carrier gas. Both liquid injections (1 μl) and SPME fibre desorptions were operated in the splitless mode at 250 °C. The column effluent was split (1:1, all-glass press-fit splitter, Supelco) to the FID and a heated outlet port (200 °C). The port effluent was sniffed by three panellists asked to describe odour-active peaks and estimate their intensity from 1 (low) to 3 (high).

2.5. GC-mass spectrometry
Broccoli volatiles were analysed on a HP 5973 with SPME desorptions, liquid injections, and GC parameters run under identical conditions as for GCO analyses. The MS was operated in the electron-impact mode at 70 eV.

2.6. Sensory properties of TTH
The perception odour threshold of TTH was determined by GCO using a DB-Wax capillary column (30 m x 0.32 mm i.d., d_f 0.25 μm). Oven temperature programme: 30 °C (2 min) to 50 °C at 40 °C/min, from 50 °C to 240 °C (10 min) at 8 °C/min. Other conditions were identical to those reported above. Stepwise dilutions of TTH in diethyl ether were injected in the splitless mode (1 μl) until no odour could be perceived by any of the 4 panellists at the retention time of TTH. The lowest detectable amount (LDA) was defined as the amount for which a positive answer of 50% was obtained. The perception threshold of TTH in water was determined by adding known amounts of TTH dissolved in alcoholic solutions to mineral water (Vittel, Netslé Waters, France). TTH concentrations ranging from 0.1 to 100 ppb (ng/ml) were screened first for orthonasal detection by a sensory panel of 7 members. Tasters were asked to identify spiked samples (0.1 to 1 ppb) in triangle tests. The perception threshold was defined at 50% correct answers.

3. RESULTS AND DISCUSSION

3.1. Raw broccoli
To establish the volatile composition of broccoli before any treatment or cooking process, the volatile fraction of freshly harvested broccoli florets was isolated from the headspace by SPME and analysed by GCO and GC-MS. As shown in Table 1, the major odour-active notes were sulfury (MT, DMTS), green/fruity ((Z)-3-hexen-1-ol and its ester derivatives), raw vegetable/earthy (methoxypyrazines), and mushroom-like (1-octen-3-one). In agreement with recently reported data [9], we could not be detect any TTH in fresh raw broccoli by GCO or GC-MS. Traces of TTH have been reported in raw Chinese cabbage [2] after cutting into small pieces and frozen storage at -30 °C.

3.2. Vapour-cooked broccoli
In order to preserve the fresh character of vegetables during cooking, kitchen-chefs often use vapour cooking in pressurised cookers. In the case of broccoli, however, this procedure was not efficient in delivering pleasant and characteristic cooked broccoli notes. Instead, tasters described sulfury, strong cabbage, and other pungent attributes. This was confirmed by analytical results (Table 1), which showed the presence of sulfury, oxidised odorants, whereas no TTH could be detected by GCO or GC-MS. Besides MT and DMTS, methional (boiled potato-like), 2-acetyl-1-pyrroline (roasted), (*Z*)-1,5-octadien-3-one (metallic), and (*E*,*E*)-2,4-octadienal (fatty, oxidised) were formed. On the contrary, odorants associated with green and fruity notes disappeared during the heating process.

3.3. Water-cooked broccoli

Cutting or peeling of vegetables is the first step in home-style cooking or food manufacturing. A similar treatment was applied in the present work, which may favour activation of endogenous enzymes generating appropriate aroma precursors. Therefore, after cutting freshly harvested broccoli florets into small pieces and thus allowing exposure to air, additional volatile compounds were formed upon boiling in water. In particular, detectable amounts of TTH were found, as shown both by GCO and GC-MS analyses (Table 1). In addition to the key odorants formed during vapour cooking of broccoli, the combination of enzymatic reactions followed by heating processes resulted in diacetyl and phenylacetaldehyde, which reinforced the sweet-like overall sensory perception. On the other hand, oxidised-like compounds as well as MT and DMTS showed a reduced impact.

3.4. Frozen storage and TTH level

As demonstrated earlier [10], enzymatic reactions were still taking place in *Brassica* vegetables kept at -20 °C, although at a much reduced pace. Cut pieces of broccoli florets were stored at -20 °C for 6 weeks, and then boiled in water. The formation of TTH was clearly enhanced as compared with the experiment without frozen storage (Table 1). Besides TTH, other sulfur-containing compounds were also reinforced, which support the hypothesis of combined enzymatic and thermally-induced formation pathways. Cutting broccoli florets and storing the cut parts in the frozen state allowed the development of enzymatic reactions, while limiting oxidation processes. The postulated accumulation of sulfide precursors may then lead to higher yields of poly-sulfide formation during the heating step.

3.5. Formation of TTH in broccoli

The main enzymatic reaction responsible for the formation of sulfide precursors in *Brassica* species involves in the presence of pyridoxal-5'-phosphate the transformation of methylcysteine sulfoxide into methylsulfenic acid, pyruvic acid, and ammonia (Figure 1a). The subsequent boiling in water induces a series of chemical reactions leading to dimethyl sulfide (DMS) and DMTS, which may recombine to TTH as a secondary reaction product (Figure 1b).

The thermal degradation of *S*-methyl cysteine and its sulfoxide to TTH has been reported [11]. This strongly suggests a combined enzymatic and thermal origin of TTH and other poly-sulfides. TTH might also be generated by thermal degradation of sulforaphane in aqueous solutions. Photolysis of DMDS [12] and DMS [13], thermal degradation products of methionine, also lead to TTH. Naturally occurring unusual amino acids have also been suggested as precursors of TTH formed by C-S lyase-mediated enzymatic conversion without any heat treatment [14].

Table 1. Selected odorants of the aroma composition of broccoli samples.

Compound	RI	Fresh, raw[c]	Vapour-cooked[c]	Cut + cooked[c]	Cut + frozen + cooked[c]
Methanethiol (MT)[b]	675	9	9	7.5	9
2,3-Butandione (diacetyl)[a]	978			8	7.5
Hexanal[a]	1082				6
Dimethyl disulfide (DMDS)[a]	1088				7.5
1-Octen-3-one[b]	1304	8	9		
(*Z*)-3-Hexenyl acetate[a]	1320	9			
2-Acetyl-1-pyrroline[b]	1345		7.5	8	7.5
(*Z*)-1,5-Octadien-3-one[b]	1386		9	5	9
(*Z*)-3-Hexen-1-ol[a]	1395	9			
(*Z*)-3-Hexenyl propanoate[a]	1406	6			
Dimethyl trisulfide (DMTS)[a]	1412	9	9	6	9
2-Isopropyl-3-methoxypyrazine[b]	1425	9	7.5	9	
Methional[b]	1476		9	9	9
2-Isobutyl-3-methoxypyrazine[b]	1525	5	3		
(*E*,*E*)-2,4-Octadienal[a]	1616		7.5		
2,3,5-Trithiahexane[a]	1630	-	-	5	9
Phenylacetaldehyde[a]	1644			8	7.5

[a]Compounds were identified on the basis of retention index (RI) on DB-Wax, odour quality, and MS of reference compounds. [b]Tentatively identified (too weak MS signal). [c]Sum of perceived intensities from 3 panellists using 1 (weak), 2 (medium), and 3 (strong).

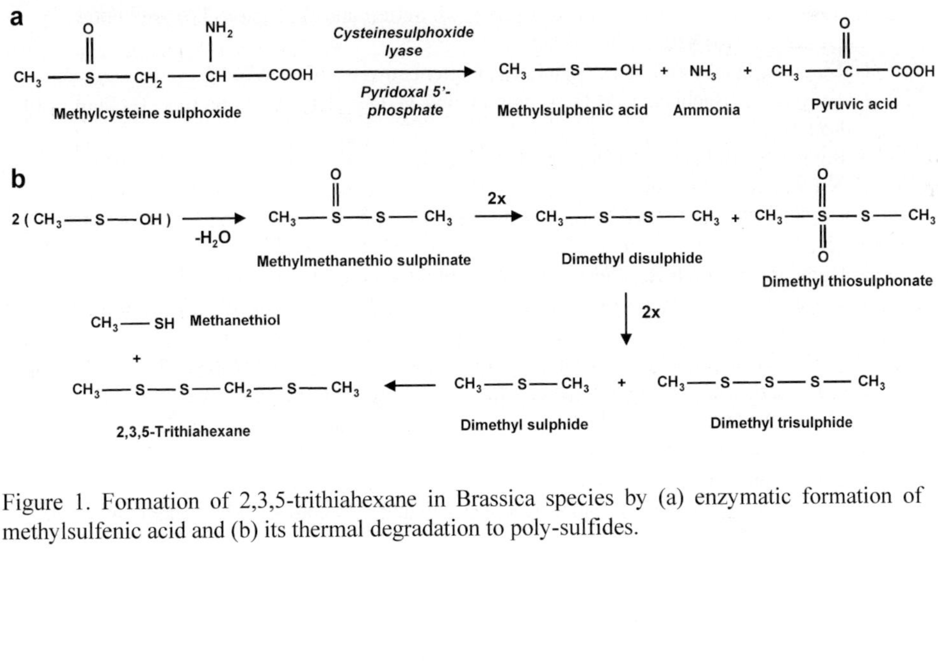

Figure 1. Formation of 2,3,5-trithiahexane in Brassica species by (a) enzymatic formation of methylsulfenic acid and (b) its thermal degradation to poly-sulfides.

3.6. Sensory significance of 2,3,5-trithiahexane

This was assessed by two approaches: (i) lowest detectable amount (LDA) by GCO and (ii) orthonasal perception threshold in water by triangle test. Panellists were also asked to describe the odour properties of TTH. Taking into account that only half of the injected amount reached the sniffing port, GCO measurements indicated an LDA of about 40 pg for TTH. Under the experimental conditions (make-up air 50 ml/min, TTH peak width of 5-8 s), this was equivalent to a minimum detectable concentration in air of 7 pg/ml.

The olfactory threshold of TTH was determined as 0.8 ng/ml on the basis of triangle tests. The tasters were also asked to describe the olfactory attributes of TTH. Around the threshold value, the major descriptors were sulfury, cabbage-like, rubbery. At higher concentrations, TTH was also perceived as onion-like, garlic-like, pungent. Therefore, it is not surprising that the odour recognition threshold in water reported to be 40 ppb [5].

4. CONCLUSION

The formation of TTH in broccoli required the combination of an enzymatic process to generate precursors followed by heat-induced reactions. The enzymatic part of the formation pathway was associated with cutting of broccoli florets into small parts, which activated endogenous enzymes to form methylsulfenic acid. Upon boiling this precursor forms DMS and DMTS, which react to TTH. Both odour thresholds and odour descriptors of TTH confirmed its potential role in the aroma of freshly cooked broccoli. Careful processing of vegetables allowing enzymatic reactions followed by thermal treatment may give desirable flavours known from home-style preparations.

References

1. R.G. Buttery, D.G. Guadagni, L.C. Ling, R.M. Seifert and W. Lipton, J. Agric. Food Chem., 24 (1976) 829.
2. H. Siegl, A. Hanke and W.H. Schnitzler, Gartenbauwiss., 62 (1997) 17.
3. L. Jirovetz, G. Buchbauer, M.B. Ngassoum and M. Geissler, Eur. Food Res. Technol., 214 (2002) 212.
4. S. Rapior, S. Breheret, T. Talou and J.-M. Bessiere, J. Agric. Food Chem., 45 (1997) 820.
5. K. Kubota, S. Ohhira and A. Kobayashi, Biosci. Biotechnol. Biochem., 58 (1994) 644.
6. J. Lin, D.H. Welti, F. Arce Vera, L.B. Fay and I. Blank, J. Agric. Food Chem., 47 (1999) 2822.
7. P. Duby and T. Huyh-Ba, European patent application, EP 545085 (1992).
8. T.R. Schultz, R.A. Flath, T.R. Mon, S.B. Eggling and R. Teranishi, J. Agric. Food Chem., 25 (1977) 446.
9. D. Ulrich, A. Krumbein, I. Schonhof and E. Hoberg, Nahrung, 42 (1998) 392.
10. C.E. Hansen, J.-C. Spadone, I. Meyer, J. Cloke and C. Hall, Unpublished data (1994).
11. R. Kubec, V. Drhova and J. Velisek, J. Agric. Food Chem., 46 (1998) 4334.
12. R.G. Buttery and R.M. Seifert, J. Agric. Food Chem., 25 (1977) 434.
13. L. Horner and J. Dörges, Tetrahedron Lett., 12 (1963) 757.
14. K. Kubota, H. Hirayama, Y. Sato, A. Kobayashi and F. Sugawara, Phytochem., 49 (1998) 99.

Discrimination of virgin olive oil defects - comparison of two evaluation methods: HS-SPME GC-MS and electronic nose

Sonia Esposto[a], Maurizio Servili[a], Roberto Selvaggini[a], I. Riccò[b], Agnese Taticchi[a], Stefania Urbani[a] and GianFrancesco Montedoro[a]

[a]*Food Science Department, Section of Food Technologies and Bio-technologies, University of Perugia, Via San Costanzo Perugia, Italy;*
[b]*Sacmi Industry, Imola, Italy*

ABSTRACT

The headspace analysis of olive oil volatile compounds can be performed using solid phase microextraction (SPME) GC-MS and the electronic nose. This paper presents a comparison of these methods to evaluate their ability to detect the most important olive oil defects such as 'sludge', 'heated' and 'vinegary', and to classify the olive oils according to off-flavour intensities. The research was carried out on blends of refined olive oil and virgin olive oils (I.O.O.C. certification) with the defects indicated above at adequate concentrations to reach four different odour thresholds. Furthermore, by SPME GC-MS, the most abundant volatile compounds were identified for each defect considered.

1. INTRODUCTION

The substances responsible for olive oil aroma belong to the following chemical classes: esters, aldehydes, ketones, aliphatic alcohols, hydrocarbons, oxygenated terpenes, fatty acids, furans and thiophene derivatives [1]. The positive descriptors indicate sensations such as 'cut grass', 'almond', 'tomato', or 'hay-like' while the negative descriptors suggest defects like 'sludge', 'vinegary', 'metallic', and 'rancid'. The sensory quality of virgin olive oil is determined according to the European Union Regulation or the International Olive Oil Council trade standards [2,3]. Both official methods prescribe sensory evaluation using panels of trained assessors. However, panel tests are costly and slow procedures that are not always at the disposal of small producers or cooperative societies; only large retailers and sellers may be able to afford such tests. Furthermore, insufficient training may lead to the subjective opinion of assessors to undermine the final overall evaluation. Alternative solutions to the panel test for the identification of the defects are based on the detection of volatile compounds that can be evaluated by HS-SPME GC-MS and electronic nose which have been widely used in analytical chemistry. The HS-SPME GC-MS is an innovative and very efficient method [1] but it cannot be applied in online processing of virgin olive oil. The second one, made of an array of chemical sensors, is a new technology in the quality assessment of foods [4]. Few investigations have been made to relate off-flavour volatile substances to rancid and vinegary defects [5]. As consequence the goal of this research was to evaluate the capability of the electronic nose and the SPME GC-MS to discriminate the olive oil samples according to the vinegary, heated and sludge defects at different intensities.

2. MATERIALS AND METHODS

2.1. Materials

The research was carried out using 3 different defective virgin olive oils with I.O.O.C. (International Olive Oil Council) Certification defects ‘sludge’, ‘heated’ and ‘vinegary’, respectively. From these oils a number of blends were produced by dilution with a refined olive oil having an odourless oily matrix.

2.2. Sensory testing

Using a trained panel of assessors as described by ASTM (E 679-91) [6], three blends where the defect had an intensity corresponding to the detection threshold were selected; one for each defect considered. Furthermore, after the same panel had carried out alignment tests, three other blends were selected for each off-flavour, having the defect present at the following levels: saturation odour threshold, recognition threshold and a level under the detection threshold. The quantity of defective VOOs present in each blend was as indicated in Table 1.

2.3. Instrumental testing

The 12 blends described in Table 1 were analysed by SPME GC-MS in duplicate and by electronic nose in quadruplicate. The HS-SPME GC-MS analysis was conducted according to Spanier *et al.* [1] and the electronic nose analysis was carried out using an Electronic Olfactory System EOS^{835} consisting of six different metal oxide sensors, built by SACMI S.c.a.r.l. To evaluate the results obtained, the feature $\Delta R/R0$ (R=Resistance measured in ohm) was calculated.

2.4. Data analysis

A PCA was conducted on all data collected to classify the objects: the chemometric package SIMCA-P v. 8.0 (Umetrics AB, Umeå, Sweden) was used for HS-SPME GC-MS results and the NPE (Nose Pattern Editor) was used for feature extraction and feature elaboration of the electronic nose results.

Table 1. Composition of the blends containing refined olive oil (RO) mixed at different concentrations, with defective virgin olive oils (V.O.O.) certificated by the International Olive Oil Council (I.O.O.C.).

	Blend 1 Saturation's odour threshold	Blend 2 Recognition threshold	Blend 3 Detection threshold	Blend 4 A level under the detection threshold
I.O.O.C. 'vinegary' VOO : RO	1:10	1:20	1:40	1:80
I.O.O.C. 'sludge' VOO : RO	1:20	1:40	1:80	1:120
I.O.O.C. 'heated' VOO : RO	1:40	1:80	1:160	1:320

3. RESULTS AND DISCUSSION

As illustrated by the scores plot in Figure 1, showing the SPME GC-MS results, the second component separated the objects in three different clusters according to the off-flavour, while the first component, within each cluster, discriminated the oil samples in order of the different defect intensities. The relative loadings plot (data not shown) allowed the identification of the most abundant volatile compounds for each defect such as phenylethanol for 'sludge', propionic acid for 'heated' and acetic acid for 'vinegary'.

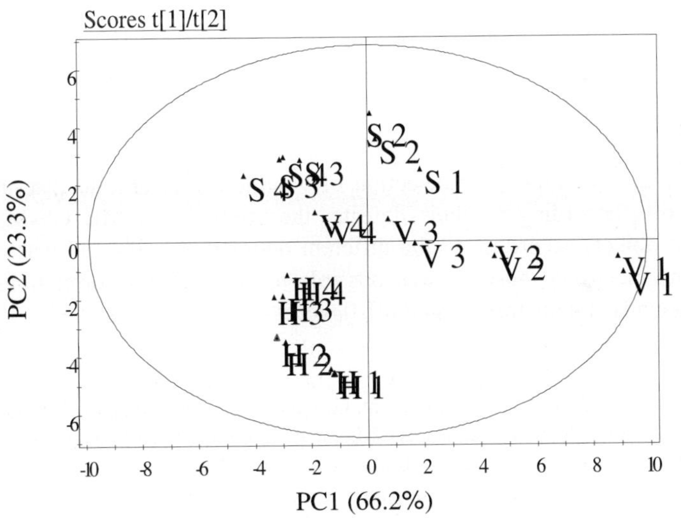

Figure 1. Scores plot of PCA model of HS-SPME GC-MS results on the blends described in Table 1. Explained total variance 90%. Legend: S: sludge; V: vinegary; H: heated. 1: saturation defect's threshold; 2: recognition defect threshold; 3: detection threshold; 4: a level under the detection threshold. Each point represents one measurement.

Figure 2 shows the score plot built from the electronic nose results on the same blends. Also in this case, separation of the oils with respect to the defect class, was possible.

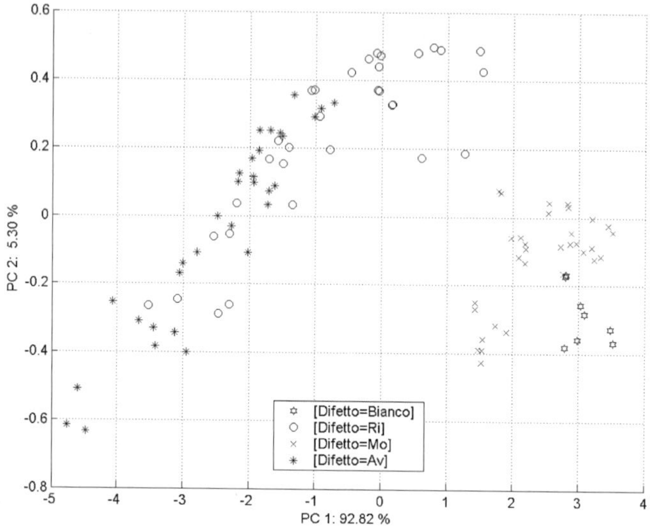

Figure 2. Scores plot of the PCA model of EOS results on the blends containing refined olive oil as matrix. Each point represents one measurement.

4. CONCLUSION

Both instrumental techniques for detection of volatiles were able to discriminate the virgin olive oils according to their defects; the study by SPME GS-MS allowed separating the objects according to the different odour thresholds and was considered better than the electronic nose. It was possible to find several correlations between volatile compounds and the three tested off-flavours.

References

1. A.H. Spanier, F. Shahidi, T.H. Parliament, C. Mussinan, C.-T. Ho and E.T. Contis (eds.), Food flavors and chemistry: advances of the new millennium: proceedings of the 10th international flavor conference, Cambridge, UK (2001) 236.
2. E.C. Off. J. Eur. Communities, Regulations 2568/1991 and 1513/2001.
3. IOOC, International Olive Oil Council, COI/T.20/Document 15/ Rev. 1. Organoleptic assessment of olive oil. Resolution RES-3/75-IV/96, Madrid, Spain, 20 November 1996.
4. H.V. Shurmer and J.W. Gardner, Sensor Actuators B, 8 (1992) 1.
5. R. Aparicio, M. Silvia Rocha, I. Delgadillo and M.T. Morales, J. Agric. Food Chem., 48 (2000) 853.
6. ASTM, Standard practice designation, E 679-91, American society for testing and materials, Philadelphia, PA (1997); 34–38 OR (2002).

W.L.P. Bredie and M.A. Petersen (Editors)
Flavour Science: Recent Advances and Trends

Characterisation of volatile compounds in selected citrus fruits from Asia

Jorry Dharmawan[a], Philip J. Barlow[a] and Philip Curran[b]

[a]*Food Science and Technology Programme, Block S3, Level 6, Science Drive 4, National University of Singapore, Singapore 117543;*
[b]*Firmenich Asia Pte Ltd, 10 Tuas West Road, Singapore 638377*

ABSTRACT

In this study the volatile compounds present in Pontianak orange (*Citrus nobilis* var microcarpa) from Indonesia, Mosambi (*Citrus sinensis* var mosambi) from India and Dalandan (*Citrus reticulata*) from Philippines were characterised. The flavour compounds of the hand-squeezed juices were extracted by headspace solid-phase microextraction (SPME) and solvent extraction methods. More than 70 compounds could be identified and most of them are commonly found in citrus fruits. However, some less common volatile compounds were identified the citrus samples, such as 4-ethenyl cyclohexene, dibutyl sulfide, β-chamigrene, eremophilene and isopiperitenone.

1. INTRODUCTION

The consumption of citrus fruits shows tremendous growth [1]. Citrus fruits are largely processed for the juice as well as for the essential oil. USA and Brazil are the main producers of citrus fruits, but south-eastern Asia is believed to be the place of origin of citrus fruits [2]. Many varieties of citrus fruits in Asia have distinct characteristics and some of these varieties have great potential to be explored further. It is the aim of this study to unveil the aroma profiles of selected citrus fruits from this region, namely Pontianak orange (Indonesia), Mosambi (India) and Dalandan (Philippines).

2. MATERIALS AND METHODS

Fruits: Ripe Pontianak oranges, Dalandans and Mosambis were obtained from fruit farms in Indonesia, Philippines and India, respectively. The freshly squeezed juice was obtained by careful hand-squeezing of the citrus fruits with a citrus juicer, while the peel oil was extracted by hand-pressing of the citrus peels.

SPME analysis: Five grams of freshly squeezed orange juice in 2 M $CaCl_2$ (1:1) was transferred into a 10 ml vial containing a magnetic stirrer. NaCl (26%) was added to the sample matrix and the vial was sealed air-tight. The SPME fibre coated with 2 cm 50/30 μm Divinylbenzene/Carboxen/Polydimethylsiloxane (Supelco) was inserted into the headspace of the sample vial. The extraction was carried out at 50 °C for 30 min.
Solvent extraction: Freshly squeezed orange juice was immediately poured into aqueous saturated $CaCl_2$ solution in order to inhibit the enzymatic reactions [3]. The aqueous mixture was extracted with diethyl ether for 6 h in a separation funnel. The extract was dried over Na_2SO_4 and was finally concentrated with a Vigreux column at 40 °C.
GC-MS analysis: Shimadzu's GC-MS QP5050A system and BPX-5 capillary column (SGE, 30 m x 0.25 mm I.D., 0.25 μm film thickness) were used to analyse the volatile compounds. The temperature was programmed from 35 °C to 200 °C at 4 °C/min and then ramped to 300 °C at 30 °C/min, with a 3 min final temperature hold. The identification of the compounds was performed by matching the mass spectra of the target compounds with those of the NIST library, and by using the standard compounds.

3. RESULTS

3.1. Volatile compounds in freshly-squeezed citrus juices

The volatile compounds detected in citrus juice samples by solvent extraction method and by SPME are shown in Table 1. The volatile compounds were arranged according to their retention indices on non-polar column in ascending order (values not shown).

Table 1. Volatiles found in freshly-squeezed citrus juices by solvent extraction and SPME.

	Solvent extraction			SPME		
Compounds	Pontianak	Dalandan	Mosambi	Pontianak	Dalandan	Mosambi
Acetaldehyde	-[a]	-	-	+	+	+
Ethanol	-	-	-	+	+	+
Methyl acetate	-	-	-	-	-	+
Ethyl acetate	-	-	-	+	+	+
Ethyl propanoate	-	-	-	+	-	-
Ethyl isobutyrate	-	-	-	+	-	-
Ethyl butyrate	+	-	+	+	+	+
Hexanal	-	-	+	+	-	+
(*E*)-2-Hexenal	-	-	-	+	-	+
Heptanal	-	-	-	+	-	+
4-Ethenyl cyclohexene	+	-	-	-	-	-
α-Thujene[b]	-	+	-	+	+	+
α-Pinene	+	+	+	+	+	+
Sabinene	+	+	+	+	+	+
β-Pinene	+	+	+	+	+	+
Myrcene	+	+	+	+	+	+
Butyl butyrate	-	-	-	-	-	+
Ethyl hexanoate	-	-	-	-	-	+
α-Phellandrene	-	+	-	-	+	-
3-Carene	-	-	+	-	-	-
Octanal	+	-	+	+	+	+

Table 1. Volatiles found in freshly-squeezed citrus juices by solvent extraction and SPME (con).

	Solvent extraction			SPME		
Compounds	Pontianak	Dalandan	Mosambi	Pontianak	Dalandan	Mosambi
D-Limonene	+[a]	+	+	+	+	+
Ocimene	+	+	+	+	+	+
γ-Terpinene	-	+	-	+	+	+
cis-Linalool oxide	-	-	-	+	+	-
Octanol	-	-	-	+	-	-
Terpinolene	+	+	+	+	+	+
Dibutyl sulfide	+	-	-	-	-	-
trans-Linalool oxide	-	-	-	+	+	-
Dehydro-*p*-cymene	-	-	-	+	+	+
Linalool	+	+	+	+	+	+
Nonanal	+	-	-	+	+	+
Ethyl-3-hydroxyhexanoate	-	-	-	-	-	+
2,6-Dimethyl-1,3,5,7-octatetraene[b]	+	-	-	+	-	-
1-Terpineol	-	-	-	-	+	-
Citronellal	+	-	-	+	-	+
β-Terpineol[b]	-	-	-	-	+	-
4-Terpineol	-	+	-	+	+	+
3,9-Epoxy-*p*-mentha-1,8(10)-diene[b]	-	-	-	+	-	+
p-Cymen-8-ol	-	-	-	-	+	-
α-Terpineol	+	+	+	+	+	+
Decanal	+	+	+	+	+	+
Dihydrocarvone	-	-	-	-	-	+
trans-Carveol	-	-	-	+	+	+
Citronellol	-	-	-	+	-	-
cis-Carveol	-	-	-	+	+	+
Neral	-	-	-	+	+	+
Carvone	-	-	-	+	+	+
Geranial	-	-	-	+	+	+
Isopiperitenone[b]	-	-	-	-	+	-
Perilla aldehyde	+	-	-	+	+	+
Thymol	-	-	-	-	+	-
Undecanal	+	-	-	+	-	-
p-Vinylguaiacol	+	-	-	-	-	-
δ-Elemene	+	-	-	+	-	-
Neryl acetate	+	-	-	+	+	-
6-Tert-butyl-m-cresol	-	-	-	+	+	+
α-Copaene	-	-	+	-	+	+
Geranyl acetate	+	-	-	+	-	-
β-Damascenone	-	-	-	-	+	-
β-Elemene	+	-	+	+	-	+
Dodecanal	+	-	-	+	+	-
2,6-Dimethyl naphthalene	+	-	-	-	-	-
β-Caryophyllene	-	-	+	-	-	+
Aromadendrene	-	-	+	-	-	+
α-Caryophyllene	+	-	-	+	-	+
γ-Selinene[b]	-	-	-	-	-	+
β-Selinene[b]	-	-	+	-	-	+
β-Chamigrene	-	-	+	-	-	+
Germacrene D[b]	+	-	-	+	-	-
Valencene	-	-	+	-	-	+
A-Farnesene	-	-	-	-	+	-

Table 1. Volatiles found in freshly-squeezed citrus juices by solvent extraction and SPME (con).

Compounds	Solvent extraction			SPME		
	Pontianak	Dalandan	Mosambi	Pontianak	Dalandan	Mosambi
Eremophilene	-	-	+	-	-	+
δ-Cadinene[b]	+	-	+	+	+	+
Hexadecanal[b]	+	-	-	-	-	-
α-Sinensal	-	+	-	-	+	-
Nootkatone	-	-	-	-	-	+

[a]+: detected; -: not detected. [b]Tentative identification by mass spectral matching with library spectra.

3.2. Volatile compounds in citrus peel oil

The results obtained indicated that the peel oil had a different profile of volatile compounds from that of freshly squeezed juice for the same citrus cultivar. Although most volatile compounds detected in the peel oil were also found in the juice, there were some compounds only detected in the peel oil. In Pontianak oranges, they were camphene, α-phellandrene, limonene oxides, 2-decenal, citronellyl acetate, α-copaene, β-caryophyllene, germacrene B and γ-elemene. It was only camphene in Dalandan while they were dodecanal, β-cubebene, β-farnesene, germacrene D, α-farnesene and 4-methyl-3-(1-methylehtylidene)cyclohexene in Mosambi.

4. DISCUSSION AND CONCLUSION

Each citrus cultivar displayed a distinct profile of volatile compounds. Although most compounds are commonly found in citrus fruits, there were some compounds like 4-ethenyl cyclohexene, dibutyl sulfide, 2,6-dimethyl naphthalene, β-damascenone, β-chamigrene and eremophilene, which were less in citrus fruits. Furthermore, the peel oil also displayed a fairly different profile of volatile compounds from that of the juice for similar citrus cultivar. This is illustrated by the detection of some volatile compounds only in the juice, such as carveol and dehydro-*p*-cymene, and by the presence of some compounds only in the peel oil, like camphene and limonene oxide. While the species-specific compounds may be the major contributors to the difference in juice flavour, this has yet to be confirmed. As for most citrus products, the flavour is a result of complex interaction between many, if not all, of the compounds present [4]. Further work on this aspect continues with a view to the possibility of developing an artificial flavour to mimic these unique Asian citrus fruits as well as of creating new flavours.

References

1. FAOSTAT agricultural data as currently available at http://faostat.fao.org/ .
2. F.S. Davies and L.G. Albrigo, Citrus, Wallingford, UK (1994).
3. A. Hinterholzer and P. Schieberle, Flavour Fragrance J., 13 (1998) 49.
4. H. Maarse (ed.), Volatile compounds in food and beverages, New York, USA (1991) 305.

Flavour and colour changes during processing and storage of saffron (*Crocus sativus* L.)

M. Bolandi[a] and H.B. Ghoddusi[b]

[a]Department of Food Science and Technology, Faculty of Agriculture, Islamic Azad University, Damghan Branch, Damghan, Iran;
[b]Department of Food Science and Technology, Faculty of Agriculture, Ferdowsi University of Mashhad, Mashhad, Iran

ABSTRACT

In the present research, saffron (*Crocus sativus* L.) was dried using three methods namely; traditional (25 °C), modified Spanish (55 °C) and microwave oven (300 watts). Changes in chemical indices of saffron according to ISO 3632 were studied during 6 months storage both in dark and exposed-light. The results indicated that the time of storage and drying method had significant effects ($p<0.05$) on chemical properties such as colouring strength, aroma and bitterness values. Samples dried in microwave oven had the highest colouring strength, aroma and bitterness values. Samples dried by a modified Spanish method had higher colouring strength than samples dried by a traditional method, but for aroma, traditional samples had significantly higher values. Regarding bitterness, samples dried by microwave oven and modified Spanish method were the same but traditional samples had significantly lower values than. The colouring strength of saffron decreased but aroma increased during storage and the bitterness did not follow an established pattern. It was also noticed that in samples exposed to artificial light (20 watts) the colouring strength was decreased while, bitterness values were unchanged and aroma values increased at first but remained almost intact after 6 months.

1. INTRODUCTION

Saffron (*Crocus sativus* L.) is one of the most expensive agricultural products [1]. It has been named 'red gold' because of its value. Iran is the major producer accounting for 150 to 170 tons per year; more than 80% of world's production and 50,000 ha is already under cultivation with this crop [1]. Saffron is appreciated for its unique colour (crocin), bitter taste (picrocrocin) and aroma (safranal) [2]. These are the main characteristics showing its quality. The drying method of saffron is very important because it determines the quality and economic value of the final product [3]. The traditional method for drying saffron, in which the stigmas are spread on a piece of cloth indoor or outdoor, avoiding direct sunlight exposure is the most common method in Iran [1]. Since it is a long process, some degradation of pigments occurs and the poor quality of final product is inevitable [4]. Since the method of drying strongly affects the sensory properties of saffron [5], the aim of the present study was mainly to assess the effects of processing condition on major characteristics of final product.

2. MATERIALS AND METHODS

2.1. Saffron samples

Flowers were hand-picked from a farm at Torbat Heidarieh during early morning. Stigmas with small parts of styles (about 2 mm) were separated by hand, mixed and placed in polyethylene containers to minimise moisture loss before drying.

2.2. Drying methods

All samples were dried to 8% moisture, using three different methods. (a) Traditional, in which the stigmas were spread in a layer 1-2 cm thick over a piece of paper at ambient temperature (20-25 °C) in the shade. (b) Microwave oven, in which 20 g stigma were spread in a layer 1cm thick on a Pyrex tray. The power was set at 300 Watts and 5-6 min was required to reach 8% moisture content. (c) Modified Spanish method in which silk screen sieves with 40 cm internal diameter were used. Stigmas were spread in a layer 2-3 cm thick on sieves and placed in an electrical heater, 40 cm above the heating source [6,7]

2.3. Packaging and storage

Dried samples were packed in a three-layer package (aluminium, paper, polyethylene) which prevented entrance of light and moisture. All packages were stored at ambient conditions (20-25 °C) for 6 months. In order to study the effect of light, a part of the sample dried by the modified Spanish method was packed in tiny glass jars and then placed in a light chamber. The light source was a 20 watt florescent lamp. Jars were put at a 10 cm distance from the lamp and were turned 180° every 15 days to have a uniformly light-induced sample.

2.4. Chemical analyses

The chemical properties including the amount of crocin, picrocrocin and safranal, were measured after storing the samples at 20-25 °C for 6 months following ISO-3632-2 [8,9]. In this standard method the absorbance of the saffron solution was measured at 440 nm, 330 nm and 257 nm by a UV-vis spectrophotometer after preparation and dilution. The colouring strength ($E^{1\%}_{440}$), aroma ($E^{1\%}_{330}$) and bitterness ($E^{1\%}_{257}$) were calculated using following formula: $E^{1\%}_{\lambda max} = A/C(g/cm^3)$. Data was analysed using a completely randomised factorial design during the 6-months storage and the means were compared with the multiple comparison LSD test.

3. RESULTS AND DISCUSSION

For all methods, the colouring strength changed with time (Table 1). The microwave-dried sample scored the highest followed by the Spanish and traditional method. This was mainly because of the different drying time giving rise to different value of crocin degradation [4,5]. Also, it was concluded that the colouring strength decreased significantly after 6 months in each sample.

Table 1. Mutual effect of time and drying method on colouring strength.

	Time (months)						
Drying method	0	1	2	3	4	5	6
Traditional	249	246	255	257	250	249	235
	hi	i	fgh	efgh	ghi	hi	j
Microwave	264	269	272	271	266	262	255
	abcde	abc	a	ab	abcd	cdef	fgh
Modified	256	270	262	258	259	255	249
Spanish	efgh	abc	bcdef	defg	defg	fgh	hi

Values marked with the same letter are not significantly different ($p<0.05$).

The mutual effect of time and drying method on volatiles is illustrated in Table 2. The longer the storage time, the more safranal was produced due to the degradation of picrocrocin during storage [10]. We observed no significant difference in volatiles among the three methods after 6 months.

Table 2. Mutual effect of time and drying method on aroma volatiles formation.

	Time (months)						
Drying method	0	1	2	3	4	5	6
Traditional	29.3	31.2	31.6	34.4	33.8	34.6	33.6
	fg	de	d	a	ab	a	abc
Microwave	29.9	32.1	31.0	33.4	34.0	33.9	34.7
	ef	bcd	de	abc	a	a	a
Modified	28.1	31.5	28.3	31.5	31.7	32.0	34.8
Spanish	g	d	g	de	d	cd	a

Values marked with the same letter are not significantly different ($p<0.05$).

The analyses of bitterness during storage (Table 3) showed no consistent differences in the levels between methods. As can be seen the bitterness fluctuated over time. A probable explanation is that picrocrocin is an intermediate compound, which is produced by zeaxanthin degradation and at the same time hydrolysed to safranal [10], resulting in variable values. Also, picrocrocin may have an inhibitory effect on zeaxanthin degradation, which is being investigated.

Table 3. Mutual effect of time and drying method on bitterness.

Drying method	Time (months) 0	1	2	3	4	5	6
Traditional	83.4	86.4	87.5	95.6	90.3	94.5	88.4
	h	fgh	efgh	ab	cdef	abc	defg
Microwave	85.2	90.5	88.0	96.7	93.0	95.4	93.0
	gh	cdef	efg	a	abc	ab	abc
Modified	88.1	93.5	86.5	90.8	91.8	93.2	92.6
Spanish	efg	abc	fgh	cdef	bcde	abc	abcd

Values marked with the same letter are not significantly different ($p<0.05$).

4. CONCLUSIONS

Drying using microwave oven appears to be a suitable way to achieve a high quality product but the cost, using potential and farmer interest should be further evaluated.

A significantly lower colouring strength was observed in the light-induced sample compared with the blank one after 4 months. The volatile compounds data showed that, although volatile development in the control sample was not as fast as the light-induced sample, they reached to the same level of aroma after 5 months. Also, the light had no significant effect on the bitterness.

Overall, at a relatively short storage (up to 3 months) the light does not significantly affect the quality of saffron, but at a longer storage period transparent packaging for saffron should not be recommended.

References

1. M. Kafi, Saffron production and processing, Mashhad, Iran (2002).
2. M. Tsimidou and C.G. Biliaderis, J. Agric. Food Chem., 45 (1997) 2890.
3. S.R. Sampathu, S. Shivashanker and Y.S. Lewis, CRC Crit. Rev. Food Sci., 20 (2) (1984) 123.
4. B.L. Raina, S.G. Agarwal, A.K. Bhatia and G.S. Gaur, J. Sci. Food Agric., 71 (1996) 27.
5. M. Hosseininejad, F. Shahidi and G. Malekzadeh, J. Agric. Sci. Technol., 16 (2) (2002) 51.
6. M. Negbi (ed.), Saffron *Crocus sativus* L., Amsterdam, The Netherlands (1999) 57.
7. P. Winterhalter and M. Straubinger, Food Rev. Int., 16 (1) (2000) 39.
8. ISO, Saffron (*Crocus sativus* L.) Part 1. Specification, International Organization for Standardization, ISO 3632-1 Case Postale 56, CH-1211 Geneve 20, Switzerland (1993).
9. ISO, Saffron (*Crocus sativus* L.) Part 2. Test methods, International Organization for Standardization, ISO 3632-2 Case Postale 56, CH-1211 Geneve 20, Switzerland (1993).
10. H. Pfander and H. Schurtenberger, Phytochem., 21 (5) (1982) 1039.

6

Flavours generated by thermal processes

W.L.P. Bredie and M.A. Petersen (Editors)
Flavour Science: Recent Advances and Trends

Investigation of the key flavour precursors in chicken meat

Michel Aliani[a] and Linda J. Farmer[a,b]

[a]Department of Food Science, Queen's University Belfast, Newforge Lane, Belfast BT9 5PX, UK; [b]Department of Agriculture and Rural Development, Agriculture, Food and Environmental Science Division, Newforge Lane, Belfast BT9 5PX, UK

ABSTRACT

This paper reports a study on the role of natural flavour precursors in the generation of odour and flavour in cooked chicken meat, with reference to the natural concentrations of these precursors in muscle. Addition of varying concentrations of ribose, ribose-5-phosphate, cysteine and ribose, thiamine and IMP to raw meat prior to cooking caused significant changes to the odour and flavour of the meat. However, only ribose showed a significant effect of small additions at close to the natural concentrations of this compound. In addition, chicken breast samples with higher natural concentrations of ribose gave significantly more 'roasted flavour'. Time after slaughter is one source of natural variation in the concentrations of ribose in chicken meat.

1. INTRODUCTION

The volatile compounds responsible for contributing to meat flavours and odours are formed during cooking by the Maillard reaction, lipid oxidation and other reactions between natural components present in raw meat. These components include reducing and phosphorylated sugars, amino acids, ribonucleotides, thiamine and lipids. Although many of the chemical reactions leading to formation of volatile compounds are known, the relative importance of each precursor for flavour is still unclear. This question has been addressed in lamb, beef and pork by monitoring the changes in volatile compounds and/or sensory quality following the addition of precursors into meat [1-3]. It is proposed that small added quantities of sugars and nucleotides, relative to the published natural concentrations, may be sufficient to increase meaty and roasted odours.

The relative importance of Maillard and related precursors for the formation of chicken flavour has received less attention [4]. Furthermore, the information available in the literature on the natural concentrations of these precursors is often sparse and/or inconsistent, making it difficult to determine the effect of 'small' additions. It has, therefore, remained unclear which of the potential Maillard precursors is most important for the formation of chicken flavour, whether the natural variation in the concentrations of these precursors in chicken is enough to cause differences in flavour and what the causes of any natural variation might be.

This paper summarises work in which the role of selected precursors for odour and flavour formation in cooked chicken have been investigated with reference to the natural quantities of these compounds in chicken muscle. The natural variation in the concentrations of these precursors within and between commercial sources is considered, together with the importance of time post-slaughter on the concentrations of these compounds.

2. MATERIALS AND METHODS

Chicken breast fillets (*M. pectoralis major*) without skin or bone from standard chickens were either purchased from local supermarkets (for analysis of precursors) or were supplied by a local poultry meat company (for sensory studies). Sugars and sugar phosphates were determined by HPLC with post-column derivatisation as described previously [5]. Procedures for the analysis of thiamine, amino acids, nucleotides, inosine and hypoxanthine by HPLC have been described in detail elsewhere [6]. Post-slaughter changes in ATP breakdown products have been studied [7] in breast meat from six individual chickens at intervals post-slaughter, prior to chilling, by the methods described previously [5,6]. These products were also monitored in two individual chickens during chilled storage after chilling.

Sensory evaluation of cooked chicken to which precursors had been added was conducted using profiling and paired comparison tests as described previously [8]. The precursors were dissolved in water at the required concentration and homogenised with meat at 10% wet weight. Breast meat from twelve individual chickens that had been analysed for ribose and ribose phosphate were subjected to paired comparison tests using 30 panellists to compare breasts with naturally high concentrations of ribose to those with lower concentration of ribose [8].

3. RESULTS

3.1. Which precursors are limiting for odour and flavour formation?

Figure 1 shows the odour profile for cooked chicken obtained by trained panellists following the addition of selected precursors. Added thiamine (340 mg/100 g) gave significant increases in 'chicken', 'roasted' and 'vegetable soup' odours compared to control. Ribose (150 mg/100 g) significantly increased 'nutty', 'roasted' odours and, interestingly, decreased 'rancid' and 'sour' aroma of cooked chicken. The effect of

added cysteine (120 mg/100 g) significantly increased the odours, 'sour' and 'rancid', but when added together with ribose the effects of ribose were repeated. IMP (400 mg/100 g, not shown) did not significantly give different odour scores to the control.
Thiamine, pentose sugars + cysteine, ribose 5-phosphate or IMP + cysteine are all potential precursors of 2-methyl-3-furanthiol and related compounds, which are important for meat flavour [2,9-11]. This experiment confirms the results of previous studies on chicken (Farmer and Nolan, unpublished data) and other meats [1-3,12] that many of these precursors have the potential to contribute to meaty odour and flavour.

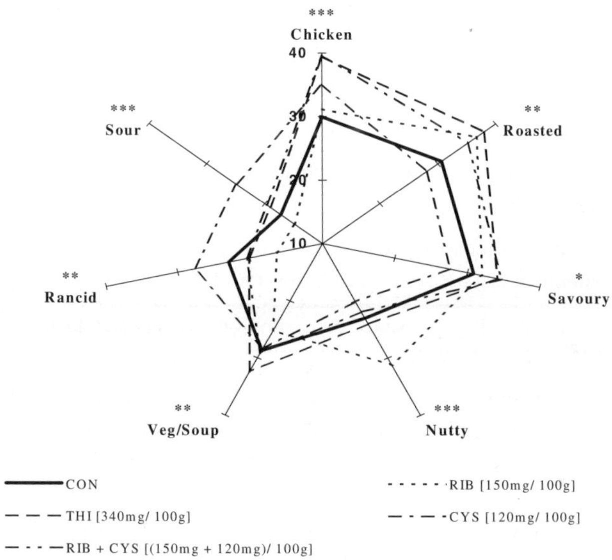

Figure 1. Effect of added precursors on mean profiling scores for selected odour descriptors [8].

However, the concentrations of precursors added in this experiment were based on literature reports and were found to be considerably higher than the concentrations since determined in raw chicken [5,6]. Thus, an understanding of which of these precursors is limiting for flavour requires determination of the natural concentrations of these precursors in chicken muscle.
Figure 2 shows the analytical results for the concentrations of selected flavour precursors in raw chicken breast, together with the standard deviation as an indication of variation [5,6]. On the basis of these data, further sensory studies were conducted to determine the effect of low added concentrations of precursors.
Table 1 shows that when thiamine, IMP and ribose-5-phosphate were added at approximately 2-4 fold of their natural concentration in chicken breast no significant increase in odour scores were observed. This is in contrast to the results shown in Figure 1 where much higher levels of added precursors were used. On the other hand, significant results were obtained with 2-4 fold addition of ribose compared to control.

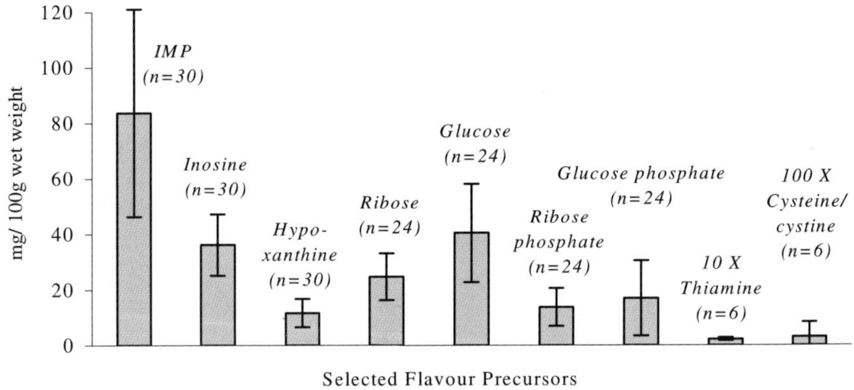

Figure 2. Mean concentrations of selected precursors in chicken breast (showing standard deviation as error bars) [5,6].

Table 1. Results of paired comparison test of the odour of cooked homogenised chicken breast samples with and without added precursors (control versus treated).

Precursor	mg added[a]	Added concentration	Chicken odour	Roasted odour
RIB	50	2X	**8 : 22***	11 : 19
RIB	100	4X	10 : 20	**8 : 22***
CYS	50	70X	16 : 14	11 : 19
RIB+ CYS	100 + 50	4X + 70X	11 : 19	**8 : 22***
R5P	56	4X	11 : 19	12 : 18
IMP	150	2X	14 : 16	12 : 18
Thiamine	0.50	2X	17 : 13	12 : 18
Thiamine	0.90	4.5X	15 : 15	19 : 11

[a]Amount (mg) of compound added to 100 g minced meat in 10 ml water. RIB: ribose; CYS: cysteine; R5P: ribose 5-phosphate. *Significant results, numbers of panellists selecting control versus treated samples as possessing more of this attribute; for 30 panellists. Critical value is 21 ($p<0.05$).

Cysteine alone, even when added at approximately 70 fold the natural concentration, did not affect the ‘chicken odour’ and ‘roasted odour’ but, in the presence of ribose, a similar result was observed as for ribose alone. These results and others [6,8] suggest that the precursor that is most limiting for increasing desirable odour is ribose.

It is evident from Figure 2 that the variation between individual chickens was considerable. In fact, the coefficient of variation for most precursors was higher than 30%. This corresponds to a range of 2-3 fold for ribose, 2 fold for thiamine, 3-4 fold for ribose phosphate, while cysteine was only present at very low concentrations. The sensory studies indicate that the natural variations observed in thiamine, ribose phosphate and cysteine would be insufficient to affect the odour and flavour of the cooked meat. However, the observed 2-3 fold differences in ribose concentration could cause a detectable difference in odour and flavour.

3.2. Can natural variations in flavour precursors cause flavour differences?

It is impossible to add precursors into the same biochemical environment that they would naturally occupy in raw meat. Therefore, a sensory experiment was designed to compare chicken breasts with naturally high concentrations of ribose (23-32 mg/100 g) to those with lower ribose concentrations (7-12 mg/100 g), but with similar concentrations of ribose phosphate. This 2-3 fold natural difference in ribose concentration was sufficient to significantly increase 'roasted chicken flavour', as determined by paired comparison tests [8]. These results offer additional evidence that small changes in ribose concentration can affect chicken flavour and indicate that the observed changes were not due to differences in the concentration of ribose phosphate.

3.3. Potential causes of variations in ribose concentration

The variations in concentration of ribose observed (Figure 2) could be due to several factors, including genetic differences between individual chickens, pre- and post-slaughter production factors or the time available after slaughter for enzymatic processes to occur. An initial study was conducted to examine the evolution of these pathways with time after slaughter.

The origin of ribose in meat is generally proposed to be via the breakdown post-slaughter of ATP to ADP, AMP and IMP, then to inosine and further into hypoxanthine and ribose [13]. The proportions of ribose-containing metabolites are shown at selected times post-slaughter (Figures 3a-d). The summed concentrations of these ribose-containing metabolites were 1.5, 1.4, 2.0 and 1.5 mmol/100 g wet weight, respectively, at 10 min, 115 min, 28 h and 150 h post-slaughter. These data indicate that this pathway is not complete but is influenced by other pathways. The interrelationships between these pathways are the subject of ongoing research.

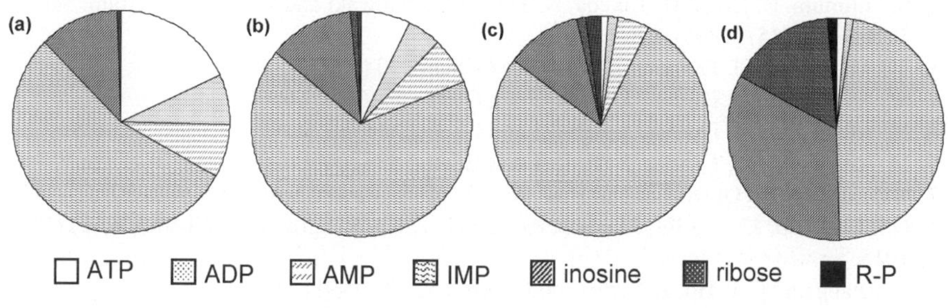

Figure 3. Percentages of ribose-containing metabolites of the ATP breakdown pathway in breast muscle at four times post-slaughter: (a) 10 min (n=6); (b) 115 min (n=6); (c) 28 h (n=2), (d) 150 h (n=2). ATP, ADP, AMP: adenosine 5'-tri-, di- & monophosphate; IMP: inosine 5'-mono-phosphate; R-P: ribose phosphate.

The proportion of ATP decreases during the early stages post-slaughter and remains low during the chilled shelf-life of the chicken (Figure 3). ADP follows a similar pattern while AMP decreases more slowly. The concentration of IMP is high even at 10 min post-slaughter (approximately 0.7 mmol/100 g), indicating a high activity of AMP breakdown by *adenylate aminohydrolase.* The concentrations of IMP present are considerably higher than any of its immediate precursors or breakdown products, suggesting that its breakdown is comparatively slow. Inosine was present at early stages post-slaughter and increased with time, especially between 28 and 150 h (2-6 days), during the expected shelf-life of the chicken. Ribose also increased during this period but was present at much lower concentrations. Hypoxanthine (results not shown) followed a similar trend to ribose, as expected if their common origin were inosine. Ribose phosphate was also present at low concentrations but its concentration decreased towards the end of the chilling period. These studies indicate that time post-slaughter and the rate of enzymatic breakdown of IMP and inosine could be important factors for ribose formation in chicken meat.

4. DISCUSSION AND CONCLUSIONS

A combination of precursor analyses and sensory investigations has indicated that ribose is a limiting precursor for chicken odour and flavour formation. This is corroborated by sensory analyses of chickens with naturally different ribose concentrations. Thus, differences in ribose concentration as small as 2-3 folds can cause perceptible differences in odour and flavour of cooked chicken. One probable cause of such concentration differences is the time that has elapsed post-slaughter.

References

1. J. E. Hudson and R.A. Loxley, Food Technol. Aust., 35 (4) (1983) 174.
2. H. Maarse and D.G. van der Heij (eds.), Trends in flavour research, Amsterdam, The Netherlands (1994) 339.
3. Y.L. Xiong, C.-T. Ho and F. Shahidi (eds.), Quality attributes of muscle foods, New York, USA (1998) 159.
4. S. Fujimura, H. Koga, H. Takeda, N. Tone, M. Kadowaki and T. Ishibashi, Anim. Sci. Technol., 67 (5) (1996) 423.
5. M. Aliani and L.J. Farmer, J. Agric. Food Chem., 50 (2002) 2760.
6. M. Aliani and L.J. Farmer, J. Agric. Food Chem., 53 (2005) 6067.
7. M. Aliani and L.J. Farmer, J. Agric. Food Chem, submitted.
8. M. Aliani and L.J. Farmer, J. Agric. Food Chem., 53 (2005) 6455.
9. G.A.M. van den Ouweland and H.G. Peer, J. Agric. Food Chem., 23 (1975) 501.
10. G.J. Hartman, J.T. Carlin, J.D. Scheide and C.-T. Ho, J. Agric. Food Chem., 32 (5) (1984) 1015.
11. Y. Zhang and C.T. Ho, J. Agric. Food Chem., 39 (6) (1991) 1145.
12. A.J. Taylor and D.S. Mottram (eds.), Flavour science: recent developments, Cambridge, UK (1996) 225.
13. R.A. Lawrie (ed.), Meat science, Oxford, UK (1985) 53.

W.L.P. Bredie and M.A. Petersen (Editors)
Flavour Science: Recent Advances and Trends

Aroma formation in beef muscle and beef liver

Jane K. Parker, Anna Arkoudi, Donald S. Mottram and A.T. Dodson

The University of Reading, School of Food Biosciences, Whiteknights, Reading RG6 6AP, United Kingdom

ABSTRACT

Beef muscle and beef liver were used as matrices in which to observe the Maillard reaction. Samples were spiked with glucose, ribose, glycine or cysteine and the effect on the formation of pyrazines and furanthiols was studied. The addition of glucose showed that glucose limited the formation of pyrazines in muscle but not in liver. The addition of glycine demonstrated how some pyrazines were reliant on glycine for their formation whereas others were not, confirming reported reaction mechanisms from in model systems. The addition of cysteine showed that in muscle the availability of H_2S was the limiting step in the formation of 3-methyl-2-furanthiol (MFT) whereas in liver, MFT could not be detected in any of the systems. This could be due to the absence of ribose in liver, but may also indicate the presence of interactions with the liver matrix.

1. INTRODUCTION

The generation of flavour during the cooking of meat is a complex process which is still a subject of interest to both flavour chemists and the food industry worldwide. Compounds responsible for the characteristic savoury meaty aroma of cooked muscle have been widely investigated and reviewed [1]. The generation of volatiles in liver has received less attention. Compared to muscle, liver contains higher levels of glycogen, free glucose and free amino acids, particularly cysteine. As a result, a more diverse range of Maillard volatiles is generated during cooking [2] and the flavour is more intense. It is well known that many of these compounds are derived from the Maillard reaction and there is a wealth of literature surrounding the formation pathways of these compounds. Aqueous model systems containing key precursors, often isotopically labelled, have been used to demonstrate major reaction pathways, in particular, for pyrazines [3-5] and furanthiols [6,7]. However, food systems are complex matrices containing many more components than a simple aqueous system. In meat model systems both lipids [8] and proteins [9] have been shown to influence the volatile profile. Very little work has been carried out to show whether the mechanisms postulated for aqueous model systems are also relevant to real food matrices. In this paper, beef muscle and beef liver, were used as matrices for observing the formation of Maillard volatiles. The addition of key precursors (glucose, ribose, cysteine and glycine) was used to probe the reactions and to evaluate the mechanisms of flavour formation.

2. MATERIALS AND METHODS

Portions (50 g) of either lean beef fillet (*M. psoas major*) or beef liver were blended with 50 ml of either water, 1% glucose, 0.2% cysteine, 2% glycine or 0.5% ribose. The homogenates were cooked in sealed jars in an autoclave for 40 min at 125 °C, cooled and rehomogenised. The volatiles were analysed using a standard dynamic headspace collection technique, followed by GC-MS as described by Elmore *et al.* [10]. The quantities of the volatiles were estimated by comparing their peak areas with the area of internal standard (1,2-dichlorobenzene). Experiments were carried out in quadruplicate.

3. RESULTS AND DISCUSSION

The quantities of volatiles produced in the liver matrix model were greater than those in the muscle matrix system, in particular the levels of pyrazines, which were 20-100 times higher in liver than in the muscle system. Over 50 volatiles were estimated quantitatively and many varied significantly on addition of precursors. This paper, however, focuses on the generation of two classes of compounds, namely pyrazines and furanthiols, for which the formation in aqueous models is well established.

3.1. Pyrazines

Addition of 0.5% glucose to muscle significantly increased the levels of pyrazines formed after cooking (Table 1), whereas in liver, it did not. This suggests that glucose was limiting the factor in the formation of pyrazines in muscle. When 1% glycine was added to liver, the pyrazines split into two groups, those which changed with glycine addition and those which remained unaffected. Pyrazines in which the entire carbon skeleton can be derived from relatively abundant sugar degradation products such as methylpyrazine and ethylpyrazine, were not affected by added glycine. For example, for methylpyrazine, aminoketones derived from methylglyoxal and glyoxal, condense to form a dihydropyrazine which is subsequently oxidised to methylpyrazine (Figure 1). Pyrazines which increased with addition of glycine were those in which the more likely formation pathway incorporates part of the glycine molecule into the carbon skeleton of the pyrazine. Mechanisms have been proposed [3-5] in which glycine undergoes Strecker degradation to form formaldehyde which subsequently reacts with the dihydropyrazine to form a methyl-substituted pyrazine (Figure 1). Where both mechanisms can occur to form 2,5- or 2,6-dimethylpyrazine, smaller increases were observed. In liver, addition of 0.1% cysteine decreased the levels of pyrazines and many other Maillard products. This is consistent with the reported loss of burnt notes on addition of N-acetylcysteine to liver sausage [12]. H_2S generated in greater quantities in presence of added cysteine may react with the low levels of sugar degradation products (deoxyosones, dicarbonyls, hydroxycarbonyls), thus reducing their availability for participation in other reaction pathways. In support of this, a decrease of some sulfur-containing compounds such as thiazoles and thiophenes (not shown), which rely on H_2S for their formation, was observed when glucose was added to these systems.

Table 1. Relative amounts (ng/25 g sample) of some volatiles collected from headspace of cooked beef muscle and liver, with and without the addition of precursor (X).

Compound	Muscle	Muscle+X		Liver	Liver+X	
X = 0.5% Glucose						
Pyrazine	4.4	11	*	251	289	
Methylpyrazine	6.0	23	*	392	378	
Ethylpyrazine	0.5	1.5	*	30	36	
2,5/6-Dimethylpyrazine	1.1	3.4	*	159	135	
X = 1% Glycine						
Methylpyrazine	13	25	*	276	338	
Ethylpyrazine	0.8	1.3	*	18	17	
2,5-Dimethylpyrazine	2.2	5.1	***	31	51	**
2,6-Dimethylpyrazine	0.8	2.8	**	33	44	*
2,3-Dimethylpyrazine	0.4	1.7	***	7	14	***
Trimethylpyrazine	0.8	5.4	***	7	32	***
Tetramethylpyrazine	-	0.5		0.04	2.6	***
X = 0.1% Cysteine						
Methylpyrazine	6.0	5.8		392	257	*
Ethylpyrazine	0.5	0.6		30	21	**
2,5/6-Dimethylpyrazine	1.1	0.6		159	45	***
2-Methyl-3-furanthiol	-	10		-	-	
2-Methyl-3-furyl methyl disulfide	-	0.4		-	-	
Bis-2-Methyl-3-furyl disulfide	-	0.4		-	-	
X = 0.25% Ribose						
2-Methyl-3-furyl methyl disulfide	-	0.4		-	-	

– Not detected. *Significant difference at $p<0.05$; ** $p<0.01$; *** $p<0.001$.

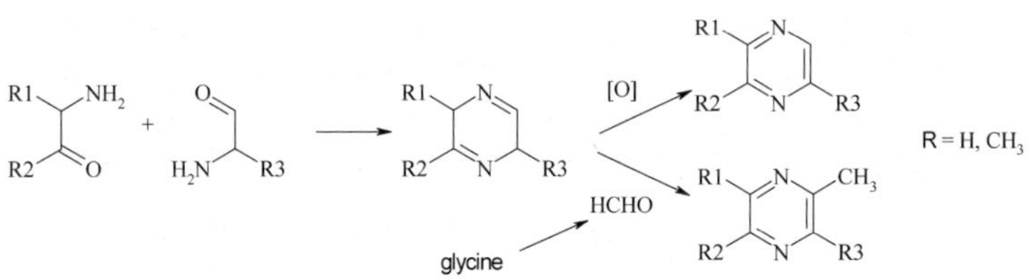

Figure 1. Two routes for the formation of substituted pyrazines [3-5].

3.2. Furanthiols

2-Methyl-3-furanthiol (MFT) has a powerful meaty aroma. Hofmann and Schieberle [7] have shown that it is predominantly formed from pentoses but can also be formed from sugar breakdown products of glucose. Neither MFT nor any of its derivatives were detected in the volatiles of cooked muscle, but after addition of cysteine, low levels of MFT and three related compounds were found. This suggests that the availability of sulfur was limiting in the formation of MFT. In liver, no MFT was generated, even after addition of cysteine, suggesting that the routes via glucose were not important. Unlike muscle, which contained low levels of ribose (Table 2), liver, being a non-contractile muscle, does not require ribose and ribonucleotides for its function and ribose was not detected in the raw material. This may explain that MFT could not be detected. However, when 0.25% ribose was added to liver (which contains much higher levels of cysteine compared to muscle), still no MFT was measurable. This leads us to suggest that any MFT and related disulfides formed in liver could be lost in the liver matrix. This is consistent with the fact that cyclic disulfides, which do not require ribose for their formation, were also found in muscle but not liver (not shown in Table 1). This apparent loss of thiols and disulfides in liver could be due to reaction with sugar degradation products such as dicarbonyls and hydroxycarbonyls. Alternatively, the mechanism may involve the disulfide interchange with protein sulfhydryl groups as proposed previously for the loss of disulfide compounds in meat [9].

Table 2. Levels of precursors (%) in liver and muscle [11] and added to muscle and liver samples.

	Muscle	Liver	Added
Glucose	0.08	0.6	0.5
Ribose	0.03	Not detected	0.25

References

1. H. Maarse (ed.), Volatile compounds in foods and beverages, New York, USA (1991) 107.
2. C.J. Mussinan and J.P. Walradt, J. Agric. Food Chem., 22 (1974) 827.
3. E.-M. Chiu, M.C. Kuo, L.J. Bruechert and C.T. Ho, J. Agric. Food Chem., 38 (1990) 58.
4. M. Amrani-Hemaimi, C. Cerny and L.B. Fay, J. Agric. Food Chem., 43 (1995) 2818.
5. V. Yaylayan and A. Keyhani, J. Agric. Food Chem., 47 (1999) 3280.
6. F.B. Whitfield and D.S. Mottram, J. Agric. Food Chem., 47 (1999) 1626.
7. T. Hofmann and P. Schieberle, J. Agric. Food Chem., 46 (1998) 235.
8. L.J. Farmer and D.S. Mottram, J. Sci. Food Agric., 53 (1990) 505.
9. R.L. Adams, D.S. Mottram and J.K. Parker, J. Agric. Food Chem., 49 (2001) 4333.
10. J.S. Elmore, D.S. Mottram, M.B. Enser and J.D. Wood, J. Agric. Food Chem., 47 (1999) 1619.
11. K. Deibler and J. Delwiche (eds.), Handbook of flavor characterization, New York, USA (2003) 463.
12. C. Hilmes, M. Gibis and A. Fischer, Fleischwirtschaft, 78 (1998) 901.

W.L.P. Bredie and M.A. Petersen (Editors)
Flavour Science: Recent Advances and Trends

Effect of baking process and storage on volatile composition of flaxseed breads

Terhi Pohjanheimo[a], Mari Hakala[a,b] and Heikki Kallio[a,b]

[a]Functional Foods Forum, University of Turku, Itäinen Pitkäkatu 4 A, FI-20520 Turku, Finland; [b]Department of Biochemistry and Food Chemistry, University of Turku, FI-20014 Turku, Finland

ABSTRACT

Volatiles of flaxseed oil, low fat crushed grains and three bakery products containing flaxseed in its different forms were analysed by GC-MS. Preliminary comparison of volatile compounds was made between freshly baked and stored products. The identified volatile compounds were mainly aldehydes, ketones and alcohols. Oxidative degradation of flaxseed oil was also monitored with rancidity tests during storage.

1. INTRODUCTION

Flax (*Linum usitatissimum* L.) is an old cultivated plant, which has been used for cloths, paints and human food since ancient times [1]. Oilseed flax contains about 45% oil, 30% dietary fibre and 20% protein. The focus in flaxseed food processing has been in the production of the oil [1]. The flaxseed oil contains exceptionally high levels of unsaturated fatty acids such as linoleic acid and α-linolenic acid, the latter usually makes up more than 50% of the total fatty acids. In general, the oil is cold pressed leaving a residual part called low-fat crushed grain. This residual part is very rich in both soluble and insoluble fibres and can be used for example in baking processes. Flaxseed being a principal source of α-linolenic acid and of lignans as well as an important source of dietary fibre, has potential in preventing and reducing the risk of some major diseases, including atherosclerosis [1].

However, applications of the highly unsaturated flaxseed oil can also be a cause for off-flavours in foods. The typical rancid off-flavour from oxidised flaxseed may be caused either by volatile compounds such as aldehydes, ketones and alcohols, or by high amounts of free fatty acids [2,3]. This may have a strong influence on acceptance and choice of flaxseed-based products by the consumer.

In our study the volatile compounds generated in three different bread products containing flaxseed were investigated just after production and after storage. The aim was to study changes in volatile composition during the product self-life. The volatiles of flaxseed ingredients were also measured. Furthermore, the oxidative stability of flaxseed oil was measured by peroxide value and free fatty acids once a week during an eleven weeks storage period. The influence of heating and storing on crushed grains has been reported earlier [4].

2. MATERIALS AND METHODS

The flaxseed oil and crushed grain samples were of a commercial brand 'Pellavainen' processed by Elixi Oil Oy (Somero, Finland) from oilseed flax variety 'Helmi'. Rye-flaxseed bread, flaxseed roll and sweet coffee bread were baked using different amounts of whole flaxseeds, oil or crushed grains by professional baker (Table 1). The bakery samples were frozen (-22 °C) in plastic bags until analysed.

Table 1. Nutrition information of the bakery samples per 100 g.

	Rye-flaxseed bread (crushed grains 5%, flaxseeds 3%)	Flaxseed roll (flaxseeds 5%, crushed grains 3%)	Sweet coffee bread (flaxseed oil 13%, crushed grains 1%)
Energy	770 kJ/180 kcal	990 kJ/240 kcal	1540 kJ/370 kcal
Protein	7.3 g	13.1 g	7.6 g
Carbohydrates	30.6 g	32.8 g	46.3 g
Fat content	3.0 g	5.4 g	16.0 g
Saturated lipids	0.4 g	1.2 g	2.2 g
Total fibre content	9.2 g	6.6 g	1.8 g

Samples were transferred into the refrigerator the day before analysis and kept at room temperature on the day of analysis. Rye-flaxseed bread was stored for seven days, flaxseed roll for four days and sweet bread for three days at room temperature before the second analysis to measure the changes during storage.

Samples of 10 g were weighted into 25 ml conical flasks. After a 30 min equilibration the volatile compounds were collected from the headspace of the closed conical flask with a SPME fibre (2 cm in length, 50/30 μm divinylbenzene/carboxen/polymethylsiloxane Stableflex, Supelco, USA) for 30 min. The headspace volatiles were analysed with a Shimadzu GC-17A gas chromatograph equipped with Shimadzu QP-5000 mass spectrometer. A column (30 m, i.d. 0.25 mm, d_f 0,25 μm) with DB-1701 as stationary phase was used. The GC temperature programming was from 40 °C to 240 °C at 7 °C/min. The volatiles were identified by comparison of their spectrum with those in a reference spectral library (Wiley 229).

Flaxseed oil for rancidity testing was stored in 100 ml glass bottles at room temperature (22 °C) in dark place. Peroxide value was performed according to AOCS method Cd 8b-90 and free fatty acids according to AOCS method Cc 5a-40. Rancidity tests were performed each time from newly opened bottles during period of eleven weeks.

3. RESULTS

A total of 47 volatile compounds were identified from the samples. The main volatiles in flaxseed oil expressed as percentage of the total volatiles were 2-butanol (18%) and hexanal (16%), and ethanol (47%) in crushed grains. Some compounds, e.g. acetaldehyde and ethanol increased. Other compounds such as 2-methylbutanal, 3-methylbutanal and hexanal decreased in all the flaxseed bakery samples during the different storage periods (Table 2).

Table 2. Changes of main volatile compounds in headspace of the samples analysed. Contents expressed as percentage (%) of total volatiles.

	Acetaldehyde		Ethanol		2-Methylbutanal/ 3-Methylbutanal		Hexanal	
	Fresh	Stored	Fresh	Stored	Fresh	Stored	Fresh	Stored
Flaxseed oil	1.9	-	3	-	9.8	-	15.6	-
Crushed grains	1.5	-	46.6	-	3.7	-	0.8	-
Rye-flaxseed bread	0.3	2.2	21.7	29.6	50.4	6.4	3.1	0.4
Flaxseed rolls	0.2	2.0	60.2	66.5	8.2	3.5	3.0	0.4
Sweat bread	0.3	0.4	56.7	54.1	5.8	3.6	3.3	1.2

The absolute content of identified compounds depended on the sample and the storage time. The number of volatiles in freshly baked products was close to the raw materials (flaxseed oil and crushed grains). The number of volatile compounds decreased during storage in all baked samples (Figure 1). During storage the most volatile compounds released from the samples entirely and were not detected in the second analysis after storage period. After the storage the number of volatiles was nearly the same in all samples.

The peroxide value and free fatty acids content showed no significant changes during eleven weeks of storage of the flaxseed oil. This strongly suggested that no oxidation of the material occurred. Even though flaxseed oil contains high levels of unsaturated fatty acids, the relatively high natural level of antioxidant active lignans may contribute to the oxidative stability of the flaxseed oil and products with incorporated flaxseed.

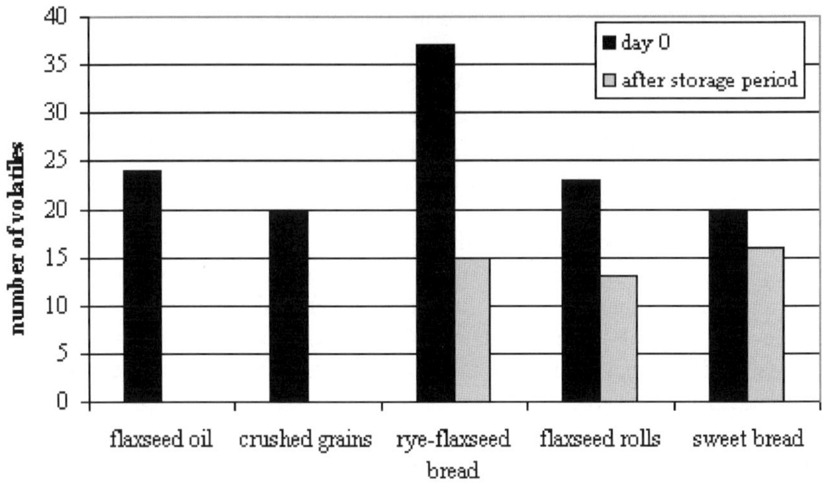

Figure 1. The number of headspace volatiles disappearing in the bakery samples during storage.

4. CONCLUSION

The volatile compounds identified in flaxseed breads, flaxseed grains and oil were mainly aldehydes, ketones and alcohols. No clear off-odours were found in the flaxseed bakery samples after relatively short storage times. The proportion of hexanal decreased during storage in all bakery samples even though it is a common indicator of oxidation of lipids [2,3]. Rancidity tests showed a high stability of flaxseed oil stored at room temperature for eleven weeks.

References

1. L.U. Thompson and S.C. Cunnane (eds.), Flaxseed in human nutrition, 2 ed., Champaign, Ill., 2 (2003) 1 and 363.
2. O. Sjövall, T. Virtalaine, A. Lapveteläinen and H. Kallio, J. Agric. Food Chem., 48 (2000) 3522.
3. M. Zhou, K. Robards, M. Glennie-Holmes and S. Helliwell, J. Agric. Food Chem., 47 (1999) 3941.
4. T. Hofmann, M. Rothe and P. Schieberle (eds.), State of the art in flavour chemistry and biology, proceedings of the 7th Wartburg symposium, Garching, Germany (2005) 190.

W.L.P. Bredie and M.A. Petersen (Editors)
Flavour Science: Recent Advances and Trends

Formation of flavour compounds in reactions of quinones and amino acids

George P. Rizzi

Procter & Gamble Company, Winton Hill Business Center, Cincinnati, OH (USA) 45224

ABSTRACT

Quinone/amino acid reactions relevant for flavour generation were studied in 2-methoxyethyl ether (MEE) (151 °C) and in aqueous (22 °C) model systems. In both systems reaction products were observed resulting from Strecker degradations. In aqueous systems containing *p*-benzoquinone (BQ) and α-amino acids visibly coloured products including 2-(4-hydroxyphenylamino)-1,4-benzoquinone (HPBQ) (λ_{max} 515 nm) were also formed. With proline and BQ a violet-coloured, highly polar substance was observed (λ_{max} 511 nm) which was tentatively identified as a BQ-induced oxidation product of the predicted Strecker intermediate 1-(4-hydroxyphenyl)-2-hydroxy-pyrrolidine.

1. INTRODUCTION

Quinones are highly reactive intermediates in the enzymatic browning reaction in which they interact with ambient nucleophilic sites on proteins to form phenolic melanoidin polymers [1]. Also, quinone-induced Strecker degradation has been invoked to explain aldehyde formation during tea fermentation [2], and in unfermented cocoa beans and green coffee beans [3]. The aim of this study was to gain further insight into the chemistry of non-enzymatic quinone/amino acid reactions and their implications for food colour and flavour.

2. MATERIALS AND METHODS

2.1. Model reaction systems

For non-aqueous reactions, equimolar amounts of amino acids (AA) and *p*-benzoquinone (BQ) each 0.30 M in 2-methoxyethyl ether (MEE) solvent, were heated to 151 °C and carbon dioxide evolved during 2 h was trapped on Ascarite® and quantified by weighing. Other volatile reaction products were isolated by a combination of steam distillation, ether extraction and preparative gas chromatography (GC). Non-volatile products were separated by ion exchange chromatography (IEC) of water-diluted reaction mixtures on a strong acid cation exchange resin. Aqueous reactions involved equimolar AA/quinone reactants at 0.05 M (1-4 h/22 °C) and Strecker aldehydes were isolated and quantified as their 2,4-dinitrophenylhydrazine derivatives.

2.2. Synthesis

4-Methyl-1,2-benzoquinone, caffeic acid quinone and catechin quinones were prepared from respective phenols by oxidation with *o*-chloranil or sodium periodate. HPBQ was obtained by reacting *p*-aminophenol with excess BQ in methanol. Quinone/ascorbic acid reactions were done at 0.015 M in methanol (22 °C/30 min) and dehydroascorbic acid (DHA) was identified as its *o*-phenylenediamine (OPD) adduct [4] formed by reaction of two moles of OPD/mole of DHA. Initial reaction of DHA forms the expected quinoxalin-lactone which undergoes lactone ring opening with a second mole of OPD to form the stable 2:1 adduct. N-Isobutylideneisobutylamine was obtained by reacting 2-methylpropanal with isobutylamine. 1-Deoxy-1-piperidino-D-fructose was prepared according to [5]. D-glucosone/OPD derivative was prepared via [6].

2.3. Identification

Paper chromatography (PC) and Silica gel (TLC) chromatography were used extensively to analyse and isolate products. Product identification was based on retention data and a combination of UV/VIS, IR and NMR spectroscopy and electrospray mass spectrometry. Named (not synthesised) compounds were commercial samples of reagent grade purity.

3. RESULTS

Reactions of BQ with amino acids in MEE solvent at 151 °C led to significant molar % yields of CO_2 indicative of Strecker degradation, namely: gly, 45; ala, 45; val, 72; leu, 86; phe, 96; ser, 21; and tyr, 14. A steam distillate of the valine reaction yielded after GC a single major product, N-isobutylideneisobutylamine. In addition, IEC and PC led to isolation and identification of isobutyl amine and *p*-aminophenol (Figure 1). Quinone/AA reactions in water led initially to chromophoric species and Strecker aldehydes (Table 1). TLC analysis of aqueous BQ/AA reactions (except β-ala) revealed a common, visibly coloured product (λ_{max} 515 nm), 2-(4-hydroxyphenylamino)-1,2-benzoquinone (HPBQ) (Figure 1). Coloured products more polar than HPBQ are currently under investigation. A violet-coloured compound (λ_{max} 511 nm) isolated from an aqueous BQ/proline reaction had a molecular weight of 177 and appears to be a zwitterionic species, $C_{10}H_{11}NO_2$ derived via a quinone-mediated oxidation of a Strecker degradation intermediate (Figure 1).

The propensity of quinones to act as oxidants was illustrated in reactions with ascorbic acid. BQ, 4-methyl-1,2-benzoquinone, catechin quinone and caffeic acid quinone all rapidly oxidised ascorbic acid to dehydroascorbic acid in 0.015 M reactions at 22 °C.

Conversely, quinones were not observed to oxidise acetol, acetoin or the Amadori compound, 1-deoxy-piperidino-D-fructose (to form D-glucosone) under similar conditions.

4. DISCUSSION AND CONCLUSIONS

The contribution of natural polyphenolic compounds to the flavour and colour of processed foods is complex and only partially understood. During processing aerobic oxidation catalysed by polyphenoloxidase produces highly reactive quinones. Our experiments in model systems proved that benzoquinones can react with ambient amino acids to form flavour compounds derived from Strecker degradation like aldehydes, amines and imines.

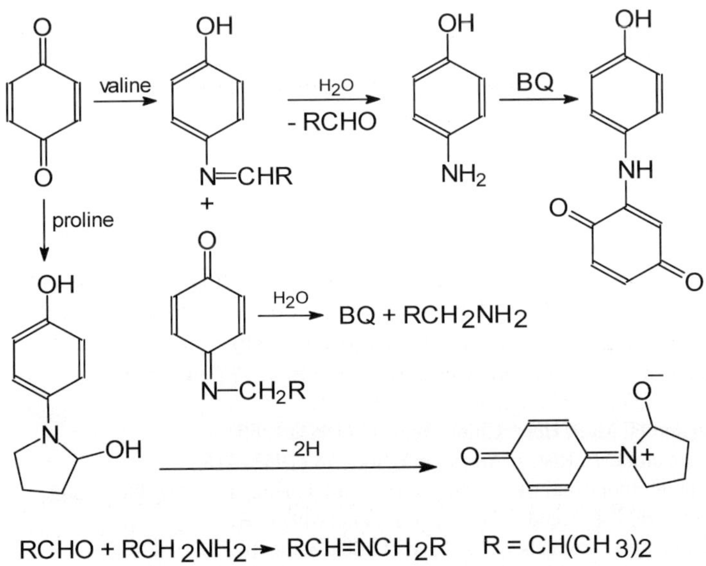

Figure 1. Strecker degradation of valine and proline with *p*-benzoquinone.

In addition, the quinone reductive amination product, *p*-aminophenol reacts further with ambient quinones to form chromogenic aminoquinones like HPBQ possibly contributing to food colours. In water BQ reacted with α-amino acids to form HPBQ and other transient coloured species that slowly decomposed to form brown polymers. Reaction of BQ with proline produced a violet coloured compound similar to one reported in proline reactions of *o*-benzoquinones [7], but whose molecular weight, polar nature, and instability suggested the dipolar zwitterionic structure shown in Figure 1. Presumably the BQ/proline Strecker intermediate shown is dehydrogenated by BQ to form the zwitterion which slowly polymerizes. Oxidation of Strecker intermediates is not unknown and has already been reported in metal ion catalysed autoxidation of Amadori compounds [8].

The chemical nature and the origin of the coloured compounds formed in reactions of quinones and amino acids will be the subject of further investigation.

Table 1. Results of 22 °C aqueous reactions of amino acids and quinones.

Quinone/amino acid	Reaction colour	(λ_{max})nm	Odour	Aldehyde (% yield)
p-Benzoquinone/glycine	Purple	(499)	None	-
α-Alanine	Orange-red	(435)	Aldehydic	-
β-Alanine	Deep red	(455, 490)	None	-
Valine	Purple	(455, 500)	Cereal-like	-
Leucine	Deep red	(455, 505)	Chocolate	2-Methylbutanal (6.9)
Phenylalanine	Red-purple	(495)	Lilac	Phenylacetaldehyde (0.6)
Proline	Deep violet	(520)	Corny	-
4-Methyl-1,2-benzoquinone/ phenylalanine	Brown		Lilac	Phenylacetaldehyde (0.5)
Caffeic acid quinone/ phenylalanine	Brown		Lilac	Phenylacetaldehyde (trace)

References

1. A.G. Mathew and H.A.B. Parpia, Adv. Food Res., 19 (1971) 97.
2. M.A. Bokuchava and N.I. Skobeleva, Adv. Food Res., 17 (1969) 215.
3. S. Motoda, J. Ferment. Technol., 57(5) (1979) 395.
4. H. Dahn and H. Moll, Helv. Chim. Acta, 47 (1964) 1860.
5. J.E. Hodge and C.E. Rist, J. Am. Chem. Soc., 75 (1953) 316.
6. N. Morita, K. Inoue and M. Takagi, Agric. Biol. Chem., 45 (11) (1981) 2665.
7. H. Jackson and L.P. Kendal, Biochem. J., 44 (1949) 477.
8. T. Hofmann and P. Schieberle, J. Agric. Food Chem., 48 (2000) 4301.

W.L.P. Bredie and M.A. Petersen (Editors)
Flavour Science: Recent Advances and Trends

Formation of 4-hydroxy-5-methyl-3(2*H*)-furanone (norfuraneol) in structured fluids

Imre Blank, Tomas Davidek, Stéphanie Devaud, Laurent Sagalowicz, Martin E. Leser and Martin Michel

Nestlé Research Center, Vers-chez-les-Blanc, 1000 Lausanne 26, Switzerland

ABSTRACT

This study investigated the influence of structured fluids on Maillard-type reactions. The decomposition of xylose and the formation of volatile compounds were affected by the type of structured fluid used as reaction medium. In model systems based on xylose and glycine or leucine, xylose was preferably degraded in the hexagonal phase, compared to other mesophases and the aqueous sample. In parallel, norfuraneol was accumulated in the hexagonal phase. The data obtained indicate that molecular organisation of the reaction medium and flavour precursors can play an important role in food systems containing ingredients that tend to form self-assembly structures.

1. INTRODUCTION

The use of structured fluids as micro-reactors for food applications has recently been highlighted in enzymatic and Maillard-type reactions [1]. Micro-emulsions are capable of solubilising considerable amounts of water-soluble and oil-soluble guest molecules. This represents an attractive approach to make better use of the potential of flavour precursors with respect to increasing the yields, guiding reaction pathways, stabilising flavour compounds generated, and controlling their delivery in the proper moment of use. Water-in-oil microemulsions have recently been reported as a means to generate and deliver novel flavours by thermal degradation of suitable intermediates during frying [2]. Reaction of pentose sugars in the presence of cysteine resulted in higher amounts of the odorants 2-methyl-3-furanthiol and 2-furfurylthiol in cubic phases based on water/monoglyceride (20/80), as compared to aqueous systems [3]. This paper deals with flavour formation using various structured fluids as reaction media. We focus on the formation of the well-known flavour intermediate norfuraneol from xylose in the presence of glycine or L-leucine.

2. MATERIALS AND METHODS

Materials. The following chemicals were commercially available: monosodium dihydrogenphosphate monohydrate, anhydrous sodium sulfate, sodium chloride (NaCl), aqueous hydrochloride (HCl) solution, diethylether (Merck, Darmstadt, Germany); D-xylose (Xyl), glycine (Gly), L-leucine (Leu), 2-hydroxy-3-methyl-2-cyclopenten-1-one (cyclotene, Fluka, Buchs, Switzerland); monoglycerides Dimodan U and Dimodan HR (Danisco, Copenhagen, Denmark).

Reactions performed in phosphate buffer. A solution of xylose (0.330 g, 2.2 mmol) and glycine (0.165 g, 2.2 mmol) or leucine (0.288 g, 2.2 mmol) in phosphate buffer (10 ml, 0.2 mol/l; pH 6.0) was heated in a Pyrex tube (30 ml, 70 °C, 7 h). Aliquots of the reaction mixture were taken at regular time intervals for analysis. Each reaction mixture (2 g) was diluted with water (8 g), cyclotene (372 μg) was added as internal standard, and the mixture saturated with NaCl (3 g). The pH was adjusted to 4 (aqueous HCl, 2 mol/l) and the volatiles were extracted with diethylether (2 x 15 ml, 45 min). The organic phase was separated, dried over sodium sulfate, and concentrated to 0.5 ml using a Vigreux column (50 cm x 1 cm) and a micro-distillation device [5]. All experiments were performed in duplicate.

Reactions performed in mesophases. The monoglyceride (8 g) in a Pyrex tube (30 ml) was heated in a silicone bath (116 °C). After 23 min, a solution of Gly (0.033 g, 0.44 mmol) or Leu (0.057 g, 0.44 mmol) in phosphate buffer (1 ml, 0.2 mol/l, pH 6.0) preheated (3 min, 116 °C), was added. After vigorous stirring (Vortex) the tube was further heated (116 °C) and stirred until a homogonous mixture was obtained (22 min). Then, a solution of Xyl (0.066 g, 0.44 mmol) in phosphate buffer (1 ml, 0.2 mol/l pH 6.0) preheated (3 min, 116 °C), was added. The mixture was vigorously stirred, heated (5 min, 116 °C), and finally cooled down to room temperature. The tube containing the reaction mixture was heated in a silicone bath (70 °C, up to 7 h). Dimodan U and Dimodan HR led to the hexagonal and lamellar phase, respectively. To obtain the cubic phase, a mixture of Dimodan HR and Dimodan U (1:1, by weight) was used.

Each reaction mixture (10 g) was transferred into a Pyrex bottle (250 ml). Diethyl ether (50 ml) and the internal standard (372 μg of cyclotene) were added and the mixture was shaken (30 min). Samples containing Dimodan HR were passed through a paper filter to separate undissolved monoglycerides. Samples containing Dimodan U were completely dissolved and thus not filtered. The resulting solution was extracted with phosphate buffer (2 x 30 ml, 0.1 mol/l, pH 8.0) and the combined extracts were centrifuged (2500 rpm, 5 min). The aqueous phase was saturated with NaCl (18 g) followed by the clean-up described above. All experiments were performed in duplicate.

Analysis of xylose. This was carried out by high performance anion exchange chromatography (HPAEC) using an electrochemical detector (ECD), as recently described [4]. An aliquot of the reaction mixture in phosphate buffer was diluted 250-times with deionised water and filtered through a PVDF filter (polyvinylidene fluoride, 0.22 μm/25 mm). The mesophase-based reaction mixture (10 g) was transferred into a Pyrex bottle (100 ml). Diethyl ether (60 ml) was added and the mixture was shaken (30 min) to break the mesophase. Water (18 ml) was added and the mixture was shaken (35 min). Samples containing Dimodan HR were centrifuged (3500 rpm, 15 min) to separate the aqueous and organic phase. Samples containing Dimodan U were completely separated without centrifugation. The water phase was diluted 25-times with deionised water, filtered through a PVDF filter, and analysed by HPAEC-ECD.

Analysis of norfuraneol. This was performed by gas chromatography-mass spectrometry (GC-MS) using a GC 6890A coupled to an MSD 5973N (Agilent, Palo Alto, USA) and equipped with a DB-Wax capillary column (J&W Scientific, Folsom, USA): 60 m x 0.25 mm, film thickness 0.25 µm. Carrier gas: helium (1.5 ml/min, constant flow). Sample injection: 1 µl, splitless, 250 °C. Oven temperature program: 2 min at 35 °C, 6 °C/min to 240 °C with a hold for 25 min. Ion source temperature: 280 °C. The electron impact (EI) MS spectra were generated at 70 eV.

3. RESULTS AND DISCUSSION

Model systems based on xylose and glycine. Various Maillard model systems were compared by heating precursors under mild reaction conditions (70 °C) in aqueous buffer solution (pH 6) and several monoglyceride-based mesophases, such as lamellar, cubic, and hexagonal phases. As shown in Figure 1, xylose was preferably decomposed in the hexagonal phase. In agreement with that, norfuraneol was accumulated in the hexagonal phase, as compared to other mesophases and the aqueous buffered sample.

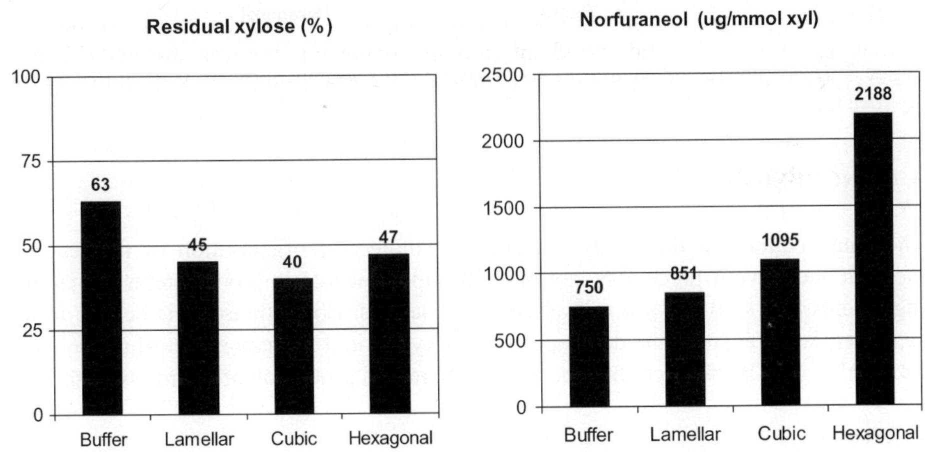

Figure 1. Decomposition of xylose in the presence of glycine at 70 °C after 7 h and concomitant formation of its degradation product norfuraneol.

Model systems based on xylose and leucine. Maillard model systems were studied based on xylose and L-leucine in various mesophases and aqueous buffer solution (pH 6) under the same mild reaction conditions (70 °C). Xylose was again preferably decomposed in the hexagonal phase (Figure 2). On the other hand, norfuraneol was readily generated in the hexagonal phase, as compared to the aqueous buffered sample.

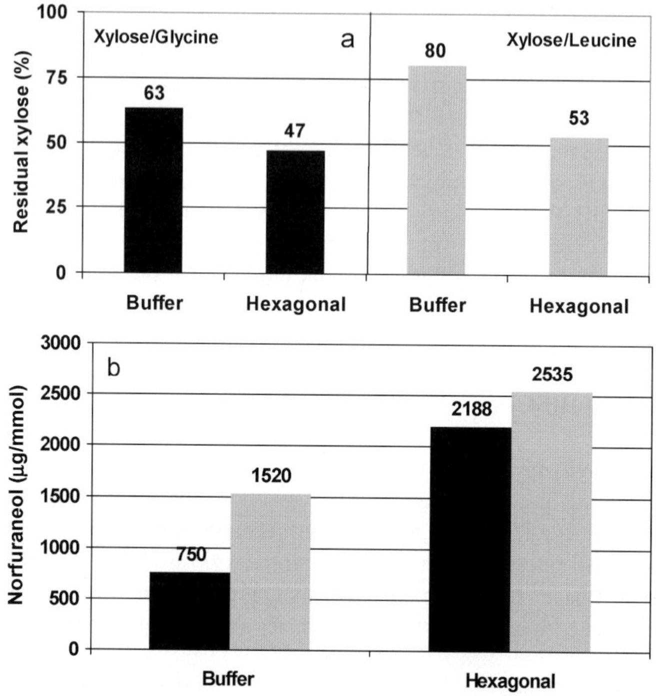

Figure 2. Decomposition of xylose (a) and formation of norfuraneol (b) in Xyl/Gly (black) and Xyl/Leu (grey).

4. CONCLUSION

The data obtained in this study indicate that molecular organisation of the reaction medium and flavour precursors can play an important role in food systems containing ingredients that tend to form self-assembly structures. Both amino acids behave in the same way with regard to the decomposition of xylose and formation of norfuraneol. The accumulation of norfuraneol may be due the increased protection of the mesophase.

References

1. N. Garti, Curr. Opin. Colloid In., 8 (2003) 197.
2. J.-L. Le Quéré and P.X. Étiévant (eds.), Flavour research at the dawn of the 21st century, proceedings of the 10th Weurman flavour research symposium, Paris, France (2003) 453.
3. S. Vauthey, Ch. Milo, P. Frossard, N. Garti, M.E. Leser and H.J. Watzke, J. Agric. Food Chem., 48 (2000) 4808.
4. T. Davidek, N. Clety, S. Devaud and I. Blank, J. Agric. Food Chem., 51 (2003) 7259.
5. D.G. Land and H.E. Nursten (eds.), Progress in flavour research, applied sciences, London, UK (1979) 79.

W.L.P. Bredie and M.A. Petersen (Editors)
Flavour Science: Recent Advances and Trends

Glycerol, another pyrazine precursor in the Maillard reaction

Christoph Cerny and Renée Guntz-Dubini

Firmenich Corporate R&D Division, Rue de la Bergère 7, CH-1217 Meyrin 2 Geneva, Switzerland

ABSTRACT

The Maillard reaction between [$^{13}C_6$]fructose and alanine was studied in glycerol/water as reaction medium using SPME-GC-MS to analyse the reaction products. Among the identified reaction products were carbonyl compounds such as 2,3-pentanedione as well as several alkylpyrazines. The isotopomer distribution of the products indicated that the ethyl group in the pyrazines stemmed largely from alanine carbons, which were also incorporated in 2,3-pentanedione. These findings are in agreement with the literature. On the other hand, a considerable amount of 2,3-pentanedione was found completely unlabelled as well as a significant percentage of carbons in the pyrazine ring. We supposed that part of the unlabelled carbons originated from glycerol used as reaction solvent. To verify this hypothesis fructose/alanine was reacted in [$^{13}C_3$]glycerol. A substantial percentage of 2,3-pentanedione (17%) and of the alkylated pyrazines (13-33%) was found [$^{13}C_3$]labelled thus confirming the integration of glycerol carbons into these molecules. Consequently glycerol not only influences the Maillard reaction by changing the physiochemical environment, but also participates as a flavour precursor.

1. INTRODUCTION

Pyrazines are important aroma compounds in a wide variety of foods and contribute roasted notes to foods like bread, cocoa, coffee, roasted meat and potato chips [1-6]. They are formed from high temperature Maillard reactions [7]. The generation of ethyl substituted pyrazines, as well as of 2,3-pentanedione, is enhanced when alanine participates in the reaction [8-9]. Labelling studies showed that two carbons from alanine are integrated into these molecules [10-11].

Glycerol is present in all natural oils, and hence in most foods, and finds its application as a solvent in the flavour industry. Glycerol influences the Maillard reaction between reducing sugars and amino acids and leads to increased browning rates, which is thought to be an effect of reduced water activity and the different physiochemical environment [12-14]. The aim of this study was to investigate the role of the solvent glycerol on the formation of volatiles formed during the Maillard reaction, in particular to determine to which extent its carbons are incorporated into the reaction products.

2. MATERIALS AND METHODS

Alanine (6.68 mg) and [$^{13}C_6$]fructose were dissolved in phosphate buffer (100 mg; 0.5 mol/l, pH 7.00), and glycerol (400 mg) was added. The solution was reacted in a closed vial at 130 °C for 2 h in a heated metal block. A similar reaction was carried out with unlabelled alanine and fructose in [$^{13}C_3$]glycerol. The samples were analysed in duplicate by headspace SPME-GC-MS using a DVB/CAR/PDMS fibre (Supelco) and 5 min adsorption time at room temperature. Desorption was at 250 °C splitless in the injector block of a GC 6890N coupled to a MSD 5973 (Agilent) equipped with a HP-5MS capillary (30 m × 0.25 mm; 0.25 μm film thickness). The oven temperature was 50 °C for 5 min, and then increased by 10 °C/min to 260 °C. Mass spectra (*m/z*) were generated in the EI mode (70 eV). The isotopomer proportions were calculated using the M^+ values (M^+-1 for 3-ethyl-2,5-dimethylpyrazine) and corrected by the naturally occurring percentages of ^{13}C and by the ratio (M^+-1)/M^+ to compensate for the loss of hydrogen observed with M^+.

3. RESULTS AND DISCUSSION

Acetaldehyde, hydroxyacetone, 2,3-pentanedione, pyrazine, 2-methylpyrazine, 2,5-dimethylpyrazine, 2-ethyl-3-methylpyrazine and 3-ethyl-2,5-dimethylpyrazine belonged to the main volatiles that were formed from the reaction of alanine with [$^{13}C_6$]fructose in glycerol/water. The corresponding mass spectra showed molecular ion signals (M^+), which were up to 8 units higher than the spectra of the unlabelled compounds. In Table 1 the compounds are listed together with the distribution of their isotopomers. Acetaldehyde (**1**) was almost completely unlabelled, and therefore the carbons do not originate from [$^{13}C_6$]fructose, but rather come from alanine, which undergoes Strecker degradation.

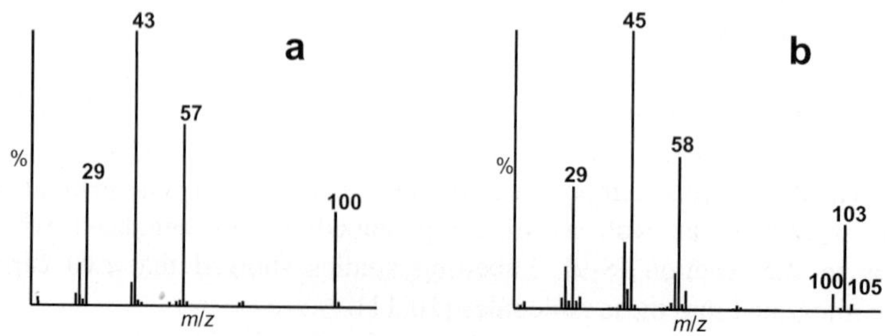

Figure 1. Mass spectra of unlabelled 2,3-pentanedione (a) and 2,3-pentanedione isotopomers (b) from the reaction of [$^{13}C_6$]fructose and alanine in glycerol/water.

Table 1. Ratio of isotopomers formed from [$^{13}C_6$]fructose and alanine in glycerol/water.

		Ratio of [13C]labelled carbons in the molecule (%)									
No.	Compound[a]	*m/z*[b]	0[c]	1	2	3	4	5	6	7	8
1	Acetaldehyde	44	98	0	2						
2	Hydroxyacetone	74	10	1	3	86					
3	2,3-Pentanedione	100	14	0	3	77	0	6			
4	Pyrazine	80	1	1	8	3	87				
5	2-Methylpyrazine	94	1	0	10	4	4	81			
6	2,5-Dimethylpyrazine	108	4	0	1	20	0	4	71		
7	2-Ethyl-3-methylpyrazine	122	1	0	11	6	3	65	14	0	
8	3-Ethyl-2,5-dimethylpyrazine	136	6	0	1	27	0	5	60	0	1

[a]Identification based on mass spectra and retention indices in the corresponding reaction between fructose and alanine. [b]Molecular ion of the unlabelled molecule. [c]Number of ^{13}C atoms.

On the other hand the majority (77%) of the 2,3-pentanedione (**3**) was triply labelled. Figure 1 shows the corresponding mass spectra (B) as well as the spectra of the unlabelled 2,3-pentanedione (A). The unlabelled fragment ion *m/z* 29 corresponding to $C_2H_5^+$ and the labelled fragments *m/z* 45 and *m/z* 58 corresponding to $^{13}CH_3$-$^{13}CO^+$ and C_2H_5-$^{13}CO^+$ suggest that C-1 to C-3 are [^{13}C]labelled and hence stem from [$^{13}C_6$]fructose. The two unlabelled C-4 and C-5 atoms are presumably originating from alanine, a presumption that is strongly supported by literature data [9,11,15]. A smaller percentage (14%) of **3** was found totally unlabelled, and consequently does not contain any carbons from fructose. It is possible that carbons from glycerol are incorporated into the molecule.

Similarly, pyrazine (**4**), 2-methylpyrazine (**5**) and 2,5-dimethylpyrazine (**6**) were found completely labelled to a level of 87, 81 and 71%, respectively, in the experiment with [$^{13}C_6$]fructose (cf. Table 1). In 8% of **4** and 10% of **5** only 2 carbons, and in 20% of **6** only 3 carbons were [^{13}C]labelled suggesting here as well an involvement of glycerol in their formation. The ethyl substituted pyrazines **7** and **8** were mainly 5-times and 6-times [^{13}C]labelled, respectively. The two ethyl carbons were supposedly being unlabelled and stemming from alanine [8-10]. A minority of **7** (20%) and **8** (28%) was only triply [^{13}C]labelled, and it would appear that, also here the carbons from a glycerol molecule were incorporated.

In order to verify this hypothesis, a second reaction with unlabelled alanine and fructose in [$^{13}C_3$]glycerol was carried out. The results in Table 2 show that 18% of **3** was triply labelled and hence unambiguously confirm that glycerol has taken part in the reaction and its carbons are recovered in **3**. Also, a significant percentage of the pyrazines contained [$^{13}C_3$]labelled isotopomers. In the alkylated pyrazines **5-8** between 13 and 27% of the molecules contained a C_3-unit from [$^{13}C_3$]glycerol. Hence, glycerol does not only influence the water activity of the reaction medium but takes an active part as a precursor for pyrazines in the reaction.

Table 2. Ratio of isotopomers formed from fructose and alanine in [$^{13}C_3$]glycerol/water.

		Ratio of [13C]labelled carbons in the molecule (%)										
No.	Compound[a]	*m/z*[b]	0[c]	1	2	3	4	5	6	7	8
1	Acetaldehyde	44	97	2	1						
2	Hydroxyacetone	74	85	4	0	11					
3	2,3-Pentanedione	100	79	2	1	18					
4	Pyrazine	80	88	1	9	0	2				
5	2-Methylpyrazine	94	77	1	7	13	0	2			
6	2,5-Dimethylpyrazine	108	65	1	1	26	0	0	7		
7	2-Ethyl-3-methylpyrazine	122	59	17	0	20	2	2	0	0	
8	3-Ethyl-2,5-dimethylpyrazine	136	62	0	3	27	0	2	6	0	0

[a]Identification based on mass spectra and retention indices in the corresponding reaction between fructose and alanine. [b]Molecular ion of the unlabelled molecule. [c]Number of ^{13}C atoms.

References

1. P. Schieberle and W. Grosch, Z. Lebensmittel Untersuch. Forsch., 185 (1987) 111.
2. P. Schnermann and P. Schieberle, J. Agric. Food Chem., 45 (1997) 867.
3. I. Blank, A. Sen and W. Grosch, Z. Lebensmittel Untersuch. Forsch., 195 (1992) 239.
4. M. Czerny, F. Mayer and W. Grosch, J. Agric. Food Chem., 47 (1999) 695.
5. C. Cerny and W. Grosch, Z. Lebensmittel Untersuch. Forsch., 194 (1992) 322.
6. R. Wagner and W. Grosch, Lebensm.-Wiss. Technol., 30 (1997) 164.
7. P.E. Koehler and P.V. Odell, J. Agric. Food Chem., 17 (1970) 393.
8. C. Cerny and W. Grosch, Z. Lebensmittel Untersuch. Forsch., 198 (1994) 210.
9. A.J. Taylor and D.S. Mottram (eds.), Flavour science: recent developments, Cambridge, UK (1996) 211.
10. M. Amrani-Hemaïmi, C. Cerny and L.B. Fay, J. Agric. Food Chem., 43 (1995) 2818.
11. V.A. Yaylayan and A. Keyhani, J. Agric. Food Chem., 47 (1999) 3280.
12. K. Eichner and M. Karel, J. Agric. Food Chem., 20 (1972) 218.
13. W.A.W. Mustapha, S.E. Hill, J.M.V. Blashard and W. Derbyshire, Food Chem., 62 (1998) 441.
14. F. Jousse, T. Jongen, W. Agterof, S. Russell and P. Braat, J. Food Sci., 67 (2002) 2534.
15. V.A. Yaylahan and A. Keyhani, J. Agric. Food Chem., 48 (2000) 2415.

W.L.P. Bredie and M.A. Petersen (Editors)
Flavour Science: Recent Advances and Trends

Influence of added carbohydrates on the aroma profile of cooked pork

Lene Lauridsen[a,b], Rikke Miklos[b], Annette Schäfer[a], Margit D. Aaslyng[a] and Wender L.P. Bredie[b]

[a]Danish Meat Research Institute (DMRI), Maglegaardsvej 2, 4000 Roskilde, Denmark; [b]Department of Food Science, The Royal Veterinary and Agricultural University (KVL), Rolighedsvej 30, 1958 Frederiksberg C, Denmark

ABSTRACT

The odour profiles of cooked pork samples with carbohydrates added at four times the estimated natural concentration were investigated. The heated pork samples were analysed by descriptive odour profiling and the aroma volatiles were measured by GC-MS. A screening of eleven different flavour precursors, which were mainly carbohydrates, showed that six precursors altered the odour compared to the reference sample without precursor addition. Glucose, glucose 6-phosphate, ribose, ribose 5-phosphate, and fructose increased the caramel and grilled odours, while lactate increased the sour odour. The remaining five precursors did not alter the sensory profile compared to the reference. GC-MS analysis of pork with added glucose, glucose 6-phosphate, and ribose, respectively, showed an increase in the formation of Maillard-derived volatiles compared to the reference. Especially samples with added ribose and glucose showed increased levels. Alkylpyrazines were the most abundant class of volatiles identified in the samples.

1. INTRODUCTION

The flavour of pork is generated primarily through the cooking process [1]. During this process Maillard reactions are perhaps the most important route to flavour formation in cooked meat [2]. Attention has been drawn to the effect of added carbohydrates on the aroma development in meat in order to both modify the aroma and to study the underlying flavour chemistry. Hudson and Loxley [3] modified the flavour of mutton meat by addition of different pentose sugars. The roasted aroma in pork and beef was significantly increased by the addition of glucose 6-phosphate and ribose at four times the natural concentration [4]. Although many important volatile compounds have been identified in meat, the role of their possible precursors and natural variations of these have received much less attention. In this study eleven carbohydrates and glycolytic intermediates were evaluated for their potential to modify pork flavour.

2. MATERIALS AND METHODS

Pork loins from seven sows were trimmed for all visible fat and grounded thoroughly. Meat samples were prepared by mixing 3 g minced pork and 250 μl of a solution containing one carbohydrate; glucose (0.6 M), ribose (0.6 M), fructose (0.6 M), glucose 1-phosphate (G1P, 0.01 M) glucose 6-phosphate (G6P, 0.3 M), glucose 1,6-bisphosphate (G1,6P, 0.01 M), ribose 5-phosphate (R5P, 0.3 M), glyceraldehyde 3-phosphate (G3P, 0.01 M), fructose 6-phosphate (F6P, 0.09 M), and fructose 1,6-bisphosphate (F1,6P, 0.01 M). A sample with added water was analysed as reference. The meat samples were equilibrated at 5 °C for 2 h (sensory analysis) and at 5 °C for 24 h (volatile analysis). Prior to analysis, samples were heated at 160 °C for 10 min in conical flask closed with a watch glass.

2.1. Sensory analysis

Descriptive odour profiling was performed using seven trained assessors. All samples were assessed just after cooking in four replicates in sensory booths under standardised conditions. Sensory responses were measured on a 15 cm unstructured intensity scale.

2.2. Aroma volatile analysis

Meat samples with added glucose, G6P, or ribose and the reference were analysed by GC-MS. Volatiles were collected on Tenax TA/Carbograph by purging with N_2 (60 ml/min) for 30 min at 50 °C. Tenax tubes were thermally desorbed on a GC-MS (Agilent Technologies, Palo Alto, USA) using a HP-Innowax column (30 m x 0.25 mm i.d. x 0.25 μm film thickness, Agilent Technologies, Palo Alto, USA). Mass spectra of the identified compounds agreed with spectra from those in reference libraries.

3. RESULTS AND DISCUSSION

3.1. Sensory analysis

The data from descriptive odour profiling were summarised by principal components analysis (PCA). The samples could be separated in three groups, where lactate constituted the first group, G3P; F1,6P; G1P; G1,6P; F6P, and the reference composed the second group, and finally R5P, G6P, fructose, glucose, and ribose constituted the third group (Figure 1). The samples with added lactate were described as very sour and unpleasant. The samples in the second group were characterised as being similar to the reference. The samples in the third group were described as more 'caramel' and 'grilled' than the reference. Figure 1 shows that the attribute 'roasted' correlated inversely with the attributes 'caramel' and 'grilled'. This attribute related more to the cooked meaty aspect and may be masked by the 'grilled' smell. Analysis of variance of the profiling data showed that fructose, glucose, and ribose had a significantly ($p<0.05$) higher score on the 'grilled' attribute than the reference, and G6P, glucose, and ribose had a significantly ($p<0.05$) higher score on the 'caramel' attribute than the reference.

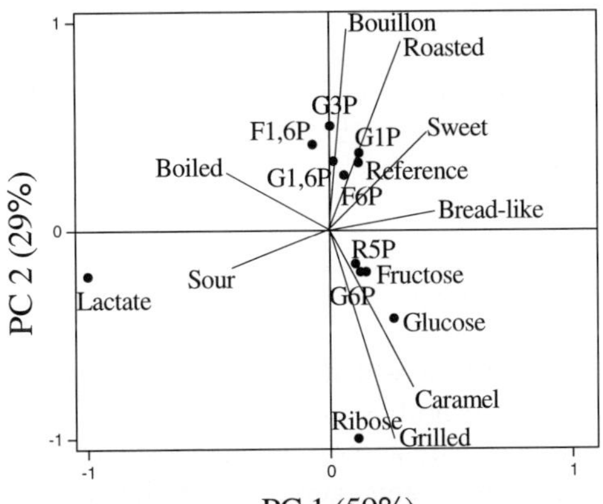

Figure 1. PCA bi-plot of the odour profiles of cooked pork (for sample codes see section 3).

3.2. Aroma volatiles

The influence of addition of glucose, G6P, and ribose on the volatile composition was investigated using GC-MS. Formation of Maillard reaction products with 'caramel' and/or 'grilled' aroma is shown in Figure 2. Pyrazines were formed in relatively high amounts and were present in large numbers compared to the other Maillard products.

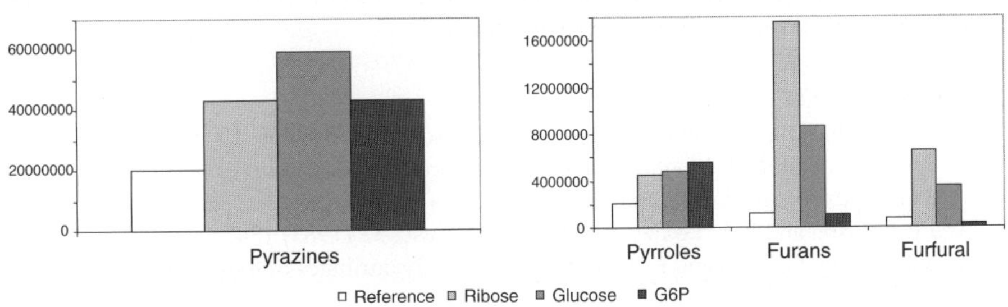

Figure 2. GC-MS peak areas (average of duplicate analysis) of Maillard-derived products.

Some variation in the level of the pyrazines in the replicates was observed, but irrespective of this variation, the ribose and glucose samples produced greater numbers of these compounds than the reference sample (Table 1). Pyrazines have previously been identified in heat-treated pork meat [5-7].

Table 1. Pyrazines identified in pork samples with added carbohydrates.

Pyrazine	Reference	Ribose	Glucose	Glucose-6P
Pyrazine	-	+	-	-
Methylpyrazine	+	+	+	+
2,3-Dimethylpyrazine	-	+	+	-
2,5-Dimethylpyrazine	+	+	+	+
2,6-Dimethylpyrazine	-	+	+	+
Trimethylpyrazine	+	+	+	+
Ethylpyrazine	-	+	+	-
2-Ethyl-3-methylpyrazine	+	+	+	+
2-Ethyl-5-methylpyrazine	+	+	+	+
2-Ethyl-6-methylpyrazine	+	+	+	+

+ Positive identification; - Not detected.

The sensory analysis clearly showed that lactate, ribose, and glucose had the greatest influence on the pork odour profile compared to the other carbohydrate flavour precursors. The sensory observations corroborated with the results from the volatile analysis. Pyrazines were the most abundant class of Maillard-derived compounds. The addition of ribose and glucose resulted in consistently higher levels of Maillard-derived volatiles compared to the reference. The addition of glucose or ribose increased the formation of all identified Maillard-derived compounds compared to the reference.

4. CONCLUSIONS

Sensory evaluation of cooked pork with added glucose, G6P, ribose, R5P, or fructose gave increased 'caramel' and 'grilled' odour notes. Since the levels of all added precursors were approximately 4 times the values reported in the literature it seems that glucose, fructose, ribose, G6P and R5P have an important contribution to modifying the odour properties of the cooked pork. The natural possible variations of these flavour precursors in pork require further investigations.

References

1. F. Shahidi (ed.), Flavour of meat products and seafoods, London, UK (1998) 61.
2. T.H. Parliment, M.J. Morello and R.J. McGorrin (eds.), Microwave, extruded and Maillard generated aromas, Washington, DC, USA (1994) 104.
3. J.E. Hudson and R.A. Loxley, Food Technol. Aust., 35 (1983) 174.
4. Y.L. Xiong, C.T. Ho and F. Shahidi (eds.), Quality attributes of muscle foods, New York, USA (1999) 159.
5. C.J. Mussian and J.P. Walradt, J. Agric. Food Chem., 22 (1974) 827.
6. D.S. Mottram, J. Sci. Food Agric., 36 (1985) 377.
7. M.L. Timón, A.L. Carrapiso, A. Jurado and J. van de Lagemaat, J. Sci. Food Agric., 84 (2004) 825.

W.L.P. Bredie and M.A. Petersen (Editors)
Flavour Science: Recent Advances and Trends

Facts and 'artefacts' in the flavour chemistry of onions

Michael Granvogl and Peter Schieberle

Technical University of Munich, Chair for Food Chemistry, 85748 Garching, Germany

ABSTRACT

Sulfur compounds play an important role in the flavour of onions. However, due to the degradation of precursors and the occurrence of transient, unstable flavour compounds it is quite difficult to define a specific type of onion odour, because a tremendous number of reactions may occur depending on the processing parameters applied. The importance of the use of appropriate analytical methods will be emphasised with respect to artefact formation, e.g. excessive generation of volatiles from precursors during the work-up procedure. The exact structure elucidation for 3,5-diethyl-1,2,4-trithiolane is shown.

1. INTRODUCTION

It is well accepted that important onion odorants are formed during cutting by a very fast enzymatic degradation of cysteine sulfoxides followed by a cascade of chemical reactions [1,2]. Thereby, significant changes in the spectrum of volatiles may occur during processing, e.g. cooking [3] or deep-frying and, additionally, during the analytical procedure. Secondary, numerous compounds have only been tentatively identified up to now in differently processed onions [3,4] only based on their mass spectra, thus producing contradictory results.

2. MATERIAL AND METHODS

In the present study, different isolation procedures and analytical techniques (solvent assisted flavour evaporation (SAFE), simultaneous steam distillation/extraction (SDE) and stir bar sorptive extraction (SBSE (Twister®)) were applied on the same batch of onions to find out differences in the amounts of certain aroma compounds, e.g. 3-mercapto-2-methylpentan-1-ol [5].

3. RESULTS

3.1. Analytical methods and artefact formation

Spanish onions were cooked as entire bulb in tap water (150 ml), using a pressure cooker (20 min) on a kitchen stove. The volatiles were isolated by SAFE distillation and the most odour-active areas were detected by aroma extract dilution analysis (AEDA). AEDA revealed 17 aroma compounds, which have been identified recently [6]. In the odour-active area 13 the onion-like, meat broth-like odorant 3-mercapto-2-methyl-pentan-1-ol (3-MMP) was identified [5]. Using 3-MMP as an example, the influence of different analytical methods on the quantitative data is specified in the following. For quantification, the isotopically labelled [$^2H_2$]-3-MMP [5] was used in stable isotope dilution assays (SIDA) applied to extracts prepared by different work-up procedures. The concentrations of 3-MMP in Spanish onions varied extremely depending on the extraction technique (Table 1).

Table 1. Concentrations of 3-MMP in Spanish onions depending on the extraction technique.

Analytical method	Concentration 3-MMP (µg/kg)
SAFE (solvent assisted flavour evaporation)	8.0 (5.4 – 10.8)
SDE (simultaneous steam distillation/ extraction according to Likens/Nickerson [7])	421.1 (406.6 – 435.6)
SBSE (stir bar sorptive extraction; Twister®)	1990 (1950 – 2030)

The higher amounts of 3-MMP analysed by SDE and SBSE in comparison to the SAFE extract have to be regarded as artefacts due to the thermal treatment in the sample preparation. The application of increasing temperatures during thermal desorption in the SBSE analyses confirmed this assumption (Table 2). Even the use of a low temperature (120 °C) resulted in an approximately fifty-fold higher concentration of 3-MMP compared to the SAFE method (Tables 1 and 2).

Table 2. Influence of the thermal desorption temperature on the amount of 3-MMP analysed by stir bar sorptive extraction (SBSE).

Temperature (°C)	Concentration 3-MMP (µg/kg)
120	383.4 (355.3 – 411.5)
160	1262 (1227 – 1297)
200	1990 (1950 – 2030)

3.2. The challenge of structure elucidation in the chemistry of onions

To illustrate this issue, in particular for compounds bearing chiral centres, the identification procedure used for the odour-active areas 12 and 13 is discussed in detail.

The same mass spectrum (Figure 1) was obtained in both areas. On the basis of the molecular ion, the following structures could be proposed: (*E*)-1-propenylpropyl-disulfide, (*Z*)-1-propenylpropyldisulfide, 4,5-diethyl-1,2,3-trithiolanes and 3,5-diethyl-1,2,4-trithiolanes.
The structure of 4,5-diethyl-1,2,3-trithiolane is not in agreement with the fragments obtained (Figure 1). In addition, the acyclic structure could be excluded by the NMR data due to the missing signals for the olefinic protons (Figure 2).
Therefore, 3,5-diethyl-1,2,4-trithiolane is the only structure which agreed with all analytical data measured. The analysis by means of an achiral GC-column yielded two peaks showing an identical mass spectrum, whereas the use of a chiral GC-column resulted in three peaks with identical mass spectra. Combining all analytical data, three different isomers of 3,5-diethyl-1,2,4-trithiolane were confirmed (Figure 3). Additionally, the analytical data of the three isomers corroborated with the structural data from the synthesised compounds [6,8].

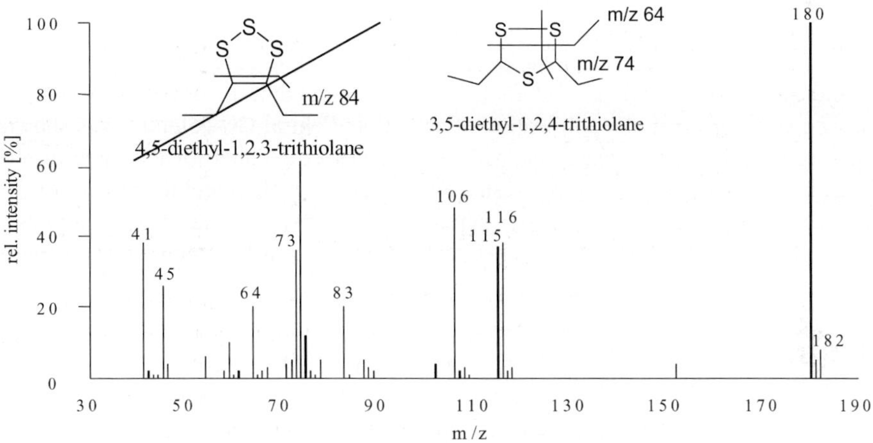

Figure 1. Mass spectrum obtained for odour-active areas 12 and 13.

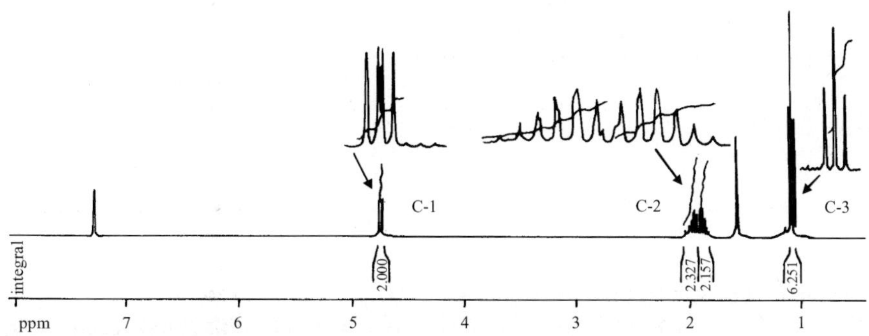

Figure 2. NMR spectrum of 3,5-diethyl-1,2,4-trithiolane.

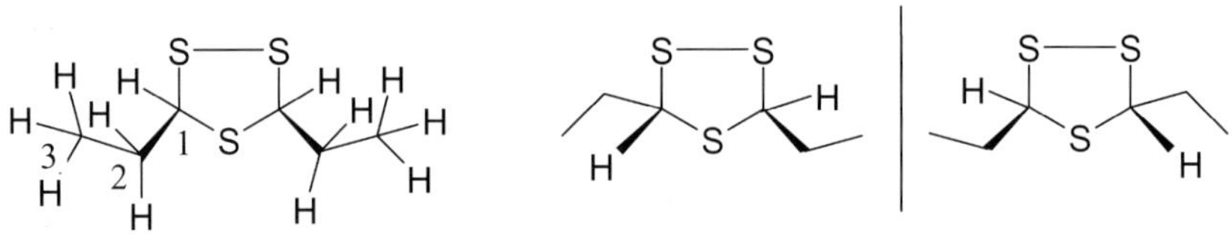

Figure 4. Different structures of the isomeric 3,5-diethyl-1,2,4-trithiolanes.

4. CONCLUSION

Depending on the method used for volatile isolation significant differences in the amounts of 3-mercapto-2-methylpentan-1-ol in the same batch of onions could be demonstrated [5]. The thermal influence was impressively shown by varying the temperature during thermal desorption in SBSE analyses. This artefact formation of 3-MMP would provoke misleading conclusions concerning its importance to the overall aroma. These results clearly emphasise the need to select an adequate analytical method to avoid excessive formation of odorants from precursors during work-up procedures.
In a second experiment, the application of achiral and chiral GC-columns was shown to be a very useful tool to identify isomeric substances as detailed for the three different 3,5-diethyl-1,2,4-trithiolanes. This result emphasises that the combination of different analytical techniques followed by syntheses of reference substances allows an unequivocal identification of the chemical nature of aroma compounds in flavour extracts.

References

1. E. Block, Angew. Chemie, 104 (1992) 1158.
2. J.F. Carson, Food Rev. Int., 3 (1987) 71.
3. Y. Tokitomo, Nippon Shokuhin Kagaku Kogaku Kaishi, 42 (1995) 279.
4. M.C. Kuo and C.T. Ho, J. Agric. Food Chem., 40 (1992) 111.
5. M. Granvogl, M. Christlbauer and P. Schieberle, J. Agric. Food Chem., 52 (2004) 2797.
6. M. Granvogl and P. Schieberle, J. Agric. Food Chem., submitted.
7. S.T. Likens and G.B. Nickerson, Am. Soc. Brew. Chem. Proc., (1964) 5.
8. G. Morel, Synthesis, (1980) 918.

Relationship between acrylamide formation and the generation of flavour in heated foods

Mei Yin Low, Donald S. Mottram and J. Stephen Elmore

The University of Reading, School of Food Biosciences, Whiteknights, Reading, RG6 6AP, United Kingdom

ABSTRACT

Strecker aldehydes and alkylpyrazines are two important classes of flavour volatiles formed together with acrylamide via the Maillard reaction in heated carbohydrate foods. The relative rates of formation of these volatiles were compared to acrylamide, using a potato model system. It was observed that the Strecker aldehydes were formed more quickly than acrylamide but consumed at a slower rate, whilst the alkylpyrazines were formed at a similar rate to acrylamide. Both Strecker aldehyde and acrylamide levels decreased on prolonged heating whilst alkylpyrazine levels continued to increase. Overall, it was found that the potato model system was able to provide consistent data amenable to kinetic modelling.

1. INTRODUCTION

Acrylamide is a neurotoxin and potential carcinogen found at relatively high levels in carbohydrate-rich foods processed at temperatures exceeding 120 °C. Researchers have highlighted the involvement of the Maillard reaction in acrylamide formation, with asparagine being the amino acid precursor for acrylamide [1,2]. The Maillard reaction is a key pathway for flavour generation and the non-enzymatic browning of heated foods. This would imply a strong correlation between acrylamide formation and flavour generation. The identification of rate determining steps or control points in the reaction, through the use of kinetic modelling, may allow desirable flavour and colour attributes to be retained whilst minimising the level of acrylamide.

This paper reports on the effect of heating time on volatile flavour formation and acrylamide in a potato model system. The objective was also to providing kinetic data for incorporating flavour formation into an existing kinetic model for acrylamide formation [3]. The aim of the research is to develop an optimal strategy for limiting acrylamide formation in food without compromising sensory quality.

2. MATERIALS AND METHODS

Model potato cakes were prepared from potato flake (McCain Foods (GB) Ltd, Scarborough, UK) and deionised water, in the ratio 1:1.3 w/w, using the method described by Elmore *et al.* [4]. These were baked at 180 °C in a band oven between 10 and 60 min and then milled to produce a fine powder for further analyses.

Flavour volatiles were collected by dynamic headspace using Tenax and by Simultaneous steam Distillation and solvent Extraction (SDE). The dynamic headspace method of Madruga and Mottram [5] was followed. The samples (7 g) were mixed with 93 g of deionised water and extracted at 37 °C for 1 h. Volatiles were collected on a glass-lined, stainless steel trap (105 mm × 3 mm i.d.) containing 85 mg of Tenax TA (Scientific Glass Engineering Ltd., Milton Keynes, U.K.). An internal standard (100 ng of 1,2-dichlorobenzene in 1 μl methanol) was added to the trap. For SDE, the method of Mottram *et al.* [6] was used with a 25 g of sample blended in 1 l of deionised water. 1,2-Dichlorobenzene (100 μg/ml in *n*-hexane) was used as internal standard.

GC-MS analyses were carried out on an HP 5972 mass spectrometer, coupled to an HP5890 gas chromatograph and a G1034C Chemstation, using a Zebron ZB-Wax column (60 m x 0.25 mm i.d., 0.25 μm film thickness; Phenomenex, UK). For the headspace samples, a CHIS injection port (Scientific Glass Engineering Ltd., UK), held at 250 °C, was used to thermally desorb the adsorbed volatiles from the Tenax trap onto the front of the column for 5 min, with a column retention gap inserted in solid CO_2. For the SDE extracts, direct injection of a 1 μl volume in the splitless mode (opening after 1 min) at 250 °C was used. The temperature program employed was: 5 min at 40 °C, a ramp of 4 °C/min to 250 °C, then 10 min at 250 °C. The helium carrier gas flow was 1.0 ml/min. The following conditions were used for the mass spectrometer: electron impact mode with source temperature of 165-175 °C, ionising voltage of 70 eV and scan range from *m/z* 29-400 at 2.05 scans/s. Linear calibration curves were prepared using the standard addition and internal standard methods for headspace concentration and SDE, respectively.

3. RESULTS AND DISCUSSION

3.1. Variation of Strecker aldehydes with cooking time

Figure 1a shows the formation of Strecker aldehydes with quantities expressed in micromoles per kilogram of potato on a dry weight basis (μmol/kg potato) to facilitate a direct comparison between these components. The data were also compared with the quantities of acrylamide produced from a similar potato model system in our laboratory [4]. It can be seen that the Strecker aldehydes were found in quantities close to or less than acrylamide. These aldehydes are reactive and hence are found at levels reflective of reaction intermediates rather than end-products.

The scale of Figure 1b has been normalised to that of percentage of the maximum amount formed for each component to allow comparison of the relative rates of formation of Strecker aldehydes and acrylamide with cooking time. All the Strecker aldehydes - except for benzaldehyde - appeared to be formed at fairly similar rates and peak at similar times, although their subsequent rates of degradation appeared to be different. Benzaldehyde, on the other hand, continued to rise with cooking time and this suggests that its formation mechanism is likely to be different and that it is also probably less reactive than the other Strecker aldehydes. It is apparent that the Strecker aldehydes are formed more quickly than acrylamide, although they appear to be consumed at a slower rate. After cooking for 15 min, levels of the Strecker aldehydes were at their maximum whilst the level of acrylamide was only about 35% of the maximum. Acrylamide reached its maximum level later at 30 min.

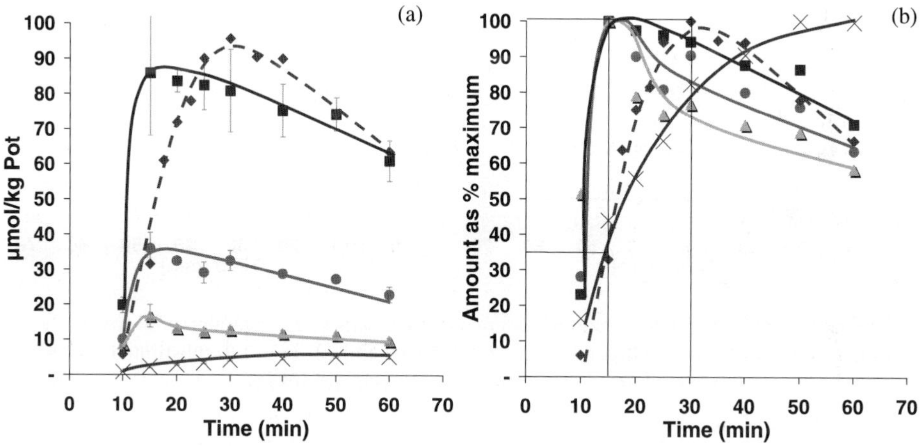

Figure 1. Trend for Strecker aldehydes with cooking time: (a) as net amount and (b) as % of maximum. 2-Methylbutanal (■); 3-methylbutanal (▲); benzaldehyde (×); phenylacetaldehyde (●); acrylamide (◆).

3.2. Variation of alkylpyrazines with cooking time

Figure 2a compares the quantities of alkylpyrazines formed, categorised by degree of substitution (i.e. total number of carbon atoms in the ring substituents). Both un-substituted pyrazine and the 4 carbons-substituted pyrazines were formed at levels similar to acrylamide. Concentrations of methylpyrazine and the 2 carbons-substituted pyrazines (i.e. dimethylpyrazine and ethylpyrazine) were found to be up to 3 to 4 times that of acrylamide, whilst the pyrazines substituted with 3 carbons (i.e. ethylmethylpyrazine, propylpyrazine, trimethylpyrazine) were approximately twice as much. The 5 carbons- and 6 carbons-substituted pyrazines were present at much lower levels than acrylamide and the other pyrazines. These observations point to α-amino ketones with 2, 3 or 4 carbon units being the most abundant precursors of alkylpyrazines in the potato model system.

It can be seen from Figure 2b that the alkylpyrazines and acrylamide appear to have been formed at similar relative rates, although the level of acrylamide peaked at 30 min while pyrazine levels continued to rise, albeit more slowly. Pyrazines are stable Maillard reaction products and therefore do not decrease with continued heating. The various alkylpyrazines were produced at similar relative rates, indicating that their kinetics of formation may be similar.

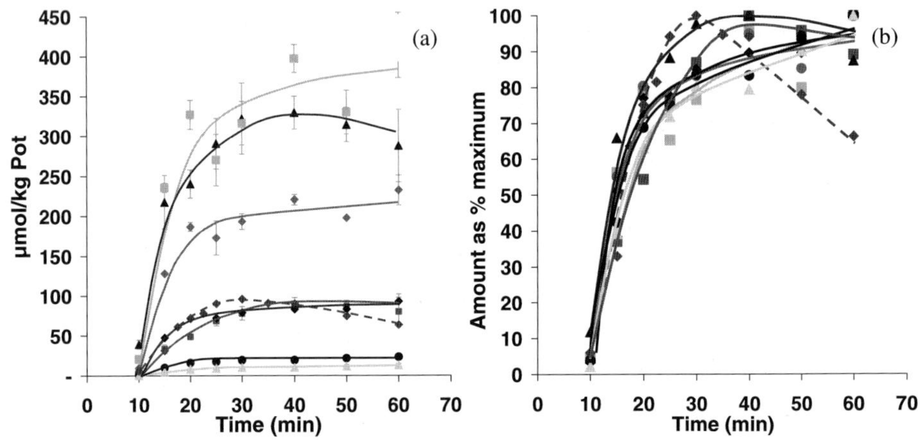

Figure 2. Formation of alkylpyrazines during cooking: (a) as net amount and (b) as % of maximum. Pyrazine (■); methylpyrazine (▲); 2C-substituted (■); 3C-substituted (●); 4C-substituted (◆); 5C-substituted (●); 6C-substituted (▲); acrylamide (◆).

4. CONCLUSIONS

The use of a potato model system with headspace concentration and SDE allowed the formation of Strecker aldehydes and alkylpyrazines with cooking time to be compared with acrylamide generation under the same conditions. A consistent kinetic data set was obtained and will be used to include flavour formation into an existing multiresponse model for acrylamide formation [3].

References

1. D.S. Mottram, B.L. Wedzicha and A.T. Dodson, Nature, 419 (2002) 448.
2. R.H. Stadler, I. Blank, N. Varga, F. Robert, J. Hau, P.A. Guy, M.C. Robert and S. Riediker, Nature, 419 (2002) 449.
3. M. Friedman and D.S. Mottram (eds.), Chemistry and safety of acrylamide in food, New York, USA (2005) 235.
4. J.S. Elmore, G. Koutsidis, A.T. Dodson, D.S. Mottram and B.L. Wedzicha, J. Agric. Food Chem., 53 (2005) 1286.
5. M.S. Madruga and D.S. Mottram, J. Sci. Food Agric., 68 (1995) 305.
6. D.S. Mottram, C. SzaumanSzumski and A. Dodson, J. Agric. Food Chem., 44 (1996) 2349.

W.L.P. Bredie and M.A. Petersen (Editors)
Flavour Science: Recent Advances and Trends

Modelling the formation of Maillard reaction intermediates for the generation of flavour

Guillaume Desclaux[a], Tahir I. Malik[b], Chris Winkel[c], D. Leo Pyle[a] and Donald S. Mottram[a]

[a]The University of Reading, School of Food Biosciences, Whiteknights, Reading, RG6 6AP, United Kingdom; [b]ICI Strategic Technology Group, Wilton Centre, Redcar, TS10 4RF, United Kingdom; [c]Quest Foods, PO Box 2, 1400 CA Bussum, The Netherlands

ABSTRACT

The Maillard reaction comprises a very complex, thermally-driven, set of reactions that leads to the formation of colour and flavour. In producing flavours, it is important to be able to quantitatively follow the course of this reaction system and thence control it by manipulating key input parameters, such as temperature, pH, type and concentration of the sugar and amino acid. To that end, the main intermediates of the reaction network have been identified and quantified so the data can be used in kinetic modelling of the reaction. Modelling should provide mechanistic insights as well as a quantitative tool to follow the progress of the complex Maillard reaction. A model developed for the glycine/xylose system is briefly reported and discussed.

1. INTRODUCTION

It is well known that the Maillard reaction involves the condensation of an amino group with a carbonyl group, usually on a reducing sugar. Although the final products of the Maillard reactions have been characterised in many papers, relatively little is known about the relative importance of the intermediates involved [1].

The reaction conditions represent a key issue in the Maillard reaction, since they influence both the nature and rate of formation of the intermediates and products. In this study, key Maillard intermediates were quantified as a function of time for selected model systems under different conditions of pH and temperature in order to get a better understanding of the reaction mechanisms and kinetics. In contrast to many earlier studies on the Maillard reaction, we here use ‘multiresponse modelling’, which utilises several state measurements simultaneously and gives insight into the wider mechanism [2]. The focus of this research was to collect data that could be used to develop and test a kinetic model of the Maillard reaction for selected amino-acid sugar systems.

2. MATERIAL AND METHODS

2.1. Analytical methods

Samples (25 g) were prepared using xylose (6 mmol), glycine (6 mmol) and buffer (0.5 M phosphate at pH 6). Reaction was carried out in stirred sealed steel tubes placed in thermostatically controlled heating block. Batch experiments were performed at pH6 and temperatures of 80, 100, and 120 °C. Measurements were taken at 30, 60, 90 and 120 min. The time to rise to the final temperature was typically 5 min.
Dicarbonyls were analysed after derivatisation with 1,2-diaminobenzene; hydroxycarbonyls were analysed after derivatisation with ethoxyamine-HCl; deoxyosones were derivatised with 1,2-diaminobenzene prior to silylation with BSTFA (bis(trimethylsilyl)-trifluoroacetamide) and heterocyclic compounds were directly analysed [3]. All analyses were carried out by solvent extraction followed by GC or GC-MS analysis.

2.2. Modelling approach

2.2.1. Proposed routes to Maillard intermediates and kinetic model

This work focused on intermediates of the Maillard reaction that are key intermediates in the routes to the formation of flavour compounds. After condensation of the sugar and the amino acid, the Amadori Rearrangement Product (ARP) which is formed breaks down to give the deoxyosones. The 1- and 3-deoxyosones (Done-1 and Done-3) are considered to be key intermediates in the subsequent network of reactions, and are the main precursors for most of the sugar-derived intermediates [1]. They act as precursors for dicarbonyl compounds such as glyoxal, methylglyoxal, diacetyl and pentanedione [4,5], but can also lead to hydroxycarbonyl compounds including hydroxyacetone and glycolaldehyde and to heterocyclic compounds such as 2-furfural and 4-hydroxy-5-methyl-3(2*H*)-furanone (nor-HDF) [6,7].
The characterisation and the quantification of individual intermediates in the Maillard reaction have been studied by a number of researchers, but only a few have reported detailed and realistic mechanisms for the formation of all the possible intermediates. To build a realistic kinetic model, we need to be confident in the routes of formation of all the species analysed. An evaluation of the published routes has led us to propose detailed mechanisms for each of the intermediates used in the model (detailed mechanisms to be published separately). The reaction network derived from this mechanistic study, which forms the basis for the kinetic model, is shown in Figure 1.

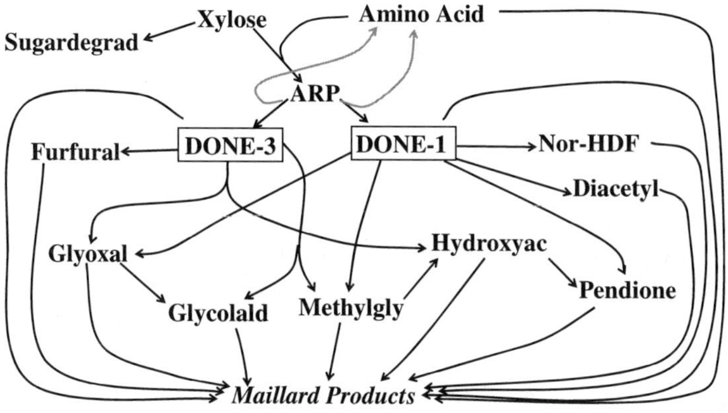

Figure 1. Proposed kinetic model pathways showing key flavour intermediates.

2.2.2. Kinetic model

The model consists of an ODE (Ordinary Differential Equation) system where the number of differential equations is determined by the number of chemical species present. Apart from the first stage reaction of sugar with amino acid to ARP, all reactions were assumed to follow first order kinetics with Arrhenius temperature dependence. In the work reported here, modelling software Athena Visual Studio (Stewart & Associates Engineering Software, Madison, USA) was used for modelling and parameter estimation; either selected or complete data sets can be used in an iterative process of model development and parameter estimation. Parameter estimation is based on non-linear least squares minimisation. The end result is a set of kinetic parameters in the form of rate constants, activation energies, pre-exponential factors or other factors.

As with any parameter estimation problem, an initial set of parameter estimates is needed. Here, values for the kinetic parameters associated with amino acid and sugar reactions were estimated from their concentrations. This provided a starting point for the development of the model. Refinement was carried out by using the parameter estimates from one round of calculations as the initial parameter set for a further iteration until satisfactory convergence and agreement with the data was achieved.

3. RESULTS AND DISCUSSION

As an example, we show the model fits obtained for a system at 100 °C and pH 6 using the model presented in Figure 1. The quality of the fits can be assessed by comparing the observed and predicted values as shown in Figure 2.

First, parameters were estimated for the initial reaction step using a simple version of the model (only including sugar, amino acid and ARP). The full model was then run, using the rate constants calculated from the simple model results as initial values, and all the reaction parameters were estimated simultaneously. The results given in Figure 2 show the fit of the model for selected species. The model fitted well the formation of hydroxyacetone and nor-HDF. 3-deoxyosone was formed rapidly in the initial reaction and then suffered a net loss over the main course of the reaction. This was also predicted well but and the prediction for the loss of xylose could be improved. The model gave rate constant of 2.79×10^{-02}/min for the formation of 1-deoxyosone from ARP and 5.50×10^{-04}/min for the formation of 3-deoxyosone, meaning that 1-deoxyosone is formed faster at pH 6 (i.e. 1-done is favoured compared to 3-done).

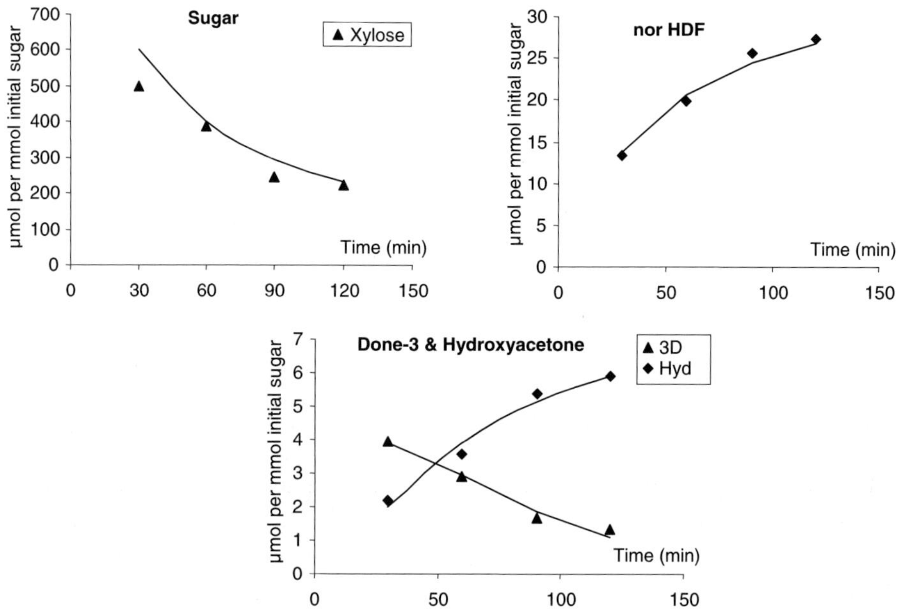

Figure 2. Fit of the model (lines) to the experimental data (points) for Xylose-Glycine pH6 at 100 °C showing xylose (sugar), nor-HDF, and 3-deoxyosone (3D) and hydroxyacetone (Hyd).

4. CONCLUSION

These results show the promise of the modelling approach to follow the formation of key flavour intermediates. We can now be confident in the proposed steps for the kinetic model which will be developed further. A subsequent paper will report the kinetic interpretation and the mechanistic insights obtained from the model.

References

1. A.J. Taylor (ed.), Food flavour technology, Sheffield, UK (2002) 27.
2. M. van Boekel, Nahrung, 45 (2001) 150.
3. T. Hofmann, Eur. Food Res. Technol., 209 (1999) 113.
4. T. Hofmann, Carbohydr. Res. 313 (1998) 215.
5. V.A. Yaylayan and A. Keyhani, J. Agric. Food Chem., 48 (2000) 2415.
6. T. Hofmann and P. Schieberle, J. Agric. Food Chem., 46 (1998) 235.
7. D.S. Mottram and I.C.C. Nobrega, J. Agric. Food Chem., 50 (2002) 4080.

W.L.P. Bredie and M.A. Petersen (Editors)
Flavour Science: Recent Advances and Trends

The effect of fatty acid precursors on the volatile flavour of pork

Annette Schäfer and Margit D. Aaslyng

Danish Meat Research Institute, Maglegaardsvej 2, DK- 4000 Roskilde, Denmark

ABSTRACT

Six fatty acids (C18:1(9), C18:2(9,12), C18:3(9,12,15), C18:3(6,12,15), C20:4(5,8,11,14) and C22:6(4,7,10,13,16,19)) were added separately to minced pork, which was subsequently heat-treated. The odour of the samples was assessed by a sensory panel. Especially the addition of the omega-3-fatty acids, C18:3(9,12,15) and C22:6(4,7,10,13,16,19), resulted in a fish-like odour, whereas the addition of C18:2(9,12) resulted in an oily odour. Three meat samples with fatty acids, C18:2(9,12), C18:3(9,12,15) or 22:6(4,7,10,13,16,19), were further analysed by gas chromatography–olfactometry–mass spectrometry, GCO-MS, using 8 assessors. The differences in odour active compounds between the three fatty acid additions were small.

1. INTRODUCTION

Raw meat has a very weak flavour. During heat-treatment, flavour compounds develop and provide the characteristic flavour of cooked meat. The flavour of fried pork is a result of the combination of compounds from lipid degradation, Maillard reactions and the interaction between these and other minor components. It is claimed that compounds from Maillard reactions are mainly responsible for the general roasted flavour, whereas the lipid degradation compounds provide the species' characteristics [1]. For that reason, the fatty acid composition has an influence on the total flavour of fried pork. The fat content of the feed can alter the fatty acid composition in the animals. Practical feeding experiments with pigs fed on various fat sources have shown that the flavour of pork meat is only slightly altered even though a high variation in the fatty acid composition has been attained (unpublished data). To understand the effect of the individual fatty acids varying in the feeding experiments, a model experiment was performed in which the pure fatty acids were added separately to minced meat.

2. MATERIAL AND METHODS

The experiment was divided into two phases. First, a sensory evaluation was conducted to compare two levels of selected fatty acids of pork. Second, three fatty acids were chosen from this study for the aroma analysis using a combination of GCO-MS.

2.1. Sensory study

A fatty acid was added to 3 g minced pork with a known fatty acid composition. Additions were either at 40 times or at 100 times the natural level present in the minced pork. For those fatty acids present in very low amounts, a theoretical amount of 0.2% was used. The meat used for the experiments had a fat percentage of approximately 1.5%. The fatty acids used were C18:1(9), C18:2(9,12), C18:3(9,12,15), C18:3(6,12,15), C20:4(5,8,11,14) and C22:6(4,7,10,13,16,19), which were present in the fat of the minced pork with a percentage of 45, 10, 2, 0.2, 0.2 and 0.2, respectively. A reference sample of meat without addition of fatty acid was also assessed.

The odour profile of the meat samples was evaluated by 5 trained assessors, all of whom where familiar with sensory assessment of pork. The following odour attributes were assessed on a 15 cm unstructured line scale from very weak to very intense: 'fried meat' (the odour of the surface of pork fried on a pan), 'sweet' (like sweet candy), 'caramel' (sweet brown), 'burnt' (like meat fried to more than well done), 'nut' (roasted hazelnuts), 'oily' (like deep frying oil), 'fried fish' (like the skin of a fried salmon), 'cod-liver oil', 'sourish' (fresh sour odour as in buttermilk)

2.2. Analysis of odour-active volatiles

From the results of the sensory study, GCO-MS was performed on a reference meat sample without addition of fatty acid and on meat samples with one of three fatty acids, each at two levels, respectively 40 and 100 times the level in the meat. The fatty acids were C18:3(9,12,15), C18:2 and C22:6. Addition of fatty acids and heat-treatment were performed as described in the sensory experiment.

Volatiles were swept onto Tenax TA. The analyses were performed on a Hewlett Packard GC-MS 6890N/5973. Thermal desorption was performed on a Perkin Elmer ATD400. The capillary column was a DB5MS (60 m, 0.25 mm ID, 0.25 μm film thickness) from J&W. The GC oven program was 40 °C for 5 min, 10 °C/min to 100 °C, 2 °C/min to 130 °C, 25 °C/min to 310 °C for 2 min. Compounds were identified using NIST/EPA/NIH Mass Spectral Database or by comparison to authentic standards. The GC was equipped with two ODP2 sniffing ports from GERSTEL.

Detection frequency (DF) method using 8 assessors for each sample was performed. The assessors were trained in the sniffing technique on artificial standards. The assessor had to place the detected odours from the samples into one of nine odour categories. The categories comprised 'chemical/synthetic' (CS), 'earthy/mushroom' (EM), 'fresh/green/nature' (FGN), 'nut/roasted/fried' (NRF), 'baked' (B), 'sourish/sweet/fruit' (SSF), 'sulfur/onion/cabbage' (SOC), 'nauseating/rancid/sour' (NRS) and 'animalic/faecal' (AF).

3. RESULTS

3.1. Sensory study

All samples to which a fatty acid had been added had less fried meat flavour than the reference (meat without additional fatty acid), independent of the level of the fatty acids (Table 1). The two fatty acids C18:3(9,12,15) and C22:6 were more intensive in 'fried fish' and 'cod liver' odours and less intensive in 'fried meat', 'sweet' and 'roasted nut' odours.

Table 1. Odour profiles of fried pork after adding 40 times or 100 times the natural level of the fatty acid. The reference sample was fried pork (meat without additional fatty acid).

	C18:3 (6,9,12)		C18:3 (9,12,15)		C18:1		C18:2		C20:4		C22:6		
Level of fatty acid	40	100	40	100	40	100	40	100	40	100	40	100	Ref
Fried meat	4.9	4.8	2.2	1.7	3.9	3.7	4.9	4.8	5.2	4.8	3.8	3.4	6.3
Roasted nut	2.9	2.8	2.0	1.7	3.5	3.3	2.9	3.0	2.9	2.8	2.3	2.1	3.3
Caramel	3.4	3.3	2.6	2.5	2.9	2.9	2.9	2.7	3.0	2.9	3.0	3.0	3.5
Burnt	4.1	4.0	3.4	3.5	3,3	3.1	3.6	3.2	3.7	3.7	4.1	4.0	3.8
Sweet	3.6	3.5	2.2	1.9	3.9	3.7	3.6	3.7	3.9	3.9	2.6	2.6	4.2
Sourish	1.6	1.7	1.7	1.7	2.1	2.1	1.3	1.4	1.6	1.6	1.6	1.7	1.7
Oily	5.2	5.4	4.7	4.7	5.1	5.2	5.9	6.2	5.1	5.4	4.5	4.4	4.4
Fried fish	0.5	0.5	4.7	5.4	0.4	0.4	0.3	0.2	0.3	0.5	3.1	3.6	0.1
Cod liver oil	0.7	0.7	4.6	5.2	1.1	1.1	0.7	0.8	0.3	0.4	2.6	2.9	0.2

There appeared to be an interaction between the fatty acids and the added level ($p=0.08$), as 'fried fish' odour only increased from 40% to 100% for the fatty acids C18:3(9,12,15) and C22:6. The intensity of 'fried meat' odour decreased as the level of added fatty acid increased for all the acids ($p<0.01$). There was a tendency towards a decrease in odour intensity with increasing addition of fatty acid for other attributes associated with fried meat such as 'burnt' ($p=0.1$) and 'caramel' ($p=0.09$), while the 'cod liver oil' ($p=0.07$) tended to be perceived more intensely.

3.2. Volatile compounds

All the aromagrams were characterised by many odour-active regions with a high detection frequency. Large quantities of 2- and 3-methylbutanal were present in all samples. At Kovats Index, KI, 830, two peaks were identified as hexanal and dihydro-2-methyl-3(2*H*)-furanone and categorised as FGN and SOC/NRF respectively. At KI 970, two peaks appeared with 100% DF in the aromagram. Dimethyltrisulfide was identified as one of the two peaks and was characterised by the category SOC by the majority of the assessors. The other peak was characterised as EM and was 1-octen-3-one. At KI 990, 2-pentylfuran was detected by the assessors. At KI 1222, the assessors agreed very much on the choice of SOC, and dimethyltetrasulfide was detected. At KI 1396, with a

DF between 85 and 100% for all three fatty acids and the reference the majority of choices of category were SSF and NRF. No compound was identified. Addition of fatty acid to the meat resulted in the presence of volatiles from oxidation or degradation of fatty acid compared with the reference. Table 2 shows the compounds identified only in the specific fatty acid sample.

Table 2. Characteristic volatile compounds from lipid degradation identified by GC-MS in fried pork with a fatty acid added before frying.

Fatty acid	Volatile compounds
C22:6	1-Penten-3-ol; 2,4-octadiene; 3-methyl-1,4-heptadiene; 2-hexenal; 1,3-(*E*)-5-(*Z*)-octatriene; 4-heptenal
C18:3	2-ethylfuran; 2-pentenal; 2-penten-1-ol; 1,3-(*E*)-5-(*Z*)-octatriene; nonane; 2-pentylfuran; 2,4-heptadienal
C18:2	pentanal; 1-pentanol; 2-hexanone; octene; 1-hexanol; 2-heptanone; methional; 2-heptenal; 2-pentylfuran; 1-octen-3-one; 1-octen-3-ol; 2,5-octanedione; 2-octenal; 3-nonen-2-one; 2-nonenal

4. DISCUSION AND CONCLUSION

Relatively high levels of several different fatty acids were added to minced pork meat, and the effect on the odour after heat-treatment was evaluated. The amount of fatty acid added to the meat had to simulate the relative amount in the actual meat, and the experiment also had to be reproducible for the small presence of, for example, C22:6. No clear differences were seen between the two levels of fatty acids, indicating that the addition of 40 times the level was enough to ensure that the fatty acids were not the limiting step in the aroma development. In practice, the variation in the content of fatty acids of pork will be smaller. The omega-3-fatty acids had the greatest effect on odour. Especially the 'fried fish' and the 'cod liver' odour were high. C18:2 gave a stronger 'oily' odour to the meat. The other fatty acids did not much alter the odour notes.
The heterocyclic compounds were only present in very small amounts, which implies that lipids or their degradation products may inhibit the formation of some heterocyclic compounds generated from Maillard reactions [2]. The assessment of the samples by GCO-MS showed many odour-active regions in the chromatograms, some of which could be linked to the identifications made. However, the variation between the aromagrams for samples from different fatty acids was rather small, even though significant differences in odours were detected. This discrepancy may be due to the efficacy of the extraction method employed. A too long sampling time resulted in overwhelming odour intensities that precluded odour prioritisation.

References

1. D.S. Mottram, R.A. Edwards and H.J.H. MacFie, J. Sci. Food Agric., 33 (1982) 934.
2. D.S. Mottram and J.E. Edwards, J. Sci. Food Agric., 34 (1983) 517.

The role of lipid in the flavour of cooked beef

J. Stephen Elmore and Donald S. Mottram

School of Food Biosciences, The University of Reading, Whiteknights, Reading, RG6 6AP, United Kingdom

ABSTRACT

The main reactions occurring during the cooking of meat are the Maillard reaction and the degradation of lipid. In particular the reaction between cysteine and ribose is a likely source of compounds with potent meat-like aromas. Compounds formed from lipid oxidation may have important aroma properties but they can also interact with products of the Maillard reaction to give further compounds. This paper reports investigations of model systems that examine the effects of linoleic acid (18:2 n–6) and α-linolenic acid (18:3 n–3) on the reaction between cysteine and ribose. These demonstrate how the breakdown products of polyunsaturated fatty acids can modify the reaction between cysteine and ribose and affect the levels of meat-like aroma compounds after cooking.

1. FLAVOUR FORMATION IN MEAT

When meat is cooked, flavour develops as a result of two main reactions. The Maillard reaction between amino acids and reducing sugars is responsible for the typical meaty flavour, and the thermal degradation of lipid provides the flavours characteristic of different species [1]. Of particular importance in the generation of meaty aromas is the reaction between cysteine and ribose, which has been widely studied. When cysteine is heated in the presence of reducing sugars, it breaks down to acetaldehyde, hydrogen sulfide and ammonia, which can react with carbonyls produced from sugar breakdown, to give a wide range of sulfur-containing heterocyclic volatile compounds, in particular thiophenes, thiazoles and sulfur-containing furans [2,3]. Many of these compounds have been reported in cooked meat and some, particularly furans and thiophenes substituted in the 3-position with sulfur, possess very low odour thresholds [1].

Compounds formed from decomposition of lipid during cooking include aldehydes, alcohols, ketones and alkylfurans. The degree of decomposition depends upon the degree of saturation of the fat. The structural lipids of the muscle, the phospholipids, are relatively unsaturated, compared to adipose tissue and intramuscular triglycerides, and are hence important contributors to the aroma of cooked meat. Lipid-derived volatiles are found at relatively high levels compared with Maillard-derived volatiles in lightly cooked meat, whereas the converse is true in pressure-cooked and well-done meat [4-6]. Many compounds, formed from the interaction of lipid with the Maillard reaction, have been identified in meat, particularly in the pressure-cooked meat of animals that have been fed supplements high in unsaturated fats [3,5,7,8] and include 2-alkylthiazoles, 2-alkyl-3-thiazolines, 2-alkylthiophenes and 2-alkyl-(2*H*)-thiapyrans.

2. EFFECT OF DIET ON BEEF FLAVOUR

Cattle diets are almost always grain-based (wheat, barley, soya) or forage-based (grass, silage, hay). Grain-based, high energy diets, contain relatively high amounts of linoleic acid (C18:2 n–6), whereas forage-based diets are low energy diets, relatively high in α-linolenic acid (C18:3 n–3). Hence animals fed on grain, when compared with grass-fed animals of the same age, are usually heavier with a higher percentage of body fat [9].
Grass- and grain-fed cattle of similar weight, reared on diets containing sugar beet pulp added to grass silage or a restricted barley/soya diet [10], were slaughtered at 14 and 24 months. Muscle lipids showed large differences due to diet, whereas age effects were small [11]. In particular, all n–3 fatty acids were higher in the muscle of the grass-fed animals, while all n–6 fatty acids were higher in the muscle of the grain-fed animals. C18:2 n–6 was over two times higher in grain-fed animals, whereas C18:3 n–3 was around 5 times higher in grass-fed animals.
Compounds derived from n–6 fatty acids, including hexanal, 1-hexanol, 1-octen-3-ol, (*Z*)-2-octen-1-ol and 2-pentylfuran, were higher in grilled steaks from grain-fed animals, whereas the n–3 derived compounds 1-penten-3-ol, (*Z*)-2-penten-1-ol and 2-ethylfuran were higher in the steaks from grass-fed animals. Although the meat from the grass-fed cattle contained the more highly unsaturated fatty acids, the quantity of lipid-derived volatiles was greater in the beef from the grain-fed animals. The higher levels of the antioxidant vitamin E in the diet of grass-fed animals, relative to grain-fed animals, have been related to improved colour retention and reduced lipid oxidation in the meat of grass-fed animals [12]. It is possible that vitamin E may also have an important role in the formation of lipid-derived flavour compounds in cooked beef.

3. MODELLING MEAT FLAVOUR FORMATION

The volatile composition of pressure-cooked meat is quite different from that of grilled meat, although many compounds, particularly lipid-derived compounds, such as straight-chain aldehydes, ketones and alcohols are common to both [6]. Pressure-cooked beef from animals fed fish oil supplements, high in C20:5 n–3 and C22:6 n–3, contained high levels of n–3 derived volatiles, such as 2-ethylfuran and 2-(2-pentenyl)furan [5], along with compounds possibly formed from the reaction of n–3 fatty acids with Maillard reaction intermediates, such as 2-ethylthiophene. Furthermore, some n–6 and n–9 derived compounds were also present at high amounts, suggesting that the fish oil n–3 acids had catalysed the breakdown of the more saturated fatty acids found in beef.

Simple reaction mixtures containing cysteine, ribose and a fatty acid (C18:2 n–6 or C18:3 n–3) were used to model flavour formation in cooked meat [3]. The reaction mixtures simulated pressure-cooking in that they were heated in a closed system at 140 °C, with no loss of water or volatile flavour precursors, such as ammonia and hydrogen sulfide. By omitting or varying the fatty acid and by heating the fatty acid on its own, the contribution of different fatty acids to meat flavour could be determined (Figure 1).

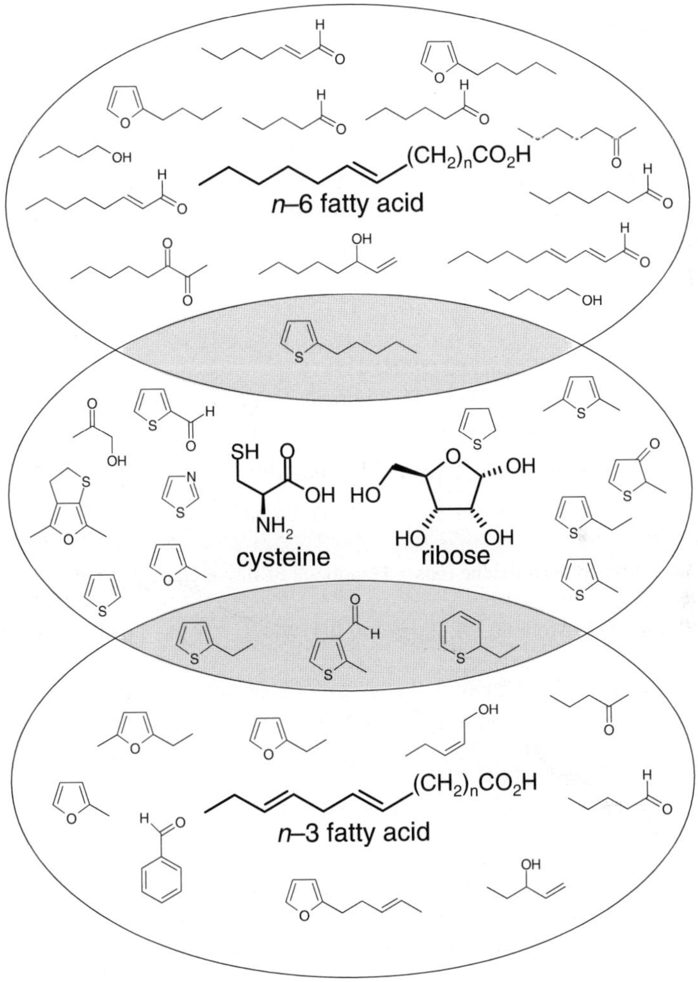

Figure 1. Volatile compounds formed in the reaction between cysteine and ribose and the decomposition of n–3 and n–6 fatty acids that have been identified in cooked beef. Compounds formed from the reaction of lipid with cysteine and ribose breakdown products are shown in grey.

Many of the compounds formed in large amounts when cysteine was heated with ribose, were absent or only formed in trace amounts when a fatty acid was present. Conversely, compounds formed from the breakdown of heated C18:2 n–6 did not change substantially in the presence of cysteine and ribose. C18:3 n–3 decomposed far more readily than 18:2 n–6 on heating, to yield compounds, which readily reacted with cysteine and ribose to give further compounds. In particular, mono- and di-unsaturated aldehydes were found at very high levels in heated C18:3 n–3, but reacted with cysteine and ribose to give 2-alkyl-3-formylthiophenes and 2-alkyl-(2*H*)-thiapyrans respectively.

4. CONCLUSIONS

Modifying the lipid composition of cattle diets will alter the lipid composition of their meat, which may affect its flavour. Lipids high in n–3 fatty acids will break down readily on heating, to give compounds that are more unsaturated and hence more reactive than compounds formed from the breakdown of the equivalent n–6 fatty acids. Such compounds could affect flavour by reacting with Maillard precursors and intermediates, as well as contributing to flavour in their own right. Furthermore, they appear to catalyse the breakdown of more saturated lipids, which may also affect beef flavour.

References

1. D.S. Mottram, Food Chem., 62 (1998) 415.
2. L.J. Farmer, D.S. Mottram and F.B. Whitfield, J. Sci. Food Agric., 49 (1989) 347.
3. J.S. Elmore, M.M. Campo, D.S. Mottram and M. Enser, J. Agric. Food Chem., 50 (2002) 1126.
4. D.S. Mottram, J. Sci. Food Agric., 36 (1985) 377.
5. J.S. Elmore, D.S. Mottram, M. Enser and J.D. Wood, J. Agric. Food Chem., 47 (1999) 1619.
6. K.D. Deibler and J. Delwiche (eds.), Handbook of flavor characterization, New York, USA (2004) 295.
7. J.S. Elmore, D.S. Mottram, M. Enser and J.D. Wood, J. Agric. Food Chem., 45 (1997) 3603.
8. J.S. Elmore and D.S. Mottram, J. Agric. Food Chem., 48 (2000) 2420.
9. P.D. Muir, J.M. Deaker and M.D Bown, NZ J. Agric. Res., 41 (1998) 623.
10. J.S. Elmore, H.E. Warren, D.S. Mottram, N.D. Scollan, M. Enser, R.I. Richardson and J.D. Wood, Meat Sci., 68 (2004) 27.
11. H. Weenen and F. Shahidi (eds.), Flavor and texture of lipid-containing foods, Washington DC, USA, in press.
12. J.D. Wood, R.I. Richardson, G.R. Nute, A.V. Fisher, M.M. Campo, E. Kasapidou, P.R. Sheard and M. Enser, Meat Sci., 66 (2003) 21.

W.L.P. Bredie and M.A. Petersen (Editors)
Flavour Science: Recent Advances and Trends

Carotenoids as flavour precursors in coffee

Andreas Degenhardt[a], Martin Preininger[b] and Frank Ullrich[a]

[a]Kraft Foods R&D Inc., Bayerwaldstr. 8, D-81737 Munich, Germany;
[b]Kraft Foods, 801 Waukegan Rd., Glenview, IL 60025, USA

ABSTRACT

Carotenoids and carotenoid-derived products were identified in green coffee beans of different origins. After extraction with acetone, the carotenoids, lutein and zeaxanthin, were identified by HPLC-Diode Array Detection. Glycosidically bound ionols were identified as carotenoid-derived products which are known to be formed by oxidative degradation of lutein and zeaxanthin [1]. These glycosides were analysed by absorption on XAD-1180 resin followed by enzymatic hydrolysis of the eluted extract with β-glucosidase, and by GC-MS analysis of the released aglycones [2].
The impact of carotenoids on aroma formation was investigated by using a model roasting system. For this purpose, green coffee beans were depleted of their naturally occurring potential aroma precursors, and subsequently enriched with β-carotene, a carotenoid similar in structure to lutein and zeaxanthin. Under coffee roasting conditions, β-carotene yielded the aroma compound, β-ionone, as its major degradation product [3]. Furthermore, the importance of 3-oxo-α-ionol as an aroma precursor was investigated in the model system.

1. INTRODUCTION

Carotenoids is a class of natural compounds which has been extensively studied due to their nutritional importance. Carotenoids are also valued as pigments for use in food products. Some powerful aroma compounds are derived from carotenoids and have been thoroughly investigated in tea, wine and fruits [3]. β-Damascenone is among the very few carotenoid-derived potent aroma compounds in coffee [4-5], and has very low flavour thresholds in the low ppb to ppt range. Carotenoids have been found in green coffee beans [4,6], and their relevance as precursors for β-damascenone has been mentioned in the literature. However, no clear identification and quantification of carotenoids in green coffee has been reported so far. Therefore, we quantified carotenoids and conducted model studies to evaluate their relevance as flavour precursors in coffee. This study on glycosidically bound flavour precursor compounds adds to the knowledge on glycoconjugated progenitors in coffee [7].

2. MATERIALS AND METHODS

2.1. Analysis of carotenoids

Ground green coffee (50 g) was extracted with acetone (2 × 200 ml; 30 min each). The extract was filtered and evaporated nearly to dryness. The carotenoids were extracted into diethyl ether (50 ml). The extract was saponified with methanolic potassium hydroxide (30 g/l; 10 ml) at room temperature for 10 min under constant shaking. The solution was washed with brine (10% NaCl) until pH 7 was reached. The ether was removed *in vacuo*, the residue taken up in acetone, and completed to 5 ml. The solution was filtered through hydrophobic filters (Sartorius Minisart SRP25, 0.45 μm), and 10 μl injected into an HPLC/Diode Array Detector system (Agilent 1100 Series). Chromatography was performed on an RP18 column (Luna; 5 μm mesh; 250 mm × 4.6 mm ID) with the eluents, acetonitrile/water (9/1, v/v; solvent A) and ethyl acetate (solvent B). A linear elution gradient (0-60% B in 16 min, 60-100% B in 16-36 min) was applied at 1 ml/min flow rate. Photometric detection was performed at 440 nm.

2.2. Analysis of carotenoid degradation products

Green coffee beans (about 200 g) were deactivated with ethanol (100 ml) and water (120 ml) for 24 h at room temperature. After evaporation of the solvents, drying and grinding, the residue (100 g) was defatted with n-hexane (300 ml).The defatted beans were extracted with methanol (3 × 250 ml) at room temperature. The methanol was evaporated, the residue taken up in water (150 ml), and cleaned up on a small XAD-1180 column (40 g resin). The column was washed with water (500 ml), eluted with methanol (about 500 ml), and the eluate evaporated to dryness (yield about 800 mg). An aliquot of the residue (400 mg) was dissolved in phosphate buffer (pH 5.0, 50 ml) and extracted with pentane/ether (1/1, v/v; 3 x 40 ml) to remove free volatile aroma compounds. After evaporation of residual organic solvent, the aqueous solution was hydrolysed with emulsin (almond β-glucosidase; 75 mg, about 7 units) at 37 °C for 72 h. After addition of 100 μL n-tridecane in ether (0.01m) as internal standard, the liberated aglycones were extracted with pentane/ether (1/1, v/v; 3 × 40 ml). The organic extract was washed with hydrochloric acid (0.1N; 2 × 40 ml), concentrated on a small Kuderna-Danish column to about 0.5 ml, and 1 μl analysed by GC-FID (HP 5890 Series II) and GC-MS (Agilent 5973, Gerstel CIS 3 injector) using a DB-WAX capillary (Agilent; 60 m x 0.32 mm ID x 0.25 μm FD). The GC oven was ramped from 40 °C to 240 °C at a 5 °C/min heating rate.

2.3. Model roasting trials

In order to remove naturally occurring potential aroma precursors, green Robusta coffee beans were extracted with water at 80 °C. The extraction was considered complete when a coffee brew from the extracted, dried, and roasted beans almost lost its typical coffee flavour, and instead tasted cereal/popcorn-like. These depleted beans were infused with β-carotene and 3-oxo-α-ionol, respectively. For that purpose, β-carotene or 3-oxo-α-ionol (10 mg each) were dissolved in ethanol (about 5 ml), diluted with water (100 ml), and infused into the depleted beans (150 g) at 80 °C for 2 h. The infused beans were dried and roasted (Neuhaus Neotec Signum roaster; 260 °C, 180 s).

2.4. Isolation of aroma compounds from model roasting trials

The volatile compounds were extracted from roast and ground (R&G) coffee (100 g) for 2 h with pentane/ether (1/1, v/v; 50 ml) using a conventional Likens-Nickerson apparatus for simultaneous-distillation-extraction (SDE). The extract was concentrated to about 2 ml on a Kuderna-Danish column, and dried over anhydrous sodium sulfate. The dried concentrate was analysed under similar GC conditions as described in section 2.2.

3. RESULTS

3.1. Identification of carotenoids in green coffee

The carotenoids, lutein and zeaxanthin, were identified in green coffee beans by HPLC-DAD analysis with photometric detection at visible wavelength. An external standard method for quantitation of carotenoids in green coffee was developed, involving alkaline saponification of bound forms. Green Robusta coffee beans contained a higher carotenoid level (sum of lutein and zeaxanthin) of 1.5 mg/kg versus Kenyan, Brazilian and Colombian Arabica with 0.2 to 0.7 mg/kg.

3.2. Identification of carotenoid-derived flavour precursors

Carotenoids are labile compounds. By oxidative degradation, primary cleavage products are formed which undergo subsequent enzymatic modification [3]. These are usually glycosidically bound, non-volatile compounds, and can be cleaved by β-glucosidase to release the respective aglycones. The aglycones can be potent aroma compounds or their volatile precursors [11]. Two glycosidically bound ionols (3-oxo-α-ionol and 3-oxo-7,8-dihydro-α-ionol) were identified in green coffee after enzymatic release from the isolated glycosidic mixture [2]. The presence of 3-oxo-α-ionol in green coffee was verified with a synthesised reference compound [8], whereas 3-oxo-7,8-dihydro-α-ionol was tentatively identified by its retention index and mass spectrum. These two ionols are known lutein- and zeaxanthin-derived aroma precursors in many foods [3,10].

3.3. Model roasting trials

The role of carotenoids as potential precursors of coffee flavour compounds was investigated in model roasting trials. Green coffee beans were depleted of their flavour precursors by extraction with hot water and were then used as a reaction matrix for added aroma precursors. β-Carotene, which has not been detected in green coffee, was added as a model compound in these roasting trials. The aroma volatiles were extracted using a Likens-Nickerson-type apparatus. The known carotenoid degradation product, β-ionone, was identified in the enriched and roasted beans by GC-MS with the help of an authentic reference compound. This finding confirms that carotenoids are degraded to aroma compounds during coffee roasting, and that carotenoids are a source of potential aroma for roasted coffee. Furthermore, synthesised 3-oxo-α-ionol [8] was infused into the depleted beans, and the bean matrix roasted. Four newly formed megastigmatrienone isomers were identified in the roasted coffee by GC-MS after SDE, utilising literature mass spectra as references [9]. A generation pathway for the megastigmatrienone isomers via 3-oxo-α-ionol has been proposed [12].

4. DISCUSSION AND CONCLUSION

The results from the present model studies prompted us to search for the occurrence of megastigmatrienones in roasted coffees. However, in R&G coffee no megastigmatrienones were detected, although they were found in green coffee. Therefore, megastigmatrienones are potentially newly identified green coffee constituents. Since glycoconjugates are labile products, which undergo cleavage and rearrangement reactions in their free forms, the content of megastigmatrienones might be altered by the sample preparation step [13]. The aroma relevance of megastigmatrienones, and their role as precursors for other aroma compounds, needs to be clarified, requiring further model reactions as well as GC-olfactometry studies. The presented data will help increasing the knowledge of flavour precursors in coffee, and elucidating further generation pathways of aroma compounds via glycosides.

References

1. R. Teranishi, G.R. Takeoka and M. Güntert (eds.), Flavour precursors - thermal and enzymatic conversions, ACS symposium series 490, Washington, DC, USA (1992) 75.
2. P. Winterhalter and R.L. Rouseff (eds.), Carotenoid-derived aroma compounds, ACS symposium series 802, Washington, DC, USA (2002) 20.
3. P. Winterhalter and R.L. Rouseff (eds.), Carotenoid-derived aroma compounds, ACS symposium series 802, Washington, DC, USA (2002) 1.
4. T.H. Parliment, M.J. Morello and R.J. McGorrin (eds.), Thermally generated flavours - Maillard, microwave, and extrusion processes, ACS symposium series 543, Washington, DC, USA (1994) 206.
5. I. Blank, A. Sen and W. Grosch, Z. Lebensmittel Untersuch. Forsch., 195 (1992) 239.
6. C. Yeretzian, A. Jordan, R. Badoud and W. Lindinger, Eur. Food Res. Technol., 214 (2002) 92.
7. B. Weckerle, G. Tóth and P. Schreier, Eur. Food Res. Technol., 216 (2003) 6.
8. A.J. Aasen, B. Kimland and C.R. Enzell, Acta Chem. Scand., 27 (1973) 2107.
9. A.J. Aasen, B. Kimland, S.-O. Almqvist and C.R. Enzell, Acta Chem. Scand., 26 (1972) 2573.
10. C.R. Strauss, B. Wilson and P.J. Williams, Phytochem., 26 (1987) 1995.
11. C. Enzell, Pure Appl. Chem., 57 (1985) 693.
12. P. Winterhalter and R.L. Rouseff (eds.), Carotenoid-derived aroma compounds, ACS Symposium Series 802, Washington, DC, USA (2002) 131.
13. P. Winterhalter and P. Schreier, J. Agric. Food Chem., 36 (1988) 1251.

7

Retention and release

W.L.P. Bredie and M.A. Petersen (Editors)
Flavour Science: Recent Advances and Trends

In vivo flavour release from dairy products: relationships between aroma and taste release, temporal perception, oral and matrix parameters

Christian Salles[a], Van Anh Phan[a], Claude Yven[b], Claire Chabanet[a], Jean-Michel Reparet[a], Jean-Luc Le Quéré[a], Samuel Lubbers[a], Nicolas Decourcelle[a] and Elisabeth Guichard[a]

[a]*Unité Mixte de Recherches INRA-ENESAD Flaveur, Vision, Comportement du consommateur, 17 rue Sully, BP86510, F-21065 Dijon Cédex, France;* [b]*Station Qualité des Produits Animaux, INRA, F-63122 St Genès Champanelle, France*

ABSTRACT

Relating flavour (aroma and taste) compounds to flavour perception experienced by a consumer is still a challenge for the flavour research. The aim of the present study was to determine to what extent the food matrix texture influences the *in vivo* release of aroma and taste compounds. Sodium release and salty perception were evaluated using four cheese analogues varying in texture, taking into account oral parameters (saliva and mastication). Aroma compound release was measured from three strawberry-flavoured yoghurts varying in viscosity. Measurement of sodium released from the cheese analogues during chewing was carried out by HPLC after swabbing saliva samples at defined chewing times, whereas salty intensity and oral parameters were evaluated simultaneously at the sampling times. The release of aroma compounds from the flavoured yoghurts was measured by SPME under shear conditions. Cheese analogues made with high shear intensity released sodium more easily in mouth. A significant release was correlated with saliva flow, chewing efficiency and electromyography (EMG) parameters. Interactions between shear intensity and fat concentration were strongly correlated to EMG parameters. For the flavoured yoghurts, the aromatic balance between strawberry note and floral note was controlled by the composition in thickeners. Thus optimal compositions for fruit preparation (starch, pectin, locus bean gum) were obtained taking into account aroma release and rheological parameters.

1. INTRODUCTION

Flavour release is an important issue in food science that has been extensively studied the last years. The systems studied were, in most cases, limited to simple poly-saccharidic or proteic matrices flavoured with one or a very limited number of compounds. In mouth, flavour release is known to be affected by several factors. During chewing and ingesting food, it is dependent on such factors as composition and texture of food, saliva and mastication parameters, and nature of the flavour compounds.
Major interindividual differences were observed for volatile and non-volatile compounds released during chewing a model cheese, but for each assessor release profiles did not differ from one compound to another [1-3]. However texture may influence flavour release, perception and oral parameters. For instance, salt release is influenced by the texture of Cheddar cheese [4] and aroma compound release is influenced by matrix structure [5].
The aim of the present study was to determine to what extent the texture of a complex food matrix influences the *in vivo* release of taste and aroma compounds. The study was conducted in two parts: For the study of taste compounds, salted model cheeses made from analogue cheese technology were used and oral parameters were taken into account. For the aroma release study, yoghurts were flavoured with strawberry aroma.

2. MATERIALS AND METHODS

2.1. Preparation and characterisation of model cheeses

Four model cheeses, called MC1, MC2, MC3 and MC4, made with anhydrous milk fat (Cormans, Belgium) and casein rennet (Eurial Poitouraine, France), were used to simulate complex model foods. The model cheeses contained proportions of volatile compounds (aroma) and non-volatile compounds (minerals, amino and organic acids) in the same concentration as described in [1]. The casein rennet and salt concentrations were 283 and 31.2 g/kg, respectively. MC1 and MC2 contained 197.2 g/kg fat (40%) while MC3 and MC4 contained 297.2 g/kg fat (50%). Aqueous flavoured solutions and water were added to complete the mixture to 1 kg. All the ingredients were incorporated in a mixer (Stephan type) and mixed at 85 °C under a stream of vapour for 6 min. The shear intensity was 2000 rpm for MC1 and MC3, and 700 rpm for MC2 and MC4. Model cheeses were kept at 4 °C in sealed food quality plastic bags. Their consume-by dates were 3 months. Rheological properties of each model cheese were determined by uniaxial compression tests using a TA-XT2 analyser (Rheo, France).

2.2. Saliva sample swabbing and sodium analysis

Saliva samples were collected from 17 panellists chewing 5 g model cheese portions and treated as described in [6]. Two training sessions were made to familiarise the panellists with the procedure. Three sessions were conducted for swabbing and EMG. For each model cheese, each saliva swab was made in triplicate. Two chewing sequences were conducted for each repetition (first sequence: 0, 20, 40, 60, 90 s and second sequence: 10, 30, 50, 70, 120 s). Collected saliva was extracted and then analysed by HPLC [2].

2.3. Oral parameter measurements

EMG, salivary flow and masticatory performances [1] were measured on each panellist and for each model cheese in triplicate using a sieve with a mesh size of 5 mm.

2.4. Data analyses

StatBox Pro software (Grimmer, Paris, France) was used for ANOVA and multiple comparison tests (Newman-Keuls test, 5%) to analyse differences in model cheeses according to their rheological properties. R software (Version 1.9.1, Foundation for Statistical Computing, Vienna, Austria) was used for MANOVA (Roy test) to test the product effects on sodium release and oral parameters. Partial Least Square regression (PLS) was used to relate model cheese texture, oral parameters and sodium release during chewing.

2.5. Yoghurt study methodology

Fat substitutes (thickeners) were introduced via the fruit preparation in the flavoured fat-free yoghurts (15 aroma compounds) to restore the changes related to the absence of fat. An experimental response surface design of Doelhert with 3 factors, 3 levels (starch, pectin, locus bean gum) was used. Thus, 13 yoghurts were analysed. The headspace composition was assessed by SPME under shear conditions in a closed vessel [7]. The flow behaviour index (***n***) and the consistency factor (***K***) were assessed using a coaxial cylinder viscometer (RM180® DIN 2:2, Rheometric Scientific, France) at 21 °C. Data 'eta64' was the apparent viscosity η measured at 10 s, at 64 s^{-1} shear rate. A sensory trained panel was asked to score odour and flavour intensity of defined attributes.

3. RESULTS AND DISCUSSION

3.1. *In vivo* sodium release from model cheeses

An analysis of variance showed a significant effect (P=0.001) for all the EMG parameters. In particular, chewing time was well correlated with the number of chews (correlation coefficient: 0.88) and mean voltage, maximum voltage and total work were correlated (correlation coefficient > 0.84).

Each model cheese could be significantly differentiated from the others for at least one of the rheological parameters (Newmann Keuls test). Whatever the shear intensity (SI) applied, the modulus of deformability increased with the fat content (FC). The increase of SI, whatever FC, leads to a decrease of fracture stress and fracture work.

For temporal flavour release of sodium as for temporal perception of saltiness, important individual differences were observed, agreeing with [2]. For statistical treatments, SI and FC and the interactions between these two factors were studied. Considering the averaged temporal release curves, the Cmax (maximum released sodium concentration) measured for model cheeses with lower FC was the highest. Moreover, for a same FC, a higher SI led to a higher Cmax.

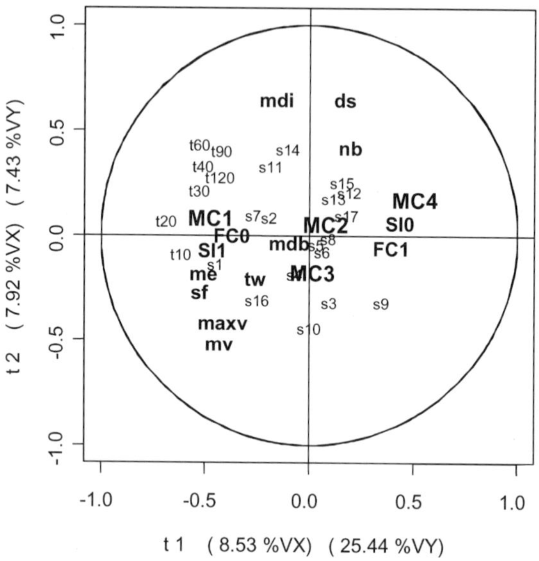

Figure 1. Partial Least Square graph: relationships between sodium release (tx) and the subjects (sx), products (MC1 to MC4) and oral parameters. SI: shear intensity (0: low, 1: high); FC: fat content (0: low, 1: high); nb: number of burst (chew); maxv: maximum voltage; mv: mean voltage; tw: total work; ds: duration of mastication sequence; mdi: mean interburst duration; mdb: mean burst duration; sf: salivary flow; me: masticatory efficiency.

The effect of the products on EMG parameters, studied by multiple analysis of variance, showed significant effects of FC and interactions FC*SI. An analysis of variance for each EMG parameter showed in particular a strong significant effect (0.1%) for the mean voltage, maximal voltage and total work parameters, and a significant effect (5%) of the interaction FC*SI for the number of chews. The effect of FC is positive if SI is high while there is no effect of FC if SI is low.

A PLS regression was conducted to explain sodium release by the variables subjects, oral parameters, and products (Figure 1). A cross validation showed that only the first axis (25% of the variance) is significant. This axis is representative of total release. In particular, it shows that high sodium release is associated to high values of SI, low FC, salivary flow, masticatory efficiency, mean and maximum voltage. Figure 1 shows also that the release of sodium is higher for cheeses made with a high SI and lower for cheeses of higher FC. As shown by the projection of MC3 on the first axis, between MC1 and MC4, the positive effect of SI and the negative effect of FC on sodium release are additive, confirming the results obtained with the analysis of variance. Concerning saltiness, the relations with texture parameters or oral parameters are more difficult to point out and need further investigation.

3.2. Flavour release and perception study of flavoured yoghurt

The results from the experiments showed significant quadratic effects and interactions between the thickeners: starch, pectin, and locust bean gum. The presence of starch and pectin decreased the quantity of flavour in the headspace of the products. Surface responses for an optimal level of strawberry aroma correspond to the high level of starch concentration and medium level of pectin concentration. Locust bean gum seemed to modulate the effect of the above thickeners. PLS analysis on the sensory and instrumental data enabled us to describe more precisely the interactions between composition, texture and flavour. Thus it appeared that the same composition in flavour compounds in the headspace of yoghurts, had an opposite influence on the perception of the 'strawberry' and 'floral' attributes. A PLS regression (PLS2) was established from significant explanatory variables obtained by a first PLS: amount of aroma compounds, rheological behaviour of the products and sensory attributes strawberry, floral, and fruity candy. The model obtained is based on three components. The variance explained by the model is 77% for Y (cumulated R2Y) and 70% for variables X (cumulated R2X). The floral note (negative character) and strawberry note (positive character) are opposed on the map presenting the components 1 and 2 from the PLS2 regression (Figure 2). The floral note was supported by a significant quantity of aroma compounds in the headspace that is unfavourable to the strawberry note. The presence of starch and locus bean gum (quadratic effect) was favourable to the strawberry note. A linear or quadratic increase in pectin concentration was unfavourable to the strawberry note, whereas it increased the floral note. The aromatic balance between strawberry note and floral note was controlled by the composition in thickeners, which like the starch and pectin tended to decrease the concentration of flavour in the headspace. The deformability of yoghurts characterised by the parameter λe180 seemed also essential to optimise the strawberry flavour of yoghurts. Thus, high level for λe180 was favourable to the floral note and conversely unfavourable to the strawberry note.

In summary, PLS analysis provided a model to explain the strawberry flavour based on 5 aroma compounds, 3 rheological parameters and the composition in thickeners.

4. CONCLUSION

Sodium release during mastication is influenced by matrix structure and composition. The use of a higher number of model cheeses and a mouth simulator should lead to better understand the role of each matrix parameter.

Experimental design and PLS analysis on the sensory and instrumental data enabled us to describe more precisely the interactions between composition-texture-flavour into a complex dairy product as yoghurt. Flavour perception in strawberry flavoured fat-free yoghurts seemed to be controlled by composition in thickeners and rheological properties of products. It appeared that an equilibrium between the strawberry and floral notes may be found for an optimal overall strawberry aroma.

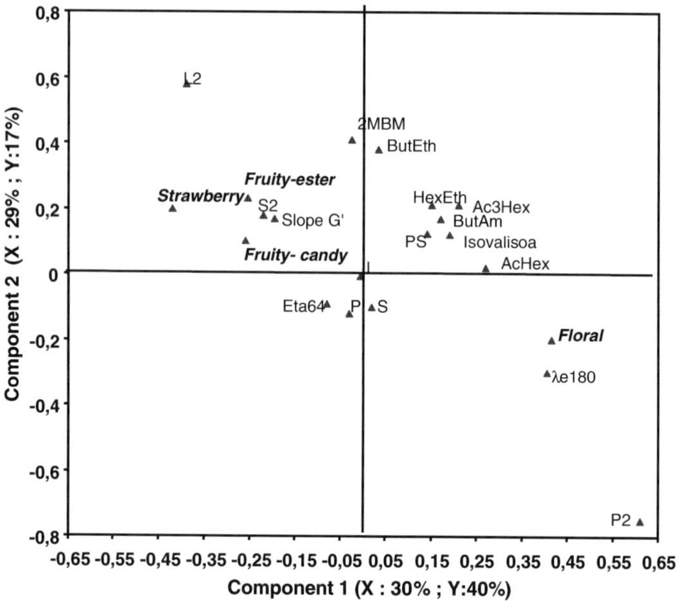

Figure 2. Plot of components 1 and 2 from a PLS regression on 16 significant physicochemical variables and 4 significant sensory variables. Variables to explain (aroma attributes): strawberry, floral, fruity candy, fruity esters. Explanatory variables (GC area): 2MBM: methyl methyl-2-butyrate; ButEth: ethyl butyrate; Ac3Hex: (Z) hex-3-en-1-yl aceate; ButAm: amyl butyrate; HexEth: ethyl hexanoate; AcHex: hexyl acetate; Isovalisoa: isoamyl isovalerate. Explanatory variables (rheological behaviour): eta64; λe180, slopeG. Explanatory variables (thickener composition): P: pectine; S: starch; L: locus bean gum.

References

1. E. Pionnier, C. Chabanet, L. Mioche, J.L. Le Quéré and C. Salles, J. Agric. Food Chem., 52 (2004) 557.
2. E. Pionnier, C. Chabanet, L. Mioche, A.J. Taylor, J.L. Le Quéré and C. Salles, J. Agric. Food Chem., 52 (2004) 565.
3. E. Pionnier, S. Nicklaus, C. Chabanet, L. Mioche, A.J. Taylor, J.L. Le Quéré and C. Salles, Food Quality Pref., 15 (2004) 843.
4. F.R. Jack, J.R. Piggott and A. Paterson, J. Food Sci., 60 (1995) 213.
5. C. Druaux and A. Voilley, Trends Food Sci. Technol., 8 (1997) 364.
6. D.D. Roberts and A.J. Taylor (eds.), Flavor release, Washington, USA (2000) 99.
7. N. Decourcelle, S. Lubbers, N. Vallet, P. Rondeau and E. Guichard, Int. Dairy J., 14 (2004) 783.

W.L.P. Bredie and M.A. Petersen (Editors)
Flavour Science: Recent Advances and Trends

How can protein ratio affect aroma release, physical properties and perceptions of yoghurt?

Anne Saint-Eve, Nathalie Martin, Cécile Levy and Isabelle Souchon

UMR GMPA, INAPG –INRA, France

ABSTRACT

Among yoghurt dairy components, the type of proteins is known to influence the behaviour of odorous volatile organic compounds (VOC). The aim of this study is to focus on the impact of the proteins on the physicochemical (rheological properties, microstructure, VOC release) and sensory (texture and flavour) properties of 4% fat flavoured stirred yoghurts. Yoghurts varied according to the ratio of milk proteins used. Large differences were highlighted concerning the structures of the different yoghurts observed by scanning electron microscopy. Yoghurts differed largely in viscosity and in texture perception. Different effects on retention of VOC and the intensity of most olfactory attributes of the strawberry flavour was observed in yoghurts.

1. INTRODUCTION

Several factors affect yoghurt quality including dairy ingredients (proteins, fat...), strains and technological parameters (thermal treatment, homogenisation...). Especially, the type and content of milk proteins is of significant importance for the physical properties and the perceived texture of the fermented product, as it has been effectively highlighted by numerous researchers [1,2]. Variations in the structure and the composition of the food matrix (due to the nature of proteins) can also be responsible for the modification of the release of VOC and the related flavour properties [3,4]. Many studies on interactions between VOC and other non-volatile constituents of food are carried out with model systems [5]. However, yoghurt is a complex and multiphasic system and to our knowledge no study has looked at the role of protein ratio on physicochemical and sensory properties of flavoured yoghurts.
The aim of the present study is to investigate and to quantify the impact of the ratio of proteins on the physicochemical (rheological properties, microstructure, VOC release) and sensory (texture and flavour) properties of 4% fat flavoured stirred yoghurts made with different milk protein ratio (caseinates on total of proteins).

2. MATERIALS AND METHODS

2.1. Product preparation

Three flavoured stirred yoghurts with equal contents of dry matter (22.5%), fat (4%) and proteins (5.4%) were used. Only the protein ratio (caseinates on total proteins) varied: ratio 86% (caseinates: CAS), ratio 81% (milk powder: MPO) and ratio 60% (whey protein: WP). MPO was the reference product. Yoghurts were flavoured with a strawberry flavour containing 17 VOCs.

2.2. Physico-chemical measurements

The complex viscosity of the three yoghurt matrices was measured at 10 °C in harmonic regime with a rheometer (details given in [6]). Sample observations and photomicrography were performed with a Scanning Electron Microscopy (SEM) (details in [7]). The VOC release in the vapour phase at 4 °C was measured by solid-phase microextraction using a gas chromatograph with an automatic headspace sampler CombiPal. Physico-chemical measurements are expressed relative to an MPO reference.

2.3. Sensory profile

A descriptive sensory analysis of the 3 yoghurts was performed by 10 trained subjects. 34 descriptors were used including visual texture with a spoon, texture in mouth, odour (orthonasal olfactory perception) and aroma (retronasal olfactory perception) and taste by a comparative descriptive analysis. The subjects evaluated the intensity of the samples (MPO, CAS, WP) compared to the reference MPO, which was anchored in the middle of the unstructured intensity scale varying from 'less intense' to 'more intense' than reference. Data acquisition was assisted by Fizz software (Biosystèmes®, 1999).

2.4. Data analysis

Data analyses were carried out using SAS software package, version 9.1 (SAS® Statistics, 1990). Factorial discriminant analyses (procedures discrim and stepdisc of SAS) were used to reveal the sensory and release differences between yoghurts.

3. RESULTS AND DISCUSSION

3.1. Physical properties and texture perception influenced by the ratio of proteins

Yoghurts differed significantly in complex viscosity. The CAS yoghurt had the highest complex viscosity: $\eta^*=80$ Pa s. On the other hand, the yoghurt WP was the least viscous ($\eta^*=40$ Pa s) and the yoghurt MPO was intermediate ($\eta^*=60$ Pa s). These results are in agreement with [8,9], who reported that, at similar protein levels, the addition of caseinate instead of MPO in the dairy mix strongly enhanced the yoghurt viscosity. Structures observed by SEM differed between the yoghurts. The protein network of WP was compact and presented a more uniform distribution of the gel. On the contrary, a heterogeneous structure with numerous pores appeared for CAS (Figure 1). These results correspond to those of [7], who observed irregular gel organisation in yoghurt from MPO and CAS and a fine network with very small pores in WP yoghurts.

Figure 1. Microstructure of yoghurts obtained from milk bases enriched with WP (a), MPO (b) and CAS (c) with ratio caseinates/total proteins: WP (60%); MPO (81%); CAS (86%). Bar-50 μm.

The sensory descriptive analyses revealed quite large differences in texture perception between the two products and the reference. The flavoured yoghurt CAS was visually more granular (+45%) and less ropy (-20.4%) than MPO. In mouth, CAS appeared more granular (+34%) and less mouth coating (-20%) than MPO. WP was perceived ropier (+25%) and less thick (-30%) than MPO. Visually, WP appeared as granular as MPO. In mouth, WP was judged mouth-coating as MPO. Thickness scores were correlated to the complex viscosity measurements. CAS was the most viscous and the thicker and WP the least. Moreover, the visual observations of gel microstructure can be related to the macroscopic evaluation of yoghurt appearance through the attribute granular. Our results corroborated those of [1].

3.2. VOC release and olfactory perception influenced by the ratio of proteins

Stepwise discriminant analysis led to the selection of a subset of 7 VOCs ($p<0.05$): ethyl hexanoate, hexanal, ethyl octanoate, methyl cinnamate, ethyl acetate, (*Z*)-3-hexenol and ethyl butyrate quoted by decreasing order of discriminating power between yoghurts (Figure 2). CAS was characterised by a low release of VOC, except for (*Z*)-3-hexenol.

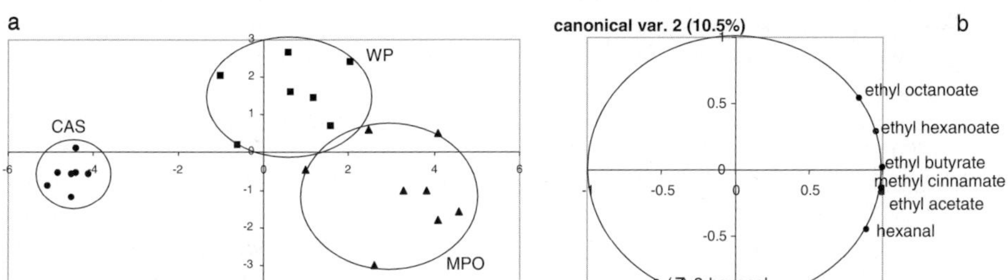

Figure 2. (a) Product mapping of the factorial discriminant analysis with CAS (●), MPO (▲) and WP (■); (ratio: WP (21.5/54); MPO (10.1/54); CAS (7.5/54)). (b) Variables (VOC) map.

The VOC release of WP and MPO was higher than for CAS, but differences between WP and MPO were revealed: the release of ethyl octanoate was higher for WP than for MPO and the contrary was observed for methyl cinnamate.
However, variations of VOC release were quite low between yoghurts: about 10% to 30% of variation between CAS and MPO and less than 10% between WP and MPO.
From the profiling, stepwise factorial discriminant analysis selected 6 orthonasal and retronasal attributes from the whole set of olfactory descriptors: pineapple in mouth, green in nose, fruity, overall intensity, peach and vanillin in mouth. CAS was not very intense in flavour, compared to WP (more intense in flavour, especially in green note). The product MPO was characterised as pineapple.
To conclude, the protein ratio modifies the behaviour of VOCs and their release. Consequently, it affects the olfactory perception. In the present study, CAS retained the majority of VOCs and was perceived less intense than the 2 other yoghurt types. The reverse was observed for WP. For ethyl butyrate and methyl cinnamate the release was highest in MPO, but no sensory effect was observed. Consequently, the sensory results could not be fully explained by physico-chemical interactions between the proteins and the VOC. This observation tends to support the hypothesis that the VOC headspace concentration can not alone predict flavour intensity. Previous studies have shown that a decrease in aroma intensity cannot systematically be explained by a variation of the concentration of VOC released in the breath when viscosity increases [3]. Sensory interactions between texture and flavour also play a role on the perception [10].

4. CONCLUSION

Proteins had principal effects on the physical properties and texture perception of flavoured yoghurt, and secondly on VOC release and olfactory attributes. There were strong relationships between rheological properties and texture attributes and also between release of VOC and olfactory attributes.

References

1. A.Y. Tamime and R.K. Robinson, Yoghurt, Paris, France (1985) 431.
2. I. Sodini, F. Remeuf, S. Haddad and G. Corrieu, Crit. rev. Food Sci. Nutr., 44 (2) (2004) 113.
3. K. Weel, A. Boelrijk, A. Alting, P. Van Mil, J. Burger, H. Gruppen, A. Voragen and G. Smit, J. Agric. Food Chem., 50 (2002) 5149.
4. N. Decourcelle, S. Lubbers, N. Vallet, P. Rondeau and E. Guichard, Int. Dairy J., 14 (9) (2004) 783.
5. A.M. Seuvre, M.A. Espinosa Diaz and A. Voilley, J. Agric. Food Chem., 48 (9) (2000) 4296.
6. E. PaçiKora, I. Souchon, E. Latrille, N. Martin and M. Marin, J. Agric. Food Chem., 52 (10) (2004) 3048.
7. F. Remeuf, S. Mohammed, I. Sodini and J.P. Tissier, Int. Dairy J., 13 (9) (2003) 773.
8. Y.H. Cho, J.A. Lucey and H. Singh, Int. Dairy J., 9 (8) (1999) 537.
9. T.P. Guinee, C.G. Mullins, W.J. Reville and M.P. Cotter, Milchwiss., 50 (4) (1995) 196.
10. A. Saint-Eve, E. PaciKora and N. Martin, Food Quality Pref., 15 (7-8) (2004) 655.

Flavour Science: Recent Advances and Trends

Role of viscosity and hydrocolloid in flavour release from thickened food model systems

Egle Bylaite and Anne S. Meyer

Food Biotechnology & Engineering Group, BioCenrum-DTU, Soeltofts Plads 221, Technical University of Denmark, DK-2800 Kgs. Lyngby, Denmark

ABSTRACT

Release behaviour of two aroma compounds, diacetyl and isoamyl acetate from viscous food model systems having different viscosities was studied. Aroma partitioning and release rates were evaluated in pure water and in hydrocolloid solutions of sodium alginate, guar gum, locust bean and gellan at different concentrations and viscosities. The results showed that the aroma compounds were released differently from viscous solutions thickened to the same viscosity by different hydrocolloids. It was concluded that the nature of aroma compounds, type and concentration of thickener, and not the viscosity of the system are determining factors in aroma release.

1. INTRODUCTION

The texture or consistency of manufactured liquid foods is often controlled by the use of hydrocolloid thickeners. The generally accepted view is that increasing the viscosity or texture of foods by addition of hydrocolloids brings about a decrease in flavour release as result of two mechanisms: a) lowering of diffusion rate in the medium due to the increased solution viscosity and b) molecular/binding interactions between flavour compounds and matrix constituents [1,2]. In order to understand and control aroma release, it is essential to clarify how the hydrocolloid matrix influences aroma release by using a fundamental approach which evaluates both thermodynamic and kinetic factors that control aroma release. The present study evaluates release behaviour of two aroma compounds, diacetyl and isoamyl acetate, from viscous food model systems. The release of aroma compounds from water and hydrocolloid solutions of two viscosity levels was evaluated by assessing the partitioning and mass transfer behaviour by static and dynamic gas chromatography methods, respectively.

2. MATERIALS AND METHODS

2.1. Sample preparation

Thickened food model systems were prepared at two viscosity levels: solutions of 0.7% sodium alginate, 0.48% locust bean gum, 0.45% guar gum and 0.065% (w/w) gellan corresponded to viscosity level I (~120 cP); solutions of 1.6% sodium alginate, 0.97% locust bean gum, 1.0% guar gum and 0.22% (w/w) gellan corresponded to viscosity level II (~ 880 cP). The solutions of locust bean and gellan were prepared by dispersing hydrocolloid into water, stirring for 30 min and then heating to 90 °C for 5 min. To prepare the solutions of guar gum and sodium alginate, a temperature of 50 °C was used. Aroma was added to viscous solutions cooled to 50 °C to give a final concentration of 20 ppm per compound. 2 ml of flavoured viscous solutions were transferred to GC vials, which were capped and stored for static headspace gas chromatography analysis (SHS-GC). An aqueous solution without hydrocolloid addition prepared under the same conditions was used as a reference.

2.2. Methods of analysis

SHS-GC was conducted after 24 h of sample preparation and equilibration at 25 °C and carried out as described previously [3]. Four concentrations of the aroma mixture were analysed in triplicate for calibration, allowing quantification of the compounds in the air phase. Air liquid partition coefficients (K) for each of the compounds in aqueous and viscous hydrocolloid solutions, were determined by dividing air phase concentration of the volatile (w/v) by the concentration in the liquid phase (w/v).
Dynamic headspace gas chromatography analysis (DHS-GC) was used to evaluate the release rates of aroma compounds in water and thickened solutions. The solutions were flushed with purified nitrogen for 15, 30, 60, 90 and 120 s; released aroma compounds were trapped onto five Tenax traps and analysed as described earlier [3]. Non-stirred conditions were applied to study the effect of different structured systems.
The viscosities of thickened solutions were matched by using Rapid Visco Analyser (RVA, Newport Scientific Instruments, Australia).

3. RESULTS AND DISCUSSION

Diacetyl is a hydrophilic diketone, strongly associated with the aqueous phase (hydrophobic constant logP = -1.8, air-water partition coefficient K_{aw}= 1.7 x 10^{-3}), whereas isoamyl acetate is more hydrophobic and volatile in water (logP = 2.3, K_{aw} = 40.3 x 10^{-3}) [4]. To estimate the effect of addition of different hydrocolloids on the thermodynamic properties of these two aroma compounds, equilibrium headspace concentrations were determined by SHS-GC and compared to those in pure water (Table 1). Expressed as the ratio $K = C_a/C_p$ between aroma concentration in the air phase above the product (C_a) and aroma concentration in the product phase (C_p) at equilibrium, partition coefficient defines the extent of aroma release from the product.

Table 1. Relative headspace concentrations of diacetyl and isoamyl acetate in the viscous solutions of hydrocolloids at equilibrium (25 °C). Data normalised against release from water.

	Viscosity I (120 cP)				Viscosity II (880 cP)			
Compound	Sodium alginate 0.7%	Locust bean 0.48%	Guar gum 0.45%	Gellan 0.065%	Sodium alginate 1.6%	Locust bean 0.97%	Guar gum 1.0%	Gellan 0.22%
Diacetyl	92.6	95.0	96.1	95.3	91.4	88.2[a]	91.2	88.1[a]
Isoamyl acetate	89.0	84.4[a]	88.4	86.6[a]	87.2	77.2[a]	78.3[a]	84.2[a]

[a]Indicates statistically significant difference from water ($p<0.05$).

The SHS-GC results (Table 1) show that the effect of hydrocolloid addition to the system differed for diacetyl and isoamyl acetate. The suppression of diacetyl release by hydrocolloid was significant only at viscosity II level which corresponded to the concentrations of 0.97% and 0.22% (w/w) of locust bean and gellan, respectively. The retention of isoamyl acetate was higher than of diacetyl as indicated by the lower relative headspace concentration values for isoamyl acetate (Table 1). This flavour retention may be due to some specific interactions occurring between the hydrophobic ester and the hydrocolloid at (relatively) high thickener concentration. The tendency of hydrocolloids to interact with hydrophobic compounds at high polysaccharide concentration was already reported in our previous papers [4,5].

At equilibrium, the effect of viscosity is nullified. Therefore, in order to evaluate the effect of viscosity on mass transfer, the release rate experiments were carried out under non-equilibrium which is the main driving force for mass transport [2]. If the viscosity was a limiting mechanism of aroma release, then the same release would occur from the solutions thickened to similar viscosities. As shown in Figure 1, that was not the case. The release of diacetyl from viscous solutions was not following an increase of systems viscosity. Gellan solutions gave significantly lower release of diacetyl compared to water, but again, the release rates were equal at both viscosities. Release of diacetyl from sodium alginate and guar gum solutions also remained unchanged with increasing viscosity and did not differ significantly from that in water (Figure 1).

Release of isoamyl acetate was affected differently by the addition of hydrocolloid (Figure 1). However, also with isoamyl acetate, solutions thickened to the same viscosity exhibited different release rates. At viscosity I, the release rates differed significantly for all hydrocolloids, even though the solutions possessed the same viscosity. Again, gellan gum solutions exhibited the same release rates independently of the systems viscosity. For these reasons the effect of hydrocolloids on the release of the flavour compound can be attributed to hydrocolloid concentration rather than to viscosity [3-5]. The data of this study complement our previously reported results on pectin, λ-carrageenan and xanthan where the effect of macroscopic (bulk) viscosity was demonstrated to be of secondary importance in determining aroma release from thickened solutions.

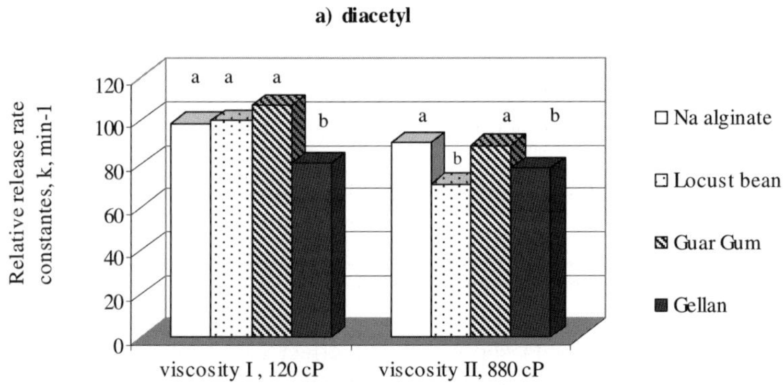

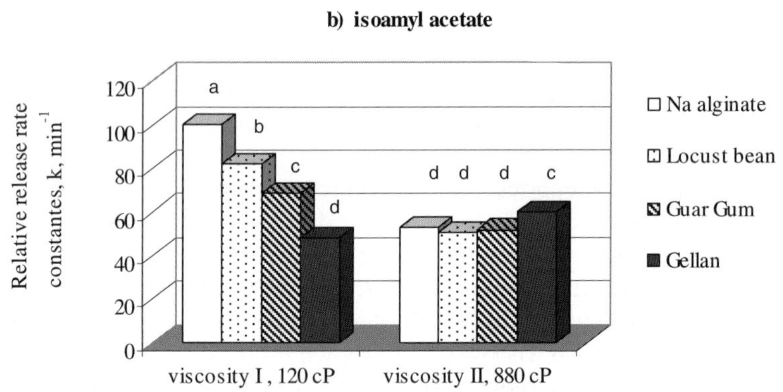

Figure 1. Relative release rate constants of diacetyl (a) and isoamyl acetate (b) in the viscous solutions of different hydrocolloids. Data normalised against release from water. Different letters indicate statistically significant difference among release rates ($p<0.05$).

Therefore, it can be concluded that the nature of aroma compounds, type and concentration of thickener, and not the bulk viscosity of the system are playing a major role in determining the effect of the thickener on aroma release.

References

1. S. Secouard, C. Malhiac, M. Grisel and B. Decroix, Food Chem., 82 (2003) 227.
2. K.B. De Roos, Int. Dairy J., 13 (2003) 593.
3. E. Bylaite, A.S. Meyer and J. Adler-Nissen, J. Agric. Food Chem., 51 (2003) 8020.
4. E. Bylaite, Ž. Ilgūnaitė, A.S. Meyer and J. Adler-Nissen, J. Agric. Food Chem., 52 (2004) 3542.
5. E. Bylaite, J. Adler-Nissen and A.S. Meyer, J. Agric. Food Chem., 53 (2005) 3577.

W.L.P. Bredie and M.A. Petersen (Editors)
Flavour Science: Recent Advances and Trends

The molecular organisation of dairy matrices influences partitioning and release of aroma compounds

Sébastien Bongard, Anne Meynier, Alain Riaublanc and Claude Genot

INRA, Biopolymers Interactions Assemblies. BP 71627, 44316 Nantes cedex 3, France

ABSTRACT

Partition coefficients of 12 aroma compounds from a strawberry aroma blend were investigated in dairy matrices containing skim-milk powder and 1.5 or 5% (w/w) of low melting point fraction or high melting point fraction anhydrous milk fat (AMF) acidified to pH~4.6 with glucono-δ-lactone. The release of aroma compounds was markedly reduced as the fat level increased, notably for the more hydrophobic compounds. The partition coefficients were significantly higher over matrices with high melting point than with low melting point fraction milk fat, probably due, in part, to the solid fat content of the lipid fraction. They were also, to a lesser extend, higher in emulsions than in stirred gels. These results demonstrate the role of structure and molecular organisation of the matrix constituents on aroma partitioning.

1. INTRODUCTION

Perceived intensity of aroma compounds is linked to their nature and concentration, and also to their physicochemical interactions with food components and modifications occurring during eating [1]. Among food components, lipids have the strongest influence on retention and release of aroma compounds due to their solvent properties [2]. Other factors are, however, also important, for instance physical state [3,4], *i.e.* not only the composition of a food product but also its structure, that is the organisation and interactions of molecules within the matrix, plays a role in perception [5]. Our objective was to sort the key factors lipid content, nature and structure of the matrix controlling the behaviour of aroma compounds in complex dairy media. Partition coefficients over emulsions and stirred gels containing 1.5 or 5% (w/w) of low melting point fraction (LMPF) or high melting point fraction (HMPF) were compared and discussed.

2. MATERIALS AND METHODS

Final composition of the dairy matrices was 4.5% (w/w) protein, 15% (w/w) carbohydrates, and 1.5 or 5.0% (w/w) fat. Initially, oil-in-water emulsions were made by pressure homogenisation (average droplet diameter: 1.4 μm, 70 °C) of 20% (w/w) LMPF or HMPF (Lactalis, Laval, France) into a solution (protein: 2% w/w) of skim-milk powder (Ingredia Dairy Ingredients, St Pol/Ternoise, France). These emulsions were then mixed with the requested amount of milk protein powder and saccharose solutions. The resulting 1.5 and 5% (w/w) fat emulsions were stored at 4 °C for 24 h. Acid gelation was carried out for 4 h at 43 °C by adding in two steps the needed amount of GDL (Roquette, Lestern, France) to achieve a final pH of 4.6. Gels were stirred with a rotor stator system for 2 min. A blend of 17 aroma compounds (Sigma Chemical Co., StQuentin-Fallavier, France) was prepared in propylene glycol (20/80 w/w) and added (0.5% w/w) to 1.5 and 5% emulsions previously equilibrated at ambient temperature or to the gels during stirring. Triplicates were prepared for each matrix.
Of each flavoured matrix, 1.5 g was poured into 22.4 ml glass vials closed with hermetic caps and equilibrated 1 h at 10 °C prior to headspace analysis. One ml of the gas phase was sampled from the headspace with a gas syringe (headspace autosampler, CombiPal) and injected into a DB624 capillary column (30 m × 0.32 mm, 1.8 μm film thickness) fitted in a gas chromatograph (HP 5890). Partition coefficients were calculated from the ratio of aroma concentration in the gas phase to the concentration in the matrix. Measurements were performed in triplicate.
Data of the nine measurements of the partition coefficients for the various matrices and aroma compounds were subjected to multivariate analysis of variance (MANOVA). The significance level was set at $p<0.05$ throughout the study. Statistical analyses were performed with Statgraphics Plus 3.0. Tested factors were the fat level, the fraction of AMF in emulsions or stirred gels and finally the type of matrix (emulsion or stirred gel).

3. RESULTS

Air-matrix partition coefficients of the 12 aroma compounds are shown in Table 1. For 6 of the 12 aroma compounds, the main factor that modified the air-matrix partition coefficient was the fat level (Table 2). The second factor was the AMF fraction and finally to a lesser extent the type of matrix. As fat level increased, the air-matrix partition coefficient of 11 of the 12 aroma compounds was significantly reduced (Table 2). Partition coefficients of hydrophilic compounds (logP < 1) such as diacetyl were weakly modified. In contrast, the retention was pronounced for aroma compounds with logP greater than 2. It ranged from 45% for γ-decalactone to 68% for limonene. Such drastic decreases of the volatility of aroma compounds in presence of a small amount of lipid have been already reported [6] and have been widely reviewed [2,7]. The significant interactions between the three tested factors show that the effect of fat also depended on the other factors.
The nature of the dispersed lipids also influenced the air-matrix partition coefficients. Retention of 8 of the 12 aroma compounds was greater in matrices containing LMPF

than HMPF. The highest differences were observed between the stirred gels containing the lowest level of lipid. Quantitatively, these modifications were especially noticeable for hydrophobic aroma compounds. For instance, emulsification of LMPF instead of HMPF resulted in a much higher decrease of concentration of ethyl octanoate in the gas phase than increasing the fat level from 1.5 to 5%. As differences between matrices containing the two lipid fractions decreased with temperature (data not shown), it can be postulated that the phenomenon was related to their solid fat content.

Table 1. Air-matrix partition coefficients of the 12 aroma compounds ($\times 10^5$) measured at 10 °C (values are mean of 9 determinations).

Matrix	Emulsion				Stirred gel			
Fat level (w/w)	1.5%		5.0%		1.5%		5.0%	
AMF fraction	LMPF	HMPF	LMPF	HMPF	LMPF	HMPF	LMPF	HMPF
Diacetyl	3.63	2.69	2.81	2.46	3.96	5.53	4.71	4.29
Maltol	0.109	0.115	0.0889	0.0790	0.0994	0.1050	0.0677	0.0675
Et-acetate	260	335	285	283	200	297	248	253
c-3-Hexen-1-ol	5.43	6.05	4.97	5.13	3.91	5.52	4.51	4.54
Hexanal	25.70	7.14	18.10	5.85	69.6	43.3	55.4	27.4
Et-butyrate	209.0	248.0	115.0	172.0	144.0	258.0	104.0	160.0
γ-Decalactone	0.230	0.619	0.126	0.332	0.337	0.827	0.174	0.301
Me-cinnamate	0.506	1.18	0.158	0.639	0.517	1.27	0.276	0.512
Et-hexanoate	45.1	69.5	16.3	34.2	27.8	74.1	13.6	28.6
Linalool	4.05	6.04	1.92	3.65	2.72	5.47	1.62	2.88
Et-octanoate	5.93	11.80	2.22	4.91	3.84	9.96	1.81	3.59
Limonene	37.3	69.6	12.1	22.8	21.2	58.5	10.1	21.1

Table 2. F values (shown if $p<0.05$) of 3 factor variance analysis (MANOVA) for the air/matrix partition coefficients measured at 10 °C of 12 aroma compounds.

	logP [a]	Fat level (A)	AMF fraction (B)	Matrix (C)	Interaction
Diacetyl	-1.34	10		189	AC, BC, ABC
Maltol	0.09	46		8	None
Et acetate	0.73		74	67	AC
3-Hexen-1-ol	1.61	26	50	82	AB, AC, BC
Hexanal	1.78	322	1529	4064	All
Et butyrate	1.81	333	249	21	AC, BC, ABC
γ-Decalactone	2.57	127	161	12	AB, AC
Me cinnamate	2.62	124	160		AC
Et hexanoate	2.83	258	175	7	AC, BC, ABC
Linalool	2.97	311	276	40	AC, ABC
Et octanoate	3.91	3817	2798	325	AB, AC, ABC
Limonene	4.57	416	240	27	AB, AC

[a]http://www.syrres.com/esc/physdemo.htm.

Indeed, it is generally considered that only liquid lipids can solubilise aroma compounds [8]. From the melting curves of the bulk fat fractions recorded by differential scanning calorimetry (DSC), we have estimated the amount of solid and liquid fractions in each matrix. Typically, matrices containing 1.5% of LMPF had liquid lipid amounts (1.17 g) close to matrices containing 5% of HMPF (1.36 g). The fact that partition coefficients between air and 5% HMPF versus 1.5% LMPF matrices were slightly, but significantly different could result from an overestimation of the solid fat content in the matrices that was based on the solid fat content of the bulk fats. Higher volatility of aroma compounds over emulsions containing high solid fat content have been reported [3,4].
Changes of organisation of matrices (emulsion or stirred gel) induced significant but smaller modifications of partition coefficients than the fat level or the nature of fat. Most aroma compounds were retained to a greater extent in the gels than in the emulsions. Retention in stirred gels did not exceed 50% and was more pronounced in the case of 1.5% fat matrices and in matrices prepared with the LMPF. It is generally accepted that increase of viscosity or gelation led to a decrease of the volatility of aroma compounds. Such tendencies have been drawn from experiments performed on fat free protein gels in presence of thickeners [9,10]. To our knowledge, no comparison of the behaviour of aroma compounds in emulsions and gels have been yet reported.

4. CONCLUSION

Extensive retention of hydrophobic aroma compounds was induced by increasing fat content from 1.5% to 5%, but greater modifications of retention, or release, were observed by modifying the nature of dispersed lipids. These results suggest that the underlying molecular organisation, which also depends on lipid content and composition, and matrix structure, takes a great part in aroma behaviour. Better knowledge on the role of lipids in the formation of interfaces and on the influence of gel network are needed to better predict the macroscopic behaviour of aroma compounds in complex matrices. If a food contains a portion of solid fat at the eating temperature, it seems essential to control its thermal history to ensure its organoleptic qualities.

References

1. P. Overbosch, W.G.M. Asterof and P.G.M. Haring, Food Rev. Int., 7 (1991) 137.
2. K.B. de Ross, Food Technol., 51 (1997) 60.
3. P. Relkin, M. Fabre and E. Guichard, J. Agric. Food Chem., 52 (2004) 6257.
4. D.D. Roberts, P. Pollien and B. Watzke, J. Agric. Food Chem., 51 (2003) 189.
5. A.G. Gaonkar (ed.), Ingredient interactions. Effect of food quality, New York, USA (1995) 411.
6. D.D. Roberts and A.J. Taylor (eds.), Flavor release, Washington, DC, USA (2000) 321.
7. L.C. Hatchwell, Food Technol., (1994) 98.
8. H. Maarse and P.J. Groenen (eds.), Aroma research: proceedings of the international symposium on aroma research, Zeist, the Netherlands (1975) 143.
9. E. Bylaite, A.S. Meyer and J. Adler-Nissen, J. Agric. Food Chem., 51 (2003) 8020.
10. N. Decourcelle, S. Lübbers, N. Vallet, P. Rondeau and E. Guichard, Int. Dairy J., 14 (2004) 783.

W.L.P. Bredie and M.A. Petersen (Editors)
Flavour Science: Recent Advances and Trends

Aroma release under oral conditions

Jacques P. Roozen[a] and Saskia van Ruth[b]

[a]*Wageningen University and Research Centre, Department of Agrotechnology and Food Sciences, PO box 8129, 6700 EV Wageningen, the Netherlands;* [b]*Department of Food and Nutritional Sciences, University College Cork, Western Road, Cork, Ireland*

ABSTRACT

The model mouth system was developed to study *in vitro* aroma release and considers sample volume, volume of the mouth, temperature, salivation, and mastication. In-nose analysis was conducted *in vivo* using real-time proton transfer reaction mass spectrometry methodology. Several experiments were carried out to determine the influence of oral conditions on the release of aroma compounds from different food products.

1. INTRODUCTION

The food matrix composition is a key feature that influences the binding and release of aroma compounds from foods. On the other hand, it is well known that individuals often perceive different aroma intensities from the same food. Although there are more factors playing a role, some of this variation may be due to differences in release of the flavour compounds in the mouth. Some authors have shown considerable differences between individuals in chewing behaviour [1]. Chewing behaviour affects the release of volatiles in the mouth by its effect on physical disruption, increase in surface area and food/saliva mixing [2]. The effect may differ for individual aroma compounds as a result of their in-mouth generation or their physico-chemical properties. Changes in the proportions of aroma compounds affect the balance of the aroma profile, in terms of release and perception [3,4]. Saliva is an important oral physiological factor for aroma release as well. Salivary flow rate and composition vary widely within and between individuals, and is related to factors such as degree of hydration, body position, olfaction, smoking, stimulation and climatological circumstances [5,6].
In order to follow aroma release real time, fast analysis with high time resolution is required. Traditional hyphenated analysis techniques are associated with low temporal resolution. High temporal resolution alternatives are techniques, which allow direct

introduction of the aroma into a mass spectrometer. This type of analysis gives fast real time analysis, however, at the expense of loss of structural information and identification. Soft ionisation techniques minimise fragmentation. They produce mainly molecular ions by addition or abstraction of a proton. Currently soft ionisation techniques such as atmospheric pressure chemical ionisation and proton transfer reaction (PTR) followed by mass spectrometry (MS) are the best options for real time aroma analysis. In PTR-MS the generation of the primary H_3O^+ ions and the chemical ionisation of the aroma compounds are individually controlled, and spatially and temporally separated processes [7].

In the present study, headspace aroma concentrations were followed real time during rehydration of dried vegetables using PTR-MS. The influence of two oral physiological factors was evaluated for aroma release from the rehydrated vegetables using PTR-MS: salivation and mastication. Furthermore, model mouth and in-nose analyses were compared for real time release measurements.

2. MATERIALS AND METHODS

2.1. Materials

Commercially dried diced leeks and French beans were supplied by Top Foods b.v. (Elburg, The Netherlands). The vegetables were packed in glass jars and stored in absence of light until sampling. In the rehydration study, dried leeks were used. For the study on the effect of mastication and salivation and for the in-nose analysis, vegetables were rehydrated by addition of 10 ml of distilled water to 1.2 g of dried vegetables in a conical flask. The conical flask was placed in a water bath at 100 °C for 10 min, and subsequently in a water bath at 20 °C for 4 min.

2.2. Rehydration experiment

2.4 g of dried leeks were placed in a headspace vial, positioned in a water bath at 37 °C, and the headspace was monitored by PTR-MS for 10 min. After an initial 10 min, 21.6 ml of distilled water was added and then the headspace monitored for another 50 min. Three replicate samples were analysed according to the method of Lindinger and co-workers [7]. The headspace was drawn in from the headspace vial at a rate of 20 ml/min, 15 ml/min of which was led through a heated transfer line into the PTR-MS for on-line analysis. Data were collected for the mass range *m/z* 20-170. The spectra were corrected for background (lab air) and transmission, and were averaged over the three replicate samples.

2.3. Mastication and saliva study

Rehydrated French beans were transferred to the flask of the model mouth [8]. The volume of saliva added was varied: 0, 2, 4, and 6 ml. These volumes refer to an addition of 0, 20, 40 and 60% (v/w) saliva, respectively. Mastication rates were varied for the 40% (v/w) saliva sample: 0, 26, 52 and 78 cycles/min. Three replicate samples of each saliva volume and mastication were analysed by PTR-MS as described in section 2.2.

The headspace was drawn in from the model mouth at 100 ml/min, of which 15 ml/min was led into the PTR-MS for on-line analysis for two minutes. Data were collected for a number of ions, based on previous Gas Chromatography-MS analyses [8]. They concerned *m/z* 69 (3-methylbutanal), 73 (2-methylpropanal), 83 (hexanal) and 87 (2-methylbutanal). Dwell time was 0.2 s per mass. Headspace concentrations were corrected for fragmentation, background and transmission. From the release curves of the individual replicates, the maximum headspace concentration (Imax) and the time to reach Imax (tmax) were calculated. These values were subsequently averaged.

2.4. Comparison model mouth/in-nose analysis

Model mouth analysis was carried out as described in section 2.3 with rehydrated leeks as sample material. In-nose analysis was carried out on rehydrated leeks using four subjects. A fork-shaped glass nosepiece was placed in the nostrils of a subject. The air was drawn in at a rate of 100 ml/min, of which 15 ml/min was led into the PTR-MS. The background was measured for 30 s, then the sample was taken, chewed and measured for 30 s. Measured concentrations were background and transmission corrected. Four replicate measurements per subject were carried out. For comparison with model mouth data, all data were normalised considering Imax = ratio x 100%.

2.5. Statistical analysis

The Imax and tmax data were subjected to analysis of variance (ANOVA) to determine significant effects of the variables. Principal component analysis (PCA) was conducted on the Imax data set. A significance level of $p<0.05$ was used throughout the study.

3. RESULTS AND DISCUSSION

3.1. Rehydration

The changes in headspace concentration of dried leeks were followed during rehydration for mass range *m/z* 20-170. Different profiles were observed for various masses. Two typical profiles are presented in Figure 1. A group of masses showed a similar profile to mass *m/z* 43 (e.g. mass *m/z* 42, 44, 47, 48, 61, 62, 63, 65, 79, 80). These masses showed that the concentration increased very rapidly, when the dry leeks were placed in the sample flask, but declined at the time water was added for rehydration. This phenomenon could relate to a compound which is released rapidly from the dry leeks, but which dissolves well in water at the time of rehydration. It could also imply that compounds are formed, but that reactions are inhibited when water is added to the leeks.

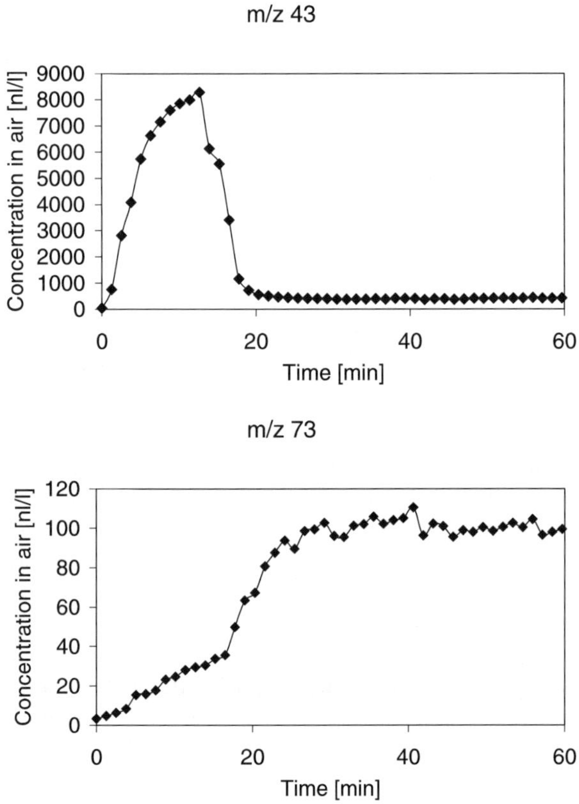

Figure 1. Change in dynamic headspace concentrations of masses *m/z* 43 (alcohols) and *m/z* 73 (2-methylpropanal) from dried leeks during rehydration - water was added at 10 min - determined by Proton Transfer Reaction Mass Spectrometry.

A second typical group shows slow release in the dry state, but an increase in release when the water was added for rehydration (mass *m/z* 73; Figure 1). This could reflect the release of more non-polar aroma compounds. For the masses *m/z* 37, 38, 39, 69, 70, 81, 83, 87, 88, 91, 95, 117, 123, 131, 132 and 145 similar profiles were observed.

3.2. Salivation and mastication

The effect of different saliva volumes on real time aroma release from rehydrated French beans was examined. The release curves of 2-methylpropanal as a function of saliva volume are presented in Figure 2. From the curves Imax and tmax values were calculated. The Imax differed significantly among the compounds [$F(3,48)=5.246$, $p=0.005$], but tmax did not [$F(3,48)=0.501$, $p=0.684$]. The Imax of 2-methylpropanal was higher than those of the other compounds. The volume of saliva did not affect the Imax of the compounds significantly [$F(3,48)=1.112$, $p=0.359$), but affected tmax significantly [$F(3,48)=4.471$, $p=0.009$]. For instance for mass *m/z* 73 (2-methyl-propanal) the average tmax was 27 s for 0% (v/w) saliva, and increased to 41,

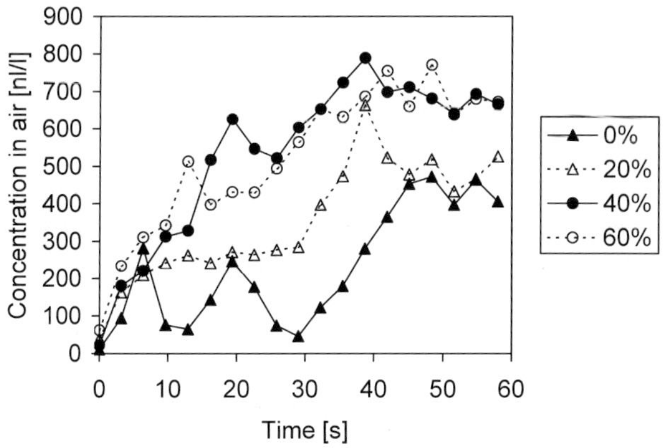

Figure 2. Real time release of 2-methylpropanal (*m/z* 73) from French beans in the model mouth, (with addition of different volumes of artificial saliva (% (v/w)) determined by Proton Transfer Reaction Mass Spectrometry (mastication rate 52 rpm).

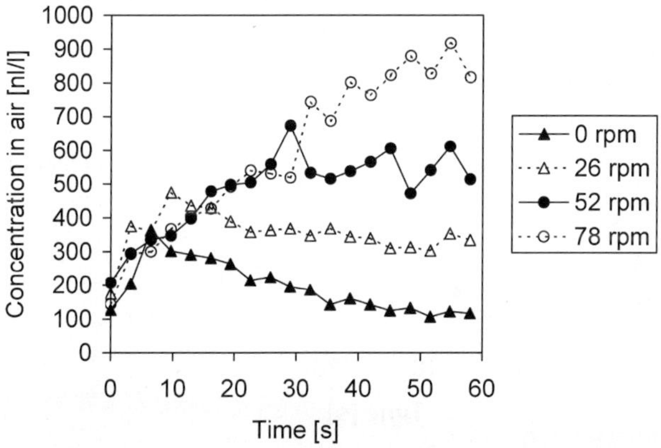

Figure 3. Real time release of 2-methylpropanal (*m/z* 73) from French beans in the model mouth, (with application of different mastication rates) determined by Proton Transfer Reaction Mass Spectrometry (40% (v/w) saliva added).

39 and 47 s for 20, 40 and 60% (v/w) saliva, respectively. Increase of tmax indicates relative greater persistence of aroma compounds. Largest differences were observed for samples with and without saliva (LSD tests, $p<0.05$). The increased persistence may be due to the breakdown of the helical structures of the amylose fraction in the starch of the beans. The dried beans contained 20.4% starch. This may have led to additional release of aroma compounds, which were retained inside the hydrophobic regions inside the polymer during the drying process.

Mastication rate had a significant effect on both Imax [$F(3,48)=19.326$, $p=0.000$] and tmax [$F(3,48)=24.769$, $p=0.000$]. The Imax values for the 0 and 26 rpm mastication rates did not differ significantly, but the 52 rpm resulted in higher values, and highest values were obtained for 78 rpm. tmax increased significantly with each mastication step (LSD tests, $p<0.05$). PCA provided an overview of the release data as a function of salivation and mastication. The samples with different mastication rates were separated along PC1, with negative scores for the lower rates and higher positive scores for the higher rates. The samples with different saliva volumes were separated along PC1 and PC2. The PCA showed that samples with relatively high saliva volumes and mastication rates were relatively similar. Both oral physiological parameters (higher saliva volume, and higher mastication rate) increased aroma release.

3.3. In-nose analysis

In-nose analysis was carried out with rehydrated leeks using PTR-MS. Mass spectral data provided real time aroma concentration profiles. The development of the concentration of mass *m/z* 73 (2-methylpropanal) both in air during consumption and in the model mouth, were compared (Figure 4). Although the curves were not identical, the concentration profiles were not significantly different ($p<0.05$).

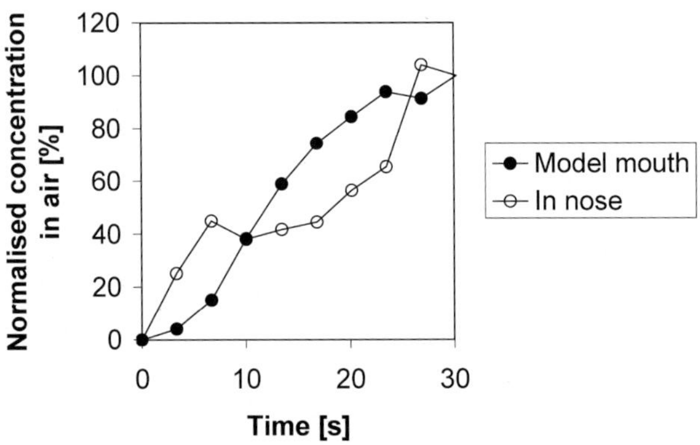

Figure 4. Comparison of model mouth and in-nose analysis of rehydrated leeks for normalised headspace concentrations of mass *m/z* 73 (2-methylpropanal, Imax = ratio x 100%).

References

1. W.E. Brown and C.E. Wilson, Trends Food Sci. Technol., 7 (1996) 444.
2. A.J. Taylor, Crit. Rev. Food Sci. Nutr., 36 (1996) 765.
3. E.N. Friel and A.J. Taylor, J. Agric. Food Chem., 49 (2001) 3898.
4. K.J. Burdach and R.L. Doty, Physiol. Behav., 41 (1987) 353.
5. C. Dawes, J. Dent. Res., 66 (1987) 648.
6. L. Wisniewski, L.H. Epstein and A.R. Caggiula, Physiol. Behav., 52 (1992) 21.
7. W. Lindinger, A. Hansel and A. Jordan, Int. J. Mass Spectrom., 173 (1998) 191.
8. S.M. van Ruth, J.P. Roozen and J.L. Cozijnsen, Food Chem., 53 (1995) 15.

W.L.P. Bredie and M.A. Petersen (Editors)
Flavour Science: Recent Advances and Trends

The role of lipids in aroma/food matrix interactions in complex liquid model systems

Celine Riéra, Elisabeth Gouézec, Walter Matthey-Doret, Fabien Robert and Imre Blank

Nestlé Research Center, Vers-chez-les-Blanc, 1000 Lausanne 26, Switzerland

ABSTRACT

The release of nine aroma compounds was investigated in complex liquid bouillon-type model systems containing various non-volatile constituents. The relative release was determined by static headspace GC-MS as a function of the bulk composition. Among the non-volatile food constituents studied, fat appeared to be very efficient in binding volatile aroma molecules. This retention by the fat was directly correlated with the intrinsic physical properties of the aroma compounds, such as the water/octanol partition coefficient. The individual effect of the major fat constituents was studied as well, indicating that, for example, even small amounts of phospholipids may effectively retain volatile compounds.

1. INTRODUCTION

The phenomena related to aroma release and retention have become a widely studied research field [1,2]. Despite the high complexity of food, most of the work published so far is dealing with rather simple systems studying the release of volatile compounds as affected by selected food biopolymers, such as starch [3], xanthan [4], carrageenan [5], pectin [6], and β-lactoglobulin [7]. Also lower molecular weight compounds of different chemical classes were studied, in particular lipids, which notably affect flavour release by lowering the vapour pressure of many odorants and changing the time scale of release with varying concentrations [8]. Reduction of fat content in food results in a drastic shift of the overall flavour profile leading to different odour sensation, even if the changes in the fat content are small [9]. The affinity of flavours to the lipid phase depends on its chemical composition, chain length, degree of saturation, and sequence of fatty acids in the triacylglycerol [10].

As the results obtained from simple model systems hardly allow prediction of the release behaviour in food systems, we studied the release of nine odorants in rather complex liquid models containing various types of macromolecules. The aim of this work was to investigate aroma release under static headspace conditions as a function of the bulk composition and to elucidate the role of different lipid constituents.

2. MATERIALS AND METHODS

Materials. The following compounds were commercially available: 3-methylbutanal, n-hexanal, n-heptanal, n-octanal, 1-hexanol, chitin (from crab cells), triacylglyceride palm oil (Fluka, Buchs, Switzerland); n-nonanal, benzaldehyde (Aldrich, Buchs, Switzerland); α-pinene (Glidco Organics, Jacksonville, USA); 1-octen-3-one (Lancaster, Cardiff, UK); cellulose gum (Aqualon, Düsseldorf, Germany); collagen (Biogel AG, Lucern, Switzerland); cured ham fat (from the local market).
Sample preparation. A stock solution composed of 3-methylbutanal (3 μl/l), hexanal (30 μl/l), heptanal (8 μl/l), octanal (5 μl/l), 1-hexanol (50 μl/l), nonanal (1 μl/l), benzaldehyde (50 μl/l), α-pinene (1 μl/l), 1-octen-3-one (2 μl/l) was prepared in mineral water (Vittel, Nestlé Waters, France) or in deionised water and adjusted to pH 7. The solution was stored at 4 °C prior to analysis. Macromolecules were used at the following concentrations: fat (5.0 g/l), collagen (2.7 g/l), cellulose gum (1.5 g/l), chitin (0.3 g/l). A fat model was composed of triacyl glycerol (85%), diacyl glycerol (3%), monoacyl glycerol (1%), free fatty acids (10%), and phospholipid (1%). Non-volatile molecules were weighted directly into a 20 ml vial and 5 ml of the volatile stock solution was added. If necessary, the samples were readjusted to pH 7 before starting with the equilibrium phase. The vials were closed with caps and equilibrated at 37 °C for 2 h. After achieving equilibration under stirring (300 rpm, 2 h), 2 ml of the headspace sample was injected using an autosampler held at 37 °C. Each sample was prepared in triplicate for GC headspace analysis.

3. RESULTS AND DISCUSSION

Effect of biopolymers. With the exception of the apolar compounds α-pinene and nonanal, of which 60% and 15%, respectively, was retained by cellulose, neither cellulose nor chitin strongly influenced aroma release into the headspace at the concentration applied (Figure 1). Collagen seemed to play a role in the release of aliphatic aldehydes and 1-octen-3-one by retaining about 20% and 70%, respectively. The release of 1-octen-3-one was much more affected suggesting a specific interaction between collagen and this aroma compound. Various chemical reactions of the carbonyl group have been reported in the literature, such as the reaction with amino and sulfhydryl groups of amino acids or proteins leading to Schiff's base formation and cysteine-aldehyde condensation, respectively [11].
Effect of fat. As shown in Figure 1, only fat produced significant differences in aroma release for all volatile components compared to the reference sample, clearly indicating aroma/lipid interactions. Relatively polar aroma compounds such as 1-hexanol and 3-

methylbutanal were readily released into the headspace reaching about 80-90% release. In contrast, lipophilic aroma compounds such as nonanal, α-pinene, and octanal were well retained by the fat (75-99%) leading to only 1-25% release compared to the reference sample. The retention of each aroma compound induced by the addition of fat is well correlated with its intrinsic physico-chemical properties, such as the octanol/water partition coeficient, thus explaining the preferred binding of lipophilic aroma compounds, in particular α-pinene (log P = 4.83), nonanal (log P = 3.27), octanal (log P = 2.78), heptanal (log P = 2.29), and 1-octen-3-one (log P = 2.13). These data confirm earlier work [12] that fat plays an important role in aroma release both by decreasing the overall headspace concentration of aroma compounds and by changing the headspace composition due to a preferred entrapment of hydrophobic volatile compounds.

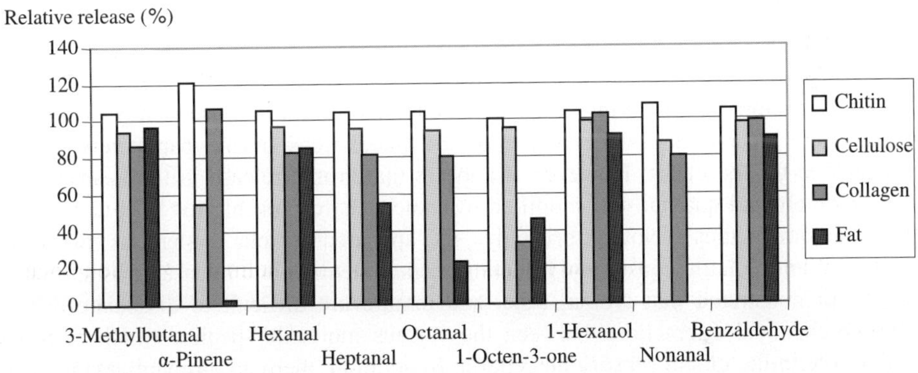

Figure 1. Individual effect of macromolecules on the release of selected volatile compounds.

Effect of the concentration. As mentioned above, neither cellulose nor chitin affected strongly, at the concentration used, the aroma release into the headspace, while collagen and in particular fat markedly modified the aroma release of most of the volatile compounds. When increasing the concentration of polysaccharides, the reduction of release into the headspace observed was clearly correlated with an increase in the viscosity (data not shown). These results confirm that the effect of hydrocolloids or proteins in aroma retention require relatively high concentrations that correspond more to a yoghurt or mayonnaise type of product rather than a bouillon application. A small increase of fat concentration resulted in drastic retention of apolar volatile compounds. On the other hand, the rather polar compounds 3-methylbutanal, 1-hexanol, and benzaldehyde were less affected by higher amounts of lipids.

Effect of lipid constituents. The headspace results obtained with the fat model were practically identical to those obtained with the natural ham fat (Figure 1). As the aroma release behaviours were similar, the simplified lipid model could be used to mimic the original cured ham fat in order to study the role of each constituent. The aroma release patterns obtained with each class of fat at the original concentration and ratio were compared with the complete system. Triacyl glycerol was the most important lipid constituent with respect to aroma retention, most likely due to its high abundance in the fat. In general, triacyl glycerol influenced the retention of the entire volatile composition. However, apolar compounds were preferably retained (50-90%) compared to 3-methylbutanal, benzaldehyde, and 1-hexanol (10-20%). These results suggest that the interactions are more driven by hydrophobic than dipole-dipole interactions at this concentration and ratio. In agreement with that, the most apolar odorants α-pinene and nonanal were retained by all lipid compounds. Despite their lower concentrations, free fatty acids and phospholipids also influenced the aroma release, in particular of apolar compounds. A possible synergistic effect of lipid constituents was checked by using various binary mixtures in aroma release experiments. None of the combinations studied showed any interaction between lipid constituents, as the measured aroma release corresponded globally to the sum of the effect obtained with each lipid species individually (data not shown).

4. CONCLUSION

As systems mimicking food ingredients are quite complex, it is consequently difficult to discuss and clearly assess the nature of the interactions or the parameters mainly responsible for the effect. However, it appears that aroma release in the headspace is quantitatively and qualitatively modified by molecules responsible for food texture (fat and macromolecules). Not surprisingly, fat appeared in our system as the most important non-volatile constituent affecting aroma retention and release. This study has clearly demonstrated that aroma, taste, and texture are difficult to dissociate due to physico-chemical interactions between the various molecule species, which certainly play a role in the global sensory perception. In addition, there are probably many other factors influencing the sensory perception as well, such as physiological and psychological interactions, which should be taken into consideration.

References

1. K. Swift (ed.), Current topics in flavours and fragrances, Dordrecht, The Netherlands (1999) 123.
2. E. Guichard, Food Rev. Int., 18 (2002) 49.
3. S. Secouard, C. Malhiac, M. Grisel and B. Decroix, Food Chem., 82 (2003) 227.
4. E. Bylaite, J. Adler-Nissen and A.S.Meyer, J. Agric. Food Chem., 53 (2005) 3577.
5. A. Juteau, J. L. Doublier and E. Guichard, J. Agric. Food Chem., 52 (2004) 1621.
6. B. Rega, E. Guichard and A. Voilley, Sci. Aliment., 22 (2002) 235.
7. E. Guichard and S. Langourieux, Food Chem., 71 (2000) 301.
8. K.B. de Roos, Food Technol., 51 (1) (1997) 60.
9. E. Graf and K.B. de Roos, ACS Symp. Ser., 633 (1996) 24.
10. H. Maarse and P.J. Groenen (eds.), Aroma research, Wageningen, The Netherlands (1975) 143.
11. A.P. Hansen and J.J. Heinis, J. Dairy Sci., 74 (1991) 2936.
12. A.-M. Haahr, W.L.P. Bredie, L.H. Stahnke, B. Jensen and H.H.F. Refsgaard, Food Chem., 71 (2000) 355.

W.L.P. Bredie and M.A. Petersen (Editors)
Flavour Science: Recent Advances and Trends

A simple model for explaining retronasal odour properties of odorants through their volatility

Vicente Ferreira[a], Jan Pet'ka[b] and Juan F. Cacho[a]

[a]Lab for Flavor Analysis and Enology. Analytical Chem. Fac. of Sciences, University of Zaragoza. 50009 Zaragoza, Spain; [b]Food Research Institute, Priemyselná 4, 824 75 Bratislava, Slovakia

ABSTRACT

A simple model able to interpret the role of the volatility of a given compound on its ortho/retro nasal odour properties and on the persistence of its aftersmell is derived. The model considers the mouth as a perfect mixing tank containing a certain amount of odorants, whose concentrations are progressively lessened by two independent processes: dilution with saliva and swallowing and stripping by the air inhaled or exhaled. Retronasal properties are related to the orthonasal odour properties of the liquid contained in the mouth. The model fits experimental data quite well.

1. INTRODUCTION

Different studies have demonstrated that the volatility of an odorant plays a key role in determining the relative intensity of the ortho/retro perceptual ways and on the intensity and duration of its aftersmell [1-4]. However, the models used are rather empirical and non-mechanistic, and so far, the parameters representing the volatility of an odorant in the proposed models are in all cases logarithms of equilibrium constants and not the equilibrium constants themselves, which are the parameters to which the actual concentration of odorant in the vapour phases are really related. In the present work a simple model derived from a general differential equation is proposed.

2. MATERIALS AND METHODS

Sensory analysis: Hydro alcoholic solutions containing different concentrations (three levels plus a blank) of eight odorants (a single odorant per solution) were evaluated by a trained sensory panel using a modified 7-point labelled magnitude scale (graphic scale with logarithmic elements described as follows: 0 = no odour, 1 = weak odour of low intensity, 2 = clear perception of odour of intermediate intensity, 3 = extremely intense odour; the intermediary values did not bear any description). First, the judges rated odour intensity of the odour by nose (orthonasal perception). This test was made successively for all cups, with time gaps to prevent adaptation. Then they were asked to sip from the liquid, keep it in the mouth swirling, savour it and rate odour intensity (buccal perception). After swallowing, the judges had to start the stopwatch and stop it when the perception of odour disappeared (aftersmell persistence). Additionally, they had to rate the maximum perceived intensity of the odour during this time (retronasal perception).

Purging constant: Synthetic wine (150 ml) containing the eight volatiles studied was poured into a bubbler flask. A stream of nitrogen was bubbled for 200 min through the solution at 100 ml/min. The volatile-enriched stream of nitrogen was passed through a trap consisting of a cartridge containing 400 mg of LiChrolut EN resin (Merck, Darmstadt, Germany). After this, volatiles were eluted with 3.2 ml of dichloromethane, 30 μl of internal standard (2-octanol, c = 700 ppm) were added to the extract which was analysed on a Fisons 8000 series gas chromatograph equipped with a flame ionisation detector. The column used was a DB-Wax (J&W Scientific, Folsom, CA; 60 m x 0.32 mm x 0.5μm). Chromatographic conditions were as follows: hydrogen as carrier gas (3 ml/min); splitless injection (splitless time 90 s); injection volume, 3 μl; injector temperature 250 °C; detector temperature 250 °C; temperature program 40 °C for 2 min, raised at 6 °C/min to 200 °C and held at this temperature for 15 min, then raised to 220 °C and held at this temperature for 20 min. Quantification was made on the basis of calibration graphs obtained for each compound.

The purging constant K_p was calculated as follows:

$$K_P = \frac{\log\left(\frac{W_0^i}{(W_0^i - W_e^i)}\right)}{t} \qquad (1)$$

where W_0^i is the initial mass of compound i in the extractor, W_e^i is its mass in the extract and t the duration of the stripping process.

3. MODEL

According to our model, the amount of volatile lost by swallowing or by transference to the gas phase (considered continuous processes) in time t can be expressed by the following differential equation:

$$-\frac{dC_i}{dt} = \left(K_{pm}^i + \frac{F}{V}\right)C_i \qquad (2)$$

where C_i is the concentration of volatile in the mouth at time t, F is the flow of saliva and the average flow swallowed, V is the volume of liquid in the mouth, K_{pm}^i is the mouth purging constant for volatile i. The equation simply expresses the idea that the amount of volatile compound remaining in the mouth after swallowing decreases due to two independent processes: transfer to the gas phase (controlled by K^{i}_{pm}) and progressive dilution with saliva and swallowing (controlled by F/V).

The integration of this expression gives:

$$\log C_{it} = \log C_{io} - \frac{1}{2.303}\left(K^{i}_{pm} + \frac{F}{V}\right)t \qquad (3)$$

where C_{it} is the concentration of volatile i at time t and C_{io} is the concentration of that volatile at time 0. Approximated parameters for this function have been estimated as follows: F has been taken as 1 ml/min, V has been taken as 1 ml and the K^{i}_{pm} values have been obtained from a dynamic headspace sampling system and have been used directly after normalisation to the volume of gas pumped from the mouth to the nostrils in average (15 x 100 = 1500 ml/min) and to the volume of liquid (1 ml). Under these conditions $K_{pm} = 1000K_p$.

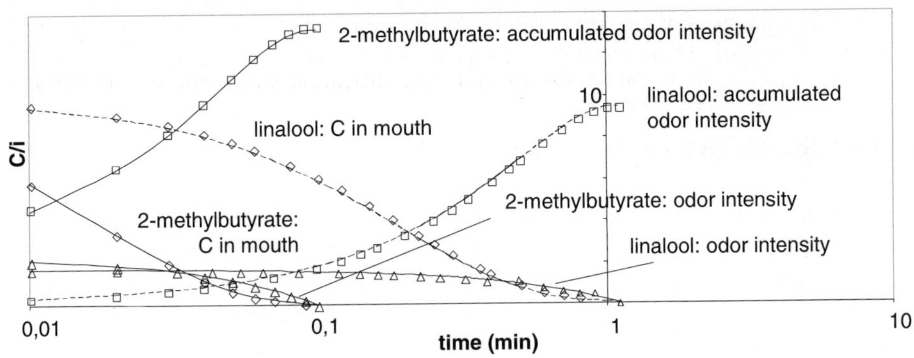

Figure 1. Practical example showing the calculation of the Accumulated Odour Intensities (AOI).

Figure 1 shows the basic ideas of the model for 2-methylbutyrate and linalool. The lines marked with a rhombus show the predicted evolution of the concentration of these compounds in the mouth (initial concentration is 10 ppm in both cases) according to equation 3. The odour intensity that the concentrations elicit are represented by the two lines marked with triangles and have been calculated from the orthonasal odour intensity/concentration plots built via direct sensory analysis of the solutions. The duration of the aftersmell of the compounds can be estimated as the point at which the intensities become zero. Finally, the two lines marked with squares give the integrals of the intensities (accumulated odour intensities). The retronasal intensities of the solutions are proportional to the square root of these latter values.

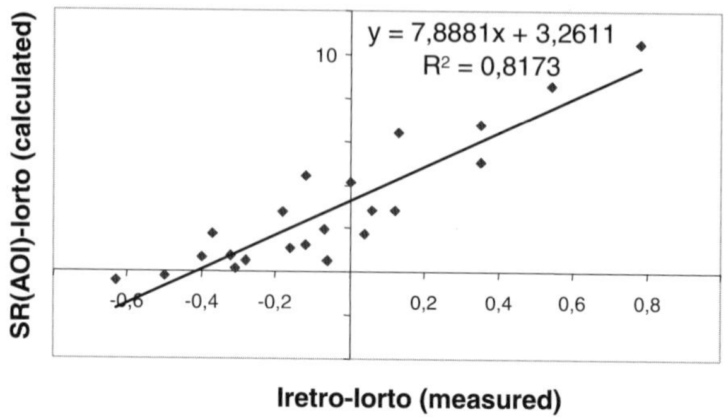

Figure 2. Relationship between the square root of the AOI and the difference between the retronasal and orthonasal odour intensities.

This simple model fits quite well the experimental results, as it can be seen in Figure 2. The figure shows the correlation between the measured and the predicted values for the difference between the orthonasal and retronasal odour intensities of a given solution for each one of the 24 solutions essayed in our experiment (8 chemicals at three different concentration levels). Similarly satisfactory correlations were found for persistence. Furthermore, the model can be applied to data presented by other authors following different techniques. For instance, the quotients mouthspace/headspace calculated by Linforth and Taylor [1] are strongly correlated to the square roots of the corresponding integrals of function 3.

4. DISCUSSION AND CONCLUSION

Once the food has been swallowed and there are no solids remaining in the mouth, the retronasal odour properties and the persistence of the aftersmell can be easily predicted by taking into account that in this simplified condition, the odorants only disappear by passing to the gas-phase and by progressive dilution and swallowing. The first process is strongly related to the volatility of the odorant and has a major influence on the time the odorant actually resides in the mouth, and therefore, on its persistence and on the intensity perceived retronasally.

References

1. R. Linforth and A.J. Taylor, J. Agric. Food Chem., 48 (2000) 5419.
2. S.M. van Ruth, C.H. O'Connor and C.M. Delahunty, Food Chem., 71 (2000) 393.
3. R. Linforth, F. Martin, M. Carey, J. Davidson and A.J. Taylor, J. Agric. Food Chem., 50 (2002) 1111.
4. M. Espinosa, Flavour Fragrance J., 19 (2004) 499.

W.L.P. Bredie and M.A. Petersen (Editors)
Flavour Science: Recent Advances and Trends

Volatile delivery under dynamic gas flow conditions

Robert S.T. Linforth and Andrew J. Taylor

Division of Food Sciences, School of Biosciences, University of Nottingham, Sutton Bonington Campus, Loughborough, Leics. LE12 5RD, UK

ABSTRACT

Under dynamic gas flow conditions (dynamic headspace or *in vivo*), the measured volatile concentrations in the headspace are very different to those expected from air-water partition data. This is due to the limited amount of the sample (effectively the immediate surface) which is involved in volatile partitioning, and restricted delivery of molecules from the bulk interior over time. Compounds with high (10^{-2}) air/water partition coefficients deliver flavour inefficiently compared those with lower ones (10^{-4}), primarily because of the proportion of molecules that have to be transferred from the solution to the gas phase as they try to reach equilibrium.

1. INTRODUCTION

The tidal volumes of air passing through the upper airway are large compared to its volume [1]. The tidal flow of breath transfers approximately 100 ml of air between the lungs and the atmosphere at velocities in the region of 150 ml/s [2]. Equally, the mouth (often thought of as a separate side chamber of the upper airway) can have over a litre of air pumped through it due to the frequent (100 chews/min) bellows-like action of chewing [2]. Consequently, the environment in which flavour is delivered during food consumption is characterised by high gas flows and is very dynamic.

Studies modelling flavour delivery *in vitro* under dynamic gas flow conditions showed that there were differences between compounds related to their air/water partition coefficients (K_{aw}) [3]. In addition, food components (e.g. emulsions) that changed the partition coefficient (air/product K_{ap}) of a compound could also influence behaviour under dynamic gas flow conditions [4]. The headspace concentration of compounds with K_{ap} values of 10^{-2} decreased rapidly and substantially during headspace dilution compared to compounds with lower K_{ap} values (10^{-4}). This is because molecules were leaving the surface faster than they could be replaced by diffusion and convection from

the bulk phase. The differences between compounds arose because, when the K_{ap} was low, only 0.01% of the molecules had to cross the interface into the gas phase to maintain equilibrium (assuming that the volume of solution actively involved in partitioning at the interface, was equal to the volume of diluting gas), whereas when the K_{ap} was high this would require 1% of the molecules present to be transported.
These experimental *in vitro* flavour delivery studies were focused on an unstirred aqueous phase with headspace dilution taking place over several minutes. The purpose of this paper was to determine whether the same differences in flavour delivery are seen *in vivo*, where time scales are much shorter, gas flows much greater and the liquid phase may undergo some stirring.

2. MATERIALS AND METHODS

Atmospheric Pressure Chemical Ionisation - Mass Spectrometry (APCI-MS) was used for both dynamic breath analyses and headspace measurements. For headspace, 500 ml aliquots of solutions of flavour compounds were prepared, placed in 1 litre flasks and allowed to reach equilibrium. The gas phase was then sampled directly into the APCI-MS at 30 ml/min and the signal intensity noted (proportional to the gas phase volatile concentration). Breath flavour measurements involved two panellists consuming 15 ml aliquots of the solutions previously used for headspace analyses. The panellists were instructed to place each solution into their mouths, swallow it, and then exhale via the mouth through a tube connected to the end of the APCI-MS sampling tube (also sampling air at 30 ml/min). Panellists consumed two replicates of each sample.
The average signal intensity when breath was sampled into the APCI-MS was divided by that for the headspace samples to show the relative concentrations of the two systems. By using the breath/headspace ratio the values obtained are indicative of the relative differences in the two gas phase concentrations and hence effectively independent of both the solution concentration and K_{aw}.

3. RESULTS AND DISCUSSION

The ratio of volatiles in the breath to the headspace concentration was compound dependent (Figure 1, Table 1). For compounds with low values of K_{aw} (10^{-4}) the concentration of the volatiles in the breath was close to that observed in the headspace, around 60%. However, the compounds with high values of K_{aw} (10^{-2}) were present at very low concentrations in the breath compared to headspace (about 10%). Overall there was a good linear correlation ($R^2 = 0.72$) between a compounds behaviour *in vivo* and its K_{aw}.
These results are consistent with studies of flavour release from emulsions, where the large differences observed between compounds dissolved in emulsions and water during headspace analysis are substantially reduced *in vivo* [4,5]. A lipophilic compound in water ($K_{aw} = 10^{-2}$) will deliver flavour inefficiently in vivo and may only achieve 1% of its headspace concentration. The same compound dissolved in an emulsion will have a far lower headspace concentration ($K_{aw} = 10^{-4}$), but, *in vivo* will deliver flavour more

efficiently reaching 50% of its headspace concentration. Hence, two systems that differ by a factor of 100 in their headspace concentration will only differ by a factor of 2 *in vivo*.

It is unlikely that these differences in flavour release are due to dynamic gas flow during swallowing. Air does not pass between the lungs and the atmosphere during swallowing and is effectively static [2], molecules pass from the solution into the gas phase as it is swallowed and are exhaled in a 'plug' that is representative of the extent of equilibration.

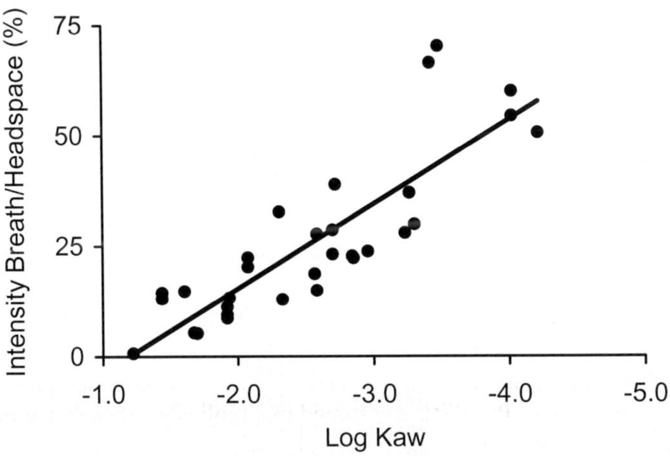

Figure 1. Relationship between Log K_{aw} and the breath/equilibrium headspace ratio. Reprinted with permission from Linforth *et al.*, J. Agric. Food Chem., 50, 1111-1117. Copyright 2002 American Chemical Society.

The fraction of the solution (and hence active volume) participating in volatile delivery could be the limiting factor. When small volumes of solutions of low K_{aw} compounds are allowed to reach equilibrium with large volumes of air (e.g. 1:100), the final concentration of the gas phase will be reasonably close to that achieved with equal volumes of the two phases. This is because the proportion of molecules required per volume of gas is low. In contrast, high K_{aw} compounds have to transfer a larger proportion of the molecules present (per volume of air) to reach equilibrium. This can deplete the amount of these compounds in the solution, resulting in a low final concentration in both phases.

The fraction of the solution delivering volatile compounds to the air may be limiting if there is limited movement of molecules from the bulk of the sample to the surface. The breath volatile concentration when consuming volatile compounds dissolved in water, or water thickened with hydrocolloids was found to be identical [6]. This would occur if the hydrocolloid did not significantly affect the transfer of bulk elements of the sample to the interface, implying limited movement for water itself.

4. CONCLUSIONS

It would appear that the restricted movement of volatile compounds from the bulk phase to the interface can be limiting for *in-vivo* flavour release (compared with static headspace analysis). This is particularly the case when achieving equilibrium requires the movement of a large proportion of the molecules from the liquid to the gas phase.

Table 1. Ratio (percent) between exhaled breath and headspace (HS) volatile concentration for a range of compounds. Each value is the mean of 4 replicates.

Compound	Breath/HS (%)	Compound	Breath/HS (%)
1-Butanol	70	Ethyl acetate	22
Ethanol	66	Octanol	22
Propan-2-ol	65	Octanone	20
Pyrazine	60	Acetaldehyde	19
Dimethyl pyrazine	51	Menthone	19
3-Hexenol	49	2-Pentanone	15
Furfuryl acetate	48	Isoamyl acetate	15
Diethyl methylpyrazine	47	2-Methylbutanal	15
2-Methylbutanol	45	Hexanal	14
Butanone	39	Ethyl butyrate	13
Carvone	39	Hexanal	13
Hexanol	38	Butanal	13
Diacetyl	37	2-Methylbutanal	9.9
Methyl acetate	33	Ethyl hexanoate	8.6
Benzaldehyde	30	Methyl propanal	7.8
2-Hexenal	29	Citronellal	7.6
Guaiacol	29	Decanone	6.3
Carvone	28	Octanal	5.4
Linalool	28	Decanol	5.2
Terpineol	28	Ethyl octanoate	2.6
2-Pentanone	28	Decanal	2.1
1,8-Cineole	26	Methyl furan	1.1
Menthol	24	Limonene	0.68
Isobutyl methoxypyrazine	23	Cymene	0.52
Methyl salicylate	23	Pinene	0.17

References

1. M. Damm, J. Vent, M. Schmidt, P. Theissen, H.E. Eckel, J. Lotsch and T. Hummel, Chem. Senses, 27 (2002) 831.
2. M. Hodgson, R.S.T. Linforth and A.J. Taylor, J. Agric. Food Chem., 51 (2003) 5052.
3. M. Marin, I. Baek and A.J.Taylor, J. Agric. Food Chem., 47 (1999) 4750.
4. K. Doyen, M. Carey, R.S.T. Linforth, M. Marin and A.J. Taylor, J. Agric. Food Chem., 49 (2001) 804.
5. D.D. Roberts, P. Pollien, N. Antille, C. Lindinger and C. Yeretzian, J. Agric. Food Chem., 51 (2003) 3636.
6. D.J. Cook, T.A. Hollowood, R.S.T. Linforth and A.J. Taylor, Chem. Senses, 28 (2003) 11.

W.L.P. Bredie and M.A. Petersen (Editors)
Flavour Science: Recent Advances and Trends

Determination of specific interactions between aroma compounds and xanthan/galactomannan mixtures

Céline Jouquand, Catherine Malhiac and Michel Grisel

Unité de Recherche en Chimie Organique et Macromoléculaire, URCOM, EA 3221, Université du Havre, 25 rue P. Lebon, BP 540, F-76058 Le Havre cedex, France

ABSTRACT

The gas/liquid partition coefficient (K) of aroma compounds was measured in pure polysaccharide solutions and in polysaccharide mixtures, using the Phase Ratio Variation method. With this method, the retention in 0.1% and 0.02% polysaccharide concentrations was determined. The results showed that in semi-dilute regime (0.1%) all solutions and mixtures induced a high retention of limonene and hydrophobic esters, thus confirming hydrophobic interactions between these compounds and the polymer chains. In dilute pure polysaccharide solutions (0.02%), a decrease in retention was noticed for the esters and a 'salting out' effect for limonene. Inversely, at 0.02%, mixtures of polysaccharides showed a good retention of aroma compounds explained by the well-known synergistic interactions between xanthan and carob gums.

1. INTRODUCTION

Polysaccharides, when used as thickeners in food formulation, are commonly known to reduce flavour release. For example, the synergistic interactions between xanthan gum (X) and galactomannans (carob gum (C) and guar gum (G)) play a major role in limonene retention according to two distinct mechanisms: (i) the decrease of limonene diffusion due to the viscosity of the matrix and (ii) the specific molecular interactions between this volatile compound and both polysaccharides [1]. The aim of this work was to determine the nature of physico-chemical interactions between aroma compounds and polymers by comparing their gas/liquid partition coefficient in water (K_{water}) and in matrices ($K_{matrices}$). Partition coefficients were measured by the Phase Ratio Variation (PRV) method. This method was chosen because it does not require external or internal calibration [2]. The differences between the interactions of pure polysaccharide solutions and polysaccharide mixtures were examined.

2. MATERIALS AND METHODS

The physico-chemical properties of aroma compounds are given in Table 1 and the concentrations of pure polysaccharide solutions and mixtures in Table 2. Limonene was first dissolved in ethanol at 2 g/l. Samples of xanthan, carob and guar gum were given by Systems Bio Industries (SKW). Polysaccharides were dispersed in pure water, in the presence of 0.01% NaCl and 400 ppm NaN_3 [1]. The solutions and mixtures concentrations were 0.02% (w/w) and 0.1% (w/w) for all solutions and mixtures. The limonene concentration was 6 ppm and the concentration of the esters was 25 ppm.

Table 1. Physico-chemical properties of aroma compounds [4] and K_{water}.

Aroma compounds	Solubility in water g/l (25 °C)	Saturated pressure mmHg (25 °C)	LogP	K_{water} at 35 °C (0.01 M NaCl)
Limonene	0.0138	1.98	4.6	1.5 ± 0.5
Ethyl butyrate	5.6	17.3	1.7	0.030 ± 0.04
Ethyl hexanoate	0.629	1.56	2.8	0.056 ± 0.03
Ethyl octanoate	0.07	0.211	3.8	0.134 ± 0.04

Table 2. Polysaccharide and mixture concentrations delimiting dilute ($c<C^*$), semi-dilute ($C^*<c<C^{**}$) and concentrated ($C^{**}<c$) regimes [1].

Polysaccharide matrices		C* (%)		C** (%)	
Xanthan		0.036		0.18	
Carob	Dilute	0.11	Semi-dilute	0.24	Concentrated
Guar	regime	0.08	regime	0.21	regime
Xanthan/carob		0.01		-	
Xanthan/guar		0.017		-	

The PRV method is based on the relationship between the phase ratio β (the ratio of the gas phase volume to the liquid phase volume) and the partition coefficient K according the following equation [2,3]:

$$C^{eq}_G = C_L / (1/K + \beta)$$

C_L: initial concentration of the compound in the liquid phase; C^{eq}_G: concentration of the compound in the gas phase at equilibrium.

Increasing volumes (1, 2, 3 and 4 ml) of polysaccharide solutions or mixtures were poured into vials (20.5 ml) to obtain different phase ratios β. Each volume was analysed in duplicate. The equilibration time was 8 h for all esters and 15 h for limonene, at 35 °C. The volume of the gas phase taken automatically by a gas syringe was 1 ml. Measurements of K were done in triplicate by GC analysis [1].

The aroma retention was determined according the following equation:

$$R = [1- (K_{matrix} / K_{water})] *100$$

3. RESULTS

3.1. Limonene

The retention of limonene in polysaccharide matrices at 35 °C is given in Figure 1. At 0.1% (semi-dilute regime for all matrices except for carob solution, Table 2), limonene was retained by all the solutions and the mixtures with a maximum of retention (> 60%) for xanthan and the mixtures, galactomannans being less efficient. Carob solutions at 0.1% (equal to C*) induced a behaviour close to that in water.
At 0.02%, limonene was only retained by the X/C mixture (semi-dilute regime, Table 2). For X/G mixtures, the 0.02% concentration which is close to C* (0.017%) induced the same behaviour as in water. At 0.02%, the other polysaccharide solutions were in dilute regime and the retention of limonene became negative, i.e. its partition coefficient was higher in matrices than in water. The aroma retention seemed to depend on the concentration regime of the polysaccharide solutions and mixtures: above C*, limonene was retained by the polysaccharides whereas below C*, there was an increase in limonene release with a high 'salting-out' effect in xanthan solutions.

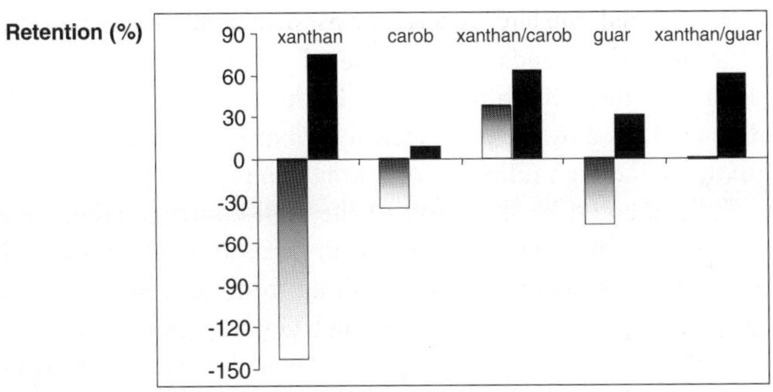

Figure 1. Retention of limonene in polysaccharide solutions and mixtures at two concentrations (0.02% ( ) and 0.1% ( )) at 35 °C.

3.2. Homologous series of esters

As for limonene, the concentration regime seemed to influence the retention of esters (Figure 2). However, the change of concentration regime (in carob and xanthan solutions) induced a 'salting-out' effect only for ethyl butyrate and a decrease in retention for the other esters. In X/C mixtures, the decrease in concentration induced a weak decrease in the retention of esters.
According to Figure 2, the retention of esters is linked to their hydrophobicity with the highest retention (>40%) for ethyl octanoate in xanthan and X/C matrices. This result

could be explained by the existence of hydrophobic interactions between esters and macromolecular chains. Because of the apolar nature of limonene, the same type of interactions can be expected for this compound.

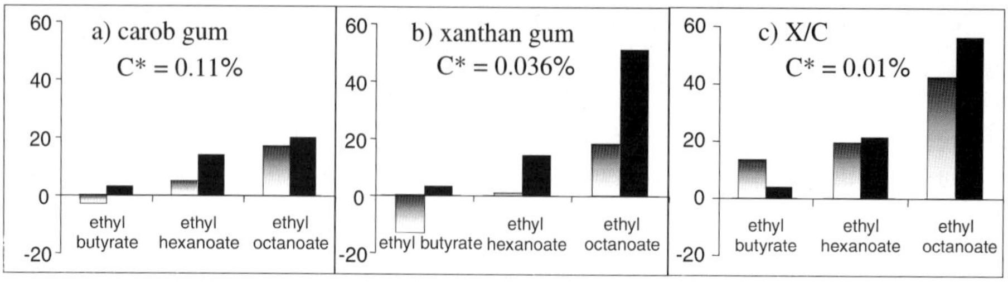

Figure 2. Retention (%) of esters in polysaccharide matrices (a) carob gum, (b) xanthan gum, (c) X/C (50/50) at 35 °C. ( ) 0.02%; ( ) 0.1%.

4. DISCUSSION AND CONCLUSION

The influence of concentration regime on aroma retention could be explained by the following mechanism: in a semi-dilute regime, the macromolecular chains begin to overlap. This could favour physico-chemical interactions with aroma compounds through steric phenomena. In a dilute regime (xanthan, carob, guar at 0.02%), the polymer chains are isolated from each other and no overlapping occurs thus limiting interactions with aroma compounds. In the case of limonene and ethyl butyrate, it is shown by an increase of their volatility, especially in xanthan solutions. This 'salting-out' effect could be explained by the polyelectrolytic behaviour of xanthan.

In 0.02% X/C mixtures, the high retention of aroma compounds when compared to pure polysaccharide solutions could be the result of the well-known synergistic interactions between xanthan and galactomannans [5-7]. It can be assumed that aroma compounds are entrapped in specific hydrophobic zones resulting from these synergistic interactions between both polymers. The formation of such hydrophobic zones was already proposed to explain the retention of apolar compounds in polysaccharide solutions [8].

References

1. S. Secouard, C. Malhiac, M. Grisel and B. Decroix, Food Chem., 82 (2003) 227.
2. C. Jouquand, V. Ducruet and P. Giampaoli, Food Chem., 85 (2004) 467.
3. L.S. Ettre, C. Welter and B. Kolb, Chromatogr., 35 (1993) 73.
4. Syracuse research corporation (2005). http://esc.syrres.com/interkow/physdemo.htm
5. R. Chandrasekaran and A. Radha, Carbohydr. Polym., 32 (1997) 201.
6. T.M.B. Bressolin, P.C. Sander, F. Reicher, M.R. Sierakowski, M. Rinaudo and J.L.M.S. Ganter, Carbohydr. Polym., 33 (1997) 131.
7. C. Schorsch, C. Garnier and J-L. Doublier, Carbohydr. Polym., 34 (1997) 165.
8. D.D. Roberts, J.S. Elmore, K.R. Langley and J. Bakker, J. Agric. Food. Chem., 44 (1996) 1321.

W.L.P. Bredie and M.A. Petersen (Editors)
Flavour Science: Recent Advances and Trends

NMR Spectroscopy study of interactions between β-lactoglobulin and aroma compounds

Céline Moreau, Laurette Tavel, Jean-Luc Le Quéré and Elisabeth Guichard

Unité Mixte de Recherche FLAVIC (INRA/ENESAD), 17 rue de Sully, BP86510, Dijon 21065 Cedex, France

ABSTRACT

Interactions between five aroma compounds (β-ionone, α-ionone, γ-decalactone, 2-nonanone and eugenol) with β-lactoglobulin in its monomeric state have been investigated by 2-Dimensional TOCSY ^{1}H NMR using the chemical shift change method. The comparison of amino acid chemical shifts of the protein alone and of each protein-flavour complex in identical experimental conditions gave information on which protein amino acids are perturbed by the presence of the aroma compound. As amino acids have been first assigned, localisation of protein binding sites has been determined for each complex and two different groups have been pointed out.

1. INTRODUCTION

Binding interactions of proteins and flavour compounds have been widely studied due to their effect on flavour release and perception [1]. Headspace or chromatographic techniques used to investigate the retention of flavour compounds in food matrices give only indirect information on aroma binding sites whereas information at a molecular level can be obtained using spectroscopic techniques. However, controversial results have been obtained for the interaction or the determination of binding sites of flavour compounds [2-4]. Among all these techniques, NMR is probably the most effective in investigating ligand/macromolecule interactions since it is a non invasive method and ligand/macromolecule complexes can be investigated in their natural environment (pH, temperature, buffer conditions). NMR spectroscopy is a powerful technique to give information at molecular level on binding site localisation and on the nature of the interactions. A first study realised in this laboratory has shown that 2-Dimensional NMR spectroscopy (2D NMR) can be successfully applied to protein/flavour compound complexes [5]. In this work, we studied interactions of five aroma compounds with β-lactoglobulin by 2D-TOCSY proton NMR using the chemical shift method. Any ligand binding event will lead to a perturbation of some amino acid chemical shifts of the protein. In a first step, amino acid chemical shifts of the protein alone have been assigned. In a second step, these chemical shifts are compared to those of the protein in the presence of a flavour compound in the same experimental conditions in order to determine which amino acids are perturbed.

2. MATERIALS AND METHODS

β-Lactoglobulin (purity ≥ 90 %, Sigma Aldrich) at 1 mM concentration was prepared in 12 mM sodium salt solution and the pH adjusted to 2.3 using diluted HCl. In these conditions the protein is in its monomeric state. Flavour solutions (Sigma Aldrich) were prepared in identical experimental conditions. From binding constant values, K_D [4], and solubility limit of each aroma compound, flavour/protein complexes were prepared to obtain at least 55% of complexed protein.

Proton NMR spectra were recorded at 315 K on a Bruker Avance DRX 500 MHz spectrometer operating at a proton frequency of 500.13 MHz and equipped with a 5 mm z-gradient probe. Temperature was calibrated using a Bruker sample temperature calibration tube (80% glycol-DMSO). In all samples, 10% of D_2O was added in order to guaranty a correct spin lock while precluding any significant H-D exchange. The residual HDO signal at 4.71 ppm was used as internal reference. Two-dimensional total correlation spectroscopy (2D TOCSY) NMR experiments using the watergate sequence for water signal suppression, were realised using a 90° flip angle impulsion length of 7 μs, a recycling delay of 3 s, a mixing time delay of 40 ms with a minimum of 32 scans and 512 x 1024 data points.

3. RESULTS

In a first step, a 2D TOCSY ^{1}H NMR spectrum of the protein alone was acquired in order to assign the chemical shifts of protein amino acids. Figure 1a shows a partial view of the entire TOCSY spectrum showing cross-peak belonging to the NH/CHα region of the β-lactoglobulin. Each cross-peak corresponds to one amino acid and their assignments have been done based on the previous work of Uhrinova *et al.*[6] .

2D TOCSY ^{1}H NMR spectra of β-lactoglobulin in the presence of each flavour compound showed a number of shift variations. As shown in Figure 1b, only eight amino acids of the protein are shifted in the presence of α-ionone, named Ile29, Ser30, Glu51, Lys101, Tyr102, Val128, Asp129 and Ala132. In the presence of β-ionone and eugenol, twelve chemical shift changes are observed. The comparison of these three spectra showed that seven perturbed amino acids are in common, i.e. Ile29, Ser30, Glu51, Lys101, Tyr102, Val128, Asp129 suggesting an identical binding site on the protein for these three flavour compounds.

a

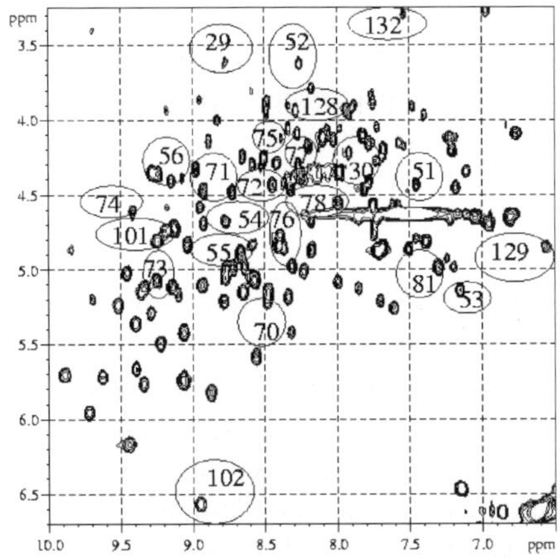

b

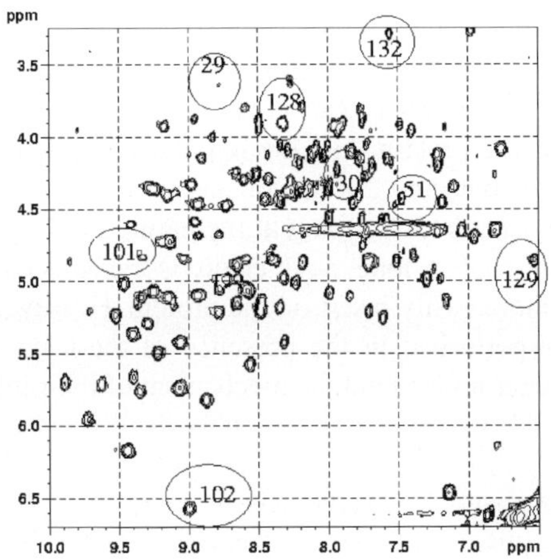

Figure 1. Zoom of the NH/CHα region of the entire TOCSY NMR spectrum of the β-lactoglobulin alone with cross-peaks of interest (a) and the β-lactoglobulin/α-ionone complex with perturbed cross-peaks (b).

In the presence of γ-decalactone and 2-nonanone, 2D TOCSY ^{1}H NMR spectra showed a different pattern of shift variations. Indeed, almost all the amino acids are perturbed whatever the molar protein/flavour compound ratio (1:1 and 1:2). However, we noted that some amino acids showed specific high shift variations which are located between **g**Glu51-Ile56 and Lys70-Val81 and these amino acids are not perturbed with β-ionone, α-ionone and eugenol.

4. DISCUSSION AND CONCLUSION

From the 2D TOCSY ^{1}H NMR spectroscopy using the chemical shift method, we showed that the localisation of binding sites of the flavour compounds to the β-lactoglobulin can be determined. The five aroma compounds could be divided into two groups depending on the chemical shift variation pattern of the protein amino acids. In a first group, α-ionone, β-ionone and eugenol lead to a variation of few amino acids which are located in the outer surface. As seven common amino acids are perturbed, these three flavour compounds interact with the protein on an identical binding site. These amino acids are principally located on the outer surface of the protein.

In the second group, γ-decalactone and 2-nonanone seem to have no specific binding sites as all amino acids are perturbed. However, the observation that some amino acids are more highly perturbed while they are not perturbed with flavours of the first group, suggests a more complex interaction. Furthermore, these amino acids are located on sheets in the hydrophobic pocket.

These results are in favour of two different binding sites of flavour compounds to the β-lactoglobulin. Contrary to some previous results, α-ionone and β-ionone have the same binding site on the protein in these conditions. Moreover, the magnitude of the NMR shift variations of the two ionones is also in agreement with the fact that the β-ionone has a higher affinity for the protein than the α-ionone [4]. The presence of a specific binding site for γ-decalactone and 2-nonanone is more controversial since all amino acids of the protein are perturbed. Looking at the nature of flavour compounds, we can suggest that the presence of a long acyl chain is responsible for the non-specific interactions as it could more easily insert on the protein. However, as some amino acids seem to be specifically perturbed in the presence of these compounds, other studies must be performed to better understand the mechanism of their interaction.

References

1. E. Guichard, Food Rev. Int., 18 (1) (2002) 49.
2. E. Dufour and T. Haertlé, J. Agric. Food Chem., 38 (8) (1990) 1691.
3. E. Jouenne and J. Crouzet, J. Agric. Food Chem., 48 (11) (2000) 5396.
4. K. Sostmann and E. Guichard, Food Chem., 62 (4) (1998) 509.
5. M. Lübke, E. Guichard, A. Tromelin and J.L. Le Quéré, J. Agric. Food Chem., 50 (2002) 7094.
6. S. Uhrinova, D. Uhrin, H. Denton, M. Smith, L. Sawyer and P.N. Barlow, J. Biomol. NMR, 12 (1998) 89.

W.L.P. Bredie and M.A. Petersen (Editors)
Flavour Science: Recent Advances and Trends

Effect of gum base and bulk sweetener on release of specific compounds from fruit flavoured chewing gum

Herdis Overgaard Fisker and Vibeke Nissen

Chew Tech I/S, Dandyvej 19, DK-7100 Vejle, Denmark

ABSTRACT

The effect of gum base composition and type of bulk sweetener used in chewing gum on release of different volatile flavour compounds, all present in an artificial fruit flavour, was investigated. Release of single flavour compounds from chewing gum to the nose space during chewing was measured by APCI-MS. Both the gum base composition - one being more polar than the other - and the type of bulk sweetener (sorbitol, xylitol and mannitol, respectively) were found to affect the release of the flavour compounds investigated. Using the more polar gum base, more of the flavour compounds were released during chewing. Using mannitol as the bulk sweetener, more of the flavour compounds were released during chewing.

1. INTRODUCTION

Long-lasting taste is the key quality parameter for chewing gum. As many volatile flavours are perceived together with sweeteners giving the overall impression [1], the challenge of producers of chewing gum is to create a matrix that releases both volatile compounds and sweeteners in a way that gives a uniform profile during chewing for up to twenty minutes.

The individual chewing gum ingredients and especially interactions between the ingredients are the overall determining factors for release of volatile flavour compounds and sweeteners from chewing gum during chewing. The composition of the gum base itself is an important factor for this release, but it has been shown in mint flavoured chewing gum [2] that also the type of bulk sweetener influences release of flavour compounds (menthone).

In this study, the effect of gum base composition and type of bulk sweetener used in chewing gum on release of different volatile flavour compounds, all present in an artificial fruit flavour, was investigated.

2. MATERIALS AND METHODS

Six chewing gums based upon two different gum bases were investigated. The two gum bases differed mostly with respect to polarity. For each gum base, the chewing gum was made with sorbitol, xylitol or mannitol as the bulk sweetener. The chewing gums were flavoured with the same concentration (3%) of an artificial fruit flavour containing, among other flavour compounds, ethylbutanoate, isopentylacetate, limonene and linalool. The concentration of the single flavour compounds in the total flavour was not equal. All chewing gums tested were without coating layer.

Release of flavour compounds from the chewing gums during chewing was measured by APCI-MS at Danish Technological Institute (Kolding, Denmark). Four persons were served the six different chewing gums in triplicate. The samples were chewed with one chew per second. Release profiles of ethylbutanoate, isopentylacetate and limonene/linalool from chewing gum to the nosespace during chewing for 7 min were followed by APCI-MS.

3. RESULTS

Release curves of ethylbutanoate, isopentylacetate and limonene/linalool from the chewing gums are shown in Figures 1-3. The different levels of release for the three flavour compounds are not only due to their different volatility but also to different concentrations of the single compounds in the total flavour.

Overall, more of the flavour compounds investigated are released from gum base 24 compared to gum base 26. This is most pronounced for ethylbutanoate and isopentylacetate, but the tendency is the same for limonene/linalool. The shapes of the release curves of the flavour compounds are only slightly affected by the type of gum base, mostly for limonene/linalool. There are no obvious trends regarding interactions between type of gum base and type of bulk sweetener in the results.

Using mannitol or xylitol as bulk sweetener in the tested chewing gums increased the release of the investigated flavour compounds. The release of ethylbutanoate and isopentylacetate was clearly higher in chewing gum with xylitol and mannitol than in chewing gum with sorbitol. Limonen/linalool was clearly released at a higher level from chewing gums with mannitol, while chewing gums with xylitol or sorbitol released limonene/linalool at equal levels.

Looking at the shapes of the release curves it is clear that limonene/linalool released at a different rate than the other flavour compounds investigated. Limonene/linalool released at a slower rate and showed an increasing concentration during the whole chewing period. The other flavour compounds reached a maximum intensity after 1-2 min of chewing and were kept at a steady or slightly decreased level during the rest of the chewing period.

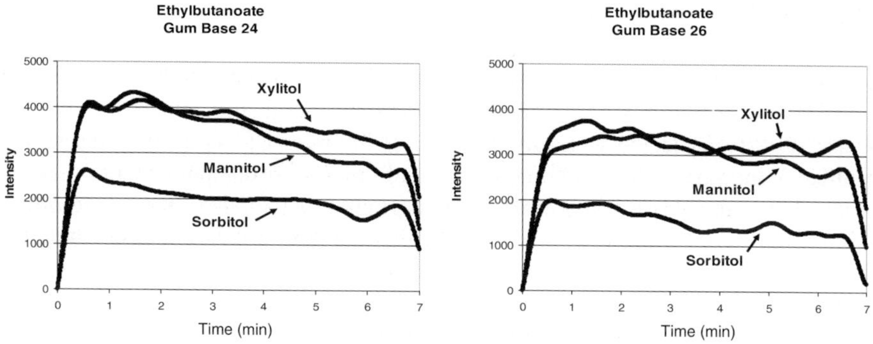

Figure 1. Release of ethylbutanoate from chewing gum made of two different gum bases and with three different types of bulk sweetener, measured by APCI-MS.

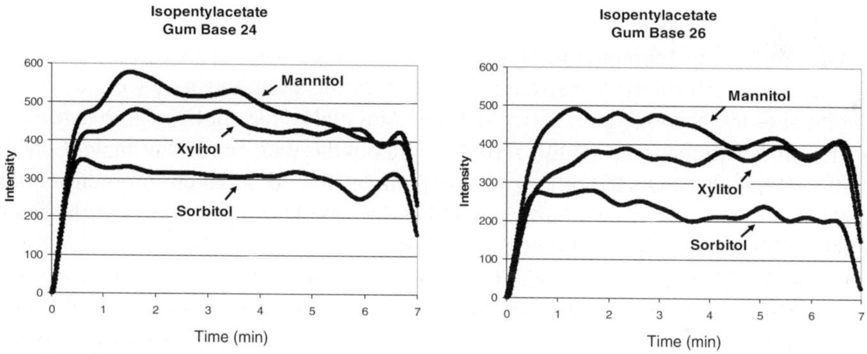

Figure 2. Release of isopentylacetate from chewing gum made of two different gum bases and with three different types of bulk sweetener, measured by APCI-MS.

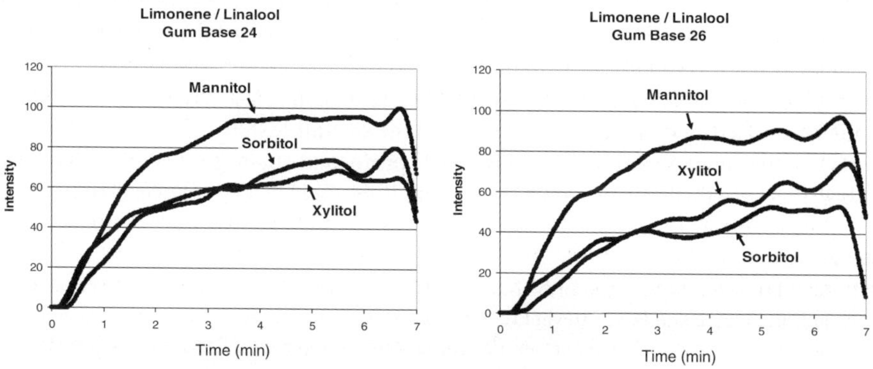

Figure 3. Release of limonene/linalool from chewing gum made of two different gum bases and with three different types of bulk sweetener, measured by APCI-MS.

4. DISCUSSION AND CONCLUSION

Overall, the main factors affecting the release of flavour compounds from chewing gum are the polarity of the matrix, reflected by partition coefficients gum-water, and mass transfer velocity, reflected by the consistency of the matrix [3,4]. In the present study, the effect of gum base composition on release of the flavour compounds was very clear. The difference in release levels from the two gum bases could be due to the different polarity of the two gum bases, gum base 24 being the most polar gum base – but also to a different consistency of the chewing gums made from the two gum bases, gum base 24 in general being softer and less elastic than gum base 26. Unfortunately, no measurements of consistency were performed in this study. Furthermore, it should be taken into consideration, that not only the gum base itself, but also the type of bulk sweetener affect the consistency of the chewing gums, so an effect of type of bulk sweetener on the consistency of the chewing gums can not be excluded.

The effect of type of bulk sweetener was very evident. Overall, the use of mannitol as the bulk sweetener increased the release of the flavour compounds investigated. The reason(s) for this phenomenon is not quite clear, but as mentioned earlier, the type of bulk sweetener affects the consistency of the chewing gum and it is well-known that xylitol makes the chewing gum softer and mannitol makes the chewing gum harder.

The non-uniform chew-out of the flavour compounds, with limonene/linalool chewed out at a slower rate than the other two compounds, should be taken into consideration when sensory analysis of the chewing gums is evaluated. The perception of the flavour could be affected by this non-uniform chew-out.

In general, a high release of flavours from chewing gums is desired in order to improve taste and for economical reasons, and based only upon these results, one should choose to work with gum base 24 and mannitol in the further development of this specific product. But in practice the situation is much more complicated. The advantages of using mannitol in relation to increased release of flavour compounds are by far overshadowed by the fact that the relative sweetness of mannitol in chewing gum is significantly lower than both sorbitol and xylitol and therefore adjustment of the level of sweetness (high intensive sweetener) in a product with mannitol is needed. Also, for economical reasons the use of mannitol as bulk sweetener compared to both xylitol and sorbitol is disadvantageous. With respect to type of gum base, the composition of the product, e.g. type and amount of flavour and the overall texture of the final product, is of more importance than the relatively small profit one could get using e.g. gum base 24, but of course this type of gum base is used, if the final product is acceptable.

References

1. J.M. Davidson, R.S.T. Linforth and A.J. Taylor, J. Agric. Food Chem., 47 (1999) 4336.
2. J.-L. Le Quéré and P.X. Étiévant (eds.), Flavour research at the dawn of the 21st century, proceedings of the 10th Weurman flavour research symposium, Paris, France (2003) 224.
3. M. Harrison, ACS Symp. Ser., (2000) 179.
4. Y. Bessiere and A.F. Thomas (eds.), Flavour science and technology, Chichester, UK (1990) 355.

W.L.P. Bredie and M.A. Petersen (Editors)
Flavour Science: Recent Advances and Trends

Influence of in-mouth aroma release on individual perception

Peter Prazeller, Nicolas Antille, Santo Ali, Philippe Pollien and Laurence Mioche

Nestlé Research Center, Vers-chez-les-Blanc, 1000 Lausanne, Switzerland

ABSTRACT

The relationship between individual in-mouth aroma release and individual aroma perception was investigated. Large variations in aroma release between subjects exist and the question to answer is to what extent these individual differences have an impact on aroma perception. To assess this, 13 subjects performed a 3-AFC test (three alternative forced choices), while their in-mouth aroma release simultaneously was measured by proton transfer reaction mass spectrometry (PTR-MS). In each test three solutions were consumed by the panellists, two of them were the reference at 6 ppmV, one the target at varying concentrations from 6.5 to 15 ppmV. The subjects had to identify the sample of the highest aroma concentration. A correlation between individual aroma release and aroma perception could be confirmed. Higher ability to discriminate between samples of low concentration difference was found for those subjects that showed a higher in-mouth release and a lower variability.

1. INTRODUCTION

During eating or drinking food aroma is released into the oral cavity and transported with the airflows associated with chewing, swallowing and breathing via the retronasal route to the receptors in the nasal cavity, where the odour perception has its initial step. Knowledge of aroma at this point is highly desirable. One way to assess this analytically is by measuring the aroma composition in the exhaled air through the nose, because that is the same air that passes over the receptors. On a personal level in-mouth aroma release measurements can aid to gain deeper insight into individual perception. Providing panellists with identical samples large individual differences of in-mouth aroma release have been reported [1-4]. Some reports link physiological parameters to olfactory function [5,6], suggestions on the influence of these differences on perception have been made [7]. This study specifically investigates whether individual differences in release have an impact on perception.

2. MATERIALS AND METHODS

2.1. Sample Materials

Samples consumed in this study were water solutions of ethyl butyrate (Sigma-Aldrich Chemie GmbH, Germany, natural, FCC FEMA 2427), presented at room temperature in a cup. Concentrations were 6, 6.5, 7.2, 8.7, 10.4, 12.5 and 15 ppmV and the volume 10 ml. Purity was checked by gas chromatography olfactometry (GCO).

2.2. Subjects

The panel consisted of 13 subjects (7 female and 6 male, ages 25-49, average 32 years) with and without prior experience in in-mouth aroma release analysis and little or no experience in sensory evaluation. In training sessions all were familiarised with nose space analysis and the samples to be consumed in the test.

2.3. Sensory evaluation

A 3-AFC test was conducted, where two of the samples given to the panellists were the reference at 6 ppmV, and one was the target at varying concentrations from 6.5 to 15 ppmV. The subjects had to identify the sample with the highest aroma concentration. For each concentration pair 20 replicates were done in a randomised order. For each panellist two sessions per day were performed over 6 days. Each day the whole set of concentrations was given to the panellists. All the tests were done approximately at the same time in the morning between 9 and 11 AM.

2.4. In-mouth aroma release analysis

The air exhaled through the nose was sampled via a tailor-made nosepiece, consisting of two glass tubings fixed on laboratory eyeglasses and smoothly fit into the nostrils. Chemical analysis was done online by PTR-MS. The method is described in more detail elsewhere [2,8]. From the time-dependent aroma release signals two parameters were extracted: 'Peak', the maximum concentration during the consumption of the sample and 'Area', the cumulated area under the concentration curve for each sample, the latter representing the amount of aroma released in the nose.

3. RESULTS

3.1. Variability of reference

Subjects differ in their reproducibility of in-mouth aroma release. Samples in all the triangles were compared to the 6 ppmV reference solution. Therefore it can be expected that the individual variability of the aroma release of the reference solution influences the overall sensory performance of the subjects. Indeed higher ability to discriminate between samples was found for people with the lowest variation in release for the reference solution. This is represented in Figure 1 by plotting for each panellist the variation coefficient (CV) of the aroma release of the 6 ppmV solution versus the percentage of correct given answers by the corresponding panellist. The general trend is that people with largest variation tend to be worst in fulfilling the discrimination task and vice versa.

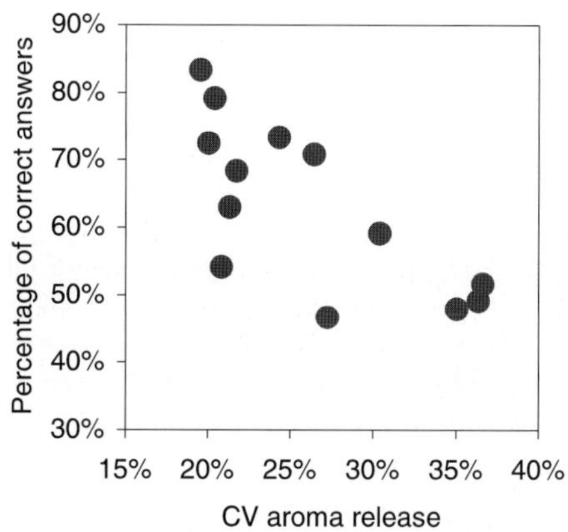

Figure 1. Relationship between variability (variation coefficient CV) of aroma release for the 6 ppmV reference solution and sensory performance; one marker per subject. Sensory performance is percentage of correct given answers to all triangles in the study; the maximum concentration in the nose during consumption of the sample was taken as value for aroma release.

3.2. Total amount release

As some people had significantly higher release than others the influence of this circumstance on the sensory performance was investigated. All the aroma release results from each person were reduced to one number, the weighted average of all areas under the curve from all samples. These values were compared to the subject's overall sensory performance. Results are shown in Figure 2. A clear trend that a larger amount released into the nose better enabled subjects to distinguish between samples of low concentration difference was observed.

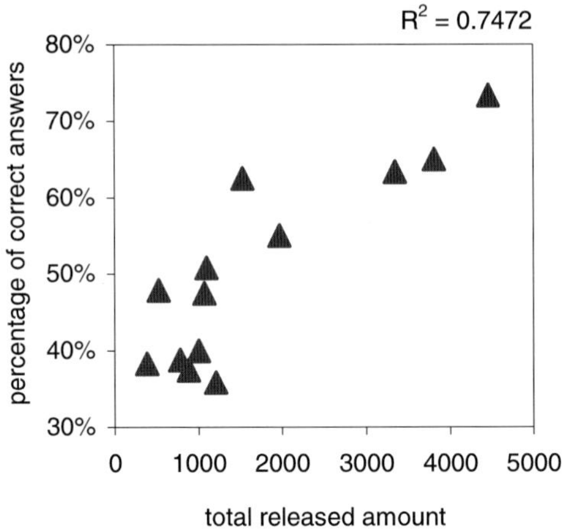

Figure 2. Relationship between subjects' sensory performance and total aroma released; one marker per subject. Sensory performance is percentage of correct given answers to all triangles in the study; total released amount is average of subjects released amount of aroma from all samples.

4. DISCUSSION AND CONCLUSION

These results prove that individual differences in aroma release have an influence on the olfactory capabilities. Perception is not only determined by the stimulus concentration, but is also affected by a variety of factors like memory, mood, expectations, age or cross modal interactions. However, findings from this study strongly support the importance of the individual aroma concentrations at the receptors. At least to a certain extent individual differences observed in perception can be explained by between-subject variability of aroma release during consumption.

References

1. K.G.C. Weel, A.E.M. Boelrijk, A.C. Alting, P.J.J.M. van Mil, J.J. Burger, H. Gruppen, A.G.J. Voragen and G. Smit, J. Agric. Food Chem., 50 (2002) 5149.
2. C. Yeretzian, P. Pollien, C. Lindinger and S. Ali, Compr. Rev. Food Sci. Food Safety, 3 (2004) 152.
3. J.B. Mei, G.A. Reineccius, W.B. Knighton and E.P. Grimsrud, J. Agric. Food Chem., 52 (2004) 6267.
4. K. Deibler and J. Delwiche (eds.), Handbook of flavor characterization, New York, USA (2003) 151.
5. K. Zhao, P.W. Scherer, S.A. Hajiloo and P. Dalton, Chem. Senses, 29 (2003) 365.
6. M. Damm, J. Vent, M. Schmidt, P. Theissen, H.E. Eckel, J. Lötsch and T. Hummel, Chem. Senses, 27 (2002) 831.
7. M. Mestres, N. Moran, A. Jordan and A. Buettner, J. Agric. Food Chem., 53 (2005) 403.
8. A. Hansel, A. Jordan, R. Holzinger, P. Prazeller, W. Vogel and W. Lindinger, Int. J. Mass Spectrom. Ion Processes, 149/150 (1995) 609.

W.L.P. Bredie and M.A. Petersen (Editors)
Flavour Science: Recent Advances and Trends

Control of aroma transfer by biopolymer based materials

Pascale Chalier, Sibel Tunc, Emmanuelle Gastaldi and Nathalie Gontard

UMR IATE, Engineering of Agropolymer and Emerging Technologies, University of Montpellier, 34095 Montpellier, France

ABSTRACT

Biopolymers such as proteins offer unique functional properties that can be used in the field of edible or biodegradable active packagings. Among the barrier properties of biopolymer based packaging, the knowledge of the aroma compound transfer is required (i) to control losses and sorption of aroma compounds which are important contributors to the sensory product quality (ii) or to emit desirable flavour from active packaging in a controlled way. Permeability of 2-heptanone in protein coated papers was investigated. It was demonstrated that coating by gluten or casein decreased the permeability of 2-heptanone and that coated papers had promising aroma barrier compared to other packaging materials such as sulphuric or paraffin papers or LDPE. The ability of biopolymer films to control aroma release was also demonstrated. An aroma compound (carvacrol) was incorporated in a gluten protein matrix, which showed adequate ability to maintain aroma compounds during film processing. The release of the aroma compound into the headspace was followed under accelerated conditions of temperature and relative humidity (30 °C and 60% RH). It was shown that the release rate was not dependent on carvacrol amount. The gluten film delayed aroma compound emission into the headspace for more than a month.

1. INTRODUCTION

Traditional packaging concepts are limited in their ability to prolong the sensory quality of food products [1]. Flavour scalping, i.e. sorption and diffusion of aroma compounds in polymeric packagings such as low density polyester film materials is well known and results in loss of flavour and taste of food products. Paper, one of the most widely used materials, has weak barrier properties against water vapour, oil, gas and aroma. To increase these barrier properties, paper is coated with thin layers of synthetic polymers but this leads to loss of biodegradability and recyclability. Proteins have already been used to develop edible food protective films with interesting functional properties such as selective properties towards oxygen and carbon dioxide [2,3]. The aim of this work is to assess the ability of proteins such as wheat gluten and casein (i) to improve aroma barrier properties of paper when used as a coating and (ii) for controlled release of aroma compounds.

2. MATERIALS AND METHODS

2.1. Coating of surfaces

Coating solutions were prepared by dissolving 20% (w/v) of wheat gluten or sodium caseinate in distilled water under stirring. For the gluten solution, acetic acid and anhydrous sodium sulphite were added. In the film preparation, glycerol and the aroma compound (carvacrol) were added. Coating solutions were either deposited on a support paper (sodium caseinate or wheat gluten coated papers) using an adjusted micrometer applicator or spread directly onto a Plexiglas surface (self-supported wheat gluten film).

2.2. Permeability measurement of coated papers

The transfer rate of the aroma compound was measured by a semi-dynamic method developed in our laboratory: the test film was sealed on the top of a permeation cell containing the pure aroma compound, this cell was placed in a diffusion chamber which was swept by a nitrogen flow. The outgoing flow charged with the aroma compound was injected at regular time intervals into a gas chromatograph using an automatic valve. The permeability P (g/(s m, Pa) was calculated at steady-state conditions from the permeate flux F using the following equation:

$$P = (F/\Delta P)\, e \qquad (1)$$

where F is the transfer rate (g/(m^2 s)), ΔP is the vapour partial pressure gradient and e the film thickness (m).

Diffusivity (D) was estimated from the half time $t_{0,5}$ at which the flux is equal to the half of maximal flux

$$D = e^2/7.199\, t_{0,5} \qquad (2)$$

2.3. Kinetic release of carvacrol from wheat gluten film

Just after processing, wheat gluten films containing carvacrol were placed in an oven at 30 °C, under an air flux and adjusted relative humidity of 60%. At different times, pieces of film were taken out from the oven and carvacrol was extracted by pentane during 4 h agitation. The residual carvacrol quantity was determined by gas chromatography analysis.

3. RESULTS AND DISCUSSION

3.1. Aroma barrier properties of coated protein papers

The permeability and diffusivity of 2-heptanone, a compound with a characteristic blue cheese aroma was determined at 25 °C and relative humidity of 10% for coated gluten and casein papers and other packagings used in cheese industry (Table 1). Whatever the nature of the protein, the coating strongly increased the barrier properties of the paper since the permeability of 2-heptanone was reduced more than 10 times. The permeability of the protein coated paper was about the same as the permeability of the sulphuric paper or a polymeric package (LDPE) and was 4 to 6 times lower than for paraffin paper, a paper coated with a hydrophobic matrix. Coating of paper with proteins had a weak effect on 2-heptanone diffusivity which was 10 times higher than that of LDPE. The barrier properties of coated papers toward 2-heptanone could be explained by a weak sorption of this compound in the protein matrix. These promising results are to be confirmed with other aroma compounds [4].

Table 1. Permeability and diffusivity of 2-heptanone to coated protein papers and other packaging at 25 °C and a relative humidity of 10%. Standard deviation is given in parenthesis.

Packaging	Thickness (μm)	Permeability (10^{-12} g/(s m, Pa))	Diffusivity (10^{-12} m^2/s)
Support paper	130	6380 (334)	12.1 (0.7)
Gluten coated paper	191	533 (44)	6.3 (0.06)
Casein coated paper	200	393 (100)	10.9 (0.9)
Paraffin paper	99	2282 (550)	2.2 (0.28)
Sulphuric paper	59	259 (5)	1.91 (0.34)
LDPE	25	670 (50)	0.65 (0.06)

3.2. Kinetic release of carvacrol from wheat gluten film

Film forming gluten solutions were used to make flavour enriched materials which emit aroma compounds in a controlled way. The losses of carvacrol during film processing including drying were determined when two different concentrations of carvacrol (8% or 16% of dry gluten) were used in the film solution. The relative losses of carvacrol were low and similar, 23.7% and 18.2% respectively. These results demonstrate the ability of gluten to retain a high quantity of carvacrol in the network.

The release of carvacrol through the processed films was followed under selected conditions at 30 °C and 60% HR (Figure 1). Whatever the carvacrol initial amount, the kinetic of release was similar and characterised by two phases: an initial fast release until day 15, and a slow release when the residual amount of carvacrol reached 40%. At day 50, the relative release reached 76.5% and 80.4% for films containing 8% and 16% of carvacrol respectively. It may be concluded from these results that the release was not affected by the amount of carvacrol included into the film and that the gluten protein matrix delayed aroma compound emission in headspace for more than a month.

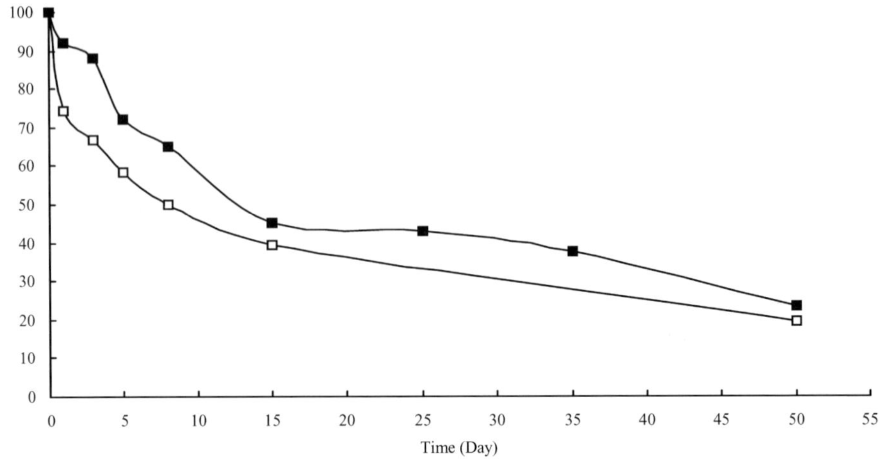

Figure 1. Retention (%) of carvacrol in wheat gluten film at 30 °C and 60% relative humidity under air flux. (■) 8% and (□) 16% of added carvacrol by dry weight gluten.

4. CONCLUSION

Papers coated by biopolymers such as proteins offered interesting and promising properties as barrier towards aroma compound transfer. Gluten films were also able to retain aroma compounds during film processing and to release them in a controlled way over time. These properties could be used advantageously for active biodegradable materials.

References

1. N. Gontard, Les emballages actifs, (2000) 1.
2. M. Miller and J. Krochta, Trends Food Sci. Technol., 8 (1997) 228.
3. H. Mujica-Paz and N. Gontard, J. Agric. Food Chem., 45 (1997) 4101.
4. A. Peyches, V. Jury, E. Gastaldi, N. Gontard and P. Chalier, J. Food Sci., submitted.

W.L.P. Bredie and M.A. Petersen (Editors)
Flavour Science: Recent Advances and Trends

Dynamics of flavour release from ethanolic solutions

Maroussa Tsachaki, Margarita Aznar, Robert S.T. Linforth and Andrew J. Taylor

Division of Food Sciences, School of Biosciences, University of Nottingham, Sutton Bonington Campus, Loughborough, LE12 5RD, UK

ABSTRACT

Aroma release from alcoholic solutions is of great interest but the presence of excess ethanol in the APCI-MS system causes non-quantitative ionisation. A recent modification to the APCI source has allowed the study of the liquid/air partitioning of volatiles from ethanolic solutions under both static equilibrium and dynamic headspace conditions. A decrease in the static equilibrium headspace concentration of aroma compounds from ethanolic solutions was noted, depending on their hydrophobicity, while, under dynamic conditions ethanol generally enhanced volatile delivery.

1. INTRODUCTION

Like in many food products, the aroma character of alcoholic beverages is influenced by the volatile compounds that are present in the gas phase. Wine is an interesting example of an alcoholic beverage where the change in perceived aroma profile with time is well-documented. In systems containing ethanol, aroma release is determined by the air-liquid partition, the presence of other solutes and other physicochemical effects such as micelle formation and surface tension effects.

Direct mass spectrometry techniques, such as APCI-MS and PTR-MS [1,2], are used for volatile release studies. Ethanol can cause significant changes in the ionisation behaviour of volatiles in these direct MS techniques and a modification of the system is required to obtain consistent data [3].

The aims of this study are to determine the effect of the presence of ethanol on volatile delivery from model solutions under static and dynamic conditions, as a first step in understanding the flavour release from wines and alcoholic beverages.

2. MATERIALS AND METHODS

A Platform LCZ mass spectrometer fitted with an MS Nose interface was used [1]. Ethanol (0-565 μl/l N_2) was added to the nitrogen make-up gas as required [3]. Solutions of 26 volatiles with a wide range of LogP values were prepared individually in either water or ethanol/water 12% (v/v). For static APCI-MS analysis, 40 ml of volatile solution was placed in 100 ml flasks fitted with a one-port lid. After equilibration for at least 2 h at 22 °C, headspace was sampled through this port into the APCI-MS (sample flow 5 ml/min). For dynamic headspace APCI-MS analysis, 100 ml of volatile solution was put into flasks fitted with a two-port lid leaving a headspace volume of 23 ml. After equilibration, N_2 was introduced through one port (at 70 ml/min) to dilute the headspace. As the gas flowed out of the second port, part of the gas flow was sampled into the APCI-MS (5 ml/min) [4].

3. RESULTS AND DISCUSSION

Ethanol was introduced into the source via the make-up gas, to act as the proton transfer reagent ion and thereby control ionisation. As ethanol concentration increased in the source, the water ions were replaced by ethanol ions above 3.2 μl ethanol/l nitrogen. Some compounds showed reduced signal (10-40%), others increased signal (150-400%) when ionised via ethanol reagent ions compared to water reagent ions. Noise also increased in most cases so there was no overall increase in sensitivity. Providing the ethanol concentration in the source was in the range from 6.5 to 11.3 μl/l N_2 and maintained at a fixed value, ionisation was consistent and quantitative [3].

3.1. Static headspace analysis

The static headspace concentration of twenty-six volatiles with a range of physicochemical properties was studied above water and ethanolic solution. For most of the volatiles the concentration in the headspace decreased when ethanol was present in the solution, by 4 to 44%. This agreed with the fact that the addition of ethanol generally increases the solubility of aroma compounds and therefore reduces their concentration in the headspace. Similar results have been found [5], although other studies only found an effect of ethanol on the partitioning of compounds when the ethanol concentration was higher than 17% [6]. The amount of a volatile compound in the headspace above ethanolic solutions is governed by its LogP value. The correlation is not linear (Figure 1) and it seems to have a parabola-like shape. This kind of correlation with LogP has been reported in biological studies on molecular mobility [7]. At high LogP values (very non-polar molecules), the presence of ethanol in the solution did not decrease the concentration of volatiles in a consistent manner. This seems to be due to changes in the hydrophobic interactions in the solution [8].

3.2. Dynamic headspace analysis

Headspace concentration of ethanol during dynamic dilution studies (Figure 2) was more stable than theoretical predictions [4]. This is probably due to its surface activity.

The variation in the ethanol trace results from the dilution of the sample gas after it left the headspace dilution flask to bring it into the working range of the APCI-MS. For most of the volatiles tested under dynamic conditions (eleven out of eighteen) the headspace dilution profile followed the profile of ethyl butyrate (Figure 2). Above water, the headspace concentration showed a substantial decrease until reaching a steady state. However, above ethanolic solution, there was a similar decrease followed by a steady state at a much higher level (at 50-97% of the initial relative intensity, depending on the volatile). This effect was so strong that under dynamic conditions the absolute volatile concentration above an ethanolic solution was greater than that above a water solution.

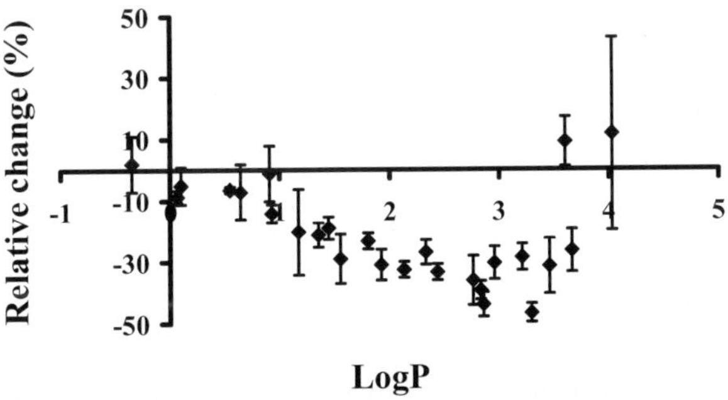

Figure 1. Mean relative change (%) of headspace concentration above a 12% ethanol solution compared to water solutions for 26 volatiles plotted against their LogP values. Error bars are +/- SD.

The effects observed during dynamic studies are direct results of the properties of ethanol. Ethanol is surface active and it adsorbs preferentially at the solution/vapour interface in aqueous solutions, thus lowering the surface tension [9]. As ethanol evaporates, some areas in the interface are depleted of ethanol, causing higher surface tension and the molecules of the adjacent low surface tension regions then move towards the high surface tension regions, carrying with them an appreciable volume of underlying liquid [10]. This phenomenon is called the Marangoni effect. In our case, both ethanol and volatile molecules are transported to the interface from the bulk phase and thus they are more available for evaporation.

Main exceptions from the ethyl butyrate type profile were compounds like limonene (hydrocarbons). Their headspace concentration profile above water and ethanolic solutions both showed a considerable decrease upon dilution (Figure 2). They adsorb preferentially to the interface, due to a favourable energy reduction as they move from the bulk to the air/water interface. So, they are immediately available for evaporation. The rest of the molecules found in the bulk phase are not able to reach the interface, after the depletion of the interfacial molecules, as they are probably found trapped in water cages, such as clathrate organisations [8].

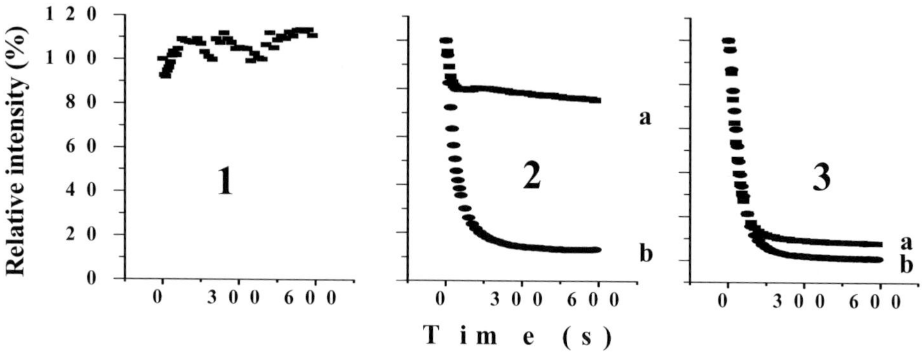

Figure 2. Average dynamic headspace dilution profile of ethanol (1), ethyl butyrate (2) and limonene (3) above ethanolic (a) and water solutions (b).

4. CONCLUSION

Now that the methodology has been proven with model systems, it is interesting to study whether ethanol has the same enhancing effect on volatile delivery in real wines and alcoholic beverages and to study the effects of other solutes on the ethanol effect.

References

1. R.S.T. Linforth and A.J. Taylor, Apparatus and methods for the analysis of trace constituents of gases, US patent no. 5,869,344 (1999).
2. W. Lindinger, A. Hansel and A. Jordan, Int. J. Mass Spectrom., 173 (1998) 191.
3. M. Aznar, M. Tsachaki, R.S.T. Linforth, V. Ferreira and A.J. Taylor, Int. J. Mass Spectrom., 239 (2004) 17.
4. M. Marin, I. Baek and A.J. Taylor, J. Agric. Food Chem., 47 (1999) 4750.
5. T. Henick-Kling, T.E. Wolf and E.M. Harkness (eds.), Proceedings for the 4th international symposium on cool climate viticulture and enology, Geneva, NY (1996) 42.
6. J.M. Conner, L. Birkmyre, A. Paterson and J.R. Piggott, J. Sci. Food Agric., 77 (1998) 121.
7. H.J. Smith (ed.), Introduction to the principles of drug design and action, Cardiff, UK, (1998) 167.
8. J. Israelachvili, Intermolecular and surface forces, London, UK (1991) 122.
9. E. Guggenheim and N.K. Adam, Proc. Roy. Soc., 139 (1933) 218.
10. A.A. Nepomnyashchy, M.G. Velarde and P. Colinet, Interfacial phenomena and convection, Boca Raton, USA (2002) 19.

W.L.P. Bredie and M.A. Petersen (Editors)
Flavour Science: Recent Advances and Trends

Active product packaging flavour interaction

Anna Nestorson[a,b], Anders Leufvén[a] and Lars Järnström[b]

[a]SIK-The Swedish Institute for Food and Biotechnology, Box5402, 402 29 Göteborg, Sweden; [b]Karlstad University, Surface Treatment Program, Chemical Engineering, 651 88 Karlstad, Sweden

ABSTRACT

The utilisation of polymer films, made from latex, onto food packaging with the aim to achieve different packaging functionalities was studied. Emphasis was given to the course of volatile compounds that affect food flavour, their adsorption, release from and distribution in hydrocolloid dispersion coatings as well as the aroma barrier properties of hydrocolloid dispersion coatings. Preliminary results indicate that the adsorption and retention during of polar and non-polar aroma compounds is different for different latex films. Furthermore, the chemical features of the aroma substances affect how they adhere to polymeric substrates.

1. INTRODUCTION

New packaging concepts that not just function as a passive, inert barrier to external conditions but play an active role in the protection of the packed food have been and are currently developed [1,2]. According to de Kruijf *et al.* [3], active packaging can be defined as a packaging solution that changes the condition of the packaged food to extend shelf life or improve food safety or sensory properties, while maintaining the quality of the packed food. Most food packaging materials interact with the food. The possible interaction processes between food and packaging material include: migration (when substances from the packaging enter the food); sorption (when substances from the food enter the packaging material) and permeation (where substances penetrate the packaging from the outside and inwards or from the inside and outwards).

By controlling food/packaging interactions, the product quality and the shelf life of the packed food can be improved [3]. Mechanisms that affect the shelf life of packed food products include physiological processes such as respiration of fresh fruits and vegetables, chemical processes such as lipid oxidation, physical processes like staling of

bread and dehydration and also microbiological aspects as spoilage from microorganisms.
In an ongoing PhD project, the influence of the packaging material on the aroma quality of the food product is considered. Dispersion barrier coatings (i.e. latex dispersions) have large accessible surface areas that will be used for desirable aroma interactions. One advantage using dispersion barriers like acrylic latex instead of conventional multilayered packaging materials is that renewable bulk material such as paper coated with barrier dispersions can be recycled as one entity in the paper mill.
The objective of the project is to study how volatile compounds that affect food flavour are adsorbed in, released from and distributed in hydrocolloid dispersion coatings as well as to study the aroma barrier properties of hydrocolloid dispersion coatings.

2. BARRIER DISPERSION COATINGS

Latex particles have a very large specific area, which is exposed to the surrounding water phase before the film is formed. After drying, an extensive three-dimensional network in the dry latex film is created. The intention is to use the interfaces that have been created to control interactions with the aroma substances that affect the flavour of the food. The film formation process from latex dispersions [4] are shown in Figure 1.

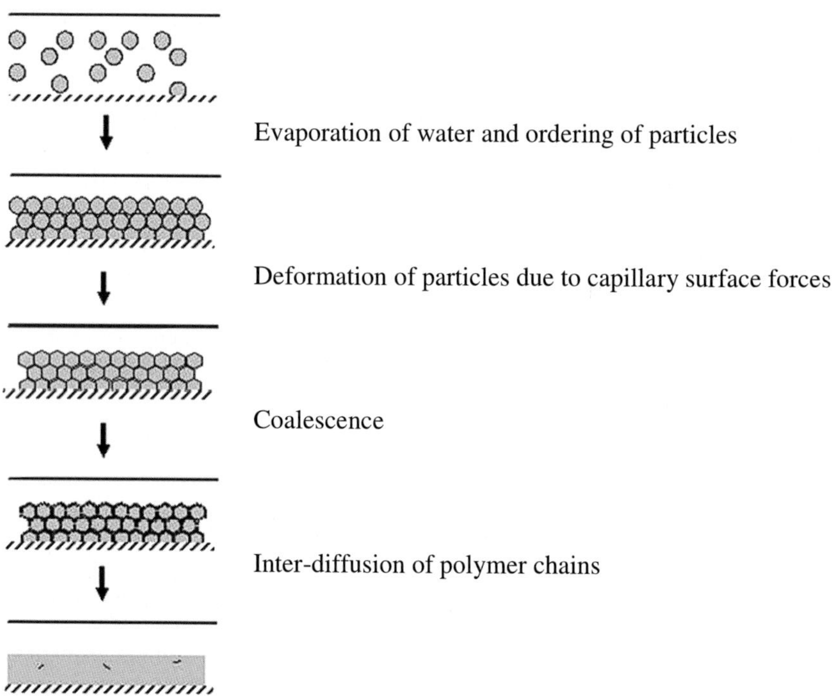

Figure 1. Schematic representation of the film formation process from a latex dispersion.

During coalescence, the polymer chains in the latex particles become tied in with each other and during inter-diffusion a uniform film is formed where it is not possible to distinguish individual polymer particles.

The most common polymers or co-polymers for barrier coating are styrene, acrylate, meta-acrylate, butadiene and vinyl acetate. The intention is to create barriers against water and water vapour, fat and gases [5]. This is needed to preserve the flavour of packaged foods.

In the initial experiments the ability of different latex qualities to absorb and retain representative aroma compounds from different chemical classes (alcohols, aldehydes, esters, ketones and terpenes) as well as partition coefficients for different aroma compounds between the latex films and gas phase have been studied. Four commercial latex grades were used (see Table 1). Aroma compounds were dissolved in propylene glycol and added to the latex dispersions. After the film formation process, the aroma concentrations in the final films were measured with multiple headspace extraction [6] and gas phase concentrations were analysed with headspace gas chromatography.

Table 1. Latex grades for formation of barrier films in this study.

Trade name	Name	Composition	T_g	Solid content	Applications	Regulatory status	Manufacturer
Latexia® 204	SA-latex1	Styrene-acrylate	~10 °C	50%	Coating binder	FDA (§176.170 and 176.180)	Raisio Chemicals
Lucidene™ 606LS	SA-latex2	Styrene-acrylate	~11 °C	47%	Coating and ink binder	FDA (§176.170 and 176.180)	Rohm and Haas
DL 950	SB-latex	Styrene-butadiene	~7 °C	50%	Coating binder	FDA (§176.170 and 176.180)	The Dow Chemical Company

3. RESULTS

Preliminary results indicate that the adsorption and retention during film formation of polar aroma compounds were greater in both SA-latex samples than in the SB-latex sample, while non-polar aroma compounds were retained to greater extent in the SB-latex sample. In addition, the chemical features of the aroma substances affect how they adhere to polymeric substrates. More details will be reported in [7].

References

1. Anon., Food Market. Technol., 16 (1) (2002) 45.

2. L. Vermeiren, F. Devlieghere and J. Debevere, Food Addit. Contam., 19 (suppl.) (2002) 163.
3. N. de Kruijf, M. van Beest, R. Rijk, T. Sipilainen-Malm, P. Paseiro-Losada and B.-D. Meulenaer, Food Addit. Contam., 19 (suppl.) (2002) 144.
4. J.H.Steward, J. Hearn and M.C. Wilkinsin, Adv. Colloid Interface Sci., 86 (2000) 195.
5. J. Brander and I. Thorn (eds.), Surface application of paper chemicals, London, UK (1997) 208.
6. B. Kolb and L.S. Ettre, Static headspace - gas chromatography: theory and practice, New York, USA (1997).
7. A. Nestorson, A. Leufvén and L. Järnström, Packag. Technol. Sci., (2005) in press.

W.L.P. Bredie and M.A. Petersen (Editors)
Flavour Science: Recent Advances and Trends

Transfer of volatile phenols at oak wood/wine interface in a model system

Daniela Barrera-García[a,b], Régis D. Gougeon[a,b], Frédéric Debeaufort[b], Andrée Voilley[b] and David Chassagne[a,b]

[a]*Institut Universitaire de la Vigne et du Vin, Jules Guyot, Université de Bourgogne, Campus Montmuzard, 21078 Dijon, France;* [b]*Equipe IMSAPS, ENSBANA, Université de Bourgogne, 1 Esplanade Erasme, 21079 Dijon, France*

ABSTRACT

In order to assess the influence of wood on the concentration of aroma compounds during ageing of wine, the transfer of volatile phenols including 4-ethylphenol, eugenol and a homologous series of guaiacols from wine to oak wood were studied in a model system at 10 °C. At equilibrium most of the volatile phenols adsorbed in the wood. The results display that the amounts adsorbed depend on the nature of the volatile phenols and the botanical origin of oak wood.

1. INTRODUCTION

During the ageing of wine in oak barrels, an aromatic complexity is acquired as a result of aroma transfer at the interface between wood and wine. Volatile compounds such as eugenol and (*Z*)- and (*E*)-β-methyl-γ-lactone are components naturally present in oak wood [1,2] which are extracted during wine ageing [1,3]

Transfer mechanisms at the interface between wood and wine were recently studied. Ramirez-Ramirez *et al.* [4] and Chassagne *et al.* [5] highlighted the sorption of aroma compounds by oak wood in conditions simulating ageing of wine. The amount adsorbed at equilibrium depended on the nature of the aroma compound. Wood sorption capacity has also been investigated for monoaromatic hydrocarbons [6], where sorption was related to the hydrophobicity and the fractional lignin content of wood.

The aim of this work was to study the transfer mechanisms of volatile phenols at the interface between wood and wine. For that purpose, a mixture of volatile phenols was adsorbed on two distinct types of oak wood, i.e. Sessile (*Quercus petrea*) and Pedunculata (*Quercus robur*) oaks.

2. MATERIALS AND METHODS

2.1. Materials

The 4-propylguaiacol and 4-vinylguaiacol were supplied from Aldrich-Sigma, 4-methylguaicol, 4-ethylguaiacol, 4-ethylphenol were purchased from TCI-EP and eugenol was purchased from Fluka, all with a minimum purity of 98%. The model wine was a 12.5% v/v hydro-alcoholic solution with a pH of 3.5 [4]. Volatile compounds were added in mixture to the model wine at concentrations between 0.5 and 30 mg/kg. These compounds were chosen because they are common in wine and exhibit a broad range of phenolic chemical structures. Dichloromethane obtained from Carlo Erba Reactives was used as solvent for extraction.

Small plates of wood (2 x 10 x 20 mm) were donated by the Office National des Forêts (ONF, France) and taken from trees of *Quercus petraea* and *Quercus robur* oak from the Forêt de Cîteaux (France).

2.2. Sorption isotherms

Experimental samples were prepared by immersing one plate of wood into 25 ml glass flasks filled with the model wine. Experimental samples and model solution without wood (control samples) were stored in triplicate at 10 °C and analysed by liquid-liquid extraction with dichloromethane as solvent. The volatiles were analysed by gas chromatography (GC TRACE ULTRA, Thermo Electron Corporation) with a flame ionisation detector and 3,4-dimethylphenol as internal standard. The column used was CP-WAX 57CB, 25 m x 0.25 id; 0.2 μm bonded phase.

3. RESULTS

The sorption isotherms for 4-methylguaiacol, 4-ethylguaiacol and 4-ethylphenol on Pedunculata oak are shown in Figure 1. For all phenol compounds, except for 4-vinylguaiacol, the same sorption behaviour is observed, i.e. a linear variation at low concentration (below 10 mg/kg) and a sharp increase above. Further detailed studies are currently carried out with the high concentration range. Due to an apparent chemical instability when in contact with wood, the sorption of 4-vinylguaiacol is not taken into account in these results. The Henry equation was fitted to the linear part of the experimental isotherm:

$$C_{wood} = K_{ww} \cdot C_{model\ wine}$$

where: C_{wood} is the quantity of aroma adsorbed per unit mass of wood;
$C_{model\ wine}$ is the concentration of aroma compound at equilibrium;
K_{ww} is the partition coefficient.

The values of K_{ww} for Pedunculata oak and Sessile oak are illustrated in Figure 2. Two distinct ranges of partition coefficients can clearly be related to the botanical origin of the wood: between 10 and 20 for the Pedunculata oak and from 40 at 70 for the Sessile oak. Furthermore, the amount of sorption is a function of the chemical structure of the volatile phenols. Although not clear for Sessile oak, sorption rises with an increase of the length of the *para*-aliphatic chain. The higher partition coefficient for 4-ethylphenol compared to 4-ethylguaiacol, also indicates that the absence of the *ortho*-methoxy substituent promotes the sorption. Eugenol which differs from 4-propylguaiacol by the saturation of the *para*-chain does not show a significantly different behaviour.

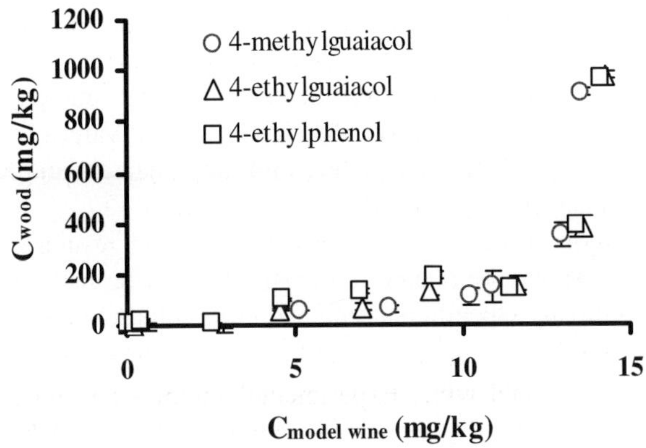

Figure 1. Sorption isotherms of 4-methylguaiacol, 4-ethylphenol and 4-ethylphenol by Pedunculata oak at 10 °C.

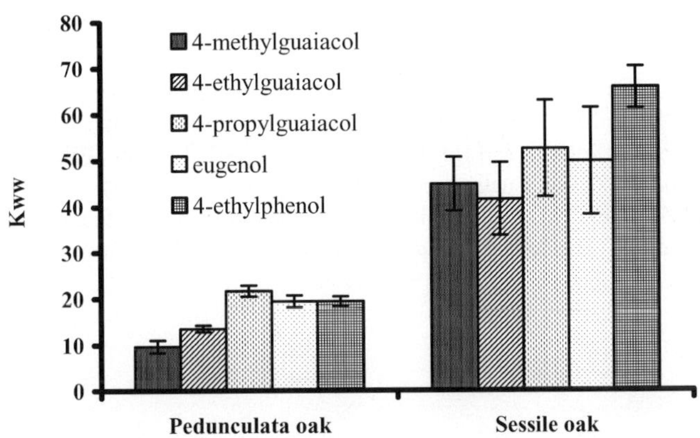

Figure 2. Influence of botanical origin of the oak on the partitioning coefficient between the wood and model wine (K_{ww}) for the different model phenols.

4. DISCUSSION AND CONCLUSION

The initial shape of the sorption isotherms suggests that partitioning is the major sorption mechanism over this range. Such mechanism can be related to that obtained for ethyl hexanoate and ethyl octanoate onto oak wood by Ramirez-Ramirez *et al.* [4]. A linear sorption of polynuclear aromatic hydrocarbons by wood fibres has already been reported [7]. Such behaviour is consistent with sorption in the presence of molecules in very low concentrations. It can be explained by the fact that there are enough active sites at the surface of wood for all the different phenol compounds to interact without competition.

For a given wood, the increased sorption with the increasing length of the *para*-chain in the phenolic compound can be related to the increase of the hydrophobic character [6]. Similarly the higher sorption of 4-ethylphenol compared to 4-ethylguaiacol may be explained by its higher hydrophobic character (logP=2.62 and 2.37 for 4-ethylphenol and 4-ethylguaiacol, respectively). The difference in logP values is consistent with the polar character of the *ortho* methoxy group in 4-ethylguaiacol and further corroborates with the sensory character of oak wood.

The higher sorption on to Sessile oak is consistent with the results obtained of Ramirez-Ramirez *et al.* [4] and may be related to the structural differences between the two oak woods. Both the porosity (specific surface and pore volume) and the surface chemistry have been shown to influence the sorption of phenolic compounds [9].

This work shows that the sorption of volatile phenols to oak wood is controlled by both the nature of the volatile compounds and the botanical origin of wood.

References

1. P. Chatonnet, D. Dubourdieu, J-N. Boidron and M. Pons, Sci. Aliment., 10 (1990) 565.
2. M.A. Sefton, I.L. Francis, K.F. Pocok and P.J. Williams, Sci. Aliment., 13 (1993) 629.
3. J.P. Towey and A.L. Waterhouse, Am. J. Enol Vitic., 47 (1996) 163.
4. G. Ramirez-Ramirez, S. Lubbers, A. Voilley, C. Charpentier, M. Feuillat and D. Chassagne, J. Agric. Food Chem., 49 (2001) 3893.
5. J.-L. Le Quéré and P.X. Étiévant (eds.), Flavour research at the dawn of the 21st century, proceedings of the 10th Weurman flavour research symposium, Paris, France (2003) 63.
6. S. Trapp, K.S. Miglioranza and H. Mosbaek, Environ. Sci. Technol., 25 (2001) 1561.
7. T. Boving and W. Zhang, Chemosphere, 54 (2004) 831.
8. G. Ramirez-Ramirez, PhD thesis, University of Bourgogne, Dijon, France (2002).
9. I.I. Salame and J.T. Bandosz, J. Colloid Interface Sci., 264 (2003) 307.

W.L.P. Bredie and M.A. Petersen (Editors)
Flavour Science: Recent Advances and Trends

Flavour release at the interfaces of stirred fruit yoghurt models

Alice Nongonierma[a], Philippe Cayot[a], Mark Springett[b], Jean-Luc Le Quéré[c] and Andrée Voilley[a]

[a]Laboratoire IMSAPS, ENSBANA, Université de Bourgogne, 1, Esplanade Erasme, 21000, Dijon, France; [b]Danone Vitapole, Centre de Recherche Daniel Carasso, R.D. 128, 91767, Palaiseau, France; [c]UMR FLAVIC ENESAD-INRA, 17, rue Sully, 21065, Dijon, Cedex, France

ABSTRACT

Model matrices consisting of a pectin and a dairy gel were investigated to study the transfer of flavour compounds between the different phases of stirred fruit yoghurt. Using a full factorial experimental design, different parameters (storage temperature, initial flavour concentration in the pectin gel and fat content of the dairy gel) were studied. The kinetics of the migration of flavour compounds between the two gels were investigated using a novel SPME method, where a Carboxen/Polydimethylsiloxane fibre was directly inserted into the pectin gel. Flavour compounds were predominantly retained in the pectin gel and thus, the presence of fat in the dairy gel did not affect their transfer. This retention depended on the hydrophobicity of the flavour compounds and the storage temperature, which modified the matrix structure and consequently the interactions between aroma compounds and matrix.

1. INTRODUCTION

Stirred fruit yoghurt is a multiphase product made of fruit pieces (8% w/w) and a mix of stirred yoghurt (80% w/w) and syrup (12% w/w). In such a food, the concentration gradients are responsible for transfer of small ligands between the different phases [1, 2]. The aim of this research was to study the different parameters which might control these transfers and to better understand the mechanisms involved.

2. MATERIALS AND METHODS

2.1. Flavour compounds

A mix of 8 strawberry key flavour compounds (ethyl acetate, ethyl butanoate, ethyl isobutanoate, ethyl hexanoate, hexanal, 2-methylbutanoic acid, linalool and (*Z*)-3-hexen-1-ol) was used. These compounds (purity > 98%) were obtained from Sigma-Aldrich (Steinheim, Germany) except for ethyl acetate, from Prolabo (Paris, France). Their logP (logarithm of the *n*-octanol/water partition coefficient) was estimated by the Hansch and Leo method [3].

2.2. Fruit yoghurt model

The dairy gels (0, 1.5 and 5% fat) consisted of stirred acidified (pH 4.4) milk gels [1]. They were made of 85.4% (w/w) water (Volvic, Danone, France), 14.5% (w/w) skimmed milk powder (Ingredia, Arras, France), 2.2% (w/w) glucono-δ-lactone (Prolabo, Paris, France) and 0.1% (w/w) potassium sorbate (Aldrich, Steinheim, Germany). The fat consisted of a mix of tributyrin (Prolabo, Paris, France) and triolein (Prolabo, Paris, France) in the ratio 2:5. The dairy gels were mixed with a syrup [1] in the proportion 1:9 syrup:dairy gel (w/w).

The fruit model [2] was a pectin gel made of 57% (w/w) water (Volvic, Danone, France), 40% (w/w) sucrose (Sucre Union, Paris, France), 2% (w/w) low amidated pectin (Grindsted™ pectin LA 110, Danisco, France), 0.2% (w/w) citric acid (Prolabo, Paris, France), 0.1% (w/w) potassium sorbate (Aldrich, Steinheim, Germany) and 0.09% (w/w) calcium chloride (Prolabo, Paris, France).

2.3. Flavour transfer at the pectin/dairy gel interface

Flavour transfers were studied with an experimental design (2^3) whose variables were the *storage temperature* (4 and 10 °C), the *fat content of the dairy gel* (0 and 5%) and the *flavour concentration in the pectin gel* (10 and 20 ppm for each compound). The central point conditions were: 7 °C, 1.5% fat, and 15 ppm added flavour.

A 50 g flavoured pectin gel and a 50 g dairy gel were placed in a cylindrical flask of 9.6 $cm^2 \times 10$ cm, resulting in a bilayer sample. The flavour compounds were extracted by direct immersion of a 75 μm Carboxen/Polydimethylsiloxane fibre (Supelco, Bellefonte, PA, USA) in the pectin gel 5 mm below the interface for 30 min (at 0, 2, 5, 8, 24, 30, 48, 72, 96, 144, 192, 504 and 672 h storage). The details of the analysis are given in [1]. After sampling, the fibre was immersed in water, dried on a paper filter, then desorbed 5 min at 250 °C and subsequently cleaned (300 °C, 45 min).

Flavour compounds were analysed with a gas chromatograph CP 3800 (Varian, Walnut Creek, USA) equipped with an HP FFAP capillary column (30 m× 0,32 mm× 0,25 μm, Agilent Technologies, Karlsruhe, Deutschland). The oven temperature was: 50 °C for 4 min, increase by 5 °C/min to 180 °C, hold for 17 min. The temperature of the injector (splitless) and the FID were 250 and 200 °C, respectively. Helium was used as a carrier gas at a velocity of 42.8 cm/s at 40 °C.

3. RESULTS AND DISCUSSION

3.1. Influence of temperature, concentration, fat content and logP on transfer

The initial rate of release (R_i) and the concentration of flavour compounds in the pectin gel at equilibrium (C_{pg}^{∞}) were correlated to the experimental design variables and the logP of the flavour compounds. The significance of these variables was evaluated by means of ANOVA (Table 1). For R_i, the significant variables were the flavour concentration and the temperature. These results are in agreement with the fact that flavour compound retention in the pectin gel is higher at 10 than at 4 °C. Furthermore, this temperature decrease affects the pectin gel structure, which might result in an expulsion of water and flavour compounds [1,2]. An increase in the flavour concentration gradient increased R_i. The logP and the dairy gel fat content did not influence R_i. This is consistent with the results of Harisson *et al.* [4], who demonstrated that the rate of release was proportional to the flavour concentration, and the mass transfer coefficient at short times.

Table 1. Effect of the storage temperature, flavour concentration, fat content and logP upon the initial rate of release (R_i) and the concentration of flavour compounds in the pectin gel at equilibrium (C_{pg}^{∞}).

	Storage temperature	Flavour concentration	Dairy gel fat content	logP
R_i	***	***	n.s.	n.s.
C_{pg}^{∞}	***	***	n.s.	***

n.s.: not significant ($p>0.05$), ***significant at $p<0.001$.

The temperature, logP and flavour concentration affected C_{pg}^{∞}. At equilibrium, the release is governed by flavour partitioning, which is affected by the physicochemical properties of the compounds (e.g. logP) and extrinsic parameters (e.g. the temperature) [4,5]. The effect of logP and temperature suggests that hydrophobic interactions [6] together with entrapment within the pectin gel structure modify flavour transfer [7,8].

3.2. Modelling flavour transfer at the pectin/dairy gel interface

The Ri and C_{pg}^{∞} were correlated with the significant variables (equations 1 and 2), except for ethyl butanoate which has an intermediate hydrophobicity (logP=1.75) and was therefore chosen to validate the models.

$$R_i = 0.055 - 0.008 * \theta + 0.006 * [C] \quad (1)$$

$$C_{pg}^{\infty} = -6.8 + 0.6 * \theta + 0.6 * [C] + 1.8 * \log P \quad (2)$$

Where θ is the temperature (°C), [C] the flavour compound concentration (ppm) and logP, the logarithm of the *n*-octanol/water partition coefficient.

The correlation between the predicted and experimental values is illustrated on Figure 1. The slopes of the curves, lower than 1, indicated that the models slightly underestimated the experimental values (by 2% for R_i and 8% for C_{pg}^{∞}). This is nevertheless below the coefficient of variation of the method (11.2%). The R^2 values show that the parameters studied are not sufficient to predict Ri and C^{∞}_{pg}. Other parameters such as the gel structure and the retention of flavour compounds in the matrices might influence flavour transfers.

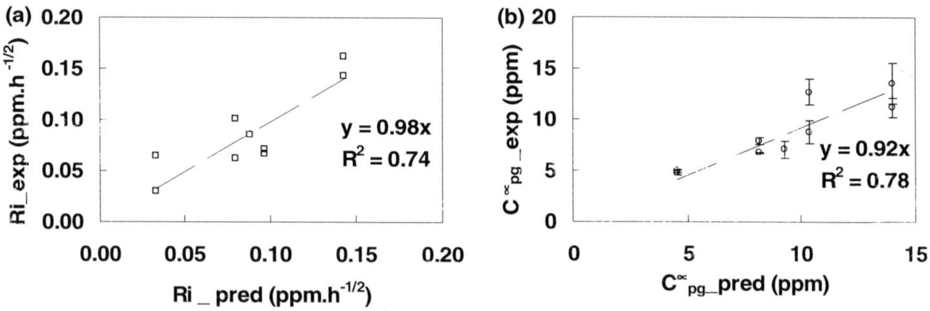

Figure 1. Relationship (a) between the predicted (R_i_pred) and experimental (R_i_exp) initial rates of release and (b) between the predicted (C^{∞}_{pg}_pred) and experimental (C^{∞}_{pg}_exp) concentrations of ethyl butanoate remaining in the pectin gel at equilibrium. R is significant for p=0.005.

4. CONCLUSION

The flavour compound concentration, the storage temperature and the logP greatly affected flavour transfer between the pectin and dairy gels. Flavour compound migration was governed by the pectin gel, which retained flavour compounds owing to hydrophobic interactions and entrapment within the gel structure.

Future studies should investigate the impact of flavour transfer upon the perception of stirred fruit yoghurts, together with the influence of the fruit preparation formulation (fruit variety, flavour compounds).

References

1. A. Nongonierma, P. Cayot, T. Saint-Denis, M. Springett, J.-L. Le Quéré and A. Voilley, J. Chromatogr. A, submitted.
2. A. Nongonierma, P. Cayot, R. Cachon, M. Springett, J.-L. Le Quéré and A. Voilley, Food Hydrocolloid., submitted.
3. C. Hansch and A.J. Leo, Substituent constants for correlation analysis in chemistry and biology, New York, USA (1979).
4. M. Harrison, B.P. Hills, J. Bakker and T. Clothier, J. Food Sci., 62 (4) (1997) 664.
5. M. Marin, I. Baek and A.J. Taylor, J. Agric. Food Chem., 47 (11) (1999) 4750.
6. A. Hansson, J. Andersson and A. Leufvèn, Food Chem., 72 (3) (2001) 363.
7. E. Bylaite, A.S. Meyer and J. Adler-Nissen, J. Agric. Food Chem., 51 (27) (2003) 8020.
8. B. Rega, E. Guichard and A. Voilley, Sci. Aliment., 22 (3) (2002) 235.

W.L.P. Bredie and M.A. Petersen (Editors)
Flavour Science: Recent Advances and Trends

Requirement for a global design to remove fat from flavoured yoghurts

Philippe Cayot, Alice Nongonierma, Guillaume Houzé, Flore Schenker, Anne-Marie Seuvre and Andrée Voilley

Department of Molecular and Sensorial Engineering of Foods and Health Products (IMSAPS). Engineer Institute of Applied Biology to Nutrition and Alimentation (ENSBANA), University of Burgundy, 1 esplanade Erasme, 21000 DIJON, France

ABSTRACT

The low-fat dairy desserts are generally not so much appreciated as the corresponding products containing fat because of altered flavour and texture characteristics. A review of the influence of fat on these aspects is proposed. Some methods used to correct for flavour imbalances also decrease the texture quality.

1. INTRODUCTION

A clear preference for sweet and fruit flavoured yoghurts compared to plain yoghurts that are judged acidic has been demonstrated [1]. The fat content of yoghurts greatly affects the perception of aromas [2]. As recently demonstrated [3], there is a strong influence of the fat added on the 2-nonanone release (logP = 2.9) whatever the state of the model matrix (gelled or not gelled β-lactoglobulin suspensions, stirred or set gels). Conversely, the release of hex-4-en-3-ol (logP = 1.7) was not influenced by the composition of the matrix. The flavour perception of yoghurt does, therefore, not only depend on its composition (fat content especially). Texture-aroma and taste-aroma interactions have been demonstrated by studying the influence of flavouring on texture or taste perception [4].
The dispersed fat in the aqueous phase behaves also as gel filler and can, therefore, modify the rheological properties of gels when the volume fraction is higher than 5% [5]. Many factors affect the rheological properties of filled gels. These are not only related to the characteristics of the gel matrix and the filler properties, but also to the interactions between filler particles and gel matrix. These interactions depend on the

volume, the deformability, the shape of the filler and also on the surface affinity of the filler for the matrix [5]. Yet no studies have been carried out with fat content below 5% despite the typical average fat volume fraction of 2% (v/v) for commercial yoghurts.
The aim of this study was to evaluate the effects of inclusion of small amounts of different kinds of fat (2%, v/v) on the visco-elastic properties of acid milk gel aged for 7 days and to investigate in parallel the effect of fat addition (1.5% and 5%) after equilibrium of 7 days upon retention of flavour molecules. The final objective was to identify solutions to replace fat globules based on the mechanisms determining the quality of fat containing yoghurt. As a result, these mechanisms should be mimicked in low-fat flavoured yoghurts.

2. MATERIALS AND METHODS

2.1. Flavour compounds

A mix of 6 strawberry flavour compounds (ethyl acetate, ethyl butanoate, ethyl isobutanoate, ethyl hexanoate, ethyl octanoate and hexanal) was studied. These compounds (purity > 98%) were purchased from Sigma-Aldrich (Germany) except for ethyl acetate, from Prolabo (France). Their logP (logarithm of the *n*-octanol/water partition coefficient) was estimated by the method described in [6].

2.2. Yoghurt model

The dairy gels (0, 2, 3.5 and 5% fat) consisted of set and stirred acidified milk gels (pH 4.4). They were made with skimmed milk powder (low heat), and 2.2% (w/w) glucono-δ-lactone (Prolabo, France). The fat consisted of a mix of tributyrin (Prolabo, Paris, France) and triolein (Prolabo, France) in the ratio 2:5 for flavour compound retention experiments. Commercial oils for rheological analysis were used: sunflower, olive and groundnut for domestic oil and anhydrous milk fat (Corman, Belgium).

2.3. Retention of flavour compounds

Flavour release from the dairy gels with 0% and 5% fat flavoured with 20 ppm of each flavour compound was studied at 4 and 10 °C with a static headspace method. The headspace (1 ml) was injected into a gas chromatograph (Shimadzu GC-14B, Japan) and analysed according to [7].

2.4. Rheological analysis of set gel

Stress relaxation tests were used to determine the firmness and the elasticity of the gels. Tests were performed using an Instron 1122 (Instron Instruments, England) directly on gels in their vials [8]. Analysis of stirred gels is perfectly correlated to set gel analysis [9] and because of the time consumption of analysis of stirred gels we just compared addition of fat with conventional analysis of set gel.

2.5. Sensory assessments of smooth/ granular character

The combination of different heating processes of milk before acidification according to [9] as well as different conditions for gel stirring, allowed us to obtain stirred gels with a different size of gel particles. The size of gel particles was analysed with a laser light scattering analyser Mastersizer MS20 (Malvern Instruments, England).

The visual sensation of smooth/granular character on a spoon was defined as: presence of particles on the back of the spoon. The oral sensation of smooth/granular character was defined as: presence of particles in mouth felt after a weak compression of the product between tongue and palate. The yoghurt sample was 'smooth' if no asperity (roughness) or surface irregularity was perceived.

3. RESULTS AND DISCUSSION

The aroma compound retention in natural yoghurts depended on logP of the compounds as shown in Figure 1. For ethyl acetate, ethyl butanoate and ethyl isobutanoate retentions in the dairy gel and in water were the same. For ethyl hexanoate, ethyl octanoate and hexanal (logP > 1.7), retention in the dairy gel was higher since their logP was higher. The addition of 5% fat (mixture of tributyrin and triolein) markedly enhanced retention of flavour compounds except for EA (logP = 0.7).

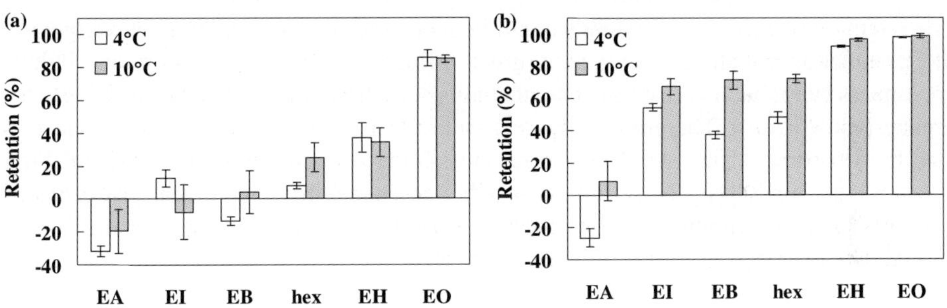

Figure 1. Retention of flavour compounds in stirred yoghurts without fat (a) and with 5% of fat (b), at 4 and 10 °C [7]. EA: ethyl acetate; EB: ethyl butanoate; EI: ethyl isobutanoate; EH: ethyl hexanoate; EO: ethyl octanoate; hex: hexanal.

Whatever the type of fat used (2% v/v), an effect of fat globules on milk gel rheological properties was also observed. It was remarkable to observe an effect on the rheological behaviour at this low volume fraction. In fact, it was previously assumed that effects could only be observed at oil volume fraction >5%. Whatever the type of fat used, both the firmness of the gel (Figure 2) and the elasticity reflected by the relaxation time were increased. Milk fat behaved differently during melting compared to other fats. A lower increase in gel firmness was observed. The temperature (especially 5 and 10 °C) had an important effect on the gel strength in the presence of fat. In the same way, the retention of flavour compounds was significantly affected by the temperature between 4 and 10 °C in presence of fat. The removal of fat in yoghurt decreases the cohesion of the gel and increases the volatility of aroma compounds, which partially explained the differences perceived between classic and low-fat yoghurts.

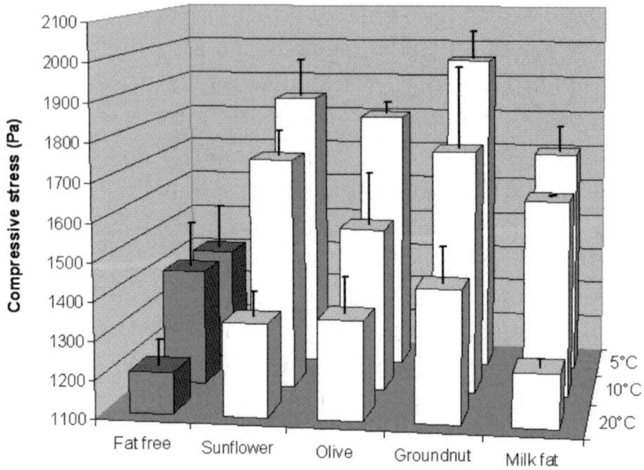

Figure 2. Influence of fat nature on acid milk gel firmness as a function of the temperature [8].

The retention of flavour compounds could be increased by encapsulation. Nevertheless, the granulation in stirred gel had a great influence on the perception of yoghurt creaminess (unpublished results). The unctuous sensation is perceived for particle sizes greater than 150 μm. Therefore, polysaccharide capsules that lie within this range will also be perceived. Using starch is another possibility to increase the retention of some flavour compounds [10]. Nevertheless it will also change the rheological properties of yoghurts [11]. No simple solutions exist, but a combination of different formulations should help to compensate flavour and texture changes.

References

1. D.L. Barnes, S.J. Harper, F.W. Bodyfelt and M.R. McDaniel, J. Dairy Sci., 74 (7) (1991) 2089.
2. M.S. Brauss, R.S.T. Linforth, I. Cayeux, B. Harvey and A.J. Taylor, 47 (5) (1999) 2055.
3. A-M. Seuvre, M-A. Espinosa-Diaz, P. Cayot and A. Voilley, Le Lait, in press.
4. A. Saint-Eve, E. Paçi Kora and N. Martin, Food Quality Pref., 15 (7-8) (2004) 655.
5. E. Dickinson, Colloids Surf. B, 20 (3) (2001) 197.
6. C. Hansch and A.J. Leo, Substituent constants for correlation analysis in chemistry and biology, New York, USA (1979).
7. A. Nongonierma, M. Springett, J.-L. Le Quere, P. Cayot and A. Voilley, Int. Dairy. J., in press.
8. G. Houze, E. Cases, B. Colas and P. Cayot, Int. Dairy. J., 15 (10) (2005) 1006.
9. P. Cayot, J.-F. Fairise, B. Colas, D. Lorient and G. Brule, J. Dairy Res., 70 (4) (2003) 423.
10. G. Arvisenet, P. Le Bail, A. Voilley and N. Cayot, J. Agric. Food Chem., 50 (24) (2002) 7088.
11. M.K. Keogh and B.T. O'Kennedy, J. Food Sci., 63 (1) (1998) 108.

W.L.P. Bredie and M.A. Petersen (Editors)
Flavour Science: Recent Advances and Trends

Phase ratio variation method as an efficient way to determine the partition coefficients of various aroma compounds in mixture

Géraldine Savary[a], Jean-Louis Doublier[b] and Nathalie Cayot[a]

[a]*UMR FLAVIC INRA-ENESAD, 17 rue Sully, 21 000 Dijon, France;*
[b]*UBIA INRA, rue de la Géraudière, 44316 Nantes, France*

ABSTRACT

The partition coefficients of aroma compounds when mixed together were determined at 30 °C in complex models of fruit preparation containing starch and pectin or carrageenan by the phase ratio variation method using a headspace autosampler. This method enabled easy determination of the partition coefficients of volatile compounds in a gas/matrix system without calibration and the results agreed well with published data obtained by other methods. Under equilibrium, an effect of polysaccharides was found only with hexanal, limonene, diacetyl and linalool. The retention was not due to interactions between aroma compounds and a polysaccharide but rather to a trapping in the complex network or to a combined effect of starch and polysaccharides.

1. INTRODUCTION

The measurement of partition coefficients has been used for many years to estimate the interactions between aroma compounds and food matrices. This was mainly applied to simple model systems in which aroma compounds were added one by one [1]. The present study is focused on complex model systems containing two polysaccharides in mixture with sucrose and an acid buffer. These complex systems are used as models of fruit preparations that are used in the dairy industry to flavour yoghurt. This work is a part of a larger study dealing with the effect of the structure of gels on aroma release. A first step was to determine the partition coefficient of each aroma compound by static headspace measurements for each system and to test the impact of each ingredient. For that purpose, the use of the phase ratio variation (PRV) method was applied. This method seems to be appropriate for the determination of partition coefficients of aroma compounds in a mixture because it necessitates neither a calibration curve nor a quantification step [2]. The method has already been used by several authors [3,4] to estimate the partition coefficient of single aroma compounds in polysaccharide solutions.

2. MATERIALS AND METHODS

2.1. Composition of model fruit preparations

The partition coefficients of nine aroma compounds were determined in different matrices. Table 1 shows the composition of these matrices (%, w/w). A modified waxy corn starch, an amidated low methoxy pectin and a mix of κ and κ2-carrageenans, were used as texture agents in a lactate buffer (pH=3.8).

Table 1. Composition of the matrices.

	Water	Buffer	Sucrose	Starch	Carrageenans	Pectin	KCl	$CaSO_4$
Matrix P	-	63.01	35.00	1.40	-	0.40	0.16	0.03
Matrix C1	-	63.36	35.00	1.40	0.05	-	0.16	0.03
Matrix C2	-	63.64	35.00	1.40	0.33	-	0.16	0.03
Matrix C3	-	63.17	35.00	1.40	0.80	-	0.16	0.03
Matrix B	-	64.81	35.00	-	-	-	0.16	0.03
Matrix W	100	-	-	-	-	-	-	-

2.2. Composition of the strawberry aroma

The blend aroma was prepared by dissolving 17 aroma compounds in propylene glycol (%, w/w): ethyl acetate (1.8), diacetyl (0.4), ethyl butyrate (2.7), hexanal (0.1), limonene (0.2), ethyl hexanoate (2.2), (*Z*)-3-hexenol (2.4), ethyl octanoate (0.1), linalool (0.19), butanoic acid (0.2), hexanoic acid (0.1), maltol (3.3), furaneol (1.9), methyl cinnamate (0.2), γ-decalactone (0.3), decanoic acid (0.1) and vanillin (1.6).

2.3. Sample preparation

The matrices given in Table 1 were obtained using the same thermal treatment (10 min at 90 °C). They were flavoured (0.8%, w/w) before cooling at 30 °C under agitation. The samples were stored 24 h at 10 °C before headspace analysis.

2.4. Gas chromatography headspace analysis

During sampling, increasing quantities (50 mg, 100 mg, 200 mg, 500 mg, 1000 mg and 2000 mg) of each preparation were poured into headspace vials (22 ml); thus each vial represented a different phase ratio β (479, 239, 119, 47, 23 and 11). After equilibration at 30 °C for 2 h, a 1 ml sample of headspace was automatically taken off (CombiPal, CTC Analytics, Switzerland) and analysed (Fison, GC 8000, MFC 800).

2.5. Phase ratio variation method

The PRV method is based on the linear relationship established by [5] between the reciprocal of the chromatographic area A at equilibrium and the phase ratio β: $1/A = a + b \cdot \beta$, with $a = K/(fi \cdot Cs)$ and $b = 1/(fi \cdot Cs)$. Cs is the initial sample concentration in the matrix and fi is the proportional factor. The *a/b* ratio corresponds to the partition coefficient K and is calculated from the values obtained by plotting 1/A against *β*. An example for ethyl butyrate is given in Figure 1.

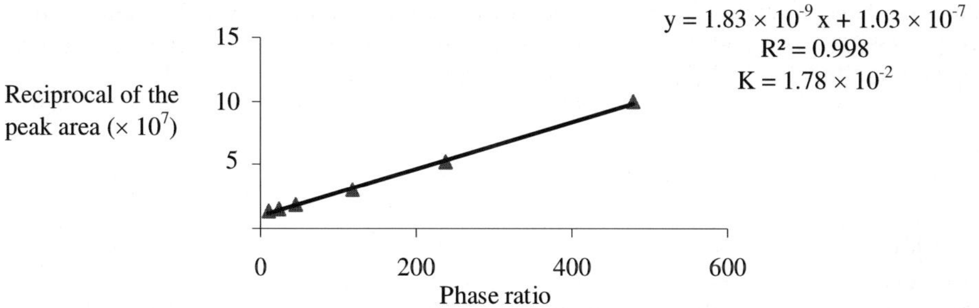

Figure 1. Graphic determination of partition coefficient gas/matrix at equilibrium by the PRV method for ethyl butyrate in the matrix P at 30 °C.

2.6. Data analysis

The K value was obtained by averaging the results of 3 series. The statistical tests (ANOVA) were performed using Statgraphics plus software.

3. RESULTS AND DISCUSSION

Nine compounds among the 17 of the strawberry aroma were volatile enough to be detected in the headspace.

Partition coefficients at 30 °C were first measured in water and then in the model fruit preparations that differed by their composition of polysaccharides (matrices P and C2) and by their texture (matrix C3: gel, matrix C1: thickened solution and matrix C2: intermediate). The partition coefficients were also measured in a simplified matrix without polysaccharide (matrix B). The overall results are reported in Table 2. Significant differences among the different measured K were observed only for diacetyl, linalool, hexanal and limonene (lower part of Table 2). For limonene, it seems that the K value was influenced by the carrageenan level. To verify this hypothesis, partition coefficients of limonene were measured in three matrices containing only water and carrageenans at various concentrations (0.05%, 0.33% and 0.80%). These matrices were flavoured by 0.8% of a solution of limonene at 0.22% in propylene glycol. We obtained the following not significantly different values $K_1 = 0.26 \times 10^{-1}$, $K_2 = 0.24 \times 10^{-1}$ and $K_3 = 0.24 \times 10^{-1}$, respectively.

Thus the retention of limonene in complex matrices was not due to simple interactions with carrageenan chains but rather to a trapping in the gelled structure or to a combined effect of the starch and carrageenans.

Table 2. Air/matrix partition coefficients of aroma compounds at 30 °C in the matrices ($\times 10^3$). The composition of the matrix types is given in Table 1.

	Matrix type					
	Water	B	P	C1	C2	C3
Ethyl acetate	16.1	16.4	16.2	16.4	16.2	16.1
Ethyl butyrate	18.3	18.4	17.8	18.6	18.4	18.2
Ethyl hexanoate	30.2	29.6	29.7	29.7	29.8	28.0
Ethyl octanoate	34.4	34.2	32.9	34.6	34.0	34.6
(*Z*)-3-Hexenol	0.5	0.6	0.6	0.5	0.6	0.6
Diacetyl	1.26^a	1.25^a	1.12^{ab}	1.03^b	1.01^b	1.14^{ab}
Linalool	3.87^a	3.82^a	3.21^b	3.21^b	3.17^b	3.27^b
Hexanal	11.6^a	11.8^a	9.68^b	9.64^b	9.64^b	9.53^b
Limonene	251^a	241^a	248^a	242^a	222^b	217^b

Values with different superscripts within a row are significantly different, ANOVA and LSD tests, $p < 0.05$.

For linalool, hexanal and diacetyl, the results revealed a significant retention effect by polysaccharides but this effect was the same whatever the quantity and the type of polysaccharides. As this phenomenon could be explained by interactions between aroma compounds and starch, the partition coefficients of linalool, hexanal and diacetyl were measured in a matrix containing only water and starch at 1.4%. This matrix was flavoured by 0.8% of a solution of these three compounds in propylene glycol.
Results indicated that the retention was higher than in water but less important than in matrices containing pectin and carrageenans. As in the case of limonene, the retention was not only due to the presence of one polysaccharide but rather to a trapping in the complex network or to a combined effect between starch and other polysaccharides.

References

1. E. Guichard, Food Rev. Int., 18 (1) (2002) 49.
2. L.S. Ettre, C. Welter and B. Kolb, Chromatogr., 35 (1993) 73.
3. C. Jouquand, V. Ducruet and P. Giampaoli, Food Chem., 85 (3) (2004) 467.
4. E. Bylaite, Z. Ilgunaite, A.S. Meyer and J. Adler-Nissen, J. Agric. Food Chem., 52 (11) (2004) 3542.
5. L.S. Ettre and B. Kolb, Chromatogr., 32 (1991) 5.
6. P. Landy, C. Druaux and A. Voilley, Food Chem., 54 (4) (1995) 387.
7. C. Yven, E. Guichard, A. Giboreau and A.D. Roberts, J. Agric. Food Chem., 46 (4) (1998) 1510.

W.L.P. Bredie and M.A. Petersen (Editors)
Flavour Science: Recent Advances and Trends

Role of mastication on the release of apple volatile compounds in a model mouth system

Gaëlle Arvisenet[a], Ludivine Billy[a], Gaëlle Royer[b] and Carole Prost[a]

[a]ENITIAA, LBAI - Rue de la Géraudière - BP 82225, 44322 NANTES cedex 3, France; [b]ESA, GRAPPE - 55, Rue rabelais - BP30748, 49007 ANGERS cedex01, France

ABSTRACT

A model mouth system was used to study the release of apple volatile compounds. Different crushing systems were designed to obtain different apple particle size. It was shown that the system giving the smallest apple particles allowed the release of the highest amount of apple volatile compounds.

1. INTRODUCTION

Dynamic headspace analysis (DHA) is commonly used to study aroma compounds of food, because it reproduces the way volatile compounds are released from food in the mouth better than other extraction techniques. Nevertheless, it is not always the most suitable technique. For example, the extract obtained by DHA from apple pieces is less representative for fresh apple aroma than a vacuum hydro-distillation extract [1,2], probably because apple is hard and crunchy, making aroma release depend of the breaking down process.

That is why we developed a model mouth system, allowing to reproduce mastication movements during extraction by DHA. Different model mouth systems have been developed by several authors to study liquid [3,5] or viscous [6] model products, but also semi-solid natural foodstuffs like vegetables [7,8]. The conditions in which volatile compounds were collected from food to headspace varied from one study to another. Generally, the temperature was maintained the same as in the human mouth, and the food sample was humidified by water or artificial saliva [4,6-8]. Depending on the apparatus and the texture of the sample, it could be either stirred [4] or crushed with a plunger [3,5-8].

The present study was undertaken to determine the most suitable intensity of mastication in our apparatus, in order to increase the extractability of apple volatile compounds.

2. MATERIALS AND METHODS

Materials: Golden delicious apples harvested in 2004 were stored at 3 °C until the day before measurements and kept at ambient temperature for 24 h before analyses. For each assay, four randomly selected fruits were peeled, cut in half and their cores were discarded. Equal longitudinal slices (0.8 mm thick) were cut from each apple. The internal standard *n*-decane (> 99%, Sigma) was used.

Mastication assays: A total of 68 g of apple and 15 ml of a solution of *n*-decane, 0.32 g/l in purified water (Millipore Corp., Molsheim, France), were introduced in the container of the artificial mouth (Figure 1), and assays were carried out with or without mastication simulation. To simulate mastication, two discs with model teeth were used, one containing flat teeth, imitating molar shape, and the second one containing pointed teeth, imitating canine shape. Simultaneously with the crushing, the headspace was purged with helium at 700 ml/min. The temperature in the device was 37 °C and the experiment lasted for 20 min.

Volatile collection and analysis: The volatiles were swept into a porous adsorbent polymer trap (Tenax 1.8 in. × 1.2 in.). After the extraction, volatiles were cryofocused at -40 °C using carbon dioxide and thermally desorbed at 195 °C. During desorption, volatiles were transferred directly into the column through the injection port in splitless mode. The volatiles were analysed with a gas chromatograph (Chrompack, CP 9001) fitted with a flame ionisation detector equipped with a capillary column (DB-Wax, 30 m in length × 0.25 mm i.d. × 0.5 μm thickness, J&W Scientific, Folsom, CA). The helium carrier gas velocity was 1 ml/min. The injector and detector were 250 °C and the oven was programmed as follows: from 40 °C to 60 °C at 1 °C/min, 60 °C held for 5 min, then from 60 °C to 100 °C at 2 °C/min and from 100 °C to 230 °C at 15 °C/min with a hold of 15 min.

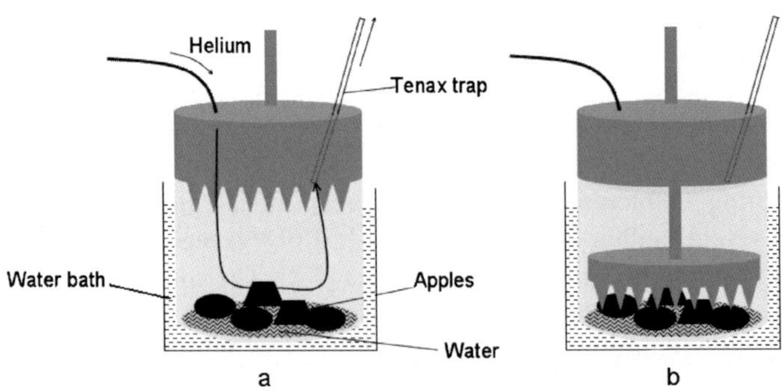

Figure 1. Model mouth system with crushing disk in up (a) and down (b) position.

3. RESULTS AND DISCUSSION

The two discs used to crush apples during extraction gave very different sizes of apple pieces (Figure 2). The slight crushing (flat teeth) only cut apples in pieces, still individualised and hydrated by water sorption. The remaining water contained few small particles of apple in suspension. With the strong crushing (pointed teeth), apples were reduced to very small particles, like a puree, and blended with water giving a suspension. These results were compared with apple masticated by a human (results not showed), and it appeared that apple pieces masticated in human mouth were a little bit larger and more homogeneous than in model mouth system with the strong crushing.

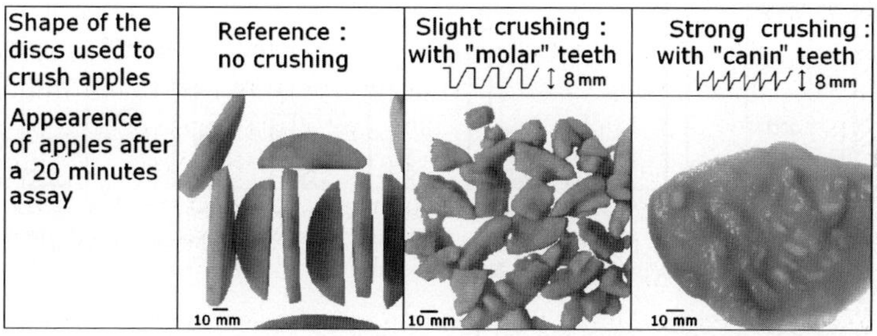

Figure 2. Appearance of apples after a 20 min assay, depending of the crushing disc used.

The extraction of *n*-decane, used as a standard, was not the same with and without crushing (Figure 3): when apple was not crushed, the release of *n*-decane was very low, and it increased by the apple crushing. This can be explained by the movement induced by the crushing, which acts as a stirring. But, surprisingly, the more the crushing was increased, the less the *n*-decane was released (Figure 3). With the strong crushing, *n*-decane may have been retained in the viscous apple suspension more than in water, were it stayed in 'slight crushing' conditions.

Because of this particular evolution of *n*-decane with crushing, the extracted amount of volatile compounds could not be expressed as a function of extracted amount of *n*-decane. Therefore it was expressed in peak areas. The total quantity of volatile compounds was significantly increased by crushing of apples. Compared to the reference without mastication, the amount of extracted compounds was increased by a factor of 6 with 'slight crushing' conditions, and by a factor of 10 with 'strong crushing' conditions.

The eight compounds giving the most important areas were studied separately. For seven of these eight compounds, stronger crushing resulted in increased release. These results confirm the importance of crushing apples to study the release of their volatile compounds. Mastication, by reducing apples into small particles, increases their contact surface with the environment and it probably enhances diffusion of volatile compounds from apple to headspace. It may also destroy the cellular structure, in the strongest crushing conditions, releasing even more volatiles. The increase of the amount of extracted volatiles seems to be particularly important for the lightest and heaviest compounds of our study. Ethyl acetate had a different behaviour from the other apple compounds studied: with slight crushing conditions, its release was seven times as high as the release in the reference without crushing, whereas with strong crushing, the release was only about 3 times higher than in the reference. Further studies will be performed in order to explain this particular behaviour.

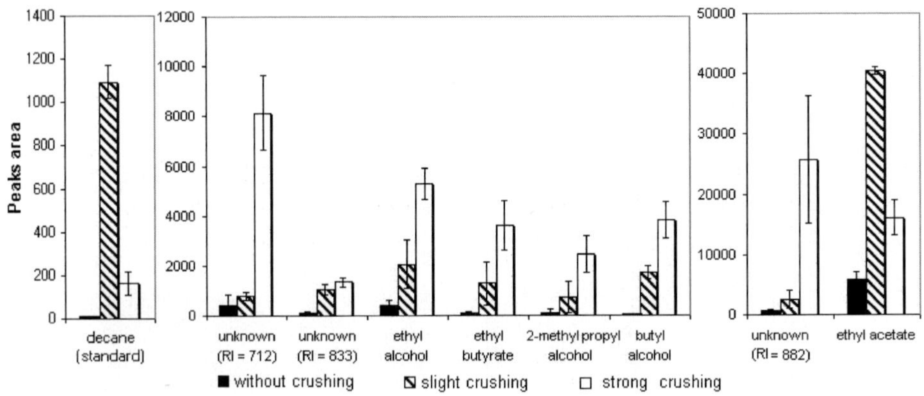

Figure 3. Influence of apple crushing on the release of apple volatile compounds in mouth conditions.

4. CONCLUSION AND PERSPECTIVES

The present work confirms how important it is to disintegrate food structure while studying flavour release. Semi-solid foods were previously studied in model mouth systems with mastication conditions [7,8] but, to our knowledge, it was never applied on foods as firm as apples. Further studies will be necessary to determine if the apple extract obtained by this method is representative for fresh apple aroma. It would be also interesting to apply this technique to other 'hard' foodstuffs and to study the release kinetics during the crushing of food.

References

1. E. Mehinagic, C. Prost and M. Demaimay, J. Agric. Food Chem., 52 (2004) 5175.
2. E. Mehinagic, C. Prost and M. Demaimay, J. Food Sci., 68 (8) (2003) 2411.
3. S.M. van Ruth and K. Buhr, Int. J. Mass Spectrom., 239 (2004) 187.
4. S. Rabe, U. Krings and R.G. Berger, Chem. Senses, 29 (2004) 153.
5. S.M. van Ruth, C.H. O'Connor and M. Delahunty Conor, Food Chem., 71 (2000) 393.
6. S. Odake, J.P. Roozen and J.J. Burger, Biosci. Biotechnol. Biochem., 64 (12) (2000) 2523.
7. S.M. van Ruth, J.P. Roozen and J.L. Cozijnsen, Food Chem., 53 (1995) 15.
8. S.M. van Ruth and J.P. Roozen, Food Chem., 71 (2000) 339.

Volatile loss from dry food polymer systems resulting from chemical reactions

Dana M. Dronen[a] and Gary A. Reineccius[b]

[a]General Mills Inc., 330 University Ave. S.E., Minneapolis, MN 55414 USA; [b]Department of Food Science and Nutrition, University of Minnesota, 1334 Eckles Ave., Saint Paul, MN 55108 USA

ABSTRACT

This study focuses on the loss of desirable flavour when encapsulated via spray drying in selected food polymers examining the relative importance of chemical reactions versus diffusion out of the polymer into the product headspace. The food polymers examined include whey protein isolate (WPI), gum acacia and 10 DE maltodextrin. Both diffusive and reactive losses of our model volatile system (18 compounds) occurred during storage and the extent of loss was dependent upon both the food polymer and volatile being considered. Losses tended to be related to the presence of protein in the food polymer although some volatiles were most stable in the whey protein isolate (e.g. acetaldehyde and pyrrole). Based on the average value of loss across all volatiles for each food polymer, volatile loss attributed to chemical reaction was approximately 35% for the 10 DE maltodextrin system, 50% for the gum acacia system and 80% for the WPI system.

1. INTRODUCTION

Information available on the diffusion of volatile compounds through dry food polymers has generally been based on long-term storage studies where the rate of loss is determined as the amount of a given compound in the dry matrix before and after some period of storage [1-3]. Most of these studies have used volatile compounds that would not be expected to undergo significant chemical reactions and thus, losses have been assumed to be due to diffusion from the food polymer into the environment [4]. We have studied a wider range of volatiles some of which may be quite reactive with the food polymer, oxygen or with other volatiles in the model system. Thus, we have found it necessary to attempt to separate losses due to chemical reactions from losses by diffusion. We have chosen to do this by encapsulating a complex model volatile system

(aroma compounds) via spray drying into various food polymers and then putting aliquots of the spray dried materials into sealed glass ampoules, where diffusional losses would be limited to those required to saturate the void volume of the sample [1]. The restricted void volume also limited (but not eliminated) any oxygen available for oxidation of volatiles.
This study reports on our observations of volatile losses over time of samples stored in closed ampoules versus those stored open to the atmosphere. While this method does not permit determination of the mechanism of loss due to chemical reaction, it limits diffusion so an appreciation can be gained for the relative importance of these two mechanisms of volatile loss.

2. MATERIALS AND METHODS

Several members of different chemical groups were included as model volatiles in this study [1]. They were (number in brackets is ppm based on carrier solids weight in the spray dryer infeed): methanol (5000), ethanol (3000), propanol (5000), acetaldehyde (1000), propanal (1000), butanal (1000), furfural (5000), acetone (2000), diacetyl (3000), 2,3-pentanedione (3000), ethyl acetate (1500), methyl formate (1500), ethyl formate (1500), propyl formate (1500), hexane (1000, internal standard), limonene (3000), pyrrole (4000) and ethyl mercaptan (1000).
Just prior to spray drying, a blend of these volatiles was homogenised into an aqueous slurry of selected food polymers: 10 DE maltodextrin (Grain Processing Corp., Muscatine, IO), gum acacia (Colloides Naturels, Bridgewater, NJ), or whey protein isolate (WPI, Davisco, LeSueur, MN). The slurries were spray dried in a Niro dryer to a water activity (A_w) of approximately 0.11.
Aliquots of each powder were placed in glass ampoules, the ampoules were sealed and then put into storage at 35 °C and -56 °C. An equivalent set of samples was placed in relative humidity controlled desiccators and also stored at 35 °C (volatiles were free to be lost from the powders into the desiccator headspace). Samples were analysed periodically over several months by gas chromatography headspace methods to determine the amount of each model aroma compound remaining in the powders. The differences in volatile concentrations between these two powders were assumed to be due to diffusion losses from the powder stored open to the environment.

3. RESULTS

3.1. Food polymer materials

The 10 DE maltodextrin provided good protection against reactive losses to all volatiles except for pyrrole (60%) (numbers in parentheses are loss over 65 days storage, see Figure 1 for all results). The only other compounds with notable losses were the aldehydes (acetaldehyde, propanal, butanal) and ethyl mercaptan. The aldehydes and ethyl mercaptan gradually decreased throughout storage with losses of 10-20%.

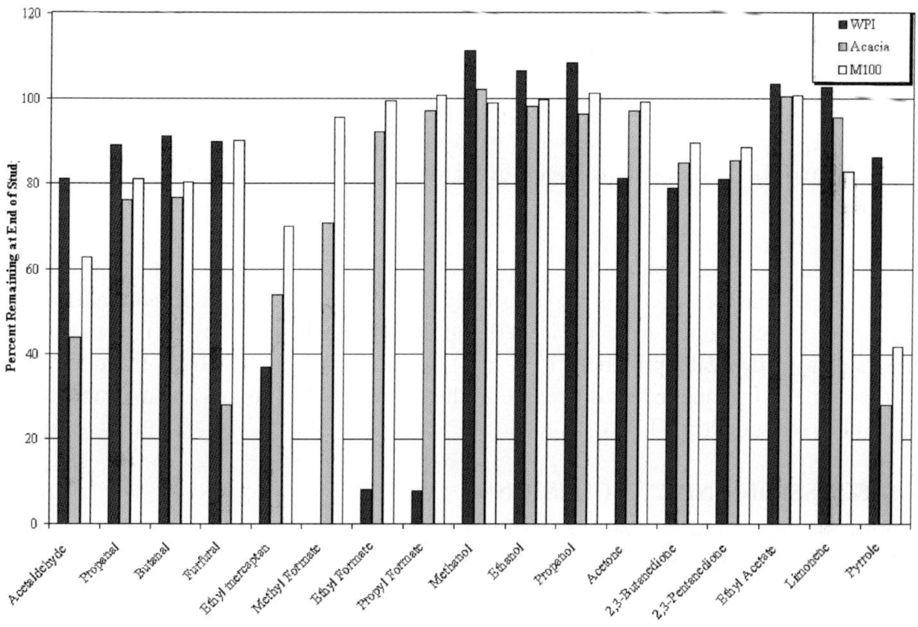

Figure 1. Losses of volatiles due to chemical reactions during storage (35 °C, A_w 0.11) when encapsulated in different matrices.

The gum acacia system demonstrated greater reactive loss of volatiles than the 10 DE maltodextrin system but less reactive losses than the WPI system. Chemical reaction was pronounced in this system for pyrrole (78%), furfural (78%) and acetaldehyde (58%). Losses of other volatiles by chemical reaction were generally small ranging from 0% (alcohols) to 20% (propanal).

WPI offered quite variable protection to our model system. Compounds such as pyrrole and furfural suffered very small losses during storage (5-15%) while the formates were nearly completely lost (about 100%). Also, nearly 50% of the ethyl mercaptan was lost. The alcohols and ketones were relatively stable in the WPI, with little or no loss of the alcohols over the storage period and 15-20% loss of the ketones and aldehydes over the entire storage period.

3.2. Volatiles

3.2.1. Carbonyls

While aldehydes tended to be lost via chemical reaction from all three food polymers during storage, losses generally tended to be relatively small (10-20% except for acetaldehyde in gum acacia). Ketones (represented by acetone, diacetyl and 2,3-pentanedione) were relatively stable with losses ranging from nearly none to slightly over 20% across all three polymers.

3.2.2. Alcohols

There appeared to be no measurable loss of the alcohols from any of the food polymers.

3.2.3. Esters

There were large losses of the formates from the WPI (90-100%) but lesser losses from the other two food polymers (0-50%). Ethyl acetate was not lost from any of the food polymers.

3.2.4. Miscellaneous compounds

The losses of these compounds varied greatly depending upon the food polymer. Ethyl mercaptan losses ranged from 65% (WPI) to 30% (10 DE maltodextrin). Pyrrole was most stable in the WPI polymer (15% loss) but 60-70% was lost from the other two food polymers. Limonene loss decreased in the order of WPI (no loss), gum acacia (6%) and 10 DE maltodextrin (17%).

4. DISCUSSION AND CONCLUSIONS

The results presented show that volatile losses attributable to chemical reactions can be an extremely important means of volatile loss from dry food systems (Aw approximately 0.11) during long term storage. Interestingly, model volatiles that exhibited large losses during open storage (where diffusional losses were permitted), showed excellent retention during closed storage (e.g. the alcohols) [1]. Thus in some cases, losses during storage are dependent upon packaging permeability and in others, losses may be excessive even with impermeable packaging systems. In summary, chemical reactions may contribute substantially to the loss of desirable flavour components, and thereby may be a limiting factor in the shelf-life of dry foods.

Our results further show that different food polymers afford differing levels of inhibition towards these chemical reactions. The effect of the polymer varied with the volatile compound and its likely mechanism of chemical reaction. Thus we would expect the food base composition to greatly influence the stability of a given flavour during storage.

References

1. D.M. Dronen, Characterization of volatile loss from dry food polymer materials, PhD Thesis, University of Minnesota, Minneapolis, MN, USA (2004).
2. B.C. Whorton, Effect of polymeric phase transitions on the controlled release and oxidative stability of flavor model systems encapsulated in traditional carbohydrate carriers, PhD Thesis, University of Minnesota, Minneapolis, MN, USA (2000).
3. Y.M. Gunning, P.A. Gunning, E.K. Kemsley, R. Parker, S.G. Ring, R.H. Wilson and A. Blake, J. Agric. Food Chem., 47 (1999) 5198.
4. H. Levine (ed.), Amorphous food and pharmaceutical systems, special publication no. 281, Cambridge, UK (2002) 98.

W.L.P. Bredie and M.A. Petersen (Editors)
Flavour Science: Recent Advances and Trends

Influence of proteins on the release of aroma compounds into polymer film

Yuichi Hirata, Mélanie Massey, Perla Relkin, Paolo Nunes and Violette Ducruet

UMR SCALE (ENSIA/CNAM/INRA), 1 Avenue des Olympiades Massy CEDEX, 91744, France

ABSTRACT

This work focused on the mass transfer between the gas phase above an aqueous mixture containing proteins and aroma compounds, and a polymer film. A permeation system equipped with a purge-and-trap injector coupled to GC-MS was optimised for this application. The permeability and diffusion kinetics of aroma compounds (from C4 to C8) through a low-density polyethylene film were tested simultaneously. Each permeation curve was obtained independently. The transmission rate for the penetrants is discussed with regard to the transport parameters such as permeability, diffusion, and solubility coefficients but also to the partition coefficient between gas phase (headspace) and liquid phase (protein solution).

1. INTRODUCTION

Barrier properties against aromas as well as gas and water vapour of polymeric packaging films are of great importance to maintain shelf-life of foodstuffs. Because of the interaction between polymer and aroma compound, high barrier gas polymers do not exhibit low aroma permeability. Moreover, standard methods (ASTM D1434-82; ASTM E96-80) used to measure gas permeability of the packaging films are not available for measuring aroma permeability. This could be due to the fact that aroma in food is a complex mixture in which the vapour pressure of each component is very low. Simultaneous measurement of the permeability of each aroma compound is required for the food packaging field, and different setups have been proposed for measurement of permeability of volatile organic compounds through packaging films at very low permeant vapour pressures [1]. In this study, a purge-and-trap injector [2] was used to enhance the rapidity of quantification of transmission rate of individual aroma compounds. Concerning release properties of aroma compounds from a food matrix, it was shown that globular proteins that are characterised by a core-structural pattern, might bind a large variety of aroma compounds, with consequences on their partition coefficient between air and aqueous phases [3].

2. MATERIALS AND METHODS

Low density polyethylene (LDPE) film was purchased by Good fellow Co., Ltd. The thickness of film was 150 μm.
Egg whites were separated from yolk and a solution was prepared in ultra pure water, at a concentration of 28% (w/w) or 0.6% (w/w). After 3 h of gentle magnetic stirring (avoiding foaming), the solution was centrifuged at 14000 rpm during 30 min. The protein solutions were then flavoured with the aroma mixture (see Table 1). The characteristics of aroma compounds used in this study are presented in Table 1. Their concentrations ranged from 1 mg/kg to 50 mg/kg. Protein solutions were stored at 4 °C prior to measurements.

Table 1. Characteristics of aroma compounds in the aroma mixture. MW: molar weight; T_{bp}: boiling point; Psat: vapour pressure.

Aroma compound	MW	T_{bp} (°C)	Psat[a] (cmHg)	Solubility in water[a] (mg/l)	LogP [b]	Content[c] (mg/l)
Ethyl acetate	88.1	77.1	9.32	80000	0.73	50
Propyl acetate	102.13	101.5	3.59	18900	1.36	20
Ethyl butanoate	116.16	121.5	1.28	4900	1.85	7
Butyl acetate	116.16	126.1	1.15	8400	1.85	7
Isoamyl acetate	130.18	142.5	0.56	2000	2.26	7
Ethyl hexanoate	144.21	167	0.156	629	2.83	1
Hexyl acetate	144.21	171.5	0.132	511	2.83	1
Cyclohexyl acetate	142.20	173	1.08	728	2.64	7

[a]At 25 °C. [b]Calculated using EPI Suite v3.12 (U.S. Environmental Protection Agency). [c]In pure aroma mixture.

The partition coefficients of the aroma compounds were determined by the Phase Ratio Variation method [4] in different matrices (water, egg white 28% and 0.6%). Increasing volumes (1, 2, 3 and 4 ml) of each aqueous solution were poured into headspace vials (22 ml); thus, each vial represented a different phase ratio β (21; 10; 6.3; 4.5). Partition coefficient of each aroma compound was determined after 3 h of equilibration at 50 °C. Headspace samples of 1.5 ml were taken by a Multi Purpose Sampler 2 (Gerstel) before injection on a DB-FFAP column (30 m x 0.32 mm i.d. x 1 μm film thickness, Agilent). Analyses were carried out in duplicate.
A permeation cell was linked to a purge-and-trap injector coupled to GC-MS system for measurement of permeation at 25 °C. The sample film was installed in the glass cell resulting in an effective film area of 28.3 cm^2. 500 ml of the sample solution was poured in the cell at the feed side and then the helium gas was periodically swept in at the permeation side by using a purge-and-trap injection system. The aroma compounds in

sweep gas were trapped in a capillary column at -110 °C and then injected to a GC-MS by heating the column to 200 °C. Calibration was carried out to relate the GC peak area of each compound to the quantity.

3. RESULTS AND DISCUSSION

At 50 °C the air-liquid partition coefficient in water was for all compounds tested higher than in the egg white solutions (Figure 1). The retention increased with the protein concentration. Furthermore, the most hydrophobic compounds were the most retained. The retention of ethyl hexanoate was almost complete in the 28% protein solution as this compound was not detected in the headspace. The binding of aroma compounds by proteins from egg white does not only depend on the nature of the aroma compound, as was shown in many studies, but also on the protein content, which was already shown for sodium caseinate [5].

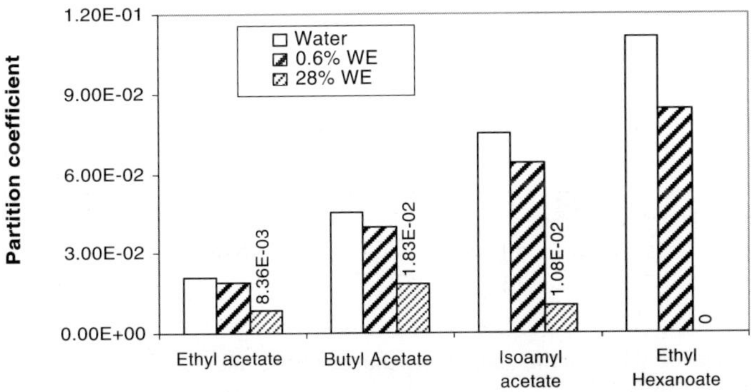

Figure 1. Partition coefficients (K) of the aroma compounds into both white of egg (WE) solution (0.6% and 28% w/w) and into water at 50 °C.

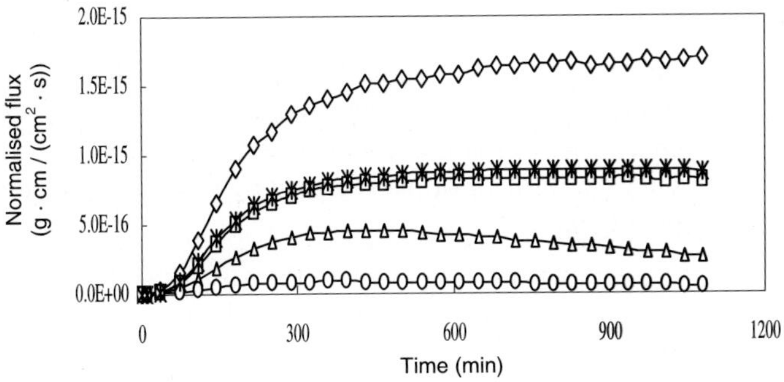

Figure 2. Permeation curves for ethyl butanoate. (◇) water solution; (□) 0.6% protein solution; (△) 28% protein solution; (*) 0.6% protein solution after one day; (O) 28% protein solution after one day.

Ovalbumin, which represents about 54% (w/w) of the protein fraction of egg white could be involved in the retention of the aroma compounds studied. Previous studies have shown that this globular glycoprotein can interact with vanillin [6].
Figure 2 shows the flux for ethyl butanoate during the permeation experiment. When ethyl butanoate was applied as water solution or 0.6% protein solution the flux was stationary at the end of permeation (1080 min). In the case of 28% protein solution, the flux increased and then decreased until the end of permeation. Preservation for 1 day depressed the flux from the 28% protein solution during the permeation experiment.
Figure 3 shows the flux for 5 aroma compounds at 1080 min in the case of one day preservation. The influence of the concentration of the protein on the flux was strong for all compounds investigated, and tended to increase with increasing logP.

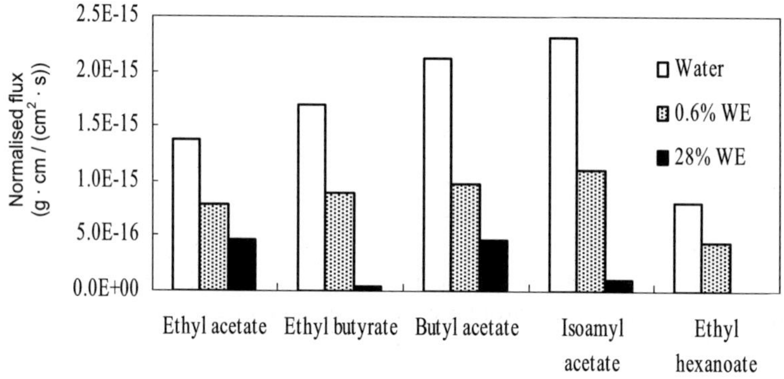

Figure 3. Influence of the protein content in the solutions on the flux of aroma compounds through LDPE.

4. CONCLUSION

Egg white proteins can interact with aroma compounds independently of the volatility of these molecules. The main effect was noticed for the more hydrophobic compounds (ethyl hexanoate) and depended on the protein concentration in the medium. The retention lead to a decreased flux of aroma compounds to LDPE. Proteins could play a positive role in the aroma stabilisation of packed food by preventing the transfer of the aroma compounds to packaging.

References

1. R. Franz, Packag. Technol. Sci., 6 (1993) 91.
2. Q. Zhou, B. Guthrie and R. Cadwallader, Packag. Technol. Sci., 17 (2004) 175.
3. S.M. Van Ruth and E. Villeneuve, Food Chem., 79 (2002) 157.
4. C. Jouquand, V. Ducruet and P. Giampaoli, Food Chem., 85 (2004) 467.
5. P. Landy, C. Druaux and A. Voilley, Food Chem., 54 (1995) 387.
6. T.V. Burova, N.V. Grinberg, V.Y. Grinberg and V.B. Tolstoguzov, Colloid. Surf. A, 213 (2003) 235.

W.L.P. Bredie and M.A. Petersen (Editors)
Flavour Science: Recent Advances and Trends

Glycosidically bound alcohols of blackcurrant juice

Camilla Varming, Mogens L. Andersen and Leif Poll

Department of Food Science, The Royal Veterinary and Agricultural University, Rolighedsvej 30, 1958 Frederiksberg C, Denmark

ABSTRACT

The release of bound volatile alcohols of blackcurrant juice from the corresponding glycosides was studied enzymatically using β-glucosidase in two different ways. The enzyme was added either directly to the juice, or to an extract of glycosidic compounds that had been isolated from the juice by chromatography using Amberlite XAD-2. The two methods resulted in the same patterns of released volatile aglyconic compounds. Fifteen aliphatic alcohols, four aromatic alcohols and one aromatic aldehyde were identified in the glycosidically bound fraction.

1. INTRODUCTION

Many fruits contain non-volatile precursors of aroma compounds in the form of glycosidically bound compounds [1]. Volatile aroma compounds can be released from the precursors by the action of e.g. heat, acid hydrolysis or enzymes. This has been a subject of interest especially in grapes and wine, whereas the presence of glycosidically bound aroma components in blackcurrant has previously only been indicated [2]. Heating and other processing of blackcurrant berries and juice cause changes in the levels of aroma components [3-6]. The possibility that some of the observed changes are caused by release from a pool of glycosidically bound aroma compounds is investigated in the present study.

2. MATERIALS AND METHODS

2.1. Materials

Commercial blackcurrant *(Ribes nigrum* L.) juice of the variety Ben Lemond was obtained from an industrial plant as described by [4] and kept at -18 °C until analysis. A 0.1 M citrate-phosphate buffer (pH 5) and an almond β-glucosidase enzyme solution (10 mg/ml, Sigma-Aldrich, Denmark) in buffer were used.

2.2. Enzymatic hydrolysis

Enzymatic hydrolysis of chromatographically separated glycosidic extract: Free and glycosidically bound aroma compounds from 100 ml juice were chromatographically separated on an XAD-2 column, according to the method described in [7]. Water soluble components were eluted with 50 ml of distilled water. The free volatile fraction was extracted with 50 ml pentane/diethylether (1:1) and discharged. The glycosidically bound aroma fraction was extracted with 50 ml methanol, concentrated under reduced pressure to approximately 1 ml and redissolved in 2.0 ml buffer. Half of this glycosidic extract was added 2.0 ml β-glucosidase solution in a small vial. The other half of the glycosidic extract was added 2.0 ml buffer, constituting the control sample.

Direct enzymatic hydrolysis: Blackcurrant juice and buffer (1:1) was adjusted to pH 5. Fifty millilitre of the mixture was transferred to a 250 ml glass flask equipped with a lid and 2.0 ml of the β-glucosidase solution was added. The juice sample not to be enzymatically treated constituted 50 ml of the juice and buffer (1:1) and 2.0 ml buffer.

All samples were incubated for 19 h in a water bath at 37 °C.

2.3. Determination of volatile compounds

Dynamic headspace sampling was carried out immediately after the enzyme treatments. The chromatographically separated glycosidic extract sample was transferred to a glass flask equipped with a purge head and diluted with water to 50 ml. The glass flask containing direct enzymatically hydrolysed juice was equipped with a purge head. One millilitre of internal standard (50 μl/l 4-methyl-1-pentanol) was added. Sample temperature was equilibrated in a 30 °C water bath for 10 min. Under magnetic stirring (200 rpm) the sample was then purged with nitrogen (100 ml/min) for 60 min. The volatiles were collected on Tenax TA traps.

Quantification of the direct enzymatic hydrolysis samples was carried out on reference aroma compounds (200 ppb) in a model solution (glucose concentration 48 g/l, fructose level 61 g/l and citric acid monohydrate concentration of 39 g/l in water, 14°B) resembling blackcurrant juice. For dynamic headspace sampling 50 ml of model solution and buffer (1:1) adjusted to pH 5 was used.

An Automated Thermal Desorber and a GC-MS system were used as described in [8]. Identification was carried out by probability-based matching with mass spectra in the G1035A Wiley library (Hewlett-Packard). Peak area calculations were based on single ions, and peak areas of aroma compounds were divided by the peak area of the internal standard.

3. RESULTS AND DISCUSSION

Glycosidically bound compounds in blackcurrant juice were isolated using an Amberlite XAD-2 column. Fifteen aliphatic alcohols, four aromatic alcohols as well as an aromatic aldehyde were identified upon enzymatic treatment of the chromatographically prepared extract with β-glucosidase (Table 1). No volatile compounds were released from a control sample of the glycosidic extract. Also, some glycosidically bound terpenoids were found which will be described in a separate paper [9].

Direct enzymatic treatment of blackcurrant juice with β-glucosidase gave thc same pattern of released volatile aglyconic compounds, except for 3-methyl-1-butanol (Table 1). However, the concentration levels are not dircctly comparable due to the different methods used.

Table 1. Free and glycosidically bound volatile alcohols in blackcurrant juice.

Compound[a]	Enzymatic hydrolysis of glycosidic extract (μg/l)[b]	Direct enzymatic hydrolysis of blackcurrant juice			
		Free (μg/l)[c]	Bound[d] (μg/l)	% Free	% Bound
2-Butanol	59	6±2	33	15	85
2-Methyl-1-propanol[e]	64	431±13	0	100	0
3-Pentanol	42	4±0	55	6	94
2-Pentanol	219	22±1	226	9	91
1-Butanol	100	954±47	67	93	7
1-Penten-3-ol	70	26±10	107	20	80
3-Methyl-1-butanol	1042	3535±100	17	100	0
3-Methyl-3-buten-1-ol	43	7±1	30	19	81
1-Pentanol	30	18±1	35	35	65
1-Hexanol	242	97±4	206	32	68
(*Z*)-3-Hexen-1-ol	117	29±4	81	26	74
(*E*)-2-Hexenol	16	4±1	10	29	71
1-Octen-3-ol	15	12±1	6	65	35
1-Heptanol	3	3±1	3	50	50
1-Octanol	14	9±0	12	42	58
Methyl salicylate	199	11±1	137	7	93
Benzyl alcohol	1313	60±65	1934	3	97
Eugenol	197	284±124	280	50	50
2,4-bis-*tert*-Butylphenol[f]	2	2±0	1	57	43
Benzaldehyde	209	117±13	111	51	49

[a]Other compounds identified in the blackcurrant juice are reported elsewhere [4]. [b]Values are not directly comparable with those of direct enzymatic hydrolysis since only one replicate was analysed. [c]Concentrations are given as average ± standard deviation (n=3). [d]Concentration of enzymatically hydrolysed juice - concentration of untreated juice. [e]Breakthrough on the adsorbent traps. [f]Quantified on the basis of thymol standard.

A large variation in the amount and proportion of the glycosidically bound volatiles in the blackcurrant juice was seen, and no obvious pattern could be recognised. For most compounds the glycosidically bound fraction was larger than the free fraction, but all compounds identified as glycosidically bound were also present in the free form. Benzyl alcohol was the glycosidically bound compound that had the highest ratio between bound and free amount (bound concentration was 32 times the free concentration) followed by eugenol, 2-pentanol, and 1-hexanol.

Existence of aromatic glycosides in blackcurrants has previously been indicated [2] and most of the glycosidically released compounds have been found as aglycones in other fruits [1].

Results of the present study confirm that the increase of aromatic alcohols observed by thermal treatment [4] is likely to be, at least partly, caused by a heat induced glycosidic release. Also, during the concentration of blackcurrant juice an increase of benzaldehyde, benzyl alcohol, eugenol, and methyl salicylate is found [5,6]. In neither of these studies did the concentration of aliphatic alcohols however increase, indicating that the applied heat does not induce a glycosidic release of these compounds. Possibly, glycosidically bound volatiles could have been released during other steps of the juice processing [3].

Some of the identified glycosidically bound alcohols are important for the odour of blackcurrant juice, i.e. 3-methyl-1-butanol, (*Z*)-3-hexen-1-ol, 1-octen-3-ol, 1-octanol, benzyl alcohol, and eugenol [3,6,8]. Glycosidic release might result in additional compounds having concentrations exceeding their odour threshold values. Hence, release from a pool of glycosidically bound alcohols by enzymatic hydrolysis can lead to changes in the sensory impression of blackcurrant juice.

References

1. E. Stahl-Biskup, F. Intert, J. Holthuijzen, M. Stengele and G. Schulz, Flavour Fragrance J., 8 (2) (1993) 61.
2. R.J. Marriott, Abstracts of papers of the American chemical society, American Chemical Society, Washington, DC, USA, 190 (1985) 33.
3. B.B. Mikkelsen and L. Poll, J. Food Sci., 67 (2002) 3447.
4. C. Varming, M.L. Andersen and L. Poll, J. Agric. Food Chem., 52 (2004) 7628.
5. H. Kollmansberger and R.G. Berger, Deut. Lebensm. Rundsch., 90 (1994) 69.
6. J.-L. Le Quéré and P.X. Étiévant (eds.), Flavour research at the dawn of the 21st century, proceedings of the 10th Weurman flavour research symposium, Paris, France (2003) 741.
7. R. Boulanger, D. Chassagne and J. Crouzet, Flavour Fragrance J., 14 (1999) 303.
8. C. Varming, M.A. Petersen and L. Poll, J. Agric. Food Chem., 52 (2004) 1647.
9. C. Varming, M.L. Andersen and L. Poll, J. Agric. Food Chem., submitted.

8

Sensory - instrumental relationships

W.L.P. Bredie and M.A. Petersen (Editors)
Flavour Science: Recent Advances and Trends

Prediction of wine sensory descriptors from GC-olfactometry data: possibilities and limitations

Eva Mª Campo, Ana Escudero, Juan Cacho and Vicente Ferreira

Laboratory for Flavour Analysis & Enology, Analytical Chemistry, Faculty of Sciences Universidad de Zaragoza, 50009 Zaragoza, Spain

ABSTRACT

A novel methodology combining a dynamic headspace technique for the preparation of wine extracts, quantitative GCO and further data treatment has been developed. This strategy makes it possible to rank the potentially most important odorants and even predict some wine sensory attributes and has been applied to two different sets of white wines. GCO data revealed that linalool, 3-mercaptohexyl acetate, guaiacol and sotolon were the potentially most discriminative odorants. PLSR and correlation analysis suggested that these compounds could explain some of the most important nuances of the studied wines, such as floral, sweet, tropical fruit or nutty-liquor. Possibilities and limitations of this strategy are discussed.

1. INTRODUCTION

The analytical chemistry aspects of aroma analysis have been well-established to determine a product's aroma compound composition in the search for components relating to a product's sensory attributes. The first stage is usually the chemical characterisation and quantification of volatiles. In order to assess the potency of isolated odorants GC-olfactometry (GCO) methods such as AEDA combined with a quantitative determination to rank the odorants by their ratio between concentration and odour threshold values (OAVs) and, finally, sensory tests on reconstructed aromas [1]. The need of new tools, in addition to GCO and OAVs for the evaluation of the potentially most important and discriminative odorants in wines encouraged us to develop a strategy where the preparation of a representative wine extract and appropriate olfactometry data [2] became the keystones. This approach may represent a practical alternative to sophisticated and expensive analytical methodology for prediction purposes.

2. MATERIALS AND METHODS

2.1. Experiment 1

2.1.1. Wine samples

Six monovarietal young white wines from 2001 vintage; Albariño, Godello, Malvasía, Parellada, Treixadura and Verdejo, were selected on the basis of their quality and representativeness of the varietal character.

2.1.2. Wine extracts preparation

Extracts for analysis were obtained by a dynamic headspace sampling technique in conditions close to retronasal olfaction: a controlled flow of N_2 (100 ml/min) passes through a mixture of 80 ml of wine and 20 ml of artificial saliva kept at 37 °C for 200 min. Volatiles are first trapped in a 400 mg bed packed with LiChrolut-EN resins and then eluted with 3.2 ml of dichloromethane. The extract is finally concentrated up to 200 µl, of which 1 µl is used for GCO analysis.

2.1.3. Olfactometry data treatment

GCO was carried out by a trained panel of eight subjects, who reported retention time, intensity (using a 7 point category scale) and descriptor for any detected odour. On this occasion, as some of the odorants in these extracts were very diluted, the olfactometry signal finally processed was not the mean of the olfactometry scores given by the different sniffers, but a mixture of both intensity and frequency of detection (labelled as 'modified frequency' – MF), which was calculated with the formula proposed by Dravnieks [3]; $MF(\%) = [F(\%)*(I)]^{1/2}$, where $F(\%)$ is the detection frequency of a given odorant expressed as percentage, and $I(\%)$ is the average intensity expressed as percentage of the maximum intensity of the odorant. Individual intensity ratings were analysed by ANOVA whereas % MF scores were analysed by chi-square test. The identification of the odorants was carried out by comparison of their odours, chromatographic retention index on both DB-WAX and DB-5 columns, and MS spectra with those of authentic compounds.

2.1.4. Wine sensory analysis

Descriptive analysis (DA) of wines (20 ml at 20 °C) was performed in duplicate by a panel of 8 trained tasters. ANOVA revealed that three sensory attributes; tropical fruit, flowery and sweet, out of a total of 10 notes assessed, showed the most discriminative power among the varieties. These three sensory attributes were selected for PLSR modelling.

2.1.5. Modelling sensory descriptors from olfactometry composition

The relationship between the olfactometry data and a single sensory attribute was explored by PLSR1. A first initial model was built for a given sensory descriptor using all the discriminating X variables (GCO scores). Different iterations excluding the least important variables were further run to look for the simplest model with the best validated prediction ability using cross validation.

2.1.6. Validation of PLSR models by sensory analysis
Validation assays were carried out by spiking a deodorised white wine with the compounds revealed as most important by PLSR. A series of triangle tests were first held to explore the potential sensory effect of an odorant in the aroma of the wine. Additional ranking and quantification tests were performed to establish relationships (positive or negative) between odorants and the intensity of a given sensory attribute.

2.2. Experiment 2
Four wines following complex elaboration and maturation procedures; Madeira, Pedro Ximénez, Sherry and Sauternes, were selected for a second experiment. Wine extracts preparation, GCO and DA were performed as indicated in 3.1. Data was analysed by simple regression.

3. RESULTS

Experiment 1. GCO revealed that wine extracts were enriched in the most volatile and least polar compounds. More than 90 different odorants were detected during the olfactometry runs, although only 20 compounds reached a minimum %MF value of 50 in at least one wine. These olfactograms were quite simple when compared with those obtained by solvent extraction in previous studies [4]. The most important odorants of the %MF ranking were the ethyl esters, the fusel alcohols and β-damascenone. Acids, phenols or aldehydes were hardly found in the extracts whereas three alkyl-methoxypyrazines were easily detected by panellists. GCO results allowed to easily rank the odorants attending to their potential importance in the aroma of wine, or to their potential ability to explain sensory differences between these samples, as is shown in Table 1.
Concerning PLSR, satisfactory models could be built for three of the most important descriptors of this set of wines, being the explained variance higher than 70% in the three cases. As shown in Table 2, a total of 5 compounds (one of them unknown) were involved in the models but only linalool and 3-mercaptohexyl acetate showed high loading weighs.
Sensory quantification tests (performed by a panel of assessors specially trained to quantify the intensity of the target descriptors) confirmed the prediction ability of PLSR models. Tropical fruit and floral notes were almost exclusively explained by the presence of 3-mercaptohexyl acetate and linalool, respectively.
Experiment 2. In the second set of wines, DA evidenced that the most important notes when describing some of these wines were 'nutty-liquor', 'aldehydic' and 'floral-sweet' (defined as raisin or dried figs). GCO revealed that linalool and guaiacol were the most discriminative odorants among varieties. Differences between maximum and minimum %MF values in these samples reached 85 for linalool and 82 for guaiacol.

Table 1. Odorants for which maximum differences between the six white wines were observed in the GCO experiment. Data is given in %MF. *, ** and *** indicates significance at $p<0.05$, $p<0.01$ and $p<0.001$, respectively. ns denotes no significance.

Compound	Max - min	Max	Mean	ANOVA
Linalool	66	66	24	***
3-Mercaptohexyl acetate	63	63	18	***
2-Methyl-3-furanthiol	48	75	45	ns
Acetic acid	46	59	44	*
3-Isopropyl-2-methoxypyrazine	46	62	49	*
Isobutyl acetate	43	63	42	*
2-Phenylethyl acetate	43	50	29	ns
3-*sec*-Butyl-2-methoxypyrazine	41	41	14	**
3-Isobutyl-2-methoxypyrazine	40	63	50	*

Table 2. Loading weights of the odorants included in the different PLSR models explaining a sensory attribute as a function of GCO scores.

Compound	Tropical fruit	Flowery	Sweet
Linalool	–	0.77	0.76
3-Mercaptohexyl acetate	0.84	-0.42	-0.50
3-*sec*-Butyl-2-methoxypyrazine	-0.37	–	–
2-Phenylethyl acetate	–	0.34	0.42
Unknown 1746	-0.41	–	–

4. DISCUSSION AND CONCLUSION

4.1. GCO

The dynamic headspace technique made it possible to obtain simpler and cleaner olfactograms than those observed in previous studies [4], in which extracts were obtained by Solid Phase Extraction of wine.

In this occasion, the recorded GCO signal takes into account not only the evaluation of intensity, but also the frequency of detection of an odorant. This can be done now because a large number of odorants are at concentrations near the threshold, and the differences in individual sensitivity between members of the tasting panel become very important. In addition to a more realistic hierarchy of the aromatic compounds, this strategy makes it possible to easily rank the odorants according to their discriminative-ability criteria. There are different indicators of such ability, such as the olfactometry range (Max-Min), or the significance of the effect of the factor wine measured through ANOVA or chi-square tests. Whichever the indicator used, there are two compounds showing outstanding potential discrimination power in wines belonging to the first set; linalool and 3-mercaptohexyl acetate. Both components have relatively low average

%MF values, but can reach very high scores (more than 60%) in some wine samples. Furthermore, there are some wines in which such compounds were not even detected. The high linalool content in the Albariño variety may explain its characteristic floral bouquet whereas 3-mercaptohexyl acetate could be responsible for the outstanding tropical fruit note of the Verdejo wine.

Other odorants with potential discriminant power in set 1 are 2-methyl-3-furanthiol, acetic acid, isobutyl and 2-phenylethyl acetates and the three alkylmethoxypyrazines.

4.2. PLSR models

The model for the tropical fruit flavour note suggested that the intensity of this nuance is positively related to the wine content in 3-mercaptohexyl acetate. Sensory experiments confirmed the significant role of the thiol in this nuance as well as the negative effect of methoxypyrazines.

Floral and sweet notes could be modelled well (correlation coefficient of 0.939, $p<0.01$). The similarity of PLSR models suggested that both attributes could be considered as equivalent for validation purposes. As shown in Table 2, both sensory notes are positively related to the presence of linalool and 2-phenylethyl acetate, and negatively affected by 3-mercaptohexyl acetate. Quantification tests confirmed the increment of the floral-sweet nuance with increasing amounts of both compounds. A significant decrease of this note was observed with additions of 3-mercaptohexyl acetate.

4.3. Simple regression

The sensory attributes of wines belonging to the second set can not be modelled following the previous strategy as a consequence of the aroma diversity of such samples. So far, these samples differ in their varietal and geographical origin, in their wine making processes and in the kind of maturation they have followed. Differences are, therefore, so intense, that it is quite easy to determine which odorants are the potential responsible for such differences, but there are not enough samples to build more complex models, through PLSR or other techniques. The existence of some paramount discriminating odorants overshadowed the role played by other odorants. This effect is illustrated in Figure 1.

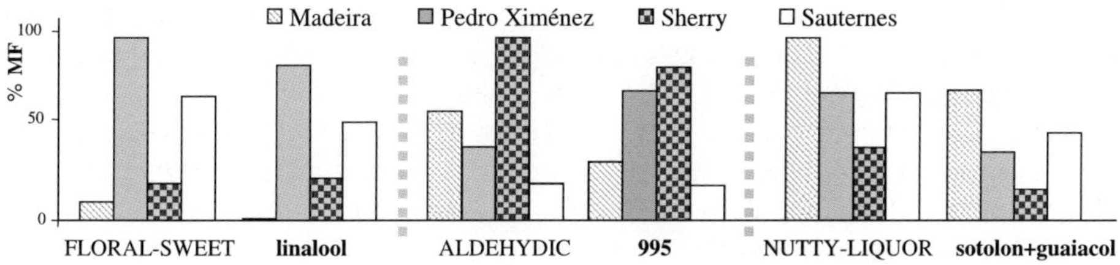

Figure 1. Aromatic notes and correlated compounds found in the four wines.

Linalool seems to dominate the floral-sweet odour note of these wines. It reaches the maximum GCO score in the Pedro Ximénez wine, while in the wine from Madeira was barely perceived. In wine from Sauternes it reaches also a high score with an excellent correlation between GCO and sensory data (Figure 1). The aldehydic note, characteristic of Sherry wine seems to be entirely related to the odorant with linear retention index 995. This odorant shows an equivalent sensory note and may be an iso-aldehyde but further identification is needed. Finally, the nutty-liquor note seems to be caused by the high levels of sotolon and guaiacol particularly in the Madeira wine.

5. CONCLUSIONS

As suggested in previous research [1], wine could form some kind of aromatic buffer (constituted by ethanol, ethyl esters, fusel alcohols, volatile phenols, β-damascenone and fatty acids) toward a wide range of aromas. This buffer can be broken only by the presence of an odorant with very different aroma properties. This is the case of compounds such as linalool, 3-mercaptohexyl acetate, sotolon, guaiacol and, perhaps, of the aldehyde with linear retention index 995. All of them are molecules with distinctive aromas and some of them were revealed as key-compounds. The detection of these key compounds has been easy to achieve by GCO of headspace extracts. These results suggest that the strategy proposed represents a practical alternative to determination of dilution factors and odour activity values, which usually need greater analytical efforts.
There are, however, some questions that such strategy is not able to explain. Why some odorants revealed as important in this study, do not actually play a manifest role in overall aroma profile? This is the case for alkyl-methoxypyrazines. None of the wines rich in these compounds presented the pepper nuances related to these components. A second question is about the role played by heavy volatiles. For example, vanillin was not detected by GCO while present at relatively high levels in the experimental wines.

References

1. A. Escudero, B. Gogorza, M.A. Melús, N. Ortín, J. Cacho and V. Ferreira, J. Agric. Food Chem., 52 (11) (2004) 3516.
2. V. Ferreira, J. Pet'ka, M. Aznar and J. Cacho, J. Chromatogr. A, 1002 (1-2) (2003) 169.
3. ASTM, Atlas of odor character profiles, Philadelphia, USA (1985) 354.
4. R. López, N. Ortín, J.P. Pérez-Trujillo, J. Cacho and V. Ferreira, J. Agric. Food Chem., 51 (11) (2003) 3419.

Interactions of basil flavour compounds in tomato soups of varying Brix and acidity

Bonnie M. King and C.A.A. Duineveld

Quest International Nederland BV, PO Box 2, 1400 CA Bussum, The Netherlands

ABSTRACT

Eight model samples of basil flavour were prepared according to a Plackett Burman design using 7 ingredients. The flavours were applied to each of four different tomato soup bases in order to study the flavour-matrix interactions by sensory profiling. The full factorial sample design for the bases included two levels each for acidity and Brix. Statistical analyses revealed significant effects for acidity (increased scores for 'sour', decreased scores for 'sweet' and 'cooked tomatoes') and Brix (increased scores for 'spices') as well as several significant interaction effects between acidity or Brix and specific ingredients in the model basil flavour. Ocimene significantly increased scores for 'umami', 'bitter', 'metallic' and decreased scores for 'spices'. Linalool significantly decreased scores for 'sour' and increased scores for 'bitter'. Hexanal increased scores for 'bitter' and decreased scores for 'alliaceous'. Eugenol increased scores for 'spices'. 2-Methyl-4-propyl-1,3-oxathiane decreased scores for 'spices' and increased scores for 'alliaceous', 'minty', 'tomato leaves/catty'. Methyl cinnamate decreased scores for 'DMS'.

1. INTRODUCTION

Taste and smell interactions in foods are well known. Several reports of fruity flavours enhancing sweetness or green notes enhancing sourness can be found in the literature, e.g. see recent review by Delwiche [1]. Tomato breeding for flavour improvement has aimed at both an increased ratio of Brix:acidity and an increase in those volatiles associated with the ripe fruit, such as *S*-Linalool [2,3]. Given the widespread use of basil in tomato-based foods, and the commonality in flavour ingredients for basil and tomato, a sensory study of their flavour-matrix interactions was undertaken.

2. MATERIALS AND METHODS

2.1. Soups

Four soup bases were prepared (two batches of 10 kg per base) and stored in 500 g plastic bags at –18 °C until used. Ingredients for all bases included tomato paste (1700 g), sugar (150 g), salt (100 g), sunflower oil (100 g) Colflo 67 (120 g), glucose syrup Malta (100 g), Monosodium glutamate (MSG, 30 g) and water (7700 g). Dry ingredients were mixed together before adding tomato paste, oil, and water. The ingredients for 10 kg soup were brought to a boil and cooked 5 min, then cooled and portioned for storage. The full factorial design included two levels each for acidity and Brix. Acidity levels were adjusted by adding either 10 g citric acid (high acidity) or 30 g sodium citrate (low acidity) to the 10 kg-batch of cooled soup. Adjustments for the high Brix level were achieved by adding 350 g Maltodextrin to the dry ingredients during soup preparation. Table 1 gives average analytical values for acidity as citric acid 0 aq (DL 70ES Mettler Toledo autotitrator), pH (Metrohm 691 Potentiometer), Brix (REUO Mettler Toledo refractometer), dry matter (evaporation at 70 °C under vacuum), refractive index (n 20/D), MSG levels (enzymatic determination) and viscosity (Brookfield LVTD, spindle 2 at two different speeds: 3 and 6).

Table 1. Analytical measurements for the four soup bases.

	Base 1	Base 2	Base 3	Base 4
Experimental design:				
Acidity	High	High	Low	Low
Brix	Low	High	Low	High
Acidity (%)	0.43	0.43	0.32	0.33
Brix (°)	10.65	14.65	10.90	15.10
Dry matter (%)	11.8	15.4	12.1	15.9
MSG (%)	0.45	0.45	0.45	0.45
n20D	1.3489	1.3552	1.3493	1.3559
pH	4.4	4.3	4.8	4.6
Visc3 (cp)	2605	3845	2940	3760
Visc6 (cp)	1915	2555	2085	2395

2.2. Flavours

Eight model samples of basil flavour were prepared according to a Plackett Burman design using 7 ingredients dissolved in MCT as shown in Table 2. Each flavour was dosed at 0.2 g/kg soup base. Soups were thawed at room temperature (approximately 4 h). Flavours were added, and the soup was brought up to serving temperature in the au-bain-marie.

Table 2. Composition (g) of the 8 Basil flavours evaluated on each base.

Flavour ingredient	Basil PB 1	Basil PB 2	Basil PB 3	Basil PB 4	Basil PB 5	Basil PB 6	Basil PB 7	Basil PB 8
Linalool (natural)	1.10	1.10	1.10	0.55	1.10	0.55	0.55	0.55
Methyl cinnamate	1.00	0.50	1.00	1.00	0.50	1.00	0.50	0.50
Eugenol	1.00	0.50	0.50	1.00	1.00	0.50	1.00	0.50
Hexanal	0.60	0.30	0.30	0.30	0.60	0.60	0.30	0.60
β-Ocimene	1.00	1.00	0.50	0.50	0.50	1.00	1.00	0.50
2-Methyl-4-propyl-1,3 oxathiane	0.60	0.30	0.60	0.30	0.30	0.30	0.60	0.60
1,8-Cineole	1.00	1.00	0.50	1.00	0.50	0.50	0.50	1.00
Medium Chain Triglycerides (MCT)	0.70	2.30	2.50	2.35	2.50	2.55	2.55	2.75

2.3. Sensory Evaluation

Panellists (members of Quest's Sensory Research Panel, N=21) were served about 50 ml soup at approximately 50 °C in 150 ml plastic cups coded with 3-digit numbers. The serving design was balanced for carry-over. Panellists evaluated the intensity of 16 descriptors using the audio method [4]. Four soups were evaluated in each of the first 8 sessions, i.e., one flavour on each of the four bases. The next 8 sessions were organised according to the design shown in Table 3.

Panel means of intensity per sample/descriptor were obtained by fitting variance components using Restricted Maximum Likelihood (REML) where panellists and all interactions with panellists were considered random effects.

Table 3. Experimental design for second series of evaluation sessions.

Day	Base	Basil flavour			
1	2	PB2	PB3	PB7	PB1
2	3	PB2	PB4	PB7	PB5
3	1	PB7	PB1	PB5	PB8
4	4	PB3	PB4	PB7	PB8
5	1	PB2	PB3	PB4	PB6
6	3	PB3	PB6	PB1	PB8
7	4	PB2	PB6	PB1	PB5
8	2	PB4	PB6	PB5	PB8

3. RESULTS AND DISCUSSION

Means for each of the 9 effects (acidity and Brix for the base, plus 7 ingredients in the flavours) were compared. In contrast to our earlier study using flavour ingredients responsible for green notes and several complete basil flavours, only a few significant effects for the base were found. It should be noted, however, that the differences between the two levels of acidity in the current study were much smaller, in keeping with realistic levels for commercial tomato soups. Increasing acidity increased scores for ‘sour’ ($p<0.001$) and decreased scores for ‘sweet’ ($p<0.001$), as could be expected. Increasing acidity also decreased scores for ‘cooked tomatoes’ ($p=0.031$) and ‘bitter’ ($p=0.110$).

Increasing Brix increased scores for ‘spices’ ($p=0.027$). Scores for ‘green’ decreased with higher Brix, but not significantly ($p=0.182$). Scores for ‘sweet’ were unaffected by the increase in Brix ($p=0.531$), which is not so surprising considering that the higher Brix came from addition of Maltodextrin. Maltodextrin increased the viscosity, as is shown by the instrumental measurements given in Table 1. Although the panel made no texture measurements, this increase (in a system that was already quite viscous) can probably be neglected. There was no effect on release of hexanal from 10% aqueous sucrose solutions thickened with pectin and calcium chloride over a much greater range of viscosity than that used in the soup bases [5]. Adding a thickening agent (pectin+starch) to low fat stirred yoghurt, however, decreased sweetness as well as green apple (hexanal) notes [6].

There were several significant interaction effects between acidity or Brix and specific ingredients in the model basil flavour. Ocimene significantly increased scores for ‘umami’ ($p=0.016$), ‘bitter’ ($p=0.033$), ‘metallic’ ($p=0.054$) and decreased scores for ‘spices’ ($p=0.003$). Linalool significantly decreased scores for ‘sour’ ($p=0.030$) and increased scores for ‘bitter’ ($p=0.038$). Hexanal significantly decreased scores for ‘alliaceous’ ($p=0.023$). Hexanal did not significantly increase scores for ‘green’ ($p=0.404$) in these experiments, although there was an increase in scores for ‘bitter’ ($p=0.061$). Bitterness perception in a model olive oil was enhanced by another green odour, (*Z*)-3-hexen-1-ol [7]. Eugenol increased scores for ‘spices’ ($p<0.001$). 2-Methyl-4-propyl-1,3-oxathiane decreased scores for ‘spices’ ($p<0.001$) and increased scores for ‘alliaceous’ ($p=0.024$), ‘minty’ ($p=0.060$), ‘tomato leaves/catty’ ($p<0.001$). Methyl cinnamate decreased scores for ‘DMS’ ($p=0.045$).

4. CONCLUSIONS

Eight model samples of basil flavour were prepared according to a Plackett Burman design using 7 flavour compounds. The basil flavours were applied to each of four different tomato soup bases. Several significant sensory effects could be observed.

References

1. J. Delwiche, Food Quality Pref., 15 (2004) 137.
2. E.A. Baldwin, K. Goodner, A. Plotto, K. Pritchett and M. Einstein, J. Food Sci., 69 (8) (2004) S310.
3. E. Lewinsohn, F. Schalechet, J. Wilkinson, K. Matsui, Y. Tadmor, K-H. Nam, O. Amar, E. Lastochkin, O. Larkov, U. Ravid, W. Hiatt, S. Gepstein and E. Pichersky, Plant Physiol., 127 (2001) 1256.
4. B.M. King, Lebensm.-Wiss. Technol., 27 (1994) 450.
5. E. Bylaite, A.S. Meyer and J. Adler-Nissen, J. Agric. Food Chem., 51 (2003) 8020.
6. E.P. Kora, E. Latrille, I. Souchon and N. Martin, J. Sens. Stud., 18 (2003) 367.
7. G. Caporale, S. Policastro and E. Monteleone, Food Quality Prefer., 15 (2004) 219.

W.L.P. Bredie and M.A. Petersen (Editors)
Flavour Science: Recent Advances and Trends

Characterisation of the flavour of infant formulas by instrumental and sensory analysis

Saskia M. van Ruth[a], Vincent Floris[a], Stephane Fayoux[a] and Margaret Shine[b]

[a]*Department of Food and Nutritional Sciences, University College Cork, Western Road, Cork, Ireland;* [b]*Sensory Research Limited, Acorn Business Campus, Mahon Industrial Estate, Blackrock, Cork, Ireland*

ABSTRACT

The flavours of seven infant formulas were evaluated by Proton Transfer Reaction-Mass Spectrometry analysis (PTR-MS) and sensory analysis. Both types of analysis indicated significant differences between the three brands, the type of formula (infant target group), and the physical form (powder/liquid) of the formulas.

1. INTRODUCTION

Food and flavour preferences in humans later in life are influenced by early flavour experiences. Flavour experiences with mothers' milk modify and serve to establish preferences [1]. Infant formulas primarily serve as a source of nutrition for those infants who are not breastfed. The type of formula consumed during infancy has been shown to affect flavour preferences of 4- and 5-years old children [2]. Furthermore, recent studies showed that preferences of children were stable from 2- to 3-years old until young adulthood [3]. Despite its importance very few studies have characterised the flavours of infant formulas. Generally, focus has been on formation of compounds in infant formulas due to oxidation [4] and through Maillard reactions [5,6].

2. MATERIALS AND METHODS

2.1. Materials

Seven infant formulas were evaluated in the study representing three brands (Aptamil, Cow & Gate, SMA), three types of formula ('first' for infants aged 0-12 months,

'hungrier' for hungrier infants aged 0-12 months, 'follow-on' for infants aged 6-24 months, and 'soy' for infants who do not tolerate cow's milk), and two physical forms (powder, liquid) of the formulas. The samples used were as follows (sample number in brackets): aptamil 'first' (1) and 'follow-on' powder (2); Cow & Gate 'first' powder (3) and liquid (4), and SMA 'first' (5), 'hungrier' (6), and 'soy' powder (7).
The powder formulas were prepared as directed on the packaging. For PTR-MS analysis, 90 ml of previously boiled water was placed in a 500 ml glass sample flask. Three scoops of infant formula were added. Total volume was about 100 ml for both prepared powder and liquid formulas. The formula was allowed to equilibrate for 45 min at room temperature while the liquid was stirred. Samples were prepared in triplicate for PTR-MS analysis. For descriptive sensory analysis, larger volumes of infant formula (500 ml) were prepared in duplicate 45 min prior to the start of the sensory analysis.

2.2. Instrumental analysis by PTR-MS

The flask containing the formula was connected to the PTR-MS transfer line. The sample was stirred continuously. The headspace was drawn at 20 ml/min, 15 ml/min of which was led into the PTR-MS (Ionicon Analytik, Innsbruck, Austria). The headspace of the sample was analysed by PTR-MS according to a method described previously [7,8]. The mass range *m/z* 20-220 was scanned. Background and transmission corrected spectra were averaged over three cycles for each individual sample replicate.

2.3. Sensory analysis using QDA

An experienced panel of 9 assessors developed the vocabulary and the quantification scale for descriptive sensory analysis. The attributes developed for odour were 'creamy' (ocreamy), 'cheesy' (ocheesy), 'vegetable' (oveg) and 'cooked oil' (ocooked), for flavour were 'sweet' (fsweet), 'bitter' (fbitter), 'cod liver' (fcod), 'full fat milk' (ffmilk) and 'skimmed milk' (fsmilk), and for after-taste 'drying' (adry). Thirty ml of infant formula was served at room temperature (20 °C) in clear wine glasses (ISO standard). All samples were evaluated in duplicate by scoring perceived intensities on a 100 mm visual analogue scale. The order of sample presentation was balanced.

2.4. Statistical analysis

Instrumental and sensory data were subjected to principal components analysis (PCA) and multivariate analysis of variance (MANOVA).

3. RESULTS AND DISCUSSION

PTR-MS analysis was carried out on the seven infant formulas. The mass spectral data were subjected to PCA (Figure 1) and MANOVA. PCA showed a separation between the infant formula brands. The SMA samples were separated from the Aptamil and Cow & Gate samples along PC1. They were generally characterised by high volatile concentrations. Samples differing in type or physical form were separated as well. MANOVA conducted previously on a larger group of samples, including present samples, showed that 69 out of the 200 masses that were measured showed significant differences for at least one of the factors brands, type, and physical form of the infant formulas. Forty-two masses showed differences between brands, 40 masses between liquids and powders, and 14 masses between different types of formula [8].

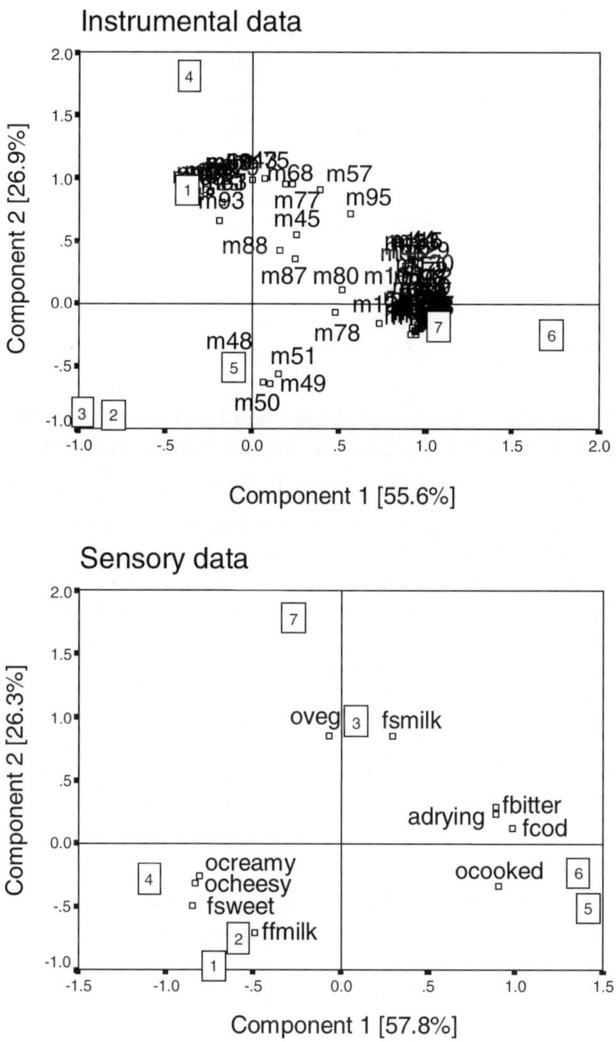

Figure 1. First and second dimensions of principal components analysis on the instrumental and sensory data of the infant formulas. Sample numbers and attribute abbreviations are explained in the materials and methods section.

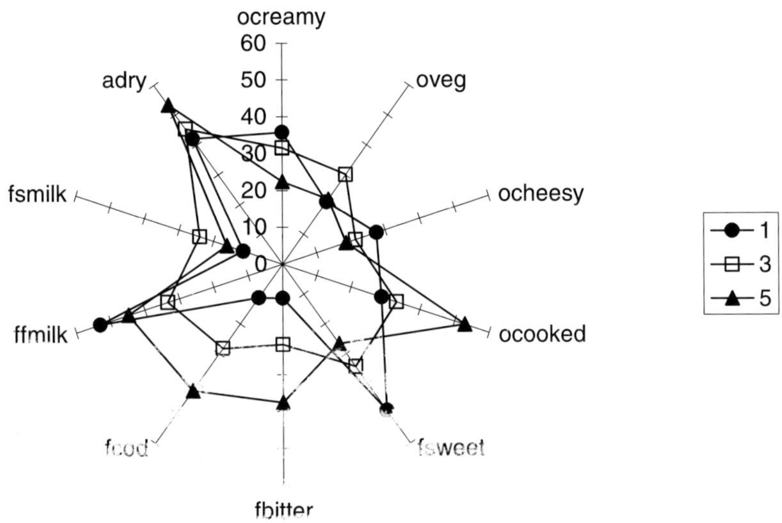

Figure 2. Spider diagram of scores for sensory attributes of three infant formulas. Sample numbers and attribute abbreviations are explained in the materials and methods section.

The sensory results for 'first' powder formula are presented as spiderweb diagram for the three brands (sample 1, 3, 5; Figure 2). The brands differed significantly (ANOVA, P<0.05) in cooked oil odour, sweet flavour, bitter flavour, cod liver flavour, full fat milk flavour, skimmed milk flavour, and drying aftertaste. PCA on the sensory data set indicated that the flavour profile of the SMA samples was quite different from the Aptamil and Cow & Gate samples (Figure 1). The cow's milk based SMA formulas were characterised by cooked oil odour, cod liver flavour, bitter flavour, and a drying after-taste. Contrarily, most of the Aptamil and Cow & Gate samples were described by the attributes creamy odour, cheesy odour, sweet flavour, and full fat milk flavour. The soy-based formula and the Cow & Gate 'first' powder were associated with vegetable odour and skimmed milk flavour.

References

1. J.A. Menella and G.K. Beauchamp, Dev. Psychobiol., 35 (1999) 197.
2. J.A. Menella and G.K. Beauchamp, Early Hum. Dev., 68 (2002) 71.
3. S. Nicklaus, V. Boggio, C. Chabanet and S. Issanchou, Food Quality Pref., 15 (2004) 805.
4. F. Fenaille, P. Visani, R. Fumeaux, C. Milo and P.A. Guy, J. Agric. Food Chem., 51 (2003) 2790.
5. E. Ferrer, A. Alegría, R. Farré, P. Abellán and F. Romero, Food Chem., 89 (2005) 639.
6. S. Albalá-Hurtado, M.T. Veaciana-Nogués, A. Mariné-Font and M.C. Vidal-Carou, J. Agric. Food Chem., 49 (1998) 2998.
7. W. Lindinger, A. Hansel and A. Jordan, Int. J. Mass Spectrom., 173 (1998) 191.
8. S.M. van Ruth, F. Vincent and S. Fayoux, Food Chem., submitted.

W.L.P. Bredie and M.A. Petersen (Editors)
Flavour Science: Recent Advances and Trends

Prediction of the overall sensory profile of espresso coffee by on-line headspace measurement using Proton Transfer Reaction-Mass Spectrometry

Christian Lindinger, Philippe Pollien, David Labbe, Andreas Rytz, Marcel A. Juillerat and Imre Blank

Nestlé Research Center, Vers-chez-les-Blanc, 1000 Lausanne 26, Switzerland

ABSTRACT

Analytical and sensory profiling were performed on different commercially available espresso coffees. Chemical information about differences in composition of the coffee headspace characterising different coffee blends was obtained by on-line analysis using PTR-MS. In addition, an expert panel trained for coffee tasting described each sample by scoring key flavour attributes on an 11-point intensity scale. The overall sensory description of each sample was correlated with the analytically obtained differences in chemical composition to develop a tool predicting the sensory profile based on analytical data. This novel and efficient approach of characterising the coffee aroma by on-line analysis may shorten the time required for the development of new products and improve quality control in a more automated and objective manner.

1. INTRODUCTION

The aroma of coffee derives from a complex mixture of volatile components whereof more than 1000 volatiles have been identified in different coffees [1]. However, only a fraction of about 50 compounds occurring in a distinct ratio is thought to contribute to the overall aroma of roast and ground coffee and the brew prepared thereof [2-4].
The correlation of analytical and sensory data and its prediction based on analytical data has been a challenge and has not been possible for coffee aroma. This may be due to the methods applied so far for data acquisition and treatment. Recently, good correlation has been reported for PTR-MS data and the sensory profile of mozzarella cheese, as both methods provided comparable sample discrimination after multivariate statistical analysis [5]. The present study aimed at correlating PTR-MS fingerprints and sensory

data of espresso coffee to achieve prediction of sensory profiles on the basis of instrumental data obtained by on-line PTR-MS analysis.

2. EXPERIMENTAL PART

Sample preparation. A total of eleven coffee samples was characterised in this study. A standard coffee machine was used for the extraction of coffee (5 g) obtaining a defined extraction volume (28 ml). The experimental conditions were maintained constant in terms of coffee quantity inside the headspace sampling vessel and water quality (Vitell) to eliminate variation due to water hardness. The extraction water temperature, coffee machine parameters, and cup temperature were standardised in the whole study to ensure identical conditions of coffee preparation. Dilution and purge gas flows, the temperatures of all tubes being in contact with the sample gas, the stirring of the coffee sample in the sample vial, and the course of analysing the samples were kept constant for all measurements.

Headspace (HS) sampling. A new HS set-up was developed to measure the volatile aroma compounds above the cup of coffee, as recently described [9]. There are basically two different approaches measuring HS concentrations. Static HS measurement (equilibrium state) is used to calculate indirectly the liquid phase concentration on the basis of Henry's law. The dynamic HS sample method (HS never gets saturated) is applicable to describe *in-vivo* situations such as aroma perception above the food. In this case the sample system is simulating an open system where the aroma molecules are readily released into the HS.

Proton transfer reaction mass spectrometer (PTR-MS). The PTR-MS gas analyser used in this study has been described in previous publications [6-8]. Soft ionisation techniques are advantageous to analyse a complex gas mixture quantitatively and on-line by MS methods. Direct quantification without the need of extensive calibration as well as long time stability and high sensitivity is recommended to achieve high quality and reproducibility of the analytic data. These requirements are fulfilled by PTR-MS. HS measurements of the coffee samples showed reproducibility higher than 95%.

Sensory profiling. A coffee panel (n=10) familiar with espresso products took part in the evaluation. Quantitative Descriptive Analysis® using the monadic approach, i.e. samples were presented one by one, was applied as sensory profiling technique. Each assessor scored the product based on the knowledge and consensus acquired during the training sessions. The samples were served according to a balanced presentation design, which was repeated twice. The products were evaluated in booths using an 11 points scale ranging from 0 (not detectable) to 10 (very intense).

3. RESULTS AND DISCUSSION

All measurements and sensory analysis were performed under well-defined conditions. One requirement for the coffee extraction procedure was to simulate consumption conditions in both sensory mapping and analytical measurements.

Time intensity profiles of espresso coffee aroma. The aroma above the cup of freshly brewed coffee is a dynamic phenomenon. In the first few seconds, a burst of volatile compounds was released into the HS followed by a continuous decrease of the HS concentration with time. In general, certain compounds decreased very quickly (within 2 min) while others were stable for long time. However, in this study we were not interested in the dynamic behaviour of the coffee HS. Only the maximum HS concentrations measured 2 min after connecting the coffee sample to the analytic set-up were used for statistical data analysis. It should be noted that the HS volume (200 ml) above the liquid was renewed within 1 min, thus simulating an open system.

Data reduction (data pre-processing). In preliminary measurements on a reduced set of different espresso samples, full mass spectra over the mass range of *m/z* 20 to *m/z* 250 were obtained. Only the most characteristic ion traces, differentiating the various espresso samples, were selected and measured more precisely using the multiple ion detection mode. 16 different ion traces were chosen representing 30 identified volatile organic compounds. The headspace intensities of these 16 ion traces were obtained for all samples with five repetitions. A pre-processing of analytic data was performed. In a first step the data set was minima-normalised by dividing the concentration value of each individual by the minimum concentration within the complete set of samples. In the next step a pseudo composition was calculated by determining the mean over all ion traces (16) and normalising the complete data set to this mean value. The result shows which ion trace is misbalanced in the whole set of espresso samples.

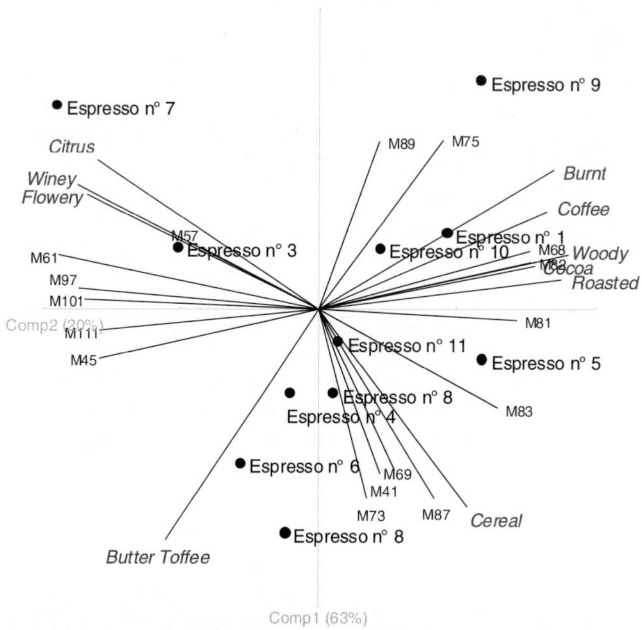

Figure 1. Principal components analysis (PCA) of selected ion traces analysed by PTR-MS and sensory profile of the coffee samples used for the model development.

A principal component analysis (PCA) based on analytical and sensory data (Figure 1) showed a good discrimination of all different espressos. To develop a sensory predictive tool on the basis of objective analytic data, a principle component regression (PCR) of analytic and sensory data was performed.
As shown in Figure 2, comparison of the sensory profiles obtained by the panel and analytical PTR-MS data are very close, indicating the potential of the predictive tool.

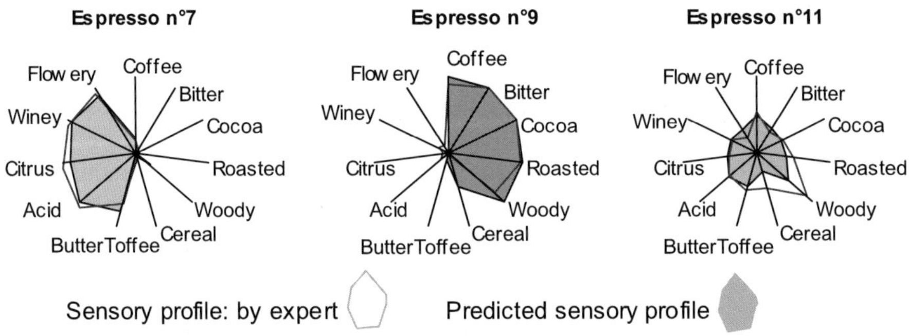

Figure 2. Development of the predicting model: sensory profiles obtained by an expert panel (open) and predicted on the basis of PTR-MS data (shadowed).

4. CONCLUSION

A predictive model was developed showing good correlation between descriptive sensory profiling and analytical differences in chemical composition obtained by PTR-MS. This novel and efficient approach of characterising the aroma of coffee blends by on-line analysis may shorten the time required for the development of new products and improve quality control in a more automated and objective manner.

References

1. L.M. Nijssen, C.A. Visscher, H. Maarse, L.C. Willemsens and M.H. Boelens (eds.), Volatile compounds in food, Zeist, The Netherlands (1996).
2. M. Czerny, F. Mayer and W. Grosch, J. Agric. Food Chem., 47 (1999) 695.
3. F. Mayer, M. Czerny and W. Grosch, Eur. Food Res. Technol., 211 (2000) 272.
4. F. Mayer and W. Grosch, Flavour Fragrance J., 16 (2001) 180.
5. F. Gasperi, G. Gallerani, A. Boschetti, F. Biasioli, A. Monetti, E. Boscaini, A. Jordan, W. Lindinger and S. Iannotta, J. Sci. Food Agric., 81 (2001) 357.
6. W. Lindinger, A. Hansel and A. Jordan, Int. J. Mass Spectrom., 173 (1998) 191.
7. W. Lindinger, J. Hirber and H. Paretzke, Int. J. Mass Spectrom. Ion Processes., 129 (1993) 79.
8. A. Hansel, A. Jordan, R. Holzinger, P. Prazeller, W. Vogel and W. Lindinger, Int. J. Mass Spectrom., 149/150 (1995) 609.
9. C. Lindinger, P. Pollien, S. Ali, C. Yeretzian, I. Blank and T. Maerk, Anal. Chem., 77 (2005) 4117.

W.L.P. Bredie and M.A. Petersen (Editors)
Flavour Science: Recent Advances and Trends

Interactions between food texture and oral processing affecting the strawberry flavour of custard desserts

Saskia van Ruth[a], Eugenio Aprea[abc] and Amaya Rey Uriarte[ad]

[a]Department of Food and Nutritional Sciences, University College Cork, Western Road, Cork, Ireland; [b]Istituto Agrario di S. Michele a/A, S. Michele, Via E. Mach 2, 38010 Italy; [c]Institut für Ionenphysik, Universität Innsbruck, Technickerstr. 25, A-6020 Innsbruck, Austria; [d]Departamento de Teconologia de Alimentos, Universidad Politecnica de Valencia, Apdo. Correos 22012, 46071 Valencia, Spain

ABSTRACT

Interactions between oral processing and food texture on flavour release during consumption of custards were examined using Proton Transfer Reaction Mass Spectrometry (PTR-MS). Significant effects of type of flavour compound, custard texture, oral processing protocol, and subjects as well as interactions were observed.

1. INTRODUCTION

The formulation of foods with controlled sensory properties remains a challenge. The composition and structure of a food system, as well as the interaction between these two parameters, determine a food's sensory flavour and texture properties. Flavour perception is for instance affected by variables such as hardness, water holding capacity, and microstructure. Texturing agents are added to food products to modify a product's viscosity. Its addition sometimes results in a significant decrease in perceived flavour [1]. This can be explained by the effect of increased viscosity which may hinder mass transfer of flavour compounds to the surface of the food product [2]. However, solutions with similar viscosity can be prepared from different texturing agents. Such solutions do not induce the same flavour perception. Furthermore, in other studies it was shown that texturing agents did not exhibit a change in in-nose measured flavour concentrations but did affect flavour perception [3,4]. Food composition, food structure as well as oral processing are likely to be important factors determining flavour release and perception from food products containing texturing agents. In the present study, two custard desserts varying in texture were examined for strawberry flavour release during consumption using PTR-MS. Two oral processing protocols were compared.

2. MATERIALS AND METHODS

2.1. Materials

A commercial strawberry flavour mixture was obtained from Givaudan (Duebendorf, Switzerland). Its composition was published previously [5]. Ethyl butyrate, ethyl iso-pentanoate, and ethyl hexanoate were present at the following concentrations 90 mg/g, 10 mg/g and 20 mg/g, respectively. High viscosity carboxymethyl cellulose (CMC, C-5013; Sigma-Aldrich Chemie, Steinheim, Germany) was used for custard preparation.

2.2. Custard preparation

Two different custards were prepared. They were composed of 0.1% and 1.0% CMC, respectively. For custard preparation, 936 (0.1% CMC custard) or 927 (1.0% CMC custard) full-fat milk was heated to 60 °C in a water bath. Sucrose (63 g; Siucra; Irish Sugar Ltd, Carlow, Ireland) was added and the mixture stirred for 3 min. The CMC was added in small increments to ensure that the CMC was fully dispersed. To obtain the custard texture, the mixture was stirred again for 5 min. The temperature of the water bath was increased to 95 °C, while stirring continued. When the custard reached a temperature of 90 °C, heating continued for another 10 min. The custard was subsequently cooled down at room temperature for 15 min and further down to 30 °C by placing the bottle in cold water. Forty gram of the custard was placed in a 100 ml glass bottle, 14 μl of the flavour mixture was injected in the custard and the bottle sealed. The mixture was stirred for 5 min and stored at 6 °C for 24 h prior to analysis. Final total flavour concentration was 56 mg/kg custard. For each type of custard duplicate batches were prepared.

2.3. In-nose analysis

For in-nose analysis, a fork-shaped glass nosepiece was placed with its two inlets in the nostrils of a subject. The air was drawn in at a rate of 100 ml/min, 15 ml of which was led into the PTR-MS. The background was measured for 30 s. During that time 7 g of custard (20 °C) was placed on a spoon. The subject transferred the custard to his/her mouth. Two different oral processing protocols were applied. Subjects were either allowed to chew and swallow freely, or they had to follow instructions (10 movements protocol). The instructions involved moving the custard ten times left to right to left in their mouths during 15 s. They immediately swallowed after the 15 s. Subjects raised their hands to indicate time of swallowing ($t_{swallow}$). Preliminary scans (mass range *m/z* 30-220) of the flavoured custards as well as the individual flavour compounds revealed that the masses *m/z* 117, *m/z* 131 and *m/z* 145 could be exclusively assigned to ethyl butyrate, ethyl iso-pentanoate, and ethyl hexanoate, respectively. Twenty-one subjects participated in the in-nose analyses. Two batches of the individual custards were analysed (2 replicates per type of custard per chewing protocol per person). The samples were analysed according to the method described by Lindinger and co-workers [6]. The spectra were background and transmission corrected. From the individual curves, maximum intensities (I_{max}) and time to maximum intensities (t_{Imax}) were determined.

2.4. Statistical analysis

The I_{max}, t_{Imax}, and $t_{swallow}$ data were subjected to multivariate analysis of variance (MANOVA) to determine significant differences between the custards, the chewing protocols, and the assessors. Principal components analysis (PCA) was conducted on the I_{max} data set. A significance level of $p<0.05$ was used throughout the study.

3. RESULTS AND DISCUSSION

The two custards varying in texture were subjected to in-nose analysis using two oral processing protocols. From the real time curves, I_{max} and t_{Imax} values were calculated (Table 1). The three volatile compounds showed significantly different I_{max} and t_{Imax} values. Highest I_{max} and t_{Imax} values were obtained for ethyl butyrate, followed by ethyl iso-pentanoate and ethyl hexanoate. This was most likely due to the concentrations of the flavour compounds in the custard as well as to flavour-matrix interactions.

Table 1. Three strawberry flavour compounds released from custards varying in texture (0.1 and 1.0% CMC) using two oral processing protocols (free and 10 movs): results from in-nose PTR-MS analysis (I_{max}, t_{Imax}, $t_{swallow}$; mean ± SD).

	Ethyl butyrate		Ethyl iso-pentanoate		Ethyl hexanoate	
	0.1% CMC	1.0% CMC	0.1% CMC	1.0% CMC	0.1% CMC	1.0% CMC
I_{max}						
10 movs	270±193	284±209	24±17	25±17	8.0±4.9	8.0±4.7
free	256±190	226±156	21±15	18±11	7.5±4.2	6.7±6.7
t_{Imax} (s)						
10 movs	30±7	31±5	32±7	33±7	40±8	42±9
free	19±3	21±5	21±4	23±5	28±5	33±12
$t_{swallow}$ (s)						
10 movs	14±4	15±5	14±4	15±5	14±4	15±5
free	4±2	6±3	4±2	6±3	4±2	6±3

CMC: carboxymethyl cellulose; 'free': free chewing and swallowing protocol; '10 movs': the 10 movements protocol; I_{max}: maximum intensity; t_{Imax}: time to reach maximum intensity, $t_{swallow}$: time to swallowing.

For the 10 movements protocol, I_{max} increased with increase of CMC concentration. For the free chewing protocol the opposite was observed. Independent of the oral processing protocol, higher values for t_{Imax} were observed for the high CMC concentration custard. Increased CMC concentrations resulted in higher variance for both oral processing protocols, which implies that differences between individuals were more pronounced for the firmer custard.

The free chewing protocol resulted in lower I_{max}, t_{Imax} and $t_{swallow}$ values compared to the 10 movements protocol. During free chewing, the custards were swallowed fairly quickly (<4 s) compared to the time to swallowing in the 10 movements protocol (15 s). Statistical analysis of the data showed significant differences between retronasal aroma delivery between the subjects for all parameters. PCA on the I_{max} data showed that the subjects could be divided into two groups: some released consistently more strawberry flavour from the high concentration CMC custard and some from the low concentration CMC custard.

References

1. R. Chandrasekaran (ed.), Frontiers in carbohydrate research, New York, USA (1992) 85.
2. A.M. Stephen, Food polysaccharides and their application, New York, USA (1995) 517.
3. T.A. Hollowood, R.S.T. Linforth and A.J. Taylor, Chem. Senses, 27 (2002) 583.
4. K.G.C. Weel, A.E.M. Boelrijk, A.C. Alting, P.J.J.M. van Mil, J.J. Burger, H. Gruppen, A.G.J. Voragen and C. Smit, J. Agric. Food Chem., 50 (2002) 5149.
5. S.M. van Ruth, L. de Witte and A. Rey, J. Agric. Food Chem., 52 (2004) 8105.
6. W. Lindinger, A. Hansel and A. Jordan, Int. J. Mass Spectrom., 173 (1998) 191.

W.L.P. Bredie and M.A. Petersen (Editors)
Flavour Science: Recent Advances and Trends

Analysis of aroma compounds from carrots by dynamic headspace technique using different purging and cutting methods

Stine Kreutzmann[a], Merete Edelenbos[a], Lars P. Christensen[a], Anette Thybo[a] and Mikael A. Petersen[b]

[a]Department of Food Science, Danish Institute of Agricultural Sciences, Research Centre Aarslev, Kirstinebjergvej 10, DK-5792 Aarslev, Denmark; [b]Department of Food Science, The Royal Veterinary and Agricultural University, Rolighedsvej 30, DK-1958 Frederiksberg C, Denmark

ABSTRACT

The aim of the present study was to evaluate which purging and cutting method that gives the best relationship between sensory quality and the release of carrot volatiles. Volatile compounds from carrots (cv. Eskimo and cv. Nairobi) were collected from shreds and half-moon cuttings by dynamic headspace sampling using dry (no water added) and wet purging (water added and blended). Headspace samples were analysed by GC-MS. Monoterpenes, sesquiterpenes and irregular terpenes accounted for more than 98% of the total mass of the volatiles. Dry purging had a significant effect on the total volatile mass compared to wet purging, resulting in higher headspace concentrations of major volatile compounds. However, sensory analysis revealed that there were no significant differences between the sensory scores for the used descriptors.

1. INTRODUCTION

Several campaigns in Denmark such as the 'Six-a-day' campaign have tried to make people eat more fruit and vegetables in order to prevent the development of cancer, cardiovascular diseases and overweight. Regarding vegetables, focus has been on increasing the consumption of different cabbages and root vegetables, like carrots, because these products contain an abundant number of vitamins, and other health promoting compounds as well as dietary fibre. Volatile compounds, sugars, polyacetylenes and phenolics are essential for taste perception of carrots and are very important for the eating quality of carrots [1,2]. More than 90 volatile compounds have been identified from carrots with mono- and sesquiterpenes being the most abundant, making up to more than 98% of the total volatile mass [2-9]. Various methods have been used to collect volatile compounds from carrots and other vegetables including static and dynamic headspace techniques [3-9]. Investigations have been performed with different purging methods, i.e. on either 'dry' or 'wet' plant material. The use of different purging and cutting methods, however, may affect the release of volatile compounds from the plant material and hence the concentration in the headspace. The aim of this study was to evaluate which purging and cutting method that gives the best relationship between sensory quality and the release of volatile compounds of carrots.

2. MATERIALS AND METHODS

2.1. Dynamic headspace sampling

Volatile compounds from carrots (cv. Eskimo and cv. Nairobi) were collected from shreds (50 g) or half-moon cuttings (50 g) by dynamic headspace sampling using dry purging (= no water added) and wet purging (= water added and blended). The volatile compounds were trapped on absorbent traps filled with 250 mg Tenax-GR (Mesh size 60-80). No breakthrough of volatiles was detected with this trap size. Samples were purged with N_2 (30 min, 50 ml/min) at 30 °C for dynamic headspace sampling. Further, wet purging was performed under stirring and 500 ppm 4-methyl-1-pentanol internal standard was added prior to collection of headspace volatiles.

2.2. Analysis of volatile compounds

Desorption of volatile compounds was done thermally by an ATD 400 automatic thermal desorption system at 250 °C with helium as carrier gas (flow: 60 ml/min in 15 min). Volatile compounds were analysed by GC-MS. Identification was done by probability-based matching with mass spectra in the G1035A Wiley library database. Identity was confirmed by comparison with mass spectra and retention indices of authentic compounds. GC-MS areas of individual volatile compounds were used as a measure for the amounts released from carrots.

2.3. Sensory evaluation and statistical evaluation

Sensory evaluation of carrots was carried out by a sensory panel consisting of nine members, who had been selected and trained according to guidelines in ISO/DIS 8596-1 and ASTM STP 758 (1993). Statistical analyses were performed using SAS software (v6.11) with $p<0.05$ as significance level. All analyses were carried out in triplicate.

3. RESULTS AND DISCUSSION

A total of forty-nine volatile compounds were repeatedly detected in carrot headspace samples. Forty-one of these were identified by comparison of their mass spectral data with those from authentic compounds and/or mass spectra suggested by the MS database as described in Materials and Methods. Most of the identified volatiles have been reported previously as constituents of carrots [2-9]. The carrot volatiles consisted mainly of compounds of terpenoid origin and included monoterpenes, sesquiterpenes, and irregular terpenes that accounted for more than 98% of the total volatile mass. Especially, the monoterpenes impart the characteristic aroma typical of carrots and they are considered to be the most important volatile compounds responsible for 'green', 'earthy', and 'carrot top' flavours in carrots [3,4,7,8]. The results showed that dry purging had a significant effect on the total volatile mass compared to wet purging, resulting in higher headspace concentrations of major volatile compounds (Figure 1). The largest differences were observed for the monoterpenes α-terpinene, β-myrcene, and *p*-cymene and the sesquiterpenes (*E*)-γ-bisabolene, β-caryophyllene, and α-humulene showing differences of 46%, 35%, 29%, 33%, 32%, and 24%, respectively, in cv. Eskimo. Furthermore, there was a clear effect of cutting method. The highest total volatile mass was observed in shreds of cv. Eskimo and cv. Nairobi.

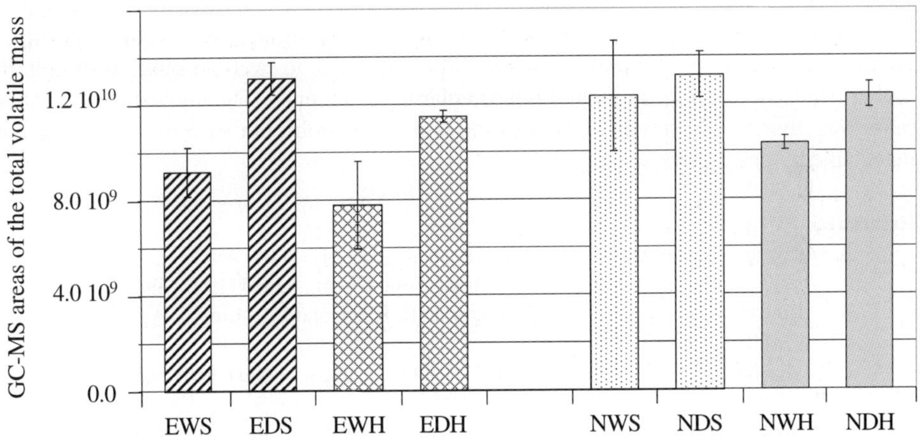

Figure 1. GC-MS areas of the total volatile mass for the two cultivars sampled by wet and dry purging, shreds and half-moon cuttings. (E: Eskimo; N: Nairobi; W: Wet purging; D: Dry purging; S: Shreds; H: Half-moon cuttings).

ANOVA of the sensory data is summarised in Table 1. The two-factor interaction was not significant for any of the sensory descriptors, expect for nutty flavour, which was not considered to be important. The sensory analysis revealed no changes in flavour compounds quality whether the carrots were served shredded or cut as half-moons.

Table 1. Sensory scores of carrots cv. Eskimo and cv. Nairobi prepared as shreds and half-moon cuttings. Data are averages from ratings of 9 panellists using a scale from 0 (low) to 15 (high).

	cv. Eskimo			cv. Nairobi		
Descriptor	Shreds	Moons	Significance	Shreds	Moons	Significance
Terpene aroma	5.4	4.1	NS	4.8	3.9	NS
Carrot aroma	4.4	3.9	NS	4.6	4.1	NS
Terpene flavour	4.4	5.1	NS	3.2	3.7	NS
Carrot flavour	5.8	6.4	NS	5.5	6.3	NS
Sweetness	4.7	5.0	NS	5.3	5.5	NS
Bitterness	2.9	3.9	NS	4.2	3.1	NS
Green flavour	3.3	4.9	NS	2.7	4.1	NS
Soapy flavour	2.8	2.8	NS	3.0	2.6	NS
Nutty flavour	3.7	3.5	NS	3.1	3.9	*
Burning aftertaste	3.3	4.7	NS	3.7	3.9	NS

NS: $p>0.05$; *: $p<0.05$.

4. CONCLUSION

The present study showed that the use of different purging and cutting methods had an effect on the amount of collected total volatile mass in fresh shredded or half-moon cut carrots by dynamic headspace sampling technique. Dry purging had a significant effect on the concentration of volatile compounds compared to wet purging, resulting in higher headspace concentrations of major volatile compounds. The highest total volatile mass was observed in shreds of cv. Eskimo and cv. Nairobi, but such changes were not detectable by descriptive sensory panel.

References

1. A. Czepa and T. Hofmann, J. Agric. Food Chem., 52 (14) (2004) 4508.
2. C. Alasalvar, J.M. Grigor and P.C. Quantick, Food Chem., 65 (3) (1991) 391.
3. R.G. Buttery, R.M. Seifert, D.G. Guadagni, D.R. Black and L.C. Ling, J. Agric. Food Chem., 16 (6) (1968) 1009.
4. D.A. Heatherbell, R.E. Wrolstad and L.M. Libbey, J. Agric. Food Chem., 19 (6) (1971) 1069.
5. F. Kjeldsen, L.P. Christensen and M. Edenlenbos, J. Agric. Food Chem., 49 (9) (2001) 4342.
6. R. Seljasen, G.B. Bengtsson, H. Hoftun and G. Vogt, J. Sci. Food Agric., 81 (4) (2001) 436.
7. M. Shamaila, T. Durance and B. Girard, J. Food Sci., 61 (5) (1996) 1191.
8. R. Habegger, B. Müller, A. Hanke and W.H. Schnitzler, Gartenbauwiss., 61 (5) (1996) 225.
9. C. Varming, K. Jensen, S. Møller, P.B. Brockhoff, T. Christensen, M. Edenlenbos, G.K. Bjørn and L. Poll, Food Quality Pref., 15 (6) (2004) 531.

W.L.P. Bredie and M.A. Petersen (Editors)
Flavour Science: Recent Advances and Trends

A study of sensory profiling performance comparing various sensory laboratories - a data analytical approach

Janna Bitnes[a,b], Per Lea[a] and Magni Martens[a,b]

[a]Matforsk AS, N-1430 Ås, Norway; [b]The Royal Veterinary and Agricultural University, DK-1958 Frederiksberg, Denmark

ABSTRACT

Sensory panels of different levels of expertise and training were compared with respect to their profiling performance. Training panellists for sensory profiling is important to get a consistent panel. Panellists should be trained to achieve good discrimination abilities and to reproduce their assessments of identical samples. Partial least squares regression (PLSR) and analysis of variance (ANOVA) were applied to analyse the sensory data. The sensory profiling performances of three sensory panels were investigated for two product categories. The results from the data analyses were displayed in signal-noise plots (PLSR) and p-MSE plots (ANOVA). Both methods showed the same tendency with respect to profiling performance.

1. INTRODUCTION

Sensory evaluation has been defined as a method used to evoke, measure, analyse, and interpret those responses to products as perceived through the senses of sight, smell, touch, taste, and hearing [1]. Sensory science is a multidisciplinary field that has received contributions from data analysis. You need to have a tool to analyse your data and to interpret the results. But it seems to be a jungle out there. There are a number of statistical methods available. How should a young scientist, non-statistician, get the knowledge of which method is the right one for your data? Should you use the method your colleges use, the one that the referee wants, the one that gives the answer you expect with most significant variables or the one easiest to understand? Should you compare several methods or is it enough to just use one that you know very well? Do the different methods give the same answer? Should you just give up and hire a statistician to do the job?

Sensory studies include several variables and one of the goals is to investigate if there are any significant differences between the products for the variables. However, one might also want information about how one product is different from another product. In the present paper we have chosen one univariate and one multivariate statistical approach to analyse the sensory data from the profiling of two product categories. By using profile methods we will be able to detect the sensory attributes that describe the difference between samples and to measure the intensity of the different attributes.

2. METHODS AND MATERIALS

2.1. Panels

Three trained sensory panels from 'The Sensory Study Group' of Norway participated in this investigation. The number of assessors in the panels varied between 8 and 11.

2.2. Stimuli

Two food product categories were profiled: product category 1 and product category 2. Four commercial varieties within each category were analysed in each experiment. Thus in total 8 different samples were evaluated. The products chosen had uniform quality within each variety.

2.3. Experiments/procedure

Each of the two product categories was profiled by three sensory panels. The descriptive analysis procedure was performed according to ISO 6564-1985 using a continuous unstructured line scale anchored with the terms 'no intensity' to the left and 'high intensity' to the right side. All the panels used the same list of attributes depending on the product category in question. The samples were evaluated in two replicates randomised across two sessions. Panel 3 registered their evaluations on a computer system (CSA, Compusense, Version 4.2, Guelph, Canada). The other two panels registered their evaluations manually on paper forms.

2.4. Data analysis

2.4.1. p-MSE plots, ANOVA

The sensory data were subjected to analysis of variance (ANOVA) using SAS Release 6.12 (SAS Institute Inc., Cary, NC, USA). A general linear model (GLM) was performed on the profiling sensory analysis data in order to identify the sensory attributes that differentiated between samples. Assessors and assessor*product interactions were taken to be random effects. The p-value was then plotted against the mean square error (MSE) where a low p-value indicated good discriminatory ability and a low MSE indicated good repeatability.

2.4.2. Signal-noise plots, PLSR

The sensory data were also subjected to partial least squares regression (PLSR) using The Unscrambler Version 9.1 (CAMO ASA, Oslo, Norway). The data analytical method consisted of generating PLSR models and retrieving residuals from these models [2]. This resulted in plots that displayed the signal and noise of the different sensory terms. The X-variables were the sensory variables and the Y-variables were indicator variables for each product. To make the signal-noise plots the sensory responses were organised in a data table, where the sensory terms represented the columns and the products*replicates*assessors represented the rows. The signal expressed the relevant variance after correcting for level differences between the assessors (in their use of the sensory scales) and between the replicates (drift over time). The noise expressed the residual variance after modelling optimal numbers of significant PCs. The total signal was then compared to the corresponding residual noise, where higher signal than noise indicated a higher profiling performance.

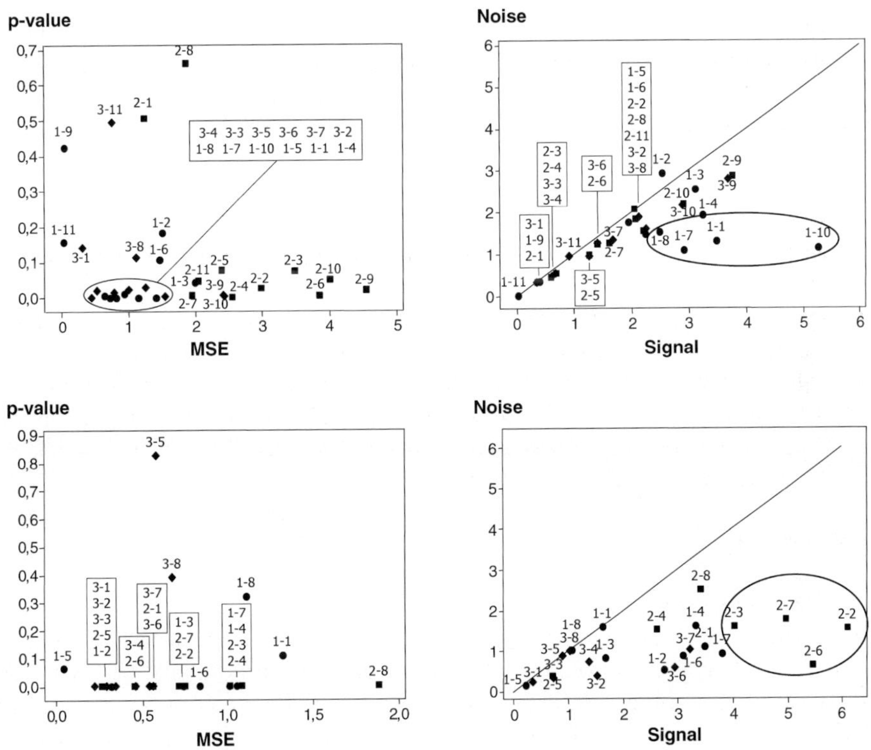

Figure 1. Left: p-value versus MSE. Right: noise versus signal. Upper: 11 attributes and 3 panels from the evaluation of four varieties within product category 1. Lower: 8 attributes and 3 panels from the evaluation of four varieties within product category 2. Coding: first digit: panel number; second digit: attribute number.

3. RESULTS AND DISCUSSION

The results from the two data analyses showed the same tendency (Figure 1). Panel 1 demonstrated a better ability to discriminate between the products than panel 2 and 3 in the profiling of product category 1. This was shown by low p-value and MSE and higher signal than noise, especially for the attributes 1-1, 1-7, 1-8, 1-10. In the profiling of product category 2 it was primarily panel 2 that had low p-values and high signals, especially for the attributes 2-2, 2-3, 2-6 and 2-7. The signal versus noise plots seemed, however, to possess a more differentiated picture of the profiling performance than the p-MSE plots. But one has to remember that these two data analytical approaches are somewhat different and the presentation of the results is aimed at visualising the panel performance data differently. The p-MSE plots are mainly used to pick out outlying observations, normally on the assessor level, but it can also, as in the present paper, be used on the panel level. The panels and attributes with low p-values are clustered close to the X-axis, but the extreme points are clearly identifiable. The signal-noise plots are mainly used to show the sensory terms that are best modelled. However, both methods showed the same tendency with respect to profiling performance.

4. CONCLUSION

The good ability to discriminate between the products was seen by both low p-values (ANOVA) and a much higher signal than noise (PLSR). Both methods showed the same profiling performance.

References

1. H. Stone and J.L. Sidel, Sensory evaluation practices, San Diego, USA (1993).
2. H. Martens and M. Martens, Multivariate analysis of quality - an introduction, Chichester, UK (2001).

W.L.P. Bredie and M.A. Petersen (Editors)
Flavour Science: Recent Advances and Trends

Influence of dehydration on key odour compounds of saffron

Marjorie Bergoin-Lefort, Christine Raynaud, Gérard Vilarem and Thierry Talou

Laboratory of Agro-Industrial chemistry, 118 route de Narbonne, 31077 Toulouse, France

ABSTRACT

Nine new compounds have been identified in saffron and six new key odorants were detected using the frequency detection GCO method. A remaining high moisture content in saffron after the drying process results in chemical and sensory modifications with the generation of 4-hydroxy-2,6,6-trimethyl-1-cyclohexene-1-carboxaldehyde (HTCC) and an 'animal' note caused by 3-methylbutanoic acid.

1. INTRODUCTION

Safranal is the major constituent of the saffron volatiles fraction, generated by hydrolysis and enzymatic reactions during the drying process [1]. The quality of the aroma, colour and flavour of saffron depends on the drying process [2] and on the final moisture content [3,4]. Currently, these properties are evaluated by international standard ISO/TS 3632 [5], measuring safranal, crocin and picrocrocin by spectrophotometry. However, compounds present in minor amounts appear to be important for the overall aroma, e.g. 2-hydroxy-4,4,6-trimethyl-2,5-cyclohexadien-1-one (lanierone) [6]. The absorption of *cis*-crocin occurs at the same wavelength as safranal and its low solubility in water has led to overestimate the total content of safranal [7]. The aim of this work was to determine the real impact of the dehydration process on the sensory and chemical properties of saffron properties using complementary analytical methodologies.

2. MATERIALS AND METHODS

Eight saffron samples (2003) from Quercy area (France) were dried by two producers, using two different drying processes: (A) electrical dehydrator, from 55 °C to 60 °C

(depending on moisture content) during 40 min or (B) ventilated oven at 60 °C during 30 min. Moisture content and safranal, crocin and picrocrocin contents were measured according to ISO/TS 3632-2 [5] by spectrophotometry using $E^{1\%}_{1cm}$.
Volatiles from 0.2 g of dried stigmas were extracted by a SPME using a carboxen-PDMS fibre [8]. Samples were analysed by a GC-MS (Agilent 6980/5973N, France) with a DB5ms column (30 m, 0.25 mm i.d., 0.25 μm f.t., JW Agilent Technologies, USA) coupled to an Olfactory Detector Port (Gerstel-ODP2, RIC, France) GC-MS/ODP [9] (split ratio 1:2). Helium was used as carried gas at a constant flow rate of 1.4 ml/min. Oven temperature started at 40 °C during 1 min and was increased at a rate of 5 °C/min to 100 °C, at 3 °C/min to 125 °C and at 6 °C/min to 240 °C. Compounds were identified by spectral comparisons with NIST 98 and Wiley libraries, retention indices and literature [1,6,10]. GC sniffing was performed with 10 subjects, who assigned a descriptor from a list of 15 odour classes to each eluting odour. The frequency detection method was used with SNIF Software (Wageningen University). The analytical and sensory data were analysed by ANOVA and Partial Least Squares (PLS) regression.

3. RESULTS

The content of moisture, crocin, picrocrocin and safranal in saffrons from different producers and drying processes are presented in Table 1. A total of 35 volatiles were isolated from saffron samples of which 32 could be identified. Nine compounds are reported fro the first time (Table 2). Three compounds coeluted with safranal: verbenone [11], eucarvone [6] and 2-hydroxy-3,5,5-trimethyl-2-cyclohexen-1,4-dione [10]. Only two volatiles could not significantly ($p>0.05$) discriminate the saffrons. The PLS1 regression on the analytical data gave good prediction for moisture content ($R_{calibration}$ 0.999, $R_{validation}$ 0.989) and crocin rates ($R_{calibration}$ 0.996, $R_{validation}$ 0.913).

Table 1. Moisture content (MC) and secondary metabolites of saffrons from 4 producers, dried by two different processes (A and B).

Saffrons	A1	A2	B1	A3	A4	B2	B3	B4
MC%	6.36	6.49	6.49	8.64	8.87	13.58	18.05	19.14
$E^{1\%}_{1cm}$Picrocrocin	93.8	114.5	97.2	106.5	105.4	95.4	84.3	82.4
$E^{1\%}_{1cm}$Crocin	270.5	274.3	235.8	256.8	255.2	221.8	129.1	84.7
$E^{1\%}_{1cm}$Safranal	11.7	27.4	24.7	31.3	26.2	27.7	26.6	32.5

GCO analysis showed 14 odorous areas in the saffron extracts. The more odorous sample (B4) and the less one (A2) had respectively 11 and 5 odorous areas. Peaks, 7, 9 and 10 have been sniffed in all the samples as peak 2 and 8 (except in A2 see Table 2). Saffron samples (Figure 1) were especially 'hay' and 'spicy' but could be 'fatty' and 'animal' when the moisture content was high. PLS1 on the sensory data confirmed that moisture content was correlated to 3-methylbutanoic acid, which has an 'animal' note.

Table 2. Volatile compounds and odours identified from saffron by SPME GC-MS/ODP.

RI	Compounds	GCO detection[c]	GCO description[d]
496	**Ethanol**[a]		
506	**Acetone**[a, b]	P1: 3%	Floral (67%), spicy (33%)
601-621	2,3-Butanedione + Acetic acid	**P2: 58%**	Fatty (68%), acidic/pungent (26%)
658	**2-Methyl-1-pentene**[a]		
667	**1-Hydroxy-2-propanone**[a]		
701	**Propanoic acid**[a]		
709	3-Hydroxy-2-butanone		
802	Hexanal		
843	3-Methylbutanoic acid	P2: 25%	Animal (59%)
854	**2-Methylbutanoic acid**[b]	P4: 6%	Fruity (40%), roasted (20%)
912	**2-[5*H*]-Furanone**[b]	P5: 5%	Roasted (20%), spicy (20%)
1040	β-Isophorone		
1063	2,2-Dimethylcyclohexane-1-carboxaldehyde		
1091	**1-(3,4-Dimethylphenyl)-ethanone**[a]		
1100	Linalool	P6: 20%	Citrus fruits/fresh (41%), fruity (24%)
1104	Nonanal		
1108	2-Methylene-6,6-dimethyl-3-cyclohexene-1-carboxaldehyde		
1115	Phenylethylalcohol + 2,6,6-Trimethyl-3-oxo-1-cyclohexenecarboxaldehyde		
1121	**α-Isophorone**[b]	**P7: 56%**	Floral (40%), hay (28%)
1143	**4-Ketoisophorone**[b]	**P8: 35%**	Hay (55%)
1156	Lanierone	**P9: 83%**	Hay (49%)
1164	**6-Hydroxy-2,4,4-trimethylcyclohex-2-en-1-one**[a]		
1171	2,2,6-Trimethyl-1,4-cyclohexanedione		
1185	**Ethylbenzaldehyde**[a]		
1199	Safranal	**P10: 92%**	Spicy (70%)
1121	**Eucarvone**[b]	P11: 3%	Floral (67%), roasted (33%)
1306	2,6,6-Trimethyl-3-oxo-1,4-cyclohexadiene-1-carboxaldehyde + β-Safranic acid		
1335	Undetected	P12: 41%	Hay (51%), floral (31%)
1379	Undetected	P13: 11%	Soft (33%), unidentified (33%)
1387	**2,4,4-Trimethyl-3-carboxaldehyde-5-hydroxy-2,5-cyclohexadien-1-one**[a]		
1417	Undetected	P14: 32%	Hay (45%)
1426	HTCC		

[a]Not previously detected in saffron. [b]Not previously detected in saffron as odorant. [c]Frequency odour detection for all saffrons. [d]Odours with $\geq$20% similar descriptors.

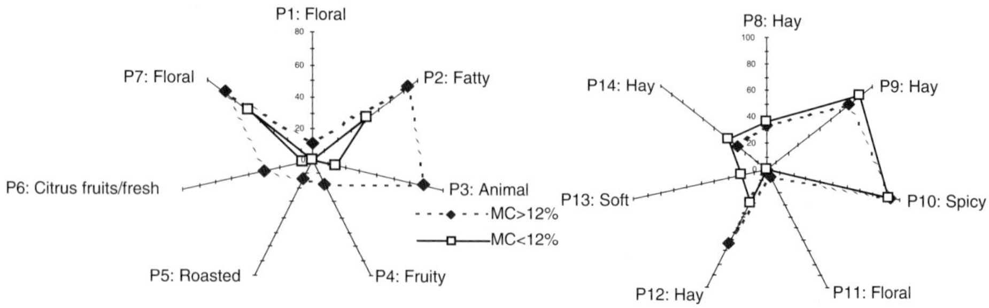

Figure 1. Sensory GCO profiles of saffrons with moisture content (MC) < 12% (A1, A2, B1, A3, A4) and MC > 12% (B2, B3, B4). The sensory scale represents the % of detection. Only the most frequently detected odours are shown.

4. DISCUSSION AND CONCLUSION

The increase of moisture content induces crocin deterioration by auto-oxidation, as mentioned by Alonso *et al.* [3], whereas safranal and picrocrocin seem to be stable. Headspace extraction shows that safranal, the product of picrocrocin hydrolysis, increased slightly with the moisture content also shown by Tsimidou [4]. HTCC is preferably generated by soft hydrolysis in a high moisture medium [1]. The presence of β-safranic acid in samples (MC > 12%) could be the result of safranal oxidation [11]. Safranal and lanierone with respectively 'spicy' and 'hay' notes were the major odorants detected by GCO. Lanierone was, however, present in a much lower amount than safranal. The present study showed that other minor volatiles also participate to the saffron aroma. The 3-methylbutanoic acid, responsible of the animal note (Figure 1), could appear by oxidation of the corresponding aldehyde in a high water content medium (MC > 13.6%). The study clearly illustrated that the drying process and moisture content have a major impact on the aroma volatiles in saffron.

References

1. W. Rodel and M. Petrizka, J. High Resolut. Chromatogr., 14 (11) (1991) 771.
2. B.L. Raina and S.G. Agarwal, J. Sci. Food Agr., 71 (1996) 27.
3. G.L. Alonso and R. Varon, Boll. Chim. Farm., 132 (4) (1993) 116.
4. M. Tsimidou and C.G. Biliaderis, J. Agr. Food Chem., 45 (8) (1997) 2890.
5. ISO/TS 3632-1/2, International organization for standardization, Switzerland (2003).
6. K.R. Cadawaller and H.H. Baek, Spices: flavor chemistry and antioxidant properties, Washington, USA (1997) 66.
7. G.L. Alonso and M.R. Salinas, Food Sci. Technol. Int., 7 (3) (2001) 229.
8. J.A. Fernandez and F. Abdullaev (eds.), Proceeding of the first international symposium on saffron biology and biotechnology in acta horticulturae, Leuven, Belgium (2004) 355.
9. A.M. Spanier, F. Shahidi, T.H. Parliament, C. Missinan, C.T. Ho and E. Tratas Cortis (eds.), Recent advances in food and flavor chemistry, in press.
10. N.S. Zarghami and D.E. Heinz, Lebensm. Wiss. Technol., 4 (2) (1971) 43.
11. P.A. Tarantilis and M.G. Polissiou, J. Agric. Food Chem., 45 (1997) 459.

W.L.P. Bredie and M.A. Petersen (Editors)
Flavour Science: Recent Advances and Trends

Determination of odour active aroma compounds in a mixed product of fresh cut iceberg lettuce, carrot and green bell pepper

Ghita Studsgaard Nielsen and Leif Poll

The Royal Veterinary and Agricultural University, Department of Food Science and The Centre for Advanced Food Studies, Rolighedsvej 30, 1958 Frederiksberg C, Denmark

ABSTRACT

The odour active compounds in lettuce, carrot, and green bell pepper, respectively, and in a mixed product of the three crops were investigated by the nasal impact frequency method (NIF). Nine judges were evaluating the odour of the four samples by GCO and seven judges performed the sensory analyses. The most important aroma compounds in order of odour impact were in lettuce: (*E*)-α-bisabolene, an unknown terpene, and α-copaene in carrots: myristicin, α-copaene, and the unknown terpene and in green bell pepper: 2-methoxy-3-isobutylpyrazine, the unknown terpene, and an unknown compound. The most important aroma compounds in the mixed product were the unknown terpene, α-copaene, myristicin, and 2-methoxy-3-isobutylpyrazine, which indicate that all of the three individual crops contributed to the odorants of the mixed product.

1. INTRODUCTION

Fresh cut vegetables are prepared in many different types. Mixture products with two or more species are commonly produced, but when mixing the crops in the packaging the aroma of one species might become predominant. The three investigated vegetables have different aroma composition. The aroma profile of iceberg lettuce is to our knowledge not reported, investigations in our laboratory have shown that the concentration of aroma compounds is very low but mainly consists of terpenes. Carrots are characterised by many terpenes [1,2] with some aldehydes are also present [2], whereas green bell pepper consist of many aldehydes, 2-methoxy-3-isobutylpyrazine

and a few terpenes, but 2-methoxy-3-isobutylpyrazine is mainly responsible for the aroma [3].
This study investigates the odour active aroma compounds in a mixed product consisting of 65% cut iceberg lettuce, 20% shredded carrot and 15% cut green bell pepper and in the three vegetables alone. This is done by GC-MS and GC-Olfactometry (GCO) to determine the important aroma compounds in the three vegetables and to investigate the impact of these aroma compounds in the mixed product by means of the nasal impact frequency method (NIF) [4]. In addition, sensory analyses of the headspace of the mixed product and the three vegetables alone are conducted in order to determine interactions of odours from the three individual species and to investigate correlations between GCO analysis and sensory analysis of the headspace of the products.

2. MATERIALS AND METHODS

Fresh iceberg lettuce (*Lactuca sativa* L.), carrot (*Daucus carota* L.), and green bell pepper (*Capsicum annuum*) were purchased from a local store. The vascular tissue of lettuce was removed from the midrib and cut into 4 mm strips. Carrots were peeled and shredded into 2 mm strips. Green bell peppers were cut into 2 mm strips.
Aroma compounds were isolated by dynamic headspace for 25 min at 30 °C from 130 g of cut iceberg lettuce, 40 g of shredded carrot, or 30 g of cut green bell pepper, respectively, or on a mixed product of the three vegetables with the same amounts. Nine judges sniffed all four samples on a HP 5890 GC equipped with an olfactory detector outlet ODO-1 from SGE, Australia. Each sniffing session continued for 40 min. The judges were instructed to note start and finish time of the odour and a description of the odour. The 9 individual aromagrams of one sample were added up to one aromagram. The NIF value was calculated as the number of judges in percentage detecting the odour at the peak as described by [4], and the SNIF value as the total of minutes the odour is detected by all judges. An odour had to be detected by at least 44% of the judges to be included in the results.
The sensory analysis was performed with 7 judges. Samples of 130 g cut iceberg lettuce, 40 g shredded carrot, or 30 g of cut green bell pepper, respectively, or a mixed product of the three vegetables with the same amount, were packed in 1567 ml glass jars, sealed with a lid, and the identity was concealed. Each judge evaluated the strength of the lettuce odour, the carrot odour, and the green bell pepper odour of each sample on an ungraduated scale from 0 to 15, where 15 indicated strong odour. Samples were evaluated at 25 °C after being sealed for 15 min.

3. RESULTS

Results from the GCO analysis are shown in Table 1 (lettuce and green bell pepper) and Table 2 (carrot and mixed product). The results of the sensory analysis of all four samples are shown in Table 3.

Table 1. Eight most important[a] odours detected by GCO analysis of lettuce and of green bell pepper.

	Lettuce			Bell pepper	
Compound	NIF	SNIF	Compound	NIF	SNIF
Total SNIF (min)[b]		13.6			21.6
(*E*)-α-Bisabolene	89	1.9	2-Meth-3-isobutylpyr[c]	100	3.2
Terpene	89	1.5	Terpene	100	1.6
α-Copaene	78	1.1	Unknown compound	89	3.7
2-Meth-3-isobutylpyr[c]	67	1.9	α-Terpinolene	89	1.8
Germacrene	67	1.2	α-Copaene	89	1.6
No peak A	67	1.1	(*E*)-α-Bisabolene	78	1.6
Valencene	67	0.7	1-Hexanol	78	1.2
α-Terpinolene	56	1.0	(+)-3-Carene	67	1.1

[a]The importance was evaluated by the NIF value (%) and, if equal, the SNIF value (min). [b]Sum of SNIF values of all of the detected odours. [c]2-Methoxy-3-isobutylpyrazine.

Table 2. Eight most important[a] odours detected by GCO analysis of carrot and of a mixed product of lettuce, carrot and green bell pepper.

	Carrot			Mixed product	
Compound	NIF	SNIF	Compound	NIF	SNIF
Total SNIF (min)[b]		21.4			44.5
Myristicin	89	4.2	Terpene	100	2.0
α-Copaene	89	1.4	α-Copaene	100	1.9
Terpene	89	1.2	Myristicin	100	4.9
(*E*)-α-Bisabolene	89	2.0	2-Meth-3-isobutylpyr[c]	89	3.1
ρ-Cymene	78	1.7	(*E*)-α-Bisabolene	89	2.1
No peak B	67	1.1	(*Z*)-3-Hexenol	89	1.8
2-Meth-3-isobutylpyr[c]	67	1.1	ρ-Cymene	78	1.6
Valencene	67	0.7	α-Terpinolene	78	1.9

[a]The importance was evaluated by the NIF value (%) and, if equal, the SNIF value (min). [b]Sum of SNIF values of all of the detected odours. [c]2-Methoxy-3-isobutylpyrazine.

Table 3. Results of the sensory analysis of lettuce, carrot, and green bell pepper odour evaluated on a scale from 0 (no odour) to 15 (strong odour). Different letters in a column indicate difference at a significance level of 5%.

	Lettuce odour	Carrot odour	Green bell pepper odour
Mixed product	1.4[b]	6.4[b]	7.1[b]
Lettuce	11.6[a]	1.0[c]	0.5[c]
Carrot	0.7[b]	12.4[a]	0.2[c]
Green bell pepper	0.8[b]	1.2[c]	12.5[a]

4. DISCUSSION AND CONCLUSION

The most important odour active aroma compounds in lettuce appeared to be mainly terpenes, but also 2-methoxy-3-isobutylpyrazine was of importance (Table 1). This compound was found to be the most important in green bell pepper, and this was also reported by [3]. Myristicin and α-copaene were found to be the most important compounds in carrot (Table 2). This is partly in agreement with [1], who found myristicin to have an odour sensation, but not α-copaene. The odour of the mixed product was mostly influenced by the green bell pepper and carrot, with both myristicin and 2-methoxy-3-isobutylpyrazine as important compounds with high SNIF values. An unknown compound was the most important aroma compound in the mixed product, and the second most important both in lettuce and green bell pepper. The compound is most likely a terpene and the description from the GCO analysis is mainly carrot-like or green. Two compounds could not be identified because no peak was present.

A total of 31 odours were detected from the mixed product, 15 odours were detected from the lettuce, 17 from the carrots, and 18 from the green bell pepper (data not shown). The total duration of these odours is displayed in Table 1 and 2 as total minutes, and a clear add up effect was demonstrated as the mixed product obtained 44.5 min, whereas the other three samples achieved 13.6 (lettuce), 21.4 (carrot), and 21.6 (green bell pepper), which indicates that the odours detected from carrot and green bell pepper were longer lasting than the odours from lettuce. Some of the compounds are present in more than one vegetable, and therefore, compounds that are not important in the individual products might become important in the mixed product, because the amounts add up. This might be the case for (*Z*)-3-hexenol which was not among the eight most important aroma compounds in any of the 3 individual crops, but was placed as the 6th most important aroma compound in the mixed product.

The results of the sensory analysis of the headspace odour of the products (Table 3) showed that the impression of the mixed product mainly consists of carrot odour and green bell pepper odour (statistically equal values). This is in accordance with the GCO results, where the most important compounds in green bell pepper (2-methoxy-3-isobutylpyrazine) and carrot (myristicin) both gave high values in the mixed product along with other important compounds found in the two vegetables. When the vegetables were mixed the perceived intensity of the individual odours of carrot and green bell pepper was almost 50% and lettuce around 10% of the intensity of vegetable evaluated as single product.

References

1. F. Kjeldsen, L.P. Christensen and M. Edelenbos, J. Agric. Food Chem., 51 (18) (2003) 5400.
2. C. Varming, K. Jensen, S. Møller, P.B. Brockhoff, T. Christiansen, M. Edelenbos, G.K. Bjørn and L. Poll, Food Quality Pref., 15 (6) (2004) 531.
3. S.M. van Ruth and J.P. Roozen, Food Chem., 51 (2) (1994) 165.
4. P. Pollien, A.Ott, F. Montigon, M. Baumgartner, R. Muñoz-Box and A. Chaintreau, J. Agric. Food Chem., 45 (7) (1997) 2630.

W.L.P. Bredie and M.A. Petersen (Editors)
Flavour Science: Recent Advances and Trends

ChemSensor classification of red wines

Inge Dirinck, Isabelle Van Leuven and Patrick Dirinck

Laboratory for Flavour Research, Catholic Technical University St.-Lieven, Gebr. Desmetstraat 1, BE-9000 Gent, Belgium

ABSTRACT

In this study the hyphenated technique of automated headspace-solid phase microextraction (HS-SPME) and quadrupole mass spectrometry (MS) as a sensing system was used, in combination with on-line pattern recognition algorithms, for classification of red wines. These ChemSensor classifications based on mass fingerprinting were compared with a time-consuming GC-MS analysis, consisting of headspace-solid phase microextraction, identification, semi-quantitative determination of the wine volatiles and principal component analysis (PCA) of the semi-quantitative data. Good correlations could be observed between both techniques.

1. INTRODUCTION

In previous work the ChemSensor system was successfully used for fast aroma characterisation of coffee [1,2]. The aim of the current study was to evaluate the suitability of ChemSensor classifications for fast objective quality evaluation of red wines. For this purpose a good relationship should be obtained between classifications based on: a) fast MS fingerprinting; b) time-consuming GC-MS profiling and c) sensory data from an expert wine panel. A first approach studied a consumer-oriented model system composed of 14 commercial red wines with important sensory differences. In a further stage the suitability of the method was demonstrated by 2 thematic comparisons of Bordeaux wines with, respectively Cabernet Sauvignon and Merlot as dominating grape. However, presentation of these results is beyond the scope of this publication.

2. MATERIALS AND METHODS

2.1. Wine samples

A consumer-oriented model system was composed of 14 commercial red wines with important sensory differences and covering a large range of regions and both

monovarietal wines and wines composed of different grape varieties (Cabernet Sauvignon, Merlot, Pinot Noir, Syrah, Grenache, Carignan, Gamay, Mourvèdre). The wines selected for the study included 13 French wines and 1 wine from Italy:
Bordeaux (BD): Château Branaire 1998 (BD/B), Plaisir de Haut-Médoc (BD/HM), Château la Grave à Pomerol 1999 (BD/P), Château Roche Guitard 2000 (BD/RG); Bourgogne (BG): Aloxe Corton 1998 (BG/AC), La Chance au Roy 2000 (BG/CR); Rhône (R): Domaine de Panisse 2001 (R/P), Crozes Hermitage 2002 (R/CH), Les Truffiers 2001 (R/LT); Roussillon (RS): Domaine de Terre Rouge 2002 (RS/TR); Loire (L): Alison 2002 (L/A); Beaujolais (BJ): Beaujolais Delhaize 2002 (BJ/B), Morgon 2002 (BJ/M); Emilia-Romagna (ER) and from Italy: Villa Giulia 2002 (ER/VG).

2.2. HS-SPME-ChemSensor analysis

The hyphenated configuration consisted of a sample preparation autosampler (MultiPurposeSampler® or MPS-2®, Gerstel) for headspace-solid phase microextraction (HS-SPME), a 6890/5973 GC-MS system (Agilent Technologies) and a workstation with ChemSensor software (Agilent Technologies) and Pirouette® pattern recognition software (Infometrix).
HS-SPME parameters were evaluated and optimised: 10 ml samples of red wine were each diluted to 10% (v/v) ethanol, 2 g of sodium chloride was added in 20 ml vials and incubated for 3 min at 40 °C in the thermostatic agitator of the MPS-2®. The sorption of wine volatile compounds was performed for 30 min on a polydimethylsiloxane (PDMS) fibre (100 μm) (Supelco).
The GC column was continuously held at 250 °C and helium was used as carrier gas (1 ml/min). The transfer lines were maintained at 280 °C. The total ion current (70 eV) was recorded in the *m/z* range from 40-230 amu (scan mode) using a solvent delay of 2 min and a run time of 5 min. For each wine sample a total mass spectrum was generated and converted by the ChemSensor software to a composite mass fingerprint, which could be easily imported into the Pirouette® pattern recognition software. Different pattern recognition algorithms were used for on-line data processing of the mass fingerprints (e.g. principal components analysis (PCA), hierarchical cluster analysis (HCA) and soft independent modelling of class analogy (SIMCA)).

2.3. HS-SPME-GC-MS analysis

The hyphenated ChemSensor configuration was also used in the GC-MS mode. Therefore, the GC column (HP-PONA 50 m x 0.2 mm x 0.5 μm, Agilent Technologies) was temperature-programmed: 40 °C for 5 min, from 40 °C to 200 °C (5 °C/min), from 200 °C to 248 °C (8 °C/min), 248 °C for 5 min. A solvent delay of 6 min was used and the total run time was 48 min. For GC-MS analysis an aliquot of the internal standard nonane was added to the vials. Each wine was analysed in triplicate. Semi-quantitative determinations of the wine volatiles were obtained by relating the peak areas of the volatiles to the peak area of the internal standard. Principal components analysis (PCA) was performed on the semi-quantitative GC-MS data using The Unscrambler® (Camo).

3. RESULTS

The GC-MS data for the red wines were analysed by PCA and Figure 1 shows the relationships between the 14 red wines the 70 volatiles identified in these products. For clarity reasons mean values of the 3 replicates for each wine were used in the PCA. Bordeaux and Bourgogne wines had negative PC1 scores and PC2 differentiated both type of wines. Bordeaux wines were characterised by higher levels of oak lactones, vitispirane and phenolic components (4-ethylphenol, 4-ethyl-2-methoxyphenol and 2-methoxy-4-propylphenol). Characterised by positive score-values on PC1 were Rhône/Roussillon and Loire/Beaujolais/Emilia Romagna. Rhône wines. These wines were characterised by high levels of flowery terpene alcohols (linalool and geraniol), while the Beaujolais wines had high levels of acetate esters.

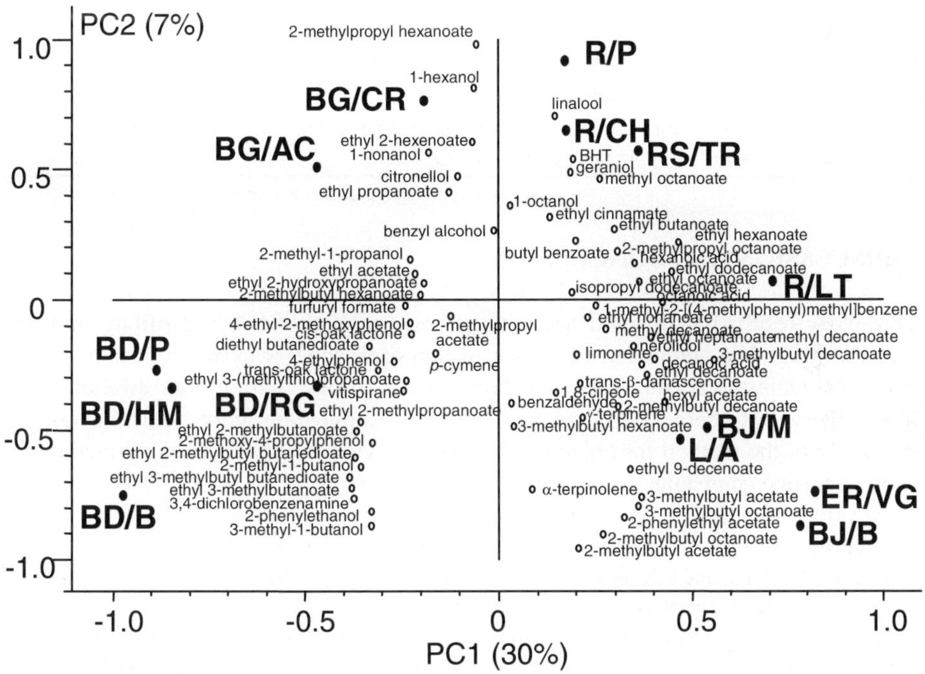

Figure 1. PCA biplot of the volatiles of the 14 red wines from different regions. For explanations on the sample abbreviations see section 2.1.

The GC-MS-PCA classification was in good correspondence with principal components analysis of the MS fingerprinting data. In Table 1 the SIMCA interclass distances of the red wines obtained by ChemSensor analysis are presented. Analogies in mass fingerprints are reflected by low interclass distances. An interclass distance lower than 4 is generally an indication for similarity. A good agreement between the MS fingerprinting and the GC-MS profiling approach was obtained.

Table 1. Interclass distances (SIMCA) between the 14 red wines obtained from ChemSensor analysis. For full names of the abbreviations used, see section 2.1.

	BD/B	BD/HM	BD/P	BD/RG	BG/AC	BG/CR	R/P	R/CH	R/LT	RS/TR	L/A	BJ/M	BJ/B	ER/VG
BD/B	-	4.50	1.60	3.03	6.80	6.19	12.57	9.57	11.65	11.41	9.30	9.37	10.76	16.25
BD/HM	4.50	-	3.21	3.56	6.73	6.15	8.66	6.07	8.35	8.10	6.24	6.69	8.47	9.99
BD/P	1.60	3.21	-	3.04	5.28	4.88	9.78	6.51	8.44	8.34	6.64	6.35	7.59	10.55
BD/RG	3.03	3.56	3.04	-	3.53	3.41	6.95	5.01	6.49	6.64	5.04	5.29	6.60	9.71
BG/AC	6.80	6.73	5.28	3.53	-	2.62	6.39	4.67	5.55	5.81	5.01	5.05	6.16	13.93
BG/CR	6.19	6.15	4.88	3.41	2.62	-	4.76	2.91	3.28	4.40	3.37	2.72	3.69	7.78
R/P	12.57	8.66	9.78	6.95	6.39	4.76	-	3.59	2.75	4.48	4.25	4.54	4.84	8.66
R/CH	9.57	6.07	6.51	5.01	4.67	2.91	3.59	-	2.34	3.39	1.97	2.31	2.65	6.46
R/LT	11.65	8.35	8.44	6.49	5.55	3.28	2.75	2.34	-	3.32	2.46	2.06	2.83	3.72
RS/TR	11.41	8.10	8.34	6.64	5.81	4.40	4.48	3.39	3.32	-	2.47	3.89	4.86	7.22
L/A	9.30	6.24	6.64	5.04	5.01	3.37	4.25	1.97	2.46	2.47	-	1.34	2.18	3.61
BJ/M	9.37	6.69	6.35	5.29	5.05	2.72	4.54	2.31	2.06	3.89	1.34	-	1.08	2.55
BJ/B	10.76	8.47	7.59	6.60	6.16	3.69	4.84	2.65	2.83	4.86	2.18	1.08	-	2.98
ER/VG	16.25	9.99	10.55	9.71	13.93	7.78	8.66	6.46	3.72	7.22	3.61	2.55	2.98	-

4. DISCUSSION AND CONCLUSION

This work resulted in a good accordance between both GC-MS profiling and MS fingerprinting. The automated ChemSensor approach appears to be promising for fast objective quality evaluation of wine. Future publications dealing with different wine types, e.g. with Cabernet Sauvignon or Merlot as dominating grape, will illustrate the suitability of the method for objective selection of wines as rapid screening method for wine importing companies.

References

1. J.-L. Le Quéré and P.X. Étiévant (eds.), Flavour research at the dawn of the 21st century, proceedings of the 10th Weurman flavour research symposium, Paris, France (2003) 572.
2. T. Hofmann, M. Rothe and P. Schieberle (eds.), State of the art in flavour chemistry and biology, proceedings of the 7th Wartburg symposium, Garching, Germany (2005) 98.

W.L.P. Bredie and M.A. Petersen (Editors)
Flavour Science: Recent Advances and Trends

Comparing predictability of GC-MS and e-nose for aroma attributes in soy sauce using PLS regression analysis

Tetsuo Aishima

Chemometrics and Sensometrics Laboratory, 1-197 Sengen-cho, Omiya-ku, Saitama 330-0842, Japan

ABSTRACT

Sensory attributes identified in soy sauce aroma by quantitative descriptive analysis were correlated with GC-MS profiles and electronic nose (e-nose) responses using PLS regression analysis. Highly predictive PLS models were obtained for every attribute based on GC-MS peaks but predictability in PLS models calculated from e-nose responses was unsatisfactory. No specific relationship between compounds and sensors was indicated by PLS2 analysis applied to the whole GC-MS and e-nose data sets.

1. INTRODUCTION

In food research and development, quantitative descriptive sensory analysis is widely used as an essential tool to characterise flavour. The basic principle of this method is similar to that of chromatographic analysis [1]. This similarity gives the rationale of handling obtained sensory data with various chemometrics techniques. Although at present no method can accurately illustrate the composition of flavour [2], GC-MS has generally been used for flavour analysis. Various successful classifications of the 'flavour' of products have been reported with the e-nose utilising metal oxide semiconductor gas sensors [3]. However, attempts to correlate e-nose responses to sensory data succeeded only partly [4]. Therefore, the predictability of GC-MS peaks and e-nose responses for sensory data were compared by PLS regression analysis.

2. MATERIALS AND METHODS

2.1. Samples and sensory evaluation

Fourteen deep-coloured type soy sauce samples (A-N) were purchased from a local market in Tokyo. The quantitative descriptive sensory analysis was performed for the soy sauce samples by 12 trained panellists using a line scale.

2.2. GC-MS and e-nose analysis

Headspace volatiles in soy sauce were purged and trapped on a Tenax TA column installed in a Tekmar LSC2000 (Tekmar Inc., Cincinnati, OH). The trapped volatiles were analysed by an HP5971 mass spectrometer (Agilent Technologies, Palo Alto, CA) connected with an HP5980 gas chromatography equipped with a DB-WAX (60 m × 0.25 mm, film thickness: 0.25 μm, J & W Scientific Inc, Folsom, CA) capillary column. Peak components were identified by matching their mass spectra with authentic ones and based on their retention indices. The integrated peak areas were used as variables.

A FOX 4000 e-nose from Alpha-MOS (Toulouse, France) installing 18 metal oxide semiconductor gas sensors was used. One ml of soy sauce sample was placed in a 10 ml volume of vial and heated at 50 °C. One ml of headspace air was injected into the e-nose and sensor responses were recorded for 120 s. The maximum responses from each of 18 sensors were used as the e-nose response.

2.3. Statistical analysis

The relationship between a sensory attribute and GC-MS profiles or e-nose responses was analysed by PLS1 regression analysis using Unscrambler version 7.01 (CAMO ASA, Trondheim, Norway) but PLS2 was applied to depicting relationships between the whole GC-MS and e-nose data sets. F(p) was calculated by ANOVA.

3. RESULTS

3.1. Sensory analysis

Among 15 attributes identified by the panel for describing soy sauce aroma, 8 attributes, 'sweet', 'fruity', 'woody', 'roasted', 'medicinal', 'earthy', 'mouldy' and 'ink' listed in Table 1, were significantly different ($p<0.05$) between samples.

3.2. Sensory versus GC-MS

Ninety-eight GC-MS peaks commonly found in all 14 soy sauce samples were used as predictor variables in PLS1 regression analysis. As Table 1 indicates, all R^2 calculated for the attributes were greater than 0.76. This indicated that attributes identified in soy sauce aroma were strongly correlated with headspace GC-MS profiles. A PLS model based on the 20 most important peaks from the initial PLS models gave a greatly improved R^2.

Table 1. Contributing proportions (R^2x100) in PLS models calculated from 98 GC-MS peaks, selected 20 peaks, and e-nose responses for 8 attributes where significant difference ($p<0.05$) between samples was indicated.

Attribute	Description	F (p)	GC-MS		E-nose
			98 peaks	20 peaks	
Sweet	Cotton candy	1.87 (0.04)	87	97	9
Fruity	Sweet fruity	1.96 (0.03)	90	97	17
Woody	Pine or cider wood	2.09 (0.02)	86	93	16
Roasted	Roasted cereals	1.88 (0.04)	90	96	2
Medicinal	Medicine-like smell	4.22 (0.00)	76	91	20
Earthy	Dusty or earthy	2.06 (0.02)	96	95	38
Mouldy	Moulded food	2.37 (0.01)	78	94	22
Ink	Smell of ink	2.15 (0.01)	76	94	7

3.3. Sensory versus e-nose

R^2s calculated from e-nose responses were greater than 0.5 only for the attributes 'alcoholic' and 'fishy' even though they were not significantly different between the samples. However, R^2 calculated for the other 13 attributes were much smaller (Table 1). According to cross-validation, the 5 components PLS model with R^2=0.863 was the best for predicting 'fishy' but R^2 from the cross-validation was 0.599. However, 100% variance contained in sensor responses had already been extracted in the first two PLS components (Figure 1). The factor loadings plot indicates that 18 sensors are simply classified into contrasting two groups, one characterised by positive and another by negative loadings on the PLS component 1, which accounted for 99% of variance.

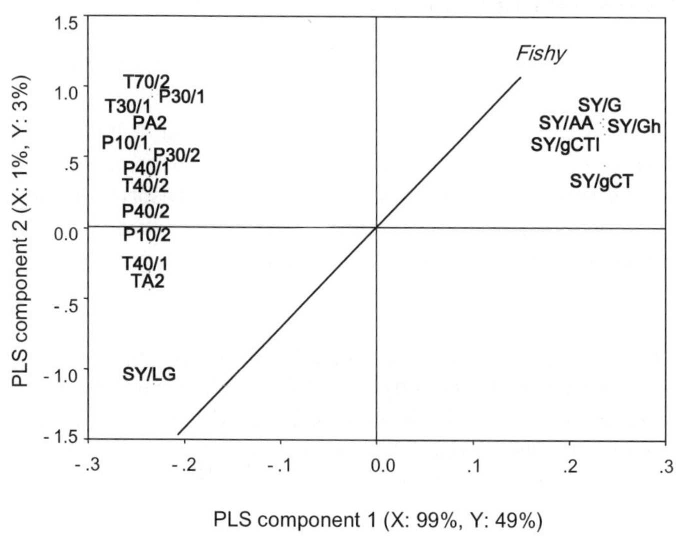

Figure 1. Factor loadings of PLS components 1 and 2 on 18 gas sensors and the 'fishy' attribute.

3.4. GC-MS versus e-nose

Relationships between e-nose responses and whole GC-MS peaks were analysed by PLS2 regression analysis. Figure 2 shows factor loadings for 98 GC-MS peaks and 18 gas sensors with samples (A-N) superimposed, where two sensor clusters locate in the opposite sides and GC-MS peaks are placed in between. The contributing proportion for 18 gas sensors attained 93% by PLS components 1 and 2 but only a few GC-MS peaks show weak relationships with gas sensors.

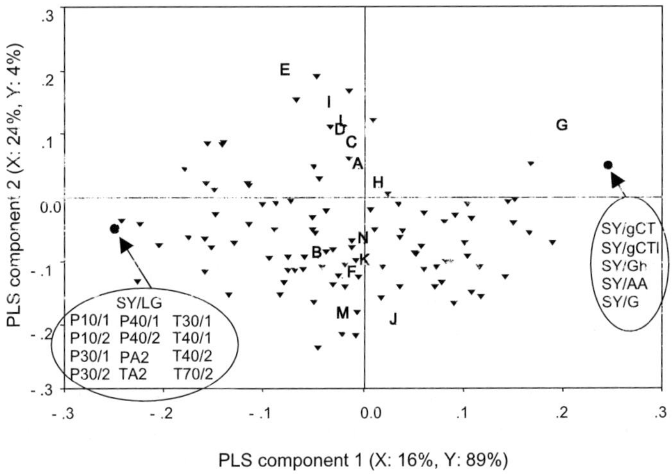

Figure 2. Factor loadings of PLS components 1 and 2 on 98 GC-MS peaks (▼) and 18 sensors.

4. DISCUSSION AND CONCLUSIONS

Every sensory attribute could be well correlated to GC-MS peaks by PLS1 analysis. On the other hand, the predictability of e-nose responses for sensory attributes was unsatisfactory. Large variance extracted in the PLS component 1 does not seem to account for any flavour attribute or volatile compounds. However, no current e-nose seems to be able to supply sufficient information relevant to sensory attributes. These results strongly suggested that novel gas sensors with higher specificity and wider diversity should be developed to construct a real e-nose that can complement and substitute sensory analysis.

References

1. T. Aishima. J. Chromatogr. A, 1054 (2004) 39.
2. A.J. Taylor (ed.), Flavour technology, Shefield, UK (2002) 210.
3. T. Aishima, Anal. Chim. Acta, 243 (1991) 293.
4. M. Morvan, T. Talou and J.-F. Cois Beziauc, Sensor. Actuat. B-Chem., 95 (2003) 212.

W.L.P. Bredie and M.A. Petersen (Editors)
Flavour Science: Recent Advances and Trends

Role of Strecker aldehydes on beer flavour stability

P. Guedes de Pinho and A.C. Silva Ferreira

Escola Superior de Biotecnologia, Universidade Católica Portuguesa, Rua Dr. Antonio Bernardino de Almeida, PT-4700-072 Porto, Portugal

ABSTRACT

In this work, attempts were made in order to measure the importance of Strecker aldehydes on flavour stability of beer by correlating chemical and sensory data. Methional and phenylacetaldehyde accumulated during storage were correlated with the aroma quality. A 'fresh beer' was spiked with methional, phenylacetaldehyde and also with (*E*)-2-nonenal, alone or in combination with these Strecker aldehydes. The degree of similarity was then determined between these samples and an aged beer. The highest similarity value was 7.2 when the three compounds were added simultaneously and the combination of the two Strecker aldehydes gave a degree of similarity of 5.4

1. INTRODUCTION

A large number of studies devoted to flavour stability of beer are available in the literature [1-7]. Several mechanisms are recognised as contributors to the undesirable 'stale character' typical of flavour deterioration [7]. Two mechanisms are widely accepted as the most relevant, namely lipid oxidation and Maillard reactions [4]. The first pathway is responsible for the presence of (*E*)-2-nonenal, one of the most undesirable aroma compounds in beer described as 'cardboard-like' with an extremely low odour threshold of 0.035 μg/l [1].

The Strecker degradation is a minor pathway of the second mechanism and is responsible for the formation of the Strecker aldehydes. In beer, 3-methylbutanal, 2-methylbutanal and 2-methylpropanal, are considered to be responsible for the 'malty' character [2-4]. However, methional was related to the flavour of aged beer [5,6]. Yeast activity plays an important role for the levels of free carbonyl compounds, in fact they are reduced to alcohols during fermentation [10-12]. On the other hand SO_2 formed during the process can bind these molecules, thus contributing to a reduction in the perceived 'aged character'. Nevertheless the possible release of aldehydes from the sulfite adducts during bottle storage may also contribute to the accumulation of these molecules in beer.

The aim of this work was to select suitable chemical substances responsible for aroma deterioration, to be monitored during beer production. This in order to provide information concerning the key steps along the process which most affect the flavour stability and thus the shelf-life of lager-type beers.

2. MATERIAL AND METHODS

2.1. Lager Beers

Group 1: In order to promote aroma degradation of beer, a 'forced aging' experiment was implemented. A total of 34 lager beer samples were divided into two groups: i) Beers stored at a temperature of 37 °C for 7 days; ii) Beers kept at 4 °C for 180 days. *Group 2*: Eight beer samples were analysed after the six-months corresponding to the shelf-life period of the product. Two were kept at 4 °C at the brewery and the other six, kept under commercial storage conditions, had been returned to the point of production after the shelf-life period had expired.

2.2. Sensory studies

A sensory panel was composed of 17 trained assessors including brewery workers, university students and laboratory personnel. They were trained every week for two months. Tests were performed individually using tulip glasses containing 30 ml of beer in a room at 20 °C.

The AFNOR NFV-09-021 [13] procedure was used to select the most important descriptors related to the typical aroma of aged beer. Methional, phenylacetaldehyde, and (*E*)-2-nonenal, were added singly and in combination to a fresh beer in the following concentrations 3 μg/l; 4 μg/l and 0.25 μg/l. The 'Similarity Value' (SV), of each sample with the aged beer was determined using a discontinuous scale from 0 to 10. An overall quality score was obtained using a continuous quality scale from +1 (no defect) to –3 (major defect), as usually used at the brewery.

2.3. Chemical Studies

GCO analysis was employed using dichloromethane extracts in order to identify the substances responsible for the aromatic notes associated with descriptors of the aged beer. Methional and phenylacetaldehyde were quantified by GC-MS/MS as in [8].

3. RESULTS

Four descriptors 'malty', 'honey like', 'cooked potato' and 'metallic', were selected using the AFNOR NFV-09-021 [13] procedure as the most relevant on the characterisation of the typical aroma of 'aged' beer. GCO analysis highlighted six odour-active zones showing aromas close to the descriptors selected. They were described as 'aged beer like', 'bread', 'cooked potato' for a retention index RI=1463, 'honey' (RI=1690), 'melted sugar' (RI=2038) and 'aged Port' (RI=2189). Using GC-MS and chemical standards,the coinciding molecules were identified as methional (RI=1463), phenylacetaldehyde (RI=1690), and Sotolon (RI=2189).

Aroma deterioration was promoted by 'forced aging' experiment group 1 samples, in order to correlate the quantities of methional and phenylacetaldehyde with the flavour quality. The quantities of aldehydes were significantly higher in samples stored at the higher temperature and a high correlation (r=0.76) was observed between them indicating a concomitant formation during Strecker degradation. These results are in agreement with other studies [8,9]. Due to the extreme conditions applied to 'forced aged' samples, which may promote reactions different from those produced during regular aging, methional and phenylacetaldehyde were also analysed in beers submitted to normal aging during the six-month shelf-life period. This was done for both for samples exposed to commercial storage temperature and a control group kept at 4 °C ('group 2' samples). The first group presented higher levels of both aldehydes, indicating that these compounds are formed in 'normal aging' conditions, with an average value of 0.5 μg/l (SD=0.2) and 2.4 μg/l (SD= 0.5) for methional and phenylacetaldehyde respectively. On the other hand, their levels observed in samples kept at 4 °C were 0.3 μg/l and 1.4 μg/l, respectively.

Strecker aldehyde formation is greatly affected by two parameters, namely temperature and the level of dissolved oxygen [9]. Since the values of oxygen in these samples were lower than 0.2 mg/l, the observed variations within the commercial storage samples were most likely related to temperature. All of these observations point toward a progressive accumulation of methional and phenylacetaldehyde during storage as 'indicators' of aroma deterioration. The flavour quality of the samples was evaluated using 'group 1' beers. Average scores for each beer were calculated and good correlations were observed with the levels of methional and phenylacetaldehyde with r=0.61 and r=0.60 respectively.

Methional and phenylacetaldehyde were added separately or in combination to a 'fresh' beer, in the concentrations close to those found in the 'aged' beers respectively 3.0 μg/l and 4.0 μg/l. Considering the high impact of (*E*)-2-nonenal on the beer aroma degradation, this compound was also included in the test at 0.25 μg/l. A simple paired comparison test was carried out in order to rate the degree of similarity between each of the supplemented samples and the 'aged beer'. Statistical differences for each pair were determined using the Tukey test. Mean rating scores (MRS) were arranged according to magnitude, and the LSD at 95% was determined [14]. The average of the similarity values and the standard error (SE) were calculated for each pair as well as the Tukey test (Figure 1).

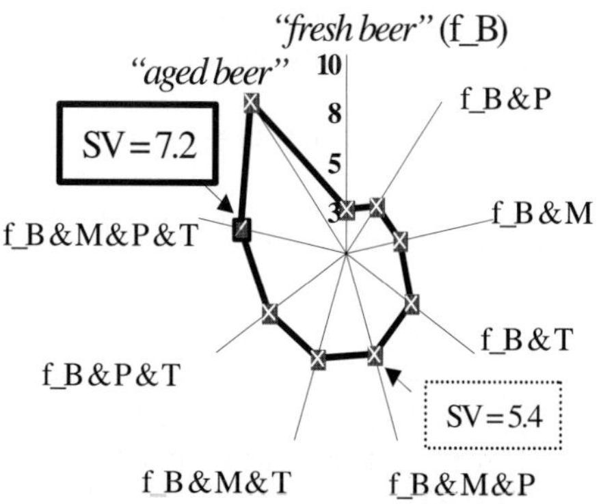

Tukey's Test

	M&T&P	T&P	M&T	M&P	T	M	P
M&T&P							
T&P	-						
M&T	-	-					
M&P	-	-	-				
T	+	-	-	-			
M	+	+	-	-	-		
P	+	+	+	+	-	-	
No Add	+	+	+	+	+	-	-

(+) different
(-) NOT different

Least Significant Difference (LSD): 2,20
Standard Error (SE): 0,49

Figure 1. Similarity Values and Tukey results, for each pair: supplemented sample-'aged beer'. f_B: fresh Beer; P: Phenylacethaldeyde; M: Methional; T: (*E*)-2-nonenal.

The ANOVA calculations for the data showed differences between samples ($p<0.001$) and no significant differences between assessors. All compounds added simultaneously contributed in a high degree to 'aroma spoilage' perception, SV ranged from 5.44 to 6.0. It is interesting to note that the SV for both 'Strecker aldehyde' additions (5.4) were close to the (*E*)-2-nonenal SV=5.0. The presence of the two Strecker aldehydes, formed during aging, has a higher impact on the perceived aroma associated with the 'aged' beer.

4. CONCLUSION

Methional and phenylacetaldehyde are key odorants for the perceived aroma of 'aged' beer. Due to the cumulative behaviour observed during aging, measuring their concentration can be useful to establish the best storage conditions as well as indicators of shelf-life period. These findings look promising in view of the possible applications in shelf-life control by minimising the Strecker aldehyde formation.

References

1. C. Liegeois, N. Meurens, C. Badot and S. Collin, J. Agric. Food Chem., 50 (26) (2002) 7634.
2. K.D. Deibler and J. Delwiche (eds.), Handbook of flavor characterization: sensory analysis, chemistry, and physiology, New York, USA (2004) 473.
3. R.K. Boccorh and A. Paterson, Food Quality Pref., 13 (2) (2002) 117.
4. P. Perpète and S. Collin, J. Agric. Food Chem., 47 (1999) 2374.
5. M.K. Visser and R.C. Lindsay, Role of methional in the development of off-flavours in beer, J Am. Soc. Brew. Chem., (1971) 230.
6. L. Gijs, P. Perpete, A. Timmermans and S. Collin, J. Agric. Food Chem., 48 (12) (2000) 6196.
7. M. Moll and N. Moll, Revue de la Brasserie et des boissons, 6 (1990) 99.
8. A.C. Silva Ferreira, P. Guedes de Pinho and T. Hogg, J. Agric. Food Chem., 51 (5) (2003) 1377.
9. A.C. Silva Ferreira, P. Guedes de Pinho, P. Rodrigues and T. Hogg, J. Agric. Food Chem., 50 (2002) 5919.
10. S. Collin, M. Montesinos, E. Meersman, W. Swinkels and J.P. Dufour, European brewery convention, proceedings of the 23rd congress, Oxford, UK (1991) 209.
11. A. Debourg, M. Laurent, V. Verlinden, L. van de Winkel, C.A. Masschelein and L. van Nedervelde, EBC-Symposium draught beer, packaging, dispense, Edinburgh, UK (1996) 158.
12. T.L. Peppard and S.A. Halsey, J. Inst. Brew., 87 (6) (1981) 386.
13. AFNOR NFV09-021, Recueil de normes françaises, contrôle de la qualité des produits alimentaires, analyse sensorielle, AFNOR (1991).
14. M. Meilgaard, G.V. Civille and B.T. Carr (eds.), Sensory evaluation techniques, Boca Raton, FL (1999) chapter 4.

Holistic taste analysis

Ben Nijssen[a], Leon Coulier[a], Eduard Derks[a], Mickaël Labbé[b] and Mark Springett[b]

[a]TNO Quality of Live, P.O. Box 360, 3700 AJ Zeist, The Netherlands; [b]DanoneVitapole Centre de Recherche Daniel Carasso, R.D. 128, 91767 Palaiseau Cedex, France

ABSTRACT

The suitability of holistic (combined GC- and LC-MC) analyses is demonstrated in the case of an investigation of the compounds responsible for the bitterness of a fermented milk product. Samples were prepared with different ingredients and fermented with two types of bacteria varying in composition. By means of multivariate data analyses of GC-MS and LC-MS data, it was possible to correlate bitterness and astringency, as evaluated by a sensory panel, to the sample composition. The compounds giving the best correlation with the sensory evaluation were selected for further identification.

1. INTRODUCTION

In metabolomics studies it is a prerequisite that all changes of metabolites within for instance a cell can be investigated. To achieve this goal, holistic methods are developed which are capable not only to detect all volatiles but also many compounds, which normally are not separated by GC methods. This is possible with derivatisation techniques which transform compounds like for instance organic acids, amino acids (including small peptides), sugars and polyphenols into compounds which can be analysed with GC-MS. For the identification of these compounds new databases are compiled. The mixtures obtained are in many cases so complicated that they can only be separated with GC or GC*GC. Of course not all compounds can be analysed this way. Therefore LC methods most complementary to the GC methods were developed.

TNO is investigating the applicability of both GC and LC methods in taste analyses; especially in the case of bitterness (e.g. milk or soy based products and vegetables). To be able to correlate taste with instrumental data, advanced pattern recognition tools are necessary. Preprocessing such as alignment and deconvolution of the complex chromatograms is indispensable. The advantage of this approach (holistic analysis

combined with data analyses) is that not only compounds are found which positively correlate with a certain taste aspect, but also compounds with a negative correlation.
The suitability of this approach is demonstrated by case of a fermented milk product in which the composition of the ingredients and the microflora was varied. Generally bitterness in milk products is attributed to peptides derived from caseins. A detailed review on this subject is already given in 1992 by Lemieux and Simard [1]. The approach in the present investigation was to find possible positive and negative correlations between bitterness and astringency, as observed by a sensory panel, at milk products with varying composition and the detection of as many as possible compounds with GC-MS and LC-MS.

2. MATERIALS AND METHODS

Twenty five samples were produced in which 8 ingredients and 2 microorganisms were varied. The samples were scored by a trained panel for bitterness and astringency using a 9-point scale with 9 as ‘very bitter’. Each sample was evaluated at least twice.

2.1. Instrumental analyses

For GC-MS analyses freeze-dried samples were derivatised, i.e. by silylation. At several steps of the sample preparation reference compounds were added to check the whole procedure for loss of compounds, the effectiveness of the derivatisation, the inertness of the injection system and the injection volume. GC-MS analysis was performed using an Agilent 5973 quadrupole mass spectrometer. A 1 μl was injected splitless on a HP-5MS column. The oven was programmed up to 325 °C.
For LC-MS analyses the freeze-dried extracts were dissolved in a small volume water/methanol (3:1). The instrument used was a Finnigan LTQ linear ion trap mass spectrometer. Samples (20 μl) were separated on a reversed phase C18 column using a linear gradient from water + 0.1% HCOOH to MeCN + 0.1% HCOOH at a flow rate of 0.2 ml/min. Compounds were detected by electospray ionisation (ESI; positive ionisation mode) in the range *m/z* 50-2000.

2.2. Data analysis

Data preprocessing was only carried out for the LC-MS data. The data were processed with in-home developed software to correct for small changes in retention time and MS response using internal standards. The GC-MS data were directly used for multivariate data analyses using deconvolution software. PCA and PLS was carried with in-home developed software.

3. RESULTS AND DISCUSSION

Bitterness in the samples, as perceived by the panel, varied between 0.14 and 3.56 (repeatability 0.76, reproducibility 0.61). Astringency varied between 0.79 and 5.60 (repeatability 0.96, reproducibility 0.73). Based on the sensory data only, a correlation was found between the ingredients and both the bitterness and astringency ratings on the basis of an unbalanced ANOVA correlation. Typical chromatograms obtained with GC-MS and LC-MS are shown in Figures 1 and 2.

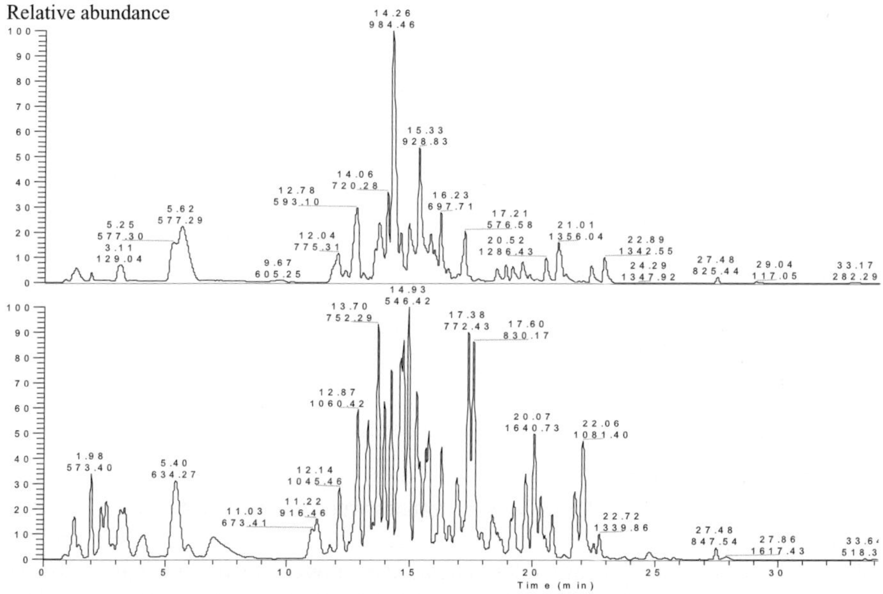

Figure 1. Typical LC-MS chromatograms (M3 and M13).

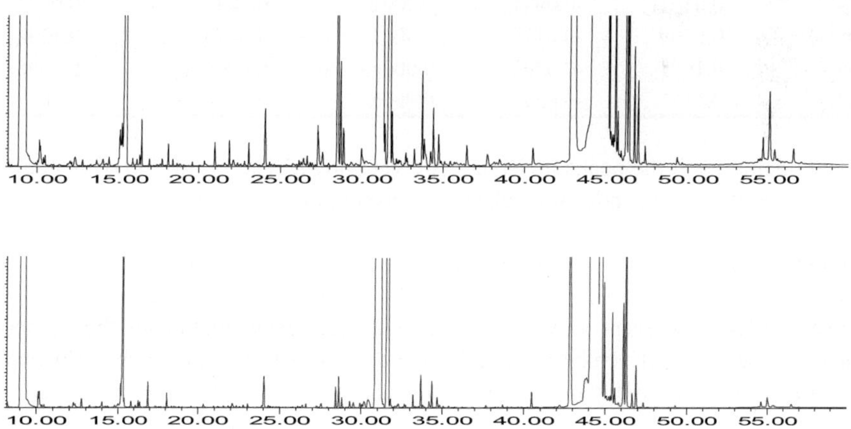

Figure 2. Typical GC-MS chromatograms (M8 and M17).

Many differences can be seen between the LC and GC chromatograms of the different samples. So far, MVDA correlations have only been investigated for the GC-MS patterns. Good correlations were found between these fingerprints and the composition, bitterness and astringency. In Figure 3 the predicted bitterness is shown against the actual bitterness.

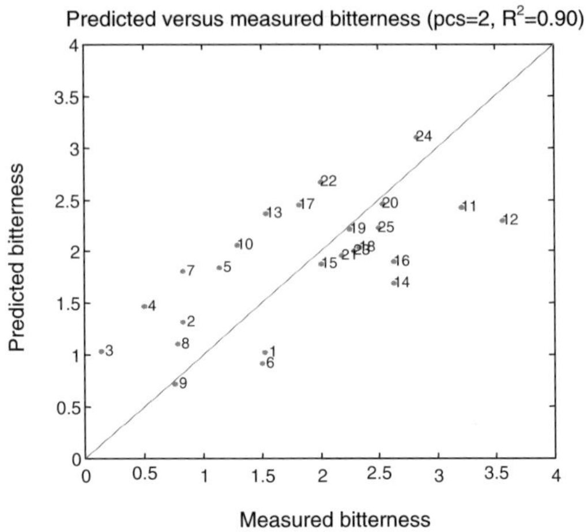

Figure 3. Correlation of predicted versus measured bitterness based on GC-MS results.

The same procedure will be performed with the LC-MS data. From the GC-MS data the five peaks with the best correlation with bitterness were selected (Table 1).

Table 1. Bootstrap PLS coefficients (with Bonferoni correction for multiple testing).

Peak no.	beta	beta bias	beta var	beta L (99%)	beta H (99%)
6	0.0736	-0.0128	0.0002	0.0121	0.1352
29	0.0395	-0.0071	0.0001	0.0043	0.0747
45	0.0709	-0.0116	0.0002	0.0074	0.1344
55	0.1005	-0.0202	0.0006	0.0004	0.2005
57	0.1155	-0.0213	0.0005	0.024	0.2071

Not all identities of the peaks in Table 1 are yet known. The same kind of selection can be made for the compounds most related to masking of bitter taste.

4. CONCLUSIONS

Holistic analysis delivers very promising results for investigation of the compounds related to bitter taste. Even better results can be obtained when the number of samples is increased and when the sensory scores are analysed at the level of individual panel members.

References

1. L. Lemieux and R.E. Simard, Lait, 72 (1992) 335.

W.L.P. Bredie and M.A. Petersen (Editors)
Flavour Science: Recent Advances and Trends

Black tea mouthfeel characterisation by NMR analysis and chemometrics

Marcial Peña y Lillo[a], Paul N. Sanderson[b], Emma L.Wantling[b] and Paul D.A. Pudney[b]

[a]Beverages Global Technology Centre; [b]Measurement Science, Unilever R&D Colworth, Sharnbrook, Bedford, MK44 1LQ, UK

ABSTRACT

The molecular understanding of black tea mouthfeel has been hindered by the difficulties of complete separation and identification of its constituents. The aim of the current work is to better understand the contribution of the whole non-volatile matrix of black tea to its in-mouth sensation, with a focus on astringency. Eight single origin black teas from different parts of the world were studied. ^{1}H NMR data were acquired directly from the tea infusion. Descriptive analysis of the aroma and in-mouth characteristics of the tea infusions was carried out by a trained sensory panel. Significant differences were found among the teas for astringency. The NMR spectra were used as predictors of the intensities of the astringency attribute using PLS models. Prediction errors calculated through cross validation were smaller for the models constructed with a single interval of the spectrum, rather than the total spectrum. In turn, models that included combinations of spectral intervals resulted in shorter prediction errors. The zones of the NMR spectra of highest contribution to the predicted astringency were those that contained signals from gallated species, caffeine and aliphatic methyls from low-level constituents.

1. INTRODUCTION

Black tea is the second most consumed beverage throughout the world (water is the first), and is prepared by infusing the fermented leaves of *Camellia sinensis* in boiling water. Black tea is a complex product, with very distinctive in-mouth sensations, caused by hundreds of non-volatile components [1,2]. Characterisation of the non-volatile composition of this product remains a scientific challenge, particularly with regard to the high molecular weight thearubigin fraction, which can constitute up to 60% of the solids in a black tea infusion [3]. Nevertheless, spectroscopic techniques (such as

nuclear magnetic resonance, NMR) provide an efficient method for gathering compositional data from the full non-volatile matrix of complex beverages. However, with such multivariate data, there are considerable statistical challenges to be met in order to extract relevant taste information from relatively few samples - a common situation in food sensory research.
The aim of the current work is to explore the capabilities of combining a spectroscopic non-biased analytical method (NMR); multivariable analysis methods (using variations of Partial Least Squares, PLS); and sensory profiling, to gain further understanding of the contribution of the non-volatile components of black tea to its in-mouth sensation, specifically astringency.

2. METHODOLOGY

2.1. Black teas

Eight single origin black teas from various regions around the world were selected by expert tea-tasters to represent a wide range of aroma and in-mouth sensory characteristics: Argentine BOP (WAB), Assam (WAS), Ceylon Good Medium Dust (WCM), Ceylon UVA Fanning (WCU), Darjeeling Whole Leaf (WDR), Kenyan BP1 (WKB), Turkey (WTB), Vietnam Dust (WVD).

2.2. Sample preparation

Infusions were prepared with 8.0 g leaf tea per litre of boiling mineral water. Tea was brewed for 4 min and filtered through muslin.

2.3. Sensory

Thirteen trained sensory panellists assessed freshly brewed infusions for a full profile of aroma and in-mouth attributes using descriptive analysis techniques. They described astringency as 'the degree to which a drying, puckering sensation is perceived in the mouth'. Infusions were presented according to a complete-block design balanced for order and carry-over effects, with each panellist assessing each infusion three times. Infusions were assessed on unmarked linear line scales, which were then transformed into a linear scale going from 0 (lowest intensity) to 10 (highest intensity). Significant differences among products were identified using ANOVA.

2.4. NMR

Samples comprised 570 µl of tea infusion + 30 µl D_2O. 400MHz ^1H NMR data were acquired at 300K with presaturation of the water resonance. NMR data pre-processing conditions were established to retain peak resolution whilst minimising data set size; phasing and baseline correction were adjusted for each NMR data set.

2.5. Chemometrics

Predictive models were built for the intensity of astringency (Y block) from NMR data (X block) with Matlab version 7.0.1 and the iToolbox freeware [4]. The data were mean centered and normalised (function 'auto'). Due to the low number of samples included in this study, all were used for cross validation, with 8 segments (function 'syst123'). No external validation was performed. The root mean squared error of cross validation, RMSECV, was used as an indication of model performance.

For the global model, PLS was applied on all spectroscopic variables (2346 vars). For interval PLS (iPLS) models, the X block variables were divided into intervals, and PLS applied to one interval at a time. The synergy PLS (sPLS) models were built with combinations of up to 4 intervals.

The models tested aimed for lowering prediction error by first reducing X block variables (from a total of 2346 for the global model) to a number comparable to the number of samples (a minimum of 7 for the iPLS and 5 for sPLS models). The aim of the sPLS models was to recover the multivariable power lost by the reduction of the X block variables. However, due to computation power limitations (restricted to not more than 10^6 models), only 2, 3 and 4 interval combinations were tested in a partial factorial design.

3. RESULTS AND DISCUSSION

3.1. Error minimisation and model fit

The global PLS model minimal RMSECV was 0.5 for 3 PLS factors. Since the number of variables (2346) was very high compared with the number of samples (8), alternative models were explored to reduce the input variables and therefore enhance the model performance. iPLS models showed better performance than the global model with an optimum error of 0.1 with 300 intervals and 3 PLS factors. The rest of the iPLS models tested consisted of 20, 40, 80, 160 intervals, with corresponding errors of 0.25, 0.23, 0.2 and 0.18. sPLS proved to be the best method tested, obtaining an optimum performance when the data was split into 600 intervals and combinations of only 2 of these intervals were used by the model at a time. The optimum number of PLS factors was 6, and the corresponding error 0.01. Different variations of interval size (from 20 to 800) with 2, 3 or 4 interval combinations were tested; these gave higher errors.

Figure 1a shows the optimal sPLS model with the predicted astringency scores being almost identical to the measured intensities. All eight samples are well predicted by the model. However this exploratory work is based on cross validation of a small set of teas, and proper external validation should be incorporated into the study.

3.2. Regions of high contribution

The three different pairs of intervals with the lowest error for the optimal sPLS model are mapped onto the NMR spectra in Figure 1b. These provide an indication of the chemical nature of the components that give the highest contribution to astringency. These interval pairs respectively indicate a polyphenolic gallate-like group paired with an aliphatic methyl, caffeine paired with a high-field methyl and an aliphatic methylene and methyl pair. More precise chemical identification remains to be undertaken.

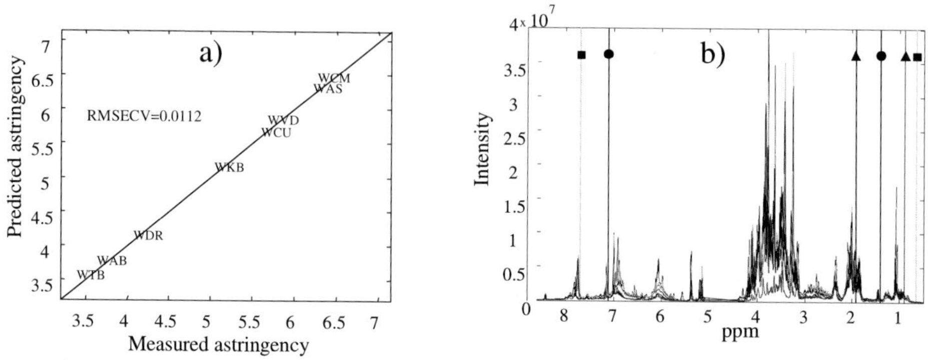

Figure 1. (a) Optimal model fit: sPLS with 600 intervals, 2 combinations and 6 PLS factors. (b) NMR regions of highest contribution to the astringency model. (●) polyphenolic gallate-like group paired with an aliphatic methyl; (■) caffeine paired with a high-field methyl; (▲) aliphatic methylene and methyl.

4. CONCLUSIONS

Chemometric analysis has been shown to effectively extract chemical information from NMR data for the prediction of black tea astringency. Regions of the NMR spectra of highest contribution to the astringency model were identified. These indicate the chemical nature of potential astringent constituents of black tea, including previously identified tastants (such as gallated groups and caffeine) as well as other yet to be identified minor components. This preliminary study demonstrates the potential for combining spectroscopic fingerprinting techniques and chemometrics, to provide an understanding of the molecular origin of black tea mouthfeel. On-going research is being extended to include FTIR spectroscopy, larger data sets and further data processing algorithms.

References

1. Y.S. Zhen, Z.M. Chen, S.J. Cheng and M.I. Chen (eds.), Tea, bioactivity and therapeutic potential, London, UK (2002) 173.
2. K.C. Willson and M.N. Clifford (eds.), Tea, cultivation to consumption, London, UK (1992) 555.
3. E. Haslam, Phytochem., 64 (2003) 61.
4. L. Nørgaard, A. Saudland, J. Wagner, J.P. Nielsen, L. Munck and S.B. Engelsen, Appl. Spectrosc., 54 (3) (2000) 413.

W.L.P. Bredie and M.A. Petersen (Editors)
Flavour Science: Recent Advances and Trends

Determination of commercial orange juice quality factors using descriptive and GCO analyses

A. Elston[a,b], C. Sims[a], K. Mahattanatawee[a,c] and R. Rouseff[a]

[a]University of Florida, Institute of Food and Agricultural Sciences, Citrus Research and Education Center, Lake Alfred, Florida, USA; [b]Jim Beam Brands, Research and Development, Clermont, KY, USA; [c]Department of Food Technology, Siam University, 235 Petkasem Rd., Phasicharoen, Bangkok 10163, Thailand

ABSTRACT

Thirty commercial orange juice samples were evaluated by descriptive sensory analysis using a 15 point scale. The juices could be separated into three groups based on quality scores; good (>8.9), fair (5.0-8.8), and poor (<5.0). Significant ($p<0.05$) correlation coefficients were observed for fruity/floral (0.99) orange juice aroma character (0.97), green/fatty/metallic (-0.98) and cooked/processed (-0.98) scores. Principal component analysis, PCA, of flavour attribute scores also indicated there were three distinct groups corresponding to good, fair and poor quality. Attributes most associated with overall flavour quality included: fruity/floral, orange juice character, green/fatty/metallic and cooked/heated/processed. Over 40 aroma active components were observed by time-intensity GCO analyses of 13 selected juices. PCA of the GCO data showed the most discriminating aroma active components to be octanal, ethyl butanoate/hexanal, 4-mercapto-4-methyl-2-pentanone, furaneol, 4-vinylguaiacol, *p*-methen-8-thiol, 2-methyl-3-furanthiol, vanillin, methional, (*E*)-2-nonenal, and 4-mercapto-4-methyl-2-pentanol.

1. INTRODUCTION

Commercial orange juice (OJ) is one of the major agricultural commodities bought and sold throughout the world. However, few studies [1] have examined the factors which determine flavour quality in commercial OJ in a modern way. Early studies quantified major OJ volatiles [2-6]. Later investigations [5,7,8] demonstrated that many key aroma components remained unidentified. Recent GCO studies [9-12] have examined self prepared OJ only. The flavour components in self prepared OJ's are affected by fruit maturity and cultivar. However, the flavour components in commercial OJ are also influenced by processing history, storage conditions, and container type. These latter factors will introduce flavour components in the juice that would not be present in freshly prepared juice. The purpose of the present study was to identify all factors in commercial OJ which differentiate juice quality independent of cultivar, processing history, storage conditions and container type.

2. MATERIALS AND METHODS

2.1. Sensory panel and techniques

Twelve panellists (6 men, 6 women, age range 25-62) evaluated 30 commercial orange juices. The Spectrum descriptive analysis method was employed to develop and quantify the nine flavour attributes. A 15 point defined overall flavour score was also assigned for each juice. More details can be found in [13].

2.2. Sample preparation

Thirty juices (12 NFC, 10 refrigerated FC, 3 shelf-stable FC, 2 frozen, 3 canned) of various quality and type were purchased from local markets. Juice volatiles (20 ml) were extracted with diethyl ether (2x2 ml) using a Mixxor-like apparatus previously described in detail [14]. However, there was no concentration of the extract.

2.3. Gas chromatography-olfactometry

The extract was separated under conditions similar to that described in [14] using a ZB-5 column, 0.5 μl injection splitless. GCO conditions and data treatment were similar to those described in [11]. Chromperfect software was used to integrate both GC-FID and human response. Each extract was sniffed by two experienced assessors in duplicate.

2.4. Statistical analyses

Correlations and multivariate analyses utilised the means from the 30 duplicate juice values for both sensory and GCO experiments. Single and multiple regressions were obtained using Statistica v5.1 (StatSoft, Tulsa, OK, USA). PCA and PLS results were obtained using the Unscrambler v7.0 (CAMO, Woodbridge, NJ, USA).

3. RESULTS

3.1. Single and multiple correlations of descriptive analysis results

Significant ($p<0.05$) correlation coefficients were observed for mean descriptive scores for: fruity/floral (0.99), sweetness (0.98), OJ aroma character (0.97), green/fatty/ metallic (-0.98) and cooked/processed (-0.98) scores with flavour quality. Correlation coefficients for bitterness (-0.92), peel oil (0.85), sourness (-0.73) and aroma strength (0.47), were lower but still significant. Principal components analysis, PCA, was employed to examine the underlying structure of the sensory data. The resulting score plot employing the sensory descriptive scores are shown in Figure 1. Three distinct groupings can be observed which contain juices of similar average overall quality scores but were formed from the nine sensory attribute scores. The distribution of the overall flavour quality scores are shown in Figure 2. A least significant difference, LSD, analysis confirmed the categories indicated in Figure 1 and are shown as dashed lines in Figure 2. The separation between good and fair juices can easily be determined from the distribution shown in Figure 2. However, the separation between fair and poor juices is not as obvious and PCA groupings were used to set the boundaries between these two groups. A PCA load plot for the nine attributes for all 30 juices is shown in Figure 3. It can be seen that PC 1 alone can explain 91% of the total variance. It can be seen that the highest loading values belong to OJ character, fruity/floral, fatty/metallic/green and cooked/ heated/processed.

3.2. GCO of juices with varying flavour quality

A representative subset of 13 commercial juices was examined by TI-GCO. A total of 47 aroma active peaks were observed and tentatively identified based on aroma quality and retention indices related to *n*-alkane.

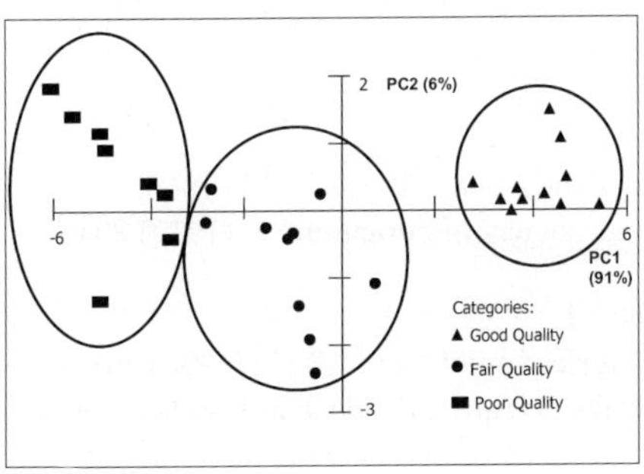

Figure 1. Score plot for 30 commercial orange juices. Highest quality juices = triangles, lowest quality juices = rectangles.

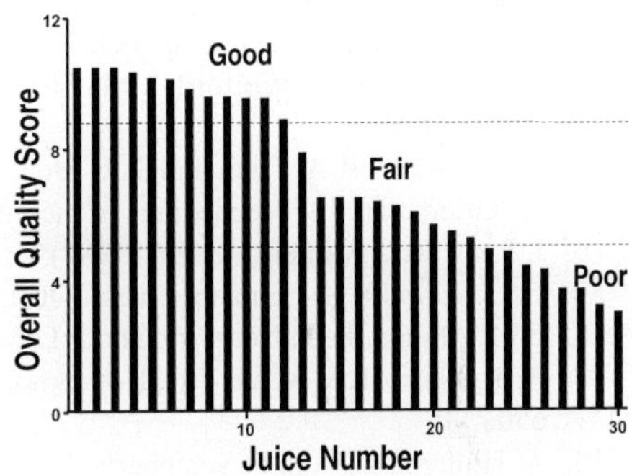

Figure 2. Overall OJ quality rankings.

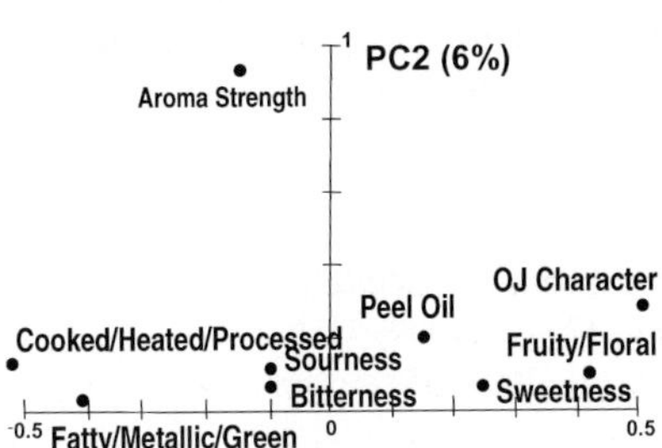

Figure 3. PCA loading plot all 30 juices.

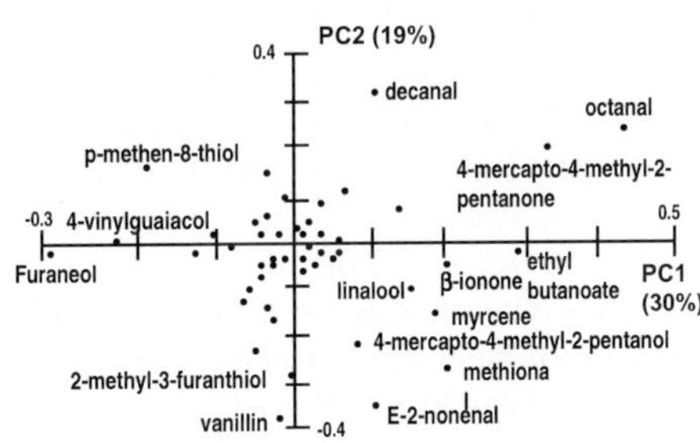

Figure 4. PCA load plot of 47 aroma active peaks found 13 representative juices.

The aroma active compounds observed are in agreement with previous GCO studies of fresh juice [9-11,15] and processed and/or stored juices OJ [16-19]. The PCA loadings of the GCO intensity values for these aroma active compounds showed some of the most differentiating aroma compounds (Figure 4). On PC1 these included: octanal, ethyl butanoate/hexanal, 4-mercapto-4-methyl-2-pentanone, furaneol, 4-vinylguaiacol, and *p*-methen-8-thiol. The most differentiating compounds on PC2 included: 2-methyl-3-furanthiol, vanillin, methional, (*E*)-2-nonenal, and 4-mercapto-4-methyl-2-pentanol.

4. CONCLUSIONS

Descriptive analysis indicated that OJ character, fruity/floral, fatty/metallic/green and cooked/heated/processed were most heavily associated with flavour quality of OJ. GCO analysis of 13 commercial juices detected a total of 47 aroma active components. Surprisingly, some of the most discriminating compounds for OJ flavour quality, are sulfur compounds reported in fresh [20] or processed grapefruit juice [14,21].

References

1. D. Tonder, M.A. Petersen, L. Poll and C.E. Olsen, Food Chem., 61 (1/2) (1998) 223.
2. R.W. Wolford, J.A. Attaway, G.E. Alberding and C.D. Atkins, J. Food Sci., 28 (1963) 320.
3. K.S. Rymal, R.W. Wolford, E.M. Ahmed and R.A. Dennison, Food Technol., 22 (12) (1968) 1592.
4. T.H. Schultz, R.A. Flath and T.R. Mon, J. Agric. Food Chem., 19 (6) (1971) 1060.
5. W. Grierson (ed.), Proceedings of the international society of citriculture, 3 (1977) 804.
6. J. Pino, Acta Aliment., 11 (1) (1982) 1.
7. E. Benk and R. Bergmann, Fluess. Obst, 42 (1975) 1.
8. E.M. Ahmed, R.A. Dennison and P.E. Shaw, J. Agric. Food Chem., 26 (2) (1978) 368.
9. A.B. Marin, T.E. Acree, J.H. Hotchkiss and S. Nagy, J. Agric. Food Chem., 40 (4) (1992) 650.
10. A. Hinterholzer and P. Schieberle, Flavour Fragrance J., 13 (1) (1998) 49.
11. R. Bazemore, K. Goodner and R. Rouseff, J. Food Sci., 64 (5) (1999) 800.
12. J. Leland, P. Schieberle, A. Buettner and T.E. Acree (eds.), Gas chromatography-olfactometry: the state of the art, Washington, DC, USA, 782 (2001) 33.
13. A. Elston, R. Rouseff and C.A. Sims, J. Food Sci., submitted.
14. P. Jella, R. Rouseff, K. Goodner and W. Widmer, J. Agric. Food Chem., 46 (1) (1998) 242.
15. J.-L. Le Quéré and P.X. Étiévant (eds.), Flavour research at the dawn of the 21st century, proceedings of the 10th Weurman flavour research symposium, Paris, France (2003) 706.
16. J.H. Tatum, S. Nagy and R.E. Berry, J. Food Sci., 40 (4) (1975) 707.
17. M.A. Petersen, D. Tonder and L. Poll, Food Quality Pref., 9 (1/2) (1998) 43.
18. Y. Bezman, R. Rouseff and M. Naim, J. Agric. Food Chem., 49 (11) (2001) 5425.
19. J.G. Dreher, R. Rouseff and M. Naim, J. Agric. Food Chem., 51 (10) (2003) 3097.
20. A. Buettner and P. Schieberle, J. Agric. Food Chem., 47 (12) (1999) 5189.
21. J.M. Lin and R.L. Rouseff, Flavour Fragrance J., 16 (6) (2001) 457.

W.L.P. Bredie and M.A. Petersen (Editors)
Flavour Science: Recent Advances and Trends

Quality of old and new carrot cultivars from ecological cultivation

Edelgard Hoberg[a], Detlef Ulrich[a], Dietrich Bauer[b] and Rolf Quilitzsch[b]

[a]Federal Centre for Breeding Research on Cultivated Plants, Institute of Plant Analysis, Neuer Weg 22-23, D-06484 Quedlinburg, Germany; [b]Association for the Promotion of the Biological-Dynamic Vegetable Seed Breeding 'Culture Seed', Dottenfelderhof, D-61118 Bad Vilbel, Germany

ABSTRACT

Open pollinating cultivars and hybrids of carrots were ecologically cultivated in the spring and autumn of 2002 and 2003. Sensory quality, aroma compounds as well as carotene content and texture were evaluated. The paper presents the different effects of genotypes and cultivation conditions on taste and aroma. The volatile compounds β-myrcene, α-humulene and caryophyllene are closely linked with unpleasant sensory sensations and can be proposed as marker for quality selection. In most cases expression of aroma compounds is stronger influenced by the year than by the cultivar.

1. INTRODUCTION

Latest results of nutrition research and psychology demonstrate the essential power of a high sensory quality for the increased consumption of fruits and vegetables, and consequently for human health [1]. The consumers often prefer sweet products. Generally carrots meet this preference. But their quality is also affected by bitter compounds, volatiles, and free amino acids [2]. Particularly carotenoids and crude fibres are responsible for the health-promoting quality. The amounts and profiles of all these compounds are influenced by the cultivar, the cultivation conditions, and the interactions between both factors. Over the past decades the number of hybrids and cultivars, which were bred for conventional farming, has been increasing. It is unknown, whether these genotypes grow also at the ecological conditions. Comparisons between the newer hybrids and those cultivars from ecological selection should be based on rigid scientific investigations in order to select hybrids suitable for ecological production.

2. PLANT MATERIAL AND METHODS

2.1. Carrots

The carrots were cultivated with ecological methods at Dottenfelderhof in Bad Vilbel. A total of 34 cultivars in 2002 and 40 cultivars in 2003 were divided into 7 groups according to the cultivation time. Group 1: early open pollinating cultivars ('Nantaise' different provenience); 2: early hybrids ('Espredo' p.e.); 3: late open pollinating cultivars ('Rodelika' different provenience); 4: late hybrids; 5: 'Lange Rote Stumpfe' - Qlb; 6: 'Cubic'- Sp; 7: 'Flakkee 2 -Trophy'. The weather conditions were extreme in both years. The year 2002 was very wet and the soil was compressed, that means disadvantageous conditions for the carrots, whereas 2003 was very dry and hot.

2.2. Sensory evaluation

The sensory evaluation was carried out by a trained panel consisting of 15 members with a maximum of 6 samples per session, one session a day. The panellists quantified the features (Table 1) on a non-graduated 10 cm linear scale. The acceptability was classified as 'very high' (5), 'high' (4), 'middle' (3), 'low' (2) and 'very low' (1). Every cultivar was tested twice during the experimental period. Statistical calculations were carried out with Excel 2000 from Microsoft® and Statistica, V6 from StatSoft®.

2.3. Instrumental analyses

The analysis of aroma volatiles was carried out using headspace SPME and GC with FID and MSD [3]. The total carotene content was analysed with spectral photometric and NIR-methods [4]. The texture was determined with a penetrometer [5]. The aroma and carotene analyses, the sensory evaluation as well as texture measurements were carried out simultaneously on fresh carrots, without storage.

3. RESULTS AND DISCUSSION

The results of the sensory and aroma analyses as well as of the carotene and texture measurements were modelled by a 2-factorial ANOVA with interaction. The significant effects are summarised in Table 1. The total carotene content and the texture were significantly distinguished by the groups, the years and the interaction. In some cases the significance was caused only by the groups 5, 6 and/or 7. 'Cubic' (group 6) was in 2003 the reason for the significant difference in 12 of 27 parameters. Especially differences of cultivars or groups were of interest. The acceptability ratings of different carrot groups decreased in the order: 1 - early open pollinating cultivars; 2- early hybrids; 3 - late open pollinating cultivars; 4 - late hybrids; 5 - 'Lange Rote Stumpfe' Qlb; 6 - 'Cubic' Sp; 7 - 'Flakkee 2 - Trophy'. The well accepted groups were the early hybrids, the late open pollinating cultivars, and 'Lange Rote Stumpfe'. The 'Cubic' cultivar had a very low acceptability rating.

Table 1. Significant influence (HSD Tukey-Test, α = 5%) of the carrot groups, the year and the interactions on the sensory parameters and the aroma compounds.

Variables	Label	Effects		
		Group (g) 1-7	Year (y) 2002-2003	Interaction (g) x (y)
Odour (orthonasal)				
Typical carrot	S1	x	n.s.	n.s.
Herbaceous	S2	x 6)	n.s.	x 6/03
Green	S3	x 6)	x	x 6/03
Sweetish	S4	x	n.s.	n.s.
Spicy	S5	n.s.	x	x
Musty	S6	n.s.	n.s.	n.s.
Chemical	S7	n.s.	x	n.s.
Flowery	S8	n.s.	n.s.	n.s.
Mushroom-like	S26	x 6)	n.s.	x 6/03
Pungent	S9	x 6)	x	x 6/03
Off-odour	S10	n.s.	n.s.	n.s.
Taste				
Sweet	S11	x	x	x
Bitter	S12	x 6)	x	x
Odour (retronasal)				
Typical carrot	S13	x	n.s.	x
Herbaceous	S14	x 6)	x	n.s.
Green	S15	x 6)	x	x 6/03
Sweetish, flowery	S16	x	x	x
Spicy	S17	n.s.	x	x
Nutty	S18	x	x	x
Chemical	S19	x 6)	n.s.	x 6/03
Musty	S20	x 6)	n.s.	x 6/03
Off-odour	S21	x 6) 7)	x	x 6/02/03
Aftertaste soapy	S22	x 6)	x	x 6/02
Mouth sensation				
Al dente	S23	x	x	x
Juiciness	S24	x	x	x
Raspy	S25	x 6)	x	x 6/03
Acceptability	S27	x	x	x
Aroma compounds				
α-Pinene	A1	n.s.	x	x
Sabinene	A2	n.s.	x	n.s.
β-Pinene	A3	n.s.	x	x
β-Myrcene	A4	x 6)	n.s.	x 6/03
Limonene	A5	n.s.	x	n.s.
γ-Terpinene	A6	x	x	x
Terpinolene	A7	x 5)	n.s.	n.s.
Caryophyllene	A8	x 6)	n.s.	x 6/03
α-Humulene	A9	x 6)	x	x 6/03
Myristicin	A10	n.s.	n.s.	n.s.
Σ Aroma compounds	A11	n.s.	x	x

n.s.: no significant differences.

Figure 1 shows the multivariate relationships of the parameters. On the right side are the negative sensory qualities and the 'Cubic' (group 6) carrots. They correlate with α-humulene, caryophyllene and β-myrcene. On the left side are the 'positive' sensory parameters together with 'acceptability'. Texture and carotene content contribute only little but positively to the group differentiation. Sweet taste correlates well with 'acceptability' (r=0.94). Groups 1, 2, 3, and 'Lange Rote Stumpfe' are in both years left. 'Lange Rote Stumpfe' has as the chemical features β-terpinene and terpinolene.

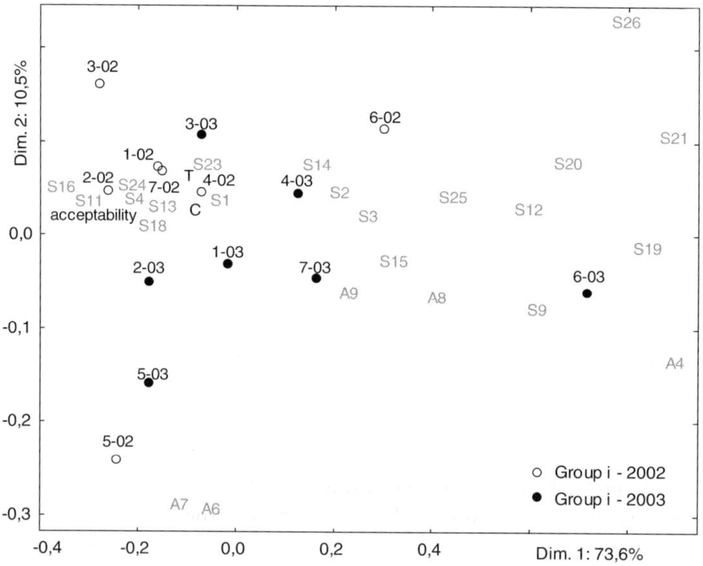

Figure 1. Correspondence analysis obtained for sensory parameters (Si), aroma compounds (Ai), carotene (C) and texture (T) of the seven carrot groups in two years.

There is a great variability among the different carrot types. The efforts of ecologically plant breeding resulted in genotypes with acceptabilities which are comparable with those of hybrids. Rodelika types possess a high acceptability. High contents of sugar and juiciness, low bitterness and low contents of α-humulene, caryophyllene and β-myrcene may characterise a well tasting carrot.

References

1. I. Possemeyer, Geo Wissen, 28 (2001) 72.
2. L.R. Howard, D. Braswell, H. Heymann, Y. Lee, L.M. Pike and J. Aselage, J. Food Sci., 60 (1995) 145.
3. J.-L. Le Quéré and P.X. Étiévant (eds.), Flavour research at the dawn of the 21st century, proceedings of the 10th Weurman flavour research symposium, Paris, France (2003) 440.
4. R. Quilitzsch and E. Hoberg, VDI-MEG Kolloquium: qualität von agrarprodukten, Potsdam, Bornimer Agrartechnische Berichte, 18 (1998) 244.
5. M. Pöschl (ed.), Proceedings of the 5th international conference on food physics, Brno, Tschechien (2002), 83.

W.L.P. Bredie and M.A. Petersen (Editors)
Flavour Science: Recent Advances and Trends

Effect on time-intensity results - comparison of time information versus no time information

Kirsten Lorensen and Line Budde Andersen

Chew Tech I/S, Dandyvej 19, 7100 Vejle, Denmark

ABSTRACT

Time-intensity analysis is used to assess the changes of menthol intensity in chewing gum. To observe whether information on the time-span on which the judgement should be based had an influence on the results, time information was given in some tests and omitted from other tests. Furthermore, the influence of testing one or multiple attributes - the perceptual dumping effect - was also evaluated. The results showed no significant difference between the test results with and without time information. The results indicated that the menthol intensity when judging with a single attribute was more sensitive than tests when assessing three attributes in one session.

1. INTRODUCTION

Time-intensity (TI) testing is a sensory method increasingly used in chewing gum research and development. The method plays an important role in the description of flavour release and changes in texture. However, the optimal procedures and possible biases for TI analysis are not clearly defined. Several authors [1,4] describe that information on how long time the assessors have to chew might cause a reflection on their evaluation modifying their final responses. Another bias is the 'dumping' effect [2,3]. Assessors may judge the intensity for one attribute too high, when presented with one intensity scale. When more attributes are available this enhancement is limited. A better understanding of these biases could provide more valid and reliable TI results.

2. MATERIALS AND METHODS

2.1. Sensory analyses

The TI method was used to assess the changes on menthol intensity in chewing gum. Twelve separate experiments were performed (Table 1). Eight trained assessors participated. All experiments were carried out with the same kind of menthol chewing gum (produced by Chew Tech I/S, Vejle, Denmark). The tests with one attribute were evaluated first and the tests with three attributes were evaluated last. The succession of the two kinds of test set-up was random. To avoid adaptation the assessors were given a 10 min break between sessions. For all evaluations computerised data acquisition software (FIZZ, Biosystemes, France) was used.

Table 1. The experimental set-up of the study.

Experiments	Attributes tested	Time between judgements
Without time information - one attribute	Menthol	15, 30 and 60 s
With time information - one attribute	Menthol	15, 30 and 60 s
Without time information - three attributes	Menthol, softness, sweetness	15, 30 and 60 s
With time information - three attributes	Menthol, softness, sweetness	15, 30 and 60 s

The intensity of menthol, softness and sweetness was indicated on a horizontal 15 cm unstructured line scale, with 'min.' anchored 1.5 cm from the left and 'max.' anchored 1.5 cm from the right. The assessors rinsed their mouths with tap water and neutralised their palates between samples by eating cucumbers and neutral biscuits. Prior to the test, the assessors were presented with a warm-up sample.

2.2. Chemical analyses

To support the sensory results and to see whether there were any differences in the flavour level, the release of menthol and sweeteners was analysed. Menthol was analysed by HS-GC-MS. General conditions: HS: Perkin Elmer Turbo Matrix 40, eq. temperature 75 °C, eq. time: 20 min. GC: Perkin Elmer Clarus 500, column: Perkin Elmer Elite 1 (60 m, 0.25 mm ID), GC program: 1 min. at 50 °C, 10 °C/min. to 280 °C, hold 5 min. MS: Perkin Elmer Clarus 500, scan mode: full scan *m/z* 25-500.

The release of sweeteners was analysed by HPLC-UV-PDA and HPLC-RI. General conditions: HPLC Dionex P680 Pump, ASI-100, column oven TCC-100 (80 °C for bulk, 40 °C for HIS). Detectors: Shodex RI-101 and Dionex UVD340U. Columns: Shodex SP-810 and Waters Symmetry C_{18} 5 μm. For further details: please contact Chew Tech I/S, Vejle, Denmark.

3. RESULTS

3.1. Sensory analyses

The results for menthol are presented in this paper. There was no significant difference between the test conditions where the assessors had received information on how long they had to chew and when they had not received this information (Figures 1-3). When comparing the first 5 min of the evaluation there was a tendency to obtain a higher score. When it comes to the single attribute compared to the triple attribute test there was little difference in the judgement.

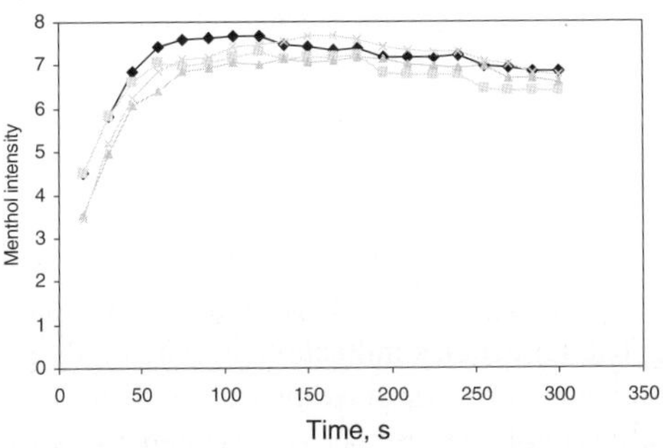

Figure 1. TI curve for 'menthol', 5 min test.

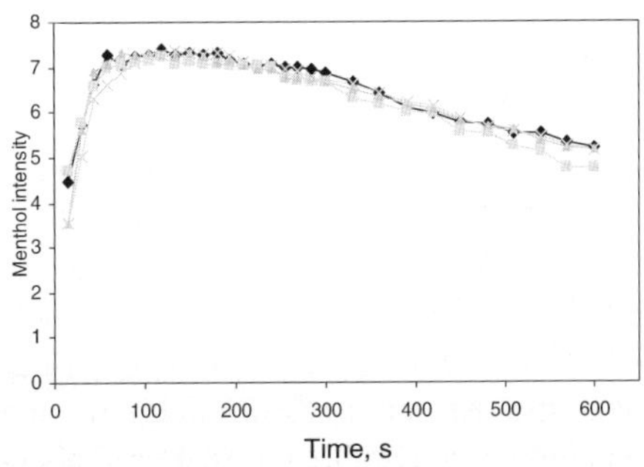

Figure 2. TI curves 'menthol', 10 min test.

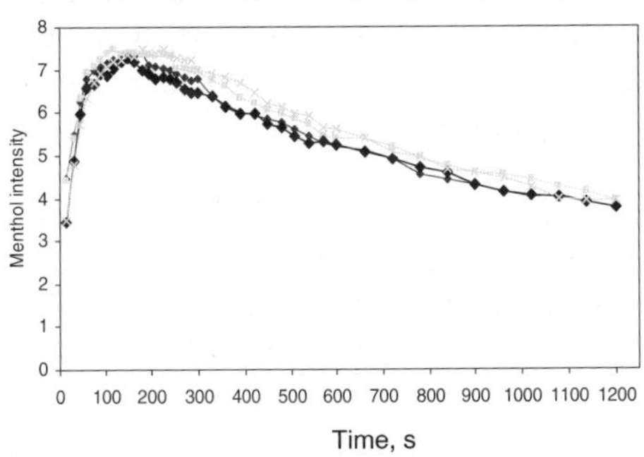

Figure 3. TI curves for 'menthol', 20 min test.

Legends for Figures 1-3:

- Without time information
- Without time information, 3 attributes
- With time information
- With time information, 3 attributes

When looking at the TI curves from individual assessors, the majority of the assessors were not influenced by the test information. However, for one assessor it was interesting to note that there was a difference in the way this particular assessor reflected over the test. Figure 4 shows the assessor's results from the 5 min test condition without time information, whereas Figure 5 shows the results when the time information was given.

3.2. Chemical analyses

The analysis done by GC-MS and HS-GC-MS showed no differences in the menthol concentration over time. As for the sweeteners, there was a decrease by approximately 97% (results not shown) in the concentration over time.

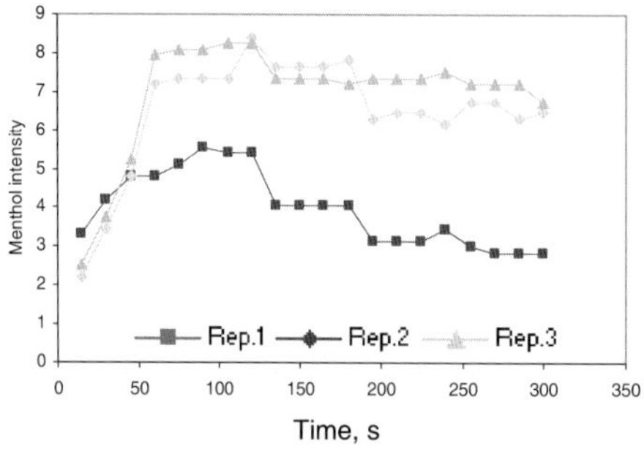

Figure 4. TI curves 'menthol', 5 min test without time information.

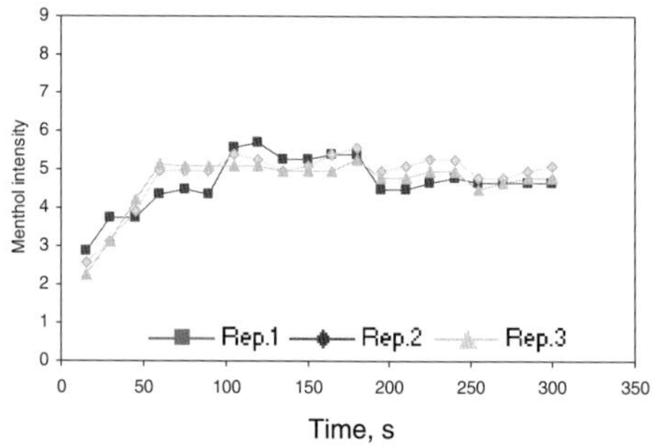

Figure 5. TI curves 'menthol', 5 min test with time information.

4. DISCUSSION AND CONCLUSION

When comparing results from TI tests with and without time information and with one and three attributes significant differences only appear when the tests with one attribute is compared with a test with three attributes. The tendencies indicate that, when judging one attribute, the assessors judge the menthol intensity higher compared with the test where they judge three attributes. This could be explained by an exaggerated focus on the one particular attribute resulting in a perception of higher intensity than in fact exists and is consistent to the findings of Frank *et al.* [2]. In additional, the abundance of time could result in intensity scores becoming systematic instead of being independent. When, on the other hand, the assessors evaluate three attributes at a time a more natural focus is applied resulting in a momentary attribute evaluation where the actual intensity score is independent of prior intensity scores.

The analysis done by HS-GC-MS showed no differences in the menthol concentration of the chewing gum over time. As for the sweeteners, there is a decrease in the concentration over time (results not shown). This concludes that there is an adaptation/ perception to menthol that leads to the lower judgement on menthol intensity at the end of the sensory test - see Figures 1-3, and that the decrease in the menthol intensity is due to the sweeteners decreasing during chewing.

Further experiments need to be performed to provide full understanding of the results in this experiment. Some of the assessors may have been aware of the test set-up and this may have influenced the results. Because of this, it would be appropriate to make a different set-up to avoid that the assessors remember from test to test.

References

1. G. Dijksterhuis and J.R. Piggott, Trends Food Sci. Technol., 11 (2001) 284.
2. R.A. Frank, N. Wessel and G. Shaffer, Chem. Senses, 15 (1990) 576.
3. H.T. Lawless and C.C. Clark, Food Technol., 45 (7/12) (1991) 81.
4. C. Peyvieux and G. Dijksterhuis, Food Quality Pref., 12 (2001) 19.

W.L.P. Bredie and M.A. Petersen (Editors)
Flavour Science: Recent Advances and Trends

The perception of strawberry aroma in milk

Zdenka Panovska, Alena Šedivá, Jan Pokorný and Dobroslava Lukešová

Department of Food Chemistry and Analysis, Institute of Chemical Technology in Prague, Technická 3, 166 28 Prague, Czech Republic

ABSTRACT

The flavour release in foods depends on the nature and concentration of flavour compounds present and their interaction with the food matrix. Studies on flavour release should also include the evaluation of food properties by using the human senses. Perceived intensity and pleasantness of strawberry aroma in milk were tested by sensory methods. The ability of assessors to distinguish fattiness of milk and small concentration of strawberry aroma was tested by a ranking test. Samples were evaluated by descriptive analysis using 9 attributes to characterising the sensory profile of the milk aroma with added a strawberry flavour mixture. The results show the difference in sweet taste among the samples. The strawberry aroma influences the feeling of fattiness. The panellists judged milk containing strawberry aroma to be fatter.

1. INTRODUCTION

Sensory science evaluates the food properties by using the human senses that react to certain stimuli from food. The sensation depends not only on the type and intensity of the stimulus but also on the person´s or group of person´s sociological and psycho-physiological conditions [1]. Moreover it is also important to choose the suitable method that allows one to determine the reliability of the data and it is also necessary to know in detail the properties and composition of compounds that form the food.

Milk can be produced to differ in fat content. The fat in milk is emulsified as micro globules and the first sensation is the aroma of fat soluble volatile flavour molecules, perceived through the nose and mouth.

Frøst and co-workers studied sensory perception of fat in milk [2]. They investigated the effects of various factors such as thickeners, cream flavours and whiteners on sensory properties and perceived fattiness. Their experiments showed that the fat does not affect the sensory properties of milk in a linear fashion and that there are larger sensory differences between milk with 0.1% and 1.3% fat than there are between 1.3% and 3.5% fat. Tepper and Kuang studied the effect of flavour addition in milk. They showed that products with a high level of added aroma compounds were perceived similar to products with a higher fat content [3]. The aim of this study was to evaluate the effect of strawberry aroma on perception of fat and intensity of sweet taste.

2. MATERIALS AND METHODS

2.1. Material

Pasteurised milk with three different levels of fat content (0.5%, 1.5% and 3.5%) was used. The strawberry flavour mixture (0.1 g/l) was kindly provided by Givaudan.

2.2. Sensory analysis

Sensory evaluation was performed in the laboratory with 6 individual booths equipped according to requirements of the international standard ISO 8589. The preparation and serving of the samples were in agreement with the requirements of ISO 6658-1985. Tap water was used for washing the mouth before the first sample and between samples. Samples were served blind coded in 20 ml portions at 15 °C and 22 °C.

2.3. Sensory panel

Sixty-five assessors (22 men, 43 women) mostly university students were experienced in sensory analysis, particularly with the sensory profiling and ranking methods.

2.4. Sensory analysis

Ranking tests were performed with two sets of milk samples (0.5%) with different flavour concentration were prepared. The first set had 8 samples (0; 0.05; 0.1; 0.15; 0.2; 0.25; 0.3; 0.35 g/l) and the second set had 5 samples (0; 0.2; 0.4; 0.6; 0.8 g/l). Sensory profiling was done using unstructured line scales of 100 mm long, constructed after the standard ISO 4121-1985. The scales were anchored by descriptors. Statistical analyses were done using Microsoft Excel 2000 and SPSS software v12.0.

3. RESULTS

3.1. Testing ability of panellist to recognise level of fat in milk

The assessors were tested for their ability to recognise differences in fat content of milk by rating on a sensory scale the fattiness of milk containing 1.5% fat (Figure 1).

3.2. Testing the ability of panellists to rank the samples with strawberry aroma

The assessors ranked the model samples of milk (0.5%) with different levels of strawberry aroma by their intensities. In the first set the step differences in concentration of strawberry flavour were too small, namely 0.05 g/l. The assessors could not distinguish the differences among samples. Therefore, a second set was prepared with step differences in concentration of strawberry flavour of 0.2 g/l. About 50% of assessors could distinguish correctly the intensity of strawberry aroma in milk. It was easier for assessors to distinguish differences at the lower concentrations (0.2 and 0.4 g/l) than at the higher concentration of 0.4 to 0.6 and from 0.6 to 0.8 g/l (Figure 2).

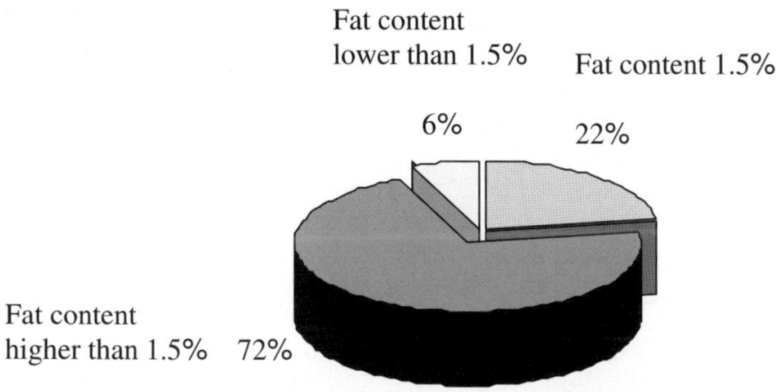

Figure 1. Results from judging the fat content of milk containing 1.5% fat. Only 22% of the panellists correctly indicated the fattiness of the milk.

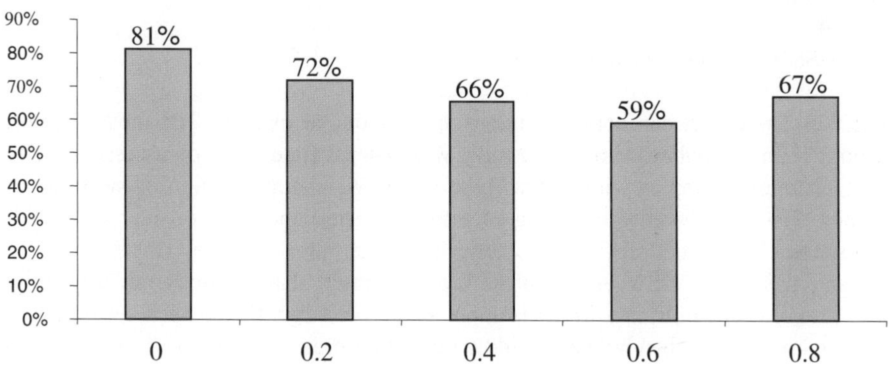

Figure 2. Result from intensity ranking of strawberry flavour in milk with different added concentrations of strawberry flavour (g/l). Percent of correct answers form the ranking test.

3.3. Profile of low fat milk with and without aroma

The profile of milk with and without aroma was evaluated by assessors for all types of milk. The main differences were observed in the profile of low fat milk (Figure 3).

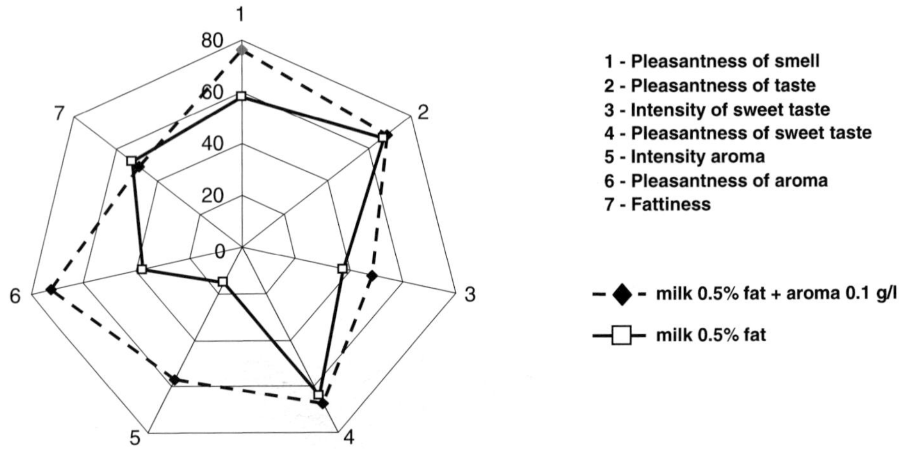

Figure 3. Sensory and hedonic profiles of low fat milk with or without added strawberry flavour.

Analysis of variance (ANOVA) using 95% of confidence intervals was run for each of descriptors to disclose possible differences among the samples. The statistical differences were in pleasantness of smell, perception of sweet taste and also the aroma influenced the feeling of fattiness. The panellists considered milk with aroma fatter. The same profile was done also for fatter milk but there were no significant differences between samples.

4. DISCUSSION AND CONCLUSION

Only 22% of assessors can correctly judge the fat content in milk. The most pleasant concentration of strawberry aroma in milk was 0.4 g/l (one third of assessors). The differences among samples with and without aroma were in perception of sweet taste. The aroma also influenced the feeling of fattiness. The panellists consider milk with aroma fatter. The same profile was done also for fatter milk but there were no significant differences between samples. We confirmed that fat does not affect the sensory properties of milk in a linear fashion. There are larger differences between low fat milk and middle fat than between middle fat and fat milk.

References

1. P. Alvarez and M.A. Blanco, J. Sci. Food Agr., 80 (2000) 409.
2. M.B. Frost, G. Dijksterhuis and M. Martens, Food Quality Pref., 12 (2001) 327.
3. B.J. Tepper and T. Kuanga, J. Sens. Stud., 11 (3) (1996) 175.

9

Advanced instrumental analyses

W.L.P. Bredie and M.A. Petersen (Editors)
Flavour Science: Recent Advances and Trends

Novel concept of multidimensional gas chromatography. New capabilities for chiral analysis and olfactometric detection

Alain Chaintreau, Frédéric Begnaud and Christian Starkenmann

Firmenich SA, Corporate R&D Division, P.O. Box 239, CH-1211 Geneva 8, Switzerland

ABSTRACT

In multidimensional gas chromatography (MDGC), the analyte transfer from the first to the second column is usually achieved using either a valve, or a pneumatic switcher. Both systems have significant drawbacks (degradation of analytes, switcher complexity, etc.). A new design, based on the double-cool-strand interface (DCSI) overcomes these limitations due to: 1) simplicity and low cost, 2) no degradation of labile compounds, 3) no dead volume, 4) optimised chiral MDGC conditions with a single oven, 5) higher sensitivity and peak resolution. Examples illustrate the suitability of this technique when various detection means are hyphenated to MDGC: FID, mass spectrometry and olfactometric detection. In addition, the use of the new interface in conjunction with a chiral column allows the odour evaluation of enantiomers with an interval of several minutes. Such a possibility that may not be achieved with previous interfaces, allows the panellist to recover from the first stimuli prior to smelling the second one. This is particularly important for long-lasting compounds, such as mercaptans, as exemplified with 3-methyl-3-mercapto-1-hexanol found in human sweat.

1. INTRODUCTION

In the area of flavours, fragrances and essential oils, the analysis of chiral volatile compounds is an important challenge, as the naturally occurring isomer may impart a specific note to the overall odour or aroma [1]. A severe limitation of conventional chiral GC is its poor compatibility with olfactive detection. Even with a baseline resolution, and especially for long-lasting odours such as sulfur compounds, correct olfactometric evaluation requires that an odour-free zone exists between both enantiomers so that the human nose recovers from the perception of the first peak. In the absence of this free zone, the second odorant may be minimised by the panellist due to adaptation or saturation.

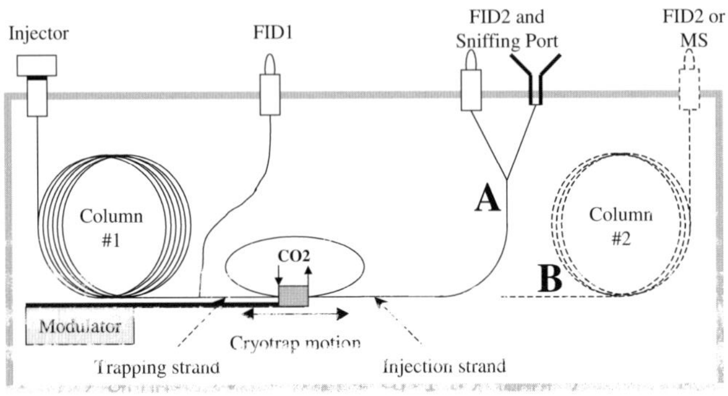

Figure 1. Schematic lay-out of the MDGC with the double-cool-strand interface. **A:** Chiral-GCO design **B:** previous design [3].

The DCSI principle is based on the controlled transfer of peaks between two strands of a capillary column owing to a cryo-trap. The target peak and the following ones are respectively trapped in the injection strand and in the trapping strands of a capillary loop (Figure 1), until previous peaks have been eluted from the second column. Then, the target peak is released and eluted. Meanwhile, following peaks are themselves released from the trapping strand and trapped again in the injection strand. This operation creates a free zone in the chromatogram where the target peak is eluted without interference. In contrast to the technique of comprehensive two-dimensional GC, that requires very rapid detectors (at least 50 Hz [2]), the DCSI interface allows using conventional capillary columns as a second dimension, and 'slow' detectors, such as conventional MS, or a sniff port in association with the human nose. Such a configuration was tested to improve the resolution of a model fragrance mixture of 6 compounds (terpinolene, nonanal, tetrahydrolinalool, Zestover, linalool and phenylethanol, see Table 1 for experimental conditions) [3]. These compounds were strongly co-eluted with a non-polar column in a single-dimension GC-MS (Figure 2). After having trapped this zone using MDGC-DCSI and the same non-polar column as the first dimension, these peaks were fully separated in the second polar dimension (Figure 3). Excellent peak shapes were observed, with short widths and maximised heights. Hence the sensitivity was increased.

A more detailed description can be found in [3] and [4], as well as numerous applications. The present work aims at extending the capabilities of the DCSI to achieve a delay of several seconds or minutes between the release of two enantiomers. The system will be applied to determine the chirality and sensory characteristics of 3-methyl-3-mercapto-1-hexanol enantiomers recently discovered in the human sweat malodour [5].

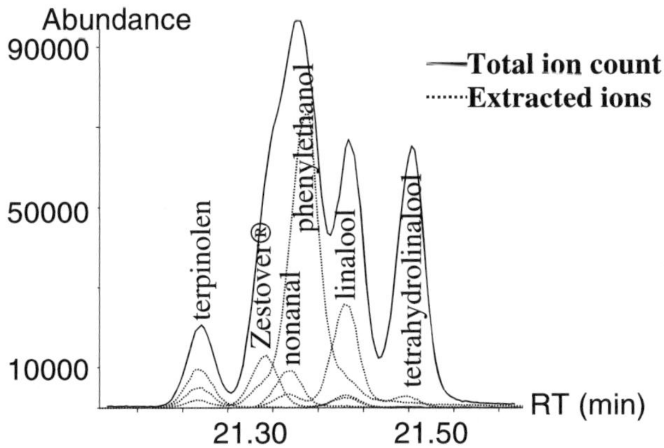

Figure 2. Single dimension GC-MS chromatogram of the model mixture using a non-polar phase.

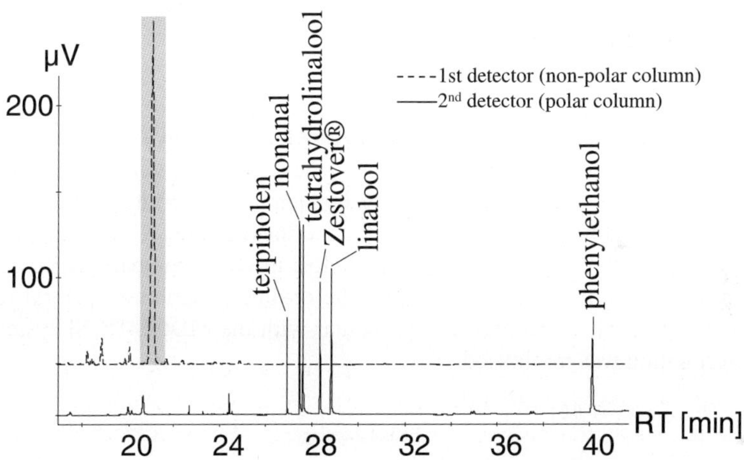

Figure 3. MDGC-MS analysis of the fragrance model using the DCSI. First column: dashed line, second column: solid line. The grey rectangle represents the cryotrapping zone.

2. MATERIALS AND METHODS

All parameters are summarised in Table 1. Detectors used after the 1st and 2nd column were FIDs, except for the characterisation of the fragrance model mixture and for the olfactive characterisation where respectively an ion trap MS (IT-MS) and a homemade sniffing port were connected to the outlet of the 2nd dimension deactivated capillary (see [3] for more details).

Table 1. Analytical conditions used for the different examples.

Product	1st column/2nd column	Oven temperature program
Resolution of a fragrance model mixture	SPB1, 30 m x 0.25 mm x 1.0 µm, Supelco/DB-Wax 30 m x 0.25 mm x 0.25 µm	5 min at 40 °C then 5 °C/min to 220 °C 5 min
Characterisation of 3-methyl-3-mercapto-1-hexanol in sweat malodour	SPB1, 30 m x 0.25 mm x 1.0 µm, Supelco/Megadex DMPβ 10 m x 0.25 mm x 0.25 µm, Mega	5 min at 50 °C then 5 °C/min to 150 °C 1 min then 25 °C/min to 130 °C 15 min then 15 °C/min to 210 °C 5 min
Olfactive characterisation of 3-methyl-3-mercapto-1-hexanol enantiomers	Megadex DMPβ 10 m x 0.25 mm x 0.25 µm, Mega/Deactivated capillary 1 m x 0.25 mm, Supelco	15 min at 130 °C then 15 °C/min to 180 °C 2 min

3. RESULTS AND DISCUSSION

3.1. Enantiomeric ratio of 3-methyl-3-mercapto-1-hexanol in a sweat extract

The potential lability of 3-methyl-3-mercapto-1-hexanol and its only trace presence in the extract (Troccaz *et al.* [5] estimated it to be close to 4 ppb) did not allow classical isolation techniques. In recent experiments, the inertness of the DCSI towards thiols was demonstrated for 2-methyl-3-furanethiol (2.3% lost in DCSI compared with 61.5% in a valve interface, to be published). Therefore the on-line isolation and chiral resolution was the most appropriate technique, in particular with the MDGC-DCSI system, as the risk of deterioration was minimised.

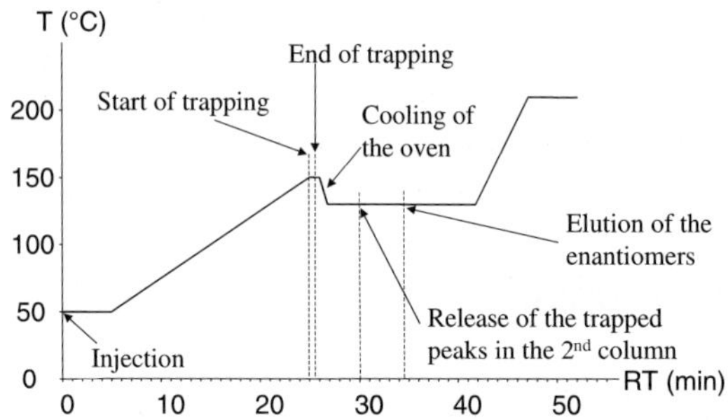

Figure 4. Oven temperature program with main events of the auxiliary malodour extract analysis.

After determining the retention time of the target on the first column by injecting the synthetic racemate, the double-cool-strand interface was programmed to trap it selectively and retain the following peaks (experimental conditions in Table 1).

Before releasing the target peak into the chiral column, the oven temperature was cooled down according to Figure 4 to reach the optimal temperature for its enantiomeric resolution. The peak identities were determined by comparison with the chromatogram of the racemate and by injecting the synthetic (*R*)-3-methyl-3-mercapto-1-hexanol in the MDGC-DCSI (Figure 5). Due to the minute quantity of material present in the extract, peak areas were small and provided only a rough estimate of the (*S*):(*R*) enantiomeric ratio of 3:1.

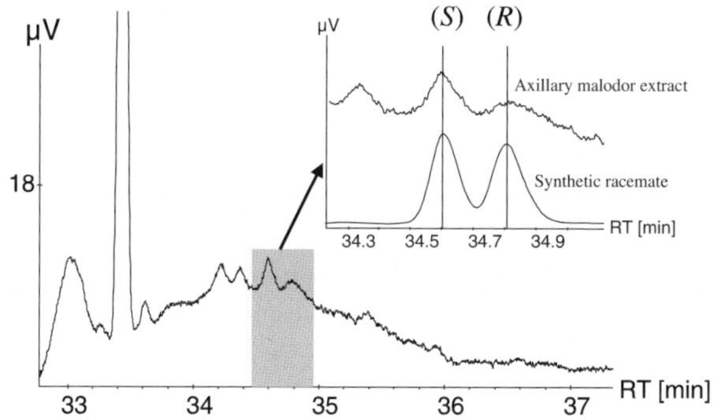

Figure 5. Chromatogram of auxiliary malodour extract and comparison with synthetic racemate.

3.2. Chiral-GC-Olfactometry of 3-methyl-3-mercapto-1-hexanol

As DCSI uses a longitudinally modulated cryogenic system, the minimum difference between the retention time of two target peaks must be at least 1 s [6] which is compatible with conventional enantiomeric resolution. By connecting the DCSI between a first chiral column and a transfer line as a second column, the second enantiomer could be trapped in the DCSI whereas the first one eluted through a sniffing port. The delay before the remobilisation of the second enantiomer would be determined for an olfactometric baseline recovery, i.e. the time required to allow the panellist's nose to recover from the first stimulus. With such a system, a panellist can independently characterise both enantiomers during a single GC run.

To evaluate the odours of 3-mercapto-3-methyl-1-hexanol, a solution of synthetic racemate was injected in the new chiral GCO system (Table 1 for chromatographic conditions). Three chromatographic zones were transferred to the sniff port under the DCSI control: release of the pure *S*-isomer (A, Figure 6), then of a mixture of both enantiomers (B, Figure 6) and then of the pure R-isomer (C, Figure 6). This product was so powerful and lingering that the time interval between the last peak (C, Figure 6) and the preceding one (B, Figure 6) had to be about 4 min long to reach the olfactive baseline for independent evaluation of the peaks. The odour of the *S*-enantiomer (4.34 min) was described as sweat, onion and animal, whereas the R-enantiomer (9.14 min) was more fruity, grapefruit and sulfury. These results agreed with the enantiomeric ratio of *S*:*R* being 3:1. The sweat malodour extract was described as sweat, sage-like [5] thus was mainly imparted by (*S*)-3-methyl-3-mercapto-1-hexanol.

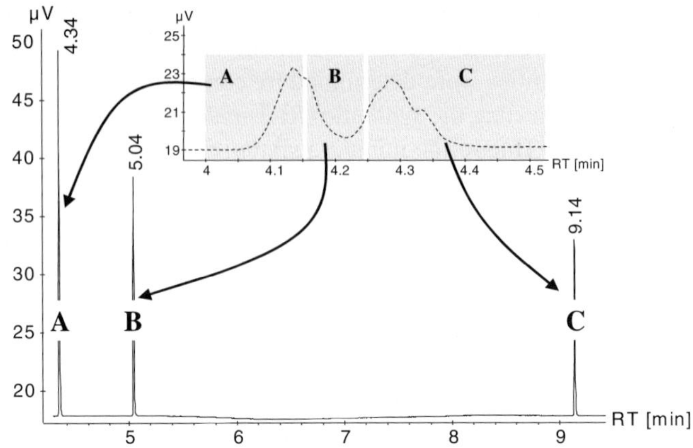

Figure 6. Chromatogram of the enantiomers of 3-mercapto-3-methyl-1-hexanol eluting from the chiral column (dashed line) and the transfer line (solid line). The grey rectangles represent the three trapping zones. (A) corresponds to 2/3 of the first peak (*S*-isomer), (B) to the racemate and (C) to 2/3 of the second peak (*R*-isomer).

4. CONCLUSION

The ability to simultaneously and independently delay a target peak and all following ones during a flexible lapse of time induced two new applications for MDGC devices: i) the modification of the chromatographic conditions during the run, first dimension separation as well as second dimension separation can then be optimally achieved without the need for a second oven, ii) the free setting of the time interval between the release of two successive peaks from the cryotrapping zones. The efficiency of MDGC-DCSI was demonstrated for the analytical and olfactometric characterisation of 3-methyl-3-mercapto-1-hexanol in sweat auxiliary malodour. The enantiomeric ratio was found to be about *S*:*R* = 3:1 and it was demonstrated that the major contributor for the auxiliary sweat malodour was (*S*)-3-methyl-3-mercapto-1-hexanol.

References

1. S. Arctander and M.G. Arctander (eds.), Perfume and flavor chemicals, Las Vegas, Nevada, USA (1969).
2. J. Dallüge, J. Beens and U.A.Th. Brinkman, J. Chromatogr. A, 1000 (2003) 69.
3. F. Begnaud and A. Chaintreau, J. Chromatogr. A, 1071 (1-2) (2005) 13.
4. F. Begnaud and A. Chaintreau, New multidimensional gas chromatography apparatus and analyte transfer procedure using a multiple-cool strand interface, PCT/1132004/001589 (2004).
5. M. Troccaz, C. Starkenmann, Y. Niclass, M. Van de Waal and A. Clark, Chem. Biodivers., 1 (2004) 1022.
6. R.M. Kinghorn, P.J. Marriott and P.A. Dawes, J. High Resolut. Chromatogr., 23 (2000) 245.

W.L.P. Bredie and M.A. Petersen (Editors)
Flavour Science: Recent Advances and Trends

The artificial throat: a new device to simulate swallowing and *in vivo* aroma release in the throat. The effect of emulsion properties on release in relation to sensory intensity

Alexandra E.M. Boelrijk[a], Koen G.C. Weel[a], Jack J. Burger[b], Maykel Verschueren[a], Harry Gruppen[c], Alphons G.J. Voragen[c] and Gerrit Smit[a,c]

[a]NIZO food research, Flavour Division, P.O. Box 20, 6710 BA Ede, The Netherlands; [b]Quest International Nederland BV, Naarden, The Netherlands; [c]Department of Agrotechnology and Food Sciences, Wageningen University, Wageningen, The Netherlands

ABSTRACT

The effects of oil content and droplet size distributions of dilute oil-in-water emulsions on release of 4 esters with different hydrophobicities were studied under *in vivo*, static-headspace, and artificial throat conditions. The effect of oil content on orthonasal and retronasal perceived intensity of ethyl hexanoate was studied using a sensory panel. The results indicate that the effect of droplet size distribution on aroma release strongly depends on the hydrophobicity of the aroma compound, the emulsion characteristics and the dynamics of the measurement. The lowest oil content that had an effect on aroma release was determined for all systems.

1. INTRODUCTION

For liquid products, it has been shown that swallowing determines the *in vivo* aroma release rather than the preceding oral processing. After swallowing, the majority of the sample disappears into the oesophagus, but a thin layer of the liquid sample remains on the surface of the pharynx [1]. During the exhalation after swallowing, a steep gradient in aroma concentration exists between the thin liquid layer on the surface of the pharynx and the exhaled air that passes over this surface. An artificial throat was developed with the aim of mimicking aroma release from this thin liquid layer in the throat [2]. In many food systems lipids are present as oil droplets dispersed in an aqueous phase. Droplet size has been reported to increase, to decrease, as well as to have no effect on aroma release for different systems. In the present study, the lowest effective oil content and the effect of oil droplet size distribution on release of a set of ester compounds have been studied under *in vivo*, static headspace, and artificial throat conditions. The effect of oil content on orthonasal and retronasal perceived intensity of ethyl hexanoate was studied using a sensory panel. Because changes in oil content changes viscosity, the effect of changes in viscosity on release was studied too.

2. MATERIAL AND METHODS

Emulsion preparation: Medium chain triglycerides (MCT) oil was slowly added to a final concentration of 5% to an aqueous solution of 5% gum arabic while stirring in a Cyclotron mixer. The mixing continued for 5 min, yielding a crude emulsion. A Rannie 250 homogeniser processed this crude emulsion at several regimes, see Figure 1 for characteristics. These three emulsions were mixed overnight with aqueous solutions of ester compounds, water, and gum arabic solution, to produce a range of emulsions with different oil contents (ranging from 0.01 to 2.0% lipid content).
Viscosity measurements: An Ubbelohde viscometer was used to measure the kinematic viscosity of the samples at 20 °C.
MS-Nose: Aroma compounds in the air releasing from the artificial throat, from the breath of panellists (P1 and P2), or from the calibration [2] were monitored by on-line sampling by APCI-MS. A strict consumption protocol [3] was used.
Artificial throat: *In vivo* aroma release was simulated by the artificial throat. For a schematic overview see Figure 2.

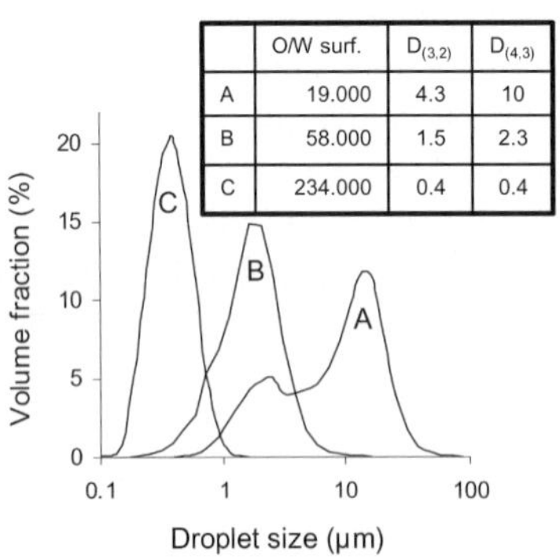

Figure 1. Characteristics of emulsions used.

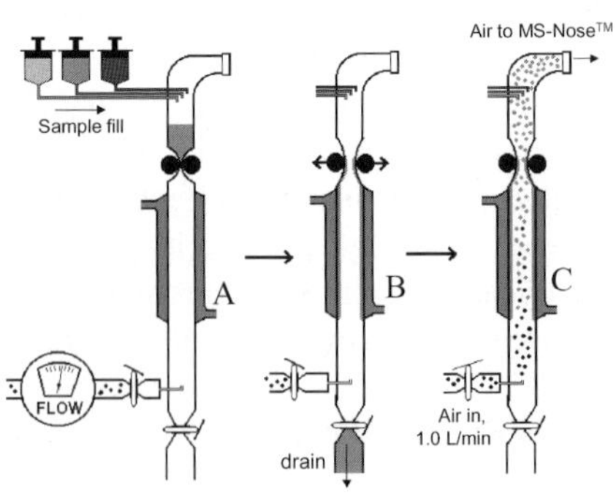

Figure 2. Schematic of artificial throat.

Measurements were done in 2 replicates and recorded for 1 min. The equilibrium headspace aroma concentrations of esters were determined by gas chromatography.
Sensory evaluation: The intensity of ethyl hexanoate was measured for six emulsions by a panel consisting of 7 trained judges.

3. RESULTS

The effect of viscosity on aroma release *in vivo* and in the artificial throat was studied by means of a range of aroma solutions with increasing carrageenan concentrations. An increase in viscosity resulted in an increase in released amount of ethyl butanoate in the artificial throat (Figure 3). The other ester compounds showed similar effects (results not shown). In contrast to the artificial throat results, *in vivo* measurements showed no effect of viscosity on the amount of aroma released (P1, P2 Figures 3 and 4). This can be caused by differences in the swallowing mechanism and by aroma dilution with saliva.

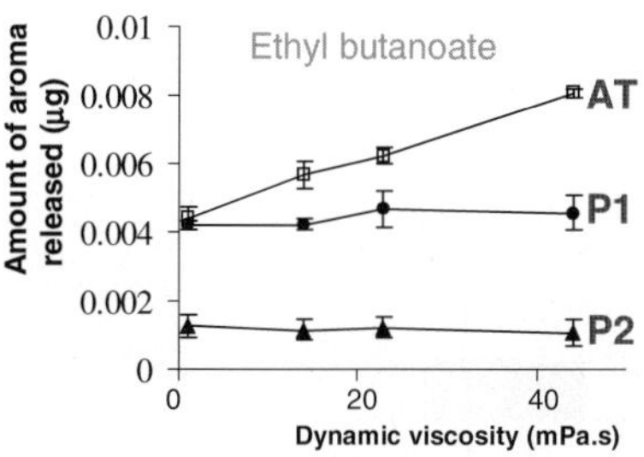

Figure 3. Artificial throat (AT) and *in vivo* results (P1 and P2).

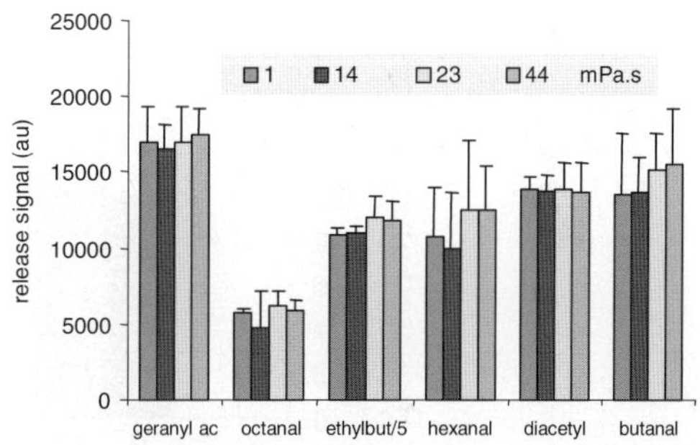

Figure 4. *In vivo* results for aromas depending on viscosity of liquid product.

Swallowing in the artificial throat is driven by gravity, while *in vivo* the liquid is forced downwards by pharyngeal peristalsis. It seems reasonable to assume that the film formed *in vivo* will be different. Future versions of the artificial throat should take peristalsis into account.

Table 1. Lowest contents of medium chain triglycerides from emulsion that differ significantly (α=0.05) from 0% oil emulsions.

	Ethyl butanoate	Ethyl hexanoate	Geranyl acetate
Panellist 1	(>5%)[a]	0.5%	0.1%
Panellist 2	(>5%)[a]	1%	0.05%
HSGC	0.5%	0.01%	0.01%
Artificial throat	0.5%	0.2%	0.05%
Panel orthonasal	n.m.	0.2%	n.m.
Panel retronasal	n.m.	0.5%	n.m.

[a]A possible significant effect at higher oil content. [b]n.m.: not measured.

If one studies the lowest MCT content from emulsions that has a significant effect on aroma release, lower effective oil contents are 1) found for esters with a higher hydrophobicity, 2) found for techniques based on partitioning, instead of release dynamics from a thin liquid. In case of ethyl hexanoate, the effective oil content of the artificial throat (0.2%), of the *in vivo* aroma release (0.5-1%), and of the retronasal perception (1%) are all within the same range, indicating the relevance of artificial throat and *in vivo* aroma release measurements for aroma perception. The effect of droplet size distribution on the release of ethyl butanoate, ethyl hexanoate, and geranyl acetate, was tested at 2 oil contents (0.1% and 1%), The HSGC and artificial throat measurements did not yield a significant effect of droplet size distribution. A significant effect of droplet size distribution was found for geranyl acetate at an oil content of 0.1% under *in vivo* conditions. The release from the crude emulsion (A) was lower than from the emulsions homogenised at low (B) and high energy (C). This might be due to a difference in diffusion time scale within the droplets which is determined by their size.

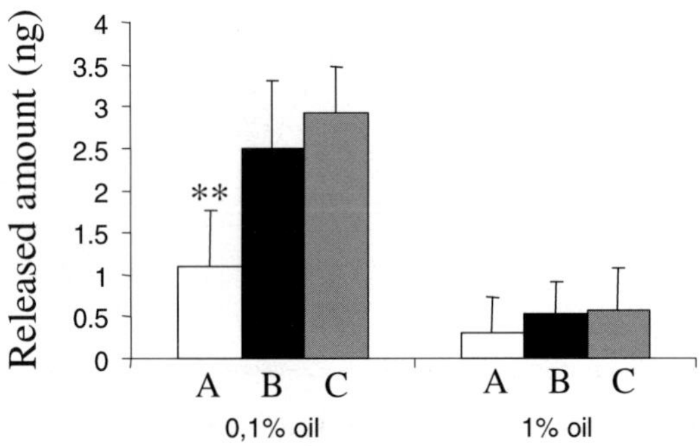

Figure 5. Effect of droplet size on *in vivo* release of geranyl acetate. A, B and C: see Figure 2.

4. CONCLUSION

In contrast to the artificial throat results, *in vivo* measurements showed no effect of viscosity on the amount of aroma released. This can be due to differences in swallowing mechanism. The effect of oil on aroma release is smaller under *in vivo* conditions than under static conditions, and smaller for retronasal perception than orthonasal perception. The effect of oil on aroma release in the artificial throat correlated well with the effect under *in vivo* conditions, providing support for the 'thin layer' hypothesis, and thus showing that the artificial throat is a suitable device to simulate *in vivo* aroma release.

References

1. A. Buettner, A. Beer, C. Hannig and M. Settles, Chem. Senses, 26 (9) (2001) 1211.
2. K.G.C. Weel, A.E.M. Boelrijk, J.J. Burger, M. Verschueren, H. Gruppen, A.G.J. Voragen and G. Smit, J. Agric. Food Chem., 52 (21) (2004) 6564.
3. K.G.C. Weel, A.E.M. Boelrijk, J.J. Burger, H. Gruppen, A.G.J. Voragen and G. Smit, J. Food Sci., 68 (2003) 1123.

W.L.P. Bredie and M.A. Petersen (Editors)
Flavour Science: Recent Advances and Trends

Thermal flavour generation: insights from mass spectrometric studies

David Cook, Guy Channell, Maaruf Abd Ghani and Andrew Taylor

Division of Food Sciences, University of Nottingham, Sutton Bonington Campus, Loughborough, Leics. LE12 5RD, UK

ABSTRACT

Fundamental studies of the reaction between glucose and L-asparagine demonstrated effects of moisture and of temperature on the generation of Strecker degradation products, exemplified here by acrylamide. Increasing the humidity of the reaction environment significantly enhanced acrylamide production. Irrespective of humidity, acrylamide was not observed until the reaction temperature was ramped from 100 to 130 °C or above. Studies using a more complex food system (wheat bread) have also demonstrated that humidifying the environment during cooking can change the resulting concentrations of key flavour compounds such as pyrazines and Strecker aldehydes.
The generation of acrylamide and of 2,5-dimethylpyrazine in a hydrated potato flake system were followed at 180 and 200 °C using an on-line gas-phase MS-MS technique. Each of these compounds can be formed subsequent to Strecker-type reactions between amino acids and a common pool of carbonylic intermediates. Production of the pyrazine preceded that of acrylamide, perhaps indicating that it began at a lower temperature.

1. INTRODUCTION

Thermally generated flavours are produced in a complex series of reactions which are under kinetic control (influenced by factors such as precursor concentrations, temperature, time, moisture, pH etc.). Here, we report results from mass spectrometric studies designed to monitor the evolution of flavour volatiles and their precursors under controlled conditions of temperature and humidity. The studies embody a 'bottom to top' approach of studying chemistry at a fundamental level using a film reactor [1], through trials using model food systems, to measurements of flavour volatiles in a real product (wheat bread).

2. FUNDAMENTAL STUDIES ON THE GLUCOSE-ASPARAGINE SYSTEM

Asparagine and glucose were reacted in a thin film formed in an on-line micro reactor [1]. A continuous stream of carrier gas of controlled humidity was configured to flow over the film so that effluent from the reactor was introduced directly to the Atmospheric Pressure Chemical Ionisation source of an Ion Trap Mass Spectrometer. Controlled temperature change within the reaction film under a stepped regime (60 °C → 100 °C → 130 °C → 170 °C at 1 °C/s) demonstrated the onset of acrylamide production at 130 °C with accelerated formation at 170 °C (dotted grey trace in Figure 1; constant humidity). When, at each temperature plateau, environment humidity was cycled through intermediate-zero-high moisture levels (shaded bars in Figure 1), acryl-amide production was shown to be enhanced by high moisture levels and suppressed in the absence of water (black trace in Figure 1). This was particularly notable at 170 °C where the signal for acrylamide dropped in intensity in the absence of humidity, but rose sharply again once moisture was re-introduced (dark humidity bar, 170 °C).

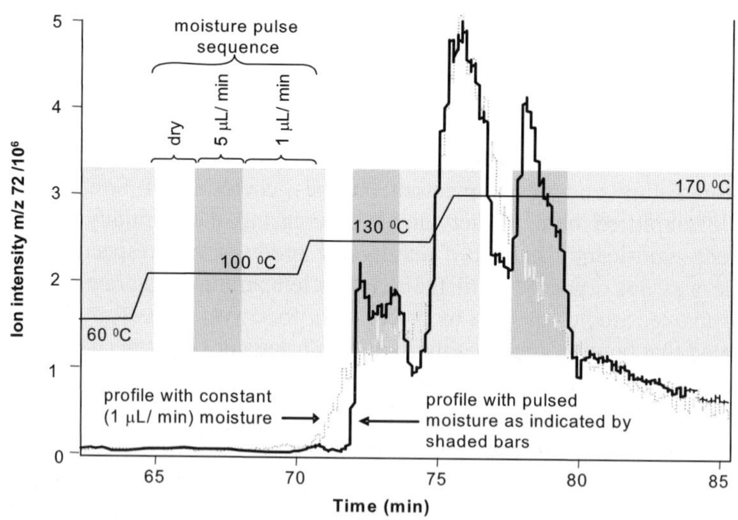

Figure 1. Effect of moisture on acrylamide formation using a 50 mM glucose/50 mM asparagine reaction system in an 'on-line Film Reactor' [1].

In parallel experiments, where constant humidity was maintained throughout the same temperature regime, acrylamide production (quantified by the area under ion trace *m/z* 72) was 2.5 times as high at the intermediate moisture level (1 μl/min water) as in the dry condition (mean of 2 replicate measurements in each condition). These results demonstrate how changes in humidity can alter carbon flux within the Maillard reaction. Tentative assignment of other molecular ions from the glucose-asparagine reaction showed a range of sugar breakdown products, amino acid/sugar adducts, transamination products, and the Strecker aldehyde, 3-oxopropanamide to be produced.

3. INSIGHTS INTO THE GENERATION OF MAILLARD FLAVOUR PRECURSORS FROM ON-LINE MONITORING OF ACRYLAMIDE

The relationship between acrylamide formation and that of key flavour impact compounds can be understood with reference to a kinetic scheme of the Maillard reaction developed by Bronek Wedzicha and co-workers (Figure 2).

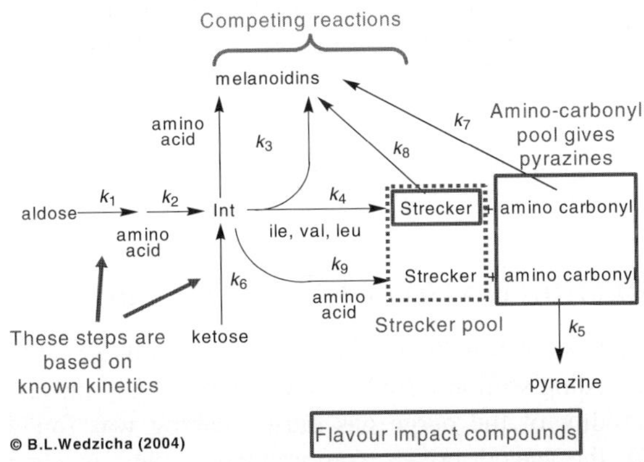

Figure 2. A kinetic scheme of the Maillard reaction indicating the key rate determining steps ('k_n') in the formation of flavour impact compounds. Reproduced with the kind permission of Bronek Wedzicha, University of Leeds, UK.

In Figure 2 'Int' represents a pool of carbonylic Strecker reagents. Asparagine competes with other free amino acids for this pool of reactive intermediates and a small fraction of asparagine is ultimately converted to acrylamide. Modelling studies suggested that, at 180 °C, these reactions were rapid and occurred in proportion to the relative amounts of each amino acid available [2]. Hence, acrylamide production might act as a 'marker' of the availability of the key flavour precursors ('Int'). This hypothesis was investigated by following the generation of acrylamide and of a pyrazine flavour compound using an on-line gas-phase MS-MS technique whilst cooking hydrated potato flakes (1 part flakes : 1.3 parts water; 450 mg). These experiments were conducted using a sample cell housed in a GC oven and purged with heated/humidified N_2 gas, as described previously [3]. Acrylamide was monitored using the characteristic MS-MS transition m/z 72→55 and 2,5-dimethylpyrazine was followed at m/z 109→82 (Figure 3). Some commonality was noted between the temporal production of acrylamide and that of 2,5-dimethylpyrazine. However, it would be simplistic to overstate the similarity. At each oven temperature, the generation of pyrazine preceded that of acrylamide by 1-2 min, possibly indicating that it commenced at a lower temperature. The decay rate from maximum of each species also differed, as might be expected from consideration of factors such as thermal stability (acrylamide is known to degrade at these temperatures) and the depletion of reactants specific to the formation of each compound (e.g. asparagine for acrylamide, aminoacetone for 2,5-dimethylpyrazine).

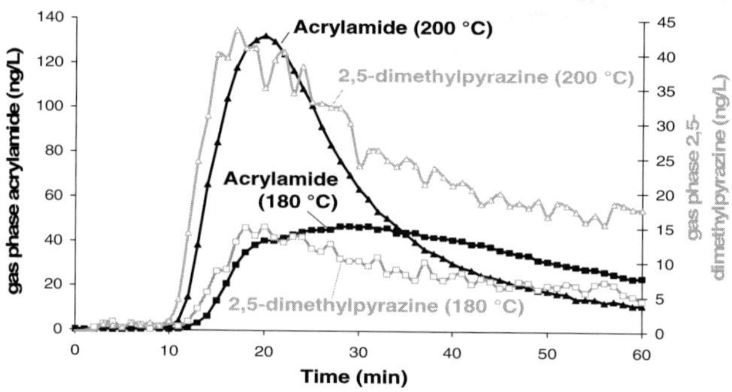

Figure 3. On-line monitoring of acrylamide and 2,5-dimethylpyrazine release from hydrated potato flake during cooking (curves are the mean of 3 replicate experiments).

4. EFFECT OF HUMIDIFICATION ON VOLATILE PRODUCTION IN BREAD

Aroma volatiles of mini wheat breads baked (120 °C for 30 min) in the apparatus described in the preceding section were extracted in to CH_2Cl_2 and analysed by GC-MS. Increasing the humidity of the purge gas during baking was found to influence the volatile profile of the baked breads (representative data is shown in Figure 4). Methylpyrazine, 2-acetylpyrazine and the Strecker aldehyde 3-methylbutanal all increased in concentration over the range of humidity investigated. However, this trend was not observed for all aroma volatiles, e.g. 2-furanmethanol and heptanal exhibited a small downwards trend in concentration as humidity was increased (data not shown).

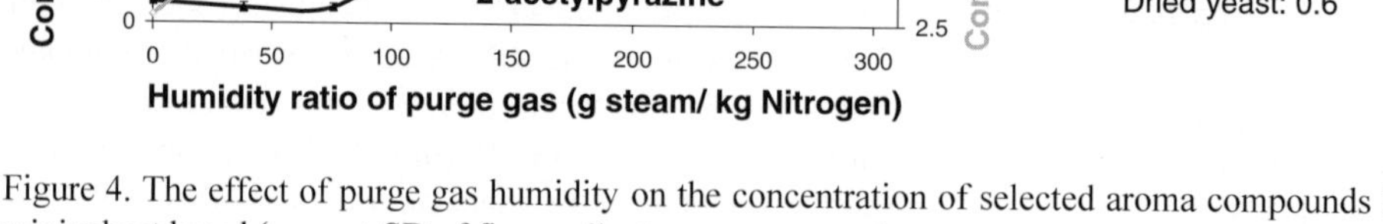

Figure 4. The effect of purge gas humidity on the concentration of selected aroma compounds in mini wheat bread (mean ± SD of five replicate measurements).

References

1. D.K. Weerasinghe and M.K. Sucan (eds.), Process and reaction flavors, Washington, DC, USA (2005) 181.
2. D.S. Mottram and M. Friedman (eds.), Chemistry and safety of acrylamide in food, New York, USA (2005) 235.
3. D.S. Mottram and M. Friedman (eds.), Chemistry and safety of acrylamide in food, New York, USA (2005) 303.

W.L.P. Bredie and M.A. Petersen (Editors)
Flavour Science: Recent Advances and Trends

Innovative mass spectrometric tools for the structural elucidation of flavour compounds

Ulrich Krings, Holger Zorn and Ralf G. Berger

Zentrum Angewandte Chemie, Institut für Lebensmittelchemie, Wunstorfer Straße 14, 30453 Hannover, Germany

ABSTRACT

The substantial contribution of combined MS approaches - accurate mass detection, simultaneous EI and EI/CI and high energy collision - to the structural elucidation and confirmation of novel flavour compounds is demonstrated for pyruvate decarboxylase derived hydroxy ketones and sesquiterpene bioconversion products.

1. INTRODUCTION

High resolution gas chromatography (HRGC) coupled to low resolution mass spectrometers is an indispensable tool in flavour analysis. Extensive spectral data libraries are available, and high performance search machines enable fast and reliable confirmation of known flavour compounds. However, structural identification or even the derivation of a presumed structure from EI mass spectra of an unknown compound is rarely achieved. Sophisticated NMR techniques, on the other hand, require amounts of pure substances larger than they are usually available. The combination of high resolution gas chromatography with high resolution mass spectrometry (HRMS) represents an alternative. Accurate mass determination together with simultaneous EI and CI ionisation as well as linked scan options contribute substantially to the structural elucidation of novel flavour compounds [1,2].

2. MATERIALS AND METHODS

HRGC-HRMS device consisted of: GC 6890, Agilent; DB-5 capillary column; column splitting (1:1); heated transfer capillaries to in-axis EI and CI sources; a double focusing sector field MS-device (AMD QuAS^{3}AR, Harpsted, Germany) equipped with a high energy collision cell working with air as collision gas. Accurate mass determination was achieved at a dynamic working resolution of R=2000 (10% valley), a scan rate of 0.1-

1.0 scans per mass decade and integrated calibration with high boiling perfluoro kerosine (PFK) prior to and lock/shift mass calibration during analysis.

3. RESULTS

3.1. Simultaneous generation and rapid alternating recording of EI and EI/CI ions

(*R,S*)-(*E*)-3-Hydroxy-5,9-dimethyldeca-4,8-dien-2-one, a pyruvate decarboxylase (PDC) catalysed condensation product of pyruvate and geranial showed no molecular ion (M^+) in the EI mode. The base peak was found at *m/z* 69 and the dominating ions present in the *m/z* range > 100 were *m/z* 153 and 151, respectively. Figure 1 demonstrates the concerted action of the two independent working ion sources in the alternating recording mode.

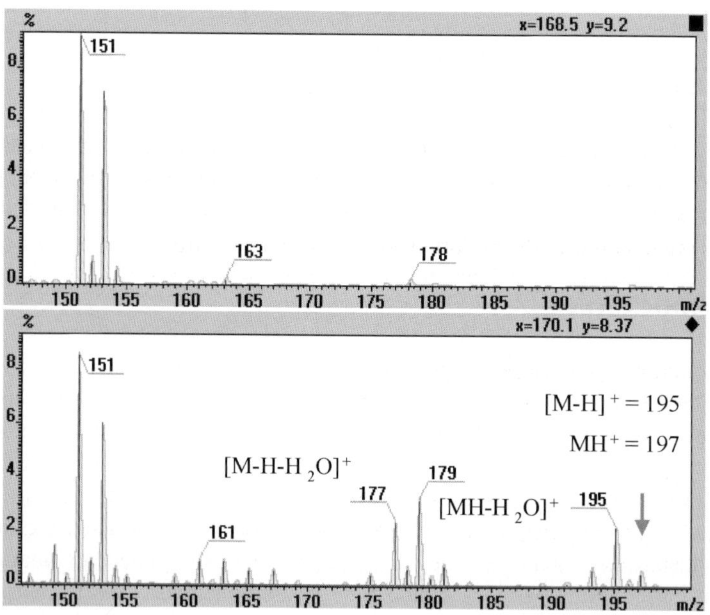

Figure 1. Alternating EI (top) and EI/CI mass spectra of (*R,S*)-(*E*)-3-hydroxy-5,9-dimethyldeca-4,8-dien-2-one (*m/z* > 146).

3.2. Accurate mass determination in GC-MS analysis

Feeding of the fungus *Chaetomium globosum* with the highly unsaturated acyclic sesquiterpene farnesene yielded a complex mixture of biotransformation and autoxidation products. Exemplarily for hydroxylated biotransformation products the accurate mass and, thus the elemental composition of significant fragment ions (*m/z* > 100) of 6-hydroxy farnesene was determined (Table 1).

Table 1. Accurate mass detection of fragment ions of 6-hydroxy farnesene.

m/z (rel. int.) (%)	Determined mass (u)	Elemental composition	Calculated mass (u) (± ppm)
220 (0.3)	n.e. (220.1819)	$C_{15}H_{24}O$	220.1827 (-3.6)
205 (4.3	205.1587	$C_{14}H_{21}O$	205.1592 (- 2.4)
202 (9.1)	202.1710	$C_{15}H_{22}$	202.1722 (-5.9)
187 (4.7)	187.1473	$C_{14}H_{19}$	187.1487 (-7.5)
162 (28)	162.1425	$C_{12}H_{18}$	162.1409 (+9.9)
159 (24)	159.1177	$C_{12}H_{15}$	159.1174 (+1.9)
137 (14)	137.0989	$C_9H_{13}O$	137.0966 (+16.8)
133 (13)	133.1032	$C_{10}H_{13}$	133.1017 (+11.3)
119 (18)	119.0877	C_9H_{11}	119.0861 (+13.4)
105 (27)	105.0726	C_8H_9	105.0704 (+17.1)

n.e.: not evaluated in the multichannel mode (determined by peak matching).

3.3. High energy collision induced dissociation - linked scan-options

The EI mass spectrum of 6-hydroxy farnesene (MW = 220 g/mol) possessed an intensive fragment ion at *m/z* 162. The accurate mass determination revealed an elemental composition of $C_{12}H_{18}$ (formal loss of C_3H_6O). Linked scan experiments with precursor ions at *m/z* 220, 205, 202, 187, and 177 revealed that the *m/z* 162 ion was derived from the molecular ion and not from a fragment ion. As a result of these linked scan experiments on 6-hydroxy farnesene, the loss of neutral acetone was claimed. The fragmentation can be explained by an oxygen shift from C6 to C2 through an intermediate six-membered ring.

4. DISCUSSION AND CONCLUSION

4.1. Impact of EI/CI ions

The most important piece of information that a mass spectrum can provide is the molecular weight of the compound. The electron ionisation (EI) mass spectra of many compounds do not contain abundant, sometimes not even detectable M^+ ions [3]. To obtain both structural information derived from molecule fragmentation in the EI mode and unambiguous determination of the molecular weight in the CI mode at least two GC runs are required. Fast alternating recording of two independent working ion sources within one GC-peak provides the information of both ionisation modes and assures anytime a distinct allocation of measured data. This gains importance especially for densely occupied GC chromatograms with badly separated or even co-eluting compounds. It is apparent from the EI spectrum (Figure 1) of (*R,S*)-(*E*)-3-hydroxy-5,9-dimethyldeca-4,8-dien-2-one that ions at *m/z* 153 or *m/z* 151 do not represent the molecular ion of the acyloin. A less abundant signal at *m/z* 178 was recorded which by mistake could be postulated as the molecular ion. But the methane CI mass spectrum of the acyloin yielded abundant MH^+ (*m/z* 197) and $M\text{-}H^+$ (*m/z* 195) ions together with the corresponding fragment ions caused by the favoured loss of water. Because of the fast switching between EI and superimposed EI/CI, contributing CI ions can be distinguished from EI ions and undoubtedly define the molecular weight of the acyloin. The switching technique was also successively applied to discover: i) further acyloins in yeast fermented foods, ii) lactones in cocoa flavour and iii) hydroxy terpenoids in transformation media of terpene fed fungi.

4.2. Accurate mass determination in flavour analysis

High resolution mass spectrometry coupled to high resolution gas chromatography adds indispensable contributions to structure elucidation in flavour analysis. Although complex flavours consist of a large number of compounds, different in functionality and size, the multi-channel recording in the fast V/E mode allows to cover of the entire mass range (1.2 mass decades) for accurate mass detection. The elemental composition of every ion (M^+ or fragment ions) is amenable, and the entire structure of molecules might be unveiled. For the sake of sensitivity and the amount of data to be processed a working resolution of R=2000 (10% valley) was chosen for accurate mass detection within an entire GC-run. As a result the relative mass deviation was found to exceed the 5 ppm range (required for accepted elemental composition) in this mode of operation.

4.3. High energy collision induced dissociation - linked scan-options

As a result of linked scan experiments with (*R,S*)-(*E*)-3-hydroxy-5,9-dimethyldeca-4,8-dien-2-one, the loss of acetone was claimed. The fragmentation can be explained by an oxygen shift from C6 to C2 through an intermediate six-membered ring. However, such a complex fragmentation mechanism is not easily derived without the confirming information gained from linked scan experiments. To explain the origin of fragment ions supplements ideally the accurate mass determination of fragment ions in the structural elucidation of unknown compounds.

References

1. H. Zorn, M. Schüler and R.G. Berger, Biocatal. Biotransfor., 21 (2003) 341.
2. A. Rueda, E. Zubía, M.J. Ortega and J. Salvá, J. Nat. Prod., 64 (2001) 401.
3. B. Munson, Int. J. Mass Spectrom., 200 (2000) 243.

W.L.P. Bredie and M.A. Petersen (Editors)
Flavour Science: Recent Advances and Trends

Optimisation of stir bar sorptive extraction (SBSE) for flavour analysis

Carlos Ibáñez and Josep Solá

Lucta S.A., Flavours, Fragrances & Feed Additives, Ctra. de Masnou a Granollers km 12,400, 08170 Montornés del Vallés, Barcelona, Spain

ABSTRACT

A model mixture consisting of twenty common flavouring substances having different chemical functional groups and solubility and volatility properties was used to optimise extraction and analysis from aqueous solutions with the stir bar sorptive extraction (SBSE) method. Stir bar extraction, desorption, and injection of the volatile components extracted were optimised for the most significant operational parameters.
The optimised SBSE method was compared with solid phase microextraction (SPME) and simultaneous distillation-extraction (SDE) for the analysis of commercial samples of flavoured foods.

1. INTRODUCTION

The stir bar sorptive extraction (SBSE) is a recently introduced technique for sample extraction. Aqueous analyte solution extraction is based on the partition of the solute between the aqueous phase and the bulk of a polymeric phase covering a stir bar. After exposure to the analyte solution, the stir bar is removed, washed and placed in a thermal desorption unit. Volatile components are thermally desorbed and analysed by gas chromatography [1].

2. MATERIALS AND METHODS

2.1. Materials

The model mixture was prepared weighing equal amounts of each flavour chemical obtained from various suppliers and tested for their purity (+98%). Acetaldehyde, isooctane, diacetyl, ethyl butyrate, limonene, phenylethyl isopropyl ether, linalool, propylene glycol, butyric acid, isoborneol, citral, benzyl acetate, alpha-ionone, phenylethylalcohol, furaneol, isopentyl salicylate, methyl anthranilate, gamma-undecalactone, coumarin, vanillin and benzyl cinnamate were selected. Elix/Simplicity (Millipore) water was spiked with the model mixture and extracted with SBSE stir bars in an SBS multipoint magnetic stirrer or an Ika stirrer. Gerstel SBSE glass/metal stir bars were 10 mm long and coated with a 0.5 mm layer of polydimethylsiloxane (PDMS).

2.2. Gas chromatography set-up

All chromatographic analysis were performed on an Agilent GC 6890 coupled with a Mass Selective Detector Agilent MSD 5973, equipped with a Gerstel MPS2 Multipurpose Automatic Sampler, a Gerstel thermal desorption unit (TDU) and a Gerstel cooled injection system (CIS4). The fused silica capillary column was a Supelcowax10, 30 m x 0.25 mm i.d. x 0.25 μm from Supelco. Liquid Nitrogen for cryogenic temperature control was kept in a 50 l pressurised Dewar tank from Messer (Germany).

2.3. Gas chromatography methods

Chromatographic conditions were the following: carrier gas helium, constant flow mode. Retention time locked for limonene at 6.9 min. Oven program: 60 °C (0 min), 4 °C/min to 80 °C (0 min), 10 °C/min to 230 °C (25 min). MSD parameters: Quadrupole temperature 150 °C, ion source: 230 °C, transfer line: 250 °C. Scan mode from 35 amu to 300 amu.

3. RESULTS

Chromatographic conditions were previously studied injecting 1 μl of the model mixture, diluted in ethanol 1/10, in a conventional split/splitless injector with a 1:20 split. The chromatogram obtained was taken as reference to compare for future differences due to changes with desorption system parameters.

The first step was the optimisation of the sample desorption and injection conditions in absence of the SBSE stir bar. This process was carried out by injecting 1 μl of the model mixture, diluted 1/10 in ethanol, in a plug of glass wool, inside the glass insert of the TDU, where the SBSE stir bars would be placed for analysis. Initial conditions were common values taken from examples [2,3] of scientific publications. Starting from these conditions the main parameters concerning desorption and injection were evaluated: desorption flow (1.3 ml/min to 30 ml/min), desorption type (split or splitless), cryogenic collection temperature (-30 °C to -150 °C) and injection type (split or splitless) were studied (see Table 1).

For the next optimisation step the model mixture was diluted 1/500000 in water. A volume of 50 ml of the final diluted sample was stirred at 1400 rpm with an SBSE stir bar for half an hour. Once extracted, rinsed with water and cleaned, the stir bar was left in a glass insert of the MPS2 automatic injector and injected.

When split desorption and splitless injection was applied bad peak shapes were observed for very volatile peaks. Therefore, it was decided to use splitless desorption and split injection for the rest of the optimisation procedure.

When re-injecting the last stir bar used, some memory effects were observed for high volatility compounds, possibly due to low desorption flows. Desorption flow was then studied. High flows increased the areas of the low volatility compounds but decreased the areas of the high volatility compounds (Table 2). Memory effects decreased with this flow and all peak shapes were acceptable.

Table 1. Changes in peak areas of compounds of different volatility. (–): significant area decrease relative to initial conditions; (+): significant area increase relative to initial conditions.

	TDU desorption flow		Desorption/injection type		Cryogenic temperature	
Volatility	1.3 ml/min	30 ml/min	Splitless/split	Split/splitless	-30 °C	-150 °C
High	–	+	+	–	–	+
Medium	–	+	+	+	+	+
Low	–	+	+	–	+	+

As expected, acetaldehyde, diacetyl, butyric acid and propylene glycol (low logP octanol/water values) were badly recovered. Phenylethylalcohol and vanillin also gave recovery problems although they have slightly higher logP. Extraction conditions were then studied. Volume of sample (5 ml, dilution 1/50000 and 50 ml, dilution 1/500000) and extraction time (30 min to 150 min) were tested.

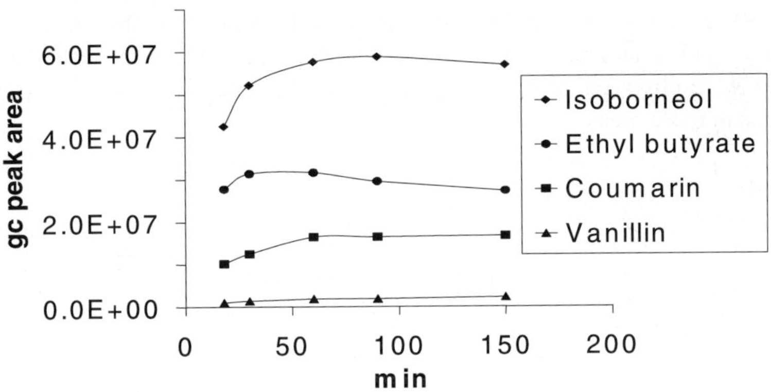

Figure 1. SBSE recovery versus sampling time of 5 ml sample.

A maximum of recovery was not achieved for any of the components of the 50 ml samples even at 60 min of extraction time. In contrast, 5 ml samples gave a maximum of recovery for most of the components between 60 min and 90 min (see Figure 1) except for ethyl butyrate and vanillin (30 min and 150 min), and gave in general, areas higher than 50 ml samples.

TDU desorption rate, cryogenic collection temperature and CIS4 injection time at maximum temperature (1 min and 3 min) were also studied (Table 2).

Table 2. Changes in peak areas of compounds of different volatility. (–): significant area decrease relative to initial conditions; (+): significant area increase relative to initial conditions; (++): very significant area increase relative to initial conditions.

Volatility	TDU desorption flow 23 ml/min	50 ml/min	70 ml/min	CIS4 Cryogenic temperature -70 °C	-110 °C	-120 °C	TDU desorption rate 30 °C/ min	70 °C/ min	90 °C/ min
High	+	+	–	–	+	–	–	+	++
Medium	+	+	+	–	+	+	–	+	+
Low	–	+	++	–	+	+	–	+	+

Optimum final conditions were taken as follows:
Extraction: 5 ml sample, dilution 1/50000, extraction time 90 min at 1400 rpm.
Desorption: splitless, flow 50 ml/min, 30 °C to 250 °C at 90 °C/min, then 10 min.
Injection: split 1/20, -110 °C to 250 °C at 12 °C/s, then 3 min.
Method repeatability was tested at the optimum conditions. Using the same stir bar most of the components exhibited coefficients of variation (CV) between 0.14% and 1.8%. With different stir bars most of the components had CVs between 0.7% and 2.4%.

4. CONCLUSION

A method for extracting flavours from water solutions using SBSE stir bars and subsequent analysis by GC-MS was optimised examining the main extraction, desorption and injection parameters. Method repeatability was calculated. Main problems were observed with very volatile and/or soluble compounds. More work has to be done in those areas.

References

1. E. Baltussen, C.A. Cramers and P.J.F. Sandra, Anal. Bioanal. Chem., 373 (2002) 3.
2. F. David, B. Tienpont and P. Sandra, LC GC Eur., 16 (7) (2003) 410 + Jul. 2003.
3. K.A.D. Swift (ed.), Advances in flavours and fragrances, (2002) 27.

W.L.P. Bredie and M.A. Petersen (Editors)
Flavour Science: Recent Advances and Trends

A novel prototype to closely mimic mastication for *in vitro* dynamic measurements of flavour release

C. Salles[a], P. Mielle[a], J.-L. Le Quéré[a], R. Renaud[a], J. Maratray[a], P. Gorria[b], J. Liaboeuf[b] and J.-J. Liodenot[b]

[a]INRA, UMR Flavic, 17 rue Sully, BP 85610, 21065 Dijon, France;
[b]Plateform'3D, I.U.T, 12 rue de la fonderie, 71200 Le Creusot, France

ABSTRACT

Flavour release during eating of a food depends upon many parameters that can hardly be managed. *In-vivo* measurements by the APCI MS-nose method allowed temporal sensory evaluation and flavour release data to be directly correlated, but several limitations have frequently been reported. These were: high inter-individual variability, low repeatability of measurements, and weak experiment throughput due to panellists' exhaustion. To overcome most of these limitations, the use of an artificial mouth for on-line measurement of flavour release is recommended. However, the systems used in previous reports were limited in terms of reproducing *in-vivo* oral functions and parameters. This paper introduces a newly designed chewing simulator, which reproduces as realistically as possible most of the physiological functions of the human mouth. The active part of the system is a dedicated cell, precisely tooled in a biocompatible and inert material, which is 3 axes fully actuated and computer driven. The cell is composed of several mobile parts that are able to reproduce accurately in real time shear and compression strengths and tongue functions, according to *in-vivo* collected data. The flavour release can be on-line monitored by either APCI-MS or chemical sensors. On-line taste sampling is under development.

1. INTRODUCTION

Perception of sensory properties of food products during eating by humans is closely related to the availability of each of the individually released compounds for the aroma and taste receptors. The influence of the different processes occurring in-mouth during eating or drinking on the flavour perception is not well understood and could lead to a better understanding of the release as a function of time.

In addition to human perception, analytical instruments can be hyphenated with the human mouth, i.e. using APCI-MS 'in-nose method' [1]. In that case, the measurement is still *in vivo*, but the panellist is only asked to masticate the sample, and its sensations are no more collected. This nosespace extraction (breath-by-breath analysis) could allow temporal flavour release and sensory evaluation data to be directly correlated. However, several limitations have frequently been reported:

- high inter-individual variability;
- high breathing noise, high amount of exhaled air, leading to a dilution of aroma;
- acceptability problems may occur, depending on personal preferences (onion, garlic, cheese…) or amount of added flavour or due to regulations on human experimentation.

In addition, flavour release during chewing by humans is related to numerous parameters assessed by panels, such as structure and rheological properties of the food matrix, and physical masticatory parameters (dentition, forces, milling torque...).

Taking into account the experimental uncertainties listed and the number of parameters involved, it is virtually impossible to model the human chewing process. To overcome most of these limitations, the use of an artificial mouth for on-line measurement of flavour release is highly suited. Some simulators were used in previous reports [2-4], but with only some of the oral function features, either temperature or chewing movements or liquid proofness. Moreover, none of these systems were at the same time gas- and liquid-tight, and fully inert towards the aroma compounds.

2. DEVELOPMENT

We designed and developed a novel chewing simulator 'Artificial Mouth', which reproduces as faithfully as possible most of the physiological functions of the human mouth (Figure 1). The specifications were as precise as possible to closely mimic operation, parameters and the most important phenomena occurring in a human mouth from the introduction of a food to the final swallowing of the bolus.

a: Air cylinder
b: Gas sampling
c: Chewing cell
d: Temperature control
e: Teeth angle motion
f: Angle motion actuation
g: Teeth linear motion
h: Tongue linear motion

Figure 1. Left: active cell, external view. Right: detail of the actuation and position control mechanism.

2.1. Specifications

This artificial mouth environment, developed to study the flavour compounds released during mastication, should in particular closely match the different conditions of human mastication and should be universal to study a large variety of foods and drinks:

- Initial introduction of food sample, and continuous introduction of artificial saliva;
- Cycle time from 800 ms, chewing duration up to 180 s;
- On-line gas sampling with embedded sampling system; at-line liquid sampling.

All the parameters imitate those of the human mouth and then were set up in accordance with human measurements by electromyography and downsized physical sensors:

- Motion, force and torque control of tongue, mandibles and teeth;
- Volume, temperature, gas and liquid flow-rates.

2.2. Active parts

The main part of the system is a dedicated cell, precisely tooled in a biocompatible and inert material, which is 3 axes fully actuated and computer driven via an intelligent 4 axes PCI card. The software was developed under LabView (National Instruments) and allows simultaneously automatic mechanical zeroing at power-on, actuation and force control, and data acquisition of the mechanical, physical and fluid values for each cycle. For the system design, all mechanical parts were divided in functional subsets, and each subset was validated after its optimisation. The active cell is composed of several mobile parts: a fixed upper mandible, an actuated lower mandible (displacement, revolution, force and torque control), and an actuated tongue (displacement and force control). The prototype is able to reproduce in real time shear and compression strengths and tongue functions (up to 250 N). Strengths, shears, forces, and torques can be accurately reproduced according to *in vivo* collected data in amplitude and frequency.

After 3D digitalisation of human molars and 3D modelling and rendering, teeth were 3D tooled. Antagonist teeth are not the 'negative' draft of molars, but fit into each other like natural human ones (Figure 2).

2.3. Sampling

Artificial saliva and neutral carrier gas are introduced in the cell at variable flow-rates (respectively 0 to 5 g/min and 10 to 50 ml/min). A dedicated sampling system together with the carrier gas reproduces the movements of air inside the mouth and through the retronasal cavity occurring during mastication. The flavour release can be on-line monitored by APCI-MS or chemical sensor array. For the parameter optimisation, a hybrid array of 4 MOS and 8 QMB sensors was used (VocMeter, AppliedSensor, Germany). The most critical study was the gas tightness and chemical inertness. The specifications were draconian: fully gas tightness should be insured with no gasket or sealant, and the use of lubricant was not considered acceptable. We succeeded in meeting these tight specifications with a very precise tooling of materials (operational tolerance in the micrometer range). A screening of available materials that were likely to be precisely tooled assessed for the inertness. Not only the chemical inertness was concerned, but also the adsorption onto the buffed material walls.

3. DISCUSSION AND CONCLUSION

Preliminary results were obtained from dairy matrices, particularly flavoured model-cheese. Deconstruction of food products using the chewing prototype was very close to that obtained in the human mouth (see Figure 2, right part). This enables on-line monitoring of aroma release from food products and comparison of the results with the mouthspace technique. Future prospects include the relation between controlled oral parameters reproduced from *in vivo* data with sensory measurements.

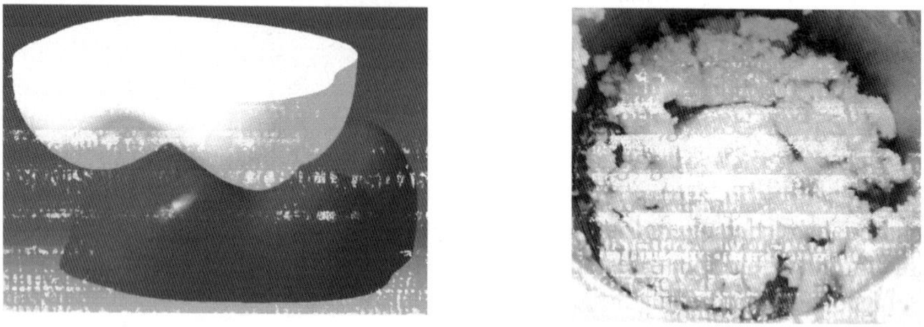

Figure 2. Left: teeth profiles 3D rendering. Right: food breakdown after artificial chewing of cheese.

The chewing forces and torques for humans are depending on both the appetite for the presented food and the bolus destruction degree. This means that all the chewing parameters and the saliva generation present a complex time-intensity profile.
Fortunately, the prototype would allow the decoupling of mixed-up actions measured in humans by electromyography in separate chewing parameters. This is expected to lead to a better understanding of the flavour release processes. Different tooth profiles corresponding to real human dentures can be tooled and replaced easily, as a rapid dismantling and cleaning of the active parts was also in the initial specifications. All the safety issues were assessed and respected. On-line taste sampling is under development.

References

1. A.J. Taylor, R.S.T. Linforth, B.A. Harvey and A. Blake, Food Chem., 71 (2000) 327.
2. J.-L. Le Quéré and P.X. Étiévant (eds.), Flavour research at the dawn of the 21st century, proceedings of the 10th Weurman flavour research symposium, Paris, France (2003) 228.
3. K.D. Deibler, E.H. Lavin, R.S.T. Linforth, A.J. Taylor and T.E. Acree, J. Agric. Food Chem., 49 (2001) 1388.
4. H. Maarse and D.G. van der Heij (eds.), Trends in flavour research, Amsterdam, The Netherlands (1994) 59.

W.L.P. Bredie and M.A. Petersen (Editors)
Flavour Science: Recent Advances and Trends

MS-nose flavour release profile mimic using an olfactometer

Peter M.T. de Kok, Alexandra E.M. Boelrijk, Catrienus de Jong, Maurits J.M. Burgering and Marc A. Jacobs

NIZO food research, Flavour Division, P.O. Box 20, 6710 BA Ede, The Netherlands

ABSTRACT

An olfactometer has been 'trained' to produce flavour release curves imitating profiles of authentic foods as measured with an MS-Nose. These sensory cues have been used to verify the effect of combining food reformulations (matrix effects) with flavour release profiles of the non-modified food. Cross-modal interactions have been proven important factors to take into account while optimising flavour intensity and quality.
Understanding and tailoring the olfactometer has been proven essential in order to be able to produce reliable and accurate controlled aroma release profiles using the olfactometer.

1. INTRODUCTION

A decade ago, flavour quality and flavour performance in foodstuffs were largely seen in the context of knowing the exact composition of flavour systems. Many studies focused on establishing the aroma compositions using state-of-the-art analytical methods. With the introduction of e.g. the AEDA [1] and Charm [2] methods, establishing the (relative) importance of compounds to the overall sensory impression became feasible. However, especially with the increasing consumer demand for reduced fat levels in foods, it became evident that not only flavour composition, but also flavour release is an important factor to take into account. MS-Nose technology provided a unique tool to measure in vivo flavour release profiles. These detailed release curves enabled the development of mathematical models to describe and predict release behaviour in foods [3]. Based on many studies in the field [4], understanding the parameters that affect release did increase rapidly. Liquid layer formation in the throat after swallowing proved important [5] and tools were developed to simulate and measure the effects [6].

Although detailed and predictive flavour release models have been developed and are available, manipulating and adapting flavour release profiles have proven difficult. As release is largely determined by the food matrix properties, experimentally verifying release profiles, and the effects variations thereof exhibit, is still lacking. Recently, Hort *et al.* [7] used a multi-channel flavour delivery system with time intensity measurement techniques while Hummel *et al.* have applied olfactometry in many studies (e.g. [8]). Using a Burghart OM4 olfactometer, odour profiles were generated closely mimicking release curves as measured using APCI-MS nose space analysis (MS-Nose). This has enabled the simulation of sensory experiences from modified foods with authentic aroma release profiles and/or (over-)compensating for food matrix effects.

2. MATERIAL AND METHODS

For this study, *in vivo* and simulated release profiles were measured using an APCI-MS spectrometer (MS-nose). The *in vivo* release was determined for ethyl butyrate in strawberry flavoured whole milk and skim-milk. The *in vivo* measurements were performed using a standard protocol for swallowing and breathing [9].

Figure 1. Burghart OM4 olfactometer.

For simulation of the *in vivo* release profiles, a Burghart OM4 four channel mono-rhinal olfactometer was used, as shown in Figure 1. The olfactometer was filled with a 60 ml solution of 10 ppm ethyl butyrate in polyethylene glycol. The sum of the dilution and flavoured flows was kept constant at 7.5 l/min and the exhaust flow was set at 8 l/min. During the release simulation the flavoured flow varied in the range 0.2-1.3 l/min for the whole milk simulation and 0.9-5.9 l/min for the skim-milk profile. Pulses duration the simulation were set at 600 ms, whereas the timing between pulses was varied according to the *in vivo* profile to be simulated. The generated odour concentrations were measured using the MS-Nose.

3. RESULTS

After swallowing a strawberry flavoured milk sample, a broad range of flavour components is released into the air stream from the throat to the nose space. From this range of components, ethyl butyrate, a sweet candy-like flavour, was selected as the model component for simulating of the release profile using the olfactometer. Hence strawberry flavoured milk- and skim milk samples were used to generate the MS-Nose time-intensity curves (not shown) for ethyl butyrate. Both samples did contain 18 ppm ethyl butyrate. Simulating the MS-Nose time-intensity curves of a particular arbitrary subject, the release profiles of skim-milk and whole milk generated with the olfactometer are shown in Figures 2a and c, respectively. The inhalation/exhalation cycles are clearly represented by peaks in the simulation. The profile for skim-milk shows high initial concentrations, which subsequently rapidly decline.

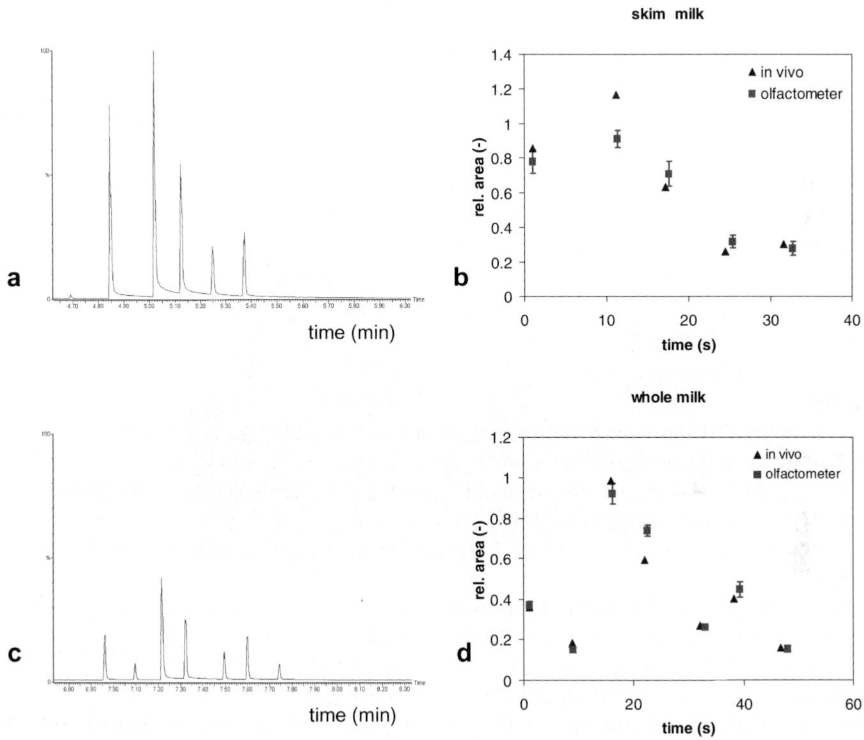

Figure 2. Olfactometer simulation of release profiles. (a) and (c) APCI-MS spectrometry (MS-Nose) chromatograms of ethyl butyrate release in skim milk and whole milk, respectively. (b) and (d) Comparison of relative peak areas (MS-Nose) for *in vivo* release and integrated olfactometer generated profile.

In the case of whole milk, containing 3% fat, the maximum concentrations in the release profile are substantially lower, due to partitioning of the hydrophilic flavour into the dispersed fat phase. As one would expect, the partitioning of the flavour results in a higher persistence of the release, as the fat phase acts as a flavour reservoir.

Figures 2b and d compare the integrated relative peak areas in MS-Nose recorded release profiles with the integrated peak areas as were obtained using the olfactometer, the latter simulating the release profiles from the MS-Nose. After careful configuration, only a single significant difference between the MS-Nose target and the olfactometer-produced load was observed for low fat skim-milk 10 seconds after the first exhalation response (second peak). In this particular instant, the olfactometer was set to generate very high flows through the flavour solution in order to simulate the maximum intense flavour release of a the lipophilic compound from a low fat system. For all lower stimuli, such as in the case of whole milk, the peak areas generated closely matched the ones obtained from the *in vivo* measurements.

4. CONCLUSION

To tailor release profiles, insight in the physics of an olfactometer and its implementation is essential. Additionally, an MS-nose is required to measure target *in vivo* release curves and for calibration of the instrument. Timings, solvent effects and dilutions of the flavours are important parameters to operate this system in order to successfully mimic release curves. Once experimental/operational parameters have been established, controlled aroma delivery is feasible. Early experiments have shown that cross-modal interactions (e.g. texture and orthonasal; aroma delivery) are significant. In order to minimise sensory effects while modifying textures, aroma delivery systems require extra corrections relative to unaltered transferring delivery profiles from the authentic foodstuff (not reported here).

References

1. F. Ulrich and W. Grosch, Z. Lebensmittel Untersuch. Forsch., 184 (1987) 277.
2. T.E. Acree, J. Barnard and D.G. Cunningham, Food Chem., 14 (1984) 273.
3. G. Lian, M.E. Malone, J.E. Homan and I.T. Norton, J. Controlled Release, 98 (2004) 139.
4. A.J. Taylor, Food Safety, 1 (2002) 45.
5. A. Buettner, A. Beer, C. Hannig, M. Settles and P. Schieberle, Food Quality Pref., 13 (2002) 497.
6. K.G.C. Weel, A.E.M. Boelrijk, J.J. Burger, M.Verschueren, H. Gruppen, A.G.J. Voragen and G. Smit, J. Agric. Food Chem., 52 (2004) 6564.
7. J. Hort and T.A. Hollowood, J. Agric. Food Chem., 52 (2004) 4834.
8. D. Krone, M. Mannel, E. Pauli and T. Hummel, Phytother R., 15 (2001) 135.
9. K.G.C. Weel, A.E.M. Boelrijk, J.J. Burger, H. Gruppen, A.G.J. Voragen and G.A. Smit, J. Food Sci., 68 (2003) 1123.

W.L.P. Bredie and M.A. Petersen (Editors)
Flavour Science: Recent Advances and Trends

Nosespace with an ion trap mass spectrometer - quantitative aspects

Jean-Luc Le Quéré, Isabelle Gierczynski, Dominique Langlois and Etienne Sémon

Institut National de la Recherche Agronomique (INRA), UMR ENESAD-INRA Flaveur, Vision, Comportement du Consommateur (FlaViC), 17 rue Sully, BP 86510, F-21065 Dijon, France

ABSTRACT

A new APCI source and interface connected to an ion trap mass spectrometer have been designed in order to allow introduction and analysis of a gaseous flow within the source. In vitro detection limits, linear response ranges, and repeatability (daily and day-to-day) have been determined for a set of flavour molecules of various chemical classes. Detection limits and linear response ranges have been found compatible with aroma compounds concentrations generally found in foodstuffs. Repeatabilities were found within the values already published for another interface connected to an ion trap mass spectrometer. During breath-by-breath data acquisition it appeared clearly that some competition between volatiles towards the chemical ionisation process occured. This point has been addressed specifically. Mixtures of 2-heptanone, ethyl hexanoate, 2,3,5-trimethylpyrazine and 2-heptanol in varying concentration ranges have been analysed and their responses compared to pure compound responses. In the case of 2-heptanol, it clearly appeared that its APCI-MS response decreased when ethyl hexanoate concentration increased in the mixture.

1. INTRODUCTION

Effective flavour release measurement on-line in human breath by mass spectrometry (MS) has been achieved using Atmospheric Pressure Chemical Ionisation (APCI) as a soft ionisation technique [1-3]. Protonated water clusters formed from moisture in the expired air are used as reagent ions. Breath-by-breath analysis of expired air from the nose or the mouth of a consumer during food consumption allows the determination of flavour compounds that are released from the food matrix while the consumer is chewing and ingesting [1,2]. Correlation with simultaneous time-course sensory evaluation (Time-Intensity) should provide basic knowledge on parameters affecting flavour perception.

An ion trap mass spectrometer has the unique advantage to allow tandem mass spectrometry (MS-MS) to be undertaken in a rather compact instrument. The APCI technique giving rise essentially to pseudo-molecular ions, MS-MS experiments are valuable tools for flavour compound determination as isobaric peaks frequently occur. As a first step it was necessary to design a source and an interface for connection to the ion trap mass spectrometer in order to sample human breath [4]. The aim of the present work was to validate the interface by obtaining quantitative data and breath-by-breath flavour release profiles.

2. MATERIALS AND METHODS

2.1. Materials

The mass spectrometer used was an ion trap mass spectrometer (Bruker Esquire LC, Bremen, Germany). A low dead-volume APCI source had been specially constructed, with the corona needle facing the ion entrance capillary [4]. Using the implemented auxiliary gas, a Venturi effect was created to allow introduction of vapours in the source via a fused silica capillary tubing (0.53 mm i.d.) inserted into a heated transfer line maintained at 150 °C to avoid condensation of water. The fused silica tubing was inserted near the Venturi region in a capillary adjustment device that allows the inlet flow rate to be precisely adjusted from 0 to 100 ml/min. The vapour inlet flow rate was optimised for signal-to-noise ratio and set to 30 ml/min.

2.2. Methods

In vitro measurements were obtained from the headspace generated from 250 ml aqueous solutions of volatiles contained in 20 l Teflon bags inflated with 17 l of nitrogen. The advantage of this Teflon bag method is a constant headspace concentration delivery for several minutes. The bags were maintained at room temperature to equilibrate the headspace for at least 12 h and then connected to the APCI source via the capillary transfer line. Mass spectra were acquired and averaged for approximately 1 min after stabilisation of the signal and maximum intensities of the pseudo-molecular ion profiles recorded. Headspace concentrations in the bags were determined by a gas chromatography method using Tenax trapping of a defined volume from the same Teflon bags and an external standardisation.

Detection limits and linear response ranges were determined from triplicate measurements for six volatile compounds belonging to various chemical classes (diacetyl, butyric acid, 2-heptanone, 2-heptanol, ethyl hexanoate, and 2,5-dimethylpyrazine, all obtained from Sigma-Aldrich, StQuentin Fallavier, France).

Daily and day-to-day repeatabilities were determined for 2-heptanol and ethyl hexanoate (5 ppm solution in water). Competition data were obtained using the same Teflon bag method and from two repetitions. Ion intensities resulting from a fixed medium-concentration of ethyl hexanoate, 2-heptanone, 2-heptanol and 2,3,5-

trimethylpyrazine (Sigma-Aldrich) one by one in solution were compared to ion intensities resulting from the same molecules at the same concentrations but present in different mixtures. Mixtures were produced containing three concentrations (low, medium, high) within the linearity ranges previously determined of the other volatiles, for instance: 0.2 mg/l ethyl hexanoate compared to 0.2 mg/l ethyl hexanoate in a mixture also containing 0.06 mg/l 2-heptanone (low), 1.93 mg/l 2-heptanol (medium) and 0.32 mg/l 2,3,5-trimethylpyrazine (medium). Results were also compared to those obtained using standard gas chromatography.

3. RESULTS

3.1. Detection limits, linear response ranges and repeatability

Detection limits and linear response ranges of the complete interface and APCI source assembly were determined for the six compounds mentioned above. The results presented in Table 1 were found compatible with aroma compound concentrations generally encountered in foodstuffs. Moreover, they are comparable with those found by Taylor *et al.* [5] with a quadrupole mass spectrometer.

Table 1. Detection limits and linear response ranges of the nosespace interface (reported values are means of triplicate measurements).

Volatile compound	Detection limit in headspace, ng/l air (in solution, ppb)	Linear response range in headspace, μg/l air (in solution)
Diacetyl	500 (1000)	4 – 125 (5 – 112 ppm)
Butyric acid	1.7 (1000)	0.018 – 2 (5 – 500 ppm)
2-Heptanone	27 (10)	0.166 – 4 (48 – 1200 ppb)
2-Heptanol	32 (53)	0.09 – 3 (100 – 3000 ppb)
Ethyl hexanoate	46 (11)	0.177 – 7 (42 – 1100 ppb)
2,5-Dimethylprazine	12 (110)	0.09 – 8 (1 – 86 ppm)

Daily (n = 6) and day-to-day (5 consecutive days, n = 10) repeatabilities were evaluated from the relative standard deviation calculated for the intensities measured using a 5 ppm solution of 2-heptanol and ethyl hexanoate. The values of about 5% daily and 15% day-to-day, respectively, were found within those already published for another interface connected to an ion trap mass spectrometer [6], and considered satisfactory taking into account the whole source of uncertainty (concentration variation during storage, measurement temperature and instrument response).

3.2. Volatiles in mixture. Competition within the ionisation process

During various breath-by-breath data acquisition assays, it clearly appeared that within the chemical ionisation process some competition between volatiles may occur. This point was addressed by comparing the headspace response of pure compounds to the headspace of the same compounds found in mixtures of various composition and concentration. From the results presented in Figure 1, despite a rather poor repeatability, it clearly appears that the responses of APCI-MS for 2-heptanol decreased when ethyl hexanoate concentration increased in the mixture. On the contrary, the responses of GC-FID detector for 2-heptanol were not affected by increasing amounts of ethyl hexanoate.

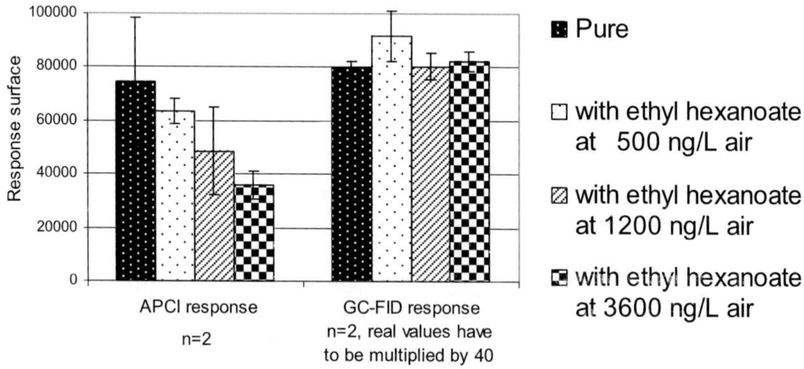

Figure 1. Surface response of APCI and GC for 2-heptanol, pure or with ethyl hexanoate.

As the composition of the mixtures has very little influence on the partition at the concentration used in this study, the observed results were tentatively explained by the respective estimated proton affinities. However, no clear explanation could be obtained. Despite this clear drawback of the technique if quantitative determinations are expected, *in vivo* breath-by-breath measurements have been obtained for various foodstuffs.

References

1. D.D. Roberts and A.J. Taylor (eds.), Flavor release, Washington, DC, USA (2000) 8.
2. P. Schieberle and K.H. Engel (eds.), Frontiers of flavour science, Garching, Germany (2000) 261.
3. G. Zehentbauer, T. Krick and G.A. Reineccius, J. Agric. Food Chem., 48 (2000) 5389.
4. A.E. Ashcroft, G. Brenton, J.J. Monaghan, Advances in mass spectrometry, Amsterdam, The Netherlands, 16 (2004).
5. A.J. Taylor, R.S.T. Linforth, B.A. Harvey and A. Blake, Food Chem., 71 (2000) 327.
6. A.M. Haahr, H. Madsen, J. Smedsgaard, W.L.P. Bredie, L.H. Stanke and H.H.F. Refsgaard, Anal. Chem., 75 (2003) 655.

W.L.P. Bredie and M.A. Petersen (Editors)
Flavour Science: Recent Advances and Trends

The specific isolation of thiols using a new type of gel

Ian Butler[a], Andrew Smart[b] and Neville S. Huskisson[b]

[a]Danisco (UK) Ltd, Denington Road, Wellingborough, Northants, NN8 2QJ, England; [b]Severn Biotech Ltd, Unit 2, Park Lane, Kidderminster, Worcestershire, DY11 6TJ England

ABSTRACT

A new type of agarose gel has been produced that enables the selective isolation and recovery of thiols from a model system. Using the approach of column chromatography the method is simple to use and can be applied on a routine basis with very little work up.

1. INTRODUCTION

Thiazoles, thiophenes, thiazolines and thiols are found in cooked foods, various fruits, vegetables and dairy products and in many cases these compounds have been found to characterise the profiles of these foods [1]. Thiols in particular have been found to be important odorants for foods such as coffee [2], beef [3] and wine [4]. The low odour thresholds of these compounds and their presence in many foods at low ppb levels make their identification difficult especially when part of a complex flavour isolate.

To facilitate their detection a number of approaches can be adopted, e.g. using some form of detection that allows for increased sensitivity and selectivity for sulfur compounds. The use of sulfur specific detectors such as chemiluminescence, AED (Atomic Emission Detection), FPD (Flame Photometric Detector) help qualify the presence of sulfur containing compounds in foods. However this approach is not suitable when identification of unknown compounds is required. An alternative approach is to reversibly and selectively isolate thiols from a flavour extract, after which they can be recovered and analysed by standard techniques such as GC-MS. *p*-Hydroxymercuric acid, has been used to isolate and identify 4-mercapto-4-methyl-2-pentanone a character impact component of Sauvignon blanc wines [5].

Another approach using affinity chromatography was developed by Full and Schreier [6] using a commercially available affinity gel manufactured by Bio-Rad consisting of a cross linked agarose gel terminated with phenylmercuric chloride groups. The thiol compounds are isolated by displacement and substitution of the chloride ion with the thiol group resulting in liberation of HCl and the formation of an Hg-S-R bond. The thiols can then be recovered by eluting with 1,4-dithiothreitol (DTT). The gel was later used to identify the key odorants in a thermally treated solution of ribose and cysteine [7].

Unfortunately, this gel is no longer commercially available. Therefore, the aim of this work was to develop an equivalent thiol binding gel and validate its efficiency in selectively isolating and recovering thiol compounds from a model system consisting of *p*-mentha-8-thiol-3-one, 2-furfurylthiol and 1-*p*-menthene-8-thiol dissolved in a solution of single fold Florida sweet orange oil.

Orange oil was used to replicate the complexity of a flavour isolate as it contained a range of volatile compounds with different functionalities such as alcohols, aldehydes and terpenes but no thiols.

2. MATERIAL AND METHODS

2.1. Chemicals

Pentane, dichloromethane, diethyl ether, dithiothreitol, 2-furfurylthiol, *p*-mentha-8-thiol-3-one were purchased from Sigma-Aldrich (Dorset, England) and 1-*p*-menthene-8-thiol from Natural Advantage. The mercuric agarose gel was purchased from Severn Biotech Ltd (Worcestershire, England). Single fold sweet Florida orange oil was obtained from Danisco (Florida, USA).

2.2. Gas chromatography-mass spectrometry

GC-MS analysis was performed on a Agilent 5973 MSDN fitted with a DB-WAX fused silica column (internal diameter: 0.25 mm; film thickness: 0.25 μm; length: 30 m). The oven temperature program was: 30 °C to 220 °C at 3 °C/min and held at 220 °C for 10 min. The injection volume was 1 μl and split 30:1.

2.3. Model system

A 100 ml stock solution containing 0.1 g 2-furfurylthiol, 0.1 g 1-*p*-menthene-8-thiol and 0.1 g *p*-mentha-8-thiol-3-one was prepared in diethyl ether. The model system was then prepared by diluting the stock solution (100 μl) into 15 ml of pentane containing 0.05 g sweet Florida orange oil.

2.4. Isolation of thiols

The mercuric acetate agarose gel was converted to mercuric chloride using a saturated sodium chloride solution and added (1 g) to a glass column fitted with a tap and sintered disk and filled with 20 ml pentane-dichloromethane (2:1 v/v). After passing through one volume (20 ml) of the pentane-dichloromethane mixture, pentane was added to the column followed by the model system. The gel was then washed with 8 x 15 ml aliquots of pentane-dichloromethane to remove the orange oil; the bound thiols were then eluted with 15 ml 10 mM dithiothreitol in dichloromethane. GC-MS analysis was carried out on the model system before and after passing through the gel, the 8 x 15 ml aliquots of pentane-dichloromethane as well as the final elution using dithiothreitol.

3. RESULTS

The chloride conditioned gel consisted of an agarose backbone containing a 6-spacer atom side chain terminated with a phenylmercuric chloride functionality. Referring to Figure 1 the GC-MS chromatogram shows that both 2-furfurylthiol [**2**] and *p*-mentha-8-thiol-3-one [**4**] are selectively isolated from the orange oil and are later recovered from the gel by elution with 10 mM DTT. The flow rate through the column has to be carefully controlled to prevent breakthrough of the thiols in the feed solution. In this work a flow rate of about 0.1 ml/min was used. Unfortunately the grapefruit smelling compound 1-*p*-menthene-8-thiol [**3**] was not recovered. Similar observations have been made in thiol binding experiments using the Bio-Rad Affi-Gel 501[8]. Based on this work it appears that this new type of gel has binding efficiencies similar to the Affi-Gel 501. However it has yet to be established whether this new gel can isolate aliphatic thiols. This will be the focus for future work.

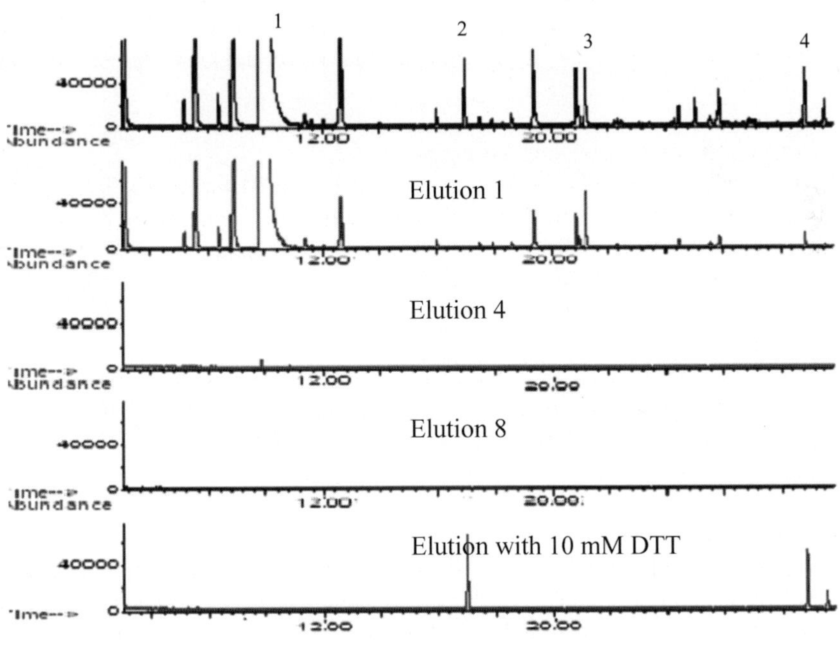

Figure 1. GC-MS profile of the thiol model system before (top) and after passing through the mercuric agarose gel, limonene [**1**]; 2-furfurylthiol [**2**]; 1-*p*-menthene-8-thiol [**3**]; *p*-mentha-8-thiol-3-one isomers 1+2 [**4**].

4. CONCLUSION

The newly developed gel appears to have thiol retention properties similar to the Bio-Rad Affi-Gel 501 when applied to a model system. By using column chromatography the method is easy to use and can be applied on a routine basis.

References

1. C.J. Mussinan and M.E. Keelan (eds.), Sulfur compounds in foods, Washington, DC, USA (1994) 1.
2. A.J. Taylor and D.S. Mottram (eds.), Flavour science: recent developments, Cambridge, UK (1996) 200.
3. U. Gasser and W.Z. Grosch, Z. Lebensmittel Untersuch. Forsch., 186 (1988) 489.
4. P. Bouchilloux, P. Darriet, D. Dubourdieu, R. Henry, S. Reichert and A. Mosandl, Eur. Food Res. Technol., 210 (2000) 349.
5. P. Darriet, T. Tominaga, V. Lavigne, J.-N. Boidron and D. Dubourdieu, Flavour Fragrance J., 10 (1995) 385.
6. G. Full and P. Schreier, Lebensmittelchem., 48 (1994) 1.
7. T. Hofmann and P. Schieberle, J. Agric. Food. Chem., 43 (1995) 2187.
8. G.A. Reineccius and T.A. Reineccius (eds.), Heteroatomic aroma compounds, Washington, DC, USA (2002) 2.

W.L.P. Bredie and M.A. Petersen (Editors)
Flavour Science: Recent Advances and Trends

Prediction of gas-chromatographic retention indices as a tool for identification of sulfur odorants

Jean-Yves de Saint Laumer and Alain Chaintreau

Firmenich SA, Corporate R&D Division, P.O. Box 239, CH-1211 Geneva 8, Switzerland

ABSTRACT

The positive identification of aroma compounds requires the use of two criteria [1], most frequently a mass spectrum and a retention index. However, as the retention indices of many compounds are unknown, a theoretical prediction of these values, from chemical structures, would be a useful tool for confirmation of identity. Firstly, a model was generated using the classical Multiple Linear Regression (MLR) from a data set containing 580 compounds. Unlike previously published studies the model was developed for compounds having a wide chemical diversity. The model was then refined using a specific data set of sulfur-containing compounds using molecular descriptors related to the sulfur functions. The predictive validity was checked by measuring indices of novel authentic compounds. This new model has been used to restrict the number of re-injections and/or syntheses of authentic compounds to the most probable candidates within the huge number of proposals generated by comprehensive two-dimensional GC-TOF-MS analyses [2].

1. INTRODUCTION

Gas chromatographic separation combined with mass spectrometry is the most widely used separation technique of volatile organic compounds prior to their identification. In this case, the retention index is a key parameter to confirm the identity of a compound [1]. Because many indices have never been determined, their prediction would help the analysis of mixtures that contain a great number of unknown compounds. QSPR (quantitative structure-properties relationships) methods have been used to predict either gas chromatographic retention times or directly the retention indices [3-5]. The prediction quality obtained with various methods can be excellent for families of structurally-related compounds. Here, we developed a QSPR model from a data set of compounds used in the flavour and fragrance industry and which have a wide chemical diversity. As this work particularly focuses on the identification of sulfur compounds, the predictive model was extended to take them into account.

2. EXPERIMENTAL

The QSPR methods can be summarised by the three following steps. First a large number of molecular descriptors is calculated for a set of compounds for which the property is known. Then a regression method is applied to establish a linear relation between a subset of the descriptors and the properties. In the last step, the predictive power of the equation is evaluated by a validation procedure.

2.1. Data set

The data set used in this study initially included 580 organic compounds from our internal database. The Kovats retention indices were measured with a DB1 column (Agilent Technologies, Wilmington, DE), 30 m length, 0.25 mm i.d., 0.25 μm phase thickness. The oven temperature was programmed from 60 up to 240 °C. Helium was used as carrier gas at a flow of 1.0 ml/min. The database was divided into two randomly chosen sets (a learning set of 480 compounds and a test set of 100 compounds). In a second phase, additional sulfur compounds were measured to constitute a data set of 169 sulfur compounds.

2.2. Molecular Descriptors

Following a preliminary study where we compared the predictive ability of various descriptor systems, we selected the Molconn-Z program [6], which gives predictions comparable to published results. The descriptors available in this software are based on the 2D structure of the molecule. They are directly calculated from the molecular structure entered in the SMILES format. The descriptors encoded various topological descriptors such as path lengths, connectivity indices and electrotopological states. Details concerning these descriptors can be found on the Edusoft web site [6]. A total of 96 descriptors were calculated using this program. We added to those descriptors a first set of functional descriptors representing chemical functions frequently found in the database: (primary alcohol; secondary alcohol, tertiary alcohol, ketone, ester, etc.). In a second part of this work, we added supplementary functional descriptors for functions and fragments involving sulfur atoms. All these functional descriptors were calculated using an internally developed program.

2.3. Stepwise Multiple Linear Regression

An important tool in a QSPR model building is the regression analysis. The regression methods are used to produce one or several equations to relate a property to descriptors. In this work we applied the Stepwise Multiple Linear Regression, as its application and interpretation are simple, and because it allows the selection of important variables from a large set of descriptors for the model. In the stepwise MLR, a multiple-term linear equation is produced, but not all variables are used. Each variable is added to the equation in turn. At each step, a new regression is performed. The new term is retained only if the equation passes a test for significance (Probability value < 0.001).

3. RESULTS

Equation (1) was obtained frcm the training set (480 compounds). Sixteen variables were iteratively introduced in the model by the stepwise MLR algorithm. The selected parameters are connectivity descriptors, Kappa shape indices, electrotopological indices, hydrogen bond indices (see [6] for a complete description) and our functional descriptors. The large number of variables selected by this method can be explained by the large diversity of the training set which could be described only by many parameters.

$$RI = 54.83 - 340.02\,X0 + 240.34\,X1 + 222.07\,Xv0 - 241.28\,Xch5 - 33.17\,Ka1 + 66.47\,Sumi - 41.26\,Sumdell - 10.13\,Sharom + 28.29\,Shhbd + 165.78\,Nhbint4 - 5.07\,Shbint4 - 45.37\,Aldehyde - 89.59\,Ester - 30.99\,Ether - 69.25\,Furan + 91.8\,\gamma\text{-}Lactone \quad (1)$$

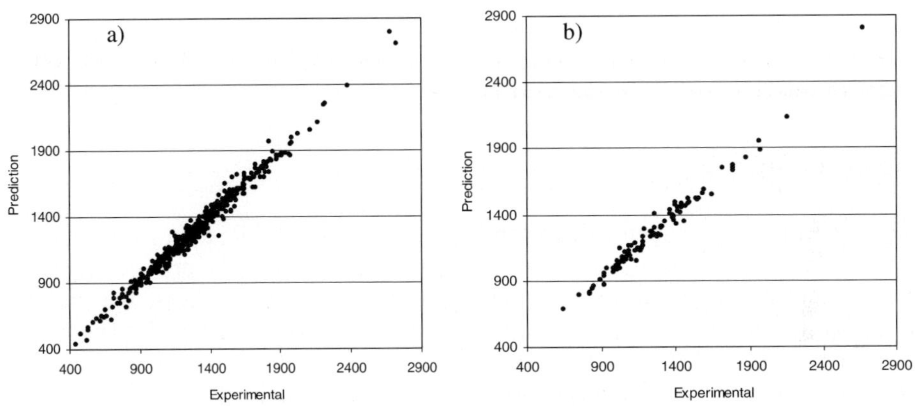

Figure 1. Plot of the calculated retention indices versus experimental value for (a) the training set (R=0.991; N=480; S=44.2) and (b) the prediction set (R=0.989; N=100; S=45.9).

When applying equation (1) to the 169 sulfur compounds data set, we observed a satisfactory correlation (R=0.989; N=169; S=41.2) between the calculated and measured values. However, for some functional classes systematic differences were observed between predictions and measurements. We then introduced into the model corrective factors related to chemical function of sulfur compounds: disulfide, isothiocyanate, thiol, thiazole, thioether, thiophene and thioester. These factors were obtained by performing a descending stepwise MLR on the residuals (the differences between prediction using equation (1) and measured values). In this regression the intercept was fixed at 0, and only parameters having a probability value < 0.005 were introduced into the model. The advantage of this approach is that the prediction of the non-sulfur compounds is not modified by the addition of the corrective factors.

RI = Equation (1) + 21.75 Disulfide + 62.02 Isothiocyanate + 16.06 Thiol - 21.83 Thiazole + 22.71 Thioether - 49.82 Thioester (2)

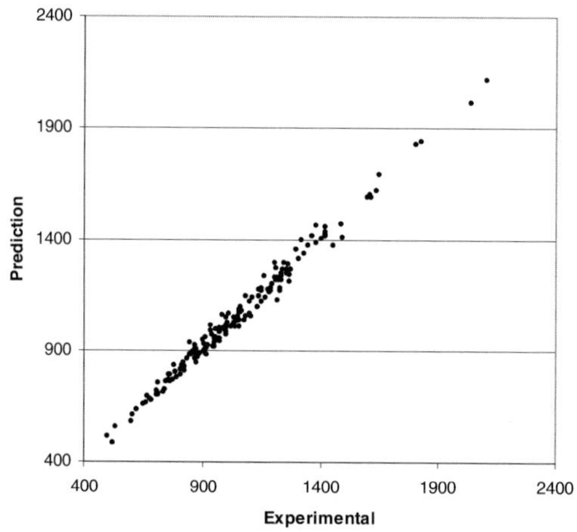

Figure 2. Plot of the calculated retention indices against experimental values for the sulfur compounds data set (R=0.993; N=169; S=31.7).

4. CONCLUSION

Good estimation of retention indices can be obtained using MLR, even for a database with large chemical diversity. This investigation has shown that the prediction quality can not only be improved owing to the number of reference indices, but also by increasing the number of functional descriptors of training set compounds. The model obtained has been used to restrict the number of re-injections and/or syntheses of authentic compounds to the most probable candidates within the huge number of proposals generated by comprehensive two-dimensional GC-TOF-MS analyses [2].

References

1. I.O.F.I., Z. Lebensmittel Untersuch. Forsch., A 204 (1997) 400.
2. W.L.P. Bredie and M.A. Petersen (eds.), Flavour science: recent advances and trends, proceedings of the 11th Weurman flavour research symposium, Amsterdam, The Netherlands (2006) 601.
3. R. Kaliszan, Quantitative structure-chromatographic retention relationships, New York, USA (1987) 69.
4. C.G. Georgakopoulos, J.C. Kiburis and P.C. Jurs, Anal. Chem., 63 (1991) 2021.
5. T.F. Woloszyn and P.C. Jurs, Anal. Chem., 64 (1992) 3059.
6. MolconnZ v. 3.5 user´s guide, Hall Associates Consulting, Quincy, Ma. Also presently available on-line at http://www.eslc.vabiotech.com.

W.L.P. Bredie and M.A. Petersen (Editors)
Flavour Science: Recent Advances and Trends

Re-investigation of sulfur impact odorants in roast beef using comprehensive two-dimensional GC-TOF-MS and the GC-SNIF technique

Alain Chaintreau, Sabine Rochat and Jean-Yves de Saint Laumer

Firmenich SA, Corporate R&D Division, P.O. Box 239, CH-1211 Geneva 8, Switzerland

ABSTRACT

Comprehensive two-dimensional gas chromatography (GC x GC) allows the detection of thousands of compounds from a food aromas such as roast beef. As the identification of this huge number of constituents would be time consuming and not useful, strategies must be developed to extract pertinent information. In this work, more than 60 sulfur compounds were found by GC x GC hyphenated to time-of-flight mass spectrometry (TOF-MS). Due to the non-availability of many reference compounds, all their retention indices could not be compared to authentic samples. The missing reference indices were simulated using a multiple linear regression to confirm peak identities [1]. In parallel, impact odorants were determined by GC-olfactometry, using the detection frequencies of the GC-SNIF technique. From a total of 18 impact sulfur odorants mentioned in the literature concerning cooked beef, only 7 of them are considered to play a significant role in the roast top note.

1. INTRODUCTION

Among the 110 volatile sulfur compounds found in cooked/boiled beef aroma [2], 18 play an impact role [3-9], and only 3 to 5 in the roast or grilled beef aroma [3,4]. This work investigates the sulfur odorants of an oven-roasted beef, in the absence of any other ingredient. This requires a sensitive and/or an enrichment technique, as sulfur compounds often occur as components. But, if many volatiles are detected, which of them really influence the overall aroma? The first challenge is overcome by using comprehensive two-dimensional GC, hyphenated to a time-of-flight mass spectrometer (GC x GC-TOF-MS) that simultaneously offers an unrivalled peak resolution and a much greater sensitivity than conventional GC-MS. The second challenge is solved by GC-olfactometry (e.g. GC-SNIF) [10]. However, GC x GC generates thousands of peaks, whose complete interpretation is beyond the analyst's capabilities. In this preliminary work a strategy is tested to focus on, and to identify the targeted sulfur compounds.

2. EXPERIMENTAL

Sirloin beef was roasted in a tubular oven (Figure 1) under a stream of pure air, at 200 °C for 18 min. Vapours were either directly extracted by an SPME fiber (PDMS 100 μm), or condensed, percolated trough an Affi-Gel 501 column (BioRad) to trap thiols that were released by dithiotreitol and trapped by SPME. SPME samples were analysed with a Pegasus 4-D GC x GC-TOF-MS system (Leco), equipped with DB1 x DB225 columns (20 m x 0.18 mm x 0.18 μm and 1 m x 0.10 mm x 0.15 μm, respectively). The sampling of the roast beef aroma volatiles and the GCO experiment are described in [11].

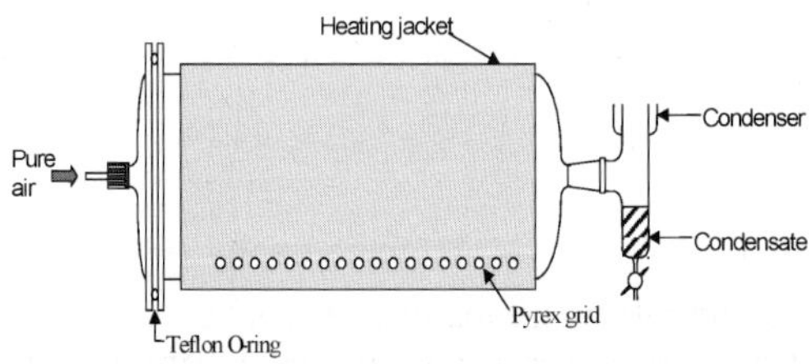

Figure 1. Schematic overview of the roasting of the beef in a tubular oven.

3. RESULTS AND DISCUSSION

In SPME extracts, 4,700 peaks (over 15,000 apexes in the vapours, Figure 2) were tentatively identified by the MS libraries. S-containing compounds were selected (100 compounds) and their retention indices (RI) calculated. Among them 57% did not have a reference RI in the initial databases. To limit the injection (and the synthesis) of authentic standards to the best candidates, the use of simulated indices was tested.
From the existing RI database, compounds were categorised according to their functional classes, and a model was generated by Multiple Linear Regression (MLR) with a standard deviation of 32 index units [1]. For an index difference between the MRL calculation and the unknown of less than SD, the proposal was considered to be 'likely', and 'possible' between SD and 2SD.
The identification of thiazoles (Table 1) exemplifies that many library proposals agreed with the simulated indices, and were confirmed by the re-injection of the standard. However, the simulation accuracy does not allow choosing between several position isomers (e.g. 2- and 4-methylthiazole, Table 1).

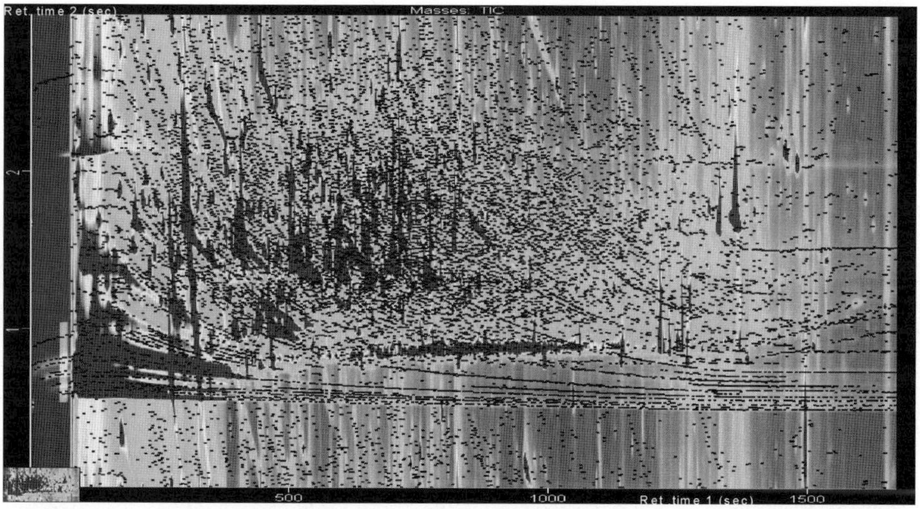

Figure 2. GC x GC-TOF-MS chromatogram of roast beef vapours trapped with a SPME (black points represent the peak apexes).

Table 1. Identification of thiazoles in the roast beef top note.

	Similarity	Ret. Index	Ret. Index	Ret. Index	Occurrence
	GC x GC-TOF	GC x GC-TOF	MLR	Authentic	
Thiazole	834	705	696	709	Confirmed
2-Methyl-	835	777	783		Likely
4-Methyl-	836	786	781	**786**	Confirmed
2-Ethyl-	837	869	883	867	Confirmed
5-Ethyl-	838	917	885	914	Confirmed
2-Isobutyl-	839	1010	1033	**1010**	Confirmed
4-Ethyl-	840	880	880	873	Confirmed
2,4-Dimethyl-	841	857	868	**858**	Confirmed
4,5-Dimethyl-	842	910	876	907	Confirmed
4-Methyl-5-ethyl-	843	991	977	988	Confirmed
2-Ethyl-4-methyl-	844	948	968	948	Confirmed
2,4,5-Trimethyl-	845	974	963	973	Confirmed
5-Et-2-*iso*-Pr-4-Me-	846	1179	1221	**1179**	Confirmed
Benzothiazole	847	1187	1144	1190	Confirmed
2-Acetyl-	848	981	1037	986	Confirmed
2-Propanoyl-	849	1083	1125	1085	Confirmed
2-Butanoyl-	850	1174	1213		Possible

In bold: experimental values (under present GC conditions) initially absent from the index database.

Using this approach, seven compounds exhibiting a high impact in the GC-SNIF analysis were positively identified (NIF > 55%, Table 2). Only derivatives of 2-methyl-3-furanethiol (MFT) were found (e.g. methyl 2-methyl-3-furyl disulfide) as it rapidly forms disulfides with other thiols. Therefore, the corresponding olfactive peak was considered to be positively identified. 2-Methyl-3-mercaptopropanol could not be confirmed by GC x GC-TOF-MS. However, its RI in the olfactogram, and its sensory descriptors fitted those of the authentic sample well. Its impact remained very low.

Table 2. Impact sulfur odorants found in the GC-SNIF aromagram of the roast beef top note (9 panellists) and confirmed by MS and retention indices.

Compound	NIF (%)	MS Similarity	Ret. ind. Beef	Ret. ind. MLR	Ret. ind. Authentic
Methanethiol	55.5	702	87 s[a]	444	90 s[a]
2-Methyl-thiophene	77.7	711	745	791	748
2-Methyl-3-furanethiol	77.7	n.d.	n.d.	880	844
Dimethyl trisulfide	99.9	923	943	919	943
Methional	88.8	916	863	847	861
2-Methyl-3-mercaptopropanol	11.1	n.d.	n.d.	896	921
2-Acetylthiazoline	55.5	870	1061	1065	1069

[a]Retention time (s) because the retention index was less than 500.

4. CONCLUSION

The simulation of RI saves time by focusing the analyst's effort (standard injection and/or synthesis) on the most probable MS proposals. However, its accuracy is insufficient to choose between different isomers. The GC x GC-TOF-MS and GC-SNIF analyses allow a positive identification of 6 sulfur compounds in the roast beef top note.

Reference

1. W.L.P. Bredie and M.A. Petersen (eds.), Flavour science: recent advances and trends, proceedings of the 11th Weurman flavour research symposium, Amsterdam, The Netherlands (2006) 597.
2. VCF00, BACIS, Boelens aroma chemical information service (1999).
3. C. Cerny and W. Grosch, Z. Lebensmittel Untersuch. Forsch., 194 (1992) 322.
4. K. Specht and W. Baltes, J. Agric. Food Chem., 42 (1994) 2246.
5. U. Gasser and W. Grosch, Z. Lebensmittel Untersuch. Forsch., 186 (1988) 489.
6. D. Machiels, S.M. Van Ruth, M.A. Posthumus and L. Istasse, Talanta, 60 (2003) 755.
7. U.C. Konopka and W. Grosch, Z. Lebensmittel Untersuch. Forsch., 193 (1991) 123.
8. H. Guth and W. Grosch, J. Agric. Food Chem., 42 (1994) 2862.
9. R. Kerscher and W. Grosch, Z. Lebensmittel Untersuch. Forsch., 204 (1997) 3.
10. P. Pollien, A. Ott, F. Montigon, M. Baumgartner, R. Munoz-Box and A. Chaintreau, J. Agric. Food Chem., 45 (1997) 2630.
11. S. Rochat and A. Chaintreau, J. Agric. Food Chem., submitted.

W.L.P. Bredie and M.A. Petersen (Editors)
Flavour Science: Recent Advances and Trends

Searching the missed flavour: chemical and sensory characterisation of flavour compounds released during baking

Barbara Rega, Aurélie Guerard, Murielle Maire and Pierre Giampaoli

ENSIA- UMR Scale, 1 av. des Olympiades, Massy 91300, France

ABSTRACT

An original 'on-line' system coupled with SPME (solid phase microextraction) and GC-MS analysis was developed in order to extract and analyse aroma compounds released during baking. Four different SPME fibres were used in order to investigate their extraction efficiency and how well they represented the odour produced during baking. A similarity test showed that the global odour emerging from SPME extracts was close to that of the reference (the odour of the baking cake). The hedonic test on SPME extracts and gas-chromatography coupled to olfactometry allowed to identify key odorant zones related with the 'tasty' odour of cake baking.

1. INTRODUCTION

The baking process of amylaceous products leads to the formation of numerous volatile compounds which determine the sensory quality of final products and increase consumer acceptance. One of the most important reactions taking place is the Maillard reaction which leads to the formation of coloured compounds as well as a large number of volatile aroma compounds like pyrazines which are responsible for typical nutty and crust odours [1]. Although the baking process causes flavour development, it is also responsible for a strong loss in pleasant aroma compounds which are released in steam during the thermal treatment. Concerning baked products, the typical flavour of bread, crackers, and cookies was largely studied [2,3], and a few studies have been made on flavours produced during extrusion cooking [4]. The aim of this study was to develop an on-line device to analyse aroma compounds released during the baking of a model sponge cake in order to gain knowledge in which volatile compound contributes to the pleasant odour diffusing out of the oven.

2. MATERIALS AND METHODS

Ingredients and making procedure were established following the CANAL Arle Project protocol [5].
During cake baking, vapours were carried to a flask (flow = 170 cm^3/s) containing the SPME fibre. SPME extraction of volatile compounds was performed dynamically in two ways: (i) during the whole time of baking (25 min) and (ii) at different baking intervals (5-10 min, 10-15 min, 15-20 min and 15-25 min). Four different SPME fibres were used: 100 μm PDMS, 65 μm PDMS/DVB, 75 μm CAR/PDMS and Stableflex 50/30 μm DVB/CAR/PDMS (Supelco Bellfonte, PA). The extraction temperature was optimised to 15 °C by preliminary tests. All samples were immediately analysed in triplicate by GC-MS. SPME fibres were desorbed into a Fisons gas chromatograph equipped with a Trio 1000 mass detector and a DB-Wax column (J&W Science, i.d. 0.32 mm, 30 m, film thickness of 0.5 μm). Mass spectral matches were determined by comparison with NIST (Gaithersburg, MD), ENSIA (France) and INRAMASS (INRA, France) mass spectra libraries. Linear Kovats' Indices of authentic compounds were used to confirm identifications.
A similarity test was made using a direct olfactometry device according to Rega *et al.* [6] and working with eight trained panellists. In order to familiarise subjects with the odour of the sponge cake baking, they were allowed to memorise this odour during the cake making. The similarity test was performed in duplicate on the five samples presented in Latin square: four SPME extracts (25 min) and a sponge cake aroma. A dummy product (the sponge cake aroma) was always presented first. Sniffers were first asked to smell and memorise the reference (the sponge cake just taken out of the oven and put in a sealed box). Then they evaluated samples rating similarity to the reference using a 10 cm scale ranging from 0 to 10 (close to/far from the reference). ANOVA and Newman-Keuls test were performed on similarity rates ($p<0.05$).
In order to find interesting aroma compounds released from the cake during baking, thirty eight panellists were asked to hedonically evaluate the SPME extracts obtained at different cooking times by the DVB/CAR/PDMS fibre. Iso-intense extracts were obtained at 5-10 min, 10-15 min, 15-20 min and 15-25 min of baking. The hedonic test was performed using direct olfactometry. Samples were presented in a Latin square. Panellists were asked to rank samples according to their preference and then to rate them using a 10 cm scale ranging from 0 (dislike) to 10 (like). They were also invited to give free semantic descriptions. Friedman test and Page test were performed on ranking, ANOVA and Newman-Keuls test were performed on hedonic rates ($p<0.05$).
Gas chromatography-olfactometry analysis (GCO) was performed on the most pleasant fractions using experienced sniffers in order to identify volatiles involved in the 'tasty' odours of cake baking. Chromatographic conditions were the same as previously mentioned but the GC effluent was split 1:1 between the FID detector and the sniffing port (250 °C). For each odour stimulus, panellists recorded the detection time and gave a free semantic description.

3. RESULTS AND DISCUSSION

The on-line device was able to efficiently extract a great number of volatile compounds, e.g. a large number of alcohols, heterocyclic compounds, aldehydes and ketones. Different SPME fibres led to very different chromatographic profiles. The DVB/CAR/PDMS fibre led to the highest number of extracted compounds, whereas the CAR/PDMS fibre allowed to extract very volatile compounds like the Strecker aldehyde 3-methylbutanal. The PDMS fibre was less effective. Table 1 shows that among the heterocyclic compounds extracted, a large number of mono- and dimethylpyrazines (which are known to be key odorants in bread crust) were identified as well as maltol and its precursor 2,3-dihydro-3,5-dihydroxy-6-methyl-4(*H*)-pyran-4-one.
The similarity test showed that the odour of all total SPME extracts (25 min) were very close to that of the reference with similarity rates ranging from 3.8 to 4.3 despite the sponge cake aroma (7.7 and 8). Among SPME fibres, DVB/Car/PDMS gave the best similarity results (3.8), even though the Newman-Keuls test showed no significant differences among the SPME extracts.

Table 1. Heterocyclic compounds extracted by on-line SPME.

Identification	RI	DVB/Car/PDMS	Car/PDMS	PDMS
2-Pentylfuran	1215	X	X	
Methylpyrazine	1249	X	X	
2,5-Dimethylpyrazine	1305	X	X	X
2,6-Dimethylpyrazine	1312	X		
2,3-Dimethylpyrazine	1329	X	X	
2,3,5-Trimethylpyrazine	1389	X	X	X
Acetylpyrazine	1605	X	X	
Furfural	1441	X	X	X
5-Hydroxymethylfurfural	2461	X	X	X
Maltol	2022	X	X	
Furaneol	2066	X	X	X
2,3-Dihydro-3,5-dihydroxy-6-methyl-4(*H*)-pyran-4-one	2244	X	X	X

By means of the DVB/Car/PDMS fibre flavour compounds produced at different baking times (5-10 min, 10-15 min, 15-20 min and 15-25 min) were extracted. The direct olfactometry hedonic test showed that preference increased gradually with baking time (Page test significant at 5%, Table 2). The analysis of the free semantic descriptors contributed to explain hedonic results: the global odours of 10-15 min, 15-20 min and 20-25 min extracts are related to 'baking cake', 'sweet' and 'crust' descriptors (Figure 1) whereas the 'not tasty' descriptor was only used for the 5-10 min extract.

Table 2. Preference rating and ranking test of the on-line SPME extracts during baking.

	5-10 min	10-15 min	15-20 min	20-25 min
Preference rating	4.08	4.68	5.03	5.16
Rank	2.89	2.47	2.28	2.36

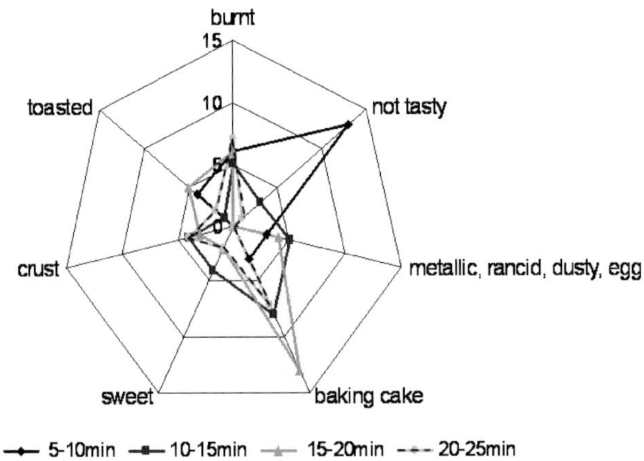

Figure 1. Sensory profile of SPME baking extraction.

GC-Olfactometry was applied to a 15-25 min baking extract. Six very interesting odour zones related with the 'tasty' flavour were identified. The main odour zone (relative to intensity and frequency of detection) was related to 'praline', 'baking cake' and 'very pleasant' descriptors and it is most probably due to the presence of acetylpyrazine.

4. CONCLUSION

The 'on-line' system coupled with SPME was able to extract volatile compounds released during the baking of a model cake. Instrumental as well as sensory analyses coupled with the direct olfactometry device allowed us to identify the vapour fraction mainly responsible for the 'tasty' odour released during the baking process.

References

1. H. Maarse (ed.), Volatile compounds in foods and beverages, NY, USA (1991) 41.
2. C. Prost, C.Y. Lee, P. Giampaoli and H. Richard, J. Food Sci., 58 (1993) 586.
3. G. Zehentbauer and W. Grosch, J. Cereal Sci., 28 (1) (1998) 81.
4. R.L. Heiniö, K. Katina, A. Wilhelmson, O. Myllymäki, T. Rajamäki, K. Latva-Kala, K.-H. Liukkonen and K. Poutanen, Lebensm. Wiss. Technol., 36 (2003) 533.
5. T. Hoffman, M. Rothe and P. Schieberle (eds.), State of the art in flavour chemistry and biology, proceedings of the 7th Wartburg symposium, Garching, Germany (2005) 408.
6. B. Rega, N. Fournier and E. Guichard, J. Agric. Food Chem., 51 (2003) 7092.

Workshops

W.L.P. Bredie and M.A. Petersen (Editors)
Flavour Science: Recent Advances and Trends

Gastronomy: the ultimate flavour science?

Thorvald Pedersen (chairman)[a], Claus Meyer[a], Harry Nursten[b] and René Redzepi[c]

[a]Department of Food Science, The Royal Veterinary and Agricultural University, Frederiksberg C, Denmark; [b]School of Food Biosciences, The University of Reading, Reading, United Kingdom; [c]Restaurant Noma, Copenhagen, Denmark

1. INTRODUCTION

Inspiration for creating novel flavours and sensorily attractive foods and meals is nowadays increasingly sought at the interface between science and the art of cooking. Gastronomy understood at the molecular level has also encouraged interest in the Scandinavian countries, but has a longer tradition elsewhere in Europe. In the workshop at the symposium, senior chemists and renowned Michelin-starred awarded Danish cooks were brought together to discuss and illustrate with real samples how chefs and flavour scientists can cooperate in a constructive way by creating ever more satisfying sensory experiences, based both on artistic and scientific skills. A brief report on the past, present and some future considerations for molecular gastronomy in Denmark are presented together with the recipes of the treats from the workshop.

2. ORIGIN OF MOLECULAR GASTRONOMY (MG)

Nicholas Kurti (1908-1998) was of Hungarian descent, eventually becoming Professor of Physics at Oxford University and specialising in low-temperature physics. Kurti took an interest in the history of physics and was particularly impressed by Count Rumford (1753-1814), who wrote a 400-page treatise, '*On the construction of kitchen fireplaces and kitchen utensils together with remarks and observations relating to the various processes of cookery and proposals for improving the most useful art*' (1794). Count Rumford was the founder of 'The Royal Institution' in London, famous especially for its Friday lectures (1799). In consequence, 1969 was the 170th anniversary of this innovation and Kurti was asked to give one of these Friday lectures. The title of his talk was '*The physicist in the kitchen*'.

Kurti used the following quotation from Rumford's treatise: 'The advantages that would result from an application of the late brilliant discoveries in philosophical chemistry and other branches of natural philosophy and mechanics to the improvement of the art of cookery are so evident that I cannot help flattering myself that we shall soon see some enlightened and liberal minded person of the profession to take up the matter in earnest and give it a thoroughly scientific investigation. In what art or science could improvements be made that would more powerfully contribute to increase the comforts of mankind?' How right he should have been in the last observation, but he was clearly ahead of his time. Kurti himself observed: 'It is a sad reflection that we know more about the temperature inside the stars than inside a soufflé' - and promptly proceeded to measure the temperature inside a soufflé during the lecture.

Out of his talk came an interest in the fundamental processes of cooking that was to remain with him for the rest of his life. He wrote a book together with his wife, Gianna, '*But the crackling is superb*', a number of Fellows and Foreign Members of The Royal Society contributing to this so-called '*Anthology on food and drink*'.

In about 1990, Kurti got together with the congenial French physical chemist, Hervé This-Benckhard, and started a series of workshops on different aspects of MG held at The Majorana Centre in Erice, Sicily.

International Workshops on Molecular Gastronomy

- International School of Molecular and Physical Gastronomy (1992).
- International School of Molecular and Physical Gastronomy (1995). 'Sauces or dishes made from them'
- International School of Molecular and Physical Gastronomy (1997). 'Cooking'
- International School of Molecular and Physical Gastronomy (1999). 'Flavours, how to get them, how to distribute them, how to keep them'
- International School of Molecular and Physical Gastronomy (2001). 'Textures of foods: how to create them'
- International workshop on Molecular Gastronomy 'Nicholas Kurti' (2004). 'Interaction of food and liquids'

Hervé This expressed his view on the fate of Rumford's wish, as follows: 'The food industry flourished, but the individuals who cook at home did not benefit from the advances of science. Cooking at home or in restaurants remained almost the same activity as in the Middle Ages: the same tools were used, the methods did not evolve and the ingredients changed little. In other words, while it is true that Rumford's wish has been to some extent fulfilled, good basic science and engineering has greatly helped the development of the food industry in the last 50-100 years, but it still seems to be very rare to see the professional scientist cum amateur cook using his physics, his chemistry, his mathematics to explain, to explore, to improve everyday processes in the domestic kitchen and, in doing so, perhaps even to create new dishes.'

It was the aim of the workshops to address the cook at home and in the restaurant kitchens as well as the scientists and indeed a fair percentage of the (about 40) participants of each workshop were cooks or food writers, while the rest were scientists.

3. CURRENT DEVELOPMENTS

Many of the current developments are centred on Hervé This but MG is rapidly attracting interest in other countries. The project on the introduction of innovative technologies in modern gastronomy for modernisation of cooking (INICON) [1] is a good example. This is a collaboration between four restaurants and their chefs (au Crocodile, Strasbourg, Emile and Monique Jung; el Bulli, Barcelona, Ferrán Adrià; Grashoff, near Bremen, Oliver Schmidt; and the Fat Duck, Bray, UK, Heston Blumenthal), an educational institute (École Gregoire-Ferrandi, École Supérieure de Cuisine Française, Paris), three firms operating in the food area, and two research and development establishments (Technical Transfer Zentrum, Bremerhaven; College de France, Paris, Hervé This). The web-site gives considerable detail of recipes for some 40 of their creations.

The available literature of molecular gastronomy is well covered by Martin Lersch [2] with frequent updates.

3.1. MG in Denmark

In Denmark the head of the Department of Food Science at The Royal Veterinary and Agricultural University (KVL), Grete Bertelsen, announced at a departmental meeting in January, 2004, that Molecular Gastronomy was to become part of the research activities during the next couple of years. Thorvald Pedersen (TP), a retired chemist from the University of Copenhagen, was hired for one year to catalyse the process. His background from a food perspective was 15 years of monthly essays on the chemistry of cooking plus a book '*Kemien bag gastronomien*' (2002) [3].

One of his first initiatives was a workshop in august 2004 aimed at coming to grips with MG. During this workshop, TP proposed a simple and operational definition of MG. He looked up the word 'Gastronomy' in the 'Oxford Learned Dictionary of Current English' and found the following: '*The art and science of choosing, preparing and eating good food*'. He then proposed that Molecular Gastronomy be defined as: '*The science of choosing, preparing and eating good food*'. Most likely Kurti would have accepted this definition. Starting from this definition, the subject and syllabus to be taught develop naturally, as follows:

The science of choosing the food

- Vegetables (freshness, storage, deterioration, preservation)
- Meat (structure, maturation)
- Fish (freshness, deterioration)
- Oxidation processes
- Microbial processes
- Enzymatic processes
- Preservation

The science of preparing food

- Creation/composition of dishes and foods
- Cooking procedures (including, for example, slow cooking)
- Baking of breads and cakes

The science of eating good food

- Sensory science

The cooks' skills enter especially during the process 'creation/composition' and one chef in particular, Heston Blumenthal of The Fat Duck, has been at the forefront of implementing such compositional ideas as 'flavour pairing' and 'nostalgia'. Flavour pairing means putting raw materials together that match each other flavour-wise. According to Blumenthal, the stone of a mango smells like green pepper while the pulp smells of terpenes, hence some of the pairings in the dishes described below. The sensory aspects of appreciating and eating the food emerge during the last point. In this connection, the term 'good' in the description of the food (*cf.* the definition) becomes open for discussion - and contemplation.

3.2. The future of MG in Denmark

Claus Meyer, a nationally renowned gastronomic entrepreneur and innovator, was appointed as external lecturer at KVL at the start of 2005 and from 2006 as affiliated professor. Claus has excellent connections with Danish cooks and has already attracted some activities. A research professorship at KVL will be launched as soon as funding becomes available. A PhD student in MG has started under the supervision of Professor L. Skibsted (Food Chemistry) in collaboration with the sensory science group and Claus Meyer. A new course on gastronomic meal design as part of the B.Sc. in Food Science is being jointly developed by the sensory and food chemistry groups. With the current high international interest in MG, the field in Denmark may also develop into a new interdisciplinary research area linked to education of food scientists, nutritionists and cooks.

4. THE RECIPES FROM THE WORKSHOP

The following dish is served fully at The Fat Duck and was served partly by René Redzepi during the workshop:

- Mango and Douglas fir purée
- Bavarois of lychee and mango
- Blackcurrant sorbet
- Blackcurrant and green pepper jelly

Four recipes were prepared by René Redzepi and his colleagues for tasting by the participants of the workshop.

1. Parsley sorbet and smoked bone marrow
The sorbet is made of a base with chicken stock, flour, and lots of parsley. Frozen and then turbined in the Pacojet (Pacojet A.G., Zug, Switzerland). Ryebread is dried in the oven. The bone marrow is salted for 3 days, then smoked over oak.

2. Yoghurt powder, lobster gelée, fresh green aniseed (wild Danish seeds)
Yoghurt is made of water and yoghurt. Frozen in a Pacojet cup, then spun in the Pacojet machine. Lobster bouillon is made into a gel with agar agar. Fresh raw aniseed is cut fine (julienne).

3. Ryebread 'surprise' (øllebrød)
Ryebread is soaked in milk for a day and then cooked to a light porridge, which is flavoured with sugar and dark beer. Put into a chiffon bottle (a metal bottle with a small tap at the top) and add gas. Raisins are soaked 1 month in aquavit.

4. Bavarois recipe for Weurman Workshop
The Bavarois is made of whipped cream, gelatine, and whipped egg whites, sugar being added as well as mango purée. The green pepper jelly is made of green pepper seeds (Madagascar) boiled with blackcurrant juice. Stiffened with gelatine after the pepper seeds have been removed.

4.1. Discussion

After they had been tasted, René Redzepi described each of the four dishes that had been prepared and asked for comments. The fact that bone marrow had been used led to a discussion of the ethics of serving animal products to vegetarians. It was made clear that proper descriptions were always available to customers in his restaurant and that most chefs endeavour to adapt their menus to their customers' requirements.
Overall, the four dishes were received with interest and appreciation. In discussing them, sensory descriptors need inevitably to be used. A useful list of descriptors selected after considerable research is that of Harper *et al.* [4]. Such a list also constitutes a palette of flavours, useful in creating/constituting new recipes. Some of the descriptors could well be refined. This applies to 'nutty', which had been used several times during the Symposium [5].

New ideas and inspiration for future recipes can come from:
- Consideration of the palette of flavours
- New flavour science, such as some of that presented at the Symposium
- Sociological diversity (ethnic foods)
- Chefs' observations (e.g., coffee with garlic)
- Traditional practices (old wives' tales)
- Research (e.g., on feeding regimes for animals and cultivation of food plants)

There is still much on which to ponder.

References

1. As currently available from 'www.inicon.net'.
2. As currently available from 'http://folk.uio.no/lersch/mat/index.e.html'.
3. T. Pedersen, Kemien bag gastronomien, Copenhagen, Denmark (2002).
4. R. Harper, D.G. Land, N.M. Griffiths and E.C. Bate-Smith, Brit. J. Psychol., 59 (1968) 231.
5. R.C. Clark and H.E. Nursten, Int. Flavours Food Add., 8 (1977) 197.

W.L.P. Bredie and M.A. Petersen (Editors)
Flavour Science: Recent Advances and Trends

Methods for artificial perception: can machine replace man?

Wender L.P. Bredie[a], Christian Lindinger[c], Gunnar Hall[d], Anne-Maria Hansen[e], Gerald Reinders[f] and Magni Martens[ab] (chairman)

[a]Department of Food Science, The Royal Veterinary and Agricultural University, Frederiksberg C, Denmark; [b]Norwegian Food Research Institute, Matforsk, Aas, Norway; [c]Nestle Research Centre, Lausanne, Switzerland; [d]Technological Institute, Kolding, Denmark; [e]Swedish Institute for Food and Biotechnology (SIK), Göteborg, Sweden; [f]Symrise, Holzminden, Germany

SUMMARY

The previous 'Scandinavian' Weurman Flavour Symposium held in Oslo, Norway, in 1987 had as the main themes chemistry, biotechnology, sensory science and data analysis in flavour research [1]. Since then, many advances have been made, however, when comparing with the Weurman Symposium in 2005, the subject areas and topics of research have not changed very much. Flavour scientists still work with identification of aroma and taste components in foods, their formation, stability and release as well as sensory and data analytical aspects. The knowledge about flavour components in foods and the understanding of the way we perceive them has though considerably expanded. Advances have also been made in the way one can analyse flavours in foods and beverages, and measure responses from and in human subjects. One can say that the toolbox with methods and techniques for flavour analysis has expanded remarkably offering many new possibilities to understand flavour from different perspectives and levels of details.

The workshop presented some state-of-the-art applications of modern *in vivo* and *in vitro* flavour analysis as well as visual sensory assessments of meals made by machines equipped with sensors and artificial networks processing capability. Also, new ways of studying multisensory processes by stimulating sensory subjects with defined stimuli were discussed. Intelligent 'artificial perception' systems may replace some routine sensory analysis and monotonous production tasks in the future, but the development of such systems still require sensory assessments by humans.

1. PREDICTION OF COFFEE ODOUR PROFILES BY PTR-MS

Proton transfer reaction mass spectroscopy (PTR-MS) has become a valuable tool for real-time monitoring of changes in the composition of aroma volatiles in vapour phases. The instrument has been successfully employed in retronasal flavour release studies by analysing the breath of humans during the eating process. In the workshop a new application of the PTR-MS for quality judgements of coffee was presented. Descriptive odour profiles were well predicted from normalised PTR-MS signals for 16 masses. The model was with success used for the prediction of new espresso varieties.

2. *IN VIVO* AND *IN VITRO* VOLATILE ANALYSIS

Modern techniques for the measurement of volatile compounds in foods and subjects were presented. At the Technological Institute, the GC-sniff technique has been useful in analysis of off-flavours. The APCI-MS using the MS-nose interface is typically used to study the effect of food matrix and the eating process on flavour release. The membrane-inlet-MS (MIMS) system linked to a chewing machine can simulate *in vivo* release. The greatest challenge for new instrumental techniques in flavour research is to accommodate detection relevant to perception - which is directed to measure changes.

3. CONTROLLED MULTISENSORY EVALUATION

A great number of flavour stimuli in foods are known. Understanding the perceptual and physiological roles of these stimuli in human subjects requires controlled stimulation. New ways of measuring flavour perception, e.g. by dynamic olfactometry, gustometry and audiometry combined with instruments aquiring brain activity may give insights in the perceptual and physiological attributes of flavour stimuli. When developing new foods and meals, correlations between neurological activity and measures of sensory quality including affective and satiety judgements should be explored.

4. ARTIFICIAL APPEARANCE-BASED EVALUATION OF MEALS

A new method for measuring meal appearance by an artificial vision system [2] based on a self-organising map (SOM) trained by sensory data has been developed. The SOM discriminate meals by considering the similarity between test samples and a sensory model. Artificial features are extracted from meal images to match to the sensory attributes and are processed by artificial neural networks intensities of the sensory attributes as output. The artificial vision system evaluated the meals (72 attributes) and gave similar results as the sensory panel in 82% of the attributes.

References

1. M. Martens, G.A. Dalen and H. Russwurm Jr. (eds.), Flavour science and technology: proceedings of the 5th Weurman flavour research symposium, Chichester, UK (1987).
2. P. Munkevik, Artifical sensory evaluation - appearance-based analysis of ready meals, Licentiate thesis, Örebro University, Department of Technology, Örebro, Sweden (2005).

W.L.P. Bredie and M.A. Petersen (Editors)
Flavour Science: Recent Advances and Trends

Challenges for data analysis in flavour science

Rasmus Bro (chairman)[a], Per M. Bruun Brockhoff[b] and Thomas Skov[a]

[a]Department of Food Science, Royal Veterinary and Agricultural University, Copenhagen, Denmark; [b]Informatics and Mathematical Modelling, Technical University of Denmark, Kongens Lyngby, Denmark

1. INTRODUCTION

The analysis of data from instrumental and sensory flavour analyses and related data types often pose special challenges for the data analyst. This may be due to the richness of information in the data or due to special artefacts that need to be handled. In this workshop, some of the basic and more advanced tools for handling various types of data were illustrated. First, an overview of multivariate methods was given, followed by a description of analysis of sensory data. Finally, new methods for handling GC and electronic nose data were described.

2. MULTIVARIATE ANALYSIS

Rasmus Bro started to point out why we need multivariate analysis. Multivariate data analysis uses all available data simultaneously - exactly as in human pattern recognition. For most interesting problems the information is in the relation between variables. Examples were given on how Principal Component Analysis can make interpretation of data tables easier. Even simple questions as ‘Which of the samples are most similar?’ can be very difficult to answer by looking at a data table in a univariate way.
Multivariate data analysis also enables an exploratory approach to data. No previous hypotheses are needed - all data can be analysed and new information can be found.
The techniques can be taken one step further: calibration can be extended to multivariate calibration and relations between complex data matrices can be described. One type of variables can be predicted from other types - typically from measurements that are more easily available, more unspecific and more complex. The following examples were given: analysis of water in product from infrared spectra, consumer preference from sensory panel data, off-flavour in beer from fluorescence spectra, and concentration of active components in tablets using near-infrared spectra.

A very important issue using these techniques is to distinguish between causality and correlation. Correlation is an observed relation. For instance sales of sunglasses correlate to sales of ice-cream. This is a true and valid relation, but the real cause is an underlying (latent) factor - the sun. Causality includes an explicit relation. For example absorbance depends on amount of analyte in a solution. This is a true and valid relation and the cause is direct: the analyte absorbs.

When processes are to be optimised, designed experiments are required, and causal relations is a prerequisite. Using the above examples: Change sales of ice/amount of analyte and see what happens to sales of sunglasses/absorbance.

It was concluded that multivariate analysis provides a window into the multivariate space enabling investigation of similarity, grouping, trends, outliers, variables, influence, importance, correlations and relation between blocks. A list of relevant web-sites was given [1-6].

3. ANALYSIS OF SENSORY DATA - WHAT IS SENSOMETRICS?

Per Bruun Brockhoff introduced the area of sensometrics, and defined sensometrics as the scientific area that applies mathematical and statistical methods to model data from sensory and consumer science. Examples of sensometrics research were given which both included ANOVA type data and regression/multivariate type data. Also the Sensometric Society which arranges meetings every second year was shortly presented [7].

Different types of relevant statistical methods were then presented. This included binomial statistics and related issues (classical, Thurstonian modeling, random effects/mixed model versions), rank based statistics, design of experiments and analysis of variance (classical, mixed model extensions) and multivariate analysis (analysis of profile data, relating sensory properties to instrumental and/or consumer data, consumer segmentation, multivariate quality control, time intensity data).

A typical sensory experiment includes a number of panelists, a number of samples and a number of replications. Simple 3-way mixed model ANOVA is typically carried out for each variable, the main issue being a test for product differences. Assessor/panel performance and monitoring may, however, also be an issue. For this, either the basic ANOVA model or the basic assessor model may be used, and these will investigate differences in level, scaling, disagreement, variabilities and sensitivities [8,9].

The methods mentioned until now are univariate, but since most sensory data is multivariate and at least 3-way, research is carried out to combine ANOVA and multivariate techniques. The combination of univariate models for scaling differences and bi-linear models for the multivariate product differences leads to 3-way models of the PARAFAC type. Finally it is often a wish to relate the multivariate sensory profile data to other data, like chemical/instrumental data on one hand and consumer preferences on the other. This calls for the classical bias regression techniques used extensively in chemometrics, such as Principal Component Regression (PCR), Partial Least Squares Regression (PLSR), and for the consumer data also for segmentation/clustering techniques to identify relevant consumer segments.

4. ANALYSIS OF GC-MS AND ELECTRONIC NOSE DATA

Thomas Skov outlined the basic differences between GC-MS and electronic nose, where the electronic nose is characterised by low specificity and by being most suitable for more rough 'good/bad' analysis. Electronic nose data are typically 3-way (Sample×Time×Sensor). The traditional approach to this type of data is simplification of data by feature extraction resulting in 2-way data suitable for PCA or PLS analysis. An advanced approach is to keep the internal data structure, treating the 3-way data with multi-way models. This has the advantage that there is no initial loss of information.

Shifted data will however give problems. This will occur when elution times in GC-MS or sensor time profiles in electronic noses are shifted. This results in obscured data, and the common bi/tri-linear model does not work. A method to overcome this is warping. Warping eliminates shift-related artefacts from measurements by correcting the time axis of a sample profile towards a reference. Successful warping increases explained variance with the use of fewer principal components. Furthermore the width of peak shaped loadings decreases and the shape of secondary artifactual loadings (derivatives) is improved. An example was given on PCA versus PARAFAC modelling (licorice with off-flavour), and improved clustering and separation could be demonstrated using three-way models.

It was concluded that data structure and dimensionality is a key factor when analysing GC-MS and electronic nose data. Shifted data needs attention, but can be taken care of before modeling or in the modeling step, and advanced chemometric models can give a more efficient modeling of data.

References

1. www.models.kvl.dk (KVL; software, courses etc.).
2. www.spectroscopynow.com (Links on chemometrics).
3. www.models.kvl.dk/ris/risweb.isa (Database with 8000 papers).
4. www.camo.no/ (Unscrambler).
5. www.umetrics.com/ (Simca).
6. www.eigenvector.com (PLS Toolbox for matlab).
7. www.sensometric.org (the Sensometric Society).
8. P.M. Brockhoff and I.M. Skovgaard, Food Quality Pref., 5 (1994) 215.
9. P.B. Brockhoff, Food Quality Pref., 14 (5-6) (2003) 425.

Author index

Keyword index

The page numbers of the keywords refer to the first page of the respective chapter.

A

B

C

D

E

F

G

H

I

K

L

M

N

O

P

Q

R

S

T

U

V

W

X

Y

Z